changing the way the world learns

To get extra value from this book for no additional cost, go to:

http://www.thomson.com/wadsworth.html

thomson.com is the World Wide Web site for Wadsworth/ITP and is your direct source to dozens of on-line resources. *thomson.com* helps you find out about supplements, experiment with demonstration software, search for a job, and send e-mail to many of our authors. You can even preview new publications and exciting new technologies.

thomson.com: *It's where you'll find us in the future.*

PLANT BIOLOGY

THOMAS L. ROST / MICHAEL G. BARBOUR

C. RALPH STOCKING / TERENCE M. MURPHY

University of California
Davis, California

WADSWORTH PUBLISHING COMPANY

I(T)P® An International Thomson Publishing Company

Belmont, CA • Albany, NY • Bonn • Boston • Cincinnati • Detroit • Johannesburg • London • Madrid • Melbourne
Mexico City • New York • Paris • Singapore • Tokyo • Toronto • Washington

BIOLOGY PUBLISHER: Jack C. Carey

PROJECT DEVELOPMENT EDITOR: Kristin Milotich

EDITORIAL ASSISTANT: Michael Burgreen

DEVELOPMENTAL EDITORS: Mary Arbogast, Elizabeth Zayatz

PRODUCTION EDITOR: John Walker

PRINT BUYER: Karen Hunt

PRODUCTION: Mary Douglas, Rogue Valley Publications

DESIGNERS: Carolyn Deacy; Gary Head, Gary Head Design

ART COORDINATOR: Myrna Engler

ARTISTS: Page Two; Precision Graphics; Illustrious Interactive

PHOTO RESEARCH: Roberta Broyer, Stephen Forsling

COVER DESIGNER: Gary Head, Gary Head Design

COVER PHOTOGRAPH: Carr Clifton/Minden Pictures

COMPOSITION AND COLOR PROCESSING: American Composition & Graphics, Inc.

PRINTING AND BINDING: Courier/Kendallville

Library of Congress Cataloging-in-Publication Data
    Plant biology / Thomas L. Rost . . . [et al.].
        p.   cm.
    Includes bibliographical references and index.
    ISBN 0-534-24930-2
    1. Botany.   I. Rost, Thomas L.
    QK47.P57   1997
    580—dc21   97-44578

BOOKS IN THE WADSWORTH BIOLOGY SERIES

Biology: Concepts and Applications, Third, Starr

Biology: The Unity and Diversity of Life, Eighth, Starr/Taggart

Human Biology, Second, Starr/McMillan

Laboratory Manual for Biology, Perry and Morton

Perspectives in Human Biology, Knapp

Plant Biology, Rost/Barbour/Stocking/Murphy

Introduction to Biotechnology, Barnum

General Ecology, Krohne

Introduction to Microbiology, Ingraham/Ingraham

Living in the Environment, Tenth, Miller

Environmental Science, Seventh, Miller

Sustaining the Earth, Third, Miller

Environment: Problems and Solutions, Miller

Environmental Science: A Systems Approach to Sustainable Development, Fifth, Chiras

Introduction to Cell and Molecular Biology, Wolfe

Molecular and Cellular Biology, Wolfe

Cell Ultrastructure, Wolfe

Marine Life and the Sea, Milne

Essentials of Oceanography, Garrison

Oceanography: An Invitation to Marine Science, Second, Garrison

Oceanography: An Introduction, Fifth, Ingmanson/Wallace

An Introduction to Ocean Sciences, Segar

Plant Physiology, Fourth, Salisbury/Ross

Plant Physiology Laboratory Manual, Ross

Plants: An Evolutionary Survey, Second, Scagel et al.

Fundamentals of Physiology: A Human Perspective, Second, Sherwood

Human Physiology: From Cells to Systems, Third, Sherwood

Human Heredity, Fourth, Cummings

Psychobiology: The Neuron and Behavior, Hoyenga/Hoyenga

Sex, Evolution, and Behavior, Second, Daly/Wilson

Dimensions of Cancer, Kupchella

Evolution: Process and Product, Third, Dodson/Dodson

For more information, contact Wadsworth Publishing Company, 10 Davis Drive, Belmont, California 94002, or electronically at

http://www.thomson.com/wadsworth.html

International Thomson Publishing Europe
Berkshire House 168-173, High Holborn
London, WC1V7AA, England

Thomas Nelson Australia
102 Dodds Street
South Melbourne 3205, Victoria, Australia

Nelson Canada
1120 Birchmount Road
Scarborough, Ontario, Canada M1K 5G4

International Thomson Editores
Campos Eliseos 385, Piso 7
Col. Polanco, 11560 México D.F. México

International Thomson Publishing GmbH
Königswinterer Strasse 418
53227 Bonn, Germany

International Thomson Publishing Asia
221 Henderson Road, #05-10 Henderson Building
Singapore 0315

International Thomson Publishing Japan
Hirakawacho Kyowa Building, 3F
2-2-1 Hirakawacho, Chiyoda-ku, Tokyo 102, Japan

International Thomson Publishing Southern Africa
Building 18, Constantia Park
240 Old Pretoria Road
Halfway House, 1685 South Africa

CONTENTS IN BRIEF

DETAILED CONTENTS

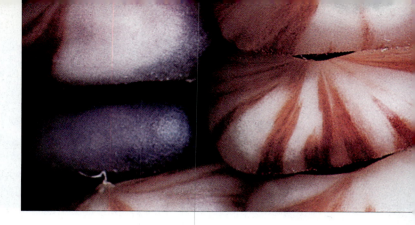

PREFACE

There has never been a time when the importance of plants has been clearer and our understanding of them has been more urgent. This century's history of regional famine and global population growth has had a profound impact on the course of politics, research, and land-use management. In the 1970s, a "green revolution" resulted in the development and promotion of astonishingly productive new varieties of grains, using the principles of plant physiology, pathology, and genetics. For a while, the result was indeed an increase in food supply, a reduction of land devoted to agricultural use, and an avoidance of famine. These benefits, however, came at a cost—an increasing dependence on mechanized farming and use of fertilizers, a decreased diversity of crop genotypes, and a (still) increasing human population. Ultimately, the rising population brought per-capita food production back down to pre-revolution levels. Now we are entering a "gene revolution," which no doubt will result in another round of food production increases. Will we pay the same costs again, or can we instead learn more about plants, the environment, and ourselves to avoid repeating this unprofitable loop? At the same time, now late in the twentieth century, we face additional serious problems of pollution as it alters the world's atmosphere and climate. Will the Earth's carrying capacity for humans (and all other species) decline or expand as a result of these changes? We who are writing this book do not have the answers; we hope and trust that you, who are about to read it, will find the solutions.

This book is the capstone to a series of plant biology textbooks that share a lineage of related authors. That lineage is probably lengthier than for any other existing biology textbook in the world. The ancestral book, *A Textbook of General Botany*, was co-authored in 1924 by Richard M. Holman, at the University of California, Berkeley, and Wilfred W. Robbins, in the Botany Division of what was then called the University Farm, located in the small town of Davis. The book was widely adopted through its four editions. After Richard Holman passed away, Professor Robbins invited a young faculty member in the Farm's Botany Division to be co-author with him of a 1950 book, *Botany: An Introduction to Plant Science*. That faculty member was T. Elliot Weier, a cytologist and pioneer in the use of the electron microscope. After Wilfred Robbins' death, Dr. Weier invited a plant physiologist colleague in the department, C. Ralph Stocking, to become co-author of the second and third editions. Michael Barbour, a plant ecologist, joined Weier and Stocking on the fourth (1970) and fifth (1974) editions and Tom Rost, a developmental anatomist, joined for the sixth edition (1982). Plant physiologist Robert Thornton helped them write two editions of a smaller book called *Botany: A Brief Introduction to Plant Biology* (Rost et al., 1979 and 1984). Now, Terence Murphy, a molecular plant biologist, has become co-author number eight in a story that has wound through 13 versions/editions, 74 years, hundreds of thousands of books printed, two publishers, and four generations of readers. Welcome to you, the fourth generation!

What is new and characteristic in this edition? Although the coverage of material is complete, its level of detail will not overwhelm students taking a one-quarter or one-semester course. An entirely new art program provides pedagogically useful and attractive illustrations, including more than 70 full-color micrographs in the plant anatomy chapters alone. Throughout the text many interesting boxed essays introduce applied topics ranging from bee pollination, harvesting peat bogs for the generation of electricity, and the use of botany in forensic medicine, to the making of oak wine barrels.

We have included a new chapter on the chemistry of life which integrates and spans basic inorganic chemistry and biological molecules. The four chapters on plant anatomy are now arranged from meristems and tissues to organs in a sequence that better captures plant function and facilitates comprehension by the reader. Metabolic chapters immediately follow anatomy chapters and continue to focus on the links between form and function. Three new chapters—plant life cycles, genetics, and biotechnology—provide an important conceptual framework for the last third of the book. Also, the classification system we follow is more modern, taking advantage of recent research about plant phylogeny.

What else is new? The text has been thoroughly updated and some sections have been expanded. The new coverage of positional information theory explains cell differentiation on the basis of position. The discussion of meristem structure and function is expanded. We include new information on rhizospheres and on the relationships between roots, soil, and microorganisms as well as information on root function and specializations. We have expanded the discussion of monocot stems and updated our treatment of leaf specializations as related to environmental adaptations and responses to stress. The presentation of photosynthesis shows both the historical development of research and our current understanding of the intracellular locations of various steps in this complex process. This edition includes expanded coverage of the various strategies to cope with environmental stresses, such as trapping CO_2 at night by succulent plants, and variations in the carbon cycle exhibited by decreasing the water potential of their guard cells. Recent principles of membrane transport are applied to the question of how roots take up minerals. The treatment of reproduction and pollination, as well as seed dispersal, is all new. The topic of stress signals as newly recognized plant hormones, along with the relation of hormones and signal cascades to gene expression, is thoroughly updated. The coverage on the critical comparison of two hypotheses to explain cell wall expansion—microfibril breakage and re-attachment and "extendin" is newly revised. The new chapter on biotechnology has a thorough discussion of the methods of transforming plant cells, the uses of transformation in plant improvement (actual and contemplated), and questions regarding the dangers of transformation. The presentation of endosymbiosis as a source of evolutionary variety, with an outline of the molecular evidence for endosymbiosis is expanded. There is an updated discussion of the cellular aspects of the establishment of the *Rhizobium*/legume symbiosis. The text includes new material on the hypersensitive response as one method by which plants defend themselves against parasites. It also includes major expansion of the plant ecology chapters, with a unique emphasis on the ecological and adaptive value of fungal traits. The chapters on plant diversity and the evolutionary connections of the plant groups have been upgraded with many new photographs.

Several pedagogic practices pervade each chapter of the text. (1) Topics are intimately integrated so that form is immediately linked with function, environment, and evolutionary significance. (2) Summary statements are used as headings to alert the reader to the main point of the section to come. (3) We have reduced jargon to those terms we consider absolutely essential, and each term is defined the first time it is used. (4) Only illustrations that complement and clarify the text are used. (5) Connections between basic science and applied science (technology) are explicitly made, and boxed essays focus on topics of general interest. (6) At the end of each chapter, major conclusions are highlighted in summary and question sections. We have tried to choose the most reader-friendly literature to recommend for further reading, rather than merely the most recent—but less accessible—references. (7) Our common goal has been to devote space to each topic in proportion to its complexity and difficulty.

We have also listened to our reviewers regarding the breadth and sequence of topics they require to support the general botany/plant biology courses they teach. Consequently, structure, function, reproduction, behavior, and biology of flowering plants precede chapters on other organisms because the flowering plants are more familiar to readers, making the new information more easily absorbed. Our survey of the plant world includes detailed excursions into the Protista, Fungi, Monera, and viruses as well as into the Kingdom Plantae. The overall sequence of major subjects in the book builds from metabolism to whole plant function to reproduction within flowering plants, then from primitive to advanced plants in a series of survey chapters, concluding with an ecological segment that ties together the entire book.

SUPPLEMENTS

InfoTrac College Edition. This online library is available free with each copy of *Plant Biology*. It gives students access to full articles—not abstracts—from more than 600 scholarly and popular periodicals dating back as much as four years. The articles are available through InfoTrac's impressive database that has such periodicals as *Discover, Science, BioScience, Horticulture: The Magazine of American Gardening, Agricultural Research, Journal of Soil and Water Conservation*, and *The American Midland Naturalist*.

Student Guide to InfoTrac's College Edition. This guide is on the Wadsworth Biology Resource Center site on the World Wide Web. It has an introduction to InfoTrac and a set of electronic readings for each chapter, updated frequently.

Biology Resource Center. It contains hyperlinks and practice quiz questions for each chapter. It also includes flashcards for all glossary terms, botanical clip art, and links to botanical organizations. The address for the Wadsworth Biology Resource Center is

http://www. wadsworth.com/biology

Introduction to the Internet. This 80-page booklet helps students learn how to get around on the Internet when using a browser such as Netscape, search engines, e-mail, setting up home pages, and related topics. It lists useful sites on the net that correspond to book chapters.

The Botanical Society of America Website. This site will be maintained on the Wadsworth web page at

http://www. thomson.com

BioLink 2.0. With this presentation tool, instructors can easily assemble art and database files with lecture notes to create a fluid lecture that may help stimulate even the least-engaged students. It includes almost all diagrams from the text, animations and films from the general biology CD, and art from other Wadsworth biology textbooks. BioLink 2.0 also has a Kudo Browser with an easy drag-and-drop feature that allows file export into such presentation tools as Power Point. Upon its creation, a file or lecture with BioLink 2.0 can be posted to the Web, where students can access it for reference or for studying needs.

Instructor's Manual with Test Items, written by Drs. John Jackson and Joseph McCulloch, has an outline, objectives, key terms, detailed lecture outline, presentation suggestions, and 40 test questions per chapter. The test bank is also available in electronic form for IBM and Macintosh in a test-generating data manager.

Laboratory Manual, written by Dr. Deborah Canington, University of California, Davis, contains 22 experiments and exercises, with numerous labeled photographs and diagrams. Many experiments are divided into parts for individual assignment, depending on available time. Each consists of objectives, terminology, discussion (introduction, background, and relevance), a list of materials for each part, procedural steps, and laboratory quizzes.

Instructor's Manual for the Lab Manual, also written by Dr. Deborah Canington, contains lists of materials and equipment, instructions for preparations of lab materials, the sources of botanical materials, planting schedules, and suggested lab schedules to fit different academic calendars and lab lengths.

Study Guide and Workbook: An Interactive Approach, written by Dr. Cherie Wetzel, lets students write answers to questions and label diagrams, which are arranged by chapter section with references to specific text pages.

Transparencies. Full-color acetates and black-and-white masters are available.

Photo Atlas for Botany. In full-color, it includes more than 600 labeled photographs and photomicrographs utilizing light, transmission, and scanning electron microscopy, allowing students to recapture what they have seen in lab.

ACKNOWLEDGMENTS

We offer our profound gratitude to many individuals who helped this project reach completion. First on the list is Jack Carey of Wadsworth, who enthusiastically adopted our project several years ago and who relentlessly (yet tactfully) pushed us to complete it according to a vision we jointly shared. Mary Arbogast of Wadsworth and Mary Douglas and Myrna Engler of Rogue Valley Publications oversaw the difficult and complex process of meshing illustrations with text in the galleys and proofs. Never once did they get rattled or cranky with the authors. We thank them for their thoroughness, professionalism, and good humor. We thank Dr. Robert Thornton, Senior Lecturer in the Section of Plant Biology at the University of California, Davis, for his preparation of two chapters in this text: Fungi and Evolution. Dr. Thornton has received several honors for his excellence as a teacher, and his work here illustrates that excellence.

We acknowledge the reviewers who consistently gave positive, detailed advice, which led to numerous significant improvements. They are: Joseph Ammirati, University of Washington, Seattle, WA; Rolf W. Benseler, California State University, Hayward, CA; Maynard Bowers, Northern Michigan University, Marquette, MI; Richard G. Bowmer, Idaho State University, Pocatello, ID; James Dawson, Pittsburg State University, Pittsburg, KS; Roger del Moral, University of Washington, Seattle, WA; Stephanie Digby, St. Cloud State University, St. Cloud, MN; H. W. Elmore, Marshall University, Huntington, WV; Michael Gardinar, University of Puget Sound, Tacoma, WA; John Green, Nicholls University, Thibodaux, LA; William Harris, University of Arkansas, Fayetteville, AR; John D. Jackson, North Hennepin Community College,

Tom Rost, Ralph Stocking, Michael Barbour, and Terry Murphy

Brooklyn Park, MN; Michael Marcovitz, Midland Lutheran College, Fremont, NE; Robert Mellor, University of Arizona, Tucson, AZ; Harvey A. Miller, University of Central Florida, Orlando, FL; H. Gordon Morris, University of Tennessee at Martin, Martin, TN; Loyd Ohl, Eau Claire, WI; P.C. Pendse, California Polytechnic State University, San Luis Obispo, CA; James Raines, North Harris College, Houston, TX; Professor Barbara Schumacher, San Jacinto College-Central, Pasadena, CA; Nancy Smith-Huerta, Miami University, Oxford, OH; John Stucky, North Carolina University, Raleigh, NC; Cherie Wetzel, City College of San Francisco, San Francisco, CA; Terry F. Werner, Harris-Stowe State College, Saint Louis, MO.

Our families exhibited gracious patience and support—especially our wives Ann Rost, Valerie Whitworth, and Judith Murphy. Many colleagues who contributed illustrations, information, or other resources are unfortunately too numerous to list individually here, but we trust that they know we deeply appreciate their help.

Meeting our objective for this book was not easy. We've been refining the text for half a dozen years, and before that we tried out our ideas in books brought out by other publishers, and before that we began the long process of developing our own approaches to plant biology in the classroom and in our research. Collectively we represent 110 years of teaching, and a portion of virtually every one of those years was spent individually or team-teaching an introductory plant biology class. This book is a distillation of our experiences and efforts over that long time. We think that our separate areas of research have helped us write with a knowledgeable, personal voice. We hope that reading our book will make your own learning and teaching as rewarding and exciting as those same activities have been for us.

We dedicate this book to T. Elliot Weier (1903–1991), our wise, kind, and energetic mentor. We miss him and we wish he were here to celebrate this moment with us.

1

ABOUT PLANT BIOLOGY

1. Plants include a quarter of a million species of mosses, ferns, conifers, and flowering plants. Plantlike relatives include another quarter of a million species of certain bacteria, fungi, and algae.

2. Plants and their plantlike relatives are of vital importance to all life on Earth, including humans. Economic wealth is largely dependent on plant products. A major challenge of the 21st century will be to attain sustainable use of our plant resources, which means no loss in the carrying capacity of the Earth (the number of organisms the Earth can support) and no loss in diversity among the millions of species that coexist on our planet.

3. The scientific method is a repeating process of formulating predictions (hypotheses) about the world and then testing those hypotheses. This is only one method humans have for explaining their environment, and it has both positive and negative consequences.

4. Plant biology is the study of organisms classified either as plants or plantlike near relatives.

1.1 THE PERVASIVE IMPORTANCE OF PLANTS

Seen from the perspective of the moon, 385,000 kilometers (km)—240,000 miles (mi)—away, Earth has a blue color (Fig. 1.1a). Blue oceans cover more than two-thirds of the planet's surface, and a blue atmosphere extends like a fuzzy halo from the entire globe, merging eventually into black space. Seen from the perspective of 10 km (6 mi) above the surface, however, the Earth looks green (Fig. 1.1b). The Earth's unique green color originates from its plant life. The surface of the moon is not green, nor is the surface of Mars. Sunlight contains green wavelengths, but it does not look green to the human eye.

Earth is green because of the enormous number of plants that carpet the ground, extend into the air, occupy the soil, and float below the surface of lakes and oceans. This tangle of plant tissue selectively absorbs red and blue wavelengths of sunlight, but it reflects green wavelengths or allows them to pass through. Consequently, sunlight reflected from a forest, a meadow, or a field of corn is green, as is the dappled shade beneath a tree.

The red and blue wavelengths of light are transformed by plant tissue into chemical energy in a process called *photosynthesis*. Technically, a *plant* is an organism that is green and photosynthetic, producing organic sugar from inorganic carbon dioxide, water vapor, and light. Plants grow attached to one place and their *cells* (the basic units of which all organisms are composed) are surrounded by a rigid wall made of *cellulose* molecules rarely found in other organisms. Plants have multicellular but simple bodies lacking obvious circulatory, digestive, or nervous systems. Their bodies, however, are

a

b

Figure 1.1 The Earth from space. (**a**) The Earth is blue from the distance of the moon. (**b**) At closer range, the Earth is predominantly green because of the prevalence of plants.

adapted to the stresses of life on land: They can regulate water loss; they have strengthening tissue that keeps them upright, they can regulate their temperature; and they can reproduce with microscopic, drought-tolerant cells called *spores*.

This technical definition of *plant* includes about 250,000 species of trees, shrubs, herbs, grasses, ferns, and mosses. These different types of plants exhibit an incredible diversity in habitat, shape, life history, human use, and ecology. Most of this book is about them. They are formally classified in the kingdom *Plantae* (Fig. 1.2).

This technical definition is really too narrow, however, because there are another 250,000 species of plantlike organisms in three other kingdoms of life, and this book is about them as well. These plantlike organisms are

Figure 1.2 Examples of plants. (**a**) scarlet oak tree, (**b**) tree ferns in Hawaii, (**c**) moss, (**d**) the broad-leaved herb mule ears (a relative of sunflower), and (**e**) wheat (a grass).

a

b

c

d

e

a

Figure 1.3 Examples of plantlike organisms in other kingdoms. (**a**) Photosynthetic bacteria (Monera). (**b**) Mushroom (Fungi). (**c**) Algae (Protista) in a pond.

b

c

in the kingdoms **Monera** (bacteria), **Fungi** (molds), and **Protista** (algae; Fig. 1.3). Many of them are green and photosynthetic, but some are not; instead, they engulf living food or feed on dead organic remains. These organisms cannot regulate water loss, nor do they possess strengthening tissue. Many of them grow in aquatic habitats rather than on land. They, too, exhibit enormous diversity and have great economic and ecological importance to humans. They are also part of the science of **botany** or **plant biology**, the study of plants and plantlike organisms.

Plants and plantlike organisms are well known to us as sources of food, fabric, shelter, and medicine. They are the source of atmospheric oxygen and organic nitrogen. They build new land and inhibit erosion, control atmospheric temperature, decompose and cycle essential inorganic nutrients, and ultimately supply food for every living organism. Civilizations have risen and fallen throughout recorded time, depending on their access to and control of plants: for lumber to make warships, as fuel to smelt vital metals, cure pottery, and generate power and heat; and as sources of wealth in the form of

Figure 1.4 Conifer forests like this coast redwood forest continue to have enormous economic value. The wood from a single old-growth redwood tree has a retail value of tens of thousands of dollars.

Figure 1.5 Forest decline in central Europe. Beginning in the late 1970s, many hectares of forest died for unknown reasons. Ozone from automobile exhaust and acids from coal burning are among the suspected causes.

spices and industrial products such as rubber. Anthropologists often describe human cultures and history in such terms as Stone Age, Bronze Age, and Iron Age, but truly humans have always lived in a Wood Age because wood has been the ultimate measure and creator of wealth (Fig. 1.4).

Our earliest written myths and legends feature plants. The *Epic of Gilgamesh*, dated to 4700 years ago from the city of Uruk in southern Mesopotamia (now Iraq), highlights the importance of forests to humans and conveys an understanding of the ecological consequences of logging. Gilgamesh was ruler of Uruk and wanted to make a name for himself by building a great city. He required lumber, but the vast cedar forest that covered mountain slopes to the east had never been entered by humans. A demigod named Humbaba protected the forest for nature and the gods and against the needs of civilization. Nevertheless, Gilgamesh and his companions traveled to the forest with adzes and axes. They briefly lost themselves in contemplation of the forest's beauty and holiness but then set to work felling trees. When Humbaba challenged them, they killed him. Then the trees wailed in fear, and "you could hear the sad song of the cedars" in the air. In retribution, the gods cast curses and promises of fire, flood, and drought. These promises later came true when Mesopotamia's growth stripped it of natural resources and degraded the environment.

Ironically, just as we have come to recognize the critical, pervasive importance of plants, we also recognize that we are losing natural plant cover to agriculture, urbanization, overgrazing, pollution, and extinction at a faster rate than ever before. We are finding the truth behind four environmental laws described in 1961 by plant biologist Barry Commoner: (1) Everything is connected to everything else; (2) everything must go somewhere; (3) nature knows best; and (4) there is no such thing as a free lunch.

In the last section of the last chapter of this book, we write that the term *conservation* once meant "wise use"—consumption of a natural resource at a rate that would result in its sustained, continued existence far into the future. But the history of our farms, forests, pastures, and fisheries suggests that technological, growth-oriented human cultures have not been able to determine what that ideal level of resource use is. We have consistently overexploited resources beyond the balance point, degrading the landscape. In addition, human activities interact with soil, water, and air in unexpected ways, resulting in such potential global catastrophes as acid rain, ozone depletion (see sidebar, "Unexpected Links Between CFCs, Climate, and Plants," p. 5), climate change, and forest decline (Fig. 1.5). *Conservation biologists* are now asking how growth-oriented, technological

UNEXPECTED LINKS Among CFCS, CLIMATE, AND PLANTS

In the 1930s, refrigeration became widespread in homes, trains, trucks, and agriculture. The basic component in refrigerators is a gas that contains chlorine or bromine, fluoride, and carbon and that absorbs heat when it is allowed to expand. Freon was an early trade name for such gases, but they are more technically called *chlorofluorocarbons* (CFCs). They are very efficient, nontoxic, and nonreactive. They are also used as foam-blowing agents and as cleansers in computer industries. A related compound, methyl bromide, is commonly used to sterilize soils.

Four decades later, however, scientists began to realize that when CFCs leaked from refrigerators or were emitted into the atmosphere from other industrial sources, they rose high into the stratosphere (25 to 50 km—16 to 31 mi—above the Earth's surface) and destroyed ozone there. Ozone is a form of oxygen molecule made up of three atoms instead of the usual two; it is a natural component of the stratosphere, and it is ecologically important because it absorbs ultraviolet radiation from the sun. Chlorine atoms act as catalysts in this destruction of ozone. Since a catalyst is a molecule that enhances a chemical reaction without being used up itself, this means that a single chlorine atom can destroy thousands of molecules of ozone. The concentration of CFCs has reached a point where the loss of ozone is 2–20%, depending on the latitude and the season.

Ozone depletion has several serious consequences. One is the likely increase of skin cancers, cataracts, and immune deficiency diseases among humans. Cancer is expected to increase by 15% over the next several decades in the heavily populated north temperate zones of the world. Ultraviolet radiation, in particular UV-B, is known to derange plant metabolism. We may expect yields of sensitive crops, such as soybeans, to be depressed by 25%. The impact on natural vegetation is unknown, but teams of scientists that have erected UV-B lamps over grassland and tundra vegetation are studying the radiation's long-term effects (Fig. 1).

The danger of ozone depletion was convincing enough to have led to the most comprehensive international agreement ever established: a protocol signed in Copenhagen in 1992 by nearly all the countries of the world, agreeing to dramatically reduce further production of CFCs and to eliminate all production by 2030. The catalytic nature of CFCs, however, means that their impact on the global environment will linger far into the future, well beyond the year 2030.

Ironically, the elimination of CFCs will itself have a negative effect on agriculture because agribusiness has come to depend on industrial chillers and the use of methyl bromide to sterilize soils and eliminate plant pathogens. Methyl bromide is used extensively in fruit, vegetable, cotton, cocoa, coffee, and grain production. Imaginative and possibly

Figure 1 Researchers at an English grassland, directed by Philip Grime of the University of Sheffield, are currently investigating the ecological impact of 5–20% more UV-B—in addition to higher concentrations of carbon dioxide in the atmosphere, higher soil temperatures, and lower rainfall.

costly substitutes will be needed to replace the CFC-based technology.

The impacts of technology are now global. Even the upper atmosphere is changed. As one environmentalist concluded, this is surely the signal of the death of wild nature. Human influence has become so pervasive that no place on the globe is unaffected, even if that place is far distant from human occupation. We are learning that the first two laws of ecology formulated by Barry Commoner are true: Everything is connected to everything else, and everything must go somewhere.

cultures like ours can coexist with the natural environment while still preserving biological diversity. It is a question worth asking, and a field of study worth encouraging.

1.2 THE SCIENTIFIC METHOD

For most of you, this textbook and the class it accompanies represent the only formal education in plant biology you will acquire in college. You are heading toward careers that, on the surface, have little to do with plants: engineering, medicine, chemistry, climatology, marine science, history, political science. For you, we wanted to write a book that would efficiently and interestingly present a modern survey of plant biology. We also wanted to show you the many interrelationships between plants and your central field of interest—connections that are important but that probably would remain unknown to you without this book.

The linkages among different fields of knowledge may be as important as the narrower information within each field. As each area of expertise expands in its own direction, the gulf between areas widens—and the importance of people who can see connections becomes enormous. The cause of an environmental problem may lie far from the place where symptoms appear. We hope you become one of those who recognizes connections because such links warn of potential ecological disasters (see sidebar).

Scientists in every field—plant biology, physics, behavior, anthropology, geology—have a set of methods in common. They observe, ask questions, make educated guesses about possible answers, base predictions on those guesses, and then devise ways to test their predictions. If a predicted result actually occurs, that is evidence that the possible answer could be correct. More formally, the guesses are called *hypotheses* or *theories*, and the set of procedures is called the *scientific method*.

In science, a theory or hypothesis is never seen as absolute truth. It is merely the best approximate answer, and we can expect it to be modified in the future as evidence from new predictions and tests becomes available. Sometimes, the new information calls for a complete rejection of the original hypothesis and the creation of a new one. This is the scientific equivalent of a revolution.

The external world, not internal conviction, should be the testing ground for scientific beliefs. We often assume that the scientific method is completely objective and free of personal bias on the part of any scientist who practices it. But this is not always so. The questions asked, the answers imagined, and the tests used to check predictions are all products of the culture that surrounds the scientist.

In addition, one's beliefs may subconsciously influence one's science. Historian Greg Mitmann studied a group of biologists at the University of Chicago whose ideas about ecology were very powerful during the middle of the 20th century. In essence, their research showed that nature was organized around cooperative behavior and community. Mitmann pointed out that these scientists were socialists and pacifists in their private lives. Did their personal beliefs lead them to interpret their scientific findings in a certain light? Today, by contrast, the Chicago School's ideas have been turned upside down: Most biologists now hypothesize that nature is organized around competition and individuality.

The scientific method was codified and encouraged in the 17th century by René Descartes and Sir Francis Bacon, among others. It has proven to be a very powerful method for advancing human understanding. At the same time, it tends to separate humans from nature. We can become isolated observers who impersonally manipulate nature in order to test assumptions. A scientific outlook is only one of many that exist among the array of cultures in our world today, and other outlooks may lead humans to tread more softly on the Earth. Carolyn Merchant, a professor at the University of California, Berkeley, has argued strongly that the scientific method is a peculiarly masculine invention that has fostered exploitation of natural resources. The ever-giving Earth Mother image has been replaced with one of an inanimate globe whose treasures must be forcibly extracted. Merchant writes that our scientific subjugation of nature has paralleled a time of subjugation of women, and she implies that one has caused the other.

Technology is the application of information to industrial or commercial objectives. It does not require an understanding of how or why a given process functions. Science and technology are ultimately linked; but they are not the same thing, and they each can progress at independent rates in different directions. For example, consider the development of antibiotics, the wonder drugs of the mid-20th century. Penicillin, the first antibiotic, was widely and effectively prescribed by doctors for the cure of many bacterial infections beginning in the late 1940s. A scientific understanding of how the drug actually worked, however, was not attained until the 1960s. Only then did researchers discover that penicillin interferes with how a bacterial cell builds a wall around itself—and that if the wall is absent the cell contents burst open, and the cell dies. Another example of technology leading science was the development of plant-killing chemicals (herbicides) called *auxins* in the 1940s. To this day, plant biologists do not understand exactly how auxins derange plant growth and lead to death, yet the pragmatic use of auxin-based herbicides continues. Throughout this book, and especially in the sidebars, we will link the science of plant biology with important applications, trying to do justice to both areas.

We now turn to a description of the specific disciplines that constitute the modern field of plant biology.

1.3 STUDYING PLANTS FROM DIFFERENT PERSPECTIVES

Any plant that we see is a product of two interacting components: the genetic (inherited) material, which every cell in that plant carries, plus the environment in which that plant grew (Fig. 1.6). If the genetic potential of the plant is to reach a height of 2 meters (m)—6 feet (ft)—it will attain that height only if the environment is moderate enough. If certain nutrients are lacking, or if shade is too deep, or if nighttime temperatures are too cold, or if too many animals browse the plant, then it will not reach 2 m in height, no matter how much time goes by.

Two major disciplines within plant biology deal with these two components: (1) *Plant genetics* (including evolution and systematics) is the study of plant heredity, and (2) *plant ecology* is the study of how the environment affects plant organisms. Knowledge of genetics is a prerequisite for the study of *plant evolution* and classification (*plant taxonomy* or *plant systematics*). Similarly, knowledge of ecology is essential for reconstructing past climates and landscapes and for understanding how plants have come to be distributed around the world as they are (the disciplines of *paleoecology* and *biogeography*).

One of the many ways these two components—genes and environment—interrelate is through metabolism. *Metabolism* is the process by which plants perform photosynthesis, transport materials internally, construct unique molecules, and use hormones to affect their behavior. The study of metabolism is the discipline of *plant physiology*.

The combination of genes, environment, and metabolism work together to produce an individual plant (Fig. 1.6). The plant can then be studied in several basic ways. The study of how a plant develops from a single cell into diverse tissues and organs and an array of outer surfaces and shapes is the discipline of *plant morphology*. The study of a plant's internal structure—the diversity of cell types and the structure of the cells themselves—is called *plant anatomy* or *cytology*. Plants can also be studied according to their classification: *microbiology* is the study of bacteria, *phycology* is the study of algae, *mycology* is the study of fungi, *bryology* is the study of mosses, and so on (Fig. 1.6).

We've chosen to write about each of these disciplines in a certain order. We start, in Chapters 2–15, with a combination of physiology, morphology, and anatomy. How do the billions of cells, dozens of tissues, and several major plant parts cooperate to achieve a complex, functioning organism? Why do roots grow downward but stems upward? How is it possible for leaves and flowers to track the sun's path during the day, maximizing the amount of light they receive? How can a species regulate where its seeds germinate, increasing the chances of survival of offspring? How does a plant repair injuries? How

Figure 1.6 Interrelationships among several plant biology disciplines.

does a plant tell time and season, so that its many life cycle events occur at the appropriate times? How do plants acquire and transport energy, carbohydrates, water, and nutrients? How do plants solve the problems of reproduction, drought, competition, and a changing environment—despite being rooted to a single location?

In chapters 16 and 17 we describe the rapidly growing science of plant genetics and its technological implications for breeding new crops with increased yields (resulting from increased tolerances to diseases and weeds). Can our understanding of genetics reach a point where we are capable of designing and engineering new species by moving bits of genetic material around—or by creating the genetic material ourselves? Can we create genetic bank accounts for all rare and endangered species, as insurance against extinction? There is a great deal of promise in these developments, but our exploitation of them has been rather limited so far.

In Chapters 18–25 we focus on the taxonomic diversity of plants and plantlike organisms, highlighting the unique way each group's morphology, anatomy, physiology, and reproduction are suited to its particular range of habitats. We also try to reconstruct the evolutionary history of each group: To what is it most closely related? When did it begin to evolve separately? Is the group still actively evolving, or has it stagnated in some way?

Today, most biologists adopt a five-kingdom classification that was first developed in detail by the plant ecologist Robert Whittaker in 1969. The names of the five kingdoms are Monera, Fungi, Protista, Plantae, and Animalia. In our book, all but the last kingdom are surveyed sequentially and compared to each other.

The Kingdom Monera Includes Photosynthetic Bacteria

Organisms in the kingdom Monera have *prokaryotic* cells. Such cells lack an organized nucleus with a bounding membrane. They have no specialized compartments (organelles) within the cell sap (cytoplasm), their cell volume is very small, and the genetic material (DNA) in their chromosome is circular rather than linear. Sexual reproduction is unknown, although these organisms do exchange bits of genetic material occasionally across cellular bridges.

Bacteria, which exist as single cells or as filaments of cells, are in this kingdom (see Fig. 1.3). The earliest forms of life, as revealed by fossils 3 to 4 billion years old, apparently were bacteria. Bacteria are microscopic, but collectively their numbers and weight (biomass) are astronomical. They occur in some of the most extreme environments on Earth, such as in hot springs, in oxygen-deficient lakes, deep in soil, on the surface of snow and ice, and floating high in the atmosphere. Bacteria are enormously beneficial as *decomposer* organisms, able to break down complex organic material into simpler nutrient molecules that can then be taken up by other plants. Bacteria also have detrimental effects as *pathogens* (disease-causing agents). Human diseases caused by bacteria include botulism, bubonic plague, cholera, diphtheria, gonorrhea, leprosy, meningitis, syphilis, tetanus, tuberculosis, and typhoid fever. Other bacteria cause diseases of plants.

One group of bacteria—called *cyanobacteria* or blue-green algae—has the pigment *chlorophyll* in its cells; hence it is green and capable of conducting photosynthesis. Another characteristic that cyanobacteria share with plants is that their cells are surrounded by a rigid wall.

Many cyanobacteria species also have the unique capacity to incorporate inorganic nitrogen gas (N_2) from the atmosphere into organic nitrogen in the form of ammonium ion (NH_4^+). They reproduce with spores.

The Kingdom Fungi Contains Decomposers and Pathogens

Fungi are nonphotosynthetic organisms with *eukaryotic* cells; that is, their cells have a membrane-bounded nucleus, special organelles within the cytoplasm, linear DNA, and much larger cell volumes than prokaryotes. Most of them are known to reproduce sexually. Fungi include molds, mildews, mushrooms, rusts, and smuts (see Fig. 1.3). They are typically microscopic and filamentous, and their cells are surrounded by a rigid wall made of *chitin*—a substance more commonly found in certain animals than in plants or plantlike organisms. They reproduce sexually in a variety of complex life cycles, and they reproduce asexually with spores.

Fungi are widely distributed throughout the world, but they are mainly terrestrial. They are ecologically important because many of them are decomposers, like the bacteria. Others form intimate associations with the roots of plants, improving the absorbing capacity of the plant's root system. Some fungi, such as mushrooms and morels, are important foods for animals and humans. The decomposing action of yeast fungi creates flavored cheeses, leavened bread, and alcoholic drinks. Fungi such as *Penicillium* have been used commercially to produce antibiotic drugs. Other fungi are pathogens, invading the tissue of animals or plants, causing illness, and reducing crop yields by billions of dollars annually.

The Kingdom Protista Includes Grasses of the Sea

Protists are simple, often microscopic, and usually aquatic organisms. Their cells are eukaryotic, and they reproduce both sexually and asexually. This kingdom contains some photosynthetic organisms, called *algae*, and some nonphotosynthetic organisms, called *slime molds*, *foraminiferans*, and *protozoans*. It is a catchall kingdom, containing such widely divergent groups as the ancestral forms of members of the plant, animal, and fungal kingdoms. As biologists learn more, protists will no doubt be classified in a different way in the near future.

Algae occur as single cells or as clusters, filaments, sheets, or three-dimensional packets of cells (see Fig. 1.3). Every cell in most multicellular algal bodies can carry out photosynthesis and obtain water and nutrients directly from its liquid environment. Algae are important because great numbers of them float in the uppermost layers of all oceans and lakes. There they are grazed upon by small animals, and these in turn provide food for larger fish and ultimately for humans. These microscopic algae are called *phytoplankton* and are commonly referred to as *grasses of the sea* because they form the base of vast natural *food chains*, which transfer energy to all organisms in an ecosystem. They also produce half of all the oxygen in the atmosphere as a by-product of photosynthesis.

The Kingdom Plantae Contains Complex Plants Adapted to Life on Land

The kingdom Plantae contains mosses, ferns, pine trees, oak trees, shrubs, vines, grasses, and broadleaved herbs (see Fig. 1.2). Organisms in this kingdom are adapted to life on land, and they are among the most recent forms of life to evolve and appear in the fossil record. They are large and abundant and they give the landscapes of our world their characteristic appearance, making up most of the biomass of forests, meadows, shrublands, deserts, marshes, woodlands, and grasslands. Members of the kingdom Plantae also share certain unique biochemical traits. They have eukaryotic cells with walls made of cellulose, they accumulate *starch* as a carbohydrate storage product, and they have special types of chlorophylls and other pigments. Only a few protists have these same traits.

Plants have more complex bodies than monerans, fungi, or bacteria. This complexity is visible as differences from cell to cell and from region to region within the body. Some cells and tissues are specialized to transport fluids, to store reserves, to perform photosynthesis, or to add strength. Regions of the plant form such unlike structures as leaves, stems, roots, flowers, and seeds.

Plants have major ecological and economic importance. They form the base of terrestrial food chains, they are the principal human crops, and they provide building materials, clothing, cordage, medicines, and beverages. Virtually all of our modern terrestrial ecosystems are dependent upon organisms in the kingdom Plantae.

1.4 A CHALLENGE FOR THE 21ST CENTURY

The last two chapters, 26 and 27, focus on the discipline of ecology. They take us through past and modern landscapes, describing the distribution of plants and their elegant solutions to environmental stresses. How do plants meet the challenges of a changing environment for continued existence? They must partition time and energy, within an individual life span, in such a way that their kind continues for another generation. What can plants tell us by their presence, vigor, or abundance about the past, present, and future of their habitat? Can the presence of certain plants predict the success or failure of our hopes for managing the land? If, as Barry Commoner says, "Nature knows best," can we ever understand the environment well enough to permit us to restore endangered vegetation and degraded landscapes?

Can we continue to increase our human population while retaining natural biological diversity and developing a sustainable use of the world's forests, grasslands, and cropland? As a species, we have not yet been successful in achieving this objective. It will remain as a major challenge into the 21st century, a century in which we expect readers like you to contribute new insights for books yet to be written and wise actions yet to be taken. We wish you every success.

SUMMARY

1. Plant biology is the study of plants, organisms formally classified in the kingdom Plantae, and of an equal number of plantlike organisms formally classified in the kingdoms Monera, Fungi, and Protista.

2. These half-million organisms share at least some of the following traits: They are photosynthetic, nonmotile, and have cells surrounded by a rigid wall; they reproduce asexually by spores; and they have relatively simple unicellular or multicellular bodies lacking obvious digestive, circulatory, nervous, and skeletal systems.

3. Plants and plantlike organisms not only pervade nearly every habitat on earth, they are of central importance to every ecosystem and to human populations.

4. Plants and plantlike organisms are at the base of natural food chains. They produce oxygen, incorporate nitrogen, and stabilize land surfaces. They provide food, fabric, pharmaceuticals, and structural products for humans.

5. Plant biologists use the scientific method to test hypotheses about plant behavior. They study plant form and development (morphology), plant metabolism (physiology), plant genetics and evolution, plant structure (anatomy and cytology), and plant ecology.

6. Plantlike organisms include cyanobacteria in the kingdom Monera. These organisms are prokaryotic (lack a cell nucleus).

7. Other plantlike organisms are molds in the kingdom Fungi. Although they do possess rigid cell walls and reproduce with spores—as do plants—they are nonphotosynthetic.

8. Bacteria and molds are ecologically important as decomposers. They also are important pathogens (disease-causing agents).

9. Algae are simple, generally aquatic plantlike organisms in the kingdom Protista. Some algae share important biochemical traits with true plants, but their bodies lack the complex adaptations to life on land exhibited by members of the kingdom Plantae.

10. A major challenge of the 21st century will be to achieve sustainable use of our plant natural resources in the face of increasing human population size and increasing pollution from human societies.

Questions

1. Why does the color green in nature typically signify the presence of plants?

2. Give some examples of plants in the kingdom Plantae and of plantlike organisms in each of the kingdoms Monera, Protista, and Fungi. Is there any single trait that all half-million species share?

3. List at least one way in which plants and plantlike organisms are ecologically important to the world's ecosystems. List at least one way in which they are economically important to humans.

4. What is the difference between science and technology? How does the story of our use of CFCs illustrate the difference?

5. What is the subject matter of the disciplines of genetics, ecology, physiology, morphology, and anatomy?

6. In what way are bacteria in the kingdom Monera different from organisms in all other kingdoms? How are molds in the kingdom Fungi and algae in the kingdom Protista different from organisms in the kingdom Plantae?

7. Why will sustainable use of natural resources be an important challenge for humans in the 21st century?

Further Readings

Barbour, Michael G., Jack H. Burk, and Wanna D. Pitts. 1987. *Terrestrial Plant Ecology.* 2d ed. Menlo Park, Calif.: Benjamin/Cummings. An unusually readable college textbook that concludes with a chapter summarizing the major vegetation types of North America.

Carey, Stephen S. 1994. *A Beginner's Guide to the Scientific Method.* Belmont, Calif.: Wadsworth. Excellent book for students and instructors, written by a philosophy instructor.

Carson, Rachel L. 1962. *Silent Spring.* New York: Houghton Mifflin. This slender book—304 pages in paperback—was the midwife to the ecological movement. Carson was given the Conservationist of the Year award by the National Wildlife Federation, and the then Secretary of Interior Stewart Udall said, "A great woman has awakened the nation by her account of the dangers around us. We owe much to Rachel Carson."

Commoner, Barry. 1975. *Making Peace with the Planet.* New York: Pantheon Books. A pioneering book, written for a lay audience, about environmental problems caused by our technological society and possible cures for them. It includes Dr. Commoner's famous four laws of ecology.

Fiedler, Peggy L., and Subodh K. Jain, eds. 1992. *Conservation Biology.* New York: Chapman and Hall. A sampling of modern thinking about how biologists can study, measure, and then work to maintain biological diversity, especially in threatened habitats.

Makofske, William J., and E. F. Karlin, eds. 1995. *Technology and Global Environmental Issues.* New York: HarperCollins. A fine summary of current major environmental issues. Concise but readable.

Merchant, Carolyn. 1980. *The Death of Nature: Women, Ecology, and the Scientific Revolution.* New York: HarperCollins. A feminist history of science and the consequences of a scientific viewpoint for human interactions with nature.

Mitmann, Greg. 1992. *The State of Nature: Ecology, Community, and American Social Thought, 1900–1950.* Chicago: University of Chicago Press. The author describes the rise and fall of a school of animal ecology that dominated scientific thought for several decades in the mid-20th century. He shows how the political and social beliefs of several ecologists probably colored their scientific research and the conclusions they reached (and preached) from it.

Perlin, John. 1989. *A Forest Journey.* Cambridge, Mass.: Harvard University Press. A fascinating history of the exploitation of forests and the central value of wood to the rise and fall of Western civilizations, over the past 1,000 years.

Whittaker, Robert H. 1969. "New Concepts of Kingdoms of Organisms." *Science* 163: 150–60. This was the first time Dr. Whittaker proposed his five-kingdom classification scheme.

2 THE CHEMISTRY OF LIFE

1. Living organisms are made from chemical compounds, and everything they do must obey all the laws of chemistry and physics. In order to understand organisms, one must have a basic appreciation of the principles of chemistry.

2. Most of the substances in organisms are based on carbon-containing compounds. Carbon atoms can form complex chains held together by stable covalent bonds. The large number of possible structures provides a basis for the complex activities of organisms.

3. The chemical properties of water make it uniquely suited as a milieu for living beings.

4. The functional chemical units of living organisms—proteins, carbohydrates, lipids, and nucleic acids—are large polymeric molecules, formed from simpler monomers: amino acids, sugars, fatty acids, and nucleotides. The information that directs the assembly of the polymers is itself encoded in the structure of the nucleic acid polymers.

2.1   CHEMISTRY AND PLANTS

There are many ways to study plants. Poets and painters, anthropologists and agriculturalists all have their own ideas about the importance and usefulness of the Earth's flora (see sidebar, "Plants as Pharmacists," p. 13). In recent years, biologists have made notable strides in understanding how plants function. Much of this understanding has come from deducing the chemical basis of life, and particularly of plant life. To fathom the most important discoveries about the life of plants, one must grasp the basic concepts of chemistry. What are plants made of? Why are the stems of some plants (like grasses) soft, whereas those of others (like trees) are hard? What are the substances that plants need to grow? How is it possible to manipulate the characteristics of plants through biotechnology? The answers to all these questions are based in chemistry. This chapter provides an overview of the concepts needed to understand plant cell structure and function as discussed in later chapters.

2.2   THE UNITS OF MATTER

Every type of matter—including all the components of living cells and organisms and all the nonliving materials on which living things depend—is built from very small units. **Molecules** are the smallest particles that retain the chemical characteristics of their type of matter. There are many thousands of kinds of molecules, which differ in their sizes, shapes, and behaviors. The largest molecules in a plant cell carry hereditary information; they are slender threads over 1 millimeter (mm) long when extended. Molecules of table sugar (sucrose) are closer to the average size; about 3 million molecules of sucrose placed end

hydrogen   carbon

Figure 2.1   The basic structures of the hydrogen and carbon atoms. Hydrogen has one proton in its nucleus, surrounded by one orbital with one electron. Carbon has six protons and six neutrons in its nucleus, surrounded by one inner orbital with two electrons and four outer orbitals with one electron each.

to end would span the printed word *cube*. A plant builds all of its own molecules by rearranging the parts of simpler molecules taken from the environment.

Molecules Are Made of Atoms

A single molecule is formed from component parts arranged in specific positions relative to one another. These parts are the **atoms** (Fig. 2.1), and they in turn are formed from three types of particles: **protons**, each of which has one unit of mass and one unit of positive electrical charge; **neutrons**, which have one unit of mass but no electrical charge; and **electrons**, which have 0.0005 units of mass and one unit of negative electrical charge. The protons and neutrons form the atomic **nucleus**, a small kernel at the center of an atom. If an atom were magnified to the size of a house, the nucleus would be about as large as a pinhead. Electrons move around the nucleus in **orbitals**, ill-defined regions in the space outside the nucleus. The size of the atom—as measured by how many can be packed into a given volume—is determined by the distance of the outer electrons from the nucleus. The weight of the atom is determined mainly by the number of protons and neutrons, since each is about 1840 times heavier than an electron.

The number of protons in the nucleus determines the chemical characteristics of an atom and its identity. Matter made from only one type of atom is called an **element**. Some of the most prominent elements in living organisms are listed in Table 2.1, along with the letter symbols that chemists use to indicate their presence in a molecule.

Although neutrons affect the weight of an atom, they influence its chemical characteristics only slightly. It is not uncommon in nature for an element to consist of atoms that differ in their number of neutrons; such atoms are called **isotopes** of the element. For instance, the atoms of the element carbon may have 6, 7, or 8 neutrons. Because every atom of carbon has 6 protons—and because neutrons weigh as much as protons—the relative weights of these three isotopes are 12, 13, and 14; so these isotopes are identified as ^{12}C, ^{13}C, and ^{14}C. The balance in the number of neutrons and protons seems to be a factor in holding the nucleus together. Nuclei with about the same number of neutrons and protons tend to be stable.

PLANTS AS PHARMACISTS

Plants are superb chemists. They synthesize a great variety of chemicals beyond those needed to perform the basic functions of cells. For many years these chemicals were known as *secondary compounds* because they were not needed for the functioning of all cells. The term suggested that the compounds had no significance in the life of the plant. Now it is recognized that these compounds often play important roles in the interactions between plants and the other organisms with which they associate. For instance, some of the chemicals are attractants, promoting the transfer of pollen from one plant to another of its species or the dispersal of seed. Some of the chemicals are antibiotics or toxins, restricting the ability of pathogens and herbivores to feed off the plant.

The subject of ethnobotany, the use of plant extracts by peoples of many cultures, has recently become popular. An understanding of these uses may help us control disease bacteria that are becoming resistant to present-day antibiotics. They may also help cure conditions like AIDS, for which there is as yet no successful treatment, and give us new ways of alleviating pain, anxiety, or neurological disorders.

Many of the chemicals in plants are used to prevent or to fight cancer. A large class of compounds, called antioxidants, prevents damage that seems to lead to cancer. These compounds include ascorbic acid (vitamin C), alpha-tocopherol (vitamin E), and beta-carotene (which is converted into vitamin A), all of which are accumulated in many fruits and vegetables. These work by reacting with (and thus taking out of circulation) chemicals with unpaired electrons (free radicals) that would otherwise combine with,

oxidize, and inactivate DNA, RNA, proteins, or membrane lipids. They are accumulated in plants to protect the plants' own cells, but they probably do the same job in our bodies. Some plants have special compounds that may perform the same function: Broccoli contains sulforaphane, garlic and onions have allyl sulfides, and teas have catechins. Many of these chemicals have been found to prevent cancer in experimental animals. The evidence that they prevent human cancers is less well established and is a subject of current research.

Chemicals used to fight cancer are generally toxic to cancer cells. Because cancer cells reproduce rapidly, the most effective compounds are those that interfere with cell division. Such compounds include vincristine and vinblastine, which come from the periwinkle plant (*Vinca rosea*), and taxol from yew trees (*Taxus brevifolia*). Psoralens from celery, parsley, and citrus leaves make cells sensitive to ultraviolet light; this treatment allows physicians to kill cells in localized regions by irradiating them with a UV laser. Gossypol (from cottonseed), which has been touted as a potential male contraceptive, also exhibits anticancer activity, possibly by stimulating the formation of free radicals that oxidize membrane lipids in cancer cells.

Even though chemists are becoming more proficient at designing and synthesizing complex molecules, the use of plant-derived chemicals will probably continue and expand. The specificity of biological catalysts in plant cells (enzymes—see Chapter 8) is the reason. There are many three-dimensional molecular structures, formed within plant cells, that are impossible for even the most skilled chemists to duplicate.

Table 2.1 The Twelve Most Common Elements in Living Organisms*		
Name of Element	Symbol	Number of Protons
Hydrogen	H	1
Carbon	C	6
Nitrogen	N	7
Oxygen	O	8
Sodium	Na	11
Magnesium	Mg	12
Phosphorus	P	15
Sulfur	S	16
Chlorine	Cl	17
Potassium	K	19
Calcium	Ca	20
Iron	Fe	26

* Listed in order of size (number of protons).

Nuclei with the number of neutrons much different from the number of protons may be unstable and may spontaneously decompose, a process known as radioactive decay. ^{12}C and ^{13}C are stable, but atoms of ^{14}C are radioactive, giving off energetic electrons—called beta (β) particles—as they decay into a more stable state.

Radioactive atoms tend to be rare, especially among the elements that make up living organisms; so radioactivity is seldom a factor in the chemistry of life. However, high concentrations of radioactive material can damage cells because the high-energy particles of radioactive decay can destroy the complex molecules they hit. On the other hand, low concentrations of radioactive atoms are very useful in biological studies because their decay products are like spotlights, enabling researchers to track the fate of these atoms in chemical reactions and to trace their movements through cells and tissues.

Matter made from atoms of two or more elements is called a **compound**. Chemists indicate the composition of a compound molecule by using the elemental symbols and subscripts to show the kind and number of atoms present. Thus, the symbol CH_4 represents a molecule that has one carbon atom (the subscript 1 is assumed if no number is supplied) and four hydrogen atoms. The five atoms in this molecule are held together by interactions among their electrons and nuclei.

Electrical Forces Attach Electrons to Nuclei

Electrical forces attach the electrons to the nuclei to form atoms and molecules. Each electron carries a unit of negative electrical charge, and each proton carries the same amount of positive electrical charge. Two particles attract one another if they carry opposite charges, which is why electrons are attracted to the nuclei. Particles repel one another if they carry charges of the same sign, so two or more electrons avoid one another as they move around the atom. An atom or molecule that has equal numbers of electrons and protons is said to be electrically neutral because any force that its electrons exert on a distant object is countered by an opposite force exerted by the protons.

Although nuclei take up fairly definite positions in a molecule, electrons are more difficult to locate. The orbitals that they occupy are not fixed lines, like the orbits of planets around the sun. Rather, they are fuzzy areas around the nuclei where a probability exists that an electron might be found at any particular time. In some parts of the orbital, the probability is higher (which means the electron spends more time there); in other parts, the probability is lower (the electron spends less time). There is a general rule that applies to all orbitals: An orbital may have zero, one, or two electrons—no more. Once two electrons have occupied the orbital closest to the nucleus (where the attractive forces are greatest) no other electrons can join them. Other electrons must move to orbitals farther away. An atom or molecule is most stable when it has exactly two electrons in each orbital.

The potential energy of an electron depends on the position of the orbital it occupies. It takes energy to move an electron away from the nucleus against the attractive electrical force. That energy is given up if the electron falls back toward the nucleus. Thus, an electron in an orbital close to the nucleus has less potential energy than one in an orbital farther from the nucleus. Orbitals in which electrons have the same potential energy are said to be in the same **shell**. Electrons can change shells by losing or gaining energy. As an electron drops into an orbital of a shell closer to the nucleus, it gives up energy in the form of a photon (a unit of light or heat radiation), which leaves the atom and can sometimes be detected as a flash. Conversely, for an electron to move into an orbital of a shell farther from the nucleus, it must absorb a photon from the environment. Such changes in electron energy are at the heart of photosynthesis (Chapter 10). An electron can also leave its home orbital and take up residence in the orbital of another atom, or it can move to an orbital shared by its atom and another atom. These movements lead to the formation of ionic and covalent bonds, respectively, and they result in the formation of new molecules.

Ionic Bonds Consist of Electrical Forces

It is possible for atoms and molecules to have more electrons than protons or vice versa. When they do, the atoms or molecules are called **ions**. The excess charge of an ion is denoted with a superscript. For example, K^+ indicates the potassium ion, with one more proton than electrons; SO_4^{2-} indicates the sulfate ion, with two more electrons than protons. Ions with a net positive charge are called **cations**; those with a net negative charge are called **anions**. Ions are formed when an electron moves from one originally neutral atom to another. For instance, the transfer of an electron from a sodium atom to a chlorine atom (Fig. 2.2) results in the formation of a sodium cation (Na^+) and a chlorine anion (called chloride, Cl^-). The loss of an electron is called **oxidation**; the gain of an electron is called **reduction**. In this transfer, the sodium is oxidized and the chlorine is reduced (that is, the chlorine atom's charge is reduced from 0 to -1). Because the oxidation and reduction occur together, they are often referred to as an **oxidation–reduction** (redox) **reaction**.

Ions with the same sign (both positive or both negative) repel each other, while ions with opposite signs attract each other. When two ions of opposite charge are held close together by electrical forces, chemists often say they are joined by an **ionic bond**. A compound containing anions and cations held together by ionic bonds is called a **salt**. As one example, sodium and chloride ions, connected by ionic bonds, form sodium chloride crystals, also known as table salt.

Covalent Bonds Consist of Shared Electrons

The strongest type of chemical bond is a *covalent* bond. Most types of molecules are formed from covalently bonded atoms. While some of the orbitals in such a molecule surround a single nucleus and are said to be **nonbonding** orbitals, other orbitals are distributed between two or more nuclei. These are said to be **bonding orbitals**, and the electrons occupying them are **bonding electrons**, shared by the nuclei. The nuclei that share the bonding electrons are said to be joined by a **covalent bond**. The covalent bond is strong because the electrons in the bonding orbital are in relatively stable positions—that is, their attraction to the nuclei is strong and their mutual repulsion is minimized. It would take a lot of energy to separate the nuclei and move the electrons into different orbitals. Covalent bonds within a molecule can be shown in different ways (Fig. 2.3).

It is important to know how orbitals form around the nuclei of hydrogen (H), carbon (C), nitrogen (N), and oxygen (O) because these make up the bulk of the molecules of the cell. H is the simplest nucleus, consisting of just one proton. The charge on the proton is so weak that

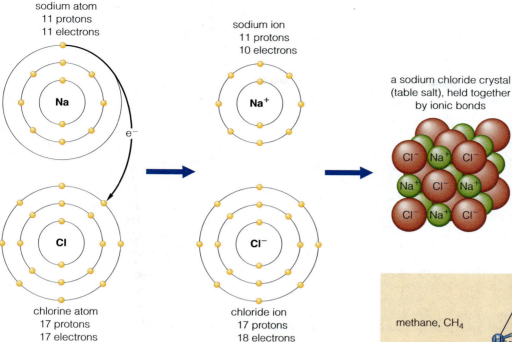

sodium atom
11 protons
11 electrons

sodium ion
11 protons
10 electrons

e^-

Na

Na$^+$

a sodium chloride crystal
(table salt), held together
by ionic bonds

Cl

Cl$^-$

chlorine atom
17 protons
17 electrons

chloride ion
17 protons
18 electrons

Figure 2.2 The oxidation of sodium and reduction of chlorine to form ions. In crystals, ions are held to-gether by the mutual attraction of opposite electrical charges.

Number of bonds to central atom

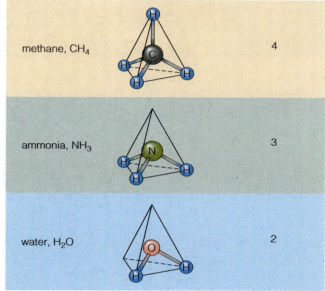

methane, CH$_4$ 4

ammonia, NH$_3$ 3

water, H$_2$O   2

Figure 2.4 The three-dimensional structures of methane, ammonia, and water and their relationship to a tetrahedron.

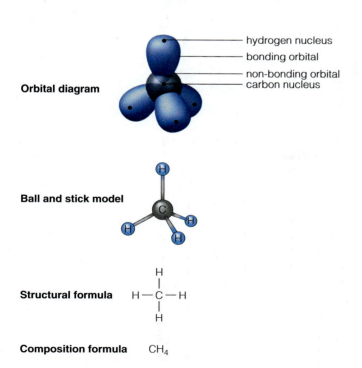

Orbital diagram

hydrogen nucleus
bonding orbital
non-bonding orbital
carbon nucleus

Ball and stick model

Structural formula

$$H - \overset{\overset{\displaystyle H}{|}}{\underset{\underset{\displaystyle H}{|}}{C}} - H$$

Composition formula   CH$_4$

Figure 2.3 Several ways to symbolize a molecule of methane, which is held together by covalent bonds. Methane, a major component of natural gas, is produced when certain bacteria decompose organic matter in the absence of oxygen.

only one orbital forms around the nucleus. C, N, and O nuclei have 6, 7, or 8 protons, respectively, and they exert a much stronger attraction for electrons. Each of these nuclei is the center for five orbitals. The first orbital is in a spherical region very close to the nucleus. The two electrons in this orbital are nonbonding electrons. The four remaining orbitals are oblong and tend to orient themselves so that they are as far apart as possible, because the negative charges of the electrons repel one another. They are said to point to the vertices of a tetrahedron (Fig. 2.4).

When atoms come together to make molecules, bonding orbitals are formed. You can think of them as representing an overlap, or hybrid, of two orbitals from adjacent atoms (although they have a shape of their own). If possible, these orbitals form in such a way that each contains exactly two electrons.

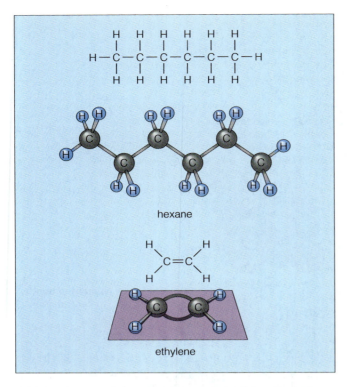

Figure 2.5 Two hydrocarbon compounds: hexane, with single bonds between the carbon atoms, and ethylene, which contains a double bond.

Carbon has six electrons. Two of these are in the inner, nonbonding orbital; four are available for sharing in the outer, bonding orbitals. Thus, all four of the outer orbitals can participate in the formation of bonding orbitals. This is seen most simply with CH_4, or methane. The C nucleus with its inner electrons forms the center of the molecule. Each of the outer, oblong orbitals forms a bonding orbital with an H nucleus embedded in one end and the C nucleus in the other (Fig. 2.4). Each of these orbitals contains two electrons. If we think of an H nucleus as getting a half-share in the electrons of a bonding orbital, then one unit of negative charge is available to balance its unit of positive charge. Similarly, there are six units of negative charge available to balance the six protons in the C nucleus: two in the inner orbital and one from each of the four bonding orbitals (again, a half-share of the two electrons in each orbital). Overall, the molecule is neutral. This illustrates the general rule that each C nucleus forms four covalent bonds. Of course, C can form bonds with other nuclei besides H, such as N and O. Most importantly, it can form bonds with other C nuclei, to produce chains of carbons. In a **hydrocarbon compound**, each C nucleus is bound to one or more other C nuclei and to hydrogens (Fig. 2.5).

Nitrogen has one more proton than carbon and thus has one more electron than carbon. The extra electron fills up one of the outer orbitals, so that each N nucleus can participate in only three bonds. A good example is NH_3, ammonia (Fig. 2.4). Similar considerations apply to the behavior of oxygen, as illustrated by H_2O (water). Oxygen has two more protons than carbon and thus attracts two more electrons. These fill up two of the outer orbitals, so that an O nucleus can form only two covalent bonds with hydrogens.

It is possible for two nuclei to form two or three bonding orbitals and thus to share four or six electrons. The sharing of four electrons is known as a **double bond**. Chemists represent double bonds by two lines: C=C, for example. Ethylene, a very short hydrocarbon chain with two carbons and four hydrogens, contains a double bond (Fig. 2.5). The two C nuclei are connected by two bonding orbitals, one occupying an oval region in the center and the other consisting of two sausage-shaped regions on either side. The double bond resists being twisted, so that the six nuclei of this molecule sit rigidly in a single plane. In longer hydrocarbon chains, occasional double bonds produce kinks in otherwise flexible molecules. In the molecules found in cells, double bonds such as C=C, C=O, and C=N are common.

2.3 THE MILIEU OF LIFE

Water is the environment of life—even for terrestrial organisms—at the cellular and molecular levels. The physical and chemical properties of water are essential to cellular reactions and processes. Perhaps the three most important characteristics of water are its ability to dissolve a great many compounds, to remain a liquid over a wide range of temperatures, and to form weak *hydrogen bonds* with itself and other molecules. These characteristics derive from the arrangement of its atoms.

Water Owes Its Unique Properties to Its Polarity

The amount and distribution of electrical charge in a molecule are very important in predicting its behavior. Some molecules have their electrons spaced evenly throughout the orbitals: these molecules are called **nonpolar**. Other molecules have local regions of positive and/or negative charge; these are said to be **polar**.

Polarity is established within a molecule because electrons that are shared between two unlike nuclei may spend more of their time closer to one nucleus than the other. A measure called **electronegativity** expresses the tendency of a nucleus to attract electrons. Nuclei of oxygen and nitrogen are more electronegative than those of carbon, sulfur, hydrogen, and phosphorus. Because of these differences, a molecule tends to be more negative near N and O nuclei and positive near C, H, P, and S nuclei.

Water is the quintessential example of a polar molecule (Fig. 2.6). The high electronegativity of the O nucleus attracts electrons—not only the electrons in the nonbonding orbitals, but also the electrons in the bonding orbitals. As a result, there is a deficiency of electrons near the

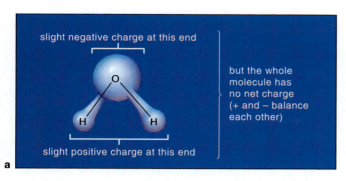

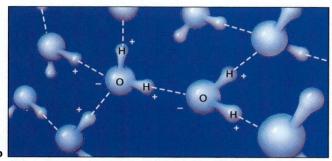

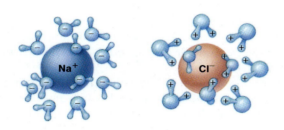

Figure 2.6 Properties of water. The polarity of water, shown in (**a**), is the reason why water can (**b**) form hydrogen bonds and (**c**) dissolve substances, such as the salt ions shown here.

hydrogen nuclei. This means that there is a slight negative charge at the O nucleus and slight positive charges at the two H nuclei. This property of water is extremely important in determining the structures of the molecules that make up a living organism. These molecules—proteins, carbohydrates, lipids, and nucleic acids (introduced later in this chapter)—assume three-dimensional shapes, determined in part by their interaction with water. And their three-dimensional shapes are critical for their functions. This is why all life exists in a watery environment, and although some organisms can survive being dried and rehydrated, no life—so far as we know—can function without water.

A strong polarity in an orbital involving H leads to a new type of bond. The H of one water molecule can be attracted to the electrons around the O of a neighboring water molecule because of their opposite charges. In fact, the H can bounce between the electronegative Os of the two water molecules. This is a stable situation, so it is said that a **hydrogen bond** has formed between the molecules. A hydrogen bond is represented by a dotted line: for example, O···H. Each of the two Hs in a water molecule can participate in a hydrogen bond. Even though this bond is only about 1/16 as strong as a covalent bond,

the large number of hydrogen bonds among water molecules makes water more stable (less volatile) than we might otherwise expect. This means that a water molecule can evaporate (leave the liquid state and move into the gaseous state) only by simultaneously breaking all the hydrogen bonds that connect it to other molecules in the liquid. As a consequence, water can absorb a large amount of heat as it evaporates, and it releases the same amount of heat as it condenses. The absorption of heat removes heat that otherwise would raise the water's temperature; the condensation provides heat to keep the water warm. Thus, evaporation and condensation tend to stabilize the temperature of liquid water. This fact is important to living systems, which are damaged by large swings in temperature.

Other molecules besides water can form hydrogen bonds. This happens when a hydrogen nucleus that is sharing electrons with O or N comes close to another O or N, with the three nuclei approximately in a straight line, like this: O—H···N. Many biological molecules that contain O—H or N—H groups participate in hydrogen bonds with water by donating their H to the bond; other molecules with groups like C=O participate by attracting an H from a water molecule. Biological molecules may also form hydrogen bonds with each other.

The polarity of water and its ability to form hydrogen bonds make it an excellent **solvent** for many other substances. A pure substance in the solid form consists of many molecules of the same kind packed regularly (a crystal) or irregularly (an amorphous solid or glass) and held together with chemical (but not covalent) bonds. On contact with water, the molecules at the surface of the solid can leave to become surrounded by and bound to water molecules. The mass is said to **dissolve**; the molecules that become intermixed in the water are the **solute** molecules. This homogeneous mixture is called a **solution**. The more hydrogen bonds that can form between the solute and the water, the more easily soluble is the solute. Polar, nonionic molecules, like ammonia (NH_3, see Fig. 2.4), are very soluble for that reason. Ions are also soluble, but for a different reason: Water molecules surround each ion with their oppositely charged poles pointing at the ion (Fig. 2.6c). Most biological molecules either form hydrogen bonds with water or are ionic—and thus dissolve in water. This allows the molecules to move, mix, and react chemically with one another; it is one of the principal reasons why all life exists in a water solution.

Nonpolar molecules such as hydrocarbons (in which the carbon and hydrogen nuclei have similar electronegativities and the electrons are evenly distributed) cannot form hydrogen bonds with water or anything else. If tossed into water, these molecules force the water to form a sort of cage around them—an energetically unfavorable (unstable) situation. Over time, the nonpolar molecules will tend to move together and stick together to minimize the amount of water that is used to form cages: this ten-

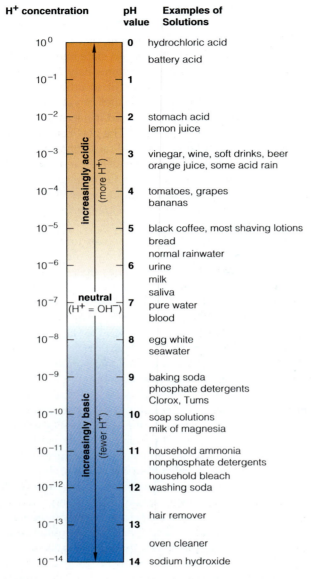

H$^+$ concentration	pH value	Examples of Solutions
10^0	0	hydrochloric acid battery acid
10^{-1}	1	
10^{-2}	2	stomach acid lemon juice
10^{-3}	3	vinegar, wine, soft drinks, beer orange juice, some acid rain
10^{-4}	4	tomatoes, grapes bananas
10^{-5}	5	black coffee, most shaving lotions bread normal rainwater
10^{-6}	6	urine milk saliva
10^{-7}	7	pure water blood
10^{-8}	8	egg white seawater
10^{-9}	9	baking soda phosphate detergents Clorox, Tums
10^{-10}	10	soap solutions milk of magnesia
10^{-11}	11	household ammonia nonphosphate detergents household bleach
10^{-12}	12	washing soda
10^{-13}	13	hair remover oven cleaner
10^{-14}	14	sodium hydroxide

increasingly acidic (more H$^+$)

neutral (H$^+$ = OH$^-$)

increasingly basic (fewer H$^+$)

Figure 2.7 The pH scale. Acidic solutions (pH < 7) are associated with a sour taste. Basic solutions (pH > 7) are often bitter and have a slippery feel, rather like soap.

dency is called a **hydrophobic bond** (because the nonpolar molecules in sticking together act as though they had a fear of or an aversion to water). In contrast, nonpolar solutes will dissolve in nonpolar solvents (like oil or kerosene), but polar solutes will not. In general, we can predict that polar or **hydrophilic** molecules will dissolve in polar solvents (water), and nonpolar or **hydrophobic** molecules will dissolve in nonpolar solvents (oil).

Acids Donate—and Bases Accept— Hydrogen Nuclei

In water solutions, both within and outside living cells, there is a rapid exchange of hydrogen nuclei among solutes and the water solvent. Molecules that contain an

H nucleus bonded to a strongly electronegative atom (the same molecules that might donate an H to a hydrogen bond) can lose the H nucleus entirely. The H is not really lost, of course; in general it is transferred to a molecule of the surrounding solvent (water) forming a *hydronium* ion. The original molecule retains the two electrons in the bonding orbital and thus acquires an extra unit of negative charge. Molecules that donate an H nucleus are called **acids**. Molecules that accept an H nucleus are called **bases**. For instance:

$$H-Cl \text{ (hydrochloric acid)} + H-O-H \Leftrightarrow H-\underset{\underset{H}{|}}{O}-H^+ \text{ (hydronium)} + Cl^- \text{ (chloride)}$$

(acid)   (base)   (acid)   (base)

Water can be both an acid and a base:

$$H-O-H + H-O-H \Leftrightarrow H-\underset{\underset{H}{|}}{O}-H^+ \text{ (hydronium)} + O-H^- \text{ (hydroxyl)}$$

(acid)   (base)   (acid)   (base)

These are reversible reactions: the hydronium and chloride can react together to form hydrochloric acid and water, and the hydronium and hydroxyl can react to form two molecules of water. In these examples (considering the reactions in both directions), hydrochloric acid, water, and hydronium are the acids, and chloride, hydroxyl, and water are the bases.

The concentration of hydronium ions determines the acidity of the solution. More hydronium means a more acidic solution. Chemists express the hydronium concentration by a measure known as pH (Fig. 2.7). A change of 1 unit on the pH scale is a 10-fold change in hydronium concentration. The lower the pH, the higher the hydronium concentration. So, a solution with pH 4 has 10 times as much hydronium as a solution with pH 5. Acids and bases can be defined as substances that lower and raise, respectively, the pH of a solution, and solutions can be classified as acidic or basic according to their pH. The acidic juice squeezed from a lemon has a pH of about 2 to 3. Pure water, which is neither acidic nor basic, has a pH of 7. Household bleach, a basic solution, has a pH of about 12.

Instead of talking about hydronium, chemists often abbreviate the concept by referring to the H$^+$ ion itself. From this viewpoint, pH refers to the H$^+$ concentration, although free H$^+$ never occurs in water solution. Most of the figures in this book will indicate hydronium by the symbol H$^+$.

THE SUBSTANCE OF LIFE

Living Organisms Are Made of Chemicals

All plants (and other organisms) are made of chemical compounds. These compounds influence the shapes that plant cells and their subunits take and determine their functions. Compounds that contain primarily carbon are

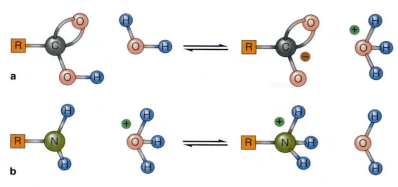

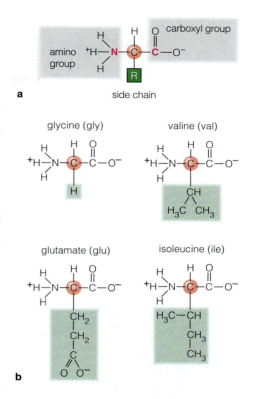

a

b

Figure 2.8 Acid–base relations in functional groups. (**a**) The release of an H+ nucleus by the acidic group, R—COOH. Note that the functional group ends up with a negative charge. (**b**) In accepting an H+ nucleus, the basic amino group, R—NH$_2$, acquires a positive charge.

Figure 2.9 Amino acid structure. (**a**) The general formula for amino acids. (**b**) Four of the 20 possible amino acids found in proteins. Notice that these molecules differ only in their side chains.

called **organic** compounds, because they were first associated with living organisms. As the techniques of laboratory carbon chemistry have advanced, it has become useful to distinguish the molecules that actually occur in cells—we often refer to these as **bio-organic**. Bio-organic molecules are based on a carbon skeleton and generally include oxygen and hydrogen. They may often contain nitrogen, phosphorus, and/or sulfur in their structures. Other elements that may be associated with bio-organic molecules (generally attached with ionic bonds) include iron, calcium, potassium, and magnesium—and less frequently sodium, boron, zinc, manganese, molybdenum, chlorine, and copper.

Biological molecules are often large and complex, but they are easier to understand when we use the concept of **functional groups**. A functional group is a small part of a larger molecule, but one that can participate in chemical reactions (Fig. 2.8). Some functional groups are acids and bases. For instance, the functional groups carboxyl (R−COOH) and phosphoryl (R−OPO$_3$H$_2$) are acids and can donate H+ nuclei to water ("R" stands for the remainder of the molecule and is not part of the functional group). The amino group (R−NH$_2$) acts as a base and accepts H+ nuclei from hydronium. Other functional groups may participate in hydrogen bonds, hydrophobic bonds, or oxidation–reduction reactions.

Most of the organic compounds found in living systems can be classified into four families: proteins, carbohydrates, lipids, and nucleic acids. Together these molecules speed up biological reactions, act as structural and fuel molecules, and serve as libraries of genetic information. The following sections discuss these compounds.

Proteins Have Diverse Shapes and Functions

Proteins are large molecules formed by stringing together between one hundred and several hundred **amino acids** into a long, unbranched chain. The amino acids are the **monomers** (individual units) that form the protein **polymer** (multiunit molecule). Each amino acid has a backbone containing an amino group, a central carbon, and a carboxyl (carboxylic acid) group (Fig. 2.9). The amino and carboxylic acid groups give the amino acid its name. Twenty different types of amino acids are found in proteins. They all have the same backbone, but each has a different side chain attached to the central carbon.

The amino acids can be linked together at their backbones: an amino group of one amino acid attached to the carboxyl group of a second amino acid, the amino group of the second amino acid attached to the carboxyl group of a third amino acid, and so on (Fig. 2.10). The bonds linking the amino acids are called **peptide** bonds, so the chain is called a **polypeptide** chain. Any one of the 20 different types of amino acids might be placed in any of the positions of the chain. Thus, there are 20^{100} different possible types of chains that are 100 amino acids long—and even more possibilities with longer chains. Every particular type of protein has its own specific arrangement of amino acids. This arrangement promotes the folding of the protein into a very specific three-dimensional shape, with different parts of the protein held together with ionic, hydrogen, and hydrophobic bonds between the side chains of the different amino acids (Fig. 2.11).

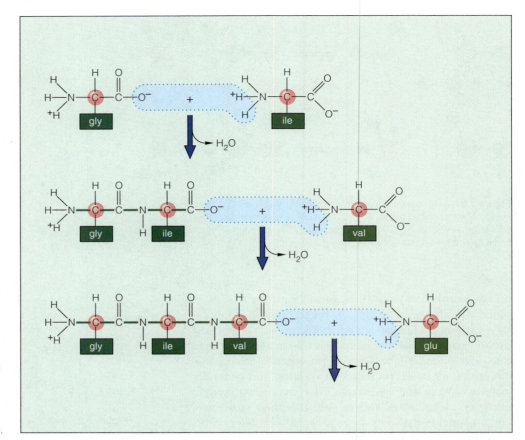

Figure 2.10 Peptide bonding in protein formation. Different amino acids can polymerize via peptide bonds to form a polypeptide chain. The length of the chain and the order of the amino acids can vary according to the type of protein.

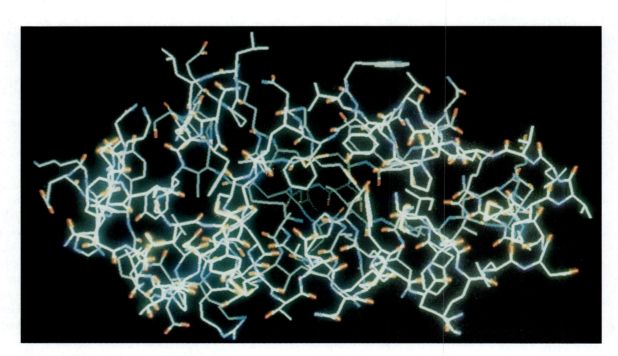

Figure 2.11 A model of a protein's structure. In this computer-generated picture, the lines represent covalent bonds; there are atoms at the angles and ends of the lines. Different colors mean different atoms: For instance, green represents carbon; blue, nitrogen; and red, oxygen. The three-dimensional organization of the atoms is specific to this type of protein.

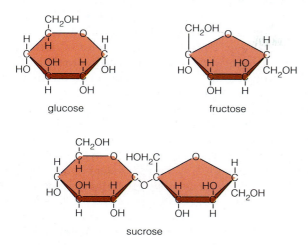

Figure 2.12 Structural formulas of two simple sugars, glucose and fructose (the most common sugar monomers in plants), and of the oligosaccharide sucrose (also known as common table sugar).

The functional properties of the protein depend on its shape. Individual protein molecules may associate with each other or with other types of molecules to form larger structural complexes that give form to the cell, direct movement within the cell, or provide a scaffold for chemical reactions. All **enzymes**, which catalyze the chemical reactions of the cell (see Chapter 8 for a discussion of catalysis), are proteins. Proteins are high-energy compounds (that is, free energy for running chemical reactions can be recovered when protein bonds are broken—again, see Chapter 8 for an explanation), so some proteins may be used for storage of energy (in seeds, for example).

Heating a solution of protein to a temperature of 50°C or more breaks the relatively weak ionic, hydrogen, and hydrophobic bonds between the side chains and un-wraps the protein; this changes its three-dimensional shape, even though it does not alter the order of the amino acids in the polypeptide chain. Protein molecules that have been heated or otherwise treated to change their three-dimensional shape lose their function. These protein molecules are said to be **denatured**. Denatured proteins often form large masses and become insoluble. The proteins in a cooked egg white are a good example.

Cells depend on functioning proteins to live. The function of a protein depends on its three-dimensional shape. And a protein's shape depends on the arrangement of amino acids in its polypeptide chain. Thus the information that guides the arrangement of amino acids into a polypeptide chain is a key element of life. This information is carried by other biological molecules, as is explained in the section on nucleic acids (p. **22**).

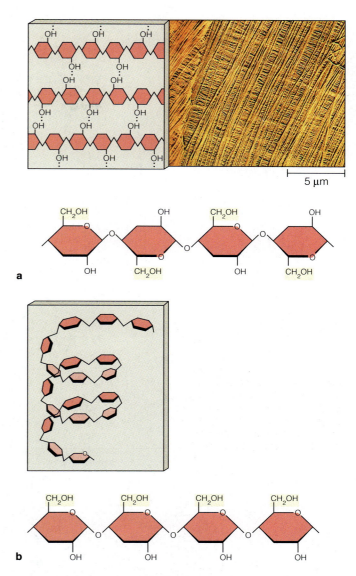

Figure 2.13 Models of the polysaccharides (**a**) cellulose and (**b**) starch. In both molecules, the monomer is the sugar glucose. The two molecules differ in the ways the adjacent glucose monomers are connected. This allows the cellulose molecule to form straight chains, whereas the starch molecule forms helices in solution.

Carbohydrates Include Sugars and Polysaccharides

Carbohydrates include simple sugars, oligosaccharides, and polysaccharides. Simple sugars are formed from carbon atoms with associated oxygens and hydrogens in the proportions $C_nH_{2n}O_n$ (n = 3 to 7). The carbons in a simple sugar are often arranged in a ring, with each carbon bonded to two other carbons, one H and one O (Fig. 2.12). **Oligosaccharides** are small chains of two or more simple sugars. **Polysaccharides** are long chains of simple sugars (Fig. 2.13). Some polysaccharides—for instance, **cellulose** and **pectin**—contribute to structure, especially

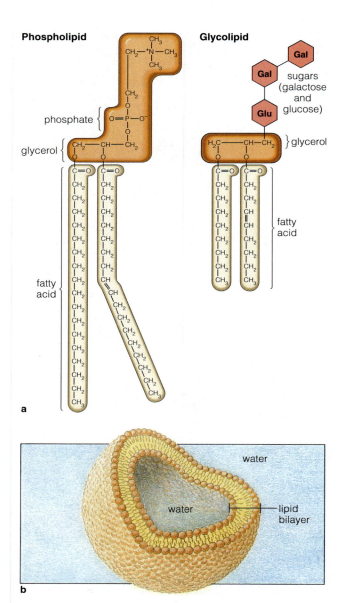

Phospholipid

CH₃

CH₂—⁺N—CH₃

CH₃

CH₂

phosphate { O=P—O⁻

O

glycerol { CH₂—CH—CH₂

O   O

C=O C=O

...

fatty acid

Glycolipid

Gal

Gal — sugars (galactose and glucose)

Glu

H₂C—CH—CH₂ } glycerol

O   O

C=O C=O

fatty acid

a

b

water

water — lipid bilayer

Figure 2.14 Phospholipids, glycolipids, and membranes. (**a**) Models of a phospholipid and a glycolipid. Both have hydrophobic hydrocarbon tails and a hydrophilic head group that may include elements of glycerol, phosphate, sugars, or other compounds. (**b**) Phospholipids in solution form a bilayer complex that keeps water away from the tails and in contact with the heads. The bilayer is flexible, and the individual molecules can move relative to one another, as long as they stay in the plane of the bilayer.

in the cell wall. Cellulose forms cables that are very strong and can keep a cell from bursting under pressure (see Chapter 3); pectin acts as a glue to hold cells together (see Chapter 4). Other polysaccharides—for instance, **starch**, a polymer of the simple sugar glucose, and **inulin**, a polymer of fructose—are used for storage of energy; free energy for running chemical reactions can be recovered when these polysaccharides are broken down into their component sugars and the sugars are oxidized.

Lipids Are Insoluble in Water

The name **lipid** applies to any oil-soluble (nonpolar) substance in the cell; however, two classes of lipids—the **phospholipids** and **glycolipids**, which associate to form thin sheets (Fig. 2.14)—play special roles in cells. These lipids do have water-soluble (polar, hydrophilic) functional groups, which may be phosphate groups (in phospholipids) or sugars (in glycolipids). They also have water-insoluble (nonpolar, hydrophobic) components, which are the hydrocarbon ends of fatty acids. A fatty acid is a molecule with a hydrocarbon chain attached to a carboxylic acid. The carbons of the hydrocarbon chain may all be connected by single covalent bonds, in which case the chain is flexible but generally straight; or they may have one or more double bonds, which insert a bend in the chain. The phosphate or sugars and the fatty acids are connected by a three-carbon molecule called *glycerol*. By forming sheets two molecules thick, phospholipids and glycolipids are able to bury their hydrophobic components in the middle, away from the solvent water, and expose their hydrophilic components to the solvent. These sheets provide the structural basis for the **membranes** that are found throughout cells.

Sterols are another type of lipid that contributes to membrane structure. Sterols are large, multiringed hydrocarbons that dissolve in the hydrophobic part of a membrane and keep it flexible, preventing it from developing cracks. In animals, the primary sterol is cholesterol; in plants (including the vegetables and fruits that you eat), there is no cholesterol; a related molecule, ergosterol, has the same function.

Other lipids, the **triglycerides**, are used as energy storage compounds because the oxidation of their components releases a great deal of useful energy. Triglycerides have three fatty acid molecules attached to a glycerol connector. They are very insoluble in water and form discrete lipid bodies inside plant cells. Extracted from plant organs, they form oils (liquid at room temperature) or fats (solid at room temperature), depending on the lengths of the fatty acid hydrocarbons and whether or not they have double bonds. (Fatty acids in oils are shorter and/or have more double bonds.) Most salad oils are triglycerides extracted from seeds of maize, soybean, sunflower, sesame, cotton, peanut, or other plants. Three commercially important fats from plants are palm "oil", coconut "oil", and cocoa butter.

Nucleic Acids Store and Transmit Information

There are two types of **nucleic acid: deoxyribonucleic acid (DNA)** and **ribonucleic acid (RNA)**. Both DNA and RNA are polymers, long unbranched chains of **nucleotide** monomers. Each nucleotide is formed from a simple five-carbon sugar attached to a phosphate group and to a one- or two-ringed molecule called a *base*. (The word

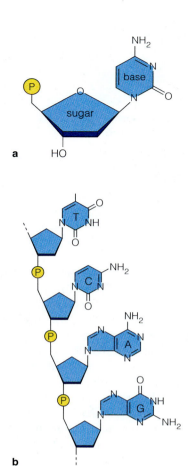

a

b

Figure 2.15 Structures of a nucleic acid polymer. (**a**) The basic nucleotide subunit has a phosphate (P), sugar, and base. DNA and RNA nucleotides are distinguished by their sugars (the one shown here is deoxyribose from DNA). (**b**) Nucleotides are connected through their sugars and phosphates.

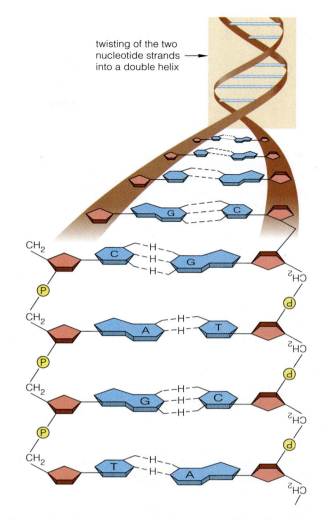

twisting of the two nucleotide strands into a double helix

Figure 2.16 The complementary bases A and T, and G and C can form base-pair complexes, connected by two and three hydrogen bonds, respectively.

base in this context has a different meaning from that used on p. **18**. There, *base* referred to any compound that accepted H⁺. Here, it refers to a few specific molecules.) The sugars give the nucleic acids their names: the sugar of DNA nucleotides is deoxyribose; the sugar of RNA nucleotides is ribose. Chains of nucleotides are formed by attaching the phosphate group of one nucleotide to the sugar of another (Fig. 2.15).

Different nucleotides are distinguished by their different bases. There are four types of nucleotides in DNA, which have bases called A, T, G, or C. There are four corresponding types of nucleotides in RNA, with A, U, G, or C bases. The nucleotide bases can bind together with hydrogen bonds to form complementary pairs. A binds only to T (or U); G binds only to C (Fig. 2.16).

DNA is formed from two chains of nucleotides wound around each other in a formation called a *double helix* (Fig. 2.17). At each position, the base of one chain is hydrogen-bonded to the corresponding base of the other chain. Therefore, the two chains are called *complementary*,

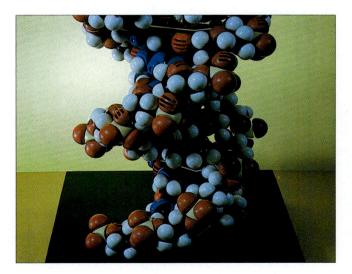

Figure 2.17 A model of DNA, the genetic material, showing two chains of nucleotides wound tightly around each other in a helix. Each base (part of a nucleotide) on a chain binds to a complementary base on the opposite chain.

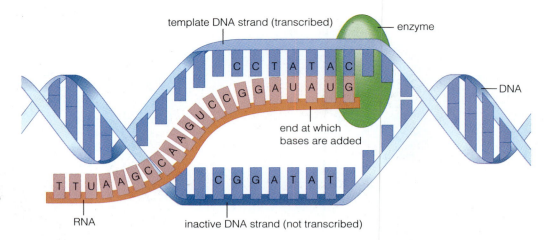

Figure 2.18 Transcription, the synthesis of RNA. One of the two DNA strands serves as a template to specify the order of bases in the growing RNA chain.

just as the individual bases are complementary. DNA stores **genetic information** in its sequence of nucleotide bases. Thus, a sequence of bases such as ATGCCC has a different meaning from another sequence, such as AAGTTA. Consequently, the order of the bases in DNA is critical to its function, as the order of amino acids is critical to the function of proteins. Although there are not as many types of bases as amino acids, the potential variation for DNA is much greater than for proteins because DNA molecules are so much longer. A single DNA molecule generally has several million nucleotides.

Genetic information is used to specify the order of amino acids in proteins. A set of three bases on a DNA molecule specifies one amino acid according to a **genetic code**. So a section of DNA containing 100 sets of three bases, all in order, can specify the amino acid sequence of a protein of 100 amino acids. Such a section of DNA (together with extra bases controlling when the genetic information is expressed to form proteins) is a **gene**.

RNA is made from a single strand of nucleotides. There are various types of RNA. Some are extended; others may loop back on themselves and be wound into different shapes. The different types of RNA molecules have different functions, but almost all are involved in some way in the use of genetic information to synthesize proteins.

GENE TO PROTEIN   The story of how information locked in the sequence of bases in DNA produces the enzymes needed for a cell's life and growth is one of the most exciting discoveries of modern biology. The process has several steps and many components. It is amazing both for the simplicity of the basic principles behind it and the complexity of their implementation.

An early step is to transcribe the genetic code of DNA onto an RNA molecule (Fig. 2.18). **Transcription** involves using the base sequence of a section of DNA as a **template**. The two strands of DNA separate, and an enzyme moves along one of the strands, assembling an RNA molecule with a base sequence that is complementary

to that of the DNA strand. *Complementary* means that at each position the RNA base fits with the base on the DNA: A fits to T, C to G, G to C, and U to A. Thus the base sequence of the DNA template specifies the base sequence of the RNA strand that is produced.

There are several types of RNA synthesized, each with its own base sequence specified by different sections of the template DNA. One type is **ribosomal RNA (rRNA)**. Three separate rRNAs, in combination with several proteins, form the basic machinery (**ribosomes**) for making proteins. A second type is **transfer RNA (tRNA)**. Transfer RNA serves as a decoding molecule, translating a base sequence into an amino acid sequence. The RNA molecules that specify the amino acid sequences of particular proteins are called **messenger RNAs (mRNA)**. As described in Chapter 3, in plant cells mRNAs carry the code (message) for the protein from the nucleus, where the genetic information is stored, to the cytoplasm, where the protein is synthesized.

The mRNA is translated to make a protein by interacting with ribosomes and tRNAs (Fig. 2.19). In **translation**, the ribosomes bind to the mRNA and then move along the mRNA three bases at a time while binding the appropriate tRNAs. A sequence of three mRNA bases is called a *codon*. There are 64 different codons, each representing a particular amino acid, except for three codons that are "stop" signals. When the ribosome reaches a particular codon for an amino acid, it finds a tRNA with the complementary set of bases, known as the *anticodon*. This tRNA is carrying the amino acid specified by the codon on the mRNA. The ribosome connects the tRNA's amino acid to the preceding amino acid with a peptide bond. More than one ribosome may work in this fashion on one mRNA, each ribosome forming one polypeptide chain. In some cases, the polypeptide chain automatically coils into the three-dimensional structure specified by its amino acid sequence. However, some polypeptide chains must be modified before they become active. Whatever the modifications, it is the genetic information in the DNA that provides the main information for producing each enzyme and other functional proteins in every cell.

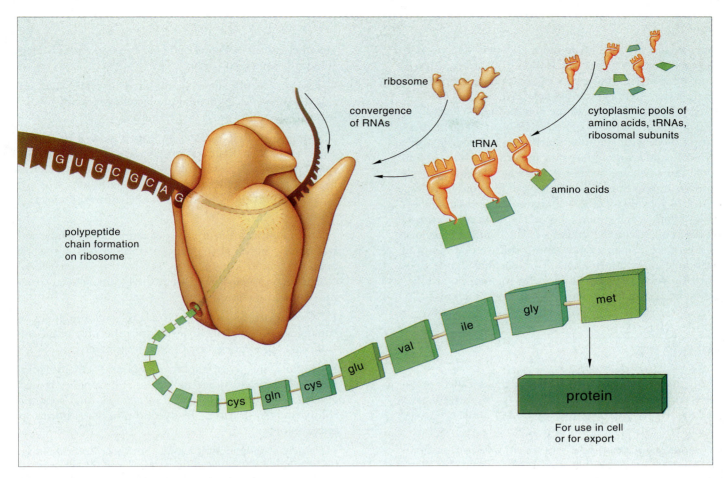

Figure 2.19 Translation, the synthesis of a protein. The ribosome serves as the central point where tRNA molecules match amino acids to the sequence of codons on the mRNA. Through enzymatic action by the ribosome, the amino acids are joined to form a polypeptide chain.

DNA REPLICATION   Reproduction is one of the most important and complex characteristics of living organisms. It is axiomatic that when a cell or organism reproduces, it must pass its genetic information to its progeny. This means that new copies of the information, and thus new copies of DNA base sequences, must be synthesized (*replicated*). The ability of a DNA strand to serve as a template (as we described for the synthesis of RNA) explains how these new copies may be formed.

The basic process of **DNA replication** involves the separation of the two original complementary strands of DNA. Each strand can serve as a template for the assembly of a new complementary strand (Fig. 2.20). An

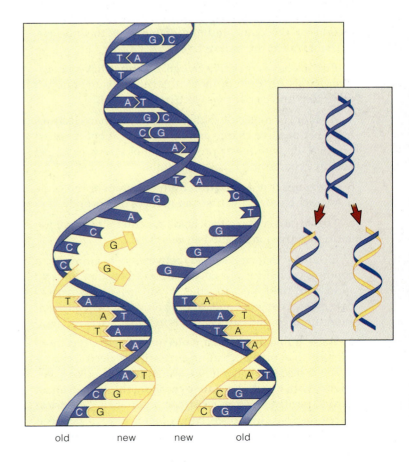

Figure 2.20 The replication of a DNA molecule. Shown is a DNA molecule that has been partially unzipped by the breaking of the hydrogen bonds between the bases. Each strand serves as a template for creating a new, complementary DNA strand. The template strand and the new strand stay together after replication is complete.

old   new   new   old

enzyme attaches the nucleotides together one by one, choosing at each step a nucleotide with a base that is complementary to the opposite base on the template. The result is two identical double helical molecules of DNA, each with one original and one new strand.

The actual synthesis of a new DNA molecule is more complicated, with synthesis starting in the middle of the molecule and working outward in both directions. Additional enzymes are needed to separate the strands and promote unwinding and rewinding of the double helices, as well as to weld together separate, adjacent segments of newly synthesized strands. It is amazing that the process works at all, given the millions of nucleotides that must be joined in proper sequences to produce one new full copy of genetic information. It is essential that it does work, however, because DNA forms the molecular foundation for all organisms and most of their functions.

SUMMARY

1. All matter is made up of simple units, called molecules. Molecules are made of atoms. An atom contains a nucleus (with protons and neutrons) and electrons, which move around the nucleus in orbitals.

2. Matter made from one type of atom is an element; matter made of more than one type of atom is a compound. The number of protons in the nucleus defines the type of atom; the number of neutrons contributes to the weight of the atom and the stability of the nucleus. Unstable nuclei are radioactive.

3. Protons have a positive charge. Electrons have a negative charge. An atom or molecule with the same number of protons and electrons is electrically neutral.

4. An atom or molecule with more protons than electrons has a net positive charge and is called a cation. An atom or molecule with fewer protons than electrons has a net negative charge and is called an anion.

5. The transfer of an electron from one atom or molecule to another oxidizes the first and reduces the second.

6. The attractive force that holds together a positive ion and a negative ion is called an *ionic bond*.

7. Electrons that are shared between two atoms in a bonding orbital form a covalent bond between the two atoms and hold them together. Hydrogen atoms participate in one covalent bond, oxygen atoms in two, nitrogen atoms in three, and carbon atoms in four.

8. Some atomic nuclei in a molecule pull electrons more strongly than others, creating a polarity of electric charge in the molecule. Water is an example of a polar compound, with a partial negative charge at the oxygen and a partial positive charge near the hydrogens. Electrical forces result in hydrogen bonds being formed between two water molecules or other molecules where a hydro-

gen nucleus can be positioned between two strongly electronegative nuclei (O or N).

9. Polar molecules tend to dissolve in water and are called *hydrophilic*. Nonpolar molecules tend not to dissolve in water and are called *hydrophobic*.

10. Molecules that donate hydrogen nuclei are called *acids*; molecules that accept the hydrogen nuclei are called *bases*. Excess hydrogen nuclei in water are attached to the water molecules to form hydronium ions. The pH scale measures the concentration of hydronium ions in a solution.

11. A protein is a linear chain of amino acids connected by peptide bonds. Each type of protein has a specific order of amino acids and a specific three-dimensional structure. A protein that loses its three-dimensional structure cannot function.

12. Carbohydrates are simple sugars and polymers of sugars. Some contribute to structure, especially of the cell wall, and others are a form of energy storage.

13. Lipids are hydrophobic compounds made from hydrocarbons and other molecules. Phospholipids and glycolipids form sheets, called *membranes*, that separate compartments in a cell. Triglycerides serve as forms of energy storage.

14. Nucleic acids include DNA and RNA. Both are long, linear chains of nucleotides. DNA, a double-stranded molecule, stores genetic information in the sequence of its nucleotide bases. RNA, a single-stranded molecule, is an intermediate in the use of genetic information to make functional proteins.

15. The process of making proteins starts with transcription, the synthesis of RNA using DNA as a template. Three types of RNA are needed: ribosomal RNA, transfer RNA, and messenger RNA. Ribosomal RNA (together with proteins) forms the machinery that uses transfer RNA to translate the sequence of codons of messenger RNA into a sequence of amino acids in a polypeptide chain.

16. The presence of two complementary chains in DNA and their use as templates form the chemical basis for the replication of genetic information.

Questions

1. How many protons and how many electrons are present in a molecule of (a) methane, CH_4; (b) ammonia, NH_3; (c) water, H_2O?

2. When an atom of ^{14}C carbon undergoes radioactive decay, its nucleus emits an electron. Since the electron carries one negative charge, the nucleus has gained one positive charge. One of the neutrons in the nucleus has turned into a proton! Is the atom still an atom of carbon? Explain your answer.

3. Sketch a molecule of methane, showing the positions of the nuclei of the five atoms and the general positions of the bonding and nonbonding electrons. Illustrate that the total number of elec-

trons in the molecule equals the total number of protons in the five nuclei.

4. Sketch a molecule of water, showing the positions of the nuclei of the three atoms and the general positions of the bonding and non-bonding electrons. Illustrate that the total number of electrons in the molecule equals the total number of protons in the three nuclei. Do all the outer orbitals of the oxygen form chemical bonds? Show how the tendency of electrons to move to the larger nucleus gives the molecule a polarity.

5. Sketch the following hydrocarbon molecules, showing the positions of the C and H atoms (as well as you can on a flat piece of paper) and the numbers of covalent bonds between adjacent atoms. Remember that all carbons should have four bonds and that some carbons may have double bonds between them. Also, some carbon chains can form rings. (a) ethane, C_2H_6; (b) butane, C_4H_{10}; (c) hexane, C_6H_{14}; (d) cyclohexane, C_6H_{12}; (e) ethylene, C_2H_4; (f) benzene, C_6H_6.

6. The pH of pure water is 7. What will the pH be if (a) you increase the hydrogen (hydronium) ion concentration 1000-fold; (b) you add a compound that forms complexes with 99% of the hydrogen (hydronium) ions?

7. Explain why salad oil and vinegar separate after you shake them up together.

8. Rearrange the following list so that the largest items are at the bottom and the smallest at the top:

carbon atom

water

neutron

protein

proton

DNA

amino acid

electron

sugar

triglyceride

9. Match the monomers and polymers in the following lists:

amino acid	DNA
fatty acid	cellulose
nucleotide	enzyme
sugar	RNA
	phospholipid
	starch
	protein

10. (a) Assume that one strand of DNA contains bases in the following order: AATGCTACGTTAA. Write the order of the bases in the complementary strand of the DNA molecule.

(b) Assume that a strand of DNA contains bases in the following order: TTTGCACTAAAA. Write the order of the bases in a strand of RNA that would be formed using this DNA as a template.

Further Readings

Miller, G. T. 1991. *Chemistry: A Contemporary Approach*. 3d ed. Belmont, Calif.: Wadsworth.

Olson, A. J., and D. S. Goodsell. 1992. "Visualizing Biological Molecules." *Scientific American* 267 (November): 76–81.

Richards, F. M. 1991. "The Protein Folding Problem." *Scientific American* 264 (January): 54–63.

Scientific American. October 1985. "The Molecules of Life." This issue contains many articles on the chemical basis of living organisms.

3

THE PLANT CELL AND THE CELL CYCLE

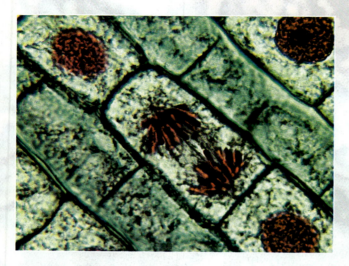

1. Every plant is constructed from small compartments called *cells*. Each cell is a living individual, possessing the basic characteristics of life, including movement, metabolism, and the ability to reproduce. Some cells in a plant develop specialized capabilities that contribute to the life of the whole organism.

2. Cells contain organelles with specialized functions. The nucleus, ribosomes, and endomembrane system participate in the synthesis of proteins; the plastids and mitochondria capture and convert energy into useful forms; the cytoskeleton directs the movement of other components around the cell. Learning the anatomy of a cell helps one understand its activities.

3. Cells reproduce by dividing. Cell division is the most complicated process that any cell can undergo. Specific genes and proteins cooperate to regulate the timing of the events in cell division.

Figure 3.1 Plant cells through the microscope. (**a**) A drawing of cell walls from the cork tissue of an oak (*Quercus* sp.) tree, published in 1665 by Robert Hooke in his *Micrographia*. (**b**) A light micrograph of leaf tissue from the aquatic plant *Elodea*, showing how the tissue is divided into cells.

3.1 | CELLS AND MICROSCOPY

Cells Are the Basic Units of Plant Structure and Function

In the late 1600s, an English experimentalist named Robert Hooke used his improved version of a microscope to look at shavings of cork tissue (the dead outer bark of an oak tree). He described "little boxes or **cells** distinct from one another . . . that perfectly enclosed air." Later Nehemiah Grew, an English clergyman, recognized that leaves were formed from collections of cells filled not with air, but with fluid and green inclusions (Fig. 3.1). It took many years for the ubiquity of cells to be realized, but in 1838 the Belgian botanist Matthias Schleiden and zoologist Theodor Schwann proposed that all plants and animals are composed of cells. Later, in 1858, Rudolf Virchow suggested that cells possess a characteristic of life ascribed by earlier observers only to organisms; that is,

a

b

cells reproduce themselves, and all cells arise by reproduction from previous cells. This set of propositions, now known as the **cell theory**, is one of the key principles of biology.

Almost all plant cells have certain similarities in structure, reflecting the fact that they share the same activities and the same problems. Most cells grow—that is, they get larger, and they divide to form new cells. In mature plants, many cells stop growing, but even these continue to synthesize new components. All these cells must accumulate chemicals that they need for the synthesis of new components. They must find sources of energy that promote the chemical reactions needed for synthesis. They must store and interpret the genetic instructions that direct the synthesis of these components at the right times and places. They must get rid of worn-out components and exclude toxins from sensitive reactions. Cells must control their own size, which means controlling the amount of water that moves into or out of them. All these functions are important to all cells—bacteria, animal cells, and protists as well as plant cells. But plant cells as a group have unique methods of carrying out some of these functions.

With very few exceptions, each cell in the plant body plays a role in the health and activities of the whole plant. To be effective, some cells have specialized structures or chemicals. Certain cells are specialized for rapid growth and cell division. Other cells have a protective function: Cells on the outer layer of a stem, for instance, secrete water-impermeable chemicals, such as waxes; these keep water vapor from diffusing out of the plant and thus keep the interior moist. Still other cells have a structural role, such as stiffening large organs so that they can support their own weight. Some cells are responsible for the transport of compounds from one part of the plant to another. Certain cells play key roles in sexual reproduction. In each case, the cell forms specialized structures that allow it to accomplish its mission in the life of the plant.

The specialized structures within cells are called **organelles** ("little organs," by analogy with the organs contained in the body of a multicellular organism). These are associated with some of the general and specialized functions that the cells must perform. The next section explains one of the most effective methods for studying organelles.

Microscopes Allow One to See Small, Otherwise Invisible Objects

Plant cells have been studied with a **light microscope** ever since Robert Hooke looked at cork. In fact, the development of the microscope was the technical breakthrough that led to the discovery of cells, and improvements in microscopy continue to contribute to our understanding of cell structure and function. Light microscopes use lenses, which bend light rays so that the object looks larger (Fig. 3.2). With a good compound microscope (one that has many lenses arranged in series), you can easily see cells that are 20 to 200 micrometers (μm) in diameter, and you should be able to see components as small as 1 μm in diameter (1 μm is 10^{-6} meter or one millionth of a meter or about four hundred-thousandths of an inch).

A light microscope has the advantage of being usable with live specimens, but it also has limitations. One involves **contrast**. Many of the organelles of a cell do not absorb light well, and so light rays coming from them look the same as rays from adjacent parts of the cell. This means that you cannot tell that the organelle is there. Microscopists partially solve this problem by staining the cells: Certain stains color particular organelles and thus increase their contrast. However, even in stained samples the scattering of light from other parts of the sample tends to wash out the image, reducing the contrast. The thicker the slice of tissue, the more serious the scattering problem. One solution to this problem is to cut a thin slice of the sample, but the soft substance of a living cell cannot withstand the chemical treatment needed for making very thin slices; so this treatment kills the specimen.

Even if the contrast of a sample in a light microscope is good, the microscope's **resolution**—its ability to distinguish separate objects—is limited by several factors. Because different colors of light are affected differently by lenses, it is impossible to focus an image perfectly when it is illuminated with white light (which contains rays of all colors). Furthermore, the resolving power of a microscope is limited to one-half the wavelength of the light being used. Since the shortest wavelength seen by the human eye is about 0.4 μm, the absolutely smallest object that can be resolved in a light microscope is about 0.2 μm in diameter.

New techniques of microscopy have minimized many of these limitations. Contrast can be dramatically increased by **confocal microscopy**. In this system, the illumination (a laser) and the detecting lens are both focused on one point in the sample at a time, scanning across the sample to assemble a whole picture. Since only one point is illuminated, there is no reduction in contrast from scattered light from other parts of the sample. Even in a relatively thick sample, the focal point of illumination can be very exact, which means that the light can be focused on different levels of a sample. If one takes separate pictures of different levels, one can assemble three-dimensional pictures of a cell.

Resolution can be improved by using **transmission electron microscopy**. Instead of light, electron microscopes use beams of electrons. Quantum theory tells us that electrons, although normally thought of as particles, also behave like light waves, with wavelengths about 1 million times shorter than those of visible light. Electron

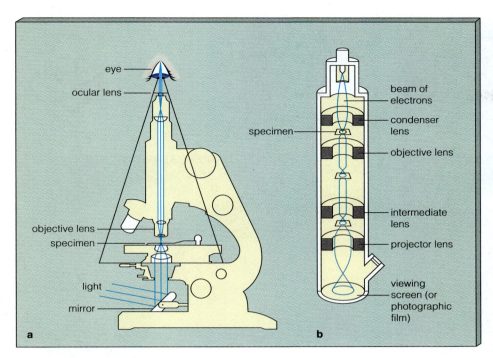

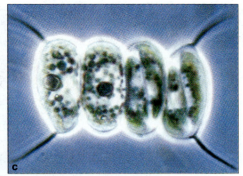

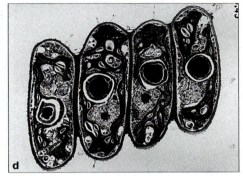

Figure 3.2 A comparison of the light microscope (**a**) and the transmission electron microscope (**b**) and the images they produce, (**c**) and (**d**). A light or electron beam is focused on the sample with glass or magnetic lenses, respectively. From each part of the sample the beam radiates toward the objective lens, forming a larger image on the other side. A series of lenses remagnifies this image, which eventually is focused on the eye (light microscope) or a photographic film (electron microscope). Both micrographs show the same organism—the green alga *Scenedesmus*—at the same size. Notice that the electron micrograph is black and white; only light micrographs can show natural color. Some electron micrographs have color added later to highlight important elements of the picture.

beams, having a negative charge, are bent by magnets; so in a transmission electron microscope magnets serve as lenses. Since the human eye cannot see electrons, the final image is made visible by using the electrons to excite a fluorescent plate or to expose photographic film. These electron microscopes have limitations, too. Electron beams cannot pass through air or through a whole cell. Therefore, the sample must be sliced ultrathin and examined in a vacuum. This technique clearly cannot be used while the samples are alive and functioning. Nevertheless, transmission electron microscopy has been responsible for the discovery of most of the smaller organelles in the cell.

A second type of electron microscopy, called scanning electron microscopy (SEM), works on an entirely different principle. A very narrow electron beam is scanned across a sample in a series of lines. The electrons that bounce off the sample are collected and used to modulate a similar beam that is forming a picture in a television picture tube. The result is a television picture of the sample. The resolution of the picture can be very high, although only the surface of the sample is shown. Like transmission electron microscopy, scanning electron microscopy must be conducted in a vacuum. In the past, this technique has required complex and careful pre-

parative techniques that kill the cells. New versions of scanning electron microscopes, however, operate in a relatively low vacuum; so live plant cells and insects can be viewed without killing them. Movements and other functions that require living cells can now be studied.

Throughout human history, advances in technology have led to discoveries in basic science. So it has been with the microscope. From the first discovery of cells, the continuing development of microscopy has led to a long series of breakthroughs in our understanding of plant cell structure and function.

3.2 THE PLANT CELL

Living cells are found throughout the plant body. They make up the internal, energy-producing (photosynthetic) cells of the leaf. They make up the pith and cortex of the stem and the cortex of the root. You find them making up the bulk of fleshy fruits. These cells all have similar organelles (Fig. 3.3), those needed for general growth and maintenance of cell function. Some also have specialized organelles for specific functions. The next section describes the components found in a generalized living plant cell.

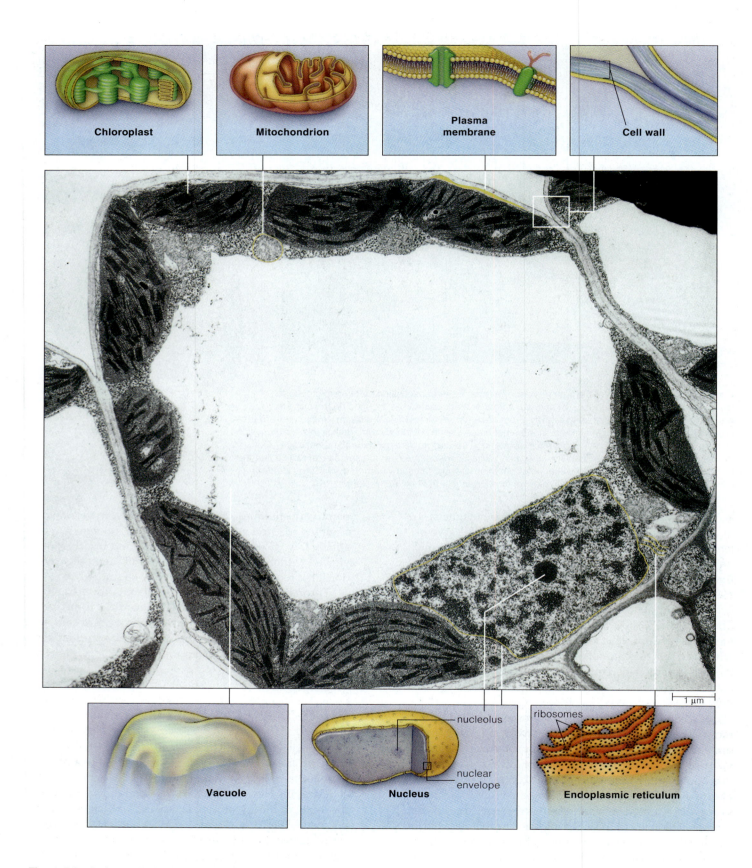

Chloroplast

Mitochondrion

Plasma membrane

Cell wall

1 μm

Vacuole

Nucleus

nucleolus

nuclear envelope

ribosomes

Endoplasmic reticulum

Figure 3.3   A plant cell, showing the major organelles, as seen in a transmission electron microscope. The cell is from a blade of timothy grass (*Phleum*).

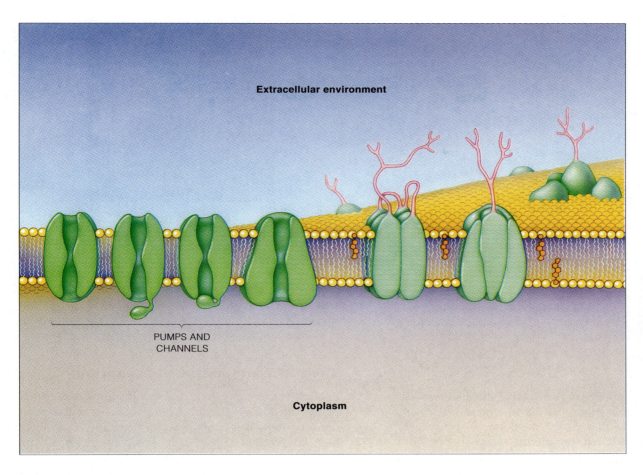

Extracellular environment

PUMPS AND
CHANNELS

Cytoplasm

Figure 3.4   A model of the plasma membrane, showing the phospholipid bilayer and various types of proteins floating in the bilayer.

3.3 **THE BOUNDARY BETWEEN INSIDE AND OUTSIDE**

The Plasma Membrane Controls Movement of Material into and out of the Cell

A thin membrane, the **plasma membrane**, surrounds each cell. Membranes are composed of approximately half phospholipid and half protein (Fig. 3.4). The phospholipids provide a flexible, continuous, hydrophobic (water-excluding) sheet two molecules thick (called the phospholipid **bilayer**). This separates the aqueous solution inside the cell (called the **cytoplasm**) from that outside the cell. The phospholipid bilayer prevents ions, amino acids, proteins, carbohydrates, nucleic acids, and other water-soluble compounds inside the cell from leaking out—and prevents those outside the cell from diffusing in. That means that for one of these compounds to move in or out, there must be a special pathway or carrier. These pathways are provided by special proteins in the bilayer.

Each type of protein in a membrane performs a different function. Some provide the special pathways by which compounds can move into or out of a cell in a highly regulated manner. For instance, some proteins function as **ion pumps**: one very important type (the **proton pump**) pumps H^+ ions from the inside to the outside of the cell (see Chapter 11 for more details). Another pumps Ca^{2+} ions to the outside of the cell. The characteristic of a pump is that it can move ions from lower to higher concentration by utilizing cellular energy in the form of ATP (see p. 127 and Chapters 9 and 10). Some proteins form **channels** for certain substances (such as K^+ ions, sucrose, or even water) to diffuse across the membrane (only from higher to lower concentration), either alone or in combination with another substance (such as H^+). The combination of phospholipids (which form a relatively impermeable sheet) and proteins (which pass specific materials through the sheet) allows the plasma membrane to control transport into and out of the cytoplasm.

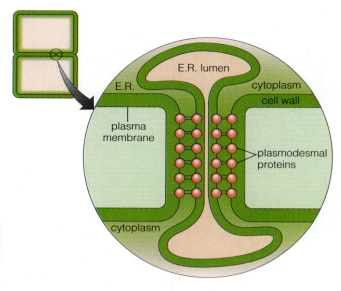

Figure 3.5 Model of a plasmodesma, showing the E.R. (see p. 36) and proteins that are thought to control the flow of materials through the channel. (Redrawn from a diagram provided by William J. Lucas.)

Figure 3.6 A small section of cell wall, as seen in a transmission electron microscope. The filaments are cellulose.

An unusual property of plant cells is the existence of connections, the **plasmodesmata** (singular, *plasmodesma*) (Fig. 3.5), between the plasma membranes of adjacent cells. At a plasmodesma, the plasma membrane on one cell extends outward through the cell wall (see next section), forming a tube and then connecting to the plasma membrane of the next cell. The continuous cytoplasm in a set of cells connected by plasmodesmata is sometimes called the **symplast**. A plasmodesma forms a passageway that may allow materials to move from the cytoplasm of one cell to the next. It is a striking idea that plant cells are so interconnected by plasmodesmata that a plant might be considered a single super-large cell, partially divided into the compartments we call cells. However, a plasmodesma is not simply an open tube. Inside is another membrane tube and proteins that control the transport of materials (especially large molecules) through the plasmodesma. Small molecules such as sugars and amino acids seem to pass easily. Larger molecules (such as proteins) and organelles generally cannot pass, although it has recently been discovered that plant viruses can open plasmodesmata so that the virus particles can spread their infection from cell to cell.

Does the plasma membrane define the limits of the cell? Many traditional cell biologists consider the plasma membrane to be the outer limit of the cell. According to this view, everything outside the plasma membrane is extracellular: something that may have come from the cell but is not really part of it. Others point to recent research that shows there is considerable metabolic activity outside the plasma membrane. They consider the space outside the cell, next to the plasma membrane and within the fibrils of the cell wall, to be part of the cell. In

either case, there is a continuity of the space around adjacent cells. This space is called the **apoplast**. It is clear that the apoplast is an important space in a plant, but it is an open question whether it is part of the plant's cells.

The Cell Wall Limits Cell Expansion

A plant cell (or any cell) that is surrounded *only* by a plasma membrane and placed in pure water will expand until it bursts. This is because a cell contains a relatively high concentration of solutes, held in by the plasma membrane. The solutes lower the effective concentration of the water inside the cell. Chemicals tend to move from a region of high concentration to one of low concentration. The plasma membrane prevents solutes from moving out, but it does allow water to move in. The flow of water from a relatively dilute solution to a relatively concentrated solution is called **osmosis**. While the inflow of water dilutes the solution inside the cell, that solution never becomes as dilute as pure water. Thus the inflow continues until the plasma membrane, which is not infinitely expandable, ruptures.

For plant cells, this problem of rupturing in a dilute solution, such as rainwater, is solved by a **cell wall** (Fig. 3.6). The cell wall is a relatively rigid structure that surrounds the cell just outside the plasma membrane. It is made of **microfibrils** formed from a polysaccharide, **cellulose**. Cellulose is a linear, unbranched chain of the sugar glucose (see Fig. 2.13). Two of these chains can line up side by side connected by hydrogen bonds. About 40 of these chains form a microfibril, which is flexible but very strong, like a cable or a nylon cord. Among the microfibrils are other, less highly organized polysaccharides (hemicelluloses and pectins) and proteins.

The microfibrils and other components form a porous network, so that water and solutes easily penetrate to the plasma membrane. If, however, an inflow of water expands the cell, it forces the plasma membrane against the cell wall. The plasma membrane cannot penetrate the cell wall, so the pressure from the water tends to expand

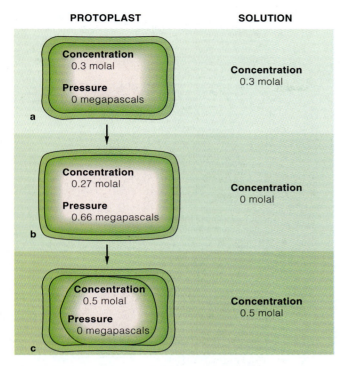

| PROTOPLAST | SOLUTION |

a
Concentration
0.3 molal

Pressure
0 megapascals

Concentration
0.3 molal

b
Concentration
0.27 molal

Pressure
0.66 megapascals

Concentration
0 molal

c
Concentration
0.5 molal

Pressure
0 megapascals

Concentration
0.5 molal

Figure 3.7 The effect of osmosis on cell size. (**a**) The cell in this example is assumed to have an initial internal concentration of solutes, which we will designate as 0.3 molal (molal is a unit of concentration). The cell is in a solution of the same concentration. The cell protoplast (volume inside the plasma membrane) is exactly the size of the cell wall in its resting (unstretched) state. (**b**) The cell is transferred to pure water. Water moves into the protoplast by osmosis, and the protoplast expands, pushing against the cell wall. The wall exerts a back pressure on the protoplast, which inhibits further influx of water. Also, the expansion of the protoplast dilutes the solutes inside the protoplast. Water stops entering the cell when the pressure exerted by the cell wall equals the osmotic pressure forcing water into the cell. (**c**) The cell is transferred to a solution of 0.5 molal. Water moves out of the protoplast by osmosis, and the protoplast shrinks until the concentration of solutes in the protoplast equals 0.5 molal. The plasma membrane pulls away from the cell wall. This effect is called *plasmolysis*.

the cell wall. The cell wall is tough but elastic, so the more it expands, the more it resists further expansion. At some point the resistance of the cell wall exactly balances the pressure of osmosis, stopping the flow of water into the cell. This keeps the plasma membrane, and consequently the cell, from expanding further (Fig. 3.7).

When the cell stops expanding, the osmotic forces pulling water into the cell do not disappear. They are simply balanced by the pressure exerted by the cell wall. The internal hydrostatic pressure in a cell can be very high. Typical cell walls maintain pressures of 0.75 megapascal (7.5 times atmospheric pressure at sea level). Cells with such internal pressure (**turgor** pressure) become stiff and incompressible. This means that they can hold heavy weights. Thus even plant cells with a thin cell wall can form and support large plant organs, so long as the solu-

tion in the apoplast is sufficiently dilute that a large turgor pressure is generated. When the supply of water is cut off, a leaf wilts. This occurs because water is lost from the cell and turgor pressure drops. Thus, besides preventing the cell from bursting when it is surrounded by a dilute solution, the cell wall is also responsible for maintaining the high turgor pressure that gives a plant organ much of its strength.

Osmosis works in both directions. If the concentration of solutes outside the cell is greater than that inside—such as is found in salt flats—then water will flow out of the cytoplasm. The volume of the **protoplast** (the space inside the plasma membrane) will shrink, and the plasma membrane will pull away from the cell wall (Fig. 3.7). This effect is called **plasmolysis**. A plasmolyzed cell has no turgor pressure. Thus an accumulation of salt in the soil can cause a plant to wilt.

The cell wall that forms while the cell is growing is the *primary cell wall*. After they stop growing, some cells deposit an additional cell wall layer between the primary cell wall and the plasma membrane—that is, a *secondary cell wall*. The secondary cell wall generally contains cellulose microfibrils plus a strong, inextensible, water-impermeable substance called **lignin**. Lignin is formed from subunits that are highly cross-linked, forming a three-dimensional network. The cross-links make these cell walls especially rigid and much more able to resist compression. Other specialized types of cell walls have **cutin** or **suberin** embedded in them. Cutin and suberin are waxy compounds that are especially impermeable to water. Cutinized cell walls are found on the surface of leaves and other organs that are exposed to the air. They retard the evaporation of water from the cells into the air. They also form a barrier to potential pathogens like bacteria, fungi, and viruses. The adaptations of specialized plant cells often involve chemical changes to their cell walls.

3.4 THE ORGANELLES OF PROTEIN SYNTHESIS AND TRANSPORT

Much of the cell is made of protein, and most of its activities depend on specialized proteins. Plant cells have specialized organelles that are responsible for producing proteins and for making sure that each is placed in its proper position in the cell.

The Nucleus Stores and Expresses Genetic Information

One of the larger and more prominent organelles is the **nucleus**. It is ovoid or irregular in shape and up to 25 μm in diameter. It stains densely with many of the stains used for light or electron microscopy. It is surrounded by a double membrane, the **nuclear envelope**. The inner and

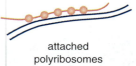

free
polyribosomes

attached
polyribosomes

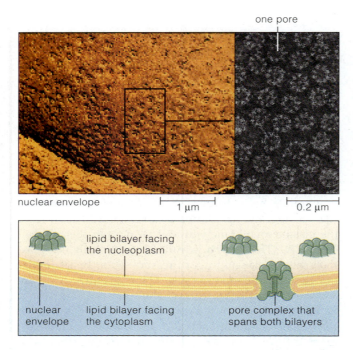

one pore

nuclear envelope

1 μm   0.2 μm

lipid bilayer facing
the nucleoplasm

nuclear
envelope

lipid bilayer facing
the cytoplasm

pore complex that
spans both bilayers

Figure 3.8 Surface view and sketch of the nucleus, showing the nuclear envelope with pores.

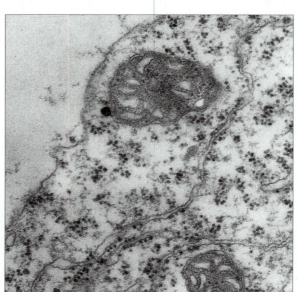

Figure 3.9 Polyribosomes as observed by transmission electron microscopy. Some of the polyribosomes in this cell from a wheat root tip are free in the cytoplasm; others are attached to the endoplasmic reticulum forming rough E.R.

outer membranes of the nuclear envelope connect to form pores, through which molecules may pass (Fig. 3.8). In the nucleoplasm (the portion inside the nuclear envelope) are found the **chromosomes**, which are made of DNA and protein. The proteins form nucleosomes (round bodies around which the DNA winds) and scaffolds (which hold loops and helices of the DNA molecules in place).

As described in Chapter 2, the DNA molecules in the chromosomes store genetic information in their sequences of nucleotides. This information is used to direct the synthesis of proteins. The first step in the process is transcription—the DNA is used as a template to direct the synthesis of RNA. Much of this RNA, for unknown reasons, stays in the nucleus or is rapidly broken down. However, a small fraction of the RNA, messenger RNA, carries the genetic information of its DNA template out of the nucleus and into the cytoplasm.

Also in the nucleoplasm are **nucleoli** (singular, *nucleolus*), seen in the light or electron microscope as densely staining regions. These are accumulations of RNA–protein complexes. The RNA–protein complexes are ribosomes (see next section), and nucleoli are the sites where the ribosomes are being formed. At the centers of the nucleoli are DNA templates, which guide the synthesis of ribosomal RNAs.

Ribosomes and Associated Components Synthesize Proteins

In the cytoplasm are the **ribosomes**, small dense bodies formed from ribosomal RNA and special proteins. In combination with other molecules, these synthesize pro-

teins. The active ribosomes are found clustered together in **polyribosomes** (Fig. 3.9). The ribosomes in a polyribosome are held together because each is attached to the same messenger RNA. Because the messenger RNA carries the information for the particular type of protein to be synthesized, all ribosomes in one polyribosome are making the same type of protein.

Although a polyribosome looks like a fixed object in an electron micrograph or a diagram, in a living cell the ribosomes move very rapidly along the messenger RNA, reading its base sequence and adding amino acids to a growing protein chain. At the end of the messenger RNA, the ribosomes fall off, releasing the completed protein into the cytoplasm.

The Endoplasmic Reticulum Packages Proteins

In a plant cell, the **endoplasmic reticulum (E.R.)** is generally a branched, tubular membrane, often near the periphery of the cell. In three dimensions, it is a closed structure (or several closed structures), so that solute molecules cannot move from the cytoplasm into the inside (the **lumen**) except by passing through the membrane.

There are many places in the cell besides the cytoplasm that need new proteins. These include the membranes (plasma membrane, nuclear envelope, and other mem-

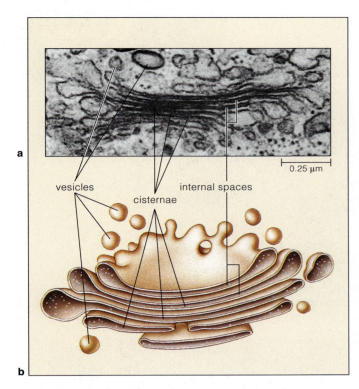

Figure 3.10 A Golgi apparatus and how it moves substances through a cell. (**a**) Transmission electron micrograph. Note the cisternae and associated vesicles. (**b**) A model of a Golgi apparatus.

vesicles
cisternae
internal spaces

0.25 μm

a

b

branes to be described), certain membrane-surrounded spaces (vacuoles, vesicles), and the apoplast. One of the functions of the E.R. is to serve as the site where proteins are synthesized and packaged for transport to these locations. Ribosomes in a polyribosome attach to the surface of the E.R. In an electron micrograph, this forms a structure known as **rough E.R.** (see Fig. 3.9), as opposed to the **smooth E.R.**, which does not have ribosomes attached. As the ribosomes synthesize a protein, the protein is injected through the membrane into the lumen. Carbohydrates are often attached to proteins in the E.R. This helps protect the proteins from breakdown by destructive enzymes. The proteins in the lumen may be considered to be packaged—that is, separated from the cytoplasm by a membrane. A small sphere of membrane containing proteins (called a *vesicle*) may bud off from the E.R. and can carry the proteins to other locations in the cell.

The Golgi Apparatus Guides the Movement of Proteins to Certain Compartments

A **Golgi apparatus** looks like a stack of membranous, flattened bladders, called **cisternae**, which often are swollen toward the edges. (It is also called a **dictyosome**.) Although it is difficult to see in an electron micrograph,

there is a direction to the organelle, with the cisternae on one side differing from those on the other.

The Golgi apparatus directs the flow of proteins and other substances from the E.R. to their destinations in the cell. For example, cell wall proteins, hemicellulose, and pectin in E.R. packages pass through the cisternae of a Golgi apparatus and then move to the plasma membrane inside a small membranous sphere. When the sphere joins the plasma membrane, its membrane becomes part of the plasma membrane. Its protein, hemicellulose, and pectin contents are then released to the outside of the cell (Fig. 3.10).

The trafficking between the Golgi apparatus, the E.R., and other organelles of the cell is rapid and continuous, so much so that even though the organelles usually appear to be separate in electron micrographs and can be isolated from one another by biochemical techniques, they can be considered parts of a complex network called the **endomembrane system**.

3.5 **THE ORGANELLES OF ENERGY METABOLISM**

The synthesis, packaging, and transport of proteins—and many other functions—require energy. Plant cells have two specialized organelles that provide this energy: plastids and mitochondria.

Plastids Convert Light Energy to Chemical Energy

Plastids are complex organelles found in every living plant cell. One cell may have 20 to 50 plastids, each 2 to 10 μm in diameter. Characteristically, they are surrounded by a double membrane. They contain DNA and ribosomes—a full protein-synthesizing system similar but not identical to the one in the nucleus and cytoplasm. Some of the proteins of the plastid are made by this system; others are made in the cytoplasm and transported into the plastid across its outer membranes.

Dividing plant cells always contain some small plastids, called proplastids. These have a few short internal membranes and some crystalline associations of membranous material, called *prolamellar bodies*. As cells mature, their plastids develop and acquire special characteristics. In this process the components of prolamellar bodies apparently are reorganized and combined with new lipids and proteins to form more extensive internal membranes.

Many cells in leaves contain plastids called **chloroplasts** (*chloro-* is derived from a Greek word for a yellow-green color; Fig. 3.11a–c). Chloroplasts contain an elaborate array of membranes, the **thylakoids**. Incorporated in the thylakoid membranes are proteins that bind the green-colored compound **chlorophyll**. It is this compound that

Figure 3.11 Plastids. (**a**) Model of a chloroplast. (**b**) A maize (*Zea mays*) leaf chloroplast, showing dense thylakoid membranes. (**c**) A maize proplastid, with only few internal (prolamellar) membranes. (**d**) A small leukoplast from an inner white leaf of endive (*Cichorium endiva*). (**e**) An amyloplast from a bean (*Phaseolus vulgaris*) seedling. (**f**) A chromoplast from a mature red pepper (*Capsicum* sp.).

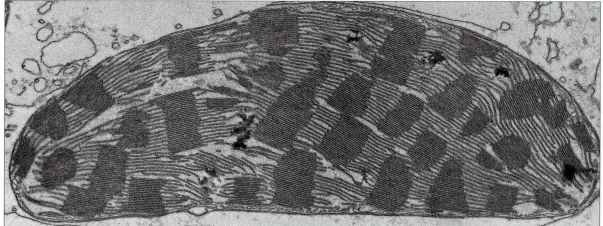

two outer membranes

a thylakoids

stroma

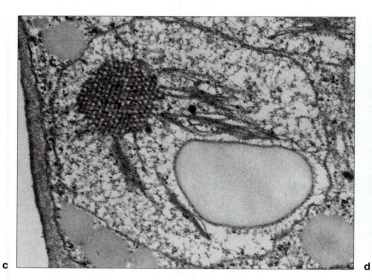

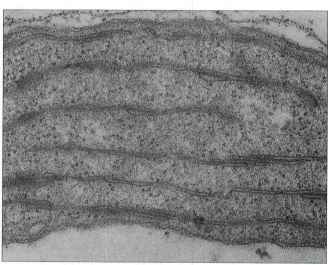

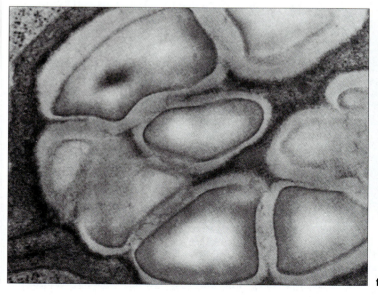

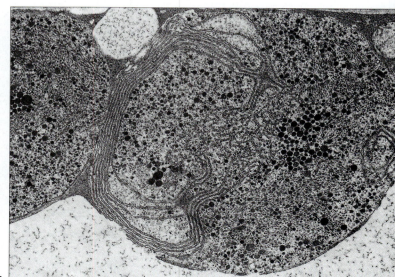

gives green plant tissues their color. Surrounding the thylakoids is a thick solution of enzymes, the **stroma**. Together, the proteins in the thylakoids and the stromal enzymes perform photosynthesis (see Chapter 10), during which light energy is converted to chemical energy. These plastids can also store carbohydrates, the products of photosynthesis, in the form of starch grains.

In roots and some nongreen tissues in stems, the plastids are **leukoplasts** (Fig. 3.11d—*leuko-* is derived from the Greek word for "white"). In these plastids, the thylakoids are missing, although there may be some less organized membranes. These plastids are also able to store carbohydrates (imported from photosynthetic tissue) in the form of starch because they have the enzymes for starch synthesis. Under a light microscope, one often sees white, refractile, shiny particles: starch grains. When leukoplasts contain large granules of starch they are often called **amyloplasts** (Fig. 3.11e—*amylo-* comes from the Greek word for "starch").

In certain colored tissues (for instance, tomato fruits and carrot roots) the plastids accumulate high concentrations of specialized lipids—carotenoids and xanthophylls. The orange-to-red color of these lipids gives the plastids their name, **chromoplasts** (Fig. 3.11f—*chromo-* meaning "color").

Mitochondria Make Useful Forms of Chemical Energy

In plant cells, as in the cells of all eukaryotes, there are **mitochondria** (Fig. 3.12). Mitochondria are made of two membrane sacs, one within the other. The inner membrane forms folds or fingerlike projections called **cristae**. These folds increase the surface area available for chemical reactions taking place on the membrane. In the center of the organelle, inside the inner membrane, is a viscous solution of enzymes, the **matrix**. Like plastids, mitochondria contain DNA and ribosomes, and they are able to synthesize some types of proteins.

Mitochondria are best known as the sites of oxidative respiration, the places where organic molecules are broken down into CO_2 and H_2O (see Chapter 9). During this process, some of the energy that is released is used to synthesize the energy-rich molecule ATP. ATP powers many of the important chemical reactions in the cell. Mitochondria are the source of most of the ATP in any cell that is not actively photosynthesizing.

3.6 OTHER CELLULAR STRUCTURES

Cells have several types of organelles besides those already described. Some of these organelles seem rather passive, serving mainly as storage compartments. Others are hard to see (even by microscopy), but they are very active, moving many of the other organelles around the inside of the cell in a continuous dance.

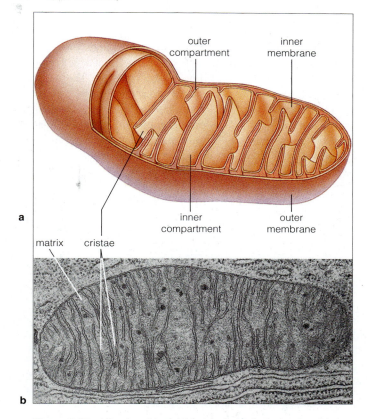

Figure 3.12 Mitochondria. (**a**) Model of a typical mitochondrion. (**b**) Transmission electron micrograph of a mitochondrion from the young leaf of a safflower (*Carthamus tintorius*) seedling.

Vacuoles Store Substances

In each living plant cell, a large compartment, bounded by a single membrane, makes up a large fraction of the cell volume (see Fig. 3.3). This is the **vacuole**. The membrane surrounding it is called the tonoplast. Like the plasma membrane, the tonoplast has a number of embedded protein pumps and channels that control the flow of ions and organic molecules into and out of the vacuole.

Depending on the cell, the vacuole may have any of several functions. It may accumulate ions (or other substances), increasing the osmotic pull of water into the cell and raising the turgor pressure inside the cell. It may store nutrients such as sucrose (common sugar). (Our commercial sugar comes from extracting sucrose from the vacuoles of sugarcane stem or sugar beet root storage cells.) The vacuole may also accumulate compounds that are toxic to predators. Finally, it may serve as a dump for wastes that the cell cannot keep and cannot excrete. Sometimes substances accumulate in the vacuole to such a high concentration that they form crystals.

Other Organelles Transport and Store Substances and Compartmentalize Reactions

The cytoplasm has many types of small, round bodies surrounded by a single membrane (see Fig. 3.3), as is the vacuole. These include **vesicles**—both the small spheres of membrane that bud off the E.R. and carry proteins (E.R. vesicles) and those that travel from the Golgi apparatus to the plasma membrane (Golgi vesicles). There are also similar but more permanent organelles in the cell. In some specialized cells, protein bodies and lipid bodies store proteins or lipids. **Peroxisomes** and **glyoxysomes** serve as compartments for enzymatic reactions that need to be separated from the cytoplasm (because some products of those reactions would be toxic to enzymes in the cytoplasm). **Lysosomes** contain enzymes that break down proteins, carbohydrates, and nucleic acids. The lysosomes may have some function in removing wastes within the living cell. In addition, when they break they release their enzymes, which dissolve the cell. In the life of a plant, there are several times when it is important to get rid of certain cells; the lysosomes accomplish this.

The Cytoskeleton Controls Form and Movement Within the Cell

In high-resolution electron micrographs, one can sometimes see a collection of long, filamentous structures within the cytoplasm. These are evidence of a higher order of organization of the organelles within the cell. They form the **cytoskeleton**, which sometimes keeps the organelles in particular places and sometimes directs their movement around the cell (the movement is called **cyclosis** or **cytoplasmic streaming**).

There are several different types of structures in the cytoskeleton. One type is relatively thick (0.024 μm in diameter)—the **microtubule**. It is assembled from protein subunits called *tubulin*. Tubulins assemble in a helical manner to form a hollow cylinder. Once assembled, microtubules are fairly rigid, although they can lengthen by the addition of tubulin molecules to one end (or shorten by the loss of tubulins from the same end); so the length of a microtubule can change rapidly.

Certain *motor proteins* can move along microtubules. These proteins—one is called kinesin, another dynein—move by stepping along the microtubule strand, making and breaking connections with adjacent tubulin subunits. The motor proteins may be attached to other organelles, such as vesicles. In this way, microtubules guide the movement of organelles around the cytoplasm. The power generated by motor proteins comes from the breakdown of a high-energy molecule, either ATP or GTP (see Chapter 8 for a description of these high-energy molecules).

Microtubules are key organelles in cell division (see p. 43). They also form the basis of *cilia* and *flagella*, motile organs that move cells by beating back and forth. Cilia and flagella are never found in the cells of flowering plants, but they are important to some algae and to male gametes of lower plants.

Another type of structure in the cytoskeleton is the **microfilament** (Fig. 3.13). The microfilament, thinner (about 0.007 μm in diameter) and more flexible than the microtubule, is made of protein subunits called *actin* that fit together in a long, helical strand. Microfilaments are often found in bundles. Like microtubules, they can serve as guides for the movement of organelles. However, myosin, rather than kinesin and dynein, is the motor protein that moves along a microfilament. The energy for this movement comes from the breakdown of the high-energy molecule ATP.

Many types of specialized proteins connect microtubules and microfilaments to other organelles. These connections are thought to coordinate many of the processes of the cell. For instance, they might direct vesicles to move to the plasma membrane and deposit new wall material when triggered by the presence of hormones. This coordination is most important, and most spectacular, during the complex process of cell reproduction. As you read the next section, notice how microtubules appear to direct several of the events of cell division.

3.7 THE CELL CYCLE

What Are the Phases of the Cell Cycle?

In higher plants, cells form aggregates—tissues and organs—that have specific structures and functions (see Chapter 4). The production of new cells allows the plant body to grow by forming new organs and by increasing the number of cells in an existing tissue or organ. The following section examines the beginning of the tissue- and organ-forming process, describing how cells divide and what mechanisms control the process.

The process of division occurs in special regions of the plant called **meristems**. (Meristems are found at the tips of roots and shoots and in some other regions of a plant.) When cells divide they must progress through a series of steps called the **cell cycle**. One cell cycle is the interval of time between the formation of a cell and its division to form two new cells.

The cell cycle has four phases: G1, S, G2, and M (Fig. 3.14). Traditionally, the G1, S, and G2 phases of the cycle are grouped together and called **interphase**. The remaining phase, cell division (M) is actually composed of two parts—**mitosis** (nuclear division) and **cytokinesis** (cytoplasmic division).

Specific Metabolic Events Occur in Each Cell Cycle Phase

Cells in each phase of the cell cycle are unique in structure and molecular composition. During G1, the cell prepares itself metabolically for DNA synthesis. These preparations include both the accumulation and synthesis of specific enzymes to control DNA synthesis and the

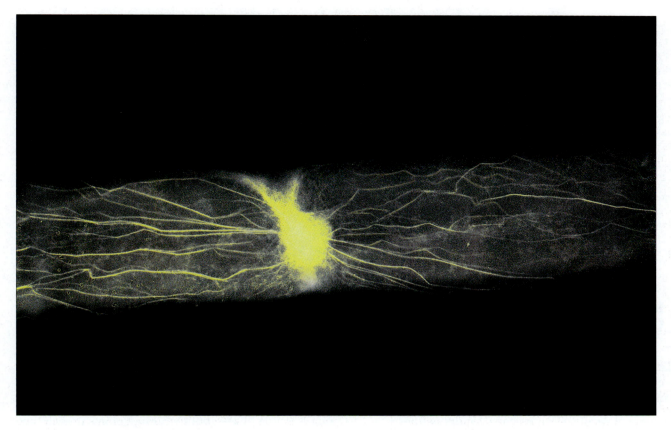

Figure 3.13 Microfilaments—one of the structures in the cyto-skeleton—in a cotton (*Gossypium hirsutum*) suspension cell. These microfilaments have been complexed with a fluorescent compound that glows yellow when irradiated with ultraviolet light.

production of the DNA subunits so that a supply is on hand when synthesis begins.

The second phase of the cell cycle is the S phase (*S* stands for synthesis of DNA). During this portion of the cycle, the cell duplicates its DNA molecules (see Chapter 2). The time needed for a cell to progress through the phases of the cell cycle depends upon the amount of DNA per nucleus. Plants with more DNA in their nuclei have longer cell cycle times than plants with less DNA. The time for a complete cycle ranges from a few hours to days.

Unless cells are metabolically active, they cannot progress through the phases of the cell cycle. Cells must respire to produce energy to stay alive (see Chapter 9), and they must be able to synthesize enzymes and other proteins needed for specific cell cycle events. DNA polymerase, the key protein in the synthesis of new DNA, is an example of a cell-cycle–specific protein. The amounts of this protein and of histones—proteins that form the nucleosomes around which DNA winds—increase at the end of G1 and during the S phase. The abundances of other proteins—such as tubulin, which forms microtubules during mitosis—can be correlated with specific periods in the cell cycle.

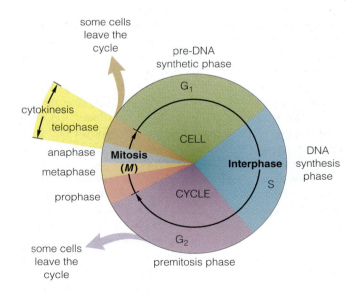

Figure 3.14 The cell cycle, consisting of the G1, S, G2, and M phases. Approximately 50% of new cells formed by cell division (mitosis) leave the cycle at either G1 or G2 and begin to differentiate; the remaining cells will cycle again.

Figure 3.15 Cell cycle control points. (**a**) Two control points located in G1 and G2 regulate the cell cycle in plant cells and in fission yeast. Consequently, a plant cell population stressed by starvation will show cells arrested in G1 and G2 because the cells lack critical proteins needed for DNA synthesis (S phase) and mitosis (M phase). The G1 control points are regulated by the activity of a gene called *cdc2*. (**b**) Budding yeast and animal cells have only one control point, in G1, which is also regulated by a *cdc2* gene.

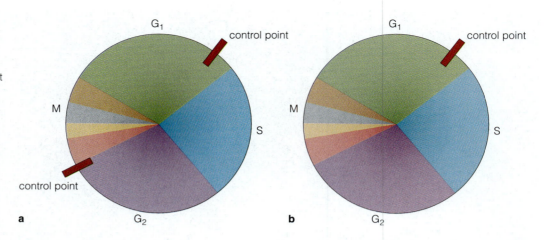

3.8 REGULATION OF THE CELL CYCLE

The Principal Control Point Hypothesis Identifies the Control Points in the Cell Cycle

The high-energy molecule ATP is essential for the synthesis of DNA and proteins, and for other processes. Cycling cells in meristems require a constant supply of carbohydrates in order to make ATP. If root tips are removed from seedlings and placed into sterile culture, their cells will progress through the cell cycle only if carbohydrate (sucrose) is added to the culture medium. Without sucrose, the cells stop cycling, but only at specific places in the cell cycle. Jack Van't Hof and his collaborators at Brookhaven National Laboratory observed that such cells became arrested in G1 and G2. The ratio of cells arrested in G1 and G2 in such experiments varies among different species. For example, in peas (*Pisum sativum*) 50% stop in G1 and 50% stop in G2, whereas in sunflowers (*Helianthus annuus*) 90% stop in G1 and 10% stop in G2. Cell cycle progression resumes after sucrose is added back to the culture medium.

These experiments led Van't Hof to develop the Principal Control Point Hypothesis, in which he proposed that control points exist in the plant cell cycle at G1 and G2 (Fig. 3.15a). If a cell progresses past the G1 or G2 control point, it then automatically progresses through DNA synthesis (S phase) or cell division (M phase), respectively. Metabolically this means that during G1 phase a cell synthesizes certain proteins in preparation for DNA synthesis in the S phase. If the cell forms these critical macromolecules, it progresses through the S phase; but if it does not, the cell is arrested in G1. The same interpretation would apply to the G2 control point, except that at that time a dividing cell would need proteins that act during the M phase.

Special Genes Regulate the Cell Cycle

Work with yeast has enabled researchers to understand how the cell cycle control points work. There are two different kinds of yeast. *Fission yeast (Schizosaccharomyces pombe)* simply divide into two new cells, whereas *budding yeast (Saccaromyces cerevisiae)* multiply by budding off a new daughter cell. In both cases the cells pass through all four cell cycle phases, but the control points in each type of yeast differ. In the case of fission yeast (Fig. 3.15a), the cell cycle is regulated at two points in the cycle—G1 and G2—just as in higher plant cells. In the case of budding yeast, only the G1 control point exists (Fig. 3.15b), as in animals. Yeast researcher Paul Nurse in England has identified a particular sequence of nucleotides in the DNA, called the *cdc2* gene, which must be active for a budding yeast cell to pass the control point. It is now known that a similar gene must be active at the G2 control point in fission yeast.

The *cdc2* gene is known to code for (to provide the genetic information for) an enzyme involved in *phosphorylation* reactions; in these, the enzyme (a protein) *adds* a phosphate functional group ($-O-PO_3H_3$) to another protein. Another kind of enzyme catalyzes *dephosphorylation* reactions; this *removes* phosphate from proteins. Proteins that have phosphate added or removed will change their three-dimensional shape and therefore their biological activity. In some cases the phosphorylated form will be the active form of the protein, but in other cases the dephosphorylated form will be the active form. This kind of phosphorylation/dephosphorylation reaction seems to be the key to regulating the cell cycle (Fig. 3.16).

Two key proteins, the **cdc2 protein kinase (C-PK)** and **cyclin**, seem to be the regulators that control the progression of cells through the cell cycle. By working together, C-PK and cyclin are responsible for triggering the G1 control point (so that cells enter into S phase) and the G2 control point (so that cells enter into M phase). The new molecule formed by the combination of

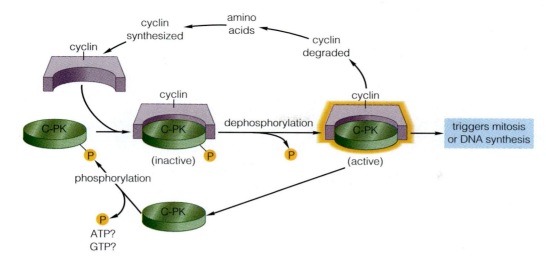

Figure 3.16 Diagram showing how the cdc2 protein kinase (C-PK) and cyclin trigger cell cycle progression in cycling cells. Starting in the upper left, cyclin has joined to C-PK in a phosphorylated form to make a new, but inactive, protein. An enzyme removes the phosphate (Ⓟ in figure) to activate this protein. This form of the protein acts as the trigger to switch on S phase and M phase. It is recycled by separating C-PK, which is phosphorylated again and recycled. The cyclin is broken down into amino acid subunits and must be resynthesized.

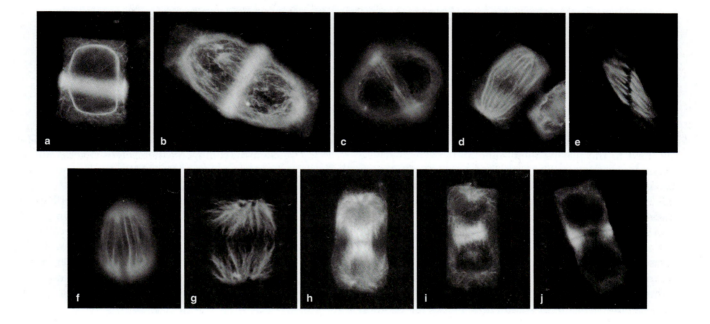

Figure 3.17 The role of microtubules in mitosis. Onion (*Allium cepa*) root tip cells were treated with tubulin antibodies tagged with a fluorescent dye and photographed by fluorescent microscopy. Thus, the photographs show only the microtubules. (**a–c**) Early prophase, showing the preprophase band of microtubules plus some microtubules along the outside of the cell just inside the cell wall. (**d, e**) The spindle apparatus in early metaphase. (**f**) Metaphase chromosomes are located in the center of the cells. (**g**) Anaphase chromosomes migrate back toward the poles, and the spindle apparatus microtubules will be dispersed. (**h, i**) The dense accumulation of microtubules in the center of the cell shows the early formation of the cell plate. (**j**) Late telophase showing microtubules only at the periphery of the cell plate.

C-PK and cyclin has a phosphate attached in its inactive form. The enzyme phosphatase removes the phosphate, making the active form of the protein. This final form of the combined protein also acts as a protein kinase, and it is this protein that is the actual trigger to start the S and M phases. After doing this, it is recycled in a three-step process. In the first step the molecule is broken into its two parts. The cdc2 protein kinase is reused after having phosphate added to it. The cyclin protein is degraded into component amino acids, and new cyclin must be resynthesized in order to restart the process.

Microtubules Set the Plane of Cell Division

Microtubules, discussed earlier in this chapter (p. 40), are intimately involved in many aspects of the cell cycle, from the regulation of cell wall formation and cell shape to the movement of chromosomes and control of the plane of cell division (Fig. 3.17). During G1 and S of the cell cycle,

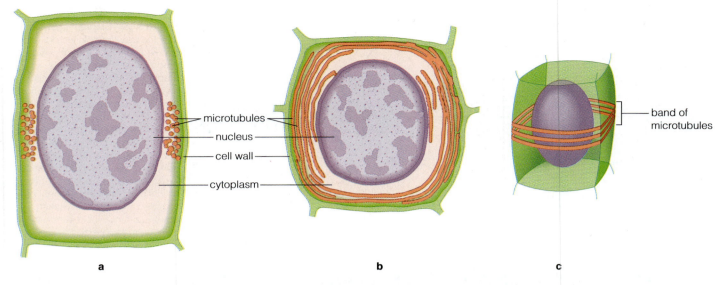

microtubules
nucleus
cell wall
cytoplasm

a

b

band of
microtubules

c

Figure 3.18 Formation of the preprophase band (PPB) of microtubules in meristematic cells of a young leaf of tobacco (*Nicotiana tabacum*). (**a**) Section at right angles to the plane of the future cell plate shows a cross section of microtubules. (**b**) Section in the plane of the future cell plate shows microtubules encircling the nucleus. (**c**) Three-dimensional drawing of (**a**) and (**b**). (Redrawn after K. Esau and J. Cronshaw. *Protoplasma* 65 (1). © 1966 by Springer-Verlag. Reprinted with permission of the publisher.)

the microtubules are located around the periphery of the cell, next to the plasma membrane and cell wall. These microtubules are involved in the deposition of cellulose in the cell wall. The cell will elongate at right angles to the axis of the microtubules and cellulose, a process described in more detail in Chapter 15.

During G2 the microtubules move into a new location and form a band surrounding the nucleus, but pressed close to the cell wall (Figs. 3.17a–c and 3.18). This organization is called the **preprophase band (PPB)** of microtubules. Its formation precedes mitosis by hours. The orientation of the PPB somehow marks the position of the new cell wall that will form at the end of mitosis. The new cell wall defines the *plane of cell division*.

Several factors may influence the position of the PPB. Within tissues, gradients of chemicals such as plant hormones regulate the start of differentiation events such as mitosis. The hormone gradients may induce the movement of the microtubules to the PPB as a first step in the overall induction of cell division.

Mitosis Occurs in Stages and Is Followed by Cytokinesis

The purpose of mitosis is to separate the DNA, which was doubled in amount during S phase, in such a way as to produce two complete sets of genetic information. The process of mitosis is generally divided into four phases to make it easier to understand.

PROPHASE During G1, S, and G2 (interphase), the DNA molecules are long and apparently tangled in the nu-

cleus. At the start of **prophase** the DNA molecules thicken by coiling on themselves several times to form dense chromosomes (Fig. 3.19a, b). The nucleolus gradually disappears.

It becomes apparent during late prophase that each chromosome is composed of two intertwined DNA molecules connected by a constricted region called a **kinetochore** (Fig. 3.19b, c). The nuclear membrane breaks down during late prophase, but the nucleoplasm does not mix with the general cytoplasm; there is a clear zone between them, devoid of organelles.

METAPHASE In **metaphase**, the chromosomes now arrange themselves on the equatorial plane of the cell, usually with their kinetochores aligned (Fig. 3.19c). This position may be in the middle of the cell, or it may be off center, resulting in an unequal cell division. The spindle fibers, composed of bundles of microtubules, can now be seen. The spindle extends from the poles near the ends of the cell to their attachment point on each chromosome, the kinetochore. All of the spindle fibers collectively are called the *spindle apparatus*.

The chromosomes are now distinct bodies of two closely associated halves, each half known as a **chromatid**. In plants such as corn (*Zea mays*), which have been intensively studied, each chromosome can be recognized (based on its shape) and numbered. There are 20 chromosomes in corn, but only 10 different types that can be distinguished by their size and form. The 20 chromosomes of corn may be arranged in 10 pairs. Geneticists have mapped corn chromosomes, showing the relative positions of many genes along them. Because each gene is

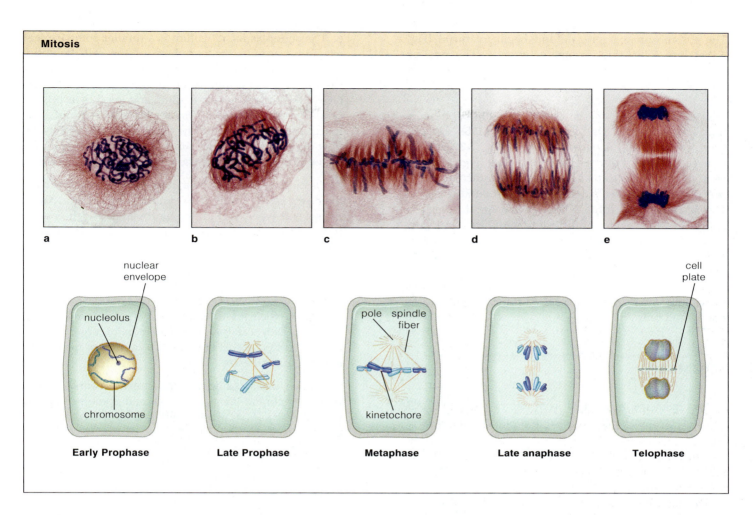

Figure 3.19 The stages of mitosis, shown in light micrographs of cells from the endosperm of the blood lily (*Haemanthus katherinae*): proteins (including microtubules) are red; chromosomes are blue. These cells are unusual in that they do not have cell walls. The lower row diagrams mitosis in the more typical meristem cell, with a cell wall. Cells from G2, with duplicated chromosomes, form the starting material for mitosis. (**a**) Early prophase: The chromosomes start to become thicker and shorter by coiling tightly. The nuclear envelope is still present. (**b**) Late prophase: The nuclear envelope and nucleolus have disappeared. On each chromosome, a kinetochore becomes visible. Spindle fibers are starting to appear. (**c**) Metaphase: The coiled chromatids are distinct. The two chromatids moving to the upper pole have separated, and the spindle has organized. (**d**) Anaphase: The chromatids move to opposite poles of the cell. (**e**) Telophase: Chromatids have aggregated at the opposite poles of the cell. The nuclear envelopes have not yet formed. A cell plate has started to form.

replicated, when the chromosomes split longitudinally, each chromatid contains a full set of genes.

ANAPHASE The chromosomes do not remain long in the equatorial plane. The individual chromatids of each pair soon separate from each other and move to opposite poles of the cell (Fig. 3.19d). This period is called **anaphase**.

The mechanism of chromosome movement and the functional role of the spindle fiber microtubules are not fully understood. There are two major theories to explain chromosome movement. The *assembly–disassembly theory* contends that microtubules move chromosomes by losing tubulin molecules from one end of the spindle. The removal of tubulin subunits from both poles of the spindle would shorten the spindle and pull the chromosomes apart. The *microtubule sliding theory* contends that the microtubules slide over each other, thereby effecting chromosome movement. Motor proteins may provide the sliding force.

TELOPHASE When the divided chromosomes have reached the opposite poles of the cell, **telophase** begins.

The chromosomes aggregate and begin to uncoil (Fig. 3.19e) into long, thin chromatin strands. The nuclear envelope and the nucleolus re-form, and the cells return once again to G1 of the cell cycle.

CYTOKINESIS Cytokinesis begins before telophase is finished. A new cell wall, called the **cell plate**, starts to form between the separated nuclei (Fig. 3.19e). This process involves both the **phragmoplast** (Fig. 3.17i), a band of microtubules that re-forms perpendicular to the cell plate, and many small membrane-bound vesicles. The small vesicles contain cell wall material, which is deposited in the center of the cell first and then outward until the new wall attaches to the side wall. The point of attachment, you will recall, was previously "marked" by the PPB (Fig. 3.18). Some of the new cells that are formed by mitosis and cytokinesis will divide again in plant meristems, and others will begin to differentiate to form specialized cells.

SUMMARY

1. All organisms are formed from one or more cells. All cells demonstrate the basic functions of life, including controlling their size, accumulating nutrients, extracting energy from the environment, reproducing, and withstanding or tolerating environmental stresses. Cells are produced only by reproduction from previously existing cells.

2. Because cells are very small, an essential tool for their study is microscopy. Light microscopes can be used to visualize live samples, but contrast and resolution are major problems in light microscopy. These limitations can be minimized or circumvented by confocal microscopy, transmission electron microscopy, and scanning electron microscopy.

3. The plasma membrane surrounds the cytoplasm of the cell. It is formed from a lipid bilayer and proteins. The lipid bilayer prevents the diffusion of water-soluble molecules into or out of the cell. Proteins in the plasma membrane control the passage of molecules through the plasma membrane, among other functions.

4. The cell wall, formed from carbohydrates and proteins, prevents excess water from entering the cell through osmosis and provides the turgor pressure that gives most plant organs a firm structure.

5. The nucleus contains chromosomes made of DNA and protein. DNA serves as a template for messenger RNA synthesis; messenger RNA directs protein synthesis. The nucleolus (or plural, nucleoli) inside the nucleus is responsible for the synthesis of ribosomal RNA. Ribosomal RNA plus specific proteins from ribosomes leave the nucleus and, together with messenger RNA in the cytoplasm, form the polyribosomes, where proteins are actually synthesized.

6. The endoplasmic reticulum is a tubular compartment extending around the periphery of the cell. Polyribosomes that are synthesizing proteins destined for export from the cell or for import into vesicles attach to the surface of the endoplasmic reticulum and inject their protein products into its lumen. Proteins leave the endoplasmic reticulum inside vesicles. The Golgi apparatus directs the flow of proteins and other substances in vesicles from the endoplasmic reticulum to their destinations in the cell.

7. Plastids are bounded by a double membrane and have their own DNA- and protein-synthesizing systems. There are several types, including chloroplasts, which enclose thylakoid membranes and catalyze the reactions of photosynthesis; amyloplasts, which store starch; and chromoplasts, which contain high concentrations of colored carotenoids and xanthophylls.

8. Mitochondria also are bounded by a double membrane and have their own DNA- and protein-synthesizing systems. They catalyze the oxidation of organic compounds to CO_2 and H_2O and use a portion of the energy released to synthesize ATP.

9. Among the specialized compartments bounded by a single membrane are the vacuole and some specialized vesicles. Vacuoles store salts, sugars, defensive compounds toxic to predators, and wastes. Some vesicles store energy-rich lipids or proteins (liposomes, protein bodies); some compartmentalize enzymatic reactions (glyoxysomes, peroxisomes); some control the release of self-destructive enzymes (lysosomes).

10. The cytoskeleton, formed from protein filaments, directs the organization and movement of organelles within the cell. Motor proteins (dynein, kinesin, myosin) move along filaments, dragging other organelles with them.

11. Cells reproduce by passing through a series of events called the cell cycle. The four phases of the cell cycle are: G1, the pre-DNA synthesis phase; S, the DNA synthesis phase; G2, the premitotic phase; and M, the cell division phase.

12. Higher plant cells and fission yeast have control points at G1 and G2. If critical metabolic events do not occur at these points, DNA synthesis and mitosis will not occur. Animal cells and budding yeast have only a G1 control point. The *cdc2* gene in yeast codes an enzyme that is essential for regulating the yeast cell cycle at the control point.

13. Microtubules have several roles in plant cell division, including control of cell plate location and regulation of chromosome movement. The preprophase band of microtubules (PPB) forms in the G2 phase and marks the plane of cell division.

14. The stages of mitosis are prophase, metaphase, anaphase, and telophase.

Mitotic Stage	Summary of Events
Prophase	Chromosomes coil; the nuclear envelope and the nucleolus break down.
Metaphase	Chromosomes move to the equatorial plane; the spindle apparatus forms and attaches to the chromosomes at the kinetochores.
Anaphase	Chromosomes move to the poles by some mechanism involving the spindle fibers.
Telophase	Chromosomes uncoil; the nuclear envelope and the nucleolus re-form.

15. During cytokinesis the cell plate, containing cell wall materials, forms first in the middle of the cell and then moves to the periphery, dividing the cytoplasm in two.

Questions

1. Explain why it is necessary to use a microscope to see a cell. Describe the difference between the image of a cell seen at high magnification but low resolution and one seen at the same magnification and high resolution.

2. Assume a cell is a cube 10 μm on a side. What is the volume of the cell in cubic μm? What is the volume in cubic meters? How many of these cells would fill a teaspoon (approximately 5 cubic centimeters)?

3. List the organelles that are needed to synthesize a protein and move it to the outside of a cell.

4. Which of the following organelles is (are) absolutely essential for the synthesis of a protein?

nuclear envelope

ribosome

microtubule

endoplasmic reticulum

Golgi apparatus

Which will be needed for treatment of the protein after it is made?

5. Almost every living plant cell has mitochondria. Explain why the presence of mitochondria is essential for a cell's well-being.

6. In what organ(s) of the plant would you find the most chloroplasts?

7. Assume that a cell must reproduce itself exactly. List (as completely as you can) the events that must occur in G1, S, G2, and M phases for the reproduction to be successful.

8. An investigator starved the cells of a yeast for sucrose, and after a few hours she found that they were all in G1 phase of the cell cycle. Were these cells more likely to be fission yeasts or budding yeasts?

9. Imagine a cell that loses the ability to make cdc2 kinase. What would happen to its ability to undergo cell division? In which phase of the cell cycle would you be likely to find it?

10. List the phases of mitosis and describe what happens in each. If you add a chemical that blocks the formation of microtubules to a cell that is starting mitosis, mitosis eventually stops. In what phase will it stop?

Further Readings

Francis, D., and N. G. Halford. 1995. "The Plant Cell Cycle." *Physiologia Plantarum* 93: 365–74. A rather advanced-level article.

Jensen, William A. 1970. *The Plant Cell*, 2d ed. Belmont, Calif.: Wadsworth. An excellent collection of transmission electron micrographs of plant cell components. Out of print, but may be available in libraries.

Murray, A. W., and M. W. Kirschner. 1991. "What Controls the Cell Cycle?" *Scientific American* 264: 56–63. A review of the role of specific genes in regulating the cell cycle in yeast and certain animal cells.

Ormrod, J. C., and D. Francis, eds. 1993. *Molecular and Cell Biology of the Plant Cell Cycle*. Dordrecht, Netherlands: Kluwer.

Wolfe, Stephen L. 1993. *Molecular and Cellular Biology*. Belmont, Calif.: Wadsworth. An advanced treatment of cell structure and function.

4

THE ORGANIZATION OF THE PLANT BODY: CELLS, TISSUES, AND MERISTEMS

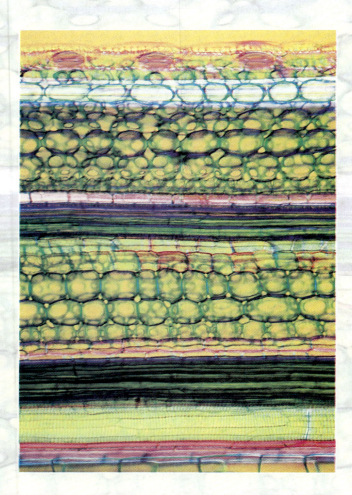

1. The plant body is composed of individual cells that are organized into aggregates of cells called *tissues*. The cells of each tissue function as a unit. Not all functional cells are alive at maturity.

2. Simple tissues are composed of cells that are all of the same type. Complex tissues are composed of more than one cell type. Tissues may function as structural supports, protective coverings, or transporters of water and nutrients; or they may be multifunctional.

3. Meristems are the sites of cell division and differentiation in the plant body. A hierarchy of meristems exist in the plant body, each with a specific role in plant development.

4.1   ORGANIZATION OF THE PLANT BODY

The next time you're outside, notice the amazing variation in the form that plants take. Despite the really big differences in the forms of plants, they all have adopted a basically common mechanism for development and a similar internal structure. In this and the next few chapters we'll examine the internal structure of the plants that are most noticeable: the vascular plants. These include ferns, cone-bearing plants (gymnosperms) like pine trees, and flowering plants (angiosperms) like rose bushes and grasses.

The plant body of most vascular plants consists of an aboveground part—the **shoot system**, which includes stems, leaves, buds, flowers, and fruit—and a belowground part—the **root system**, which includes the roots (Fig. 4.1). This plant body is constructed from millions of tiny cells, each having a characteristic shape and function (see sidebar, "Plant Anatomists as Detectives?"—p. 52). In this chapter we will examine several different cell types, tissues (aggregates of cells), and their origins in unique parts of the plant body called *meristems*.

4.2   PLANT CELLS AND TISSUES

Around each cell is a cell wall. Living cells filled with water exert force (turgor pressure) against their walls, making each cell a rigid box. Plant cells are glued to each other by a material called *pectin*, and collectively they form a very strong yet flexible plant body.

Plants consist of many different types of cells, and the cells are organized into aggregates called **tissues**. As is true in animals, organs are composed of tissues arranged in different patterns. Tissues in the plant body are made up of both living and dead cells. The dead cells, often with thick, strong cell walls, are retained as strengthening cells. Knock on your wood table right now, and you'll see firsthand how strong and hard these dead cells are.

Table 4.1 Vascular Plant Tissues and Cell Types	
Simple Tissues	**Cell Types**
Parenchyma tissue	Parenchyma cell
Collenchyma tissue	Collenchyma cell
Sclerenchyma tissue	Fiber
	Sclereid
Complex Tissues	**Cell Types**
Xylem	Vessel member
	Tracheid
	Fiber
	Parenchyma cell
Phloem	Sieve-tube member
	Sieve cell
	Companion cell
	Albuminous cell
	Fiber
	Sclereid
	Parenchyma cell
Epidermis	Guard cell
	Epidermal cell
	Subsidiary cell
	Trichome (hair)
Periderm	Phellem (cork) cell
	Phelloderm cell
Secretory structures	Trichome
	Laticifer
	(There are many other examples.)

The main tissues of plants may be grouped into three systems (Fig. 4.1). The **ground tissue system** is the most extensive, at least in leaves (mesophyll) and young green stems (pith and cortex). The **vascular tissue system** contains two types of conducting tissues that distribute water and solutes (xylem tissue does this) and sugars (phloem tissue does this) through the plant body. The **dermal tissue system** covers and protects the plant surface (epidermis and periderm).

Some of the tissues are composed mostly of a single cell type; these are called **simple tissues**. Tissues made from aggregates of different cell types are called **complex tissues**. Tissues, simple or complex, act together as a unit to accomplish a collective function. Table 4.1 lists the plant tissues described in this chapter and their cell types.

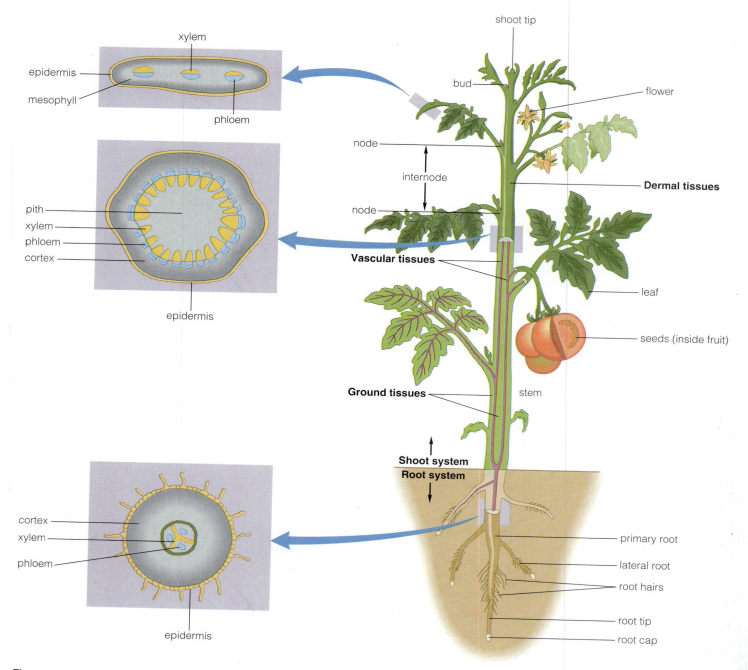

Figure 4.1 The plant body consists of the shoot system (leaves, buds, stems, flowers) and the root system (roots). Each organ is made up of cells organized into tissue systems: dermal, vascular, and ground. One way the vegetative organs (leaves, stems, and roots) differ from each other is in the distribution of the tissues.

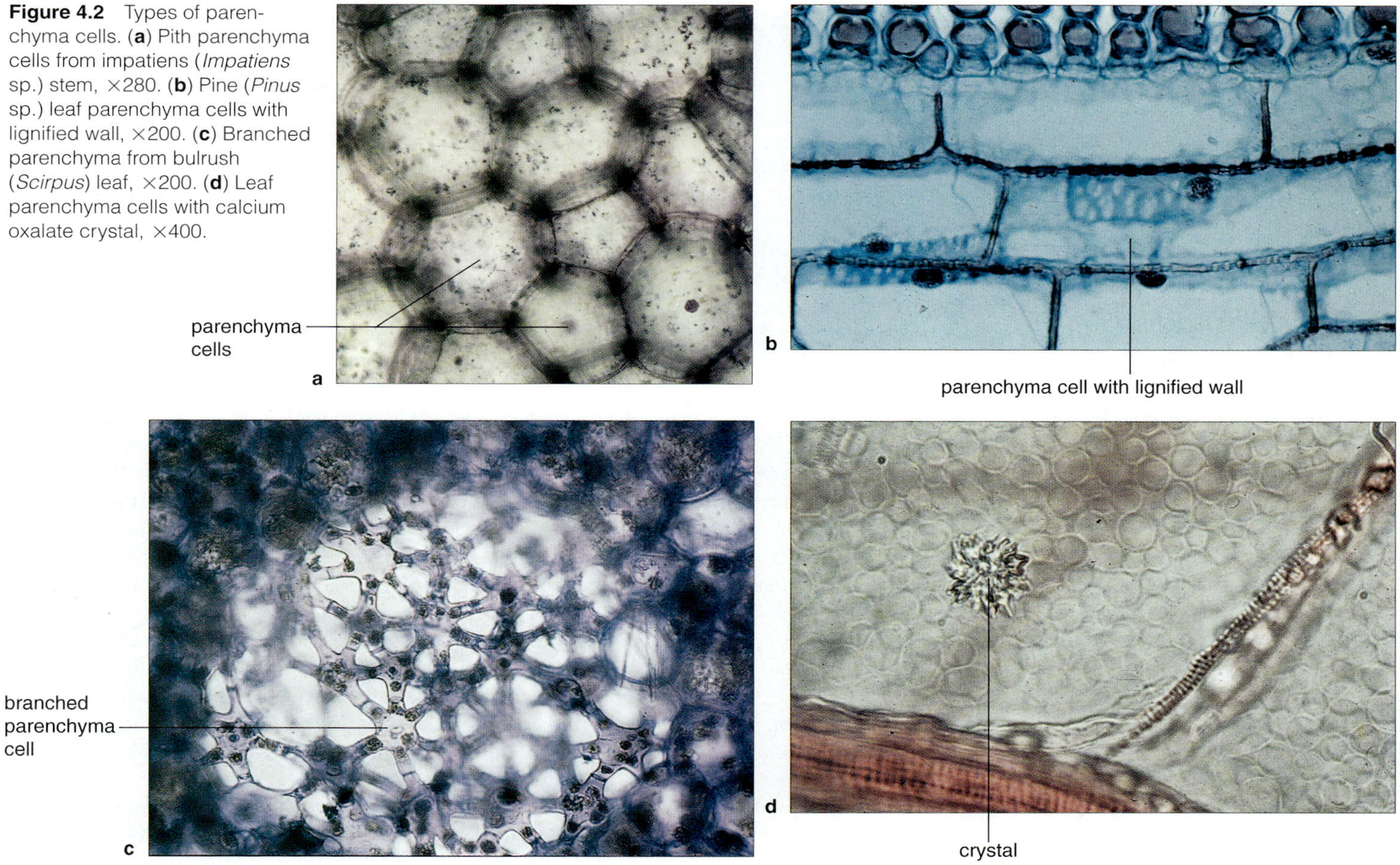

Figure 4.2 Types of parenchyma cells. (**a**) Pith parenchyma cells from impatiens (*Impatiens* sp.) stem, ×280. (**b**) Pine (*Pinus* sp.) leaf parenchyma cells with lignified wall, ×200. (**c**) Branched parenchyma from bulrush (*Scirpus*) leaf, ×200. (**d**) Leaf parenchyma cells with calcium oxalate crystal, ×400.

parenchyma cells

parenchyma cell with lignified wall

branched parenchyma cell

crystal

There Are Three Types of Simple Tissues: Parenchyma, Collenchyma, and Sclerenchyma

The simple tissues are made of cells that are the real working cells of the plant body. They do the photosynthesis, load things in and out of the vascular system, hold up the weight of the plant, store things, and generally conduct the important business and housekeeping chores needed to keep the plant body healthy and functioning.

PARENCHYMA **Parenchyma** cells are usually somewhat spherical or elongated, but they may have diverse shapes (Fig. 4.2). They usually have a thin primary cell wall, but they may have a secondary wall, which is sometimes lignified (Fig. 4.2b). Lignin is a noncarbohydrate polymer that is embedded between the cellulose microfibrils. It renders the wall impermeable to water, so that water movement occurs only through openings called **pits**. Lignin is also quite hard and strong, making lignified cells rigid and supportive even when the cells are dead.

Living parenchyma cells found in all plant organs perform the basic metabolic functions of cells: respiration, photosynthesis, storage, and secretion. Parenchyma cells

usually live for one to two years but have been known to live hundreds of years in some plants, such as cactus.

Crystals of many different shapes and sizes, usually made of calcium oxalate, are commonly found inside the vacuoles of parenchyma cells (Fig. 4.2d). They may be involved in regulating the pH of cells by crystallizing oxalic acid (which is a way of taking it out of solution).

Parenchyma cells may occur as aggregates forming **parenchyma tissue**; the cortex and pith of stems and roots and the mesophyll of leaves are composed of parenchyma tissue (Fig. 4.3a–c). The **cortex** is the region between the plant's epidermal and vascular tissues in most stems and roots. The **pith** is usually composed of storage parenchyma cells and lies at the center of many stems, inside the cylinder of vascular tissues. In some stems, such as corn (Fig. 4.3d), the vascular tissue is dispersed randomly. In this case the parenchyma tissue making up the bulk of the stem is simply called *ground tissue*, and the terms *cortex* and *pith* are not used. The **mesophyll** makes up the bulk of most leaves and is the site of most photosynthesis and water storage in leaves.

When parenchyma cells contain chloroplasts, the tissue they form is referred to as chlorenchyma (Fig. 4.3a). Air spaces usually form between parenchyma cells.

PLANT ANATOMISTS AS DETECTIVES?

FBI agents, police, lawyers, medical and veterinary doctors, agricultural extension specialists, and members of the public have all had occasion to contact me over the past several years concerning the identification of plant material. Usually this material was chewed, digested, burned, dried, or otherwise distorted, so that the identity of the plant or plant organ was no longer apparent. Yet every plant species contains cell types with special structures, shapes, sizes, and staining reactions—characteristics that make it possible to identify the plant. Regardless of the kind of case, similar microtechnique tools and methods are used, and the application of basic knowledge of plant anatomy is required.

Plant anatomists have been involved in many legal cases, but mostly we are asked for opinions dealing with medical or agricultural problems. For example, agricultural extension specialists need help in identifying root specimens from clogged sewers. Veterinarians who suspect that sick animals have eaten poisonous plants will call plant anatomists. Anatomists have also worked with anthropologists interested in ancient basketry or the identification of wood and charcoal pieces found in ancient fire pits.

Probably the most famous legal case involving a plant anatomist was the Lindbergh kidnapping. In 1932 the infant son of celebrated aviator Charles Lindbergh and his wife, Anne Morrow Lindbergh, a noted writer and poet, was kidnapped out of a second-story nursery. The kidnapper sent a ransom note, but the baby was later found dead. The only evidence left at the crime scene was a crude wooden ladder leaning against the second-story window. The police and FBI gave the ladder to the Forest Products Laboratory in Wisconsin for analysis. A plant anatomist there, Arthur Koehler, was able to specifically identify the wood used. After a man named Bruno Hauptmann was arrested for the crime, police found that several boards were missing from the floor of the attic in his house. Pieces of this wood were sent to the Forest Products Laboratory, which positively identified the wood as a match to the ladder. This testimony was used to convict Hauptmann, who was later executed.

Most cases are less poignant. In 1984 I was contacted by the Sheriff's office in Calaveras County, California (the part of California's gold country made famous by Mark Twain's story of the jumping frog contest). In this instance, a man was suspected of growing marijuana (*Cannabis sativa*) in an elaborate hydroponic setup in his attic. The suspect was warned of an impending raid just before the sheriff arrived. He hid the stems of the plants outside his house and tried to burn the stem stumps and roots in his fireplace. The sheriff arrested him and took the partially burned material as evidence. The suspected grower reportedly told a sheriff's deputy that he would get off because it wasn't possible to prove that he was burning marijuana in his fireplace. I examined the material taken as evidence and compared it with known specimens of marijuana from herbarium sheets and from identified marijuana stems taken in other arrests. The charred evidence and the known specimens showed similar pitting patterns in vessel members and similar wood anatomy. After I testified in a pretrial hearing that, based on the comparison, the partially burned material was most likely marijuana, the defendant accepted a plea bargain. The case was never brought to trial.

Several of my cases have come from the San Diego Zoo. One of them involved a group of Hanuman langur monkeys, a rare Asian species. For two years the monkeys had been fed mostly *Acacia* leaves. Suddenly three monkeys died after bouts of weight loss, diarrhea, and vomiting. An autopsy revealed intestinal lesions and plugging with masses of fibrous plant material. I examined some of this material, along with samples of *Acacia* leaf browse and other plant materials within reach of the monkey enclosure. Microscopic examination of this material revealed partially digested vascular strands with attached thick-walled cells and small epidermal fragments. By making polarized light images and comparing them to known specimens, I was able to identify the material as *Acacia* leaf vascular bundles. The keepers changed the monkeys' feed composition, and no further problems have occured.

The point of these stories is that the study of plant anatomy has uses far beyond just knowing what's inside the plant. Basic information and a few simple tools—such as a razor blade, some common dyes, and a microscope—can go a long way toward solving criminal and medical puzzles. So study hard; you never know when your knowledge of plant anatomy might come in handy.

Sometimes the air spaces are very large, especially in the stems and leaves of plants that grow in the water of marshes. This type of parenchyma tissue is called **aerenchyma** (see Fig. 4.2c).

Parenchyma cells are unique in that mature ones can be developmentally reprogrammed to form into different cell types, especially after wounding. For instance, within several hours after a *Coleus* stem is wounded, the parenchyma cells immediately around the wound start to divide. After two days or so some of these cells differentiate into xylem cells, which can transport water around the wound (Fig. 4.4).

Transfer cells are modified parenchyma cells that have many cell wall ingrowths (Fig. 4.5), which greatly increase the internal membrane surface area. This enables these cells to improve the transport of water and minerals over short distances between themselves and attached cells. Transfer cells are found at the ends of files of vascular cells, where they help load and unload sugars and other substances.

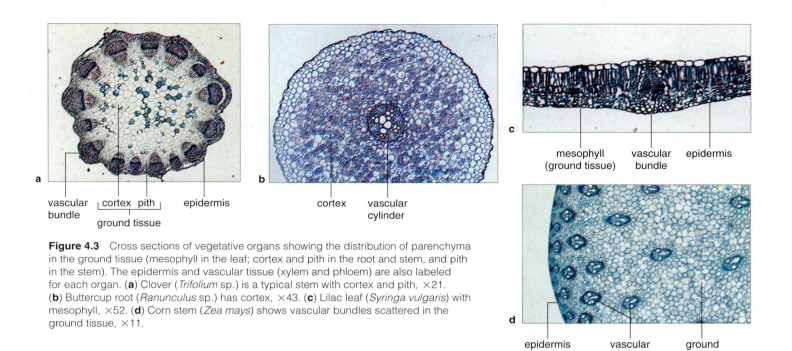

a

vascular
bundle   cortex pith   epidermis

ground tissue

b

cortex   vascular
cylinder

c

mesophyll   vascular   epidermis
(ground tissue) bundle

d

epidermis   vascular   ground
bundles   tissue

Figure 4.3 Cross sections of vegetative organs showing the distribution of parenchyma in the ground tissue (mesophyll in the leaf; cortex and pith in the root and stem, and pith in the stem). The epidermis and vascular tissue (xylem and phloem) are also labeled for each organ. (**a**) Clover (*Trifolium* sp.) is a typical stem with cortex and pith, ×21. (**b**) Buttercup root (*Ranunculus* sp.) has cortex, ×43. (**c**) Lilac leaf (*Syringa vulgaris*) with mesophyll, ×52. (**d**) Corn stem (*Zea mays*) shows vascular bundles scattered in the ground tissue, ×11.

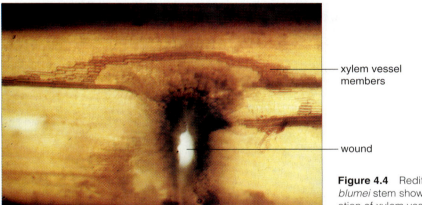

xylem vessel
members

wound

Figure 4.4 Redifferentiation of parenchyma cells. This *Coleus blumei* stem shows a severed vascular bundle and the regeneration of xylem vessel members cells around the wound, ×17.

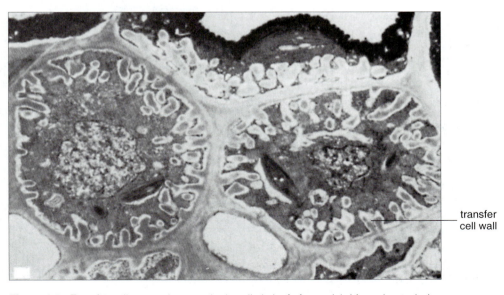

transfer
cell wall

Figure 4.5 Transfer cells around a vascular bundle in leaf of sea-pink (*Armeria corsica*), ×2900. Note the many ingrowths of the cell wall.

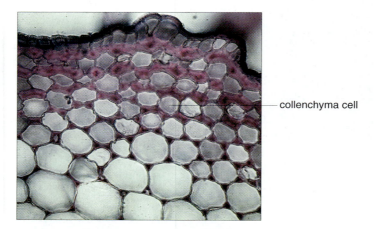

— collenchyma cell

Figure 4.6 Section of marigold stem (*Calendula* sp.) showing pink-stained collenchyma cells with thickened corners, ×170.

COLLENCHYMA **Collenchyma** is a tissue specialized to support young stems and leaf petioles (this is the leaf part that holds the leaf blade to the stem). Its cells are often the outermost cells of the cortex, being just inside the epidermis in young stems and the petioles of leaves. Collenchyma cells are elongated, often contain chloroplasts, and are living at maturity. The walls of collenchyma cells are composed of alternating layers of pectin and cellulose. In the most common type of collenchyma, the cell walls are thickened at the corners (Fig. 4.6). These thickenings are quite flexible and will stretch without snapping back (they are said to be *plastic*). Collenchyma is, therefore, an ideal strengthening tissue in young stems because it also allows for tissue growth. Collenchyma cells occur as aggregates (collenchyma tissue), forming a cylinder surrounding a stem, or as strands, such as make up the ridges of a celery stalk.

SCLERENCHYMA The cells making **sclerenchyma tissue** are rigid and function to support the weight of a plant organ. There are two types of sclerenchyma cells: **fibers** and **sclereids**. These cells tend to have thick, lignified secondary cell walls. They are dead at maturity.

Fibers can occur as a continuous cylinder around stems, as multicellular strands acting like strengthening cables embedded in concrete, or as a component of vascular tissues. They are long, narrow cells with thick, pitted cell walls and tapered ends (Fig. 4.7a, b). Fibers are sometimes very elastic and can be stretched to a degree, but they will snap back to their original lengths.

Sclereids sometimes occur as sheets (an example being the hard outer layer of some seed coats), but they usually occur in small clusters or as solitary cells. Sclereids have many striking shapes, from elaborately branched cells, to star-shaped cells, to the simple stone cells that give a gritty texture to pear fruits (Fig. 4.7c, d). Sclereid cell walls are often thicker than the walls of fibers.

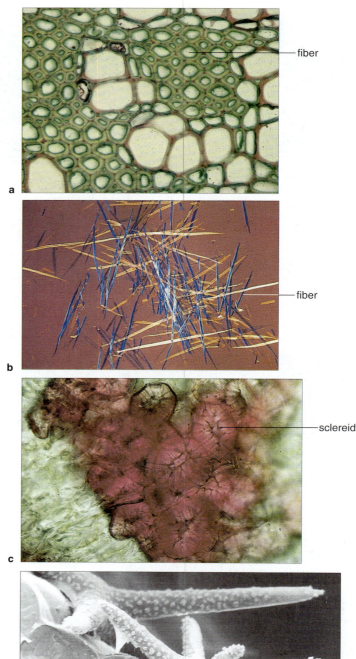

a — fiber

b — fiber

c — sclereid

d

Figure 4.7 Sclerenchyma. (**a**) Cross section of geranium (*Pelargonium* sp.) stem showing clusters of green-stained fibers, ×200. (**b**) Photograph taken with polarized light, showing cells from the wood of a tulip tree (*Liriodendron* sp.). The long, thin cells are fibers, ×100. (**c**) Stone cells are a type of sclereid found in pear (*Pyrus* sp.) fruit, ×200. (**d**) Star-shaped sclereid in water lily (*Nymphia* sp.) leaf, ×380.

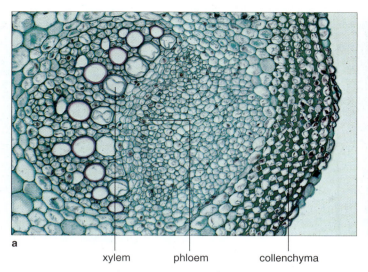

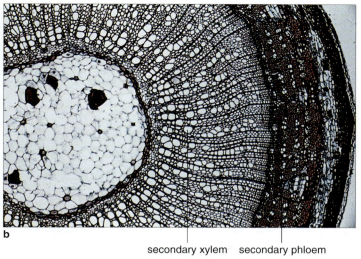

a | xylem | phloem | collenchyma

b | secondary xylem | secondary phloem

Figure 4.8 Xylem tissue and cells. (**a**) Cross section of primary vascular bundle of a sunflower (*Helianthus annuus*), ×72. Primary xylem occurs in young stems. (**b**) Cross section of a one-year-old basswood stem (*Tilia americana*) showing a ring of secondary xylem, ×50.

Complex Tissues Make Up the Plant's Vascular System and Outer Covering

THE VASCULAR SYSTEM: XYLEM The vascular system consists of an interconnected network of bundles that traverse the entire body of the plant. All cells of the plant require minerals and water, which are absorbed by the roots and transported by the **xylem**. Sugars are manufactured in the leaves and transported by the **phloem**.

The xylem is a complex tissue made up of different kinds of cells that work together to transport water and dissolved minerals. The cell types are: the water-conducting cells called **tracheids** and **vessel members** (the latter join together end to end to make **vessels**); fibers, for strength and support; and parenchyma cells, which help load minerals in and out of the vessel members and tracheids.

In leaves and young stems the xylem is found in discrete bundles called **vascular bundles** (Fig. 4.8a), and in young roots the xylem occurs in groups of cells at or near the center of the root known as the **vascular cylinder** (Fig. 4.3b). Xylem in those locations is called **primary xylem**. It is formed in the root and shoot apex very early in organ development (this will be discussed later in the chapter). Xylem that forms later in the development of stems and roots is organized in cylinder patterns and is called **secondary xylem** (Fig. 4.8b). Usually, leaves don't have secondary growth and thus have only primary xylem.

Parenchyma cells are the only living cells found in xylem. They usually have a thin primary cell wall, but in secondary xylem they often have a lignified secondary wall. Fibers function to support the xylem tissues and hold it rigid, rather like a steel rod would be used to hold up a plastic water pipe.

Vessel members and tracheids have many structural and functional characteristics in common, so the term **tra-**

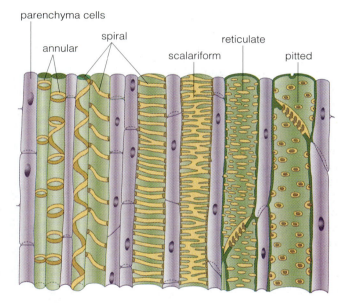

Figure 4.9 The different patterns of secondary cell walls in tracheary elements. Vessel members that form in the early primary xylem have secondary wall thickening as rings or spirals. As the organ grows, these cells can stretch. Note the two cell files labeled *annular*: The one on the left has stretched, but the one on the right has not. The xylem vessel members and tracheids that form after growth is finished tend to have pitted secondary cell walls. Intermediate forms of secondary cell walls, scalariform and reticulate, can also be observed.

cheary element is used to refer to them generally. Tracheary elements are nonliving at maturity. Before the cells die their protoplast degenerates, and the cell wall becomes thickened with cellulose and lignin.

The secondary cell wall, deposited after the cell stops enlarging, forms in one of several patterns: annular (ring-shaped), spiral, scalariform (ladderlike), reticulate (forming a network), or pitted (Fig. 4.9). These patterns relate to the timing of their formation. Tracheary elements that form and function in primary xylem of organs that are still elongating have annular or spiral secondary cell

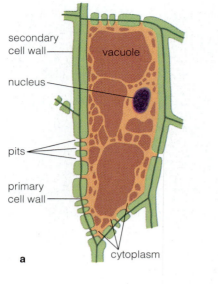

secondary
cell wall

vacuole

nucleus

pits

primary
cell wall

cytoplasm

a

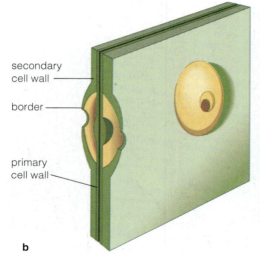

secondary
cell wall

border

primary
cell wall

b

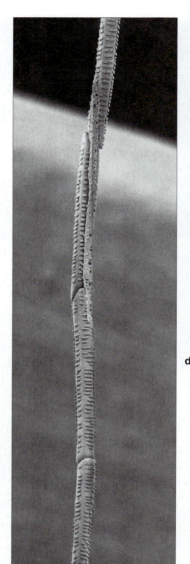

c

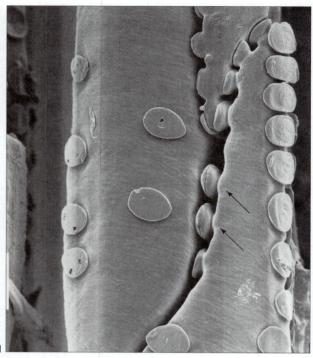

d

Figure 4.10 Pits and their locations. (**a**) Parenchyma cell with a thick secondary cell wall, showing the location of simple pits. Note that the primary cell wall (shown as a black line) remains across the pit opening. (**b**) Drawing of a bordered pit with a raised border. This type of pit is found in the walls of most tracheary elements. (**c**) and (**d**) These are plastic casts made by injecting resin into xylem tissue. After the resin hardens the walls are digested away. The hardened resin indicates the pathway of water and shows how the bordered pits (arrows) apparently fill with water; **c** ×232; **d** ×1400.

walls. Such cells are able to stretch as the surrounding cells elongate and still function in water transport. Vessel members that form after organ elongation has ended (in late-forming primary xylem or secondary xylem) tend to have pitted cell walls. These cells are quite rigid and very strong but are unable to stretch.

Since lignified secondary cell walls are impermeable to water, the only way that water can be exchanged between cells is through tiny openings called **pits**. A pit is not actually a hole in the cell wall; the primary cell wall remains intact, rather like a loose membrane across the pit opening. There are two types of pits: simple pits and bordered pits. **Simple pits** (Fig. 4.10a) are openings in the secondary walls of fibers and lignified parenchyma cells. **Bordered** **pits** occur in tracheids, vessel members, and some fibers (Fig. 4.10b–d). These pits have an expanded border of secondary wall that extends over a small pit chamber.

Tracheids and vessel members have some structural and functional characteristics in common, but there are also differences (Fig. 4.11). A vessel member is a cell with an oblique, pointed, or transverse end. The ends of mature vessel members are partially or completely digested away during their development to form a **perforation plate**. There are several different types of perforation plates; simple and scalariform are the most common (Fig. 4.12).

A vessel is a series of vessel members connected end to end (see Fig. 4.11a). Vessels are often several centime-

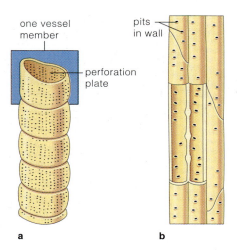

one vessel member

pits in wall

perforation plate

a   b

Figure 4.11 Tracheary elements compared. (**a**) Vessel members join end to end, but they digest out the end walls (perforation plate), forming a longish tube called a *vessel*. (**b**) Tracheids also join end to end.

ters long, and in some vines and trees they may be many meters in length. Since a vessel has a specific length, the terminal cell of a vessel will have a closed end wall containing bordered pits. Vessels connect to other vessels end to end and laterally through pits.

A tracheid is an elongated cell with more or less pointed ends (Fig. 4.11b). Tracheids are joined at overlapping ends through bordered pits, and they do not have perforation plates. Both tracheids and vessel members may be present in a single flowering plant. Most cone-bearing plants (gymnosperms), however, have only tracheids. Chapter 11 discusses the mechanism of water transport through the xylem and the advantages and disadvantages of tracheids and vessel members.

THE VASCULAR SYSTEM: PHLOEM   Phloem is the tissue that transports sugar through the plant. **Primary phloem** occurs in vascular bundles near the primary xylem in young stems and leaves and in the vascular cylinder in roots. **Secondary phloem** occurs outside the secondary xylem in older stems and roots, usually in plants that live more than one year. In flowering plants (angiosperms) the phloem is made up of several different types of cells (Fig. 4.13): sieve-tube members, companion cells, parenchyma and sometimes fibers and/or sclereids. **Sieve-tube members** are cells that join at their ends, to form long sieve tubes. Sieve tubes are the conducting elements of the phloem (Fig. 4.14a, b), which transport sugars produced by photosynthesis in the leaves to other plant parts.

A young sieve-tube member contains a nucleus, plastids, mitochondria, and Golgi stacks. As the cell matures, its nucleus disintegrates, the plastids and mitochondria

perforation plate

Figure 4.12   Scalariform perforation plate is nicely shown in this scanning electron microscope view of a group of xylem cells from tulip tree wood (*Liriodendron* sp.), ×300.

shrink, and the cytoplasm becomes reduced to a thin peripheral layer. The central part of the cell becomes occupied by a mass of dense material. This mass, which was called *slime* in the early literature, can be seen with the light microscope. It's now called **P-protein** (Fig. 4.15a) because it is actually composed of a complex of proteins that may be involved in moving materials through the sieve tubes.

When the sieve-tube member is mature, one or more **companion cells** lie connected to it by plasmodesmata (Fig. 4.13). Since companion cells have a full complement of organelles, it is thought that they regulate the metabolism of their adjacent sieve-tube member. Companion cells are also known to play an important role in the mechanism of loading and unloading the phloem.

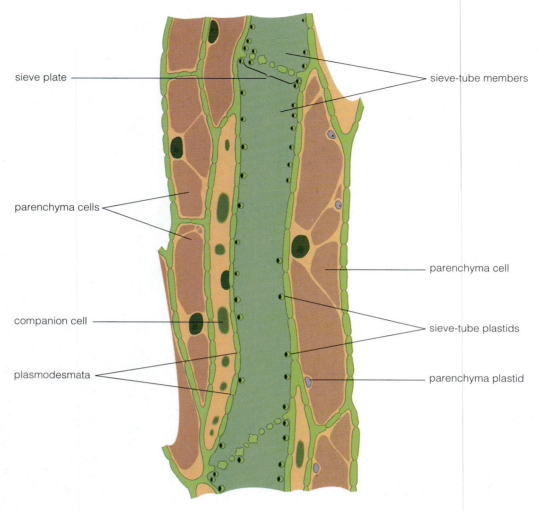

sieve plate

parenchyma cells

companion cell

plasmodesmata

sieve-tube members

parenchyma cell

sieve-tube plastids

parenchyma plastid

Figure 4.13   Phloem tissue from a stem, showing cell types.

The walls of mature sieve-tube members contain aggregates of small pores called **sieve areas**. One or more sieve areas on the end wall of a sieve-tube member is called a **sieve plate** (Fig. 4.13). The end walls of two connecting sieve-tube members are thickened, and strands of cytoplasm and P-protein pass through pores adjoining them. Sieve-tube members live and function from one to three years except in a few trees such as palms, where they live much longer.

In many studies of the structure of mature sieve-tube members, a carbohydrate known as **callose** was seen to surround the margins of the pores in the sieve areas (Fig. 4.15a). In some instances, protein is also collected at the sieve plate. It has been demonstrated in other instances that callose can form very rapidly in response to aging, wounding, and other stresses (Fig. 4.15b), and the

result is to limit the loss of cell sap from injured cells.

In gymnosperms and ferns, **sieve cells** rather than sieve-tube members are the conducting elements in the phloem. These cells are quite long, with tapered ends (Fig. 4.14c). They have sieve areas but no sieve plates at their ends. Sieve cells apparently function similarly to sieve tubes and usually lack nuclei at maturity. Adjacent **albuminous cells** are short, living cells that act as companion cells to these sieve cells.

Phloem fibers are usually long, tapered cells with lignified cell walls. Mature fibers from the ramie plant (*Boehmeria nivea*) have been reported up to 55 cm long. Such fibers are valued for their commercial uses, especially for rope and fabric. Phloem parenchyma cells are usually living cells that function in phloem loading and unloading.

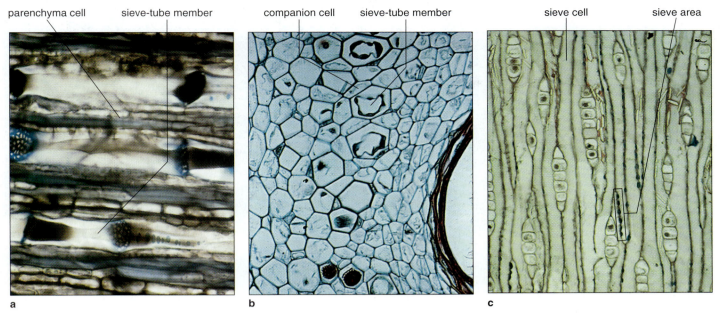

parenchyma cell   sieve-tube member   companion cell   sieve-tube member   sieve cell   sieve area

a   b   c

Figure 4.14 Phloem. (**a**) Elm (*Ulmus* sp.) tangential longitudinal section (TLS) showing sieve-tube members and parenchyma cells in secondary phloem, ×375. (**b**) Primary phloem in cucumber (*Cucurbita pepo*) stem shown in cross section, ×130. (**c**) Pine (*Pinus* sp.) secondary phloem in TLS, showing sieve cells. Note that the callose in the sieve areas is stained blue in this photograph, ×228.

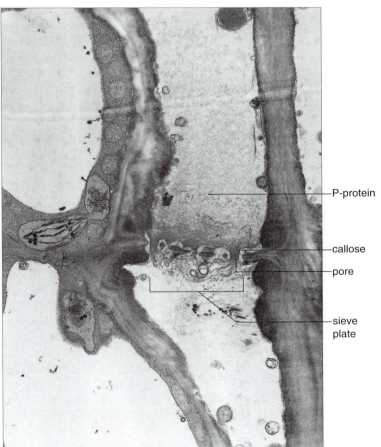

P-protein

callose

pore

sieve plate

a

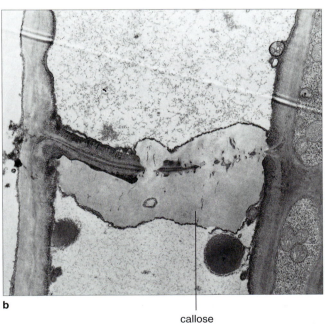

b

callose

Figure 4.15 P-protein and callose. (**a**) Beet (*Beta vulgaris*) petiole sieve-tube members with P-protein and callose surrounding the pores at the sieve plate, ×10,600. (**b**) Beet petiole one hour after being stressed by exposure to cold temperature. Note the massive amount of callose that now plugs the pores, ×10,200.

a

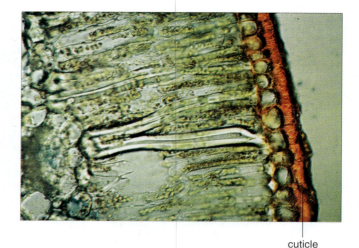

cuticle

Figure 4.17 Photograph of pincushion tree (*Hakea* sp.) leaf shown in cross section. Note the thick cuticle and the small channels that cross the cuticle, ×184.

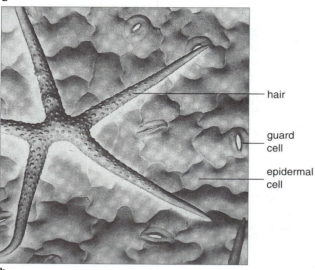

hair

guard cell

epidermal cell

b

Figure 4.16 Epidermis and epidermal cells. (**a**) Upper leaf surface of shepherd's purse (*Capsella bursa-pastoris*), ×70. (**b**) Higher magnification showing epidermal cells, star-shaped hairs, and stomata, ×350.

THE PLANT'S OUTER COVERING: THE EPIDERMIS AND PERIDERM Epidermis is the outer covering of the plant. It is a complex tissue composed of epidermal cells, guard cells, and trichomes (hairs) of various types. The epidermis is usually one layer of cells, but.it may be as many as five or six layers in the leaves of some succulent plants and in the aerial roots of certain orchids. The epidermis protects the inner tissues both from drying and from infection by some pathogens. It also regulates the movement of water and gases out of and into the plant.

Epidermal cells are the main cell type making up the epidermis. These cells are living, lack chloroplasts, are usually somewhat elongate, and often have walls with irregular contours (Fig. 4.16). The outer walls of all cells in the epidermis are often thicker than the inner and side walls. The outer epidermal wall is coated with a waxy substance (cutin) forming an impermeable layer called the **cuticle** (Fig. 4.17). All parts of the plant, except the tip of the shoot apex and the root cap, have a cuticle. In roots, the cuticle is often very thin (and has been reported to be totally lacking on the surface of root hairs).

Young stems, leaves, flower parts, and even some roots in exceptional instances have specialized epidermal cells called **guard cells**. Between each pair of guard cells is a small opening, or pore, through which gases enter and leave the underlying tissues. Two guard cells plus the pore constitute one **stoma** (plural, **stomata**; Fig. 4.18). Guard cells differ from other epidermal cells by their crescent shape and the fact that they contain chloroplasts. Another type of epidermal cell, the **subsidiary cell**, forms in close association with guard cells and functions in stomatal opening and closing. The physiological role of stomata and the mechanism of their opening and closing will be discussed in Chapter 11.

Trichomes are epidermal outgrowths and may be a single cell or multicellular (Fig. 4.19). In roots, for example, root hairs are extensions of single epidermal cells that increase the root surface area in contact with soil water. In some leaves, very elaborate multicellular trichomes may form (Fig. 4.16b). Trichomes will be discussed again in Chapter 7.

The **periderm** is a protective layer that forms in older stems and roots after those organs expand and the epidermis splits and is lost. It is a secondary tissue. This tissue is several cell layers deep and is composed of **phellem** (cork) cells on the outside, a layer of dividing cells (**phellogen** or **cork cambium**), and the **phelloderm** toward the inside (Fig. 4.20). Phellem cells are dead at maturity, and have a waxy substance (suberin) embedded in their cell walls. Phelloderm cells live longer than phellem cells and are parenchymalike.

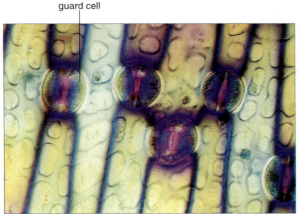

guard cell

a

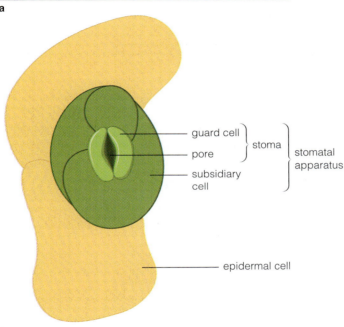

b

guard cell
pore } stoma
subsidiary cell

} stomatal apparatus

epidermal cell

Figure 4.18 Stomata. (**a**) Surface view of *Iris* sp. leaf as seen through polarized light, showing several stomata composed of two guard cells surrounding a pore, ×415. (**b**) Diagram of stomatal apparatus consisting of two guard cells and surrounding subsidiary cells, such as would be found on leaves in a plant like *Sedum*.

Secretory Tissues Produce and Secrete Materials

There are also **secretory structures**, which form mostly in leaves and stems. These may be composed of single secretory cells or very complex multicellular structures (Fig. 4.19). Some trichomes, for example, may secrete materials out of the plant to attract insect pollinators. They may also form inside the plant body and secrete materials within the plant; cells called *laticifers* (Fig. 4.21), for example, secrete latex, which discourages herbivores from eating the plant.

Table 4.2 contains a summary of the structure and function of the common cell types found in seed plants.

Figure 4.19 Glandular hairs in tobacco (*Nicotiana tabacum*), ×22. The thickened, bulblike end of this trichome secretes sticky material.

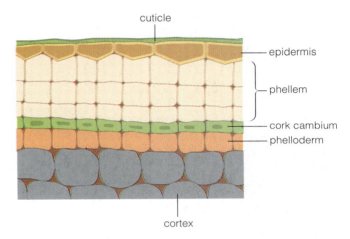

cuticle
epidermis
phellem
cork cambium
phelloderm
cortex

Figure 4.20 Diagram of periderm, showing the layers: cork cambium, phelloderm, phellem.

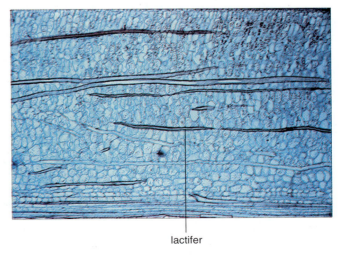

lactifer

Figure 4.21 Laticifer cells in spurge (*Euphorbia* sp.) stems, shown here in longitudinal section, ×20.

Table 4.2 Summary of Plant Types

Cell Type		Characteristics	Location	Function
Parenchyma		Living at maturity; usually more or less spherical or elongate in shape; usually has primary cell walls only	Cortex and pith of roots and stems; mesophyll of leaves; xylem and phloem	Site of basic metabolic cell functions
Transfer cell		Living at maturity; modified parenchyma cell containing cell wall ingrowths	Xylem, phloem, secretory structures	Involved in short-distance transport
Collenchyma		Living at maturity; elongate with plastic cell walls.	Leaf petioles, young stems	Support
Sclereid		Dead at maturity; elongate, star-shaped, or stone-shaped with thick, lignified secondary cell walls.	Cortex, pith, mesophyll	Strength
Fiber		Dead at maturity; long, thin cell with thick, lignified secondary cell walls.	Xylem, phloem, cortex	Most important strengthening and supportive cell type
Vessel member		Dead at maturity; lignified secondary cell walls that may be spiral, ringed or pitted; perforation plates in open end walls; members connect together to form vessels.	Xylem	Transport of water and dissolved minerals
Tracheid		Dead at maturity; similar to vessel member, except with narrower diameter and closed pitted ends; do not form vessels	Xylem	Transport water
Sieve-tube member		Living at maturity, but does not retain cell nucleus; elongate cell with aggregates of pores (sieve plates) on end walls and side walls; cells join end to end to form sieve tubes	Phloem	Transport of sugars

4.3 MERISTEMS: WHERE CELLS DIVIDE

We now know the tissue types making up the vascular plant, but how do these tissues come to be? Do plant cells specialize and group into tissues in the same manner that animal cells do? The answer is, largely, no. The growth patterns of animals and plants differ in one very significant way. Animal cells divide mostly during the embryo stage of development. After the animal reaches adult size, cells divide only in the bone marrow (to produce new blood cells), in the intestinal lining (to repair wounded tissues), and in the epithelial layers (to form new skin, nails, and hair). In vascular plants, an entirely different process exists: Cell division continues through the whole lifetime of the plant. Division occurs in special regions of the plant body called **meristems** (from the Greek word meaning "to divide").

What Is a Meristem?

A meristem is a site in the plant body where new cells form and the complex processes of growth and differentiation are initiated. **Growth** means the irreversible in-

Table 4.2 Summary of Plant Types (continued)

Cell Type		Characteristics	Location	Function
Sieve cell		Living at maturity; similar to sieve-tube members, but does not develop sieve plates or join to form sieve tubes; found only in ferns and gymnosperms	Phloem in conifers and ferns	Transport of sugars
Companion cell		Living at maturity	Connected to sieve-tube members	Metabolic regulation of sieve-tube members
Albuminous cell		Living at maturity; counterpart to companion cells, found in gynmosperms and ferns	Phloem in conifers and ferns	Metabolic regulation of sieve cells
Guard cells		Living at maturity; occur in pairs separated by a pore to form a stoma	Leaves, young stems	Gas exchange
Epidermal cell		Living at maturity; contains no chlorophyll; outer wall impregnated with cutin	Epidermis	Inhibit evaporation of water
Subsidiary cell		Living at maturity; type of epidermal cell	Occurs in contact with guard cells in epidermis	Regulate guard cell opening
Trichome		Living at maturity; elongated epidermal structure composed of one or more cells	Epidermis	Absorption, secretion, storage, and more
Phellem cell		Dead at maturity; also called cork cell; cell walls impregnated with suberin (a wax)	Outer layers of periderm	Replacement of epidermis in old stems and roots
Phelloderm cell		Most living at maturity	Inner layers of periderm	Replacement of epidermis in old stems and roots
Secretory cell		Living at maturity; several different types; may be solitary or part of a multicellular secretory structure	Most tissues and organs	Secretion of different secretory products

crease in size that comes from both cell division and cell enlargement. Cell **differentiation** refers to the changes that a cell undergoes structurally and biochemically so that it can perform a specialized function.

There are different categories of meristems, each with a specific function. Shoot and root apical meristems are at the tips of branches and roots (Fig. 4.22); they are the ultimate sources of all cells in the plant. Primary meristems, the next level of meristems, originate in apical meristems and produce the primary tissues. The secondary meristems produce the secondary tissues. These categories of meristems allow vascular plants to grow very large and

to great age. Let's now examine some details of the location and function of each meristem.

What Are the Categories of Meristems, and How Do They Differ?

ROOT AND SHOOT APICAL MERISTEMS The vascular plant body is polar, meaning that it has a shoot end and a root end. At the tip of each branch is a **shoot apical meristem (SAM)**, and at the tip of each root is a **root apical meristem (RAM**; Figs. 4.22, 4.23). These two apical

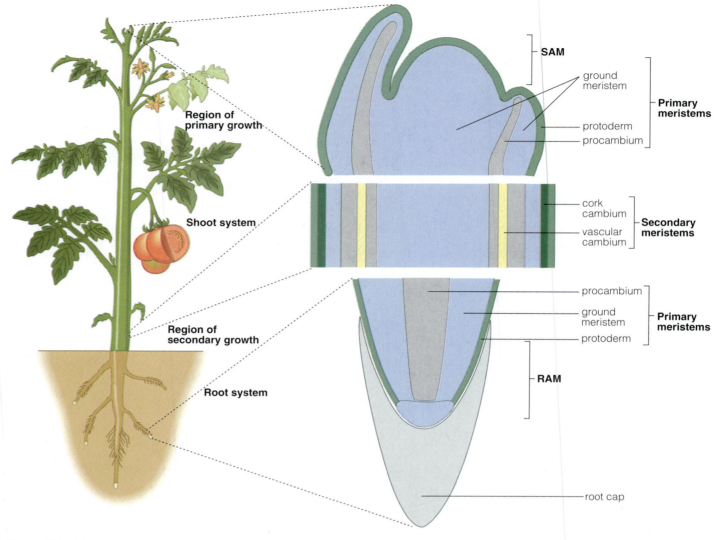

Figure 4.22 Diagram of a hypothetical flowering plant showing the relative positions of the apical meristems (RAM and SAM), the primary meristems (protoderm, ground meristem, and procambium), and the secondary meristems (vascular cambium and cork cambium) in both the shoot system and the root system.

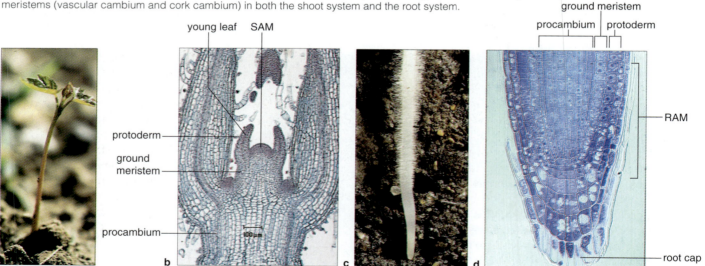

Figure 4.23 Shoot tips and root tips. (**a**) Photograph of a living plant showing the position of the shoot apex. The shoot apical meristem (SAM) is surrounded by several small leaves. (**b**) Median longitudinal section through the SAM of a *Coleus blumei* plant, ×38. (**c**) Photograph of a living root tip. The root apical meristem (RAM) is located at the very tip of the root, just inside the root cap. (**d**) Median longitudinal section through RAM of a mouse-eared cress (*Arabidopsis thaliana*) root tip, showing the root cap and RAM, ×350.

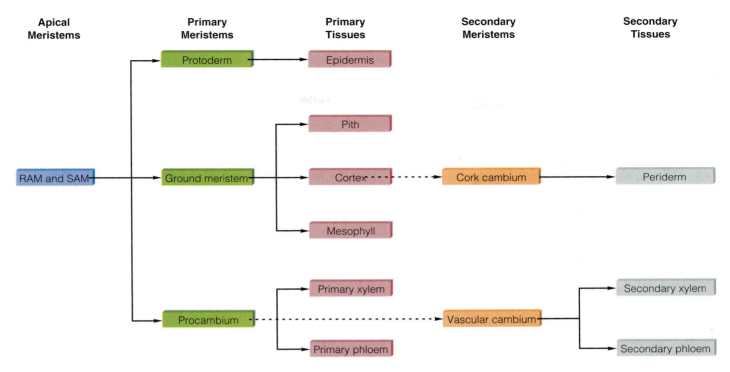

| Apical Meristems | Primary Meristems | Primary Tissues | Secondary Meristems | Secondary Tissues |

Figure 4.24 Summary of meristems and the tissues they generate.

meristems are the sites of the formation of new cells by cell division. Theoretically, apical meristems could operate forever. This does not occur, however, because some factor will always limit the size of a plant. Whether it's because of a scarcity of nutrients, or structural limitations, or heredity, eventually a plant ceases to grow. A branch, for example, can carry only a certain weight before it breaks. Also, each plant and plant organ (leaves, stems, and roots) has a system for genetic regulation of growth; every species seems to have an optimum size.

PRIMARY MERISTEMS A shoot tip and a root tip of a representative plant are shown in Figures 4.22 and 4.23. If you made a very thin longitudinal section through them (this would be done by embedding a shoot or root tip in plastic resin or paraffin and cutting a thin section), you would see that the cells of the SAM and RAM and those just basal to them (toward the more mature cells) are small, with relatively dense protoplast and large nuclei (Fig. 4.23). Cells with these characteristics are usually capable of dividing and so are referred to as *meristematic cells.*

Cells immediately basal to the SAM are ordered into distinct files of cells (Figs. 4.22, 4.23). These newly ordered cells are still meristematic; they are the *embryonic* stages of the tissues. These groups of cells are called the **primary meristems**, and they have two roles: to form the primary tissues and to elongate the root and shoot. There are three primary meristems: protoderm, procambium, and ground meristem (Fig. 4.22). The cells of the **protoderm** differentiate into the epidermis. The **pro-**

cambium cells differentiate into the cells of the primary xylem and primary phloem. The **ground meristem** differentiates into the cells of the pith and cortex of stems and roots and the mesophyll of leaves.

The primary meristems near the tips of the roots and shoots are the site of most elongation. They produce new cells, which then enlarge mostly by elongation. However, in many plants the branch or root continues to increase in girth as well. This increase in girth requires lateral growth, which involves the formation and activity of the next category of meristems, called secondary meristems.

SECONDARY MERISTEMS The **secondary meristems** are responsible for cell division, initiation of cell differentiation, and growth in a lateral direction, thereby increasing the thickness and circumference of stems and roots. The wood in trees, for example, is really secondary growth resulting from the activity of secondary meristems. Not all plants have secondary meristems. There are thousands of species that grow only one season and usually lack secondary growth. Leaves also usually lack secondary growth.

The bodies of many plants have two secondary meristems: the vascular cambium and the cork cambium (Fig. 4.22). **Vascular cambium** differentiates into secondary xylem and secondary phloem, and the **cork cambium** into the periderm (Fig. 4.24).

OTHER MERISTEMS There are several other meristems. In stems, as an example, an intercalary meristem occurs within the stem to regulate its elongation. As leaves de-

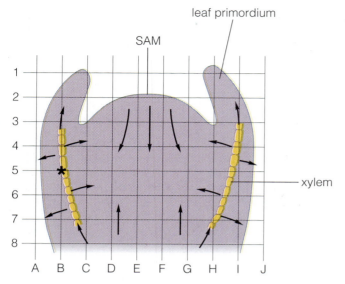

Figure 4.25 The positional information theory of cell determination. In this diagram of a longitudinal section through a shoot tip with a superimposed grid, the arrows indicate the theoretical direction of different gradients of morphogens from the base of the stem, the procambium, and the SAM. The idea is that each cell, depending on its position, would be exposed to a different combination of morphogens. Each cell would react to its cue (combination of morphogens) by turning on or off specific genes; this would direct the cell to become a xylem cell, for example, instead of an epidermal cell.

velop, there are leaf-specific meristems that regulate leaf shape. The intercalary meristem at the base of grass leaves allows the leaf to continue to grow after being grazed or mowed. Other meristems are involved in forming buds and roots in unusual places, such as at the base of trees, and also in the repair of wounds. We will discuss some of these other meristems in the following chapters.

How Is the Identity of Cell Types Determined in Meristems?

Different theories have been suggested to explain how specific cells or regions of cells are triggered to divide, enlarge, or differentiate. Perhaps the best idea is called the **positional information theory**. Basically, it states that a cell's position determines what it will become. Figure 4.25 illustrates how this is thought to happen. The diagram depicts a longitudinal section of a shoot apical meristem (SAM) of a vascular plant. The two-dimensional grid over this SAM shows that every cell within it can be identified by different 2-D coordinates.

Assume that two (or more) substances capable of affecting cell development originate at two different locations in the SAM or near it. As these substances move from cell to cell, a concentration gradient of them will be created. This means that at every position in the 2-D grid, a cell would be exposed to a different combination of the two substances, called **morphogens**. (An example of a

morphogen would be one of the hormones discussed in Chapter 15, known to be present and active in plants.)

So far, so good. But how would the cell be able to interpret these positional signals? The most likely mechanism involves genes. The activation and expression of certain genes would produce proteins that would trigger cell differentiation events.

We'll look at a hypothetical example of how this would work. We have marked a cell (*) in Fig. 4.25 at position B5. Note that this cell is positioned in a file of xylem cells (transporters of water and nutrients). Let's suppose for this example that one morphogen is diffusing down from the SAM and another substance is moving up through the developing vascular tissues. The combined effect of the two substances at the marked cell would be to trigger genes involved in the differentiation of a new xylem cell. This kind of mechanism occurring millions of times could result in all of the differentiated cells in the plant body. This means that the position a cell occupies plays a key role in determining what it will become.

SUMMARY

1. The vascular plant body is organized into a shoot system (stems, leaves, buds, flowers, and fruits) and a root system. Flowering plants (angiosperms), cone-bearing plants (gymnosperms), and ferns are all vascular plants.

2. Cells in the plant body are organized into the ground, vascular, and dermal tissue systems. There are several different cell types making up the plant body. These cells are organized into aggregates called *tissues*.

3. The simple tissues are parenchyma, collenchyma, and sclerenchyma. (Refer to Tables 4.1 and 4.2 and Figure 4.24 for summaries of cell types, tissues, and meristems.)

4. The complex tissues are epidermis, xylem, phloem, and periderm.

5. Secretory structures are specialized to secrete various substances within and outside the plant body.

6. Each cell originates from a cell that was once meristematic. The process whereby a cell changes into a mature cell is called *differentiation*.

7. Meristems are the sites of cell division, cell elongation, and the beginning of cell differentiation.

8. Apical meristems occur at the tips of stems and roots and are the ultimate source of all cells in the shoot and root systems.

9. Primary meristems (protoderm, procambium, and ground meristem) form the primary tissues (epidermis, primary xylem and phloem, and pith and cortex).

10. Secondary meristems (vascular cambium and cork cambium) form the secondary tissues (secondary xylem and secondary phloem, and phellem and phelloderm, respectively).

11. The positional information theory states that the differentiation of all cells in the plant body is regulated by a gradient of at least two substances that trigger gene expression patterns.

Questions

1. What are the tissues found in the plant body? How are they organized in each vegetative organ?

2. What are the features, cell types, and functions of each of the tissues listed below?

epidermis

periderm

xylem

phloem

parenchyma

collenchyma

sclerenchyma (fibers and sclereids)

3. Be able to define the following xylem terms:

vessel member

vessel

tracheid

bordered pit

4. Be able to define the following phloem terms:

sieve-tube member

sieve tube

companion cell

sieve plate

5. Describe the function of the following meristems:

root and shoot apical meristems

primary meristems: protoderm, ground meristem, procambium

secondary meristems: vascular cambium, cork cambium

6. Select a cell type positioned somewhere in the SAM of a hypothetical plant. Discuss the factors that would influence its differentiation in terms of the positional information theory.

7. Each tissue in the plant body has a different function. Describe these functions and discuss how the tissues and cells communicate with each other.

Further Readings

Cutler, D. F. 1978. *Applied Plant Anatomy*. New York: Longman. This is an interesting text with many examples of how plant anatomy is important to everyday life.

Esau, K. 1977. Chapters 5–13 in *Anatomy of Seed Plants*. 2d ed. New York: John Wiley. This is the classic textbook in plant anatomy. It is usually taught to third- and fourth-year students.

Perry, J. W., and D. Morton. 1996. Pages 70–82 in *Photo Atlas For Biology*. Photo atlas showing good color images of plant cells and tissues.

5 THE ROOT SYSTEM

1. The principal functions of roots are absorption of water and nutrients, conduction of absorbed materials into the plant body, and anchorage of the plant in the soil. Many roots have relationships with bacteria and fungi in the rhizosphere (soil zone near the root).

2. The root is initiated in the embryo as the radicle. It penetrates into the soil and branches.

3. The root tip is composed of the root cap, the root apical meristem, the region of cell elongation, and the region of cell maturation.

4. Roots are composed of the following tissues: epidermis, cortex, endodermis, pericycle, xylem, and phloem.

5. The endodermis regulates ion movement into the xylem. The Casparian strip embedded in the cell wall inhibits movement through the wall. The pericycle is the site of lateral root initiation and contributes to vascular cambium and cork cambium formation.

5.1 THE FUNCTIONS OF ROOTS

Although plant biologists have studied plants for hundreds of years, they have largely ignored the root system and its functions. This is probably because roots are underground, where they can't be seen. Though they are hidden from view, roots play a critical role in the everyday activities of plants. The main functions of roots are **anchorage** of the plant body in the soil (or to a surface, in the case of some vines); **absorption** of water and minerals from the soil; **storage** of foods; and **conduction** of food and water from the soil and from storage reserves into the shoot. Other, less prevalent though important functions are described on page 82, under the subhead "Some Roots Are Specialized for Special Functions."

The root system becomes more complex as the plant grows from a single root in a young seedling to a massive system of branched roots, often weighing tons in large trees. During all stages in a plant's growth cycle, there is a balance between the shoot system and the root system (Fig. 5.1). The root system must be able to supply the shoot with sufficient water and mineral nutrients, and the shoot system must manufacture enough food to maintain the root system.

The contact zone between the root surface and the soil is called the **rhizosphere** (Fig. 5.2). This region, only a few millimeters thick, is a very interesting and unique zone. Its complex chemistry includes leakage of organic material, gases, and nutrients from the roots. Consequently, the bacteria and fungi near roots are often richer and more diverse than in soil farther away. Soil fungi and bacteria form important symbiotic (mutually beneficial) relationships with roots. These will be discussed later in the chapter.

You need only walk through a streambed and observe the exposed root systems of large trees, or attempt to pull weeds, to get an idea of the anchorage function of roots. Plants with unusual roots provide anchorage in atypical ways. In ivy (*Hedera helix*), clusters of roots develop from the stems to form an unusual adhesive pad allowing these vines to cling to vertical surfaces. Parasitic plants

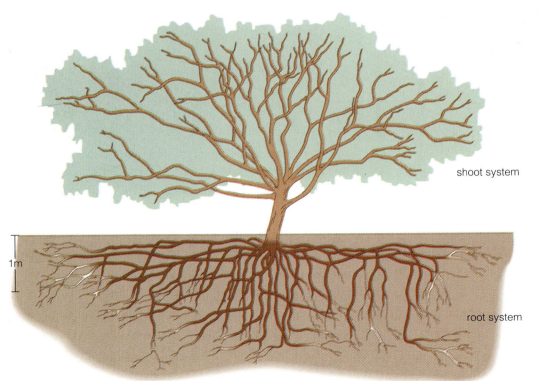

Figure 5.1 Root and shoot systems of a tree at least 10 years old. Note that the majority of the roots lie within the upper 1 m (40 in) of soil and that the total volume of roots is equal to or greater than the shoot branches.

shoot system

1 m

root system

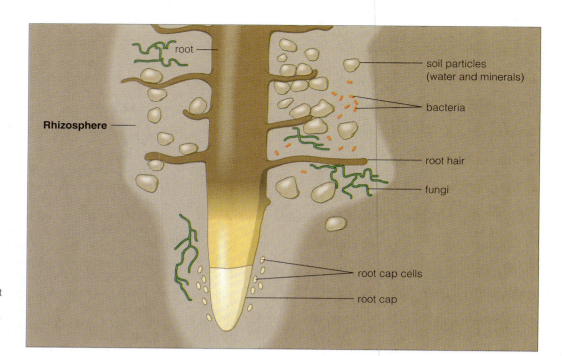

Figure 5.2 The rhizosphere, the soil area immediately around a root. This area, shown near the root tip, is rich in soil microorganisms, bacteria and fungi, and in nutrients from the root body and sloughed-off root cap cells.

like dodder (*Cuscuta* sp.) sink specialized roots into their host and then tap into its water and nutrient supply.

All plants need water. In herbaceous (nonwoody) plants, water accounts for about 90% of the plant's weight. Water is needed for all root processes and for every metabolic reaction. Plants also need the dissolved salts and minerals—such as potassium, sulfur, phosphorus, calcium, and magnesium—contained in the soil water. For these reasons, plants have developed an elaborate system for the absorption and conduction of water.

All roots, even slender ones with a primary function of absorption, may temporarily store small amounts of food. For example, when sugar moves into roots more rapidly than it can be used by growing cells, it may be converted to starch and stored. During slow growth periods, rather large quantities of starch are stored in woody roots of orchard trees. This food constitutes a reserve that is used when flowering and active growth are resumed in spring. Carrots, beets, and turnips are common root crops. Traditional cultures have also developed medicinal and other uses for roots (see sidebar, "Myths and Popular Uses of Roots," p. 71).

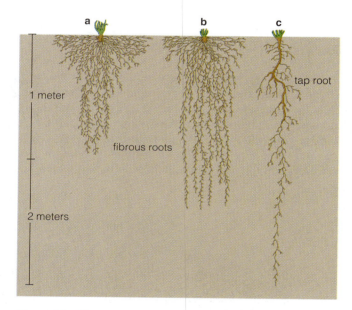

Figure 5.3 Fibrous and tap root systems. (**a**) Shallow, spreading, fibrous root system. (**b**) Fibrous root system, penetrating the soil evenly up to 1 m (40 in) in this example. (**c**) Tap root system, in which the primary root penetrates soil 2 m or more.

5.2 TYPES OF ROOT SYSTEMS

There are two basic types of root systems: fibrous and tap. They are distinguished by the way they develop and by their appearance. Many grasses and small garden plants, for instance, when pulled up bring with them a massive clump of soil. This happens because the **fibrous root system** of these plants consists of several main roots that branch to form a dense mass of roots (Fig. 5.3a, b). A typical annual grass such as corn (*Zea mays*) or rye (*Ely-*

mus cereale) will build an immense fibrous root system in one growing season. A single rye plant 50 cm (20 in) tall, with 80 tillers (shoot branches), may have a surface area of about 210 m² (1890 ft²) for its root system compared to only about 5 m² (45 ft²) for its aboveground shoot system.

Plants with a large storage root, like carrot (*Daucus carota*), have a **tap root system**, consisting of one main root from which lateral roots branch (Fig. 5.3c). Some desert plants have a rapidly growing tap root system that enables them to penetrate the soil quickly to reach deep sources of water.

MYTHS AND POPULAR USES OF ROOTS

Over the millennia, through trial and error, ancient peoples developed traditional uses—medical and otherwise—for berries, seeds, leaves, stems, and roots. Much of this plant lore has little validity in medical fact, but some traditional plants have proven to be effective remedies.

During the Middle Ages in Europe, most people believed that the way a plant looked (its shape or color), how it smelled, or some other characteristic provided clues about its uses for people. This was the so-called Doctrine of Signatures.

For example, the roots of the mandrake (*Mandragora officinarum*, Fig. 1) are thick and fleshy and tend to be irregularly branched; with a little imagination the shape of a person could be seen in a root carefully extracted from the soil. In the Middle Ages people assigned considerable importance to the manlike appearance of these roots; they believed that such roots could bring good fortune. In some cases the roots were even ground up and eaten as a love potion. Mandrake root doesn't seem to have much real medicinal value. It doesn't really give anyone good luck, and it certainly doesn't work as a love potion; but in some cases an infusion made from mandrake root may help control a cough.

Another root that looks vaguely like a person is the ginseng (*Panax quinquefolium*). In Asia, ginseng root has been considered a cure-all for many ailments. Ginseng also grows in the United States, but it has only recently become popular. Nowadays you can find it in most health food stores. Most likely, the subtle, supposedly restorative power associated with ginseng is psychological: It helps because you think it will. However, some recent evidence suggests that chemicals in ginseng roots do act to calm some people.

A root with more widely accepted utility comes from the cassava (*Manihot esculenta*), a shrubby South American plant. The whitish latex that exudes from a cut plant contains hydrocyanic acid, a deadly poison used by Brazilian Indians to make poisoned arrows. The interesting thing, however, is that when the latex is removed, the rich, starchy pulp is a good food source. Native people ground up cassava root, placed it in a sack, and hung it from a tree; when all the juice dripped out, the meal that remained could be made into cakes and eaten. It was called *farina*. Now cassava is grown commercially. You are probably familiar with the starch-rich pellets manufactured from cassava root, called tapioca.

Figure 1 These drawings of mandrake plants come from an old German herbal. Herbals were among the first books ever made. They contained drawings of plants and sometimes discussed how they could be used. (From W. H. Lewis and M. P. F. Elvin-Lewis, *Medicinal Botany: Plants Affecting Man's Health*. New York: Wiley-Interscience, 1977. Used with permission of Wiley-Interscience. Original art of "male and female" mandrake (*Mandragora officinarum*) from Peter Schöffer, *The German Herbarius*. Mencz, 1485.)

Native Americans also have a rich plant lore that includes roots. The roots of several plants were used as a source of soap. One of these—soap weed (*Yucca* sp.)—is now used as a component of popular shampoos. *Yucca* root juice was also used as glue to fasten feathers to arrows. One type of soap root was used to induce vomiting as a step in ritual purification ceremonies. Another type was thrown in streams to stupefy fish so that they could be harvested without using fishing gear. Roots are obviously important to the plant, but many also have uses for people.

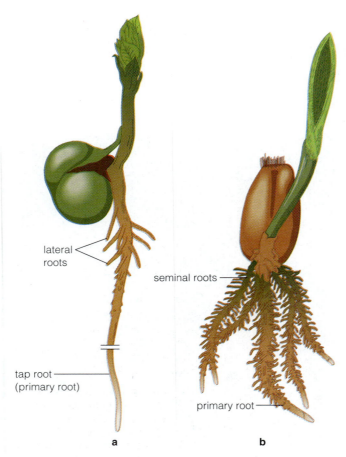

Figure 5.4 Root systems. (**a**) Seedling of pea (*Pisum sativum*), with a tap root. Short lateral roots are shown. (**b**) Seedling of wheat (*Triticum aestivum*). Seminal roots emerge from the stem (and are therefore adventitious roots because they don't emerge from another root). These create the fibrous root system. The primary root in wheat doesn't grow longer than the seminal roots.

Plants Have Different Types of Roots

The seeds of flowering plants contain a small undeveloped plant, the **embryo**. When the seed germinates, the embryonic root, the **radicle**, extends by the division and elongation of cells to form the **primary** root (Fig. 5.4a). Tap root systems develop from one primary root, which then forms **lateral roots**. Further branching results in succeeding orders of roots. Fibrous root systems develop in a slightly different way. The embryos of most grasses have a single radicle, but in addition several other embryonic roots form just above the radicle; these are called **seminal roots**. The seminal roots emerge soon after the radicle, and all of these roots branch, making a fibrous root system (Fig. 5.4b).

Roots called **adventitious roots** originate on leaves and stems. There are several common examples of adventitious roots. In a young corn plant, soon after germination **prop roots** develop on the stem just above the soil (Fig. 5.5a). Prop roots absorb water and minerals, but they also support the plant in the soil.

Banyan (*Ficus bengalensis*) trees grow in the salty mud of tropical lagoons and tidal marshes. Branches of these trees form adventitious roots—also called **aerial roots** because they are exposed to air. These extend down from branches into the soil, where they enlarge and actually hold up the large branches (Fig. 5.5b). These roots absorb water and nutrients, but their most important function is to prop up the stem. In India, merchants once held open-air bazaars among the prop roots and expansive branches of the banyan.

Mangrove trees (*Rhizophora mangle*) are native to low tidal shores and marshes in tropical and subtropical regions. In the mangrove (Fig. 5.5c) small adventitious roots called **pneumatophores**, stick up from the mud. These roots absorb oxygen and increase its availability to the submerged roots.

In *Ficus pumila*, a climbing fig, roots develop in clusters on stems just below nodes. The root cluster flattens against a surface and forms a flat adhesive pad (Fig. 5.5d). The root hairs that grow on the root cluster may secrete a sticky substance that causes the root cluster to cling.

Pieces of stem, such as a cane from a blackberry plant or a branch of willow, can be induced to make roots from their cut ends simply by placing them in moist soil. Leaves from *Begonia* and several other plants also can be rooted simply by soaking them in water. Many commercially important ornamentals are reproduced by root propagation from the leaves or stems.

Differences in the Design of Root Systems Help Plants Compete for Water and Minerals

Plants of the same and different species that grow in the same vicinity compete for soil water, mineral nutrients, and light energy. As will be discussed in Chapter 26, plants reduce the effects of competition by utilizing different parts of the environment, including the soil. This is one reason why root systems of different species also occupy different depths in the soil.

When plants growing in the prairie are carefully excavated, three general categories of roots are evident, based on how deep the roots grow (see Fig. 5.3). Some grassland species, such as blue grama (*Bouteloua gracilis*), possess a very shallow root system, most of the roots being within the top 15 cm (6 in) of soil. Other species, such as buffalo grass (*Buchloe dactyloides*), have evenly distributed roots as deep as 1.5 m (60 in). Still others, such as locoweed (*Crotalaria sagittalis*), have a tap root system, which lacks width but runs deep. By using different depths of the soil, these plants reduce competition for moisture and dissolved minerals.

In trees and shrubs, most functioning roots are localized in the upper 1 m of soil, with the majority in the upper 15 cm (6 in) (see Fig. 5.1). This depth puts them in

Figure 5.5 Types of adventitious roots. (**a**) Corn (*Zea mays*) stem showing prop roots, which emerge from the stem just above the soil. These roots help support the shoot system. (**b**) Banyan (*Ficus bengalensis*) tree with an extensive aerial root system. (**c**) Extensive adventitious root system of mangrove (*Rhizophora mangle*) growing in the tidal zone of Australia's tropical coast. Note the many air roots (pneumatophores) sticking up from the mud. (**d**) Adventitious roots of the climbing fig *Ficus pumila* forming an adhesive pad on a window.

direct competition with grasses. To help trees compete, their branched roots may extend well beyond the expansion of overhead branches. The organization of root tissues and their development will be discussed in the following section

5.3 **THE DEVELOPMENT OF ROOTS**

The tips of functional roots are thin (Fig. 5.4a) and usually white. If you were to dig up the roots of a big tree, you would see many very large roots, each of which could be followed through its branches to a thin, white tip. These tiny root tips are important parts of the root system because it is here, in just a few millimeters, where many of the important functions of roots take place.

The Root Tip Is Organized into Regions and Protected by a Root Cap

In a longitudinal section of a root tip viewed through a microscope, it is apparent that the cells are organized into three regions: the meristem, the region of elongation, and the region of maturation (Fig. 5.6a, b). The developmental events that take place in the cells of each region are somewhat specific, but the regions do overlap (Fig. 5.6c).

The **root cap** at the tip of the root apex protects the **root apical meristem (RAM)**, a group of small, regularly shaped cells, most of which are dividing. These cells are organized into two different patterns (Figure 5.7a, b).

A small, centrally located part of the RAM is called the **quiescent center (QC)** (Fig. 5.8) because its cells divide at an extremely slow rate. The function of the QC isn't exactly known, but it seems to be activated during

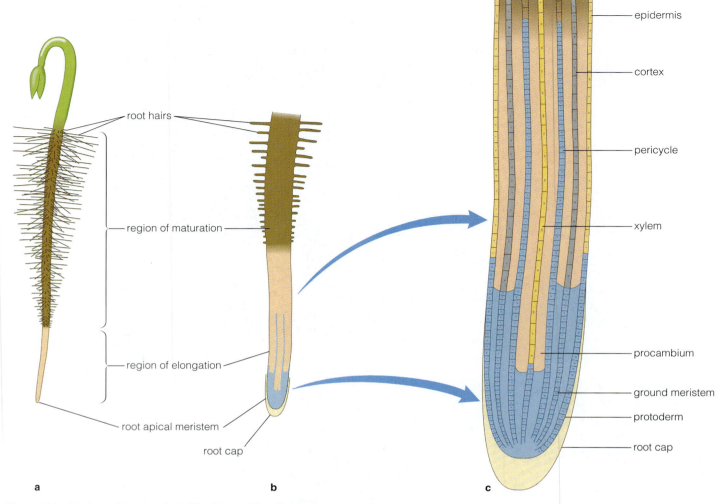

Figure 5.6 Regions of the root tip. (**a**) Radish seedling (*Raphanus sativus*), showing the root apical meristem, region of elongation, and region of maturation containing root hairs. (**b**) Closer view of the three regions. (**c**) Median longitudinal section of root tip. The precise positions of the boundaries of these regions are dependent on the tissue. Procambium cells tend to stop dividing before ground meristem cells; protoderm and pericycle cells continue dividing farther back in the root. All cells stop elongating at approximately the same point. Primary tissue cell maturation also may occur at different positions. (Dividing cells are shown in blue.)

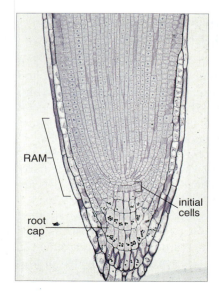
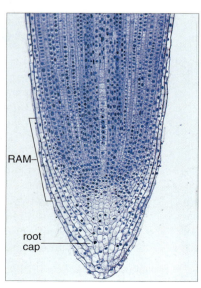

Figure 5.7 The patterns of RAM divisions. (**a**) Longitudinal section of the root tip of flax (*Linum grandiflorum*). Notice that all cell files connect directly to specific tiers of cells just above the root cap, ×200. (**b**) Longitudinal section of onion (*Allium cepa*) root tip. In this type of root apex the cell files terminate at a group of cells without any apparent organization, ×81.

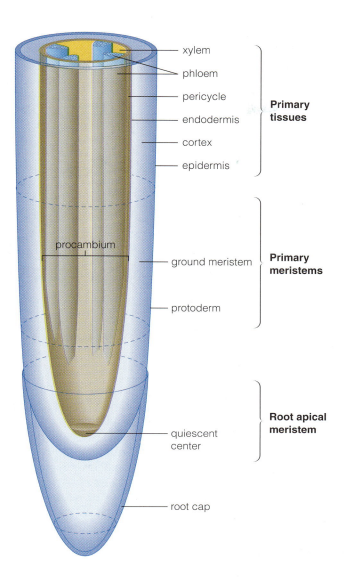

Figure 5.8 Three-dimensional diagram of a root in longitudinal view, showing the relative positions of the primary meristems and primary tissues. The quiescent center is a small group of cells that are metabolically quiescent.

xylem

phloem

pericycle

endodermis

cortex

epidermis

Primary tissues

procambium

ground meristem

protoderm

Primary meristems

quiescent center

Root apical meristem

root cap

times of acute stress. It may be a site for the synthesis of plant hormones important for controlling root development.

Cells just apical to the QC divide and produce cells to form the root cap (Fig. 5.8), the thimble-shaped layer of cells that protect the RAM as the root elongates and pushes through the soil. The root cap is also the site of gravity perception, which controls the direction of root growth (Chapter 15). Root cap cells are constantly being sloughed off at the very tip, but new cells are added by the apical meristem. The sloughed-off cells can remain alive in the soil for a time, where they provide nutrients for soil bacteria and fungi in the rhizosphere (Figs. 5.2, 5.9).

Just basal to the meristem region, toward the body of the root, is the **region of elongation**. Careful examination of a longitudinal section of a root shows that the boundary where cells stop dividing and start elongating is different for each tissue (see Fig. 5.6c). Cells in the vascular tissue, for example, stop dividing closer to the RAM than do cells in the cortex. Generally speaking, cells that are

root cap

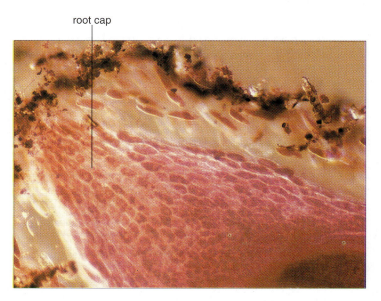

Figure 5.9 Corn (*Zea mays*) root cap, with sloughed root cap cells in the soil, ×60.

Table 5.1 Summary of Tissues and Meristems in Roots

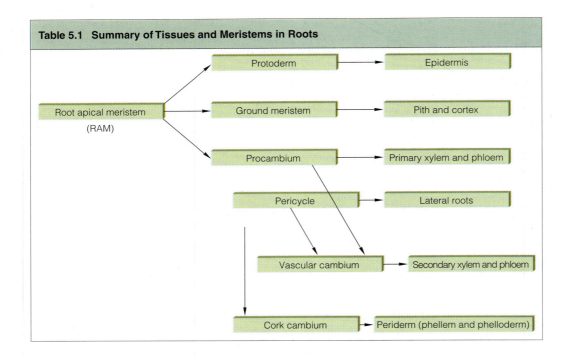

long at maturity stop dividing closer to the RAM than do cells that are short at maturity. The boundary for the termination of elongation is the same for all cell files in a root that is growing straight.

The **region of maturation** is the site of root hair formation and the maturation of other cell types. The precise position of cell maturation in different cell files is variable; cells in some files mature close to the tip, and others mature farther back (Fig. 5.6c).

The RAM Forms Three Primary Meristems

As mentioned in Chapter 4, cells change their structure according to their position, and the process of change begins in the root apical meristem. The RAM differentiates into three primary meristems: the protoderm, ground meristem, and procambium (see Fig. 5.8). These then go on to become the specialized cells and primary tissues of the root, as described in the following pages and as summarized in Table 5.1.

5.4   THE STRUCTURE OF ROOTS

The root body is composed of various tissues. Each tissue has a specific function. In the following section, each tissue will be described.

The Epidermis, Cortex, and Vascular Cylinder Are Composed of Specialized Tissues

EPIDERMIS Protoderm cells differentiate into the epidermis, which in roots is composed mostly of long epi-

dermal cells. Some cells of the protoderm develop into **root hairs** (Fig. 5.10) by the extension of epidermal cell walls into the surrounding soil. The cell walls of root hairs are thin and composed principally of cellulose and pectic substances. Root hairs tend to be sticky, so that soil particles cling to them (Fig. 5.10b). In most plants, the life of any one root hair is short; it functions for only a few days or weeks. New hairs are constantly forming at the apical end of the root-hair zone, while those at the basal end are dying. Thus, as the root advances through the soil, fresh, actively growing root hairs are constantly coming into contact with new soil particles. In the rye plant, root hairs develop at an average rate in excess of 100 million per day.

Although nearly all ordinary land plants possess root hairs, a few (such as the firs, redwoods, and Scotch pine) are apparently devoid of them. Also, many aquatic plants have no root hairs. Moreover, land plants (corn, for example) that normally develop root hairs when the root system grows in the soil may develop no root hairs when the roots grow in water. In plants devoid of root hairs, absorption is accomplished entirely through the epidermal cells.

The epidermis in roots is usually one cell layer thick; but in the aerial roots of certain plants, such as orchids, a multilayered epidermis develops that stores and possibly absorbs water from the moist air.

CORTEX The root cortex is derived from ground meristem and is composed chiefly of parenchyma cells. The innermost layer of the cortex, a single row of cells called the **endodermis** (Figs. 5.11, 5.12a, b) plays a special role in controlling mineral accumulation by the root. This is the role of the **Casparian strip**, a waxy material embedded in

a

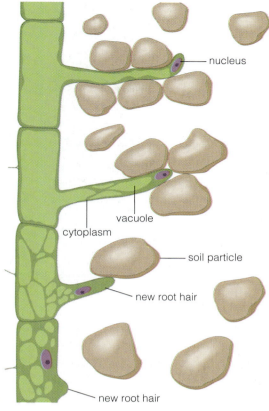

b

Figure 5.10 The development of root hairs. (**a**) Radish seedling (*Raphanus sativus*). (**b**) Stages in the development of root hairs. Note that the external epidermal cell wall protrudes and that the cell cytoplasm and nucleus move into the root hair near the tip. Root hairs are in close contact with soil particles and increase the water-absorptive surface of the root.

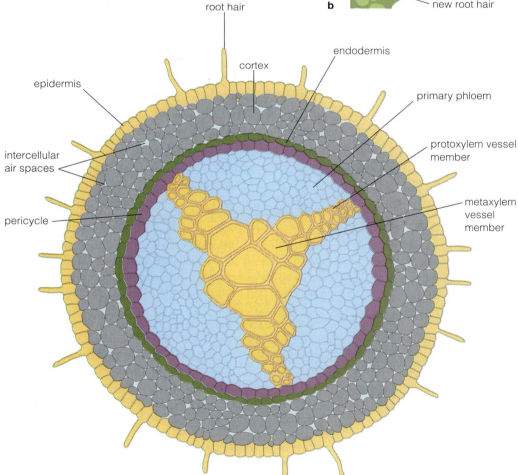

Figure 5.11 A root cross section from the area where maximum absorption occurs. The epidermis contains root hairs. The cortex has abundant air spaces between parenchyma cells. The endodermis is bounded by a Casparian strip. The pericycle is one cell layer thick. The primary xylem is distributed into three protoxylem points, with metaxylem in the middle of the root. The primary phloem alternates with primary xylem. Residual procambium occurs between the primary xylem and primary phloem.

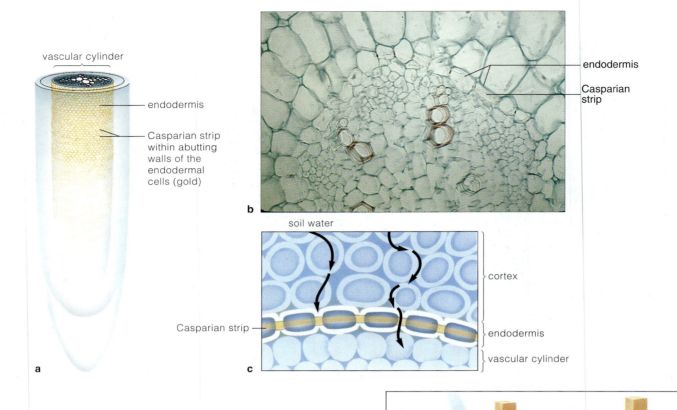

vascular cylinder

— endodermis

— Casparian strip within abutting walls of the endodermal cells (gold)

a

endodermis

Casparian strip

b

soil water

cortex

Casparian strip

endodermis

vascular cylinder

c

Casparian strip

d

Figure 5.12 Control of nutrient movement into the xylem is a function of the endodermis. (**a**, **b**). The endodermis is a single layer of inner cortex cells that have a waxy strip (Casparian strip) embedded in their transverse and radial cell walls. (b, *Nymphoides*, sp., ×300.) (**c**) The strip keeps water from moving indiscriminately through the cell walls and into the vascular cylinder. (**d**) The Casparian strip makes water move through the endodermal cells; in this way the plasma membrane can selectively control the uptake of nutrients.

the upper, lower (transverse), and side (radial) walls of endodermal cells (Fig. 5.12).

Water and dissolved minerals from the soil can move from cell to cell by two paths; they can travel freely through the porous walls of the cortex and epidermis, or they can move through the living cells (Fig. 5.12c, d). Movement through the cell wall is free movement without any constraints. Movement into a living cell, however, is regulated because it involves crossing the plasma membrane. Some minerals can move across the membrane by diffusion. (This involves a gradient in which the concentration of a mineral on one side of a semipermeable membrane will try to equalize on the other side.) Some minerals, for example potassium, can be moved across membranes through special proteins embedded in the membrane. These proteins can actually pump minerals into a living cell, even against a diffusion gradient.

The function of the endodermis is to guarantee that the minerals that finally reach the vascular cylinder can do so only by first passing across at least one plasma membrane. One reason this is important in roots is that it

provides a mechanism to increase the concentration of needed minerals through pumps in the endodermis cell membrane.

In roots of many plants, an **exodermis** containing Casparian strips occurs at the outer layer of the cortex, just inside the epidermis. This layer is present in many grass roots and in the aerial roots of orchids (Fig. 5.13). The exodermis apparently also functions to regulate ion absorption and accumulation.

VASCULAR CYLINDER The entire central cylinder of roots is composed of vascular tissue that differentiates from the procambium cells (Fig. 5.14). The primary xylem usually consists of a central core of xylem elements organized into two or more radiating points. The first xylem elements to mature, the **protoxylem**, develop at the outer points of the xylem (Figs. 5.11, 5.14b). **Metaxylem**, the last primary xylem to mature, differentiates in the center of the vascular cylinder (Fig. 5.14b, c).

The protoxylem is capable of transporting water while the root is elongating, which requires both the

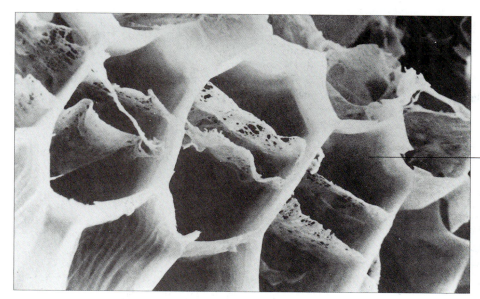

exodermal cell

Figure 5.13 The Casparian strip in exodermis cells of an aerial root of *Epidendrum* orchid. The plasma membrane has been pulled away from the cell wall by soaking the roots in salt solution. This SEM shows that the membrane is attached at the Casparian strips, ×800.

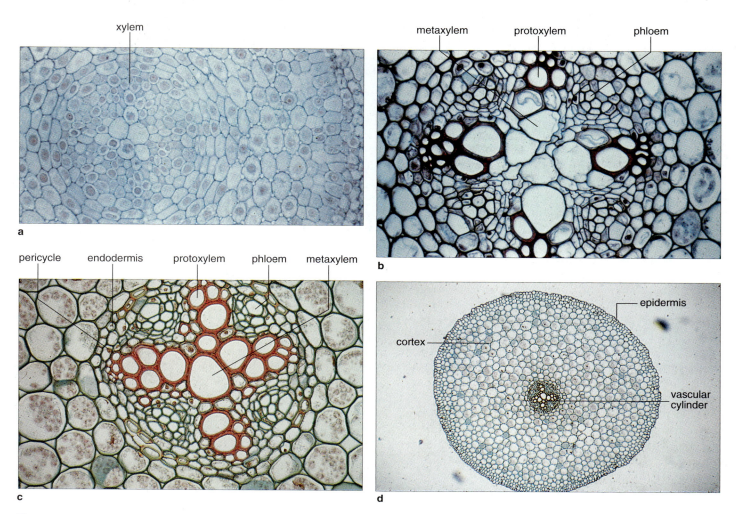

xylem

metaxylem   protoxylem   phloem

a

b

pericycle   endodermis   protoxylem   phloem   metaxylem

epidermis

cortex

vascular cylinder

c

d

Figure 5.14 Differentiation in the primary growth of buttercup (*Ranunculus* sp.) roots. (**a**) Immature region, where no secondary walls have yet formed in the xylem, ×170. (**b**) The cells of the protoxylem have developed secondary walls (shown here stained red with safranin). This particular root has four protoxylem points with phloem between, ×190. (**c**) Fully mature root with all primary tissues differentiated. Note that the endodermal cells adjacent to the protoxylem points lack secondary walls, ×170. (**d**) The same section as (**c**), but showing all tissues, ×190.

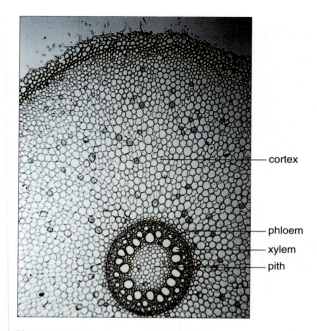

Figure 5.15 *Asparagus officinale* root in transverse section, showing all tissues and root regions. The vascular cylinder has a pith, ×50.

strength to withstand the forces that move water and still be flexible enough to stretch as the root elongates. This dual ability comes from a secondary cell wall in the shape of annular rings or spirals (see Fig. 4.9).

Metaxylem cells mature in regions of the root where elongation has been completed. Because they are no longer required to elongate, they form thick secondary cell walls with pits through which lateral exchange of water and minerals may take place. Protoxylem cells often become crushed after the metaxylem develops, but by then these cells aren't needed.

In roots of some plants (such as asparagus), a central region of parenchyma cells forms (Fig. 5.15). This region is sometimes called a *pith* (this refers to the location of ground tissue in the center of stems, which is formed from ground meristem); but in roots it is part of the vascular cylinder and originates from procambium. Xylem (both protoxylem and metaxylem) of roots consists of several other cell types, including vessel elements, tracheids, parenchyma, and fibers.

Phloem cells form in the areas between the protoxylem arms (Figs. 5.11, 5.14). The **protophloem** is actually the first part of the vascular system to become functional. These cells form at the periphery of the phloem and function primarily during root elongation. **Metaphloem** develops toward the inside and functions during the plant's adult life. Phloem of roots may consist of parenchyma, fibers, sieve-tube members, and companion cells.

The outer boundary of the vascular cylinder is the **pericycle** (Figs. 5.11, 5.14c). This tissue is unique, in that it remains capable of dividing for a long time. It has three

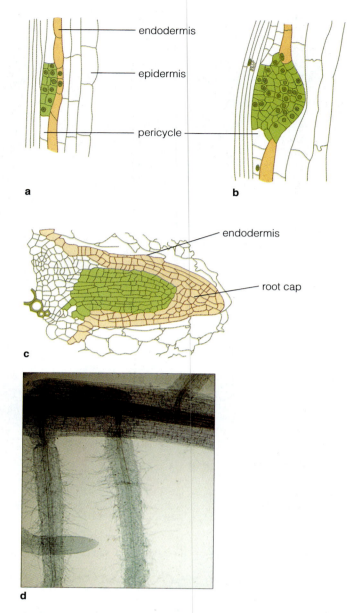

Figure 5.16 Lateral root formation. (**a**) Initiation of a lateral root of carrot (*Daucus carota*) by division of cells in the pericycle. (**b**) Formation of the lateral root primordium. (**c**) The young root pushing through the cortex. (**d**) Radish (*Raphanus sativus*) root cleared to show the continuous vascular connection to the primary root, ×12.

functions: (1) It is the site where the development of lateral roots is initiated, (2) it contributes to the formation of vascular cambium, and (3) it contributes to the formation of the cork cambium. The pericycle is usually one cell layer thick, but in some roots it has multiple layers.

Lateral Roots Begin Forming in the Pericycle

The initiation of lateral roots at particular locations is controlled by chemical growth regulators that cause pericycle cells to begin dividing at specific sites (Fig. 5.16a).

Figure 5.17 Secondary growth in a root. (**a**) At the completion of primary growth, an arc of residual procambium cells remains between the primary xylem and primary phloem. The pericycle is a complete cylinder. (**b**) The residual procambium starts to divide and joins with the pericycle cells outside the xylem arms to eventually form a complete cylinder of vascular cambium. The pericycle just outside the protoxylem points divides to form at least two cell layers. The inner layer joins with the residual procambium to form the vascular cambium; the outer layer stays part of the pericycle cylinder. (**c**) The vascular cambium forms secondary xylem internally and secondary phloem externally. The primary phloem is being pushed outward. (**d**) A root with secondary xylem and secondary phloem. The primary xylem remains in the center of the root. The primary phloem has been crushed, and the pericycle that remains will form the cork cambium. The cork cambium forms the periderm after the epidermis and cortex die. The term *bark* refers to everything outside the vascular cambium.

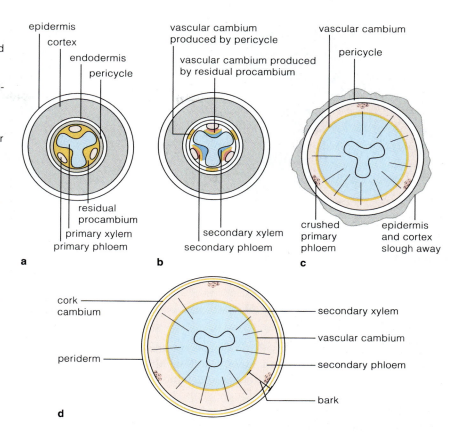

The **lateral root primordia** that result continue to form new cells, which in turn elongate. Endodermal cells outside the primordium also divide for a short time, contributing cells to the tip of the new lateral root. As it expands (Fig. 5.16b), the lateral root pushes its way through and destroys the cortical cells and the outer epidermis. The breakdown of these cortical cells is thought to be at least partly the result of digestive enzymes released from the lateral root primordium. As the lateral root emerges, its cells become organized into a root cap and root apical meristem (Fig. 5.16c, d). The wound formed by lateral root emergence is quickly healed by the secretion of mucilage and waxy substances by adjacent cortical cells.

The Vascular Cambium and the Cork Cambium Partially Form from the Pericycle

As the plant grows, the primary vascular tissues in its roots may not be able to supply the water and nutrient needs of the enlarged plant body. In long-lived dicot plants, the older regions of the root form secondary vascular tissues by activating a secondary meristem, the vascular cambium. It may be of interest to note that there is only one monocot plant, the dragon's blood tree (*Dracaena draco*), known to have secondary growth in roots. Even very tall, tree-like monocots like palms lack secondary growth in roots.

Secondary growth is initiated by the division of pericycle cells and also some leftover or **residual procambium cells** located between the arcs of xylem and phloem (Fig. 5.17a). Residual procambium cells are actually procambial cells that did not develop into primary xylem or primary phloem. They are now induced to divide, and they form secondary xylem to the inside and phloem to the outside (Fig. 5.17b). After a time, the crescent-shaped region of dividing cells joins with the pericycle, which also begins to divide, forming at least two layers of pericycle cells. The inner layer joins to the residual procambium to form an intact ring of vascular cambium (Fig. 5.17c). The outer layer remains as pericycle.

The secondary xylem or *wood* in roots with several years of secondary growth looks very much like that of woody stems. The only difference is that in young roots primary xylem occupies the middle of the root (Fig. 5.18), whereas in young stems pith occupies the middle.

Continued growth expands the root and finally causes the splitting, sloughing off, and destruction of the cortex and epidermis (Fig. 5.17c). The pressure created by this expansion apparently stimulates the remaining ring of pericycle cells to divide again. This last function of the pericycle converts it into the cork cambium, which forms the periderm (see Fig. 4.20). The **bark** in woody roots (which includes all cells from the vascular cambium outward) appears similar to that in stems; but it may be thinner and smoother on its outer surface (Fig. 5.17d).

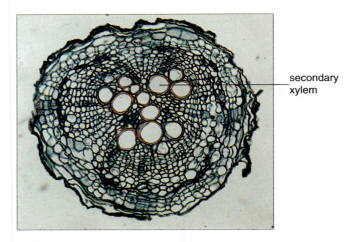

Figure 5.18 Alfalfa (*Medicago sativa*) root transverse section, with some secondary tissues and periderm on its periphery, ×125.

Some Roots Are Specialized for Special Functions

Roots of a great many plants do not have the general characteristics common to most roots. We have already discussed examples of adventitious roots which arise from nonroot origins, and the uniquely shaped roots of clinging vines. Other plants, especially those forming partnerships with microorganisms, have specialized root structures. Some important examples are haustorial roots, root nodules, mycorrhizae, and contractile roots.

Parasitic plants like dodder (*Cuscuta* sp.) anchor themselves by sinking **haustorial roots** into the vascular tissue of a host stem and thus tap the host's water and nutrient supply (Fig. 5.19).

Although nitrogen is one of the most important elements needed by plants, most plants cannot use atmospheric nitrogen (N_2) directly. Certain legumes, such as peas (*Pisum sativum*) and soybeans (*Glycine max*), are capable of *fixing* nitrogen—that is, changing N_2 that diffuses into the soil into NH_4^+ (ammonium ion), which is usable by the plant. Nitrogen fixation is the result of an unusual relationship between the bacterium *Rhizobium* and the roots of legumes (see Chapter 19; Fig. 19.15). Root cells are infected by the passage of a thin infection thread of the bacteria into root hair cells and on through to the cortical cells. The bacteria then divide and stimulate the cortical cells to divide, thereby forming a **root nodule** (Fig. 5.20). The bacteria are the actual agents for fixing the nitrogen.

Mycorrhizae are short, forked roots common to as many as 90% of seed plants. These specialized root struc-

Figure 5.19 (**a**) Haustorial roots of a plant parasite, dodder (*Cuscuta* sp.) infecting the stem of a host plant, ×53. (**b**) External view of dodder on tomato plants.

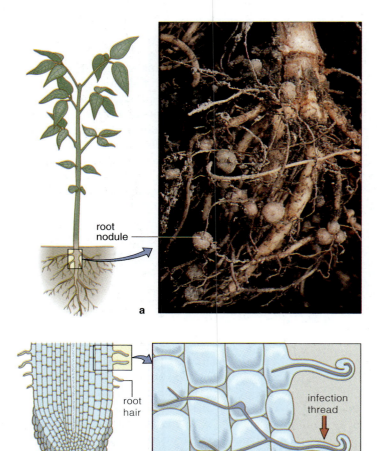

Figure 5.20 Bacterial nodules in legume roots. (**a**) Legume plants form bacterial nodules on roots. (**b**) Bacteria enter the plant by passing through a tiny infection thread, which penetrates root hairs. Once inside the host, the bacteria penetrate to the cortex of the root, forming a swollen mass of cells filled with bacteria. The bacteria, called *Rhizobium* sp., fix nitrogen, which then passes up the plant body in a usable form.

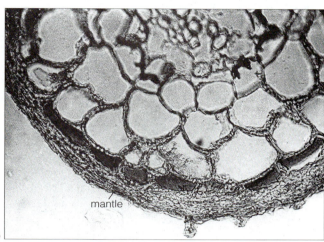

mantle

Figure 5.21 Ectotrophic mycorrhizae. (**a**) The typical Y-branched form in pine roots (*Pinus* sp.). (**b**) Transverse section through an infected mycorrhizal root, showing the mantle of fungal hyphae and the hyphae growing between the cortical cell walls.

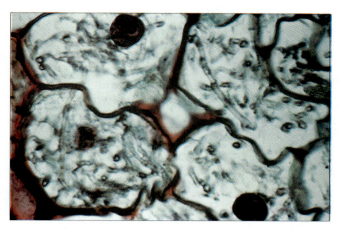

Figure 5.22 Photograph of *Coralloriza* sp. rhizome with endo-trophic mycorrhizae, showing fungal hyphae actually inside the host cells. (Note: A rhizome is actually a *stem* that grows underground. This example is used, however, because it shows the nature of the infection process very well, and it looks the same in *root* cells, ×735.)

Figure 5.23 Contractile roots of water hyacinth (*Hyacinthus orientalis*). Notice the wrinkled surface at the base of these roots, where the contraction occurs.

tures represent an association with a soil-borne fungus (see Chapter 20).

Two types of mycorrhizal roots may occur, distinguished by whether the fungus penetrates into the root cells or not. *Ectotrophic mycorrhizae* are found in roots of such trees as pines (*Pinus*), birches (*Betula*), willows (*Salix*), and oaks (*Quercus*). This type causes a drastic change in the root shape (Fig. 5.21), but the fungus does not penetrate the root cells. The fungus does, however, penetrate between the cell walls of the cortex, and it forms a covering sheath (or *mantle*) of fungal hyphae around the entire root. These mycorrhizal roots are about 0.5 cm (0.2 in) long; they lack a root cap and have a simple vascular cylinder. *Endotrophic mycorrhizae* do not form a mantle over the root, and the fungus actually enters the cortex cells (Fig. 5.22).

Mycorrhizae make roots more efficient in mineral absorption, but they are apparently not absolutely essential for the growth of the usual host plants (see Chapter 11). This is known because plants that are artificially fed adequate nutrients can grow without mycorrhizae. Mycorrhizae also may be beneficial to their host plants by secreting hormones or antibiotic agents that reduce the potential of plant disease.

Roots of the dandelion (*Taraxacum officinale*), water hyacinth (*Hyacintha orientalis*), and some other plants are capable of contracting, which keeps aboveground parts near the soil surface. This contraction is caused by the radial expansion (or, in some instances, the collapse) of cells in the root cortex. The vascular tissue in **contractile roots** forms a twisted, undulated mass (Fig. 5.23).

SUMMARY

1. The principal functions of roots are absorption of water and nutrients, conduction of absorbed materials and food, and anchorage of the plant in the soil.

2. The primary root develops from the radicle in the embryo. It generally penetrates the soil to some depth; if it dominates, a tap root system results, with a main root axis and branches.

3. Fibrous root systems are formed by seminal roots arising in the embryo in addition to the radicle. Grasses are good examples of plants with fibrous roots.

4. Adventitious roots may arise in the internodal regions of the stem.

5. The root tip is divided into overlapping zones: (a) the root cap, which protects (b) the meristematic region; (c) a region of elongation; and (d) a region of maturation, characterized externally by root hairs and internally by the formation of primary vascular tissues.

6. The epidermis forms as the outer tissue of the root. Water is absorbed through the epidermal cells and the root hairs. The next layer is the cortex. Its cells mainly store nutrients. The endodermis is the innermost cell layer of the cortex. The Casparian strip is a waxy substance found in the radial and transverse walls. It encircles endodermal cells like a ribbon. Water cannot move across the Casparian strip. Therefore, all water with dissolved nutrients must pass through the protoplasts of endodermal cells. An exodermis is present just inside the epidermis in many roots; it may also have Casparian strips and function in ion absorption and regulation.

7. In cross section, the primary xylem is star-shaped, with protoxylem at the points. Primary phloem arises between the arms of primary xylem.

8. Roots of certain grasses usually have central parenchyma and many protoxylem points.

9. The pericycle, a row of cells internal to the endodermis, represents the outermost row of cells of the vascular cylinder. Cells of pericycle may eventually initiate the differentiation of the vascular cambium and the cork cambium; they also give rise to lateral roots.

10. A vascular cambium originates from procambium cells between primary xylem and phloem and from pericycle cells exterior to the radiating points of primary xylem. The vascular cambium forms secondary xylem internally and secondary phloem externally. The resulting increase in diameter stretches and tears the endodermis, cortex, and epidermis. A cork cambium develops from the pericycle and forms the periderm.

11. Haustorial roots from parasitic plants penetrate into the host. Bacterial nodules occur in roots of nitrogen-fixing legumes. Mycorrhizae are roots infected by beneficial fungi. Contractile roots pull the shoot tight to the soil surface.

Questions

1. Be able to identify in a diagram or photograph the following root structures and tissues:

root cap

root apical meristem (RAM)

region of elongation

region of maturation

root hair

epidermis

cortex

endodermis

Casparian strip

pericycle

vascular cylinder (xylem and phloem)

2. State how each item listed in question 1 contributes to the function of the root.

3. Discuss the tissues and cells involved in mineral uptake and transport in a root.

4. Describe the differences between primary and secondary tissues in a root. Where are these located? Be able to make a labeled diagram to show both primary and secondary xylem and phloem.

5. Make a diagram to show the position of cork cambium and vascular cambium in a root.

6. Describe the structure and function of the root cap.

7. Describe two symbiotic associations involving roots (root nodules and mycorrhizae). What are the microorganisms involved, and how does the association alter the structure of the root? How do both partners benefit from the association?

8. Describe two modified roots.

Further Readings

Aikmann, L. 1977. *Nature's Healing Arts: From Folk Medicine to Modern Drugs*. Washington D.C.: National Geographic Society.

Epstein, E. 1977. "The Role of Roots in the Chemical Economy of Life on Earth." *BioScience* 27 (12): 783–87. Popular article on the function of roots.

Esau, K. 1977. Chapters 14 and 15 in *Anatomy of Seed Plants*. 2d ed. New York: John Wiley. This is the classic textbook on plant anatomy.

Fahn, A. 1990. Chapter 13 in *Plant Anatomy*. 4th ed. Oxford: Pergamon Press. A more up-to-date textbook on plant anatomy.

Feldman, L. J. 1984. "The Development and Dynamics of the Root Apical Meristem." *American Journal of Botany* 71 (9): 1308–14. A review of root apical structure, for more advanced readers.

Quinn, V. 1938. *Roots: Their Place in Life and Legend*. New York: Frederick A. Stokes.

Rost, T. L., and M. L. Sandler. 1978. "The Common History and Popular Uses of Roots." *American Biology Teacher* 40: 338–41.

Vohora, S. B., M. Rizwan, and J. A. Kahn. 1973. "Medicinal Uses of Common Indian Vegetables." *Planta Medica* 23: 381.

6

THE SHOOT SYSTEM I: THE STEM

1. The shoot system is composed of the stem and its lateral appendages: leaves, buds, and flowers. Leaves are arranged in different patterns (phyllotaxis): alternate, opposite, whorled, and spiral.

2. Stems provide support to the leaves, buds, and flowers. They conduct water and nutrients and produce new cells in meristems (shoot apical meristem, primary and secondary meristems).

3. Stems are composed of the following: epidermis, cortex and pith, xylem and phloem, and periderm.

4. Secondary xylem is formed by the division of cells in the vascular cambium and is called *wood*. The bark is composed of the periderm (formed from cork cambium) and the secondary phloem.

5. Several different types of modified stems (rhizomes, spines, and others) have important functions.

6.1 THE FUNCTIONS AND ORGANIZATION OF THE SHOOT SYSTEM

The shoot system of a typical flowering plant consists of the stem and the attached leaves, buds, flowers, and fruits. The leaves are displayed in a way that maximizes their exposure to light. Flowers and fruits are located on stems in positions that allow for pollination and the dispersal of fruits and seeds. Internally, stems provide pathways for (1) the movement of water and dissolved minerals from the roots into the leaves and for (2) food synthesized in leaves to move into roots. Some stems are modified for the storage of water and various food substances.

The stem is actually composed of repeated units called **modules**. A module is a segment of stem—an **internode**—plus the leaf and bud attached to the stem (Fig. 6.1). The point of attachment is called a **node**.

The **shoot apical meristems (SAM)** and **primary meristems** of the stem are located in buds at the ends of the branches and just above the nodes. Further down the stem in some plants are **secondary meristems**, which make secondary tissues. These meristems produce new cells and act as sites for the start of cell elongation and differentiation.

Flowering plants are divided into two groups: **dicotyledonous plants (dicots)** and **monocotyledonous plants (monocots)**. One of the bases for this division is that the dicots—such as peas (*Pisum* sp.) and oaks (*Quercus* sp.)—produce embryos with two **cotyledons** (seed leaves), whereas monocots—such as corn (*Zea mays*) and onions (*Allium cepa*)—produce embryos with only one (see Chapter 13). The stems of these two groups also have major differences in the distribution of their tissues (Fig. 6.2) and in the operation of their meristems. As we examine primary and secondary growth in stems, we will contrast the anatomy of dicots and monocots.

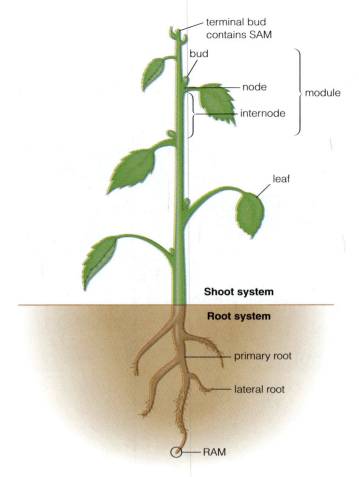

Figure 6.1 The shoot system and its parts.

6.2 PRIMARY GROWTH AND STEM ANATOMY

Primary Tissues of Dicot Stems Develop from the Primary Meristems

The shoot apical meristem (SAM) is composed of dividing cells. It is responsible for the initiation of new leaves and buds and for making the three primary meristems (Fig. 6.3). The three primary meristems—**protoderm**, **ground meristem**, and **procambium**—and their products are basically the same as in roots.

PROTODERM TO EPIDERMIS The outermost layer of cells in the shoot tip is the protoderm. This layer is called a primary *meristem* because its cells are still dividing. When the cells of the protoderm stop dividing and mature, they are called *epidermis* (Fig. 6.4). Epidermal cells, guard cells, different kinds of trichomes or hairs, and a cuticle make up the epidermis (see Chapter 4).

GROUND MERISTEM TO PITH AND CORTEX In the very center of the shoot tip and just inside the protoderm is the ground meristem (Figs. 6.3, 6.4). These cells slowly

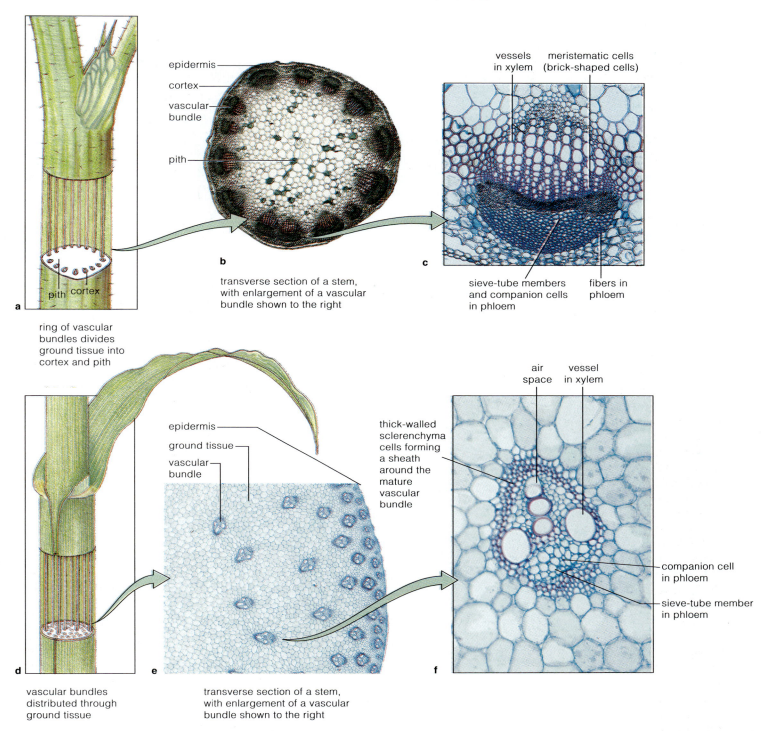

epidermis

cortex

vascular bundle

pith

a

pith cortex

ring of vascular bundles divides ground tissue into cortex and pith

b

transverse section of a stem, with enlargement of a vascular bundle shown to the right

vessels in xylem **meristematic cells (brick-shaped cells)**

c

sieve-tube members and companion cells in phloem fibers in phloem

epidermis

ground tissue

vascular bundle

d

vascular bundles distributed through ground tissue

e

transverse section of a stem, with enlargement of a vascular bundle shown to the right

air space **vessel in xylem**

thick-walled sclerenchyma cells forming a sheath around the mature vascular bundle

companion cell in phloem

sieve-tube member in phloem

f

Figure 6.2 Comparison of the anatomy of a dicot stem (top) and a monocot stem (bottom). (**a**) Typically, dicot stems have a pith surrounded by a cylinder of vascular bundles, shown in a clover (*Trifolium* sp.) stem (**b**), ×37. (**c**) These bundles usually have primary phloem toward the outside and primary xylem toward the inside, ×266. (**d**) Typically, monocot stems have scattered vascular bundles, shown in a portion of a corn stem (*Zea mays*) (**e**), ×14. (**f**) The primary phloem in these bundles is usually positioned toward the outside, ×270.

lose their ability to divide, and they differentiate mostly into parenchyma cells of the cortex (the cylinder of cells just inside the epidermis) and the pith (the core of cells in the center of the stem) (Fig. 6.2b). Along with their different positions in the stem, the cortex and pith have different functions. The parenchyma cells nearest the outside of the cortex sometimes contain chloroplasts for photo-

synthesis. Sometimes, the parenchyma cells of the cortex or pith store starch. In some instances the pith region of a stem may become hollow by the breakdown of the centrally located parenchyma cells.

PROCAMBIUM TO PRIMARY XYLEM AND PHLOEM The procambium tends to form as small bundles (Figs. 6.3,

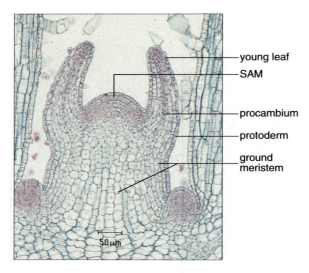

young leaf
SAM

procambium

protoderm

ground
meristem

50 μm

Figure 6.3 Shoot apex of *Coleus blumei*, a common houseplant, ×114.

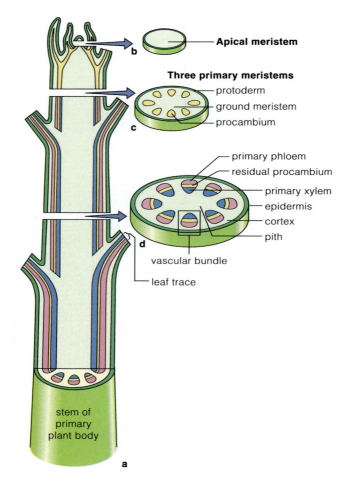

Apical meristem

b

Three primary meristems

protoderm
ground meristem
procambium

c

primary phloem
residual procambium
primary xylem
epidermis
cortex
pith

d

vascular bundle

leaf trace

stem of
primary
plant body

a

Figure 6.4 The pattern of primary vascular development in a dicot stem. (**a**) Longitudinal view showing the vascular bundles of a stem with attached leaves. (**b**) Cross section through shoot apical meristem. (**c**) Section taken slightly below the SAM, showing the three primary meristems: procambium, ground meristem, and protoderm. (**d**) Each bundle of procambium will mature into primary phloem (toward the outside) and primary xylem (toward the inside). The procambium that does not differentiate into primary xylem and phloem is called *residual procambium*. The protoderm matures to become the epidermis, and the ground meristem matures to become the pith and cortex.

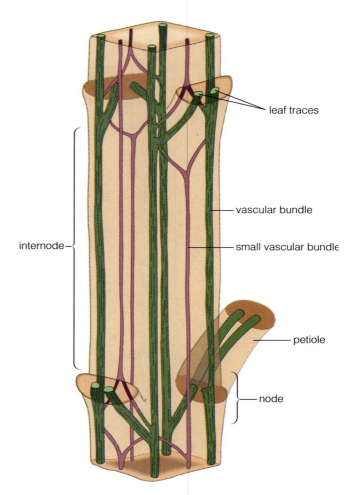

leaf traces

vascular bundle

small vascular bundle

internode

petiole

node

Figure 6.5 Three-dimensional view of the entire primary vascular system in a *Coleus* stem. In this stem there are two leaves per node, and two leaf traces pass into each leaf. A single large vascular bundle is found in each corner, with smaller bundles between the corners.

6.4) of relatively thin, long cells with dense cytoplasm. The bundles are usually arranged in a ring just inside the outer cylinder of ground meristem and below the SAM. Procambium cells divide, and then at some position down the axis they stop dividing and differentiate into primary xylem and primary phloem. Each bundle of procambium becomes a vascular bundle, with primary xylem cells toward the inside of the stem and primary phloem cells toward the outside (Fig. 6.2). In plants exhibiting secondary growth, some procambium between the primary xylem and phloem remains undifferentiated; such cells are called **residual procambium** (Fig. 6.4d).

The Distribution of the Primary Vascular Bundles Depends on the Position of Leaves

Vascular bundles in dicot stems are distributed in a **vascular cylinder**. The vascular bundles that network into the attached leaves are called **leaf traces** (Fig. 6.4a). The organization of primary vascular bundles in stems depends on the number and distribution of leaves and on the number of traces that branch and lead into the leaves (Fig. 6.5) and also into buds. The number of vascular

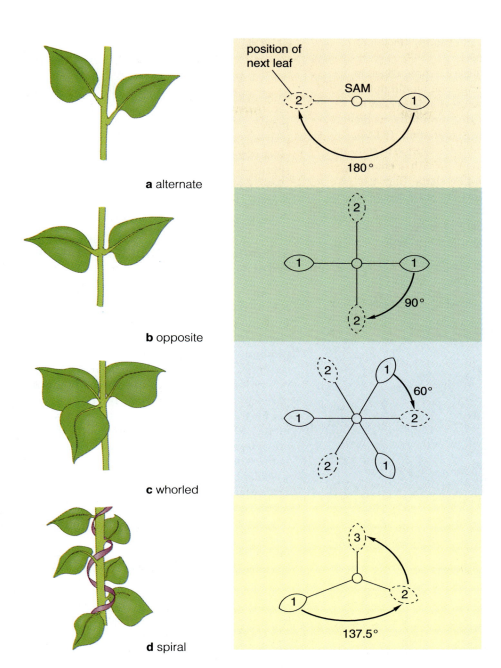

a alternate

b opposite

c whorled

d spiral

position of next leaf

SAM

180°

90°

60°

137.5°

Figure 6.6 The four basic patterns of leaf arrangement (phyllotaxis): (**a**) *Alternate* has one leaf per node and a 180° angle of divergence between leaves. (**b**) *Opposite* has two leaves per node, and 90° between sets of leaves; (**c**) *Whorled* usually has three to five leaves per node; in the case of three leaves per node, the angle of divergence between sets of leaves is 60°; (**d**) *Spiral* has one leaf per node and 137.5° between leaves.

bundles in the vascular cylinder and the number of leaf traces differ by species and are dependent on the number and arrangement of leaves.

The arrangement of leaves on the axis of the stem is called **phyllotaxis** (*phyllo* is Greek for "leaf," and *taxis* is Greek for "arrangement"). There are four basic patterns of phyllotaxis (Fig. 6.6). Plants with **alternate phyllotaxis** have only one leaf per node, and the leaves are positioned 180° from each other. The angle separating one leaf or set of leaves from another is called the *angle of divergence*. **Opposite phyllotaxis** means that the shoot has two leaves per node; the angle of divergence between leaves in successive sets is 90°. In **whorled phyllotaxis** there are three or more leaves per node, and the angle of divergence between successive sets of leaves depends on

the leaf number per set. The angle is 60° in plants with three leaves per node (Fig. 6.6c). Plants with **spiral phyllotaxis** have one leaf per node, and the angle of divergence between leaves is 137.5° (Fig. 6.6d).

Primary Growth Differs in Monocot and Dicot Stems

Monocot stems differ in several ways from the patterns described for dicot stems. One main difference is that the vascular bundles tend to be scattered throughout the stem instead of being in a ring (see Fig. 6.2). The terms *pith* and *cortex* are usually not used when the bundles are scattered; instead, the term **ground tissue** is used for all the parenchyma tissue surrounding the vascular bundles. There are some exceptions; wheat (*Triticum* sp.) stems, for example, are hollow in the stem internodes (Fig. 6.7), rather than having ground tissue there.

If you were to compare the shapes of a monocot stem and a dicot stem, one of the first things you would notice is that a monocot stem is about the same diameter near its apex as at its base. This shape is due primarily to the activity of the **primary thickening meristem (PTM)**, which is absent in dicot stems. The PTM is unique in contributing to both elongation and lateral growth, a characteristic resulting from its umbrella-like shape (Fig. 6.8). The SAM and the primary meristems are also present in these tips.

The next step in the development of stems in many plants is called *secondary growth*. This growth will increase the width of stems.

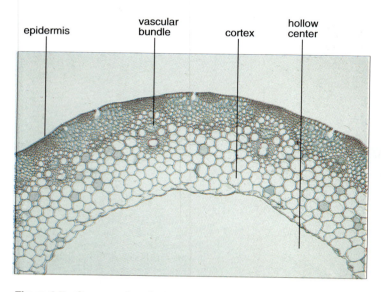

Figure 6.7   Cross section of a hollow wheat stem (*Triticum* sp.), ×38.

Most monocots and many dicots show little or no secondary growth; they are **herbaceous** (nonwoody) **plants**, which normally complete their life cycle in one growing season. By contrast, many other dicots—such as oaks (*Quercus* sp.) and maples (*Acer* sp.)—and gymnosperms—such as pines (*Pinus* sp.) and firs (*Abies* sp.)—show secondary growth starting in their first year of growth. In some plants this continues for many, even hundreds, of years; these are called **woody plants**.

Secondary Xylem and Phloem Develop from Vascular Cambium

Woody plants develop thicker, more massive stems because of the growth of secondary xylem and phloem from their secondary meristems. The first step in making secondary xylem and phloem is to form the vascular cambium (plural, *cambia*). Development of the vascular cambium involves coordinated cell division in the residual procambium inside the vascular bundles and the parenchyma cells between the bundles (Fig. 6.9). The signal for this cell division is probably given by a plant hormone. When the residual procambium cells divide, they are referred to as **fascicular cambium** (*fasciculus* is a Latin word for "bundle"). Next, the cells between the vascular bundles divide; these are referred to as the **interfascicular cambium** ("between the bundles"). The parenchyma cells near the bundles divide first, and the adjoining cells divide until a complete ring forms (Figs. 6.9, 6.10). Once the cylinder of dividing cells—the fascicular cambium plus the interfascicular cambium—is complete, it is called **vascular cambium** (Fig. 6.9).

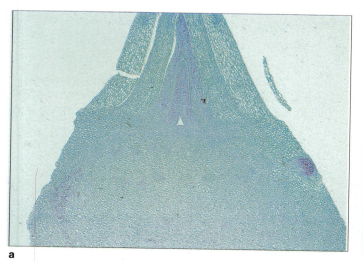

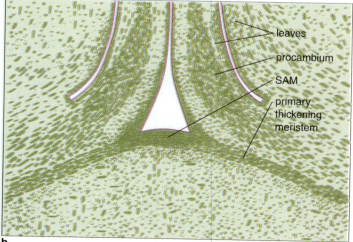

a

b

Figure 6.8   The primary thickening meristem. (**a**) Longitudinal section of *Iris* shoot apex showing the PTM, ×108. (**b**) Diagram of same apex to show the umbrella shape of the PTM more clearly.

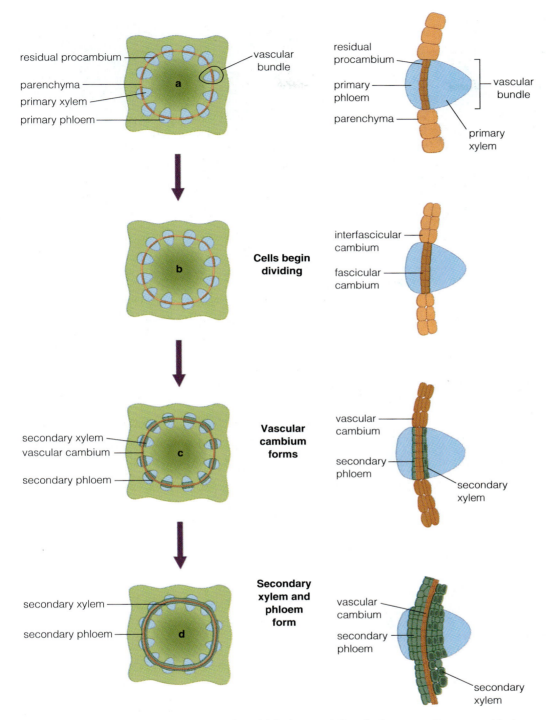

Figure 6.9 Formation of the vascular cambium. (**a**) At the completion of primary growth, some residual procambium (meristematic) cells remain between the primary xylem and primary phloem, and parenchyma cells occur between the vascular bundles. (**b**) After the cells begin dividing, the residual procambium is called the *fascicular cambium*, and the cells between the bundles are now called the *interfascicular cambium*. (**c**) When the fascicular and interfascicular cambia become connected, they are called the *vascular cambium*. (**d**) Secondary xylem (inside) and secondary phloem (outside) form from the vascular cambium.

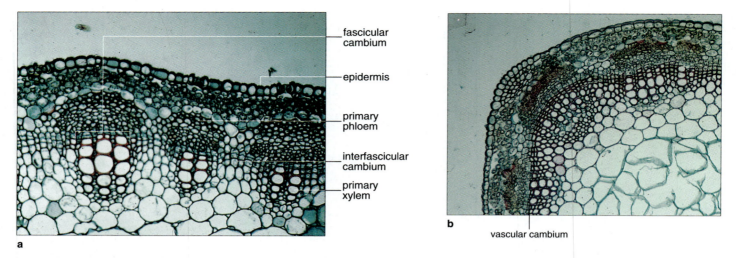

Figure 6.10 Cross sections of an alfalfa (*Medicago sativa*) stem, showing (**a**) the location of the fascicular and interfascicular cambia, ×191 and (**b**) some secondary vascular tissue, ×178.

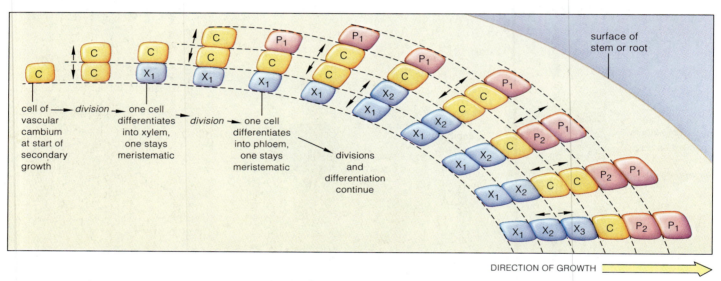

Figure 6.11 Drawing of the divisions of vascular cambium cells through a growing season. The cambium makes mostly secondary xylem cells (X_1, X_2, etc.) to the inside of the stem and secondary phloem (P_1, P_2, etc.) cells toward the outside. The result is that the stem gets wider, and the vascular cambium keeps increasing in circumference and moving outward.

The vascular cambium is an interesting meristem in that it is only one or two cells thick but divides in two directions. The cells formed to the outside become secondary phloem, and the cells formed to the inside become secondary xylem (Figs. 6.9, 6.10). Figure 6.11 summarizes the successive divisions that contribute to thickening the secondary xylem and phloem. Vascular cambium cells produce more xylem cells than phloem cells.

Some vascular cambium cells, called **fusiform initials**, form into cells of the **axial system**; vessel members are examples (Figs. 6.12, 6.13). The cells of the **ray system** are formed from vascular cambium cells called **ray initials**. The **rays** are composed of only two different cell types—ray parenchyma cells and ray tracheids. The cells of the axial system function in the longitudinal movement of water and minerals; the cells of the ray system transport water and minerals radially. (Review Chapter 4 for a list of all the cell types found in the axial system of xylem.)

Wood Is Composed of Secondary Xylem

The secondary xylem is collectively called **wood**. Examining thin sections of wood under a microscope reveals its many distinctive characteristics. Typically, these sections are made in three different planes of view: **tangential**, **radial**, and **transverse** (Fig. 6.13).

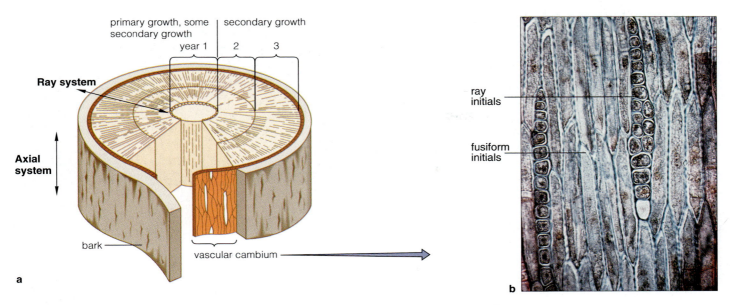

Figure 6.12 (**a**) Drawing of a three-year-old woody stem, showing the growth increments as annual rings. The lines radiating from the center are the rays; all the rays together make up the ray system. All the other cells in the wood make up the axial system. *Bark* refers to all the tissue from the vascular cambium to the outside. (**b**) The vascular cambium has two types of initial cells (inset): The ray initials make the cells of the rays, and the fusiform initials make the cells of the axial system, ×275.

Figure 6.13 Structure of red oak (*Quercus rubra*) wood. (**a**) Three-dimensional diagram showing the cells as they appear in three planes of view. (**b**) In the tangential section, the ends of the large rays are visible. (**c**) Radial sections show the sides of the rays. (**d**) In the transverse section the annual rings, springwood, and summerwood are obvious, ×20.

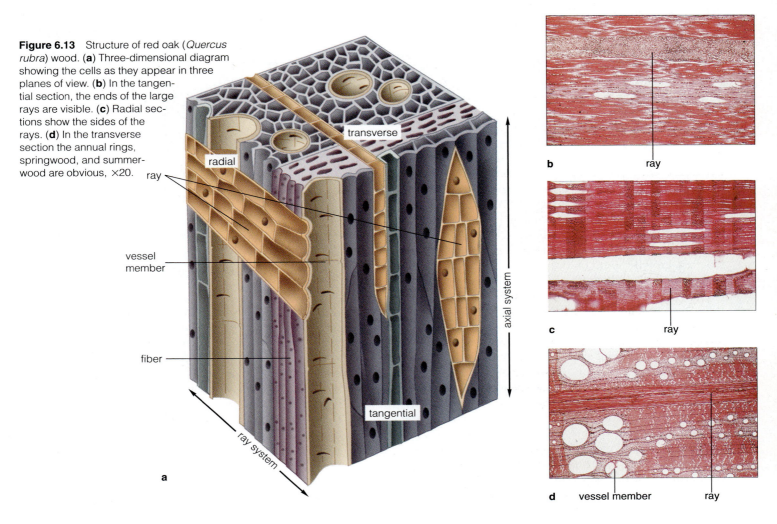

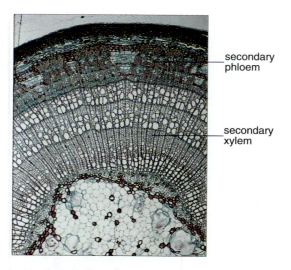

Figure 6.14 Section of a three-year-old basswood (*Tilia americana*) stem, including the secondary phloem, ×17.

secondary phloem

secondary xylem

Figure 6.15 Transverse section of diffuse porous wood in elm (*Ulmus* sp.), ×57.

vessel members

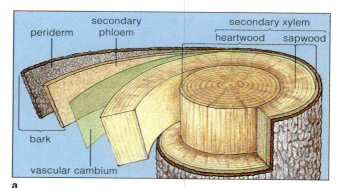

periderm

secondary phloem

secondary xylem

heartwood   sapwood

bark

vascular cambium

a

sapwood

heartwood

b

Figure 6.16 The woody stem. (**a**) Diagram showing the sapwood, heartwood, and other components of a woody stem. (**b**) Slice of a woody stem of mulberry branch (*Morus* sp.), showing the coloration difference between sapwood and heartwood.

Tangential sections of wood (Fig. 6.13b) show an end view of the rays. Wood cut in this plane makes interesting grain patterns used in furniture. Radial sections show the rays from a side view (Fig. 6.13c). Transverse sections of wood show an end view of the cells of the axial system (Fig. 6.13d).

Annual rings are concentric rings of cells in the secondary xylem (Figs. 6.12, 6.14). In trees growing in temperate climates, one annual ring forms each growing season. This means that it is usually possible to determine the age of a tree by counting the growth rings. Based on annual ring counts, the oldest known trees—the redwoods (*Sequoia*) and bristlecone pines (*Pinus*)—can be thousands of years old. Additionally, the growth rate of a tree during each growing season can be determined from the thickness of the annual rings. Ecologists use this information to speculate about climate conditions in the past.

Each annual ring has two components (Fig. 6.14). The cells in the inner part of an annual ring tend to be larger in diameter because they form in the springtime during the first growth spurt of the new season. This is the **springwood** or **earlywood**. The cells that form later in the growing season tend to have smaller diameters. This is the **summerwood** or **latewood**.

One of the structural features used to categorize wood is the distribution of large-diameter vessel members seen in transverse view. In some trees, such as Red oak (*Quercus rubra*) (Fig. 6.13), the large-diameter vessel members are located mostly in the springwood. This pattern is called **ring porous**. In other trees, like elm (*Ulmus* sp., Fig. 6.15), the large-diameter vessel members are uniformly distributed throughout both the spring- and summerwood. This pattern is called **diffuse porous**.

One last thing about transverse sections of wood: If you have ever cut down a tree, you may have noticed that the wood in the center of the tree is often darker than the wood near the periphery (Fig. 6.16). The lighter wood near the periphery is called **sapwood**. The secondary xylem cells in this part of the stem are the functioning xylem cells. This is where most of the actual transport of water and dissolved minerals takes place. The darker wood in the center is called **heartwood**. These cells are often filled with resin and other materials that block them, so that they no longer transport. In some woody plants, heartwood vessel members are blocked by structures called **tyloses**. This is true of white oak (*Quercus alba*), the wood used to make wine barrels. (See sidebar, "How Do You Make a Barrel?", p. 95.) In some wood, parenchyma cells lie adjacent to vessel members. A tylose

HOW DO YOU MAKE A BARREL?

The highest-quality red wines, as well as chardonnay (a white wine), are carefully aged in oak barrels for several months. Vintners are experts—artists, actually—at selecting the best barrel and at timing the aging process to produce the best-tasting wine. Did you ever wonder where these barrels come from and how they are made? I found out when I visited the Mendocino Cooperage, a barrel-making factory at northern California's Fetzer Winery in Hopland.

No wood is better suited for aging wine than oak—specifically, white oak (*Quercus alba*). If you look at the cellular structure of any species of oak (see Fig. 6.13, red oak) you'll see a multitude of wide and short vessel members, surrounded by fibers with thick cell walls, and large, multi-cellular rays (see Chapter 4 if you don't recall these cell types). This combination makes for a wood that is very dense and strong, yet flexible enough to make the rounded barrels.

Two other features make white oak the wood of choice. One is that the vessel members are plugged by tyloses, the bubble-shaped ingrowths from surrounding parenchyma cells. Tyloses inhibit pathogen movement in inactive wood, and they indirectly make white oak wood good for barrels because the tyloses also make the wood leak resistant (see Fig. 6.17). Nothing, including water, can move up a vessel when it is plugged with tyloses; so nothing can leak out of these barrels, either. The other feature is that the cells making up the wood contain many complex chemicals—phenolics (complex polymers of phenol units; phenols are six-carbon ring molecules), carbohydrates, and other carbon compounds—that add desirable aromas and tastes to aged wine. These aromas and tastes have become traditional and are expected by wine drinkers.

The best American barrels come from white oak trees grown in the cold climates of Minnesota and Iowa. The wood from these old trees (100 to 200 years old) has tight grain (small annual rings) because the trees grow very slowly and uniformly. The wood is clear and light colored.

Making barrels is a rare and highly valued craft. There are very few master coopers (barrel makers) in the United States. At the beginning of the coopering process, white oak wood is sawn into boards about 1m (40 in) long, a little over 2.5 cm (1 in) thick, and 7.5–12.5 cm (3–5 in) wide. The boards are stacked so that air can circulate among them, and they are aged outside for two years. The boards are then examined one at a time for imperfections, and the perfect ones are planed and trimmed into barrel staves (Fig. 1). The staves are tapered toward their ends, and their edges are beveled so that they will fit very snugly together when bent.

Each barrel is made of the same number of staves. These are collected together in a steel ring, and a hoop is pushed snugly part way up the forming barrel (Fig. 2). The barrel is then held over a small open flame (Fig. 2) to heat the wood and release whatever water is still left in the cells. At the same time, a long cable is wrapped around the bottom of the forming barrel. As the barrel continues to heat, the cooper wets the outside. The barrel slowly turns. As the cable pulls

Figures 1–4 Images of wine barrel making.

tighter, the staves bend and end up tightly pressed together (Fig. 3). After a certain amount of time, the cooper pulls the barrel off the fire and adds a second hoop on the other end. The new barrel is then placed into a machine to even all the staves, and the remaining steel hoops are pounded into position. No screws or nails are used at any step in the process.

The barrel ends are also made of oak-wood pieces. These are fired briefly, sawn to the correct length, and fitted together with wood pegs. Rush leaves are placed between the boards to seal them together. The covers are then pressed together and planed smooth, and an edge is cut all around the cover. The cooper rubs a flour-and-water paste into the groove at the lip of the barrel and hammers the cover into place very cleverly. The barrel is complete (Fig. 4). Barrels are then pressure tested for leaks with water and air. The ends of the staves are examined for water seepage and are plugged if necessary.

New barrels impart a very specific flavor and aroma to the wine. Chardonnay, as an example, will age in oak barrels for seven to eight months. Red wines may be aged longer. A typical barrel loses its characteristic flavor after being used twice. After that, the barrel may be used for several more years to store less vintage wines. A typical wine barrel has a working life of about eight years, after which you may find it cut in half and growing tomatoes in someone's garden. Making good barrels is really an art form. I'll look at them with new eyes from now on.

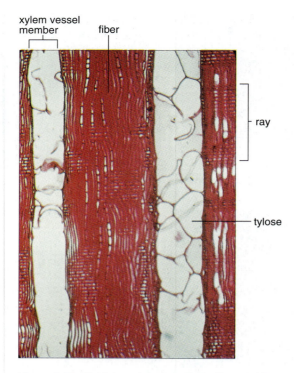

Figure 6.17 Radial section of wood of white oak (*Quercus alba*), showing tyloses plugging the large vessels in the photograph, ×48.

forms when the cell wall of the parenchyma cell actually grows through a pit and into the vessel member. Tyloses look like bubbles; they can fill a vessel member and completely block it (Fig. 6.17).

Gymnosperm Wood Differs from Angiosperm Wood

The wood we've described is angiosperm (flowering plants) wood: hardwoods such as oak (*Quercus*), elm (*Ulmus*), and walnut (*Juglans*). Gymnosperm (seed plants that do not form flowers; see Ch. 24) wood includes softwoods such as pine (*Pinus*), fir (*Abies*), and redwood (*Sequoia*). It has a simpler structure than angiosperm wood. Angiosperm wood is composed of several different types of cells, which may provide interesting and highly valued grain patterns; gymnosperm wood is composed of only a few cell types, mostly tracheids in the axial system, and very simple rays. A popular wood for building is the redwood (*Sequoia sempervirens*). This wood (shown in tangential, radial, and transverse views in (Figure 6.18) is quite simple in structure. For example, in Figure 6.18d, the annual rings of a redwood tree are apparent, and indeed they are all tracheids. Annual rings

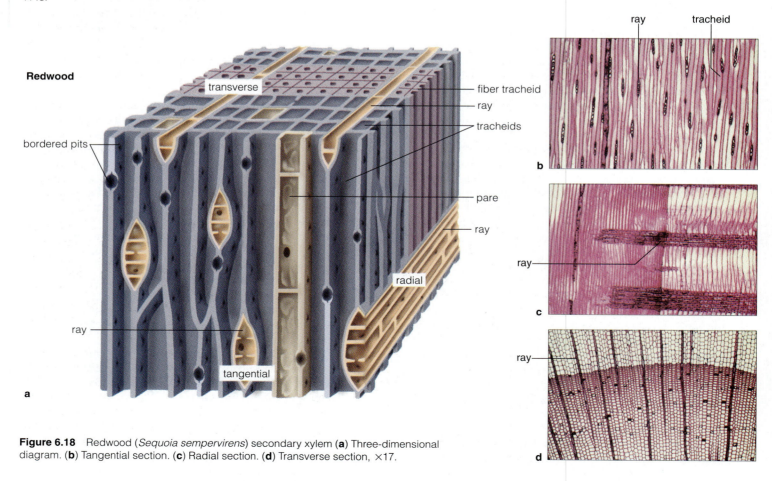

Figure 6.18 Redwood (*Sequoia sempervirens*) secondary xylem (**a**) Three-dimensional diagram. (**b**) Tangential section. (**c**) Radial section. (**d**) Transverse section, ×17.

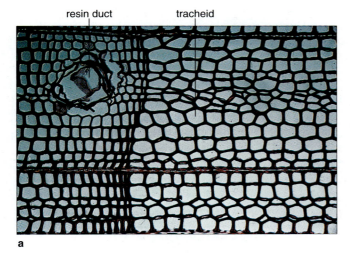

resin duct   tracheid

a

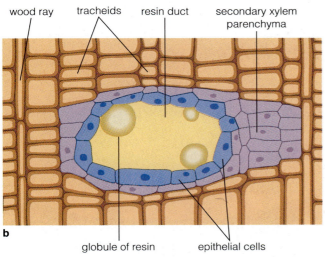

wood ray   tracheids   resin duct   secondary xylem parenchyma

b

globule of resin   epithelial cells

Figure 6.19 Resin ducts. (**a**) Pinewood (*Pinus* sp.) transverse section, showing a resin duct, ×118. (**b**) Diagram of a pinewood resin duct, showing the epithelial cells and globules of resin.

can be discerned because the springwood cells are a little wider than the summerwood cells.

Some gymnosperm wood has **resin ducts** (Fig. 6.19), which are secretory structures that produce and transport resin. The resin is synthesized and secreted by a lining of *epithelial cells*. Turpentine is one commercial product made from it. Resin has several names: It is called *sap* when it flows through the resin ducts all the way to the outside of the stem; but when it hardens, it is called *rosin*. Fossilized rosin is the *amber* popular for jewelry. Insects can get stuck in sap, which is why trapped insect bodies are sometimes found in amber.

Bark Is Composed of Secondary Phloem and Periderm

The protective covering over the wood of a tree is commonly called *bark*. In botanical terms, the **bark** is every-

thing between the vascular cambium and the outside of the woody stem. The actual composition of the bark varies a little, depending on the age of the tree. The bark of a one- or two-year-old tree includes the secondary phloem, maybe a few cells of the cortex, and one or two increments of periderm. In a very old tree, the bark would include the layers of secondary phloem plus several layers of periderm (see Fig. 6.16).

SECONDARY PHLOEM Secondary phloem (see Fig. 6.14) forms to the outside of the vascular cambium. The types of cells in the secondary phloem are sieve-tube members, companion cells, phloem parenchyma, phloem fibers, sclereids in the axial system, and ray parenchyma in the ray system. When secondary phloem is examined in radial, tangential, and transverse sections, its cellular orientation is similar to that in wood (xylem). A difference between the two in the transverse section is that secondary phloem doesn't develop in annual ring increments. You can't count phloem rings to determine tree age (Fig. 6.14). Phloem rays are composed of phloem ray parenchyma cells.

PERIDERM Usually, stems of plants that live more than one year will lose their epidermis sometime during the first or second year of growth. As the stem increases in diameter because of secondary growth from the vascular cambium, the cells of the epidermis stretch but cannot keep up with the increasing circumference. Eventually, the epidermis cracks, dries up, and flakes away from the stem.

At the same time, some cells (usually in the outer cortex) start to divide all around the periphery of the stem (Fig. 6.20). This new layer of dividing cells is the **cork cambium** (or **phellogen**). The cork cambium divides in two directions, to form **phellem cells** (also called **cork cells**) to the outside and **phelloderm cells** to the inside.

These three layers—the phellem, the cork cambium, and the phelloderm—make up the **periderm** (Fig. 6.20c). Phellem cells form in regular rows, have waxy (suberized) cell walls, and are usually dead by the time the periderm is fully functional. The phelloderm cells also form in regular rows, but these cells tend to live longer and look like parenchyma cells. The function of the periderm is to serve as an impermeable layer, inhibiting water evaporation from the protected cell layers and protecting against insect and pathogen invasion.

The cells near the surface of the stems of young woody trees are often living. These cells require oxygen to function. Consequently, a structure called a *lenticel* is present in the bark of young branches of different woody plants (Fig. 6.21). Lenticels are specialized regions of the periderm consisting of loosely packaged parenchyma cells. Their purpose is to provide a place where gases can be exchanged.

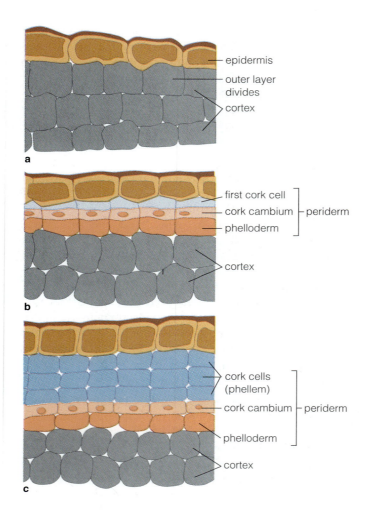

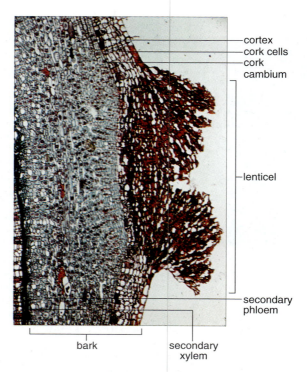

Figure 6.21 Woody stem of an elder (*Sambucus* sp.) in transverse section, ×83.

Figure 6.22 Bark of pine (*Pinus* sp.), showing the layers of periderm.

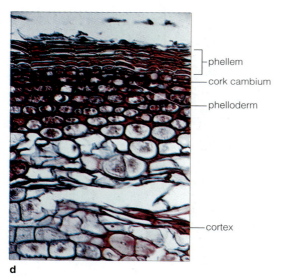

Figure 6.20 The cork cambium. (**a**) Cross section diagram showing a portion of the epidermis and outer cortex of a mature stem. (**b**) The outer cortex layer divides to become the cork cambium. (**c**) The cork cambium makes cork cells (phellem) to the outside and the phelloderm to the inside; these three layers make up the periderm. Note how these cells form in regular rows. (**d**) Microscopic view of cork cambium and periderm layers in elder (*Sambucus* sp.) stem, ×239.

At the start of each new growing season, a new increment of periderm is initiated. This means that in most trees an entirely new cork cambium is generated each spring. This new cork cambium is initiated from secondary phloem parenchyma cells, which are triggered to divide much like the outer cortex cells that divided to form the first increment of cork cambium. The consequence of this is that the one year's periderm stacks on another, forming layers (Fig. 6.22).

There are always exceptions to the rule, and cork oak (*Quercus suber*, Fig. 6.23a) is one. Instead of new cambium being generated yearly, a single cork cambium lasts the entire life of the tree. Interestingly, it divides only to the outside to form cork cells. Sheets of cork are harvested every 8–10 years for the manufacture of corks for wine bottles and other things.

Figure 6.23 External views of the bark of (**a**) cork oak (*Quercus suber*), (**b**) birch (*Betula papyrifera*), and (**c**) *Eucalyptus* sp., showing their different appearances.

In bark, the innermost periderm layer is the current active layer. *Girdling*, the removal of a continuous strip of bark around the circumference of a tree, is a sure way to kill it. That's because the nutrient-transporting secondary phloem tissue will be severed.

The external appearance of bark varies among species of trees. Most species can be recognized by their external bark texture (Fig. 6.23). There are several main patterns of bark. The ring bark of, for example, paper birch trees (*Betula papyrifera*), forms in continuous rings (Fig. 6.23b). Scale bark, which is characteristic of pine trees, forms as small overlapping scales. Shag bark, such as that found in *Eucalyptus*, has long overlapping thin sheets (Fig. 6.23c). There are also intermediate forms that are useful for tree identification.

Buds Are Compressed Branches Waiting to Elongate

The buds at the ends of woody branches are actually short compressed branches. These buds are covered with several hard, modified leaves called **bud scales**, which keep the inside of the bud fresh and moist. The bud at the end of a branch is called the **terminal bud**; those in the **axils** (at the base of the petiole) of leaves on the side of a branch are called **lateral buds** (Fig. 6.24). Some buds produce flower parts instead of leaves; these are called **flower buds**. In some fruit trees, like almonds and apricots, the first buds to open in the spring are flower buds.

When a bud elongates, as in the early spring, the bud scales open and finally fall off, leaving small scars called **bud-scale scars**. When a leaf falls, it leaves a **leaf scar** (Figs. 6.24, 6.25), and the vascular bundles located in the leaf leave **bundle scars** when they break off. The struc-

Figure 6.24 Drawing of a three-year-old twig of walnut (*Juglans regia*).

terminal bud

lateral bud

this year

last year

terminal-bud-scale scars

leaf scar

this year

lenticel

internode

flower bud

node

leaf scar

bundle scar

last year

two years ago

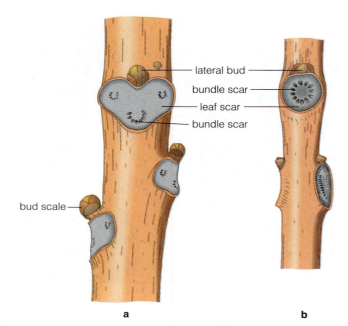

Figure 6.25 Leaf scars and buds. (**a**) Walnut (*Juglans regia*). (**b**) Catalpa (*Catalpa bignonioides*).

Figure 6.26 Treelike monocot plants. (**a**) Palms are unbranched and lack true secondary growth. (**b**) *Pandanus* sp. is branched and lacks true secondary growth. (**c**) The Joshua tree (*Yucca brevifolia*) has a branched stem and true secondary growth.

ture of these scars is characteristic of the plant and can be used to identify plants in the winter when all their leaves have fallen (Fig. 6.25).

Some Monocot Stems Have Secondary Growth

Most monocots lack a vascular cambium and do not form secondary xylem and phloem. Such plants cannot support large vertical shoot systems. Other monocots such as different types of lilies (*Lilium* sp.) can become very large by producing elaborate underground branches (rhizomes).

Treelike monocots (Fig. 6.26) such as coconut trees (*Cocos* sp.) and Joshua trees (*Yucca brevifolia*) can reach enormous size. There are actually three different types of monocot tree: (1) palms, such as the coconut tree, which are unbranched and do not have true secondary growth (Fig. 6.26a); (2) pandans (*Pandanus* sp.), which have branched stems but also lack true secondary growth (Fig. 6.26b); and (3) tree lilies (such as *Yucca*) and others (Fig. 6.26c), which have branches, a cambium, and true secondary growth.

Palm stems have a very large number of vascular bundles, most of which are leaf traces. The leaves are large, and the leaf base wraps around the stem. Each leaf has many leaf traces. Some thickening does occur at the bases of palms, but this comes from basal adventitious roots. Some thickening throughout the stem results from the division and enlargement of parenchyma cells; this is called *diffuse secondary growth*. It isn't true secondary growth because a cambium is lacking.

Several monocot plants—including *Yucca*, *Agave* (century plant), and *Dracaena* (dragon's blood tree)—have true secondary growth from a cambium. The stems of

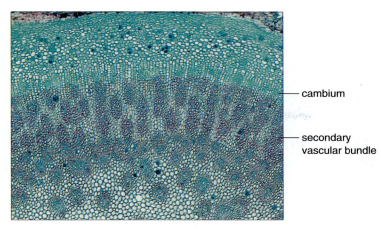

Figure 6.27 Transverse section of dragon's blood tree (*Dracaena draco*). The cambium layer produces secondary vascular bundles and parenchyma cells, ×12.

these plants are tapered: thin at the top and thick at the base. This architecture results from the pattern of cells the cambium produces. The monocot cambium is unique in forming mostly parenchyma cells with secondary vascular bundles embedded in this ground tissue (Fig. 6.27). The secondary vascular bundles are different from primary vascular bundles in that the xylem surrounds the phloem.

Some older monocot stems have a corklike layer that replaces the epidermis. This layer, called *storied cork*, isn't considered to be periderm because no cork cambium is present. Storied cork forms by the division of parenchyma cells along the stem periphery.

6.4 STEM MODIFICATIONS FOR SPECIAL FUNCTIONS

Stems of both dicots and monocots may be adapted for functions other than support, transport, and the production of new growth. They may, for instance, serve as protective devices or as attachment organs for vines, carry on photosynthesis, or store food or water.

Rhizomes are underground stems (Fig. 6.28a). They are usually light colored and burrow into the ground just below the surface. In some plants, such as bermuda grass (*Cynodon dactylon*), the rhizomes can be quite deep—40 cm (16 in) or so. Since rhizomes are stems, they have internodes and nodes. Small scalelike leaves sometimes form at the nodes, but they don't grow or become photosynthetic. The buds in the axils of these leaves elongate, producing new branches that extend to the soil surface and form new plants.

Tubers are the enlarged terminal portions of underground rhizomes. The potato plant (*Solanum tuberosum*) has three types of stems: (1) upright leafy stems; (2) underground rhizomes; and (3) swollen ends of rhizomes, the tubers (Figs. 6.28b, 6.29a). The structure of a tuber shows it to be a stem. The eyes of a tuber are actually lateral buds formed in the axil of small scale leaves at

a node. The internodes of the tuber are short, and the tuber body is filled with parenchyma cells containing starch, a storage form of sugar.

Corms and bulbs are shoot structures modified for storage of food. A **corm** is a short, thickened underground stem with thin, papery leaves. *Gladiolus* sp. corms are good examples. The central portion of the corm accumulates stored food, which is used at the time of flowering (Fig. 6.28c). New corms can form from the lateral buds on the main corm. A **bulb** differs from a corm in that the food is stored in specialized fleshy leaves (Fig. 6.28d). The stem portion is small and has at least one terminal bud (to produce a new, upright leafy stem) and a lateral bud (to produce a new bulb). Food stored in the bulb is used up during the initial growth spurt each spring. A table onion (*Allium cepa*) is an example of a bulb.

Cladophylls (also called **cladodes**) are flattened photosynthetic stems that function as and resemble leaves. They are not actually leaves in a developmental sense because they develop from buds in the axils of small, scalelike leaves. Cladophylls may bear flowers, fruits, and small leaves. Butcher's broom (*Ruscus* sp., Fig. 6.29b), asparagus, and some cacti are examples of plants with cladophylls.

Thorns originate from the axils of leaves (Fig. 6.29c) and help protect the plant from predators. Some thorns actually have leaves growing on them. From developmental evidence, we know that prickles and spines (though outwardly similar structures) are not modified stems. **Spines** are actually modified leaves; **prickles**, like those on rose stems, are really modified clusters of epidermal hairs.

Stolons, also called *runners*, are aboveground horizontal stems. Strawberry plants (*Fragaria* sp.) send out stolons (Fig. 6.29d). At each node of the stolon, a small leaf will form, and in its axil a root and a bud will sprout to initiate a new strawberry plant. The stolon helps the plant spread. Some successful weed plants, like bermuda grass, also can spread by stolons.

6.5 THE ECONOMIC VALUE OF WOODY STEMS

Trees are, without doubt, among the most valuable things on earth. Our forests are home to countless plants and animals. They are also a source of the raw material for many useful products. They purify our air, keep our soil from washing away, and affect our weather patterns. The implications of the loss of our forests are too terrible to consider. We must be extremely careful to protect and manage this natural resource even as we grow and harvest woody plants for thousands of uses.

Trees are grown for wood; for sugar (from maple trees, *Acer saccharum*); and for secondary products such as turpentine (from the pine, *Pinus palustris*), natural rub-

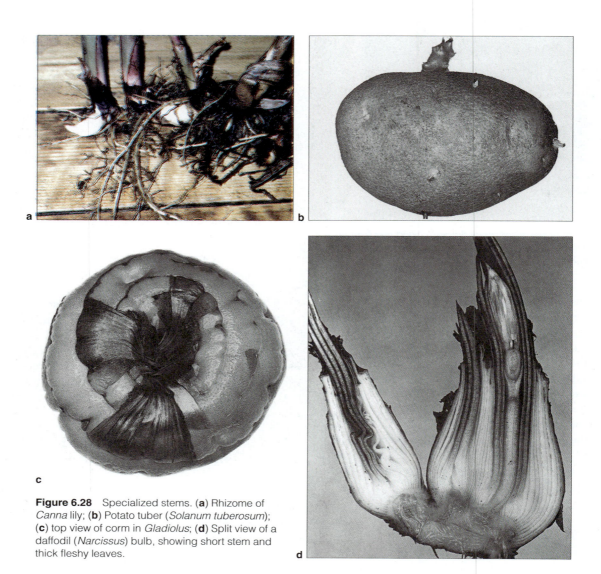

Figure 6.28 Specialized stems. (**a**) Rhizome of *Canna* lily; (**b**) Potato tuber (*Solanum tuberosum*); (**c**) top view of corm in *Gladiolus*; (**d**) Split view of a daffodil (*Narcissus*) bulb, showing short stem and thick fleshy leaves.

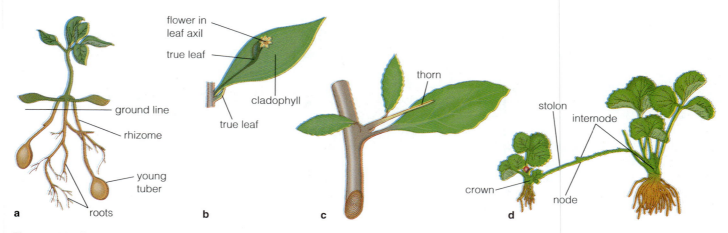

Figure 6.29 Stem modifications. (**a**) A young potato plant (*Solanum tuberosum*), showing development of young tubers at the ends of rhizomes. (**b**) Cladophyll in butcher's broom (*Ruscus aculeatus*). (**c**) Thorn growing from the axil of a *Pyracantha* sp. leaf. (**d**) Stolon in strawberry (*Fragaria* sp.).

ber (from the Brazilian rubber tree, *Hevea brasiliensis*), and chewing gum (from the sapodilla tree, *Achras sapota*). These products are harvested by making cuts in the woody stem and collecting the sap that runs out (Fig. 6.30). The raw secondary product is then processed to make commercially important materials. This is a truly renewable resource because the trees can live and continue to produce for many years.

In contrast, when trees are used for lumber they are destroyed. In the manufacture of lumber, a tree is removed from the forest with large machines and trucked or floated down a river to a sawmill. The log is moved to the mill and oriented so that tangential cuts are made along its length to make boards (Fig. 6.31a). Different types of wood grain can be obtained, depending on the type of cut made and on the quality and type of wood used.

Sheets of plywood are made by steaming the log and then placing it against a large lathe that literally peels thin sheets of wood from the slowly revolving log (Fig. 6.31b). The sheets are then cut to size, and multiple sheets are glued together to make plywood of the desired thickness.

For other industrial uses, wood is not sawed into boards or cut into sheets, but broken down by machines into individual fibers, tracheids, and vessel members. This *pulp* is then processed to make paper and cardboard products. In papermaking, the lignin in the secondary cell walls is removed. In the past such waste products have caused environmental pollution. Recently, innova-

Figure 6.30 Maple syrup being harvested from a sugar maple tree (*Acer saccharum*).

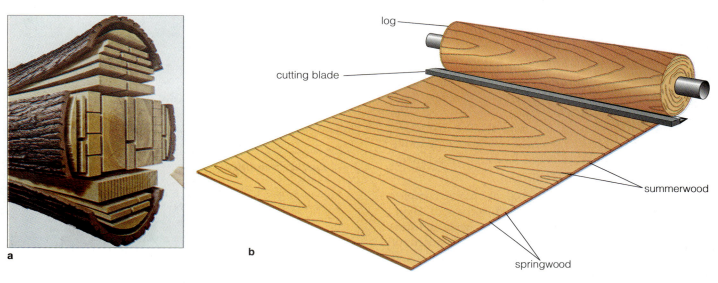

a   b

Figure 6.31 Processing of timber. (**a**) Diagram showing the different cuts of a log. (**b**) Diagram showing a sheet of plywood being peeled from a slowly revolving log.

tive products using these wastes have been developed, including soil conditioners and adhesives.

Recycling paper products is now a major activity, and this important practice means that our natural tree resources may be preserved for many generations to come. Processing recycled paper is not without its hazards; it releases vast quantities of toxic inks and chemicals used in printing. Intelligent processing, however, has resulted in the discovery of ways to detoxify and reuse these materials.

SUMMARY

1. The functions of stems are: (a) to provide the axis for the attachment of the leaves, buds, and flowers; (b) to provide pathways for movement of water and minerals from the roots and food from the leaves; and (c) to produce new cells, tissues, leaves, and buds.

2. The shoot apical meristem forms the three primary meristems: protoderm, procambium, and ground meristem.

3. Bundles of procambium form in a ring pattern between the two parts of the ground meristem (which will become the pith and cortex). Procambium cells will become primary xylem and phloem.

4. Leaves are arranged on the stem according to four basic patterns (phyllotaxis): alternate, opposite, whorled, and spiral.

5. The vascular cambium forms from the fascicular cambium (by division of residual procambium cells within the vascular bundles) and from the interfascicular cambium (by division of parenchyma cells between vascular bundles).

6. Cells in wood are arranged into axial and ray systems. The cells of the axial system come from fusiform initials in the vascular cambium; those of the ray system come from ray initials.

7. One year's increment of growth in wood equals one annual ring. Cells in springwood have relatively large diameters and form in the spring. Summerwood cells form later in the summer and have smaller diameters. Ring porous wood has large vessel members only in the earlywood; diffuse porous wood has large vessel members throughout the annual ring.

8. Angiosperm wood (hardwood) is more complex anatomically than gymnosperm wood (softwood). Gymnosperm wood is composed mostly of tracheids and simple rays, whereas angiosperm wood is composed of vessel members, tracheids, fibers, parenchyma cells, and more complex rays.

9. The bark consists of all tissues outside of the vascular cambium. Bark occurs in different patterns.

10. The periderm forms from the cork cambium by divisions of outer cortex cells. Cork cells form to the outside of the cork cambium, and phelloderm cells form to the inside. During subsequent years, new layers of periderm are added in the secondary phloem.

11. Different kinds of buds occur on woody stems: terminal buds, lateral buds, and flower buds. Bud scale scars and leaf scars form after bud scales and leaves fall from the stem.

12. Monocot stems have scattered vascular bundles and usually do not have secondary growth. Palm trees increase their thickness by diffuse division of parenchyma cells. Monocot trees, such as *Yucca* and *Dracaena*, have a cambium and true secondary growth with a vascular cambium.

13. Several different modified stems occur in flowering plants: rhizomes, corms, bulbs, tubers, cladophylls, thorns, and stolons are examples.

14. Trees are ecologically essential and provide many products for humans. It is imperative that they be conserved.

Questions

1. Be able to diagram and describe the primary meristems that develop into the primary tissues.

Table of Primary Meristems and Primary Tissues

Primary Meristem	Primary Tissue
Protoderm	Epidermis
Ground meristem	Cortex and pith
Procambium	Xylem and phloem

2. Given an illustration of a section of a stem, be able to identify the tissues and cell types present. Also, be able to make a diagram (cross section or longitudinal section) of a stem and label all of the tissues and cell types.

3. Name and describe the four types of phyllotaxis. List one plant of each type found in your area.

4. Describe the initiation of vascular cambium and the formation of secondary xylem and phloem in a stem.

5. Describe the initiation and activity of the cork cambium forming the periderm.

6. How do the stems of monocots and dicots differ anatomically?

Further Readings

Esau, K. 1977. Chapters 16 and 17 in *Anatomy of Seed Plants*. 2d ed. New York: John Wiley. This is the classic plant anatomy textbook. It has excellent illustrations.

Fahn, A. 1990. Chapter 11 in *Plant Anatomy*. 4th ed. Oxford: Pergamon Press. This is an up-to-date plant anatomy textbook, but it is rigorous reading.

Zimmerman, M. H., and C. L. Brown. 1975. *Trees: Structure and Function*. New York: Springer-Verlag. This is an excellent reference book on all aspects of tree structure.

7

THE SHOOT SYSTEM II: THE FORM AND STRUCTURE OF LEAVES

THE FUNCTIONS OF LEAVES

Leaves Are Shaped to Capture Light

The Arrangement of Leaf Cells Depends on Their Functions

Where Do New Leaves Come From?

LEAF FORM AND SPECIALIZED LEAVES

Leaf Shape May Depend on the Plant's Age and the Environment.

Leaf Abscission Is the Seasonal Removal of Leaves

■ Commercially Important Fibers from Leaves and Stems

SUMMARY

1. The most important functions of leaves are photosynthesis (creation of chemical energy from sunlight and carbon dioxide) and transpiration (evaporation of water from the leaf surface).

2. A leaf usually consists of a leaf blade and a petiole. Leaves with one leaf blade are called *simple leaves;* leaves with many leaflets are called *compound leaves.*

3. Leaves are composed of epidermis, mesophyll, xylem, and phloem. The epidermis is covered by a waxy cuticle; the mesophyll is where most photosynthesis occurs. The xylem conducts water, and the phloem transports sugars.

4. Leaf formation is initiated at the shoot apical meristem.

5. Leaf fall in the autumn involves an active process of cell division and cell breakdown at the abscission layer in the petiole.

7.1 THE FUNCTIONS OF LEAVES

Green plants, algae, and a few species of bacteria are the only organisms that use sunlight as an energy source. All other inhabitants of the earth obtain energy by consuming—directly or indirectly—what the green plants produce. It is during **photosynthesis**, the process in which plants synthesize sugars and release oxygen, that the chloroplasts within plant cells trap light energy and convert it into chemical energy.

Leaves provide large surfaces for the absorption of light and carbon dioxide. Both are the raw materials for photosynthesis. In general, most leaves are thin, so that no cells lie far from the surface. This architecture makes it easy for leaf cells and chloroplasts to absorb light and imbibe carbon dioxide gas. At the same time, however, it promotes the loss of water from plants because any opening that will permit carbon dioxide to pass will also allow water to evaporate. Evaporation, also called **transpiration** in plants, helps cool the leaf; however, excessive evaporation places the plant in danger of dehydration. To a great extent, leaf form (morphology) and anatomy are a compromise between capturing light and carbon dioxide and conserving water. Water-saving structures (for example, a thick waxy cuticle to inhibit evaporation) become important for plants that occupy hot, dry regions, such as rocky cliffs and deserts, where water loss is a great hazard.

In this chapter we will discuss the form and anatomy of leaves, describe their development, and note the modifications that help plants deal with environmental challenges.

Leaves Are Shaped to Capture Light

The leaves of dicot and monocot plants have similar functions, but their basic designs differ. Monocot plants often have leaves with a strap-shaped leaf **blade**, and

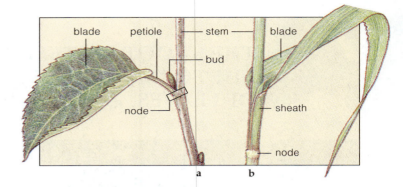

Figure 7.1 Typical dicot (**a**) and monocot (**b**) leaves compared.

their leaf bases wrap around the stem (Fig. 7.1). Dicot leaves typically are composed of a flat, thin portion (the blade) and a narrow portion (the **petiole**) (Fig. 7.1). The blade is the part that absorbs most of the light; the petiole is designed to hold the blade away from the stem. Not only does this give the blade greater exposure to light, but it allows the blade to move in the air so that gas exchange can easily take place.

THE LEAF BLADE In most dicot and monocot plants, the cells of the leaf blade contain many chloroplasts and perform most of the photosynthesis in the plant. The blade provides a broad, flat surface for capturing light and carbon dioxide. Leaves with a single blade (Fig. 7.2a) are called **simple leaves**. The form of the blade varies widely among different plants, though. For instance, the leaf edges or margins may be smooth, toothed, or lobed. Some leaves are so deeply lobed that the blade forms several separate units, or leaflets (Fig. 7.2b); leaves having this design are called **compound leaves**. There are two types of compound leaves: A *palmately compound leaf* has its leaflets diverging from a single point, and a *pinnately compound leaf* has leaflets arranged along an axis (Fig. 7.2b). These variations in leaves are useful in classifying plants.

Quite often in nature there's a practical reason for differences in structures that have similar functions. For compound leaves, the question naturally arises of how a plant might benefit by having its blade divided into leaflets. Perhaps the most likely answer is that the spaces between leaflets allow better air flow over the leaf surface. This may help cool the leaf (because of increased evaporation) while improving carbon dioxide uptake.

THE PETIOLE Dicot leaf blades are not directly attached to the stem. Instead, a petiole serves that purpose. The petiole is the narrow base of most dicot leaves. Its function is to improve photosynthetic efficiency in several ways. By extending the leaf blade away from the stem, the petiole reduces the extent to which the blade is shaded by other leaves. It also allows the blade to move in response to air currents. This is important because a layer of stagnant air, the **boundary layer**, forms at the leaf surface. During photosynthesis, the boundary layer

a Simple leaves

poplar
(*Populus*)

oak
(*Quercus*)

maple
(*Acer*)

leaflets

red buckeye
(*Aesculus*)

black locust
(*Robinia*)

honey locust
(*Gleditsia*)

b Compound leaves

Figure 7.2 Examples of simple (**a**) and compound (**b**) leaves.

becomes depleted of carbon dioxide. Leaf movements help bring fresh air containing carbon dioxide to the leaf surface and cool the leaf by increasing evaporation.

Petioles are variable in shape. They may be long or short, cylindrical or flat in cross section. They are usually attached to the base of the leaf blade (Fig. 7.1a). Leaves that lack a petiole are called *sessile*.

THE SHEATH Monocot leaves often lack petioles. Instead, the leaf base typically wraps entirely around the stem to form the **sheath** (Fig. 7.3). In grasses (for example, corn) the sheath extends almost the entire length of the stem internode. If you were to carefully examine a leaf of corn or of several other grasses, you would see a small flap of tissue, a **ligule**, extending upward from the sheath (Fig. 7.3b). The ligule apparently keeps water and dirt from sifting down between the stem and leaf sheath. In some grass species, such as barley, two additional flaps of leaf tissue, *auricles*, extend around the stem at the juncture of the sheath and blade (Fig. 7.3c).

Another interesting characteristic of grass leaves is that they grow from the base of the leaf sheath. Anyone who has had to cut the lawn in the summer has experienced this phenomenon. Have you ever wondered how the grass blades keep growing week after week? The reason for this is an **intercalary meristem**, a unique feature at the base of grass leaves. The function of this particular meristem is to make new cells at the base of the grass leaf, allowing for continued growth of the mature leaf. The intercalary meristem isn't active for the entire life of the leaf, however; it will eventually stop dividing when the leaf reaches a certain age or length.

LEAF VEINS The internal connections of the leaf to the rest of the plant are via veins. **Leaf veins** are actually **vascular bundles**. Vascular bundles (recall from Chapter 4)

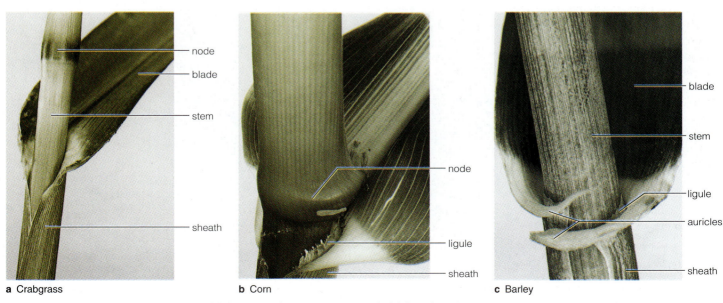

node

blade

stem

sheath

a Crabgrass

node

ligule

sheath

b Corn

blade

stem

ligule

auricles

sheath

c Barley

Figure 7.3 Examples of monocot leaves. (**a**) Crabgrass (*Digitaria sanguinalis*). (**b**) Corn (*Zea mays*), showing ligule. (**c**) Barley (*Hordeum vulgare*).

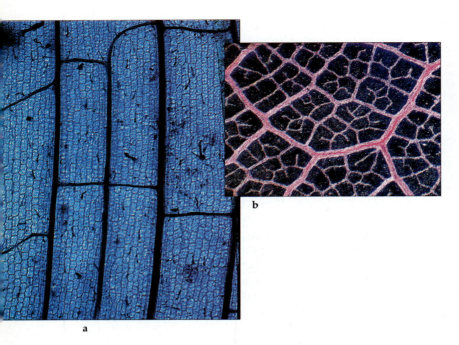

a

b

Figure 7.4 Venation patterns in leaves. (**a**) Parallel venation in a monocot leaf (*Orthoclada* sp.); (**b**) Netted venation in a dicot leaf (*Acer* sp.)

are composed of xylem (to transport water) and phloem (to transport sugars). The pattern they make in leaves is often quite beautiful (Fig. 7.4). Leaves differ from one another in many ways, including the arrangement of veins. Monocot leaves, for example, have **parallel venation**, meaning that there are several major veins that run parallel from the base to the tip of the leaf (Fig. 7.4a). In contrast, many dicot leaves have **netted venation** (Fig. 7.4b). In this pattern a major vein, the *midvein* (also called the *midrib*), runs up the middle of the leaf, and lateral veins branch out from it. The precise pattern of veins differs among species of plants and is sometimes used as an identifying characteristic.

The Arrangement of Leaf Cells Depends on Their Functions

A closer look at the cells and tissues of the leaf blade (Fig. 7.5) and petiole reveals anatomical structures that aid in photosynthesis, transport of food and water, and the transpiration of gases. The outside surface of the leaf is covered by an epidermis, the ground tissue is chloroplast-filled mesophyll, and the vascular tissue is in the form of vascular bundles (veins). The petiole has special-

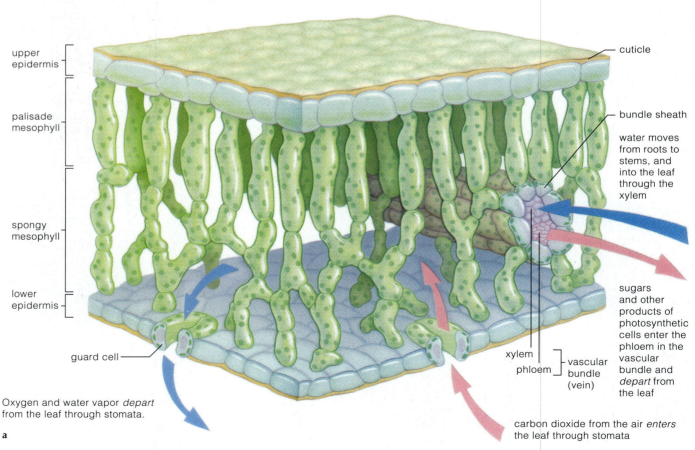

upper epidermis

palisade mesophyll

spongy mesophyll

lower epidermis

guard cell

cuticle

bundle sheath

water moves from roots to stems, and into the leaf through the xylem

xylem
phloem — vascular bundle (vein)

sugars and other products of photosynthetic cells enter the phloem in the vascular bundle and *depart* from the leaf

Oxygen and water vapor *depart* from the leaf through stomata.

carbon dioxide from the air *enters* the leaf through stomata

a

Figure 7.5a Tissues of a typical leaf blade. Shown are the upper and lower epidermis, palisade and spongy mesophyll, and a vascular bundle (vein).

ized structures that enable it to support the leaf blade and conduct food, water, and minerals. The petiole is also the point at which leaves are shed in the autumn, at the end of the growing season.

THE EPIDERMIS The epidermis covers the entire surface of the blade, petiole, and leaf sheath; it is continuous with the epidermis on the surface of the stem. In most leaves the epidermis is a single layer of cells that consists of epidermal cells, guard cells, subsidiary cells, and trichomes. Epidermal cells are flattened in a cross-sectional view of a leaf (Fig. 7.5). In surface view, epidermal cells often are irregular and sometimes look like puzzle pieces (Fig. 7.6a). The outer cell wall of epidermal cells is sometimes thickened, and its outer surface is covered by a waxy layer, the **cuticle**. The waxy cuticle inhibits the evaporation of water through the outer epidermal cell wall.

Because the cuticle blocks evaporation, special openings are needed in the epidermis for the controlled exchange of gases. These openings exist as small pores, each lying between two guard cells (Fig. 7.6b). Two guard cells plus the pore form a **stoma** (plural, **stomata**). The guard cells are connected to subsidiary cells (Fig. 7.6b), which are epidermal cells that play a special role in the mechanism to open and close the stomatal pore. The term **stomatal apparatus** describes the guard cells plus their attached subsidiary cells. The stoma permits the entry of carbon dioxide needed for photosynthesis and the loss of water vapor by transpiration. Transpiration (see Chapter 11) cools the leaf surface by evaporation and, in the process, helps pull water up from the roots. There are hundreds or even thousands of stomata per square centimeter of leaf area. Usually, there are more stomata on the bottom of the leaf than on the top (Table 7.1), an arrangement that helps limit water loss because the leaf is cooler on the underside. Stomata occur also in the epidermis of young stems and some flower parts.

Several different types of trichomes occur as part of epidermal leaf surfaces (Fig. 7.7). Some trichomes are secretory structures. These usually have a stalk and a multicellular head that does the secreting (Fig. 7.7c). The material secreted is often designed to attract pollinators to flowers.

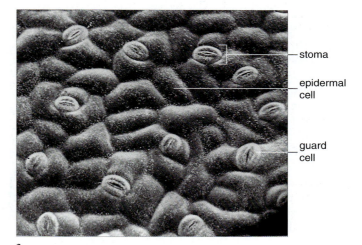

a

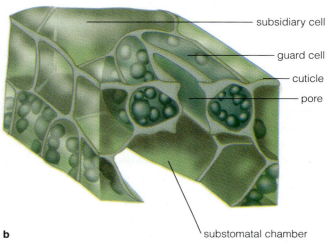

b

Figure 7.6 Features of leaf epidermis. (**a**) Epidermal surface of potato leaf (*Solanum nigra*), ×350. (**b**) Diagram of a single stomatal apparatus.

Figure 7.5b SEM view of the inside of the leaf blade of pennyroyal (*Mentha pulegium*), ×350.

Table 7.1 Average Number of Stomata in Different Plant Species per Square Centimeter (0.16 Square Inch)		
Plant	Upper Epidermis	Lower Epidermis
Alfalfa (*Medicago*)	16,900	13,800
Apple (*Malus*)	0	28,400
Bean (*Phaseolus*)	4,000	28,100
Cabbage (*Brassica*)	14,100	22,600
Corn (*Zea*)	5,200	6,800
English oak (*Quercus*)	0	45,000
Oat (*Avena*)	2,500	2,300
Potato (*Solanum*)	5,100	16,100
Tomato (*Lycopersicon*)	1,200	13,000

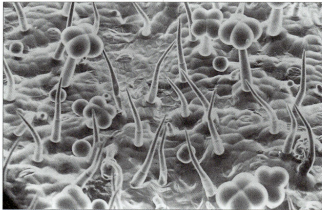

Insectivorous (insect-eating) plants have special mechanisms for snaring and digesting their prey. Some of these involve leaves with very elaborate trichomes. Venus's-flytrap (*Dionaea muscipula*), for example, has specialized leaves that look as if they are hinged along the central vein separating the two lobes of the blade (Fig. 7.8a). On the surface of the lobes are two kinds of trichomes: trigger hairs and secretory hairs. Trigger hairs are long cells with a multicellular base. When an insect moves two or more of these hairs, the trap is triggered, closing the leaf blade and entrapping the insect (Fig. 7.8b). The insect's struggles push its body against several short secretory hairs with multicellular heads. These hairs secrete digestive enzymes that break down the soft parts of the insect body into nutrients that the leaf then absorbs.

Sundews (*Drosera* sp.) capture insects by producing a thick, sticky mucilage from tentacles (Fig. 7.8c), which are multicellular structures much more complex than simple trichomes. When an insect lands on the leaf, it sticks to the mucilage, and the leaf curls to trap the insect. Secreted enzymes then digest the insect body. Pitcher plants secrete substances from glands located along the lip of the pitcher-shaped leaves (Fig. 7.8d). An insect that lands to feed and happens to crawl to the inside of the "pitcher" is likely to slip down the side. Downward-pointed sharp hairs impede the insect's progress as it tries to crawl out (Fig. 7.8e). Eventually the insect tires and falls into the digestive fluid at the bottom of the pitcher. The amino acids, lipids, sugars, and minerals released from breaking down the insect's body are used by the plant for growth. Insectivorous plants can survive without capturing and digesting insects. This highly developed specialization does, however, give them an advantage in nutrient-poor marshes and bogs.

The leaves of some desert plants such as saltbush (*Atriplex* sp.) have several short hairs. The end cell has a

Figure 7.7 Trichomes. (**a**) Branched trichome on the leaf of *Aleurites* sp. (**b**) Pointed unicellular hairs on a tomato stem, ×120. (**c**) Glandular hairs on the surface of bean stems (*Phaseolus vulgaris*), ×18.

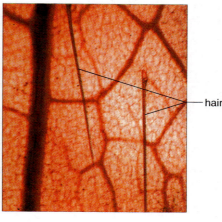

hair

Figure 7.8 Insectivorous plants that use trichomes on their leaves to attract, capture, or digest insects. (**a**) Venus's-flytrap (*Dionaea muscipula*) with open leaf lobes. (**b**) Venus's-flytrap that has captured a fly by closing leaf lobes. (**c**) Sundew (*Drosera capillaris*) captures insects by trapping them in mucilage secreted by tentacles. (**d**) Pitcher plant (*Sarracenia purpurea*) entices insects to land by secreting sweet substances, only to block with sharp hairs (**e**) the egress of those insects who wander into the leaf cup.

whitish cuticle, and the cells are filled with water. The hairs not only store water but also reflect sunlight and insulate the leaf against the extreme desert heat. Other leaves, such as those of the olive tree (*Olea europea*), have a mat of branched hairs that act as heat insulators.

THE MESOPHYLL: CAPTURING SUNLIGHT AND MAKING SUGAR A leaf's most important function is to use sunlight to manufacture food. To this end, most of the leaf blade is made up of parenchyma cells containing chloroplasts. Collectively, these cells are called mesophyll. The mesophyll in a typical dicot leaf is organized into two distinct regions (see Fig. 7.5). The **palisade mesophyll** is usually on the upper surface of the leaf. The individual cells, called **palisade parenchyma cells**, are packed together and are shaped like columns oriented at right angles to the leaf surface. The **spongy mesophyll** is made of **spongy parenchyma cells**, which are usually irregularly shaped with abundant air spaces between them.

Mesophyll cells have thin, moist cell walls that allow rapid inward diffusion of carbon dioxide and outward diffusion of oxygen and water vapor (see Fig. 7.5). Just underneath the stomata there is usually a large air space called a **substomatal chamber** (see Fig. 7.6). The mesophyll cell walls exposed to this space provide the main evaporative surfaces for gas exchange.

VEINS: TRANSPORT OF FOOD AND WATER The veins consist of vascular bundles that form a network through-

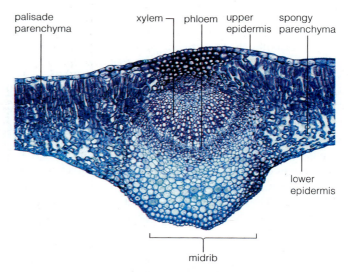

palisade parenchyma · xylem · phloem · upper epidermis · spongy parenchyma · lower epidermis · midrib

Figure 7.9 A lilac leaf (*Syringa vulgaris*) in cross section, showing the midrib, ×93.

out the leaf for transport of water and nutrients. The xylem conducts water and dissolved minerals from roots to leaves, and the phloem transports food made via photosynthesis in the mesophyll from leaves to the rest of the plant.

In dicot leaves, a large-diameter midrib (the central vascular bundle) runs the length of the leaf. In cross section the xylem is in the upper part of the bundle (Fig. 7.9). Along with the vessel members and tracheids that trans-

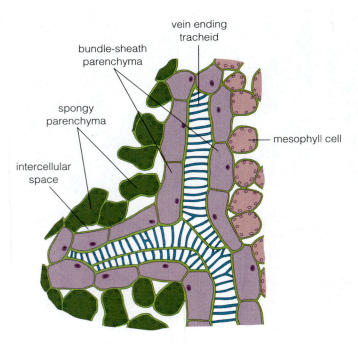

Figure 7.10 Diagram of a longitudinal section of a vascular bundle through the mesophyll and a vein ending. Shown are the mesophyll cells, bundle sheath cells, and tracheids at the vein end.

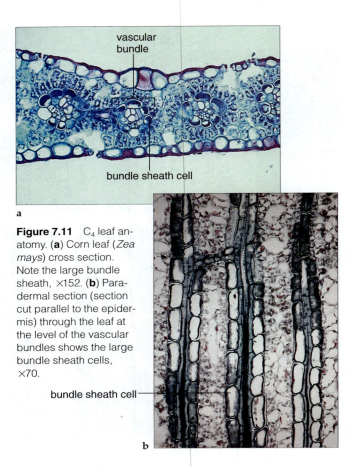

a

Figure 7.11 C_4 leaf anatomy. (a) Corn leaf (*Zea mays*) cross section. Note the large bundle sheath, ×152. (b) Paradermal section (section cut parallel to the epidermis) through the leaf at the level of the vascular bundles shows the large bundle sheath cells, ×70.

bundle sheath cell

b

port water, xylem often has fibers around the outside of the bundle, these provide mechanical support, and parenchyma cells between the vessel members. The phloem makes up the lower part of the bundle. Phloem contains sieve-tube members (STMs, which conduct sugars), companion cells, (which work with the STMs), fibers for support, and parenchyma cells (which load the sugars into the STMs). Most of the smaller vascular bundles have the same basic structure, with the smallest having only a few cells (Fig. 7.10). Usually a single layer of cells called the **bundle sheath** surrounds the vascular bundles (Fig. 7.10). The cells in the bundle sheath work to load sugars into the phloem and to unload water and minerals out of the xylem.

The main vascular system difference between dicot and monocot leaves is that monocot leaves have several main bundles running parallel along the length of the leaf rather than a single midrib with branches. In cross section, leaf vascular bundles have xylem at the top of the bundle and phloem at the bottom. In some grasses a special type of photosynthesis called C_4 *photosynthesis* occurs. This type of photosynthesis can operate at lower carbon dioxide concentrations and higher temperature than regular photosynthesis. The C_4 photosynthesis requires more steps to convert carbon dioxide into sugars, and it involves a special role for the bundle sheath cells, as described in Chapter 10. Leaves of C_4 plants have large bundle sheath cells because of this (Fig. 7.11).

Where Do New Leaves Come From?

Leaves have their origins in meristems. If you push away the older leaves of a bud, you will see several crescent-shaped bumps on the flanks of the very tip of the shoot apical meristem (SAM). These are the new leaves being formed (Fig. 7.12a). In the early stages of development they are called **leaf primordia** (singular, *primordium*). The initiation of a new leaf involves a complicated set of events. Among them is the movement of some kind of chemical signal from the vascular bundles or perhaps from the already initiated leaf primordia. The location of a new leaf depends on the plant's phyllotaxis (see Chapter 6). The signal will trigger the cells at the initiation site to start dividing. These cells become the leaf primordium (Fig. 7.12b). The shape of the new leaf will be determined by how the cells in the primordium divide and enlarge.

Recent research has shown that the cell division lineages (cells, usually in a file, that have a common origin) involved in the generation of different leaf forms are not repeated exactly in every leaf. Scientists have been able to follow the cell lineages as the cells divide by using X rays or chemicals to induce a color mutation in chloroplasts. The leaf cells containing the mutant chloroplasts are yellow instead of green. When these mutated yellow cells divide in a leaf primordium, their chloroplasts also divide and are passed on to the derivative cells (Fig. 7.13).

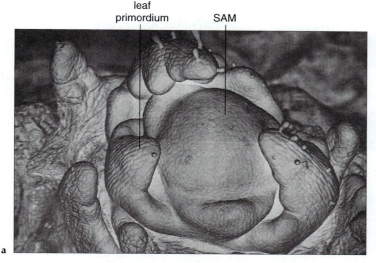

leaf
primordium   SAM

a

Figure 7.12 The development of leaf primordia. (**a**) Scanning electron micrograph of shoot apex of summer adonis (*Adonis aestivalis*), showing newly developing leaf primordia, ×80. (**b**) Diagram of a longitudinal section of the shoot apex, showing leaf primordia in different stages of development. The leader at the top indicates the position where the next leaf primordium will be initiated. The shaded cells are procambium.

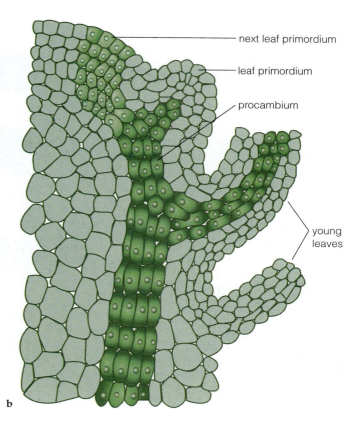

— next leaf primordium

— leaf primordium

— procambium

} young leaves

b

Figure 7.13 Yellow sectors (arrow) in this leaf were formed from cells with mutated chloroplasts.

By following the color of yellow cells, researchers have found that the cell lineages have different shapes. Also, the size of the mutant sectors depends on when the mutation is induced. Cells that are mutated early in development produce large longitudinal sectors, while those mutated late in development produce small patches. The most interesting finding is that the precise pattern of lineages is never exactly the same among leaves. This means that the development of leaf shape is controlled independently of the precise pattern of cell division.

7.2 LEAF FORM AND SPECIALIZED LEAVES

Leaf Shape May Depend on the Plant's Age and the Environment

Leaves that form during different stages of a plant's life cycle can have different shapes. The first leaves that form are part of the embryo found in seeds. These first *seed leaves*, or **cotyledons**, are slightly flattened and roundish. The cotyledons are mostly storage organs, but in some plants they enlarge and conduct photosynthesis. In bean plants the first true leaves are simple, and later-stage leaves are compound (Fig. 7.14).

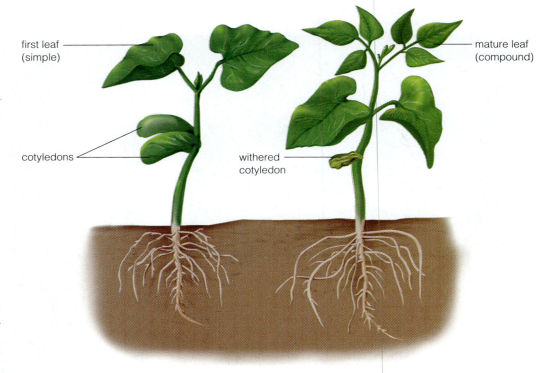

first leaf
(simple)

cotyledons

mature leaf
(compound)

withered
cotyledon

Figure 7.14 Young bean plants (*Phaseolus vulgaris*), showing the different leaf shapes that form at different stages in the plants' life cycle. Cotyledons and first leaves are simple, with smooth margin; mature leaves are compound.

a

Figure 7.15
Age-dependent leaves of ivy (*Hedera helix*).
(**a**) Juvenile. (**b**) Adult.

b

The phenomenon of different leaf shapes on a single plant is called **heterophylly**. One kind of heterophylly is related to the age of the plant. In ivy (*Hedera helix*), age-dependent changes in leaf form are related to the reproductive maturity of the plant. Leaves of juvenile ivy plants have three lobes, the plant mode of growth (or *habit*) is a vine, and juvenile plants do not flower. An adult ivy plant is upright, the leaves are not lobed, and the plant can produce flowers (Fig. 7.15).

Another kind of heterophylly is induced by the environment. In this case, the form of the leaf is influenced by the environment that the shoot apex is exposed to during leaf development. For example, in certain marsh plants, leaves that develop underwater are thin and have very deep lobes; these are called *water leaves* (Fig. 7.16). When the shoot tip extends above the water in the summertime, thicker leaves with reduced lobing are formed; these are called *air leaves*.

One last example can be seen on several tree species. Leaves that form on bottom branches, where they are mostly in the shade, tend to be thin and relatively large in surface area; these are *shade leaves*. The leaves that develop near the top of the same tree and in more direct sunlight, however, tend to be thicker and smaller. These are *sun leaves*. In each instance of heterophylly, the different leaf designs make the leaves more functionally suited to their environment.

Figure 7.16 Effect of environmental conditions on the development of leaves. Buttercup (*Ranunculus aquatilis*) leaves that form when the shoot tip is underwater are thin and more deeply lobed; those that form in the air are thicker and less lobed.

ADAPTATIONS FOR ENVIRONMENTAL EXTREMES The concept of evolution will be discussed in Chapter 18, but the fundamental idea is that some individuals in a population of plants will be better suited to the environment in which they grow because of some special characteristic. For example, of the plants growing in a dry climate, those with a thick cuticle to protect against water evaporation might be more successful. Over a period of time, the individuals best suited for a particular environment will survive to maturity and will propagate new plants that will also be adapted for that particular environment. When researchers have compared plants from different environments, they have found that the species sharing a common environment tend to share specialized structural characteristics.

Plants that grow successfully in dry climates are called **xerophytes**. These plants tend to have leaves that are designed to conserve water, store water, and insulate against heat. The oleander (Fig. 7.17a) and fig (Fig. 7.17b) are good examples. The leaves have sunken stomata and a thick cuticle to inhibit water loss. The *Crassula* plant (jade plant) has very thick leaves with few air spaces in the mesophyll (Fig. 7.17c).

Another anatomical feature common to xerophytic plants is an abundance of fibers in leaves. These long, rigid cells help support the leaf and keep it from losing its shape when it dries. These fibers often have commercial value (see sidebar, "Commercially Important Fibers from Leaves and Stems," p. 116).

Plants that grow in moist environments are called **hydrophytes**. The leaves of such plants tend to be thin, have a thin cuticle, and are often deeply lobed (see Fig. 7.16). Because these leaves always have an abundance of water, they lack characteristics to conserve water.

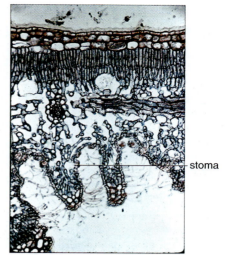
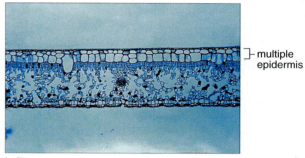

a Oleander

— stoma

— multiple epidermis

b Fig

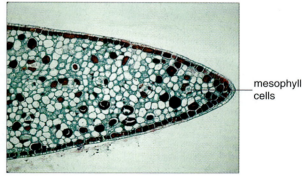

— mesophyll cells

c Jade

Figure 7.17 Xerophytic adaptations help plants survive dry and hot environments. (**a**) Oleander (*Nerium oleander*) with sunken stomata, ×225. (**b**) Fig (*Ficus* sp.) leaf with multiple epidermis that stores water and insulates against solar radiation, ×85. (**c**) Jade plant (*Crassula argentea*) mesophyll cells are tightly packed, with few air spaces, to inhibit evaporation, ×35.

COMMERCIALLY IMPORTANT FIBERS
FROM LEAVES AND STEMS

Fibers from plants have been used by people for at least 10,000 years. Cotton fibers, for example, are known to have been used in Mexico almost 9000 years ago. Cotton fibers aren't really *fibers* in terms of their cell type, however; they are epidermal hairs that grow from the cotton seed coat. The kind of fibers we will discuss here are elongated sclerenchyma fibers joined into long strands by their overlapping ends (See Chapter 4). Bundles of these fibers can be long and very strong. True fibers like these were also used in antiquity. Archaeological evidence has revealed that Native Americans were using *Agave* (sisal) to make cords about 10,000 years ago, and Stone Age Europeans were using flax plants (*Linum usitatissimum*) for linen at about the same time.

Where do fibers come from? They tend to form in strands around vascular bundles and within the ground tissues of stems and leaves. Their purpose in the plant is to strengthen stems and leaves so they don't break when moved by the wind. They act like reinforcing bars used in concrete structures. Ancient peoples figured out how to extract these strands and use them for fabrics and ropes.

There are basically two types of commercially important fibers: hard fibers and soft fibers (Table). **Hard fibers** occur in leaves of monocotyledonous plants such as *Agave* (Fig. 1). Their long, strap-shaped leaves contain thick bundles of fibers, often surrounding the vascular bundles. These fibers are very stiff and often have thick lignified cell walls (see Chapter 4). Fiber strands are extracted from the leaves, usually by crushing or beating the leaves and then scraping them to expose the fiber bundles. Once exposed, they are pulled and dried (Fig. 3). The fibers (sisal) are then woven into coarse fabric or twisted and braided into rope. Another example is Manila hemp (*Musa textilis*), which is used to make the huge ropes that tie down ships because it resists decay and breakdown by salt water. Some Manila hemp is also used to make lightweight fabric and teabags.

Soft fibers (sometimes called *bast fibers*) are found mostly in the stems of dicotyledonous plants, where they occur just outside the phloem as part of vascular bundles. Individual soft fiber cells are longer than those of hard fibers. They tend to be rather soft and flexible and are sometimes unlignified. These properties make such fibers very valuable for clothing fabric, string, rope, and even cigarette paper and money.

Probably the most famous soft fiber is linen from the flax plant, which today accounts for about 2% of the world's tex-

Table of Commercially Important Fibers			
Species	Common Name	Uses	Fiber Length
Hard Fibers			
Agave sisalana	Sisal	Rope, coarse fabric	0.8 to 8 mm
Musa textilis	Manila hemp	Marine rope, cloth	2 to 12 mm
Phormium tenax	New Zealand hemp	Rope	2 to 12 mm
Soft Fibers			
Linum usitatissimum	Flax (Linen)	Cloth, paper money	9 to 70 mm
Cannabis sativa	Hemp	Canvas, rope	5 to 55 mm
Boehmeria nivea	Ramie	Fabric	50 to 550 mm
Corchorus capsularis	Jute	Burlap, carpet backing	0.8 to 6 mm

Figure 1 *Agave sisaliana* plants (foreground) are grown as a crop in various places in the world; shown here in Malaysia. The large tree in the background is a baobab (*Adansonia digatata*). The swollen trunk of this tree stores water.

Figure 2 Flax plant, *Linum usitatissimum*.

Figure 3 Flax fibers being processed.

tiles (Fig. 2). The plant, usually about a meter (over 3 ft) tall, is pulled from the soil and bundled to dry. The fibers are extracted by a process called *retting*. This involves placing the dried plants in a watery bacterial slurry and allowing the bacteria to decay the ground tissues away, leaving the fiber bundles. These are then combed out, removed, and dried (Fig. 3). A newer way to extract them is to pass the undried plants through crushers, comb the bundles free, and dry them. The dried fibers are then spun into yarn, dyed, and woven into fabric. The elegant feel and texture of linen have made it popular for a long time. Its only problem is that it wrinkles easily. Nowadays, linen fibers are chemically treated, and the yarn is woven with cotton and/or rayon. This makes contemporary fabrics that are more wrinkle resistant but retain some of the attractive texture of the original.

Ramie (*Boehmeria nivea*) has the longest fiber cells known. They can be up to 550 mm (more than a foot) and are stronger than cotton fibers. Mercerized ramie, made by treating the fiber bundles with alkali, is among the strongest natural fibers used in clothing. The next time you see a very exotic-looking fabric, look at the label. You'll probably find linen or ramie.

a

b

c

Figure 7.18 Leaf modifications. (**a**) Cactus (*Opuntia* sp.) spines. (**b**) Tendrils from leaflets of trumpet flower (*Bignonia cap-reolata*) (**c**) Plantlets along the leaf margin of air-plant (*Kalanchoe pinnata*).

Leaves from plants in moderate climates are called **mesophytes**. These leaves have the general leaf characteristics that we have been discussing (see Fig. 7.5).

MODIFICATIONS FOR SPECIAL FUNCTIONS **Spines**, one of a handful of leaf modifications, protect against predators. Spines of various species of cactus are actually modified leaves (Fig. 7.18a). They are made of cells with hard cell walls. In the black locust, the spines are **stipules**, which are projections that form at the base of some leaves. Spines are pointed and quite dangerous to potential predators.

Tendrils are modified leaflets that wrap around things and support the shoot. In some plants the entire leaf becomes a tendril (Fig. 7.18b).

Bulbs have very thick leaves. Bulbs are modified branches with a short, thick stem and short, thick storage leaves. These leaves store food and water (see Fig. 6.28d).

Some leaves produce plantlets. In the air-plant (*Kalanchoe pinnata*, Fig. 7.18c) the leaves have notches along their margins. In the bottom of each notch, a special meristem develops that produces a new plantlet. After these little plantlets grow for a short time, they fall off the leaf and root into the soil. Similar mechanisms occur on other plants. This means of reproducing is a form of **vegetative**, or **asexual reproduction** (see Chapter 12).

Leaf Abscission Is the Seasonal Removal of Leaves

The separation of plant parts from the parent plant is a normal, continually occurring process. Leaves fall, fruits drop, flower parts wither and fall away, and even branch tips or whole branches may separate as a normal part of a plant's life. A familiar example is the fall of leaves from certain woody dicot plants in the autumn. In practically all cases, separation, or **abscission**, is the result of differentiation in a specialized region known as the **abscission zone** at the base of the petiole (Fig. 7.19a). Parenchyma cells making up the abscission zone are usually smaller than cells in adjacent tissues, and they may lack lignin in their cell walls. Xylem and phloem cells tend to be shorter in the vascular bundles at the base of the petiole, and fibers are often absent in the abscission zone. These anatomical features make this zone an area of structural weakness (Fig. 7.19b). As the zone gradually weakens, the vascular bundle cells become plugged, and the leaf finally falls. The leaf scar seals over with waxy materials so that pathogens can't enter after the leaf falls off.

Environmental cues, such as cold temperature or short days are important controls for abscission. These cues induce hormonal changes that in turn affect the formation of the abscission zone. The leaves now fall to the ground and decompose, and their nutrient components will return to the soil to be recycled.

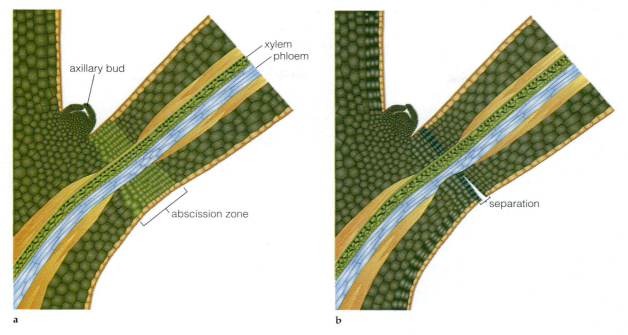

Figure 7.19 Leaf abscission. (**a**) Formation of the abscission zone. (**b**) Separation within the abscission zone and initiation of periderm.

SUMMARY

1. The main functions of leaves are photosynthesis and transpiration.

2. Leaves of flowering plants are generally flat and have thin blades that are attached to the stem by petioles or sheaths. Veins (vascular bundles) transport food and water and strengthen the leaf blade. Leaf blades may be simple or compound; leaf margins may be entire, toothed, or lobed.

3. A cross section through the blade of a typical leaf shows the following tissues: upper epidermis, mesophyll, vascular bundles, and lower epidermis. Generally, a waxy cuticle coats the epidermis. Guard cells of the epidermis form stomata that control gas exchange.

4. Mesophyll tissue is composed of palisade and spongy parenchyma and veins. Chloroplasts found in mesophyll cells convert light to energy in photosynthesis.

5. Modified leaves such as bud scales, spines, and tendrils serve to store food or water or to trap insects.

6. Young leaves are produced by a pattern of cell divisions that result in different leaf forms.

7. Environmental factors such as strong light intensity, nutrient deficiency, and too much or too little water can affect leaf form and anatomy. *Heterophylly* refers to different leaf forms induced by environmental changes—air and water leaves, for example.

8. The formation of a definite abscission zone across the petiole is responsible for leaf fall. Environmental cues and hormonal changes affect the formation of abscission zone.

Questions

1. What are some functions of leaves?

2. What are the main distinctions between monocot and dicot leaves?

3. Describe the parts of a simple and a compound leaf.

4. Describe the general anatomical characteristics of a leaf. Make a cross-sectional diagram and label the three tissue systems and the tissues in each.

5. What are the specific functions of the epidermis, mesophyll, xylem, and phloem in leaves?

6. Some leaves are modified to live in extreme environments. Identify one characteristic of each of the following leaf types: mesophyte, hydrophyte, xerophyte.

7. Leaves fall off many plants in the autumn. Describe the process of leaf abscission.

8. Describe the functions of these modified leaves: bud scales, tendrils, bulb scales, spines.

Further Readings

Esau, K. 1977. Chapters 18 and 19 in *Anatomy of Seed Plants*. 2d ed. New York: John Wiley. This is the classic plant anatomy textbook. It is an old text but is still considered an important reference.

Fahn, A. 1990. Chapter 12 in *Plant Anatomy*. 4th ed. Oxford: Pergamon Press. This is a newer anatomy text. It has up-to-date material but is rigorous reading.

Levetin, E., and K. McMahon. 1996. *Plants and Society*. Dubuque, Iowa: Wm. C. Brown.

Simpson, B. B., and M. C. Ogorzaly. 1995. *Economic Botany: Plants in Our World*. 2d ed. New York: McGraw-Hill.

8 CONCEPTS OF METABOLISM

1. Almost every chemical reaction that takes place in a cell is catalyzed (accelerated) by a specific enzyme. Each type of enzyme is formed from a unique kind of protein molecule. Some types of enzymes also require special nonprotein cofactors in order to work.

2. Chemical reactions in a cell form a network of interrelated metabolic pathways. The cell controls the synthesis and breakdown of chemicals by regulating the activities of enzymes that catalyze key reactions in the network.

3. Every reaction—and every set of interrelated (coupled) reactions—that proceeds forward spontaneously (with or without enzymatic catalysis) loses free energy. A reaction that gains free energy can be forced to proceed by coupling it to another reaction that loses even more free energy.

4. Two key reactions that lose free energy—the hydrolysis of ATP and the oxidation of NADH—power many of the activities of a cell.

8.1 SWIMMING UPSTREAM

Living organisms have a problem. They consist of highly ordered forms of matter in a universe that favors an ever greater state of disorder. Furthermore, they need a constant supply of ordered, complex molecules to stay alive. Similar to salmon swimming upstream, so life itself moves against the universe's chaotic tendencies. That it can do so stems from the ability of plants to supply not only themselves but almost all other living organisms with the basic molecules needed for life. To overcome the universal current of disorder, plants use the sun's energy to synthesize carbohydrates, hydrocarbons, and other complex molecules from much simpler molecules. They then reverse the process and use the energy stored in the new molecules to drive various life-sustaining processes. But synthesizing and destroying complex molecules poses another problem: how to orchestrate thousands of chemical reactions so that they fit together in the complex process called **metabolism**. This chapter describes some of the elegant solutions cells have devised to control the rates of chemical reactions, as well as to surmount the energy hurdles to building complex molecules and structures.

8.2 CHEMICAL REACTIONS

Living cells are made of thousands of different chemical compounds. Some of these are small, simple compounds, such as water, sugars, and amino acids; others, such as proteins and nucleic acids, are large, complex molecules. Many of these compounds, especially the complex ones, are used to construct the organelles we see inside a cell.

Some compounds are the building blocks of more complex molecules. Still others have specialized functions in the life of the cell, such as defending against pathogens or herbivores.

Virtually every plant cell synthesizes all the different complex compounds that it contains. These compounds are formed from very simple chemicals containing carbon, hydrogen, oxygen, and nitrogen, as well as smaller amounts of other elements such as sulfur and phosphorus. Water (H_2O), carbon dioxide (CO_2), ammonium (NH_4^+) or nitrate (NO_3^-) ion, sulfate (SO_4^{2-}), and phosphate (PO_4^{3-}) represent the major basic stocks for the production of all the other molecules in the cell. Water, which makes up over 90% of the weight of the protoplast (the portion of the cell inside the cell wall), is primarily a solvent, but it also is the source of most of the hydrogen atoms and some of the oxygen atoms in organic molecules. Carbon dioxide is the primary source of carbon and a major source of oxygen. Ammonium and nitrate are the primary sources of nitrogen in proteins and nucleic acids. Sulfate is the source of the sulfur found in some amino acids, and inorganic phosphate is incorporated into nucleotides.

The rearrangement of atoms from their initial positions in certain molecules to new positions in other molecules is called a **chemical reaction**. In cells of biological organisms, the reactions are often called **biochemical reactions**. It is clear that there must be thousands of biochemical reactions in plant cells to produce the thousands of bio-organic molecules found there.

Chemical Reactions Must Overcome Energy Barriers

Although every chemical reaction is different, there are concepts that describe reactions in general. Because a chemical reaction involves the rearrangement of atoms, it must also involve the breaking and re-formation of some of the covalent bonds holding the atoms together (see Chapter 2). The covalent bonds represent electrons in orbitals that are reasonably stable. To break an existing bond, it is necessary to stretch or bend it. This destabilizes the electrons, enabling them to move to other, even more stable places. Just as it takes energy to stretch or bend a spring, it takes energy to stretch or bend a covalent bond. And just as the energy used to stretch a spring is not lost—it is stored as **potential energy**—the energy used to stretch a bond is also stored as potential energy. The energy can be recovered as the stretched bond either springs back (and the electrons move back to their original positions) or breaks (and the electrons move to other stable positions).

Figure 8.1 shows the potential energy relationships of the various phases of the reaction

$$A + B \rightarrow C + D$$

Reaction: A + B ⟶ C + D

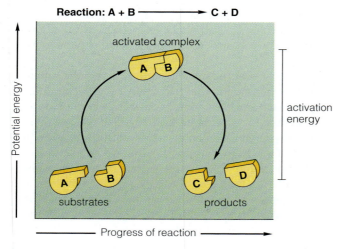

Figure 8.1 The changes in potential energy of two molecules as they exchange atoms in a chemical reaction.

where A and B, the **substrates** (or **reactants**) of the reaction, collide and rearrange their covalent bonds to form C and D, the **products** of the reaction. The two molecules A and B have a certain amount of potential energy inherent in their separate covalent bonds. As they collide, their bonds bend (that is, the electron orbitals move from their most stable positions), and their potential energy goes up. This distorted, high-energy state is called an **activated complex**. As the activated complex forms, electrons can move to new orbitals, and bonds can thus shift. For instance, the orbitals holding an atom to the other atoms of A can be broken, and new orbitals can form to attach the atom to the atoms of B. At this step, C and D form. As the new bonds shift into their most stable positions, the potential energy of C and D drops to its normal value. For A and B to react to form C and D, they must attain an increase in potential energy, called the **activation energy**. The activation energy is an energy barrier that must be scaled for the reaction to proceed.

The law of conservation of energy says that energy cannot be created or destroyed in normal chemical reactions. Where does the energy come from to bend the bonds of A and B and form the activated state? The fact that raising the temperature speeds the reaction suggests that it comes from the energy of motion—the **kinetic energy**—of the two substrates. The average amount of kinetic energy of a mixture depends on the temperature. The higher the temperature, the faster the molecules in the mixture are moving. In a solution such as the cytoplasm, molecules are closely packed and travel for only short distances before bouncing against other molecules. This means that the molecules of water and other solutes continually batter the substrate molecules; thus the kinetic energy of the substrate molecules matches the kinetic energy of the other molecules in the mixture.

Although the temperature determines the average energy of all the molecules, including the substrates, individual molecules have more or less than the average.

At a normal temperature, for instance 20°C (68°F), only a few A and B pairs will come together with enough kinetic energy to provide the potential energy needed to form the activation complex. This limits the rate at which the reaction can proceed. If the temperature rises, the reaction goes faster because a greater proportion of the A and B pairs will have enough kinetic energy.

A Catalyst Speeds a Reaction by Lowering the Energy Barrier

The rate of a chemical reaction is very important. Each of the thousands of reactions in a cell must proceed at a reasonable rate for the cell to function. There are a few ways that a chemist can speed up a reaction. Increasing the concentrations of the substrates speeds the reaction by increasing the probability that pairs of substrate molecules will meet. Increasing the temperature increases the rate at which molecules move—and thus the frequency with which pairs of substrate molecules will meet. As mentioned above, increasing the temperature also speeds the reaction by increasing the probability that substrate molecules will have enough kinetic energy to form an activated complex.

The techniques of chemists are of limited use to cells, however. Although cells can increase the concentrations of specific molecules, they can do so only up to a certain point. Raising the temperature is not a good strategy for speeding biochemical reactions in a cell because many of a cell's components are harmed by excessive heat. Instead, cells rely on **catalysts** to speed up reactions. Catalysts are not substrates or products of the reaction. They are neither used up nor formed as the reaction proceeds, although they do interact closely with the substrates. A catalyst can be a simple substance, such as the metallic element palladium, or it can be a complex biomolecule. A catalyst is thought to work by forming a temporary complex with one or both substrates, attaching to it (or them) with reversible bonds (bonds that will break after the reaction is completed). The formation of the complex distorts the bonds of the substrate, so that further bond bending or stretching requires less energy.

A diagram showing the effect of a catalyst would look similar to the original energy diagram (compare Fig. 8.2 with Fig. 8.1). Substrates plus catalyst have a certain amount of potential energy. When they come together to form an activated complex (which may be quite different in form from the original activated complex we considered), they have a high potential energy. As the products are formed and the new bonds relax, the potential energy of the trio decreases again. The difference between this case and the original, uncatalyzed one is that the amount of activation energy is less. Therefore, at a particular temperature, the group of substrates plus catalyst are more likely than the substrates alone to have enough energy to form an activated complex.

Reaction: A + B —————— enzyme ——————→ C + D

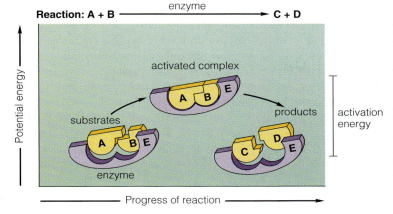

Figure 8.2 The changes in potential energy of two molecules as they exchange atoms in an enzyme-catalyzed reaction. Note that the activation energy is lower in this reaction than it was in the uncatalyzed reaction diagrammed in Figure 8.1.

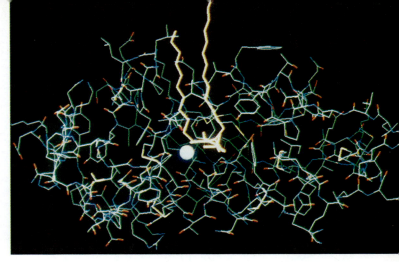

Figure 8.3 A three-dimensional model of an enzyme, phospholipase, showing how its substrate, a phospholipid, binds to its active site. The white sphere is a calcium atom that is needed for the association between the enzyme and the substrate. This enzyme catalyzes the cleavage of one of the covalent bonds of the phospholipid, breaking it into two separate molecules. Compare this picture with Fig. 2.11, which shows the enzyme without the substrate.

Enzymes Catalyze Specific Reactions

One of the features that characterizes all living cells is the presence of biological catalysts called **enzymes**. Enzymes catalyze the thousands of reactions in a cell. Enzymes differ from simple catalysts, such as palladium, in their specificity. Each enzyme works on only one set of substrates (or, at most, a very few related substrates); each enzyme catalyzes one type of reaction. That means that a cell must have thousands of enzymes to catalyze all the reactions it needs.

With a very few exceptions, enzymes are made of protein molecules. As described in Chapter 2, proteins are large polymeric molecules with complex three-dimensional shapes. Each different type of protein has a different shape. In particular, enzyme proteins have **active sites**, which are grooves or crevices in their surfaces, into which one or both of the substrates can fit (Fig. 8.3). The shape of the active site is almost complementary to that of the substrate; that is, the active site almost fits the substrate as a glove fits a hand. The substrate is held in the active site by hydrogen bonds and other forces that are weak individually but very effective as a group. As the substrate is held in the active site, its bonds are distorted in any of a number of ways. Sometimes pulling the substrate into a slightly misfitting groove distorts the substrate's shape; sometimes the active site changes shape once the substrate is present; sometimes electric charges in the active site push and pull the electrons of the substrate; sometimes functional groups (side chains of amino acids) in the active site react (temporarily) with the substrate. In any case, the distortion of the substrate makes it susceptible to the particular reaction catalyzed by the enzyme. If the reaction involves two substrates, the active site might bind both, bringing them close together in the proper orientation for the reaction to occur.

Some enzymes depend only on their protein structure for their activity; others cannot function without certain nonprotein substances called **cofactors**. Examples of cofactors include metal ions, such as Fe^{2+} and Fe^{3+}, as well as complex organic molecules without metal ions. Some cofactors are bound to enzymes by covalent bonds. Others are loosely bound and may be removed from the enzyme protein. Often, but not always, the cofactors are able to accept or donate electrons in oxidation–reduction reactions. If a cofactor (sometimes called a **coenzyme**) is loosely bound to its enzyme and is capable of accepting and donating electrons (or hydrogen atoms: electrons plus protons), it may serve as a carrier of electrons (or hydrogen atoms) from one reaction to another. Plants must be given trace amounts of certain metals that serve as cofactors (for instance, iron, copper, and molybdenum), but they make all the bio-organic cofactors they use. Many of the bio-organic cofactors that are common in plants—such as riboflavin, thiamin, niacin, and pantothenic acid—cannot be synthesized by humans. We obtain these compounds from the plants we eat.

The results of some plant enzyme activities are easy to detect. The darkening of an apple fruit after it has been cut (or bitten) results from the action of the enzyme polyphenol oxidase on chemicals released from the cells. The softening of a tomato fruit as it ripens is caused by the action of several enzymes (including cellulase and polygalacturonase) on the polysaccharides of the cell walls. Papain, an enzyme from papaya fruit, digests proteins in the fruit as it ripens. It can also be extracted from the fruit and used to tenderize meat quickly before it is cooked.

The Rates of Reactions Can Be Controlled by the Synthesis, Activation, and Inhibition of Enzymes

In a normal cell, the temperature is too low for most essential biochemical reactions to occur at reasonable

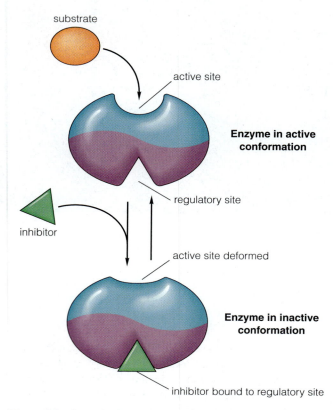

substrate

active site

Enzyme in active conformation

regulatory site

inhibitor

active site deformed

Enzyme in inactive conformation

inhibitor bound to regulatory site

Figure 8.4 A regulatory enzyme has two sites: an active (catalytic) site and a site at which a regulatory molecule may bind. In the presence of the regulatory molecule (inhibitor), the enzyme changes its shape. The substrate binding in the active site may also contribute to (or oppose) the shape change.

rates. For cells to function, all the reactions must occur fairly rapidly and in a balanced fashion. In this way, the components of the cells can be synthesized in the proper proportions, and no one material is formed (using up all the substrates) at the expense of all the others. The synthesis of the various organelles must be balanced, and various synthetic reactions must be coordinated with the reactions that break dysfunctional molecules down into their component parts.

Cells have several ways of controlling the rates of individual reactions. One method is to synthesize the enzyme that catalyzes a particular reaction when, and only when, that reaction is needed. The formation and germination of a seed provides a good example. A seed is formed from an embryo with associated tissues that store nutrients used for the growth of the embryo when the seed germinates. The nutrients include starch, and the starch is formed in the cells from simple carbohydrate subunits imported in the form of sucrose. Two enzymes (at least) are required to break sucrose down and use the products to form starch; these enzymes increase in amount as the seed is formed. Later, they disappear. When the seed germinates, the starch must be broken down into subunits, and the enzymes that accomplish this are synthesized during the early stages of germination. The synthesis of enzymes is a key aspect of develop-

ment and is treated in more detail in Chapter 15 (on the control of growth and development).

The control of reaction rates by the synthesis and breakdown of enzymes is rather slow and crude, however. A much quicker and finer control occurs through the regulation of the catalytic activity of already existing enzymes. Such regulation is possible because the catalytic activity of an enzyme depends on its three-dimensional shape, and the shape depends in turn on weak bonds (such as hydrogen bonds) that can easily be broken and re-formed. Not all enzymes can be regulated in this way; those that can generally have a special site on their surface that will bind a regulatory molecule. In some cases, a special enzyme attaches a phosphate group to that site covalently. In other cases, some other molecule (there are many possibilities) binds to the site through weak bonds, such as hydrogen or ionic bonds. Binding to the regulatory molecule changes the shape of the enzyme, and that change affects the shape of the active site (Fig. 8.4). The overall result is either (1) to lower the affinity of the active site for the substrate or its catalytic efficiency once the substrate has been bound (in which case the enzyme is inhibited), or (2) to increase the affinity of the active site for the substrate or its catalytic efficiency (in which case the enzyme is activated). The regulatory step used depends on the enzyme. It may involve a normal compound found in the cell (perhaps the substrate or product of another chemical reaction). For the regulated enzyme, this compound serves as a signal to turn off or on the reaction it catalyzes. Later in this chapter we will show how regulation of an enzyme in this way can maintain a steady concentration of a compound.

8.3 METABOLIC PATHWAYS

Most bio-organic molecules produced in a cell are formed not from a single chemical reaction but from a sequence of reactions, in which the product of one reaction becomes the substrate for the next. A series of such linked reactions in a cell is called a **metabolic pathway**.

Reactions Are Linked When One Reaction's Product Is Another's Substrate

It is not possible to make complex molecules in a single step; it takes a series of simple steps. A single step might be the addition of the components of water to the molecule, or a reduction (addition of one or two electrons), or perhaps the removal of a CO_2 group. A series of such steps, each in the correct order, can lead to quite complex compounds, as you will see in Chapters 9 and 10.

In a metabolic pathway (Fig. 8.5), the product of the first reaction is the substrate of the second reaction, the product of the second reaction is the substrate of the third, and so forth. Each reaction is catalyzed by a sepa-

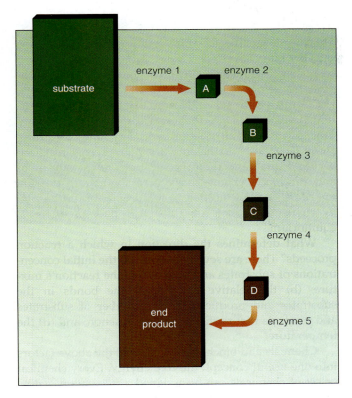

Figure 8.5 A metabolic pathway. Each letter represents a different compound. Each reaction is catalyzed by a different enzyme.

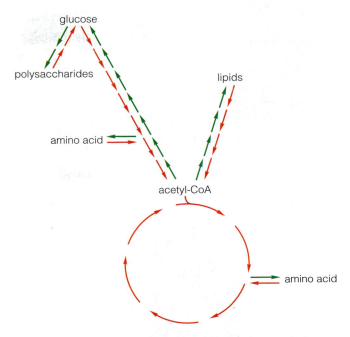

Figure 8.6 A simplified picture of some interacting metabolic pathways in a typical plant cell. Note that the anabolic reactions (green arrows) synthesizing substances and the catabolic reactions (red arrows) breaking them down generally involve different enzymes.

rate enzyme. The products of reactions, other than the final product of a series, are called **intermediates**. For this reason, the collection of all the metabolic pathways in cells is called **intermediary metabolism**. The enzymes that catalyze the reactions are linked: Each enzyme depends on the ones before it to provide substrates. Without the substrates, the enzymes cannot function. In the same way, each enzyme depends on those after it to complete the formation of the pathway's product. Without the enzymes for the later reactions, intermediates will accumulate—and no final product will be formed.

In Cells, Linked Reactions Form a Complex Network

Although Figure 8.5 shows a simple linear pathway, few if any pathways in a cell are this simple. In some cases, two or more compounds are needed to form a product. Frequently, one intermediate is used as a substrate by several separate enzymes to produce different products. In this case, the intermediate is called a **branch point** because, diagrammatically, the pathways appear to branch off in different directions (Fig. 8.6). Some intermediates are key compounds in the cell because they serve as substrates for many different reactions.

Figure 8.6 gives a very rough idea of the complexity of the metabolic reactions in a typical plant cell. This diagram shows several metabolic pathways linked together, with each arrow representing a separate reaction with a separate enzyme. Each pair of adjacent reactions is con-

nected by an intermediate, although only a few key intermediates are shown.

The reactions that produce subunits of functional or structural molecules (such as proteins or cellulose) are sometimes called **anabolic reactions**. Reactions that break damaged or unwanted molecules down into their component parts are called **catabolic** reactions. Figure 8.6 indicates these two classes of reactions, with arrows pointed in opposite directions to denote that the synthesis and breakdown of materials generally (but not always) involve different enzymes. In some cells (such as those in meristems), anabolic reactions predominate; in others (such as the cells in ripening fruit), catabolic reactions are most active. In most cells of plants and other organisms, both sets of reactions occur; there is a constant synthesis and breakdown—called a *turnover*—of cell components.

The Concentration of a Compound Often Controls Its Production in a Cell

Cells cannot afford to allow metabolic pathways to run amok. If pathways operated without any control, materials and energy would be wasted in the formation of unneeded compounds. Certain intermediates might accumulate to toxic levels; others vital to the cell might become scarce. The fact that reactions are organized into pathways immediately suggests an efficient way of controlling the formation of the end product of a pathway. Because each reaction depends on the ones before it to

oxidation–reduction reactions, for example, one compound is oxidized (loses electrons), and another is reduced (gains electrons). The oxidation can be written as a separate, but partial, reaction; and so can the reduction. Neither reaction occurs by itself, however, because electrons must be conserved. Because they must occur together, the reactions are said to be **coupled**.

Another example is the transfer of a phosphate group. One compound might lose a phosphate group through a **hydrolysis reaction** (the addition of the components of water, H$^+$ and OH$^-$, breaks the bond between the phosphate and the rest of the compound); another compound might gain a phosphate group through a **con-**

of an amino acid to a growing protein, or to run the motor proteins that attach to actin filaments. ATP is the quintessential *high-energy compound* or *energy carrier*. The fact that it is formed in the mitochondrion (by a condensation of ADP with phosphate—formally the reverse of the reaction shown in Fig. 8.9b) is the main reason that the mitochondrion is called the "powerhouse of the cell."

OXIDATION OF NADH AND NADPH The coenzymes **nicotinamide adenine dinucleotide (NADH)** and **nicotinamide adenine dinucleotide phosphate (NADPH)** are essential molecules in eukaryotic cells. The reactive part of these molecules is their nicotinamide functional

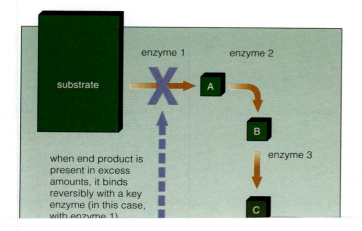

when end product is present in excess amounts, it binds reversibly with a key enzyme (in this case, with enzyme 1)

This means that the substrates and products of a reaction can be interconverted. Under arbitrary conditions a reversible reaction generally favors one direction over the other. Therefore, the rate of the forward reaction, $A + B \rightarrow C + D$, often differs initially from that of the reverse reaction, $A + B \leftarrow C + D$. As the forward and reverse reactions proceed, the overall reaction reaches a state of **equilibrium**: The rates of the forward and reverse reactions become equal, and the concentrations of the substrates and products become constant. Enzymes are thought to reduce the *time* needed to reach equilibrium but not to change the *ratio* of substrates and products at

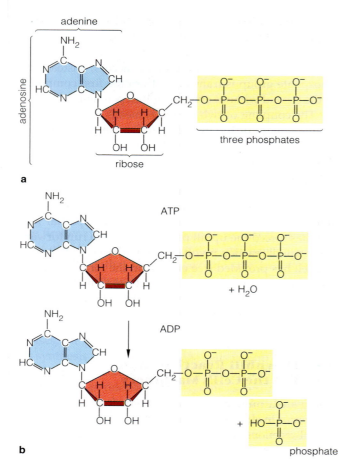

Figure 8.9 Adenosine triphosphate (ATP), the main energy carrier in cells. (**a**) The structural formula for ATP. (**b**) The hydrolysis of ATP, a strongly downhill reaction.

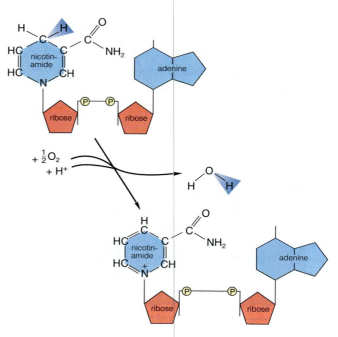

Figure 8.10 The oxidation of nicotinamide adenine dinucleotide (NADH) to NAD$^+$ by oxygen. Notice that the NADH loses one H and one chemical bond between the H and C, which represents two electrons.

SUMMARY

1. The amount of potential energy needed to force molecules to react together limits the rate of a chemical reaction. Chemical reactions can be speeded up by increasing the concentration of the substrates, increasing the temperature, or adding a catalyst.

2. Enzymes are catalysts made from protein. Most enzymes catalyze a single reaction from a specific substrate or set of substrates. In cells, the rate of a reaction is determined by the amount of enzyme and by the activity of the enzyme that is present.

3. Enzyme-catalyzed reactions can be linked together in metabolic pathways—in which the product of one reaction is the substrate for the next—to produce complex biochemicals. In a cell, the network of metabolic pathways is very complex, with a few key compounds being converted into many other compounds.

4. Feedback inhibition of the first enzyme in a pathway by the final product of the pathway enables the cell to control the concentration of that product.

5. The change in free energy of a reaction defines the direction in which it will proceed. The change in free energy is influenced by the initial concentrations of substrates and products in the reaction mixture, the relative stability of the bonds in the substrates and products, the number of independent substrates and products involved in the reaction, and the temperature. A spontaneous reaction always loses free energy.

group. (Our bodies cannot synthesize nicotinamide; we must obtain it in food, as the vitamin called *niacin*. Plant cells, however can synthesize niacin and use it to make NADH and NADPH.) NADH is the reduced form of the molecule; the oxidized form is denoted as NAD$^+$. The oxidation of NADH by oxygen gas (Fig. 8.10) is associated with the release of 52 kilocalories per mole of free energy! This is enough free energy to power the formation of several molecules of ATP; and, in fact, in the mitochondrion the oxidation of NADH is coupled to the formation of two to three molecules of ATP. The oxidation of NADPH (a parallel reaction to that of NADH) is used to synthesize fatty acids and some amino acids.

Of course, the energy locked in the bonds of ATP, NADH, and NADPH does not come without cost. The reactions that form these compounds are uphill reactions, and they occur only when they are coupled to other, downhill reactions. The metabolic pathways that provide the energy for the synthesis of these compounds are described in the next two chapters, on cellular respiration and photosynthesis.

6. If two partial reactions are coupled, the free energy change of the overall reaction is the sum of their separate free energy changes. A downhill reaction can thus force an otherwise uphill reaction to proceed.

7. Two reactions, the hydrolysis of ATP and the oxidation of NADH, are coupled to—and drive—many of the important functions of a cell.

Questions

1. Name the sources of the elements carbon, hydrogen, oxygen, nitrogen, sulfur, and phosphorus that plants use for growth. Why are enzymes needed to convert these sources to bio-organic compounds?

2. The diagram below shows the energy relations for two chemical reactions in which chemical compounds A and B are converted to C and D, and F and G are converted to H and I. Which reaction will proceed faster? Explain.

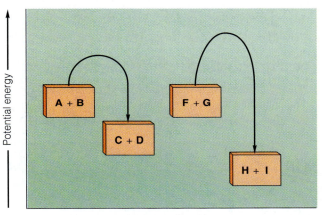

3. Would raising the temperature by 5°C (9°F) change the diagram in question 2? What would happen to the reactions?

4. Would adding an enzyme that catalyzes the first reaction (A + B → C + D) change the diagram? How?

5. Adding a weak acid to a solution containing an enzyme usually inactivates the enzyme, often irreversibly. Boiling the solution does the same thing. Suggest an explanation for these observations.

6. Give an example from your own experience (and other than the one in the text) of the action of the enzyme polyphenol oxidase. Do the same for the enzyme polygalacturonase.

7. Assume that three enzymes (E1, E2, and E3) form a metabolic pathway, like this:

$$A \xrightarrow{\text{E1}} B \xrightarrow{\text{E2}} C \xrightarrow{\text{E3}} D$$

What factors will influence the rate of formation of D?
   a. the amount of E1
   b. the amount of E2
   c. the amount of E3
   d. the temperature
   e. all of the above

8. How are catabolic reactions energetically coupled to anabolic reactions in a cell?

9. In the metabolic pathway described in question 7, are the reactions catalyzed by E1 and E2 coupled reactions? Explain your answer.

10. List five uphill reactions that occur in a living plant cell.

11. Maltose is formed from two glucose molecules bound together with a covalent bond. The enzyme maltase breaks the two glucoses apart by catalyzing the reaction

$$\text{maltose} + H_2O \rightarrow 2 \text{ glucose}$$

A solution of maltose alone produces glucose very slowly, but when maltase is added to the solution, glucose appears rapidly. Which of the following statements best explains this effect?
   a. The free energy of glucose is much less than that of maltose.
   b. The free energy of glucose is much more than that of maltose.
   c. Maltase makes the reaction irreversible.
   d. Maltase lowers the activation energy for the forward reaction.
   e. Maltase raises the activation energy for the reverse reaction.

12. No one ever detects the formation of maltose when maltase is added to a solution of glucose. Which of the statements listed as possible answers for question 10 best explains why?

13. In the presence of ATP, the enzyme luciferase (an enzyme found in the abdomens of fireflies) and its cofactor luciferin produce light. Light is energy. Energy cannot be created, so the light energy must have come from somewhere. From where?
   a. from the breakdown of the luciferase and luciferin
   b. from the hydrolysis of ATP
   c. from the heat energy in the water molecules

Further Readings

Atkins, P. 1984. *The Second Law.* New York: Freeman.

Doolittle, R. 1985. "Proteins." *Scientific American* 253 (4): 88–99.

Dressler, D., and H. Potter. 1991. *Discovering Enzymes.* New York: Freeman.

9 RESPIRATION

1. All living cells require energy to drive the reactions of life. Respiration releases energy trapped as potential energy in organic food molecules and transfers some of this energy to energy carrier molecules (ATP and NADPH). These energy carriers may then be moved in the cell and be coupled to energy-requiring reactions.

2. When a six-carbon sugar molecule is respired in a plant cell, it is broken down into pyruvate in a series of reactions called *glycolysis*.

3. In the presence of molecular oxygen, the pyruvate is broken down in an integrated series of enzymatic reactions called the *tricarboxylic acid cycle* and the *electron transport chain*. During these reactions CO_2 and water are formed; some energy is trapped in the energy carriers ATP and NADPH during their formation.

4. Enzymes of glycolysis and alcoholic fermentation are in the cell cytoplasm; enzymes involved in aerobic respiration are in the mitochondrial matrix or the inner membrane of the mitochondrial envelope.

5. The chemiosmotic theory of ATP formation proposes that the proton gradient that develops across the mitochondrial envelope during the terminal oxidation reactions drives ATP synthesis from ADP and inorganic phosphorus.

6. Some cells have developed alternate pathways of respiration. The pentose phosphate pathway transfers energy from glucose to $NADP^+$ (forming NADPH) and produces ribose 5-phosphate, necessary for the synthesis of nucleic acids. The glyoxylate pathway converts fat into intermediate compounds and ultimately into sucrose.

7. Environmental factors such as cell hydration, temperature, oxygen supply, and food availability affect the rate of respiration in a plant, as do factors such as cell type and the age and the species of a plant.

Figure 9.1 The transformation of starch to glucose, an example of digestion. Amylose, a component of starch, is a chain of sugar units (glucose) linked through oxygen. Enzymatic digestion of amylose by amylases (enzymes) involves the introduction of a water molecule between each pair of glucose units, breaking the oxygen linkage.

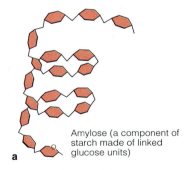

Amylose (a component of starch made of linked glucose units)

a

9.1 THE RELEASE OF ENERGY FROM FOOD

To stay alive, every living cell must obtain energy in a usable form, which it does by oxidizing food molecules. Cell membranes cannot maintain differential permeability, roots will not accumulate solutes, and cytoplasm will not flow without energy from the oxidation of food. Cells cannot synthesize amino acids, proteins, or fats without the energy and intermediate carbon compounds produced by respiration. In this chapter we will examine respiration in detail to see why it is so fundamental to life.

Early biologists applied the term *respiration* to the exchange of gases between an organism and its environment. Even today, many people consider respiration and breathing in animals as synonymous. Breathing is merely the visible indication that chemical reactions are taking place in animal cells. We define **respiration** as the oxidation of organic molecules within cells, accompanied by the release of usable energy.

Digestion Converts Complex Food into Simpler Molecules

Before large, insoluble food molecules can be respired, they must be broken down into smaller soluble components. Cells cannot oxidize even a simple soluble food molecule such as sucrose until it is broken into simpler sugars. **Digestion**, the breaking apart of complex foods (carbohydrates, fats, and proteins) into simpler compounds (sugars, fatty acids, and amino acids), occurs easily in the presence of water and specific enzymes. Large amounts of energy are not released because the transfer of electrons via oxidation–reduction reactions is not involved. Cells use water molecules in digestion, so the process is a hydrolysis reaction. Green plants differ from animals in that they normally do not ingest complex foods but synthesize them. Plants do digest foods, however, although unlike animals they have no special organs for digestion. Digestion occurs in any cell that stores complex food, even temporarily. During photosynthesis, starch often is stored as insoluble starch granules in the chloroplasts. When the time comes for this temporary starch reserve to be used by the cells or to be transported out of the cells, it must be changed back to soluble sugar. This chemical transformation of the insoluble starch into soluble glucose is one example of digestion (Fig. 9.1).

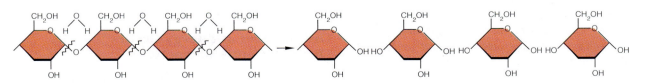

b

starch + water → glucose

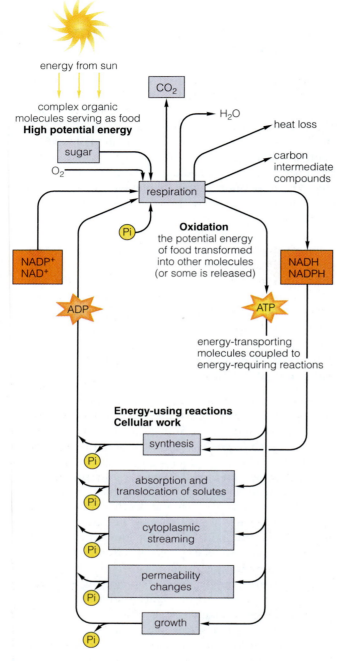

Figure 9.2 An overview of metabolism. Energy from sunlight is trapped in food during photosynthesis. Energy is released from food during respiration and is used to perform work in the cell.

Respiration Is an Oxidation–Reduction Process

Once the large food molecules have been digested, respiration results in the transfer of energy from food molecules to energy-carrier molecules in the cell. Respiration is an oxidation–reduction process because it involves the removal of electrons from electron donor molecules to electron acceptor molecules. Nevertheless, many steps do *not* involve the use of molecular oxygen. Indeed, many organisms, such as yeast and some bacteria, obtain their

energy from food without any reactions that involve molecular oxygen. Respiration that does not involve molecular oxygen is called **anaerobic respiration**. In the more common type of respiration, **aerobic respiration**, molecular oxygen plays a major role. Because of this difference in molecular oxygen usage, many biochemists restrict the definition of *respiration* to designate the final series of reactions that occur in aerobic respiration. However, we will use the term *respiration* in a broader sense to include all of the reactions involved in the oxidation of food.

The cellular organelles most involved in aerobic respiration are the mitochondria. Mitochondria produce most of the energy and the intermediate carbon compounds used as molecular building blocks in the cell. Environmental factors also play a role, for they can alter the rate and effectiveness of respiration.

Respiration Is an Integrated Series of Reactions

Respiration is more than a simple oxidation reaction; it is a series of chemical reactions that supplies the energy for most cellular processes (Fig. 9. 2). Although the reactions of respiration might seem overwhelming if we listed them all, they fall into patterns. In particular, some consume energy as food molecules are prepared for breakdown, whereas others result in the transfer of energy to molecules that act as energy carriers in the cell. It is more important for you to have an understanding of the process of respiration as a whole—and its roles in the life of a plant—than to worry about mastering each step in detail.

When we consider the total energy cycle in the living plant, we find that stored chemical energy is moved from one part of the plant to another in the form of potential energy in complex food molecules. At the cellular level, some of the potential energy in the food is trapped during respiration in the energy carriers ATP and NADH or NADPH, which may move throughout the cell.

Respiration also performs another important task in the life of the cell. During the many steps in the respiratory breakdown of food, essential intermediate compounds are formed. Living cells use many of these intermediate compounds as carbon building blocks, from which they synthesize compounds such as proteins, fats, nucleic acids, and hormones. Thus, respiration, through the oxidation of food, (1) supplies energy in a form available to do work in the cell and (2) produces intermediate carbon compounds essential to the continued growth and metabolism of cells.

The Transfer of Energy Occurs Through Coupled Reactions

To use the energy stored in food molecules, cells couple energy-yielding reactions with energy-consuming reactions. Such mechanisms transfer energy from food to the

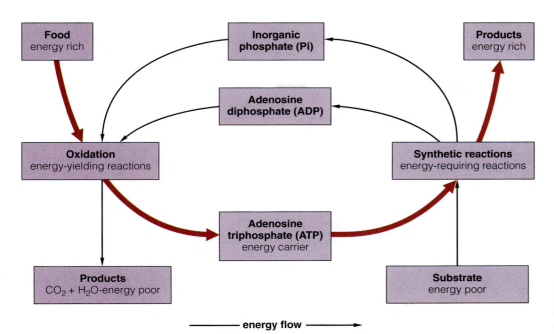

Figure 9.3 The coupling of energy-yielding reactions to energy-requiring reactions through ATP production and utilization.

energy carriers ATP and NADH or NADPH (Fig. 9.3). In some reactions, NAD$^+$ is reduced, and in others, NADP$^+$ is reduced. Whether NAD$^+$ or NADP$^+$ is reduced by accepting electrons and hydrogen depends on the enzymes catalyzing the particular reaction. When one **mole**—the molecular weight of a compound in grams (180 g in this case, about 7.7 oz)—of glucose is burned, CO_2, H_2O, and 686 kilocalories (kcal) of energy are released. This amount of energy, if released at once as heat, would destroy living cells. So instead, energy transformations in the cells occur slowly and in small steps through a sequence of many reactions. Specific enzymes catalyze each of these small reactions, allowing them to proceed at moderate rates at low temperatures. About 40% of the energy released is trapped in the bond energy of the ATP and NADPH formed. Because these molecules are in solution in the cell, they may diffuse or be carried by cytoplasmic streaming throughout a cell. Both ATP and NADPH eventually yield energy that performs work in the cell.

Because the total amounts of ADP and NADP$^+$ in a cell are quite small, a cell can never accumulate large amounts of ATP or NADPH. Consequently, cellular energy is stored as carbohydrates, fats, and protein. A cell at rest and not using energy rapidly has converted most of its ADP to ATP. At cell rest, the ATP is not used to drive work reactions; consequently, ATP is not hydrolyzed to ADP and inorganic phosphate (Pi). Some of the reactions of respiration require ADP, and because the concentration of ADP in a resting cell is very low, respiration is very slow. If such a cell is stimulated to do work—for example, to increase synthetic reactions, the rate of cytoplasmic streaming, or the amount of salt uptake—the rate of respiration will increase. When work is being done,

energy liberated from the hydrolysis of ATP to ADP and Pi helps drive these work reactions. The ADP and Pi liberated become available to the respiratory pathways, where they are synthesized again into ATP, trapping more energy and increasing the rate of respiration. This is one example of many control mechanisms that regulate the rates of various metabolic reactions in the cell.

9.2 THE REACTIONS OF RESPIRATION

If you look at the oxidation of a simple six-carbon sugar in the presence of molecular oxygen, only CO_2, H_2O, and heat (energy) are the final products. This overall process is written

$$C_6H_{12}O_6 + 6O_2 \rightarrow 6CO_2 + 6H_2O + 686 \text{ kcal}$$
$$\text{sugar} + \text{oxygen} \rightarrow \begin{array}{c}\text{carbon}\\\text{dioxide}\end{array} + \text{water} + \text{energy}$$

The equation tells us nothing about the way the reaction occurs, about the intermediate steps, or about other possible products.

It took many years for scientists to discover the many steps involved in respiration. To study respiration we separate it into phases, each of which includes reactions that produce key intermediate compounds. In the first phase, **glycolysis**, for every molecule of glucose (a six-carbon sugar) that is broken down and oxidized, two molecules of pyruvate (a three-carbon compound) are produced. No molecular oxygen takes part in the reactions, and the energy change is small. Most of the energy in the chemical bonds of the sugar being broken down still remains in the pyruvate formed. The pyruvate is then oxidized in either an aerobic or anaerobic pathway. In anaerobic respiration, the pyruvate is only partially oxidized. Alcohol and CO_2 are formed, and only a small amount of usable energy is released. In aerobic respiration, the pyruvate moves into the mitochondria. There, in the presence of molecular oxygen, it is oxidized to carbon dioxide and water. A large amount of energy is trapped in ATP and NADPH.

Glycolysis Is the First Phase of Respiration

There are three major steps in the process of glycolysis (Fig. 9.4).

1. *phosphorylation*: preparation of the six-carbon sugar for reaction by the addition of phosphate from ATP

2. *sugar cleavage*: splitting of the sugar phosphate into two three-carbon-atom fragments

3. *pyruvate formation*: oxidation of the fragments to form pyruvate, with the formation of H_2O and ATP

The six-carbon sugar, the most common sugar respired in the cell, is glucose. Although glucose is energy rich, it is stable and does not react readily with oxygen at life-sustaining temperatures; nor is it readily broken down into intermediate products. Before a cell can break glucose down and release its stored energy, glucose must react enzymatically with ATP, which donates a phosphate group and energy. This reaction occurs at two points in glycolysis, so that two ATP molecules provide energy and phosphate in the phosphorylation of the sugar molecule. (The ATP used in these reactions is replaced during later steps in respiration.) The phosphorylated sugar also undergoes a series of internal rearrangements that results in the formation of another six-carbon sugar phosphate, fructose 1,6-bisphosphate, which is split in the next reactions.

Another enzyme catalyzes the splitting of fructose 1,6-bisphosphate into two different three-carbon sugars, dihydroxyacetone phosphate and glyceraldehyde 3-phosphate. These three-carbon sugar phosphates are in equilibrium and may be converted enzymatically into one another. Glyceraldehyde 3-phosphate is oxidized through phosphoenolpyruvate to pyruvate. As it is broken down, the equilibrium shifts so that more dihydroxyacetone phosphate is converted into glyceraldehyde 3-phosphate. Thus, both three-carbon sugar phosphates are actually available for pyruvate formation.

Several enzymatic steps form pyruvate from glyceraldehyde 3-phosphate. Although molecular oxygen does not take part, the three-carbon sugar phosphate is oxidized by the transfer of two of its electrons and hydrogen to a hydrogen acceptor, NAD^+. This effectively transfers some of the energy originally in the chemical bonds of the three-carbon sugar to the newly formed NADH. Also, during the oxidation of each three-carbon sugar phos-

phate molecule to pyruvate, energy is used to form two molecules of ATP from two molecules of ADP plus two phosphates. The oxidation of the two three-carbon sugars forms a total of four ATP and two NADH molecules. However, because two molecules of ATP are used to phosphorylate a glucose molecule in preparation for its breakdown, glycolysis results in a net yield of two ATP and two NADH molecules. Glycolysis benefits the cell because usable intermediate compounds are formed. Also, although only a small amount (approximately 20%) of the energy of the glucose molecule is trapped in ATP and NADH, only about 3% is lost as heat.

The fate of the pyruvate that results from glycolysis depends upon the availability of molecular oxygen. If molecular oxygen is absent, anaerobic respiration (fermentation) may occur in most plant cells, forming alcohol and CO_2. If molecular oxygen is present, aerobic respiration may occur; the pyruvate then diffuses into the mitochondria and passes through two more phases of respiration: (1) the **tricarboxylic acid cycle**, and (2) the **terminal electron transport chain**. (Sometimes, the tricarboxylic acid cycle is called the **Krebs cycle** after the Nobel laureate physiologist Hans Krebs, whose research contributed a great deal to our knowledge of respiration.

Anaerobic Respiration Transforms Only a Small Amount of Energy

Normally, higher plants cannot live long in the absence of molecular oxygen. Under anaerobic conditions, (1) food is not oxidized completely enough by their cells to yield adequate energy for their life processes, (2) poisonous products may be produced, and (3) some necessary intermediate compounds may not be synthesized. In contrast, some fruits, notably apples (*Malus* sp.), may exist for long periods in an atmosphere containing very small amounts of oxygen and yet continue to give off carbon dioxide. Yeast may also live actively in an atmosphere with very small amounts of oxygen and still produce relatively large amounts of carbon dioxide and alcohol from the pyruvate formed during glycolysis. This type of anaerobic respiration is **alcoholic fermentation**. Yeast has a limited tolerance for alcohol, however. When the alcohol concentration of the medium in which yeast is living reaches about 12%, the yeast cells are killed. Consequently, wines and other naturally fermented alcoholic beverages do not have an alcohol content above about 12%. Yeast grows much more vigorously under aerobic conditions than in the absence of oxygen. Indeed, relatively few organisms grow only under strictly anaerobic conditions.

Some fungi, bacteria, and many animal cells may live and grow slowly under anaerobic conditions, in which they can produce many products in addition to alcohol. When oxygen is not supplied rapidly enough to vigorously exercising muscles in your body, for example, the

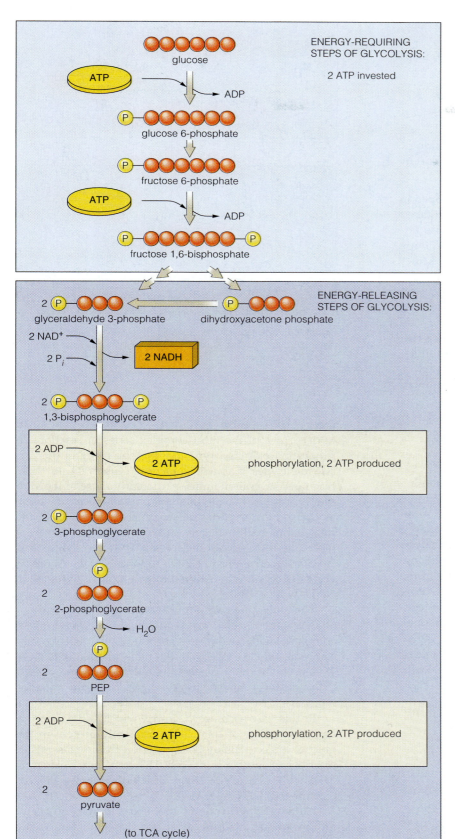

ENERGY-REQUIRING
STEPS OF GLYCOLYSIS:

2 ATP invested

glucose

ATP → ADP

glucose 6-phosphate

fructose 6-phosphate

ATP → ADP

fructose 1,6-bisphosphate

ENERGY-RELEASING
STEPS OF GLYCOLYSIS:

2 glyceraldehyde 3-phosphate   dihydroxyacetone phosphate

2 NAD⁺
2 Pᵢ → 2 NADH

2 1,3-bisphosphoglycerate

2 ADP → 2 ATP   phosphorylation, 2 ATP produced

2 3-phosphoglycerate

2 2-phosphoglycerate

→ H₂O

2 PEP

2 ADP → 2 ATP   phosphorylation, 2 ATP produced

2 pyruvate

(to TCA cycle)

Net energy yield:
2 ATP
2 NADH

Figure 9.4 Three major steps in the
process of glycolysis.

muscle cells may produce not alcohol but lactic acid, which may cause cramping.

Alcoholic fermentation takes place in two steps (Fig. 9.5):

1. One CO_2 molecule is enzymatically split off from each pyruvate. This leaves a two-carbon compound, acetaldehyde.

2. NADH, formed during glycolysis, reduces acetaldehyde to alcohol. NAD^+ is thereby released to take part in glycolysis again.

Thus, during anaerobic respiration the energy trapped in NADH is used to form alcohol (Fig. 9.5). It is significant in the life of the cell that for each mole of glucose fermented to alcohol, a net of only two moles of ATP are available to do work in the cell. This is about 14 kcal—less than 3% of the energy available in one mole of glucose. About 16% of the energy is lost as heat, and almost 84% is still locked in the two molecules of alcohol formed and is thus unavailable to the plant. In addition, the alcohol itself may be toxic to the plant. We can conclude that anaerobic respiration is not a very efficient way of utilizing the energy in food. Some plants that grow in wet, anaerobic soil avoid the toxicity of alcohol by accumulating malate, a four-carbon acid.

Aerobic Respiration Effectively Transforms the Energy in Food

When respiration occurs in an environment containing molecular oxygen, the plant cell oxidizes pyruvate more efficiently than it can under anaerobic conditions. Aerobic respiration releases adequate amounts of energy, which fuel the reactions required for life. Glycolysis releases about 20% of the energy contained in a glucose molecule, leaving most of the energy still locked in the bonds of the two pyruvate molecules formed. During aerobic oxidation of pyruvate to CO_2 and H_2O, by contrast, 40 to 50% of the energy in the pyruvate is trapped in a potentially useful form as ATP, NADH, and NADPH.

The reaction steps of aerobic respiration are divided into three parts according to the intermediate molecules and end products formed:

1. *Entry of carbon into the tricarboxylic acid (TCA) cycle of respiration.* Pyruvate releases CO_2, and the remaining two-carbon-atom fragments enter the TCA cycle—so named because several three-carboxyl (COOH) acids play prominent roles.

2. *The TCA cycle.* The two-carbon-atom fragments release CO_2 and become oxidized. ATP and NADH are formed.

3. *Electron transport and terminal oxidation.* Oxygen accepts electrons and H^+ from the reduced nucleotides (NADH, NADPH, and $FADH_2$) that function as coenzymes. ATP and water are formed (Fig. 9.6).

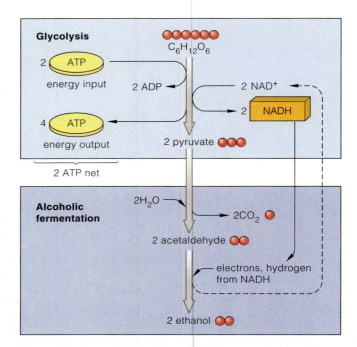

Figure 9.5 The reactions of alcoholic fermentation (anaerobic respiration). In the absence of molecular oxygen, pyruvate from glycolysis is converted to carbon dioxide and alcohol (ethanol).

ENTRY OF CARBON INTO THE TRICARBOXYLIC ACID CYCLE One of the most complex series of reactions in respiration oxidizes pyruvate, thereby forming the carbon molecules that will enter the TCA cycle. In these reactions, several vitamins, particularly those of the vitamin B complex (niacin, thiamin, pantothenic acid), serve as coenzymes or parts of coenzymes (page 123). In the final stages, each molecule of pyruvate oxidized forms a molecule of CO_2, with the transfer of two electrons and two hydrogen atoms to NAD^+ to form NADH. About 53 kcal of energy are transferred from a mole of pyruvate to NAD^+ when NADH is formed.

The remaining part of pyruvate is now a two-carbon-atom fragment called an *acetyl group*. This acetyl group complexes with coenzyme A, producing the reactive substance acetyl-CoA (Fig. 9.7), the molecule that enters the TCA cycle. In the presence of a specific enzyme, acetyl-CoA transfers (donates) its acetyl group to other acceptor molecules.

In this way each acetyl group, which still contains a high percentage of the energy originally present in the glucose molecule, may be transferred from one series of reactions to another in the cell. This transfer is analogous to the transfer of electrons and hydrogen by the coenzymes NADH and NADPH.

Although pyruvate is the usual donor of acetyl groups to coenzyme A, it is not the only donor. The breakdown of fat and protein (amino acids) forms acetyl groups that also may be donated to coenzyme A and enter the TCA cycle.

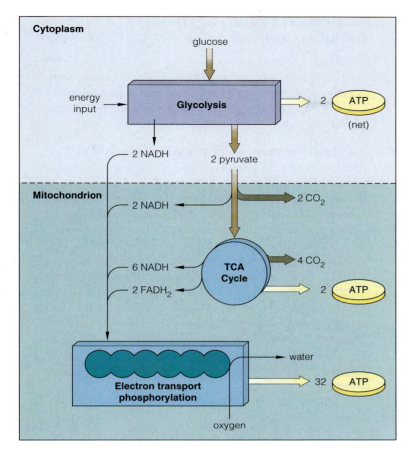

Figure 9.6 Overview of the complete oxidation of glucose during aerobic respiration.

THE TRICARBOXYLIC ACID CYCLE OF RESPIRATION In the tricarboxylic acid cycle, acetyl-CoA donates its acetyl group to an organic acid acceptor molecule, oxaloacetate, found in mitochondria. The union of oxaloacetate and the acetyl group produces citrate, which is gradually broken down in a cyclic series of reactions in which seven other organic acids (including, ultimately, a new molecule of the acetyl acceptor, oxaloacetate) are formed (Fig. 9.7). During these reactions, pairs of electrons (and hydrogen atoms are transferred to the electron carriers NAD^+, $NADP^+$, and FAD (flavin adenine dinucleotide). All of the atoms brought in with the acetyl group are gradually removed, and the carbon and oxygen are released as molecules of CO_2. Because the reactions form new oxaloacetate molecules, capable of accepting other acetyl groups, the cycle begins again.

In one turn of the cycle, some of the potential energy in one acetyl group is lost as heat to the environment, but about 66% is trapped in one molecule of ATP, three molecules of NADH, and one molecule of $FADH_2$. The ATP is free to carry its energy to other parts of the cell where work is being done. The energy in the NADH and the

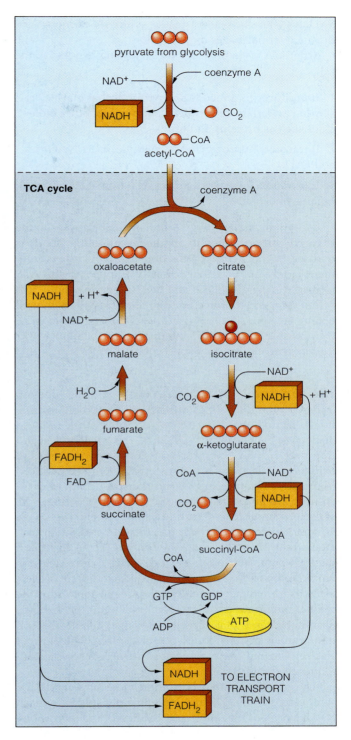

Figure 9.7 Several steps in the TCA cycle leading to the production of reduced nucleotides.

FADH$_2$ generally is used in the synthesis of more ATP during the final stages of respiration.

The TCA cycle provides the following benefits to the cell:

1. It produces intermediate compounds that may act as starting materials for synthetic pathways of other molecules such as proteins and lipids.

2. It transfers hydrogen to the energy carriers NAD$^+$ and FAD.

3. It produces the energy carrier ATP from ADP and inorganic phosphate.

Although all the energy originally associated with the reduced carbon atoms in the sugar molecule (food) is released or transferred by the time the TCA cycle is completed, only about 10% is directly trapped in the ATP formed during glycolysis and the TCA cycle. About 56% is first trapped in reduced nucleotides. How does the cell obtain the energy trapped in these reduced nucleotides (NADH, NADPH, FADH$_2$) in order to do cellular work? It utilizes an electron transport chain.

THE ELECTRON TRANSPORT CHAIN During electron transport, the final series of reactions in aerobic respiration, cells do not obtain energy directly from the reduced nucleotides. Instead they oxidize the nucleotides and trap part of their energy in ATP. In a stepwise series of oxidations, electrons are transferred from the reduced nucleotides through an electron transport chain of carriers that are alternately reduced and oxidized (Fig. 9.7). These electron carriers include a flavoprotein, coenzyme Q, and several cytochromes. **Cytochromes** contain iron atoms that are reduced to the ferrous (Fe^{2+}) form and then give up their electrons, becoming oxidized to the ferric (Fe^{3+}) form. The last step transfers a pair of electrons to an oxygen atom. Two hydrogen ions from the cellular environment then combine with the oxygen, forming water. If free oxygen were not available, this last transfer of electrons could not take place. The flow of electrons would cease, and the entire aerobic respiratory sequence would cease.

During electron transport, three molecules of ATP are synthesized from ADP and inorganic phosphate for each NADH or NADPH molecule oxidized, and two molecules of ATP are synthesized for each FADH$_2$ oxidized. This process—which couples oxidation to phosphorylation—is called oxidative phosphorylation.

Complete aerobic respiration of a glucose molecule thus involves a stepwise breakdown of the sugar molecule. It uses 6 molecules of oxygen and forms 6 CO$_2$ molecules and 6 H$_2$O molecules. The process transfers about 40% of the potential energy in the glucose to some 36 molecules of ATP, which can do work in the cell (see Fig. 9.6). About 60% of the energy is lost as heat during the various steps of respiration.

Some Plant Cells Have Alternate Pathways of Respiration

Although aerobic respiration is the most common method plant cells use to oxidize foods, alternate pathways exist that also provide energy and organic building blocks. Two common alternate pathways are the pentose phosphate pathway and the glyoxylate cycle.

THE PENTOSE PHOSPHATE PATHWAY The *pentose phosphate pathway* (PPP) occurs in the cytoplasm of some tissues. Biochemically, the PPP starts in the same way as glycolysis—that is, with the phosphorylation of a glucose molecule to produce glucose 6-phosphate (see Fig. 9.4). But instead of producing fructose 6-phosphate, the next step in the PPP is the oxidation of the glucose 6-phosphate to form two molecules of NADPH, one molecule of CO$_2$, and one molecule of ribulose 5-phosphate, from which ribose may be synthesized (Fig. 9.8). The rest of the cycle consists of a series of sugar transformations, and glucose may be regenerated in the process. Young, growing plant tissues appear to use the TCA cycle as the predominant pathway of glucose oxidation, whereas aerial parts of the plant and older tissues seem to utilize PPP as well.

The pentose phosphate pathway has two important consequences:

1. The pathway transfers energy in glucose to NADP, by forming NADPH, which is used as an energy source in synthetic reactions.

2. The pathway produces the five-carbon sugar phosphate ribose 5-phosphate, which is necessary for the synthesis of some essential compounds such as nucleic acids.

THE GLYOXYLATE CYCLE When fat-storing seeds germinate, the developing embryo requires energy and carbon compounds for growth. Because fats and oils do not dissolve in water, however, they are not transported from storage cells in the seed to cells in the growing seedling. Instead, the fats must be broken down to their constituent long-chain fatty acids and glycerol, which are then further metabolized. Although glycerol may be converted into dihydroxyacetone phosphate, thus entering the glycolytic sequence, it yields only a relatively small amount of energy or carbon units. Fatty acids, however, yield large amounts of acetyl-CoA when they are oxidized.

The disappearance of reserve fat and oil in fat-storing seeds during germination usually is accompanied by a significant increase in the sucrose content of the seeds, including the embryo. An important part of this process involves a series of reactions called the *glyoxylate cycle*. The glyoxylate cycle is an important cellular mechanism because it helps convert the immobile products of fat breakdown into soluble sugar. These water-soluble sugar molecules can then be transported to sites in the plant requiring soluble food.

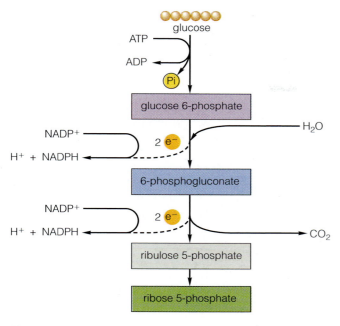

Figure 9.8 The first steps in the pentose phosphate pathway.

RESPIRATION AND CELL STRUCTURE

One cannot understand the entire process of respiration without a knowledge of the role that cell structure plays in it. Biochemists and plant physiologists isolate and purify the various cell organelles and membranes; by studying them, they can determine how cellular structures function in cell metabolism. Some enzymes are confined, during the life of the cell, to very specific locations in the cell. Two experiments show this to be true for some enzymes catalyzing respiration reactions. The first experiment demonstrates that when plant tissue is ground until all cells and organelles are broken apart—but the enzymes are preserved—glycolysis and alcohol fermentation will occur in the cell-free preparation; but complete aerobic respiration will not. This means that the enzymes of glycolysis and fermentation are soluble enzymes that do not depend upon intact organelles or membranes for their action.

The second experiment shows that if mitochondria are isolated from plant tissue and if pyruvate, ADP, Mg^{2+}, and inorganic phosphate are added, the pyruvate will break down, and the entire TCA cycle and electron transport will occur. ATP will be formed, molecular oxygen will be absorbed, and H_2O and CO_2 will be produced. The results of this experiment, in light of the first, mean that at least some steps in aerobic respiration require intact mitochondria.

In fact, biochemists have gone on to show that all of the enzymes of the TCA cycle but two are present in the mitochondrial matrix—and those two are located in the inner mitochondrial membrane. The electron transport chain enzymes and the ATP-synthesizing enzymes are also part of the inner mitochondrial membrane (Fig. 9.9). From such findings, researchers have been able to determine where in a cell specific reactions occur (Table 9.1).

Experimental findings indicate that a relationship exists between mitochondrial structure and respiration and that the membrane structure in mitochondria is important for ATP synthesis and electron transport. A mitochondrion, you'll recall from Chapter 3, consists of a matrix surrounded by a double membrane envelope made up of an outer and an inner membrane. These membranes separate the mitochondrion from the rest of the cytoplasm (Fig 9.9). The outer membrane is permeable to water and to small neutral molecules (such as urea) but not to hydrogen ions or simple sugars.

Reactions in the Inner Mitochondrial Membrane Include ATP Synthesis and Electron Transport

The final stages of respiration—electron transport and oxidative phosphorylation—occur in the inner membrane. The electron carriers NAD^+ and FAD, which were reduced to NADH and $FADH_2$ during the oxidation of organic acids in the TCA cycle in the matrix, come in contact with the inner membrane, where they become oxidized (lose electrons) to other electron carriers by being oxidized to NAD^+ and FAD.

The electron carriers making up the electron transport chain are present in aggregated clusters in the inner membrane. Each cluster appears to consist of a fixed number of molecules of each carrier. This close organization of electron carriers facilitates the rapid exchange of electrons from a reduced carrier to the next carrier in the chain as soon as it is oxidized. Electrons flow from high energy levels in the NADH or $FADH_2$ along the carrier chain down an energy gradient and are finally transferred to oxygen. Hydrogen ions are taken up from the surrounding matrix, and water is formed. Each reaction step releases energy: Some is lost as heat, but some is used to drive H^+ (protons) out of the mitochondrial matrix, through the inner mitochondrial membrane, and into the intermembrane space where H^+ accumulates. Because the inner membrane prevents the free diffusion of H^+ back into the matrix, an energy gradient is established across the membrane. This gradient has two components:

1. an osmotic component, resulting from the difference in H^+ concentration across the membrane

2. an electrical component, resulting from the accumulation of positive charges (H^+) between the inner and outer mitochondrial membrane or within the cytoplasm

This proton gradient is believed to be involved in ATP synthesis.

Figure 9.9 (**a**) A plant mitochondrion, showing the cristae, the double nature of the mitochondrial envelope, and the sites of reactions in aerobic respiration. (**b**) Section through cristae. (**c**) Detail of membranes, showing location of electron transport system and ATP synthesis.

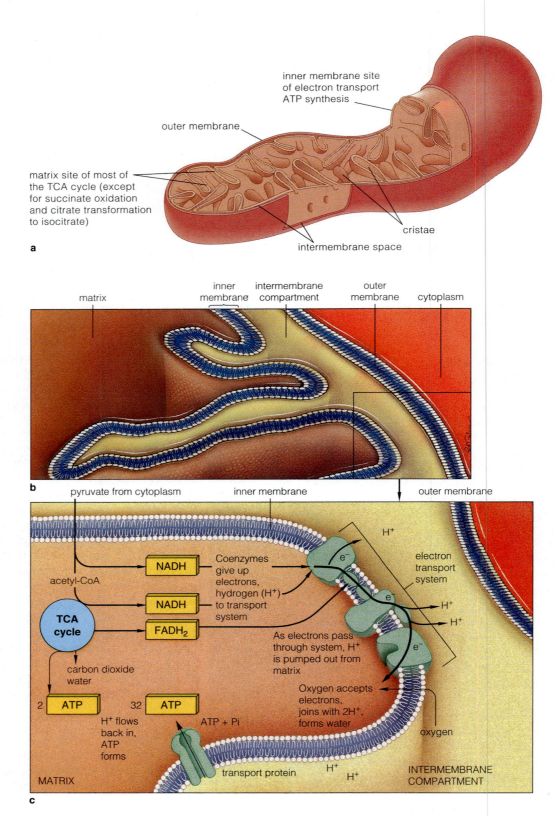

inner membrane site of electron transport ATP synthesis

outer membrane

matrix site of most of the TCA cycle (except for succinate oxidation and citrate transformation to isocitrate)

cristae

intermembrane space

a

matrix

inner membrane

intermembrane compartment

outer membrane

cytoplasm

b

pyruvate from cytoplasm

inner membrane

outer membrane

H^+

e^-

electron transport system

acetyl-CoA

NADH

Coenzymes give up electrons, hydrogen (H^+) to transport system

e

H^+

TCA cycle

NADH

FADH$_2$

H^+

As electrons pass through system, H^+ is pumped out from matrix

e^-

carbon dioxide water

Oxygen accepts electrons, joins with $2H^+$, forms water

2 ATP

32 ATP

oxygen

H^+ flows back in, ATP forms

ATP + Pi

MATRIX

transport protein

H^+
H^+

INTERMEMBRANE COMPARTMENT

c

Table 9.1 Intracellular Location of Some Respiratory Reactions

Reactions Catalyzed	Location of Reactions
Glycolysis	Cell cytoplasm
Alcoholic fermentaion	Cell cytoplasm
Tricarboxylic acid cycle (TCA)	Mitochondrial matrix (most reactions)
Citrate to isocitrate and succinic oxidation	Inner mitochondrial membrane
Electron transport	Inner mitochondrial membrane
ATP synthesis	Inner mitochondrial membrane

The Chemiosmotic Theory Explains the Synthesis of ATP in Mitochondria

In 1961, Peter Mitchell, a British biochemist, hypothesized that ATP synthesis is driven by the proton gradient that develops across the inner mitochondrial membrane during electron transport. Although the results of many experiments on oxidative phosphorylation are best explained by this theory, the details of the process are still not well understood. However, the **chemiosmotic theory** (so named because the action is both chemical and osmotic) is now generally accepted as the best explanation of the mechanism of both ATP synthesis in mitochondria and of ATP synthesis in chloroplasts during photosynthesis. Energy harvest from the proton gradient across the inner mitochondrial membrane is brought about by a complex of enzymes known as **ATP synthetase**, which is embedded in and extends across the inner mitochondrial membrane. It is believed that special channels in the ATP synthetase allow protons to move down the chemiosmotic gradient into the matrix, discharging the gradient (it is no longer present). Some of the energy released by this flow of protons is transferred to ATP when ADP and inorganic phosphate are combined during ATP synthesis. Exactly how this is accomplished is not yet clear.

9.4 THE EFFECTS OF ENVIRONMENTAL FACTORS ON RESPIRATION

Conditions inside and outside the cell affect the rate of respiration—as well as the rates of other cellular activities such as photosynthesis, water and mineral salt absorption, cell division, and growth. The rate of respiration usually is expressed as the quantity of CO_2 released or O_2 absorbed per unit of cell weight per unit of time. Many factors that affect the respiration rate act indirectly. For example, soil flooding lowers the availability of oxygen to respiring cells. Shading decreases photosynthesis and so reduces the amount of food available to respiring cells. The respiration rate also depends upon cell type and age. Cells of different ages or from different kinds of plants may respond differently to the same environmental factor. For example, at low oxygen concentrations, CO_2 is released at a much higher rate from cells in rice seedlings than from cells in wheat seedlings. This is undoubtedly because rice (*Oryza sativa*) seedlings have a greater capacity for anaerobic respiration than wheat (*Triticum aestivum* and *T. turgidum Durum group*) seedlings; rice seedlings are adapted to grow under water, where O_2 concentration is low.

In addition to its theoretical value, a knowledge of respiration and methods of its control is of great practical value (see sidebar, "Control of Respiration After Fruit Harvest," p. 142).

A few examples of the effects of environmental factors, age, and species of plant on respiration follow.

Cell Hydration May Indirectly Affect the Rate of Respiration

The water content of active cytoplasm may be as high as 90%, and small fluctuations in water content have little effect on the rate of respiration. However, in some plants starch is hydrolyzed to sugar during water stress—for example, when a plant wilts. In these plants, the rate of respiration may increase during wilting as a result of an increase in the amount of respirable food.

One of the most dramatic effects of the influence of cell water content on respiration occurs when seeds mature and begin to dry. When the water content of seeds decreases to between 16 and 17%, there is a sharp drop in the rate of CO_2 released from their cells. The water content of mature seeds frequently becomes less than 10% of the seed weight. Growth ceases, mineral salts are not absorbed, and cell division stops—as do many if not all reactions that synthesize new molecules. Respiration goes on at a very slow rate, so cells consume little oxygen, give off little carbon dioxide, and release only a very little heat. Energy is needed only to maintain a little-understood steady state in the quiescent cytoplasm. In this dormant state, seeds may be kept in large storage bins such as grain elevators for long periods.

Control of Respiration After Fruit Harvest

An understanding of the process of respiration—and of means to control and modify it—is critically important to the producers, handlers, and shippers of fruits and vegetables. High rates of respiration are associated with a complex series of biochemical changes during the ripening and maturation of fruit. If the rate of respiration is reduced, the rate of fruit maturation and subsequent senescence is frequently delayed—and storage life is extended.

One of the most effective methods for reducing the rate of respiration is to cool the produce. Most long-range shipping of fruit is done in refrigerated railcars and trucks. Because all living cells respire—and because respiration produces heat—the amount of refrigeration required to ship and store fruit is considerable. For example, 40,000 boxes of apples at 0°C (32°F) produce enough heat to melt 2.8 tons of ice in 24 hours; at 4°C (39°F), the respiratory heat loss would melt 4.8 tons of ice; and at 16°C (61°F), 19 tons of ice. Control of this heat production represents a considerable cost in marketing the product.

Fruits and vegetables vary greatly in their susceptibility to low temperature and the optimum conditions they require for successful storage. Bananas, tomatoes, and summer squash (all of tropical origin) suffer chilling injury when stored at temperatures below 10°C to 13°C (50°F to 55°F) but above freezing. (Chilling injury, which frequently shows up when the produce is taken out of cold storage, can include internal browning, scalding, and pitting.) Cold-weather fruits and vegetables such as pears, some apples, and onions, by contrast, tolerate storage at near-freezing temperatures for extended periods. The shipper's objective is to store the produce at the lowest economical temperature that still maintains the quality of the particular fruit or vegetable being shipped.

Grocery stores have tasty, crisp apples from Washington and Oregon year-round, even though the apple harvest lasts only a few weeks. If the apples are kept at room temperature after harvest, their rate of respiration rapidly increases, and production of ethylene (a gaseous plant hormone, see page 245) increases. Ethylene stimulates the aging process and decreases fruit quality. Within a few weeks, the apples become soft, mealy, and undesirable.

Cold storage will significantly prolong the storage life of the apples. Also, because oxygen is necessary for aerobic respiration, if the oxygen level in the air of the storage room is reduced from about 21% to 1–3%, storage life is increased (particularly if the CO_2 level is increased from the usual 0.03% to 1–3% at the same time). Reducing the oxygen level to zero would not be desirable because this stimulates anaerobic respiration and may result in a more rapid deterioration of the apples. A large fraction of Washington and Oregon apples are stored under refrigeration in rooms with modified atmospheres. These apples can be stored for as long as nine months and still be acceptable for eating.

In addition to the other changes that accompany respiration, there is a loss of the sugar respired. If large amounts of living material are involved, a considerable weight of sugar may be lost. As an example of this weight loss let us look at sugar beets. At harvest, the tops of sugar beets are removed, and the roots are mechanically dug up and loaded into trucks for transport to the sugar processing plant. At the plant, beets are often stored in large piles on a cement apron. If the pile loses 1% of its weight in 20 days (a reasonable figure), a 100-ton pile of sugar beets would lose 1 ton. Of course, part of the weight loss is water, but a significant portion is sugar. No practical means of reducing the rate of respiration in harvested sugar beets exists, but the faster the beets are processed after harvest the less sugar loss occurs.

If even a little water is added to viable seeds, it is imbibed. The seeds then swell, and respiration increases rapidly. If the seeds have an adequate oxygen supply, growth begins, and the seeds germinate. However, if seeds are confined too closely (as grain in a grain elevator) too high a moisture content in them may cause a rapid rate of respiration. This can increase the temperature inside the mass and kill the grain.

Living Cells Are Very Sensitive to Fluctuations in Temperature

Temperature has a marked effect upon most biological reactions, especially those that are controlled by enzymes. In the temperature range from near freezing (0°C, 32°F) to about 30°C (86°F) the rate of respiration approximately doubles for every increase of 10°C (18°F) in temperature. At temperatures above 35 to 40°C (95 to 104°F), protein molecules (which make up enzymes) progressively unfold (become denatured), resulting in the loss of enzyme activity. The longer a cell is subjected to a high temperature, the greater the loss of enzyme activity. At first, an increase in temperature of cells to 35 or 40°C may cause the respiration rate to rise; but soon the rate falls, and eventually respiration ceases.

Nevertheless, over the long time periods needed for adaptation to occur, certain organisms have evolved characteristics enabling them to survive in otherwise hostile environments. For instance, some species of algae and bacteria are adapted to respire and grow under tem-

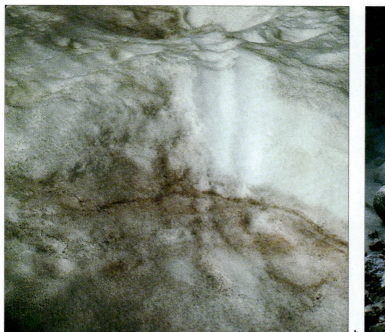

Figure 9.10   Adaptions to temperature extremes. (**a**) Cyanobacteria (formerly known as blue-green algae) growing in a snowbank. (**b**) Cyanobacteria growing in hot springs.

perature extremes that would kill most other organisms. Particularly impressive is the presence of some of these organisms in hot springs and streams, where temperatures may exceed 60°C (140°F). Others grow in snow, where the temperature remains near freezing (Fig. 9.10).

In some plants such as the potato (*Solanum tuberosum*), a decrease in temperature causes the hydrolysis of starch to sugar. Because cells cannot respire starch directly—but can respire sugar—a drop in temperature actually may increase the rate of respiration in potato cells (instead of decreasing it, as you might expect). It has been demonstrated that more sugar accumulates in potato tubers stored at 0°C (32°F) than in those stored at 4.5°C (40.1°F). At the same time, the respiration of potatoes stored at 0°C is markedly faster than it is in potatoes stored at 4.5°C, undoubtedly because of the higher sugar concentration. This accumulation of sugar in potatoes stored at or near freezing is of considerable importance to people who process potatoes because high sugar levels frequently cause undesirable browning in heat-processed potato products.

Factors That Affect Photosynthesis May Indirectly Affect Respiration

Photosynthesizing cells produce their own food in the light. Under normal conditions, they must also make enough food to supply the needs of all the other living cells in the plant. The living cells of a plant kept in the dark continue to use food, even though photosynthesis has stopped. The stored food reserves, particularly starch

and sugars, rapidly become depleted. If this condition continues, the plant will eventually starve to death. In fact, any factor that limits photosynthesis—such as light, temperature, and CO_2 level in the air—and thus reduces food availability must indirectly influence respiration.

Oxygen Gas Must Be Available for Aerobic Respiration

Cells carrying out aerobic respiration must have a continual supply of oxygen diffusing into them. They produce CO_2, which diffuses across the plasma membrane and into the surrounding environment. Most plant cells can continue for a time to oxidize foods even in the absence of gaseous oxygen by switching to anaerobic respiration. However, higher plants require molecular oxygen. The concentration of oxygen in the atmosphere rarely deviates enough from the normal 21% to appreciably affect the rate of respiration. However, underground stems, seeds, and roots may be in an oxygen-poor environment because microorganisms (as well as the plant parts themselves) may use the oxygen in the soil atmosphere faster than it is replaced from the air. Under these conditions, respiration in the cells of these underground organs may decrease. Similar conditions of low oxygen level and high carbon dioxide concentration may occur in the internal cells of bulky plant organs such as large, fleshy fruit. However, in general, the diffusion of O_2 through the intercellular spaces is rapid enough so that aerobic respiration occurs even within bulky plant parts.

SUMMARY

1. The two major functions of respiration are (a) the transformation of potential energy stored in food through the production of ATP and reduced nucleotides and (b) the production of intermediate products that are used in the synthetic reactions of the cell.

2. Energy-yielding oxidative reactions may be coupled to energy-requiring reactions.

3. Phosphorylation prepares sugar for oxidation. During phosphorylation an organic phosphorus donor, ATP, transfers phosphorus to the sugar molecule. Sugar cleavage then occurs, resulting in the production of two three-carbon sugar phosphate intermediates. These in turn are oxidized to pyruvate. This entire sequence of reactions is called *glycolysis*.

4. In the absence of molecular oxygen, pyruvate usually is metabolized to alcohol and carbon dioxide in the final stages of anaerobic respiration in plant cells. Only small amounts of energy are released in this fermentation process.

5. In the presence of molecular oxygen, pyruvate becomes further oxidized to carbon dioxide and water through the tricarboxylic acid (TCA) cycle. During aerobic respiration, large amounts of energy are stored in ATP and reduced nucleotides, which are used when work is done in the cell. Reduced nucleotides are oxidized in the electron transport chain, and ATP is produced.

6. Respiration is carried out through an integration of reactions going on in the cytoplasm (glycolysis) and mitochondria (TCA cycle and electron transport chain).

7. According to the chemiosmotic theory, the synthesis of ATP from ADP and inorganic phosphate during electron transport is driven by the proton gradient that develops across the inner mitochondrial membrane.

8. Some cells exhibit alternate pathways of respiration. One alternate pathway, the pentose phosphate pathway, transfers energy from glucose to NADP, forming NADPH and producing ribose 5-phosphate, necessary for the synthesis of nucleic acids. Another pathway, the glyoxylate cycle, which occurs in fat-storing seeds, con-verts fat into intermediate compounds and ultimately into sucrose, which may then be transported to growing cells.

9. Cell hydration, temperature, oxygen supply, food availability, and the type and age of a plant all affect the rate of respiration.

Questions

1. Define *digestion*.

2. What two major roles does respiration play in the life of a plant?

3. Why does respiration increase when a resting cell is stimulated to do work such as salt absorption or cytoplasmic streaming?

4. Name and describe the three major steps in glycolysis.

5. How would you determine whether a plant is carrying out aerobic or anaerobic respiration?

6. Why is anaerobic respiration not a very effective way of utilizing food?

7. Describe an experiment showing that some of the enzymes involved in aerobic respiration are soluble and that others are located in the mitochondrial membranes.

8. What is the chemiosmotic theory of ATP synthesis?

9. Explain how you would expect the rate of respiration in root cells to change when a plant that has been kept in the dark is moved into the light.

10. Construct a graph to show how you would expect a plant's rate of respiration to change if the temperature were increased from 10°C (50°F) to 60°C (140°F) in increments of 10°C (18°F).

Further Readings

Douce, R., and D. D. Day, eds. 1985. *Higher Plant Cell Respiration*. Vol. 18 of *Encyclopedia of Plant Physiology* (new series). Berlin: Springer-Verlag. A comprehensive treatment of plant respiration by specialists in the field.

Lehinger, A. L. 1982. Sections in Part II of *Principles of Biochemistry*. New York: Worth. An excellent and detailed treatment of the biochemistry of respiration.

Salisbury, F. B., and C. W. Ross. 1992. Sections in Chapter 13 of *Plant Physiology*. 4th ed. Belmont, Calif: Wadsworth. A useful review of plant respiration in a general plant physiology text.

10 PHOTOSYNTHESIS

1. Photosynthesis by green plants is the primary energy-storing process of almost all life, both plant and animal. The energy from sunlight is stored as chemical energy in organic compounds through a series of light-sensitive and temperature-sensitive reactions. Carbon dioxide and water are the raw materials, and the products are sugar and oxygen.

2. Chlorophyll in green plants absorbs light energy, which activates electrons in special chlorophyll *a* molecules. These electrons move along a chain of electron carriers, and some of their energy is stored in the production of ATP or NADPH. These reactions take place in association with thylakoid membranes.

3. Temperature-sensitive enzymatic reactions in the stroma use the ATP and NADPH to reduce CO_2 and produce sugar.

4. The C_3 cycle is the major path of carbon assimilation in green plants. In this cycle, two three-carbon-atom sugar phosphate molecules are formed; hence the name C_3. These molecules then combine to form a six-carbon-atom sugar phosphate, fructose bisphosphate.

5. Some plants have adapted to extreme environmental conditions by evolving variations on the carbon assimilation pathway. Succulents use crassulacean acid metabolism (CAM). CAM plants absorb CO_2 through open stomata at night and form an organic acid. During the day, when stomata are closed in these plants, CO_2 (released in the leaves from the organic acid formed) is converted into carbohydrate by the C_3 cycle.

6. Some plants of tropical origin have evolved another variation of carbon assimilation, the C_4 pathway. These plants very effectively concentrate CO_2 in their leaf mesophyll chloroplasts by forming two four-carbon-atom organic acids. The organic acids are transported into the bundle sheath chloroplasts and lose the CO_2, which is used in the C_3 cycle there.

7. Photorespiration is reduced in C_4 plants.

8. Usually only about 0.3–0.5% of light energy that strikes a leaf is stored during photosynthesis, but under ideal conditions this may increase severalfold. Photosynthesis may be limited by CO_2 concentration, light, temperature, minerals, and other environmental as well as hereditary factors.

10.1 THE HARNESSING OF LIGHT ENERGY BY PLANTS

With few exceptions, all living cells require a continuous supply of energy, which comes directly or indirectly from the sun. Although terrestrial green plants use large amounts of energy directly from the sun in both transpiration and photosynthesis, only in **photosynthesis** is a significant amount of light energy stored as chemical energy for future use. Billions of years ago, cyanobacteria (formerly known as blue-green algae) created an oxygen-rich atmosphere through their photosynthetic activity. Since that time, plants have continued to support life by being the original source of food for other organisms. The great importance of photosynthesis is twofold: the liberation of oxygen as an end product and the transformation of low-energy compounds (carbon dioxide and water) into high-energy compounds (sugars). Perhaps someday humans will use other sources of energy to drive the energy-requiring steps in the production of food and fiber. Today, however, except for a few species of bacteria (see sidebar, "Chemosynthesis," p. 147), all life is dependent on the energy-storing reactions of photosynthesis.

10.2 DEVELOPING A GENERAL EQUATION FOR PHOTOSYNTHESIS

Like respiration and other complex processes occurring in living cells, photosynthesis consists of many reaction steps. It is easier to approach photosynthesis, or any biochemical process, by looking at an overview. Indeed, this is how scientific knowledge about photosynthesis evolved; so in this section we will trace the discoveries that led to a general understanding of photosynthesis. Later sections will detail the steps of specific reaction.

Early Observations Showed the Roles of Raw Materials and Products

Until the early 17th century, scholars believed that plants derived the bulk of their substance from soil humus. A simple experiment performed by Flemish physician and chemist Joannes van Helmont disproved this idea. He planted a 2.27-kg (5-lb) willow (*Salix*) branch in 90.7 kg (200 lb) of carefully dried soil and supplied rainwater to the plant as needed. In five years it grew to a weight of 67.7 kg (169 lb), but according to van Helmont's measurements the soil had lost only 57 g (2 oz). Consequently, he reasoned that the plant substance must have come from water. This was a logical deduction, though not entirely correct. Almost two centuries elapsed before van Helmont's findings were finally correctly explained.

Our knowledge of photosynthesis begins with the observations of a religious reformer, philosopher, and spare-time naturalist, Joseph Priestley. In 1772, Priestley reported that a sprig of mint could restore confined air that had been made impure by a burning candle. The plant changed the air so that a mouse was able to live in it. The experiment was not always successful, probably because Priestley (who did not know about the role of light) did not always provide adequate illumination for his plants. In 1780, a Geneva pastor, Jean Senebier, published his own research and pointed out another important part of the process: that "fixed air," carbon dioxide, was required. Thus, in the new terminology of French

CHEMOSYNTHESIS

In 1977, scientists studying the ocean bottom near the Galapagos Islands discovered an entirely new marine community over 3000 meters (1.86 miles) below the ocean surface. Along the boundary between two plates of the earth's crust, they found fissures—hydrothermal vents—from which mineral-laden, heated sea water is pouring. Rich animal populations including large clams, mussels, and tube worms cluster around these vents, and some animals live deep within the vents themselves. Since that time, other vent communities have been discovered in various locations around the world.

The most light-sensitive photosynthesizing organism known is a green sulfur bacterium that lives 80 meters (260 feet) below the surface of the Black Sea. This bacterium utilizes energy from the very faint, pale blue rays of the sun that filter down from the surface to synthesize its food. Vent communities, by contrast, are too far from the ocean surface for sunlight to reach them, and organic material produced by photosynthetic organisms near the ocean surface does not provide adequate food to sustain these organisms far below. Instead, vent organisms obtain their food from dense microbial mats consisting primarily of bacteria associated with the vents. Many of these bacteria rely for energy on reduced inorganic compounds such as H_2S (hydrogen sulfide) that are in high concentration in the vents. These microorganisms oxidize the reduced inorganic compounds and use the energy liberated to assimilate CO_2 into organic compounds. This process is known as **chemo-synthesis**. Chemosynthesis is similar to photosynthesis in several respects, but chemical energy rather than light energy drives the synthetic reactions. Besides obtaining organic matter from the microbial mats, some of the animal species in the vents live in symbiotic association with chemosynthetic microorganisms.

A very odd and interesting observation, made in 1988, is that extremely faint infrared light is associated with hydrothermal vents (see Fig. 1). Although certain bacteria can harvest infrared energy, none is known that can use such dim infrared light. Scientists have speculated that some of the bacteria living in vents may use infrared light energy to supplement their chemosynthetic activity. Active research is currently under way to test this idea.

The discovery and research associated with hydrothermal vents illustrate how interrelated biological problems are. Many molecular studies of relationships among living organisms suggest that microbes called Archaea, that thrive at extremely high temperatures, are ancestral to all living organisms (see Chapter 18). This raises the possibility that deep-sea vent organisms were the ancestors of all life. Of course, this idea is very speculative at the present time, but it is stimulating research on hydrothermal deep-ocean vents.

Another instance in which chemosynthesis rather than photosynthesis supports the food requirements of an isolated community was found in 1986 in a limestone cave in southern Romania. Many limestone caves have areas that are devoid of light, in which are found communities of obligate, cave-dwelling organisms; however, in almost all instances, these organisms obtain their food from material of photosynthetic origin transported from the surface outside of the cave. The Romanian cave (called Movile Cave), on the other hand, supports a community of 48 species of cave-adapted terrestrial and aquatic invertebrates even though no light or outside source of organic matter is available. Water in this cave is rich in H_2S. Scientists have determined that the bacteria found floating on the water and on the walls of this cave, like those found associated with hydrothermal deep-ocean vents, use H_2S as a source of energy to synthesize organic material. Organisms that can produce their own food from inorganic materials using reduced inorganic compounds instead of light as a source of energy are **chemoautotrophs**. Just as chemoautotrophic bacteria serve as a food base for the marine life in the hydrothermal ocean vents, the chemoautotrophic bacteria in the Movile Cave serve as a food base for the large variety of cave-dwelling organisms. This is the only known cave ecosystem supported by chemoautotrophs.

chemist Antoine Lavoisier, it could be said that green plants in the light use carbon dioxide and produce oxygen.

But what was the fate of the carbon dioxide? A Dutch physician, Jan Ingen-Housz answered this question in 1796 when he found that the carbon went into the nutrition of the plant. In 1804, 32 years after Priestley's early observations, the final part of the overall reaction of photosynthesis was explained by the Swiss botanist and physicist Nicolas de Saussure, who observed that water was involved in the process. Now the experiment performed by van Helmont almost 200 years earlier could be explained:

$$\text{carbon dioxide + water} \xrightarrow[\text{green plants}]{\text{light energy}} \text{oxygen + organic matter}$$

Almost 50 years elapsed before scientists identified carbohydrates as the organic matter formed during photosynthesis in most plants. Between 1862 and 1864 a German plant physiologist, Julius von Sachs, observed that starch grains occur in the chloroplasts of higher plants and that if leaves containing starch are kept in darkness for some time, the starch disappears. If these leaves are exposed to light, starch reappears in the chloroplasts. Sachs was the first to connect the appearance of starch, a carbohydrate, both with the fixation of carbon and water and with the elimination of oxygen in chloroplasts in the presence of light.

It is easy to demonstrate in the laboratory that starch forms during photosynthesis; however, it is much more difficult to show that sugar forms *before* starch does. Proof that sugar is the first carbohydrate produced by photosynthesis had to await the availability of radioactive carbon (^{14}C). We will discuss this development in a later section.

Comparative Studies Showed That Several Molecules May Donate Hydrogen to CO_2

Glancing at the overall equation for photosynthesis,

$$6CO_2 + 6H_2O \xrightarrow{\text{light}} C_6H_{12}O_6 + 6O_2$$

you might conclude that the carbon dioxide molecule splits, liberating the oxygen molecule. Indeed, most scientists felt that the reaction proceeded in this manner until the early 1930s when Cornelis van Niel, working at Stanford University, compared photosynthesis in a number of different groups of photosynthetic bacteria. The green and purple sulfur bacteria use hydrogen sulfide instead of water to reduce carbon dioxide, for example, and van Niel found that sulfur instead of oxygen is liberated:

$$6CO_2 + 12H_2S \xrightarrow{\text{light}} C_6H_{12}O_6 + 6H_2O + 12S$$

The sulfur can come only from the hydrogen sulfide. Because the hydrogen sulfide serves the same role in these bacteria as water does in higher plants, van Niel reasoned that the oxygen evolved by higher plants comes from water, not from carbon dioxide. (Experiments using radioactive tracers have since verified van Niel's insight.)

After comparing similar reactions in other organisms, van Niel concluded that a general equation for photosynthesis should be written as:

$$6CO_2 + 12H_2A \xrightarrow{\text{light}} C_6H_{12}O_6 + 6H_2O + 12A$$
$$\text{carbon dioxide + hydrogen donor} \longrightarrow \text{carbohydrate + water + A}$$

The hydrogen donor (H_2A) can be H_2O, H_2S, H_2, or any other molecule capable of a reduction reaction, and it requires an input of energy. When H_2A gives up its hydrogen, it is oxidized to A.

10.3 LIGHT REACTIONS AND ENZYMATIC REACTIONS

Although the general equation for photosynthesis identifies the reactants and products, it tells us nothing about the individual reactions that, taken together, make up this complex process. To supply food and fiber to the increasing world population, we need to be able to increase crop yields. To increase crop yields, we need to know—among other things—the specific reactions of photosynthesis. Research spanning 100 years has shown that photosynthesis involves both light and enzymatic reactions.

Between 1883 and 1885, a German physiologist, T. W. Engelmann, in a remarkably simple experiment, demonstrated which colors of light are used in photosynthesis. The spectrum of light varies from violet to red, as we can observe when we break white light into its components by passing it through a prism. Engelmann placed together on a microscope slide a living filament of a green alga and some bacteria that would migrate toward high concentrations of dissolved oxygen. He reasoned that the bacteria would cluster near regions of the alga generating the most oxygen from photosynthesis. When he placed the alga in a spectrum produced by passing light through a prism, he found that the bacteria migrated to the sections of the algal filament exposed to the red and blue light. This demonstrated that red and blue light were trapped by the photosynthetic organelles of the alga—and that they supplied energy to drive photosynthesis and liberate oxygen (Fig. 10.1).

At approximately the same time J. Reinke, another German scientist, was studying the effect of changing the intensity (quantity) of light on photosynthesis. Reinke observed that the rate of photosynthesis increased proportionally to an increase in the intensity of light, but only at low-to-moderate light intensities. At higher light intensities, the rate of photosynthesis was unaffected by

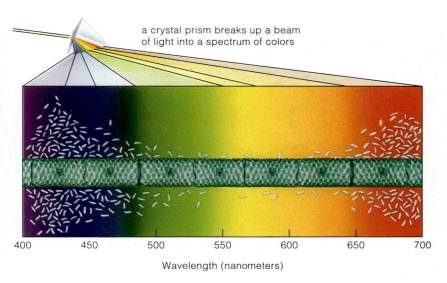

a crystal prism breaks up a beam of light into a spectrum of colors

Wavelength (nanometers)

Figure 10.1 Engelmann's experiment, demonstrating that red and blue light are effective in photosynthesis. The bacteria migrate to that portion of the algal filament where oxygen is produced.

changing the light intensity. This indicated that the reaction was then proceeding at maximum rate and was no longer affected by light; it had become light-saturated.

A further study and a more comprehensive interpretation of this phenomenon was conducted in 1905 by F. F. Blackman, a British plant physiologist. Blackman realized that the rate of light reactions is relatively unaffected by a change in temperature but that the rate of some light-insensitive reactions approximately doubles with every 10°C (18°F) increase in temperature, over a temperature range of about 10 to 25°C (50 to 77°F). He found that when photosynthesis was proceeding rapidly under adequate levels of CO_2 and high light intensities, the rate of photosynthesis more than doubled with a temperature increase of 10°C. However, if the light intensity was low, an increase in temperature had little effect on the rate of photosynthesis. Blackman reasoned that photosynthesis may be divided into two general parts: (1) photochemical reactions, which are insensitive to temperature changes and (2) temperature-sensitive reactions, sometimes called *dark reactions*. These temperature-sensitive (enzymatic) reactions are independent of light and can occur either in the light or in the dark. However, the rates of several of these reactions are affected indirectly by light because the activity of some of the enzymes involved is influenced by products of light reactions—such as changes in pH and NADPH concentration.

10.4

CHLOROPLASTS: SITES OF PHOTOSYNTHESIS

Because many of the individual reaction steps of photosynthesis are dependent on the specific cellular structure in which they occur, before we can examine photosynthesis in detail we need to understand the cellular site where

it takes place. Early studies with intact plants showed that oxygen is liberated and starch is formed in chloroplasts and that photosynthesis consists of both light and enzymatic reactions. It does not necessarily follow, however, that all of the reaction steps of photosynthesis take place in chloroplasts.

To find out, scientists isolated intact chloroplasts and parts of chloroplasts from the cell and studied the role they play in the complex process of photosynthesis. One of the earliest successful attempts to do this was by Robin Hill in Cambridge, England. In 1932, he demonstrated that chloroplasts isolated from the cell could still trap light energy and liberate oxygen. Then in 1954 Daniel Arnon, at the University of California, and others proved that isolated chloroplasts could convert light energy to chemical energy and use this energy to reduce CO_2.

Chloroplast Structure Is Important in Trapping Light Energy

Not all chloroplasts have the same shape, but they do have a universally similar structural organization critically important for photosynthesis. Chloroplasts, seen with the light microscope in living cells, appear to be homogeneously green; but in an electron micrograph the double-membrane nature of the envelope is apparent. Fig. 10.2 illustrates the chloroplast membrane systems. There are two types of internal membranes: those forming the **grana** (singular, *granum*) and those interconnecting the grana, the **stroma lamellae**. Together, these two membrane types constitute the **thylakoids** in the chloroplasts.

All biological membranes have high concentrations of both lipids and proteins. Biochemists believe that the chlorophylls and other pigments that are located in the grana and stroma lamellae are in close contact with lipids and with some of the photosynthetic enzyme systems, forming definite protein, lipid, and pigment patterns. Such a structural arrangement would improve the cell's ability to trap light energy and transform it into chemical energy during photosynthesis. Light energy initiates a series of reactions in which electrons from chlorophyll flow to other compounds (electron acceptors) in the internal membranes.

Experiments Reveal a Division of Labor in Chloroplasts

To determine what role chloroplasts play in photosynthesis, researchers need to study them independently of the rest of the plant cell. To isolate intact, functional chloroplasts, researchers cut or grind cells carefully in a dilute, buffered sugar solution or other solution of appropriate osmotic strength. They then separate the chloroplasts that remain intact in the solution from the rest of the cell contents by placing the ground material in a centrifuge tube and spinning it at moderate speeds. Large and

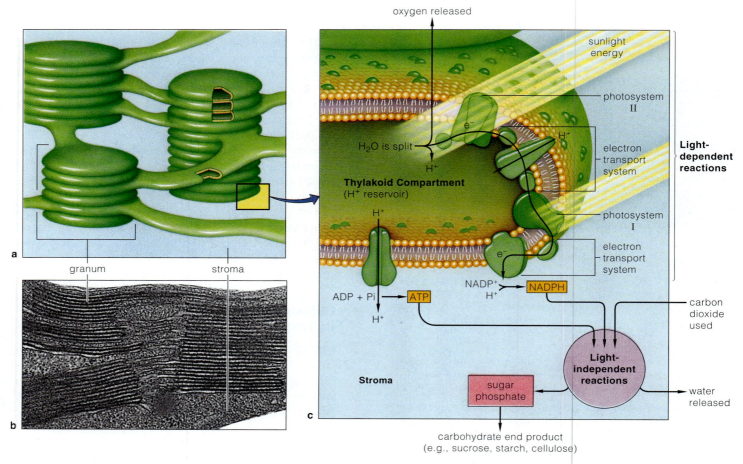

Figure 10.2 Chloroplast membrane systems. (**a**) Diagram of thylakoid membrane system of chloroplast. (**b**) Electron micrograph of chloroplast, showing membranes. (**c**) Overview of the sites where reactions of photosynthesis take place.

heavy particles—nuclei, starch grains, and cell-wall fragments—collect at the bottom of the tube. The liquid containing the chloroplasts and other, smaller cellular components is then poured into another tube and centrifuged at a slightly higher speed. The chloroplasts, free of most of the cellular debris, separate out at the bottom of the tube. Electron micrographs of these plastids show some to be complete, with internal membranes embedded in stroma and surrounded by the outer envelope. The others have lost their envelope and stroma and consist only of a system of membranes (Fig. 10.3). Research has shown that the intact chloroplasts will carry out the complete process of photosynthesis. The broken plastids, however, will carry out only part of the reactions of photosynthesis—but they will liberate oxygen, as Hill demonstrated.

A simple laboratory experiment to verify Hill's results uses dichlorophenolindophenol, which is a blue dye in its oxidized form (DCIP) and colorless in its reduced form (DCIPH$_2$). If DCIP is mixed with isolated broken chloroplasts (thylakoid fragments) in the light, oxygen is liberated. This reaction will not, however, take place in the dark.

$$2DCIP \text{ (blue)} + 2H_2O \xrightarrow[\text{thylakoids}]{\text{light}} 2DCIPH_2 \text{ (colorless)} + O_2$$

Experiments with isolated chloroplasts and with isolated thylakoids reveal a division of labor within the chloroplast. The green thylakoids capture light, liberate oxygen from water, form ATP from ADP and phosphate, and reduce NADP$^+$ to NADPH. The colorless stroma contains water-soluble enzymes, captures carbon dioxide, and uses the energy from ATP and NADPH in sugar synthesis.

10.5 CONVERTING LIGHT ENERGY TO CHEMICAL ENERGY

Knowledge of the details involved in the conversion of light energy to chemical energy during photosynthesis is not necessary to appreciate the significance of the process. However, a consideration of the major steps in the reaction, as we now know them, will help us understand how the details of such a complex reaction are

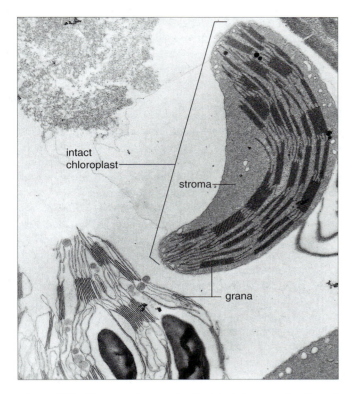

Figure 10.3 Electron micrograph of isolated chloroplasts of broad bean (*Vicia faba*), showing intact chloroplasts and chloroplasts that have lost their outer envelope and stroma, ×2500.

being unraveled. Many of the individual steps in the overall reaction of photosynthesis are the subject of current research. The more scientists know about the process, the closer humans come to producing sugar directly from the raw materials light, carbon dioxide, and water.

Light Has the Characteristics of Both Waves and Particles

First, let us consider the nature of light. Physicists have two models describing the nature of light, and both are needed to understand the role of light in photosynthesis.

The first model interprets light energy as **electromagnetic radiation** composed of waves of different lengths. The white light from the sun is composed of wavelengths ranging from red—wavelengths between about 640 and 740 nanometers (nm)—to violet—wavelengths of approximately 400 to 425 nm. White light passed through a prism or through water droplets is resolved into its separate components. The resulting colors form the visible spectrum (Fig. 10.4). We cannot see wavelengths longer than red—the infrared and radio waves—or shorter than violet—the ultraviolet, X rays, and gamma rays. The visible spectrum is only a small part of the complete electromagnetic spectrum, and only a part of the visible spectrum provides the energy for photosynthesis.

The second model of light states that in addition to acting as if it travels in waves, light also acts as if it were

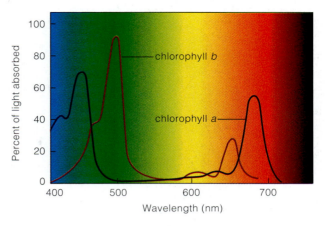

Figure 10.4 Absorption spectra of chlorophylls *a* and *b* at different wavelengths of light. The graph shows the fraction of received light that is absorbed when the pigment is exposed to various wavelengths of light. The relation between wavelength and color of light is also shown.

composed of discrete units or packets of energy called **photons**. Each photon contains an amount of energy that is inversely proportional to the wavelength of light characteristic for that photon. So the short wavelength of blue light has more energy per photon than does the longer wavelength of red light. When light is absorbed by a pigment (as in the case of photosynthesis by chloroplast pigments), *only one photon is absorbed by one pigment molecule at a time. The energy of the photon is absorbed by an electron of the pigment molecule, giving this electron more energy.*

Reactions in Thylakoid Membranes Transform Light Energy into Chemical Energy

ABSORPTION OF LIGHT ENERGY BY PLANT PIGMENTS Chlorophyll appears green because it absorbs some of the blue and red wavelengths of white light, leaving proportionally more green to be transmitted or reflected and seen. *It is the absorbed light that is used in photosynthesis.* Scientists can measure the amount of any specific wavelength of light that is absorbed by a pigment by using a **spectrophotometer**, an instrument that measures and then plots the percentage of light absorbed for each wavelength of the entire visible spectrum. The resulting graph is an **absorption spectrum**.

Vascular plants contain two major types of chlorophyll, *a* and *b*. The absorption spectra of chlorophylls *a* and *b* in solution (Fig. 10.4) show that they absorb much of the red, blue, indigo, and violet light; these are the wavelengths that are used most in photosynthesis. Part of the red and most of the yellow, orange, and green light are scarcely absorbed at all unless the chlorophyll solution is very concentrated. If, instead of a chlorophyll solution, a very thin green leaf is placed in the spectrophotometer, the absorption spectrum is quite similar to, though not identical with, those from the chlorophyll solutions. The difference can be attributed in part to the yellow pigments in the leaf, which absorb blue light; also,

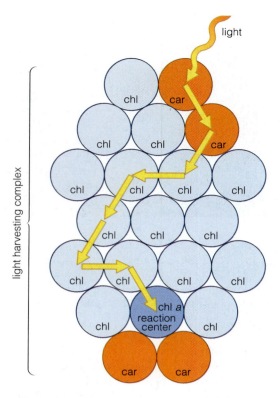

light

light harvesting complex

chl | car
chl | chl | car
chl | chl | chl | chl
chl | chl | chl
chl | chl | chl | chl
chl | chl *a* reaction center | chl
car | car

Figure 10.5 Accessory pigments in the light-harvesting complex absorb light energy and transfer it to the reaction center, where a special chlorophyll *a* molecule is excited and loses an electron with high energy (car = carotene, chl = chlorophyll).

chlorophyll in the leaf is in close association with proteins and lipids in the chloroplasts; chlorophyll in solution is not.

What happens when the chlorophyll molecules absorb light energy? How is light energy converted into chemical bond energy during photosynthesis? Recall from Chapter 2 that molecules are composed of atoms having positively charged nuclei and one or more electrons spinning around the nuclei. Energy is required to move an electron away from the positively charged nucleus. When a photon is absorbed or trapped by a chlorophyll molecule, the energy of the photon causes an electron from one of its atoms to be moved into a higher energy state. The chlorophyll molecule now is in an excited state (Fig. 10.5). This condition is unstable, and the electron tends to move rapidly, usually within a billionth of a second, back to its original energy level. As it does this, the absorbed energy usually is released as heat or as light (fluorescence). In photosynthesis, this energy also performs another function: Energy in an excited chlorophyll *a* molecule drives electrons from water to reduce NADP+. The NADPH formed subsequently reduces CO_2 in enzymatic reactions leading to sugar formation.

THE TWO PHOTOSYSTEMS Two photosystems are involved in the trapping of light energy during photosynthesis in green plants. In the 1950s, Robert Emerson at

the University of Illinois found that red light of about 680 nm or longer wavelength is very inefficient in photosynthesis, but that adding light of shorter wavelength increases the efficiency of the long wavelengths of light. This observation led to the realization that there are two light reactions and two pigment systems, designated as **photosystems I** and **II**, in thylakoids of plants that evolve oxygen. Each photosystem contains several protein molecules complexed with pigment molecules, chlorophylls *a* and *b*. Photosystem II contains **carotene** (a reddish orange pigment) as well. In addition, each photosystem has a complex of electron acceptor molecules.

Chlorophyll *a* and chlorophyll *b* are fat soluble and are found in the thylakoids together with other fat-soluble pigments such as **carotenoids**, which are red, yellow, or purple accessory pigments. These pigments act as light traps and are grouped together in functional units, **light-harvesting complexes**. The energy absorbed by these pigments is transferred to a specific chlorophyll *a* molecule in the reaction center. This chlorophyll *a* molecule is excited and loses an electron (Fig. 10.5). The reaction centers of photosystem I and II are called P_{700} and P_{680}, respectively, because they absorb light of those wavelengths. The chlorophyll *a* molecules in the reaction centers have unique light absorption properties, in part because of their association in a chlorophyll–protein complex. In addition to having P_{700} as its reaction center, photosystem I has a higher proportion of chlorophyll *a* than chlorophyll *b* in its light-harvesting complex; it is also sensitive to longer-wavelength light. In contrast, photosystem II, which is sensitive to shorter-wavelength light, contains P_{680} and has almost equal amounts of chlorophylls *a* and *b* in its light-harvesting complex.

As we will see in the next section, scientists have traced electron flow in green plants from water, through the chlorophyll *a* molecules in reaction centers, to NADP+.

NADPH FORMATION Photosynthesis is an oxidation–reduction process involving the transfer of large amounts of energy in many relatively small steps. These energy transfers are initiated with the absorption of light energy and involve the flow of electrons from reducing agents (electron donors) to oxidizing agents (electron acceptors) located in the thylakoids. The relative tendency of some of the electron carriers in the thylakoids to accept or release electrons is expressed as volts on the vertical scale in Figure 10.6. A carrier high on the scale will spontaneously transfer an electron to a carrier low on the scale, without the input of more energy into the system. Each time an electron (and energy) is transferred from one molecule to the next, some energy is lost as heat, but the rest of the energy is then in the electron acceptor. In order for electron carriers higher on the scale to gain electrons from a carrier lower on the scale, however, there must be an input of energy. Each carrier differs from the others in its ability to gain or release electrons. In addition, each

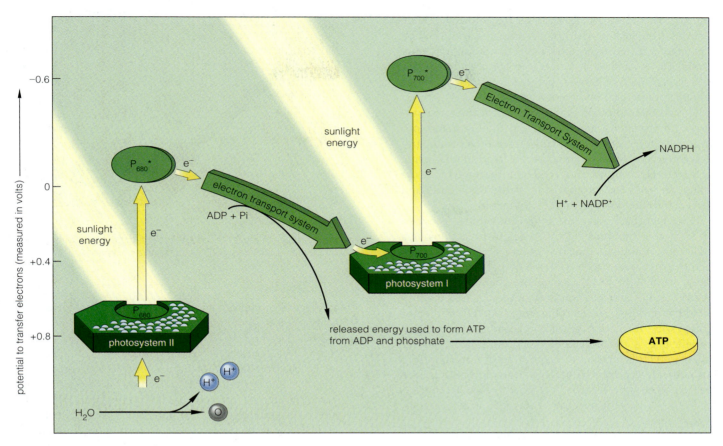

Figure 10.6 The pathway of noncyclic electron transport from water to NADPH, with the associated ATP synthesis.

carrier is located in the thylakoid membrane in a precise order; so it forms a chain with the other molecules along which electrons flow.

The absorption of light energy and subsequent transport of electrons to $NADP^+$ has two important consequences: Water molecules are split, releasing H^+ and oxygen, and NADPH is formed. Because electrons cannot move along the electron transport chain until the chlorophyll *a* molecule at reaction center I loses an electron, we start our discussion of electron flow with reactions occurring in reaction center I. As we describe the steps in this electron flow process, trace them in Figure 10.6. Observe the energy change as light energy is converted into electrical energy (electron flow) and then to the energy of chemical bonds in the NADPH molecule.

When a photon is absorbed by the light-harvesting complex associated with photosystem I, the energy is transferred to the reaction center P_{700}. An electron in this special chlorophyll *a* molecule gains energy and moves into a higher energy level, symbolized by *. (Note that the position of P_{700}^* higher on the scale in Figure 10.6 indicates that the reaction center has gained energy, not that it has physically moved in the thylakoid membrane.) In this higher-energy, excited state, the P_{700}^* has a tendency to lose an electron to an electron acceptor near it in the

thylakoid membrane. Because the electron lost by the chlorophyll *a* is in a higher energy state from the absorption of light energy, no further input of energy is needed for the electron to move down an electron transport chain. The electron is transferred quickly from one carrier molecule to another, illustrated as moving down the electron chain and finally arriving at an $NADP^+$ molecule. At this point, $NADP^+$ accepts two electrons and is reduced. It picks up a proton (H^+) from the stroma and becomes the energy-rich NADPH, which is now available for further enzymatic reactions of photosynthesis.

Meanwhile, the chlorophyll *a* at reaction center I is left as a positive ion. It holds its remaining electrons very strongly in this low-energy state. It cannot accept another photon until the lost electron is replaced. In this condition, photosystem I must await events in photosystem II.

A photon strikes photosystem II and drives an electron into a higher energy level, producing an excited state, P_{680}^*. The electron lost from P_{680}^* moves to an electron acceptor and down the electron transport chain. Finally it joins the chlorophyll *a* ion of photosystem I. Now the chlorophyll *a* molecule in the reaction center of photosystem I can absorb another photon of light. But now photosystem II has a positive chlorophyll *a* ion, which has a powerful tendency to replace its lost electron. In this highly oxidized state, P_{680} molecules gain electrons from water with the accompanying formation of oxygen and hydrogen ions, a reaction known as **photolysis**. *The exact steps in this oxygen liberation process are not known*, but manganese, chloride, and at least one enzyme appear to be involved.

This reaction probably takes place near the inner surface of the thylakoid membrane, and the hydrogen ions produced contribute to the establishment of a proton gradient across the thylakoid membrane. In addition, during the flow of electrons to photosystem I, hydrogen ions are picked up from the stromal side of the thylakoids and released into the space within the thylakoid.

If we step back from the individual steps described and view the light and electron transport reactions as a single unit, it appears that two photons have pushed an electron from water to $NADP^+$. The pathway transfers energy from light to storage in the chemical bonds of NADPH and establishes a proton gradient across the thylakoid membranes. The enzyme complex ATP synthetase catalyzes the formation of ATP in response to a proton gradient across the thylakoid membrane, in a manner similar to ATP formation in respiration in mitochondria. This pathway is **noncyclic electron transport** because the electrons from light-excited P_{700}^* flow to $NADP^+$ and do not cycle. Noncyclic electron transport involves both photosystems.

In contrast, **cyclic electron transport** (Fig. 10.7) involves only photosystem I, and electrons from the light-excited P_{700}^* do not flow to $NADP^+$. Instead they pass down the electron transport chain and back to the chlorophyll a ion in P_{700}, completing the cycle. No NADPH is produced, but a proton is removed from the stroma and released into the space within the thylakoids. ATP may be formed as a result of this proton gradient, as we will see in the next section.

ATP SYNTHESIS During electron flow, hydrogen ions (protons) are moved from the stroma and added to the fluid inside the thylakoids. The result is a proton gradient: *a difference in H^+ concentration across the thylakoid membrane*. Scientists believe that this proton gradient provides the necessary energy to form ATP from ADP and inorganic phosphate (Pi). According to this idea, the enzyme ATP synthetase, which is bound to the stromal side of the thylakoid membrane, forms ATP from ADP and inorganic phosphate in the presence of the H^+ gradient. In the process, H^+ is removed from the space within the thylakoid and is released to the stroma (see Fig. 10.2c). Consequently, the H^+ gradient across the membrane is decreased.

This light-driven production of ATP in chloroplasts is called **photophosphorylation**. There are two types of photophosphorylation, cyclic and noncyclic photophosphorylation. In **cyclic photophosphorylation** (Fig. 10.7), electrons flow from light-excited chlorophyll molecules

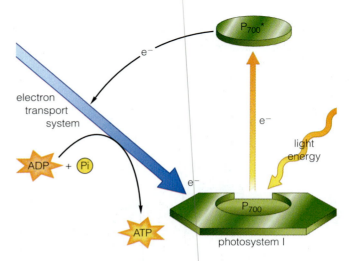

Figure 10.7 The pathway of cyclic electron transport, with the production of ATP from ADP and inorganic phosphorus (Pi).

to electron acceptors and cyclically back to chlorophyll. Only photosystem I takes part in this process. No oxygen is liberated, and no $NADP^+$ is reduced because it does not receive electrons. A proton gradient is formed during cyclic electron flow, and light energy is converted into chemical energy in the ATP molecules. However, because NADPH is not formed, cyclic photophosphorylation alone does not produce the molecules needed to bring about CO_2 reduction and sugar formation. Its significance in photosynthesis is limited to producing the H^+ gradient that leads to energy conservation in ATP.

Noncyclic photophosphorylation (Fig. 10.6) produces both ATP and NADPH. In this series of reactions, electrons from excited chlorophyll are trapped in $NADP^+$ in the formation of NADPH and do not cycle back to chlorophyll. Note that both photosystem I and photosystem II are involved in noncyclic photophosphorylation, and both ATP and NADPH are formed. Energy stored in the bonds of these molecules drives the CO_2 reduction reactions of photosynthesis.

<h2>10.6 THE REDUCTION OF CO_2 TO SUGAR: THE CARBON CYCLE OF PHOTOSYNTHESIS</h2>

In the preceding section, we dealt with the transformation of electromagnetic energy (light) into the chemical bonds of NADPH and ATP. In this section we will consider how energy stored in NADPH and ATP is used to reduce CO_2 to sugar.

Enzymes Catalyze Many Light-Independent Reactions in Photosynthesis

The complete process of photosynthesis will not take place in the dark. Many steps in the process, however, are controlled by enzymes; these steps are not directly sensitive to light but are sensitive to changes in temperature. The process of carbon dioxide reduction to carbohydrates requires many enzyme reactions. All the enzymes directly participating in photosynthesis occur in the chloroplasts. Many of them, especially those linked to the carbon cycle, are water-soluble and are found in the stroma.

One of the most extensively studied enzymes of photosynthesis—and the enzyme that probably occurs in highest concentration in many leaf cells—is **ribulose bisphosphate carboxylase (rubisco)**. This is one of the most important enzymes in plants because it catalyzes the first step in the carbon cycle of photosynthesis:

$$\text{carbon dioxide} + \text{ribulose bisphosphate} \xrightarrow{\text{rubisco}} 2 \text{ phosphoglyceric acid}$$

In this reaction, CO_2 combines with the five-carbon sugar ribulose bisphosphate (RuBP) in the plastid stroma to produce two molecules of the three-carbon atom acid, phosphoglyceric acid (PGA). This is a spontaneous reaction involving little change in energy.

The Carbon Cycle Was Identified by Use of Radioactive Carbon Dioxide and Chromatography

Although early physiologists analyzing the carbohydrate content of photosynthesizing leaves easily recognized that both starch and sugars usually accumulate during photosynthesis, they could not discover the mechanism for CO_2 reduction and sugar synthesis without learning the sequence in which carbon compounds form. The development of two sensitive analytical techniques enabled scientists to isolate the carbon compounds formed during enzymatic reactions. One method used **radioactive carbon** (^{14}C) in carbon dioxide to trace each intermediate product of the carbon cycle. The other, **chromatography**, permitted investigators to easily and accurately separate minute amounts of different organic compounds from one another. Using these techniques together, investigators could follow carbon in the series of reactions composing the carbon cycle of photosynthesis.

In the 1950s a group of scientists at the University of California under the direction of Melvin Calvin was particularly successful in using this technique to work out the early carbon pathway of photosynthesis. Calvin was awarded a Nobel prize for this achievement.

The most successful method finally adopted by this group was to expose green plant material to radioactive $^{14}CO_2$ in the light and then kill the cells in boiling alcohol after short time intervals. The alcohol extract contained essentially all of the radioactive labeled compounds formed in the cells. By varying the time to which the plant was exposed to $^{14}CO_2$, researchers determined the order in which compounds were synthesized during exposure to $^{14}CO_2$. Photosynthesis is very rapid, and the first products formed are quickly changed into other substances. Thus, Calvin used very short exposures to $^{14}CO_2$, in the range of 5–10 seconds.

If photosynthesis proceeds for an hour or so in an atmosphere containing $^{14}CO_2$, most of the labeled carbon is in carbohydrates (sugar or starch). If photosynthesis is stopped after only a few seconds, however, most of the labeled carbon is found in phosphoglyceric acid (PGA).

Phosphoglyceric acid formed during photosynthesis is thus an intermediate between carbon dioxide and sugar. By stopping photosynthesis at increasingly longer intervals (from a few seconds to several minutes), investigators were able to trace other intermediates and construct the entire pathway.

The C₃ Pathway Is the Major Path of Carbon Assimilation in Green Plants

The carbon cycle of photosynthesis is often called the **Calvin cycle** (in honor of its discoverer) or the **C₃ cycle** of photosynthesis (because the first product, PGA, contains three carbons). This cycle occurs widely in plants that photosynthesize. However, it is reasonable to assume that during the evolution of plants in a wide variety of ecological situations, adaptations and modifications in the metabolic process occurred. As we consider the C₃ pathway, we should note that memorizing the sequence of all these reactions and intermediate compounds is not particularly useful, but that understanding the roles played by this complex series of reactions in the metabolism of the plant is important. The key points to note in the C₃ carbon cycle (Fig. 10.8) are the following:

1. CO_2 enters the cycle when it combines with ribulose bisphosphate (RuBP) that is produced in the stroma. Two molecules of phosphoglyceric acid (PGA) are produced.

2. Energy to drive the cycle enters at two points: (1) when ATP donates a phosphate in the formation of RuBP and (2) when ATP is converted into ADP and phosphate (Pi) as NADPH donates electrons and hydrogen in the reduction of PGA to the sugar phosphoglyceraldehyde (PGAL).

3. Energy is stored in NADPH and ATP when these molecules are formed by light and electron-transport reactions in the thylakoids, and it is transferred into stored energy in the PGAL formed in the carbon cycle.

4. Phosphoglyceraldehyde may be converted enzymatically into another three-carbon-sugar phosphate, dihydroxyacetone phosphate. These two molecules combine to form a sugar phosphate, fructose 1, 6-bisphosphate. The plant by this process has essentially reduced one CO_2 molecule and added it to a five-carbon sugar to produce one molecule of a six-carbon sugar.

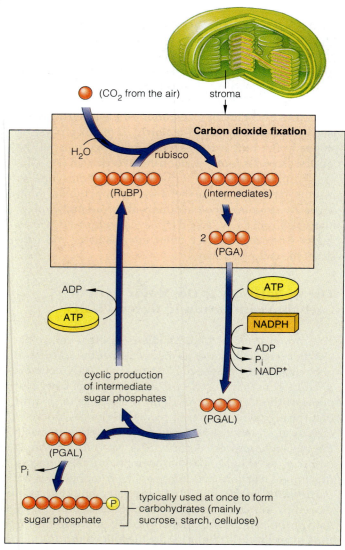

Figure 10.8 Some major steps in the C_3 or Calvin cycle of photosynthesis. Carbon dioxide enters the cycle when the enzyme rubisco combines CO_2 with RuBP to produce two molecules of phosphoglyceric acid (PGA). Carbon atoms of the key molecules are shown in red. All of the intermediates have one or two phosphate groups attached. For simplicity, only the phosphates on the resulting sugar phosphate are shown.

5. As these reactions continue, some of the fructose bisphosphate may be transformed into other carbohydrates, including starch.

6. Some of the fructose bisphosphate molecules cycle back, and RuBP is regenerated. This, in turn, accepts more CO_2. Thus, a cycle of carbon compounds exists, with CO_2 from the air and hydrogen from water entering the cycle, and various sugars being produced.

Although the C_3 cycle of photosynthesis occurs in most plants, some variations do occur. Two of these variations are found in succulent plants and in some plants of tropical origin.

Plants have adapted in various ways to extremely dry conditions existing in desert areas. To survive, plants must either endure recurrent drought or avoid drought by such means as carrying out the active part of their life cycle rapidly during brief rainy periods. Some plants, including succulents, have developed methods of storing and conserving water. The parenchyma tissue in succulent plants is highly developed, vacuoles are large, and intercellular spaces are reduced. When moisture is available, succulents absorb and store large amounts of water, and during periods of drought they resist the loss of water to the environment.

Succulents Trap CO_2 at Night

In contrast to most mesophytes, which require abundant soil and air moisture, many succulents have their stoma closed during the day and open at night. This adaptation reduces water loss during the day, when water stress is high. It could, however, be very disadvantageous for photosynthesis by reducing CO_2 uptake in the daylight, which is when photosynthesis can occur. Succulent plants minimized this disadvantage by evolving a particular type of carbon metabolism called **crassulacean acid metabolism (CAM)**. Botanists first observed it in plants belonging to the Crassulaceae family, but they now know that it is also common among 10 other large families, including Cactaceae, Euphorbiaceae, Liliaceae, and Orchidaceae.

The major features unique to CAM are:

1. During the night stomata are open, and the leaves rapidly absorb CO_2. During the day stomata are closed, preventing or greatly reducing CO_2 absorption and water loss.

2. The total amount of organic acids rapidly increases in the leaf-cell vacuoles at night.

3. At night, when CO_2 is rapidly absorbed, the enzyme PEP carboxylase initiates the fixation of CO_2. Generally, malate, a four-carbon compound, is produced.

4. Leaf acidity rapidly decreases during the day, and CO_2 is released into the leaf mesophyll.

5. Even though stomata are closed during the day, the C_3 cycle of photosynthesis usually takes place and converts the internally released CO_2 into a carbohydrate.

These features give CAM plants an effective mechanism for trapping CO_2 at night, releasing it internally during the day, and using it for photosynthesis when light is available but stomata are closed (which reduces or prevents CO_2 entrance from outside air). CAM plants use the released CO_2 in the usual way, combining it with

RuBP to yield two molecules of PGA and to complete the C_3 cycle of photosynthesis.

The C_4 Pathway Concentrates CO_2

Some plants of tropical origin exhibit a second variation of the carbon cycle, one that affects the efficiency of CO_2 absorption. We can determine how efficiently a plant absorbs CO_2 if we place it in a closed container in the light and measure the CO_2 content of the air in the chamber. At some point, the CO_2 produced by respiration will just balance or compensate for the CO_2 absorbed during photosynthesis. The percentage of CO_2 remaining in the chamber under these conditions is known as the **CO_2 compensation point**, and it varies among different plants. If a bean plant and a corn plant are placed together in a chamber in the light, the corn plant will successfully compete with the bean for the limited CO_2. Both will eventually die of starvation, but the bean will die before the corn does (Fig. 10.9). That's because plants such as corn have very low CO_2 compensation points.

In general, such plants seem to be adapted to grow in habitats with high light intensities and high temperatures, and they use water more efficiently. They also have certain structural features, such as the arrangement of large parenchyma cells in a sheath around veins (Fig. 10.10). Often, though not always, the chloroplasts of the bundle sheath parenchyma cells lack or have greatly reduced grana and frequently store starch. In contrast, the mesophyll cells between veins contain chloroplasts that have typical grana but little or no starch.

Plants that have these specialized bundle sheath and mesophyll cells rely on a special type of photosynthesis. Instead of ribulose bisphosphate combining with CO_2 (as

it does in the C_3 cycle under the influence of the enzyme rubisco), another enzyme, phosphoenolpyruvate carboxylase, catalyzes the union of CO_2 with a three-carbon-atom acid, phosphoenolpyruvate (PEP), to form the four-carbon-acid oxaloacetate (Fig. 10.11). Thus, this

Figure 10.9 Corn (*Zea mays*), a C_4 plant (right), with its low CO_2 compensation point is able to survive at a lower CO_2 concentration than bean (*Phaseolus vulgaris*), a C_3 plant (left), when they are grown together in a closed chamber in light for 10 days.

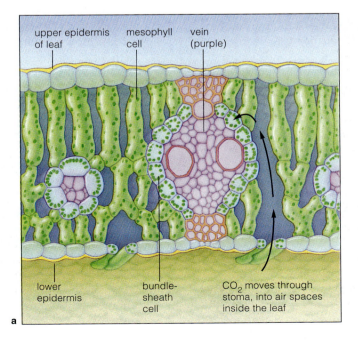

Figure 10.10 Photosynthesis in corn (*Zea mays*). (**a**) Drawing of a section through a corn leaf. (**b**) Distribution of C_3 and C_4 pathways in mesophyll and bundle sheath cells of corn, a C_4 plant.

upper epidermis of leaf

mesophyll cell

vein (purple)

lower epidermis

bundle-sheath cell

CO_2 moves through stoma, into air spaces inside the leaf

a

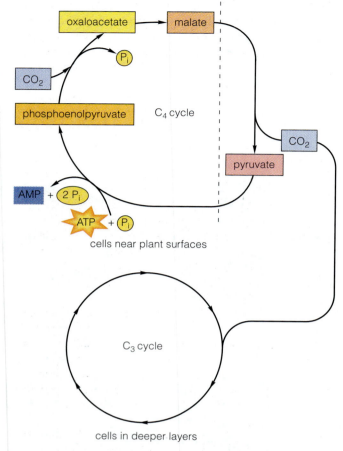

Figure 10.11 Interaction between the C_4 cycle in mesophyll cells (cells near plant surfaces) and the C_3 cycle in bundle sheath cells (cells in deeper layers) in C_4 plants.

pathway is called the **C_4 cycle**. Phosphoenolpyruvic carboxylase has a very strong affinity for CO_2. Later, two other four-carbon compounds, malate and aspartate, may form from oxaloacetate. Many plant physiologists believe that these four-carbon acids shuttle from mesophyll plastids to bundle sheath plastids, where they release CO_2 and become a three-carbon acid. This acid returns to the mesophyll, where it picks up another CO_2. The CO_2 released in the bundle sheath plastid then enters the normal C_3 cycle to form sugars.

In C_4 plants, there is a division of function between the two kinds of parenchyma cells. Mesophyll chloroplasts act as CO_2 traps. The CO_2 is then shuttled to bundle sheath plastids in the form of four-carbon acids, where it is reduced to sugars. This requires a very extensive and rapid movement of solutes over considerable cellular distances and across a number of membrane barriers. This pathway, discovered in Hawaii in 1965 by H. P. Kortschak, C. E. Hartt, and G. O. Burr and extensively studied by M. D. Hatch and C. R. Slack in Australia, is also known as the **Hatch-Slack cycle**. It differs

from the C_3 (or Calvin) cycle in that it ensures a very efficient absorption of CO_2 and results in a very low CO_2 compensation point.

Note that photosynthesis is similar in CAM and C_4 plants, in that CO_2 is first incorporated into organic acids and then is liberated to enter the usual C_3 carbon cycle. Such plants differ, however, in that CAM plants trap CO_2 at night in the same mesophyll cells that liberate and use it in the C_3 cycle during the day; C_4 plants trap CO_2 in mesophyll cells during the day and then transport the organic acids into bundle sheath cells, where the C_3 cycle takes place (also during the day). C_3, CAM, and C_4 plants all use the C_3 pathway to produce carbohydrate from CO_2.

10.8 PHOTORESPIRATION

Because respiration releases CO_2, if respiration in leaves occurs in the light then the measurement of rates of photosynthesis by measuring CO_2 uptake would be difficult. Scientists wondered for a long time whether respiration proceeds in green cells at the same rate in the light as it does in the dark. If respiration does occur during photosynthesis, it would be a wasteful process because it would release the CO_2 just fixed by photosynthesis. Studies show that light actually does stimulate respiration in some plants, in a process called **photorespiration**. However, photorespiration differs from aerobic respiration in that it yields no energized energy carriers and does not occur in the dark.

Photorespiration starts when rubisco acts on ribulose bisphosphate. The activity of the enzyme rubisco is determined by the ratio of O_2 to CO_2 in the cell. When CO_2 is high, rubisco catalyzes the addition of CO_2 to RuBP, as in the C_3 cycle of photosynthesis. When the concentration of O_2 is high and the CO_2 is low, rubisco catalyzes instead the addition of O_2 to RuBP. One of the end products of this reaction is a compound that is oxidized to CO_2, but without the formation of ATP or NADPH. Under some conditions (such as hot, sunny days), 50% of the carbon reduced during photosynthesis may be reoxidized to CO_2 during photorespiration in C_3 plants. This is a severe loss of energy.

C_3 plants have very high rates of photorespiration (particularly on hot, bright days), but C_4 plants show very little or no photorespiration. Mesophyll cells in C_4 plants have a high capacity to trap CO_2 and to present the bundle sheath cells with a high concentration of CO_2 in the form of organic acids. The high level of CO_2 that occurs in bundle sheath cells when the organic acids lose their CO_2, coupled with lower levels of O_2 that may occur at high temperatures, suppresses photorespiration. Consequently, C_4 plants may produce two or three times as much sugar as C_3 plants during the hot, bright days of summer. This is part of the reason why such C_4 plants as sugar cane (*Saccharum officinarum*) and corn (*Zea mays*) are so productive (fix high levels of CO_2). Under milder conditions, when photorespiration is less likely to occur, the C_3 plants are more efficient than C_4 plants, in part because they expend less energy to capture CO_2.

10.9 FACTORS AFFECTING PRODUCTIVITY

The photosynthetic leaf is a highly evolved unit with many features that promote an efficient capture of light and carbon dioxide. Nevertheless, only about 0.3 to 0.5% of the light energy that strikes a leaf is stored in photosynthesis. The yield may be increased by a factor of 10 under ideal conditions. In a hungry world, there is much to be gained by exploring and controlling the conditions that limit **plant productivity**, the amount of living tissue produced per unit of time by a plant or population of plants.

Greater Productivity Can Be Bred into Plants

Plant productivity is determined partly by the environment and partly by the hereditary traits of the plant. One avenue toward greater productivity is to breed more efficient plants. This approach has already been quite successful in the case of cereal grains. For instance, the productivity of corn (maize, *Zea mays*) has been increased over 200% since the 1960s. Norman Borlaug received the Nobel prize in 1970 for developing rice (*Oryza sativa*) strains so productive as to constitute a *green revolution* in tropical countries. Unfortunately, these strains require high levels of fertilizer application—an expensive and pollution-creating practice. There are also potential genetic problems associated with these grains.

Photorespiration is one example of a hereditary trait that reduces plant productivity, and the C_4 system offers a compensating hereditary advantage. Now that research has uncovered these traits, perhaps breeding programs or the use of recombinant DNA technology will lead to new C_4 varieties—as well as to C_3 plants that are less prone to photorespiration. Alternatively, chemicals may be synthesized that effectively inhibit photorespiration.

Fluctuations in the Environment Alter the Rate of Photosynthesis

Besides heredity, the environment also has many general influences on photosynthesis and plant productivity. To a certain degree some of the environmental factors are under the control of the plant grower. Crop yields can be influenced by modifying such factors, particularly in controlled growth chambers or in greenhouses. Factors generally most easily controlled are water and the mineral elements of the soil. The control of temperature, light (intensity, quality, and duration), and carbon dioxide usually require special equipment.

TEMPERATURE Plants are capable of photosynthesis over a wide temperature range. Plants native to arctic regions may photosynthesize at temperatures below 0°C (32°F). Some lichens from the Antarctic carry out photosynthesis at temperatures of −18°C (−0.4°F) and have a photosynthetic temperature optimum near 0°C. Algae in the water of hot springs may carry on photosynthesis at a temperature as high as 75°C (167°F). C_4 species generally have higher temperature optima for photosynthesis than do C_3 species, and investigators have found the majority of desert summer annuals to be C_4 plants.

Most plants function best between temperatures of 10° (50°F) and 25°C (77°F), however. If there is adequate light intensity and a normal supply of carbon dioxide, the rate of photosynthesis of most common land plants increases with an increase in temperature up to about 25°C; above this range there is a continuous fall in the rate as the temperature rises. The longer the exposure to a given high temperature, the greater the decrease in photosynthetic rate.

Under conditions of low light intensity, an increase in temperature beyond a certain minimum will not produce an increase in photosynthesis. These conditions may occur in the winter in greenhouses. If the temperature is raised too high, the plants will suffer because although the rate of photosynthesis has not changed, respiration has increased as a result of higher temperature.

LIGHT Both light intensity and wavelength affect the rate of photosynthesis. Under moderate temperature conditions and with sufficient carbon dioxide, carbohydrates produced by a given area of leaf surface increase with increasing light intensity—up to an optimum, after which their production decreases. It is not the intensity of light falling upon the leaf surface that affects photosynthesis as much as it is the intensity to which the chloroplasts are exposed. Surface hairs, thick cuticle, thick epidermis, and other structural leaf features diminish the light intensity that reaches the chloroplasts.

Intense light appears to retard the rate of photosynthesis. Many plants living in deserts and other places where the light is very bright often have structural adaptations that tend to diminish the intensity of light reaching the chloroplasts. The usual light intensity in arid and semiarid regions is well above the optimum for photosynthesis in many plants, especially crop plants introduced by humans. In these regions, the light intensity is probably nearer the optimum for photosynthesis on days when the sky is overcast than on clear, sunny days. Leaves on the surface of plants receive more intense light than those beneath, which are shaded. Therefore, some of the leaves receive light of optimum intensity, whereas others may receive light either above or below the optimum.

A special situation exists in a dense stand of plants such as a forest, where the vertical distribution of light varies dramatically—from full sunlight on the upper

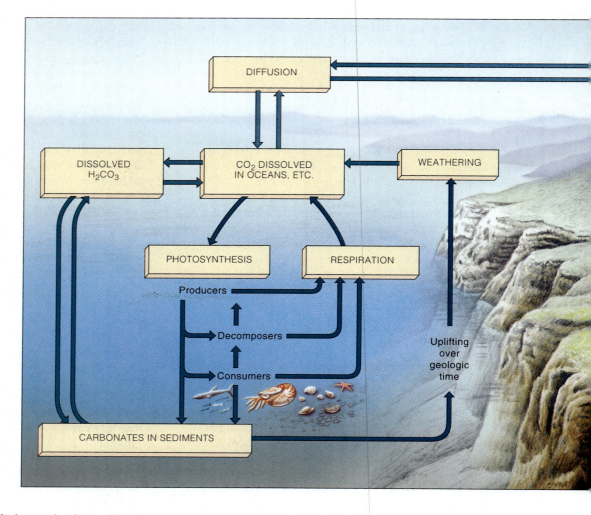

Figure 10.12 Diagram of the global carbon cycle. The portion on the left shows movement of carbon through the marine ecosystem; on the right, the movement of carbon through land ecosystems.

leaves of a canopy to very dim light on the forest floor. Plants growing below a leaf canopy receive full sunlight for relatively short periods of time, when the sun is directly overhead and pulses of light penetrate through holes in the canopy. When breezes move the canopy of leaves, these brief exposures to direct sunlight may become very short indeed. One may wonder whether plants can efficiently use light from these **sunflecks**, as they are called. Surprisingly, recent research shows that sunflecks may indeed contribute a majority of light energy utilized by understory vines, shrubs, and herbs. Research in Hawaii indicates that as much as 60% of the daily carbon gain by seedlings of such understory trees as *Euphorbia forbesii* and *Claoxylon sandwicense* comes from sunflecks.

Under many natural conditions, the quality of light striking green plants varies widely. Such variations may determine seedling survival and growth on forest floors. The leaves of the overstory canopy absorb predominantly red and blue light. Thus, lower leaves, screened from direct sunlight by upper leaves, receive light rich in green wavelengths, which are less efficiently utilized. Plants growing in deep water, such as marine algae, are subjected to light in the shorter (blue-green) wavelengths. Many such plants have developed accessory pig-

ment systems adapted to absorb blue-green light and utilize it in photosynthesis. The ability of plants to adapt to the particular quality as well as the quantity of light received is important for survival.

CARBON DIOXIDE Earth's atmosphere contains about 78% nitrogen and about 20% oxygen. The remaining small percentage is composed of traces of other gases, including approximately 0.03% carbon dioxide. Land plants absorb atmospheric CO_2 through their stomata. Once it enters the intercellular spaces of the leaf, the CO_2 dissolves in the water within the walls of the palisade and spongy parenchyma cells. During photosynthesis, carbon dioxide is removed from solution in the chloroplasts. A diffusion gradient for carbon dioxide is set up between the outside atmosphere and the chloroplasts, resulting in the diffusion of CO_2 from the atmosphere to the chloroplasts. At the same time, oxygen liberated during photosynthesis is used in respiration, or it diffuses outward—in aqueous solution through the cell walls, to intercellular spaces, and from there through stomata to the air around the leaf.

If all the chlorophyll-bearing cells of the world's plants are constantly taking carbon dioxide from the atmosphere during daylight, the quantity of this gas used

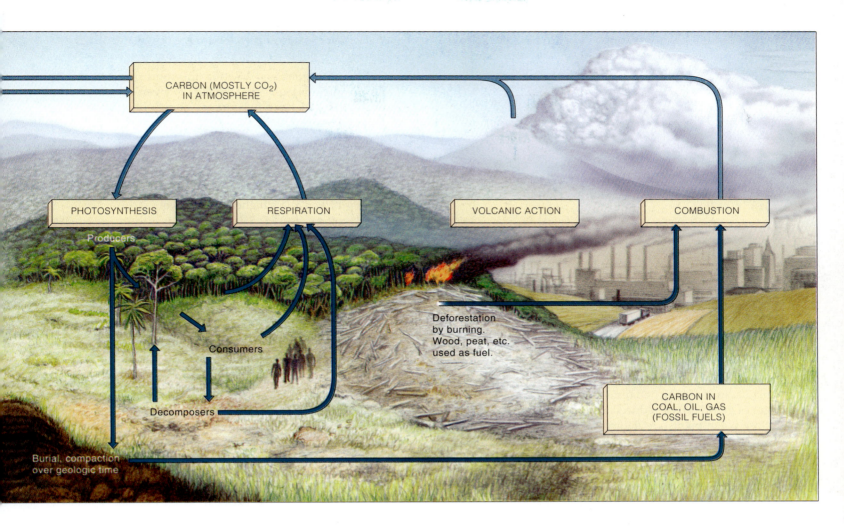

The following labels appear in the figure:

CARBON (MOSTLY CO₂) IN ATMOSPHERE

PHOTOSYNTHESIS

RESPIRATION

VOLCANIC ACTION

COMBUSTION

Producers

Consumers

Decomposers

Deforestation by burning. Wood, peat, etc. used as fuel.

Burial, compaction over geologic time

CARBON IN COAL, OIL, GAS (FOSSIL FUELS)

must be enormous, and natural processes must continually replenish the supply. A hectare (2.5 acres) of corn (*Zea mays*) contains 24,700 plants. During a growing season of 100 days, it may accumulate 6,200 kg (6.83 tons) of carbon, equivalent to 22,700 kg (25.02 tons) of CO_2. All of this carbon is derived from the CO_2 of the atmosphere.

The present atmospheric supply of carbon dioxide would be used up in about 22–30 years if it were not constantly renewed. However, it is not the possible depletion of atmospheric CO_2 but the continual rise in the level of CO_2 that is contributing to the threat of global warming, with all of its catastrophic consequences.

A number of processes result in the release of carbon dioxide into the atmosphere (Fig. 10.12).

1. The living cells of all organisms (both green and nongreen) release CO_2 in respiration.

2. The dead bodies of plants and animals, and the excretions of animals, contain large quantities of carbon and other elements in the form of organic compounds. The activities of bacteria and fungi cause these compounds to decay and release large quantities of CO_2.

3. The oceans are important reservoirs of CO_2 and may absorb or release CO_2 in response to atmospheric fluctuations of CO_2.

4. Mineral springs and volcanoes release CO_2 to the atmosphere.

5. When wood, coal, oil, gasoline, natural gas, or other carbon material burns, large quantities of CO_2 are released. It is this release of carbon dioxide that is most significant for global warming.

Even if light intensity and temperature are favorable, the amount of atmospheric CO_2 immediately around leaves limits the rate of photosynthesis. This may be particularly true in greenhouses kept closed in the winter. Under these conditions, the CO_2 may be reduced much below the average level in the air (0.03%). On the other hand, scientists have determined experimentally that (at normal temperatures and light intensities) an artificial *increase* in CO_2—up to a concentration of 0.6%—may increase the rate of photosynthesis, but only for a limited period. It appears that this high level of CO_2 is injurious to some plants after 10–15 days of exposure. Such information is of importance to greenhouse managers.

The amount of CO_2 in the atmosphere has increased during the past two centuries, partly as a result of the industrial revolution. The atmospheric CO_2 concentration is rising even faster now with the rapid destruction of tropical rain forests and the burning of fossil fuels (oil, coal, gasoline) in increasingly large amounts. Carbon dioxide tends to reduce heat loss from the earth's surface. This causes a warming trend known as the *greenhouse effect*. Many scientists and concerned citizens are worried about the consequences of the greenhouse effect on the earth's climate. Given sufficient warming, polar ice caps will melt and sea levels will rise, flooding coastal areas. Not all scientists agree that the greenhouse effect has started already. Many feel, however, that industrialized nations must act now to minimize it by reducing their dependence on fossil fuel, stopping the destruction of rain forests, and instituting massive reforestation projects to bring about a balance between the photosynthetic uptake of CO_2 and the liberation of CO_2 into the atmosphere.

WATER Most of the water absorbed by plant roots is lost as transpiration. A significant amount of the rest is retained by the highly hydrated living cells, but only 1% or less is actually used in photosynthesis. However, the rate of photosynthesis may be changed by small differences in the water content of the chlorophyll-bearing cells. Drought reduces the rate of photosynthesis in some plants, such as long pod bean (*Vicia faba*), because the turgor of guard cells is reduced and because stomata tend to close when plants are deprived of water. Then the lowered diffusion of CO_2 into the leaf limits the rate of photosynthesis. Other plants, such as the bean (*Phaseolus vulgaris*), are able to compensate for mild water deficit by increasing the solute concentration of their guard cells. This prevents a loss in turgor. Consequently the stomata remain open, and the rate of photosynthesis is less affected.

MINERAL NUTRIENTS Obviously, all of the essential elements for normal plant growth and development are required for an actively photosynthesizing plant. However, several elements are specifically required for the development of photosynthetic systems. For instance, magnesium and nitrogen are part of the chlorophyll molecule, and iron is necessary for its synthesis; the water-splitting, oxygen-evolving system requires manganese, chloride, and calcium; some electron carriers contain iron, and all proteins (including enzymes) contain nitrogen. Consequently, poor soils can result in plants with poorly developed photosynthetic capacities. In these cases, yields can be greatly increased by effective fertilizer programs.

SUMMARY

1. Photosynthesis is the primary energy-storing process of life, in which light energy is stored as chemical energy in organic compounds.

2. Carbon dioxide and water are the raw materials. The products of photosynthesis are sugar and oxygen.

3. Light reactions and electron transport occur in thylakoid membranes. Thylakoids of green plants that evolve oxygen have two photosystems, photosystem I and photosystem II.

4. Light energy absorbed by light-harvesting pigment complexes in the two photosystems is funneled to special chlorophyll *a* molecules in the reaction centers. The chlorophyll *a* molecule becomes excited as one of its electrons is driven into a higher energy level.

5. The electron that has gained energy escapes from the chlorophyll *a* molecule and passes down an electron transport chain.

6. $NADP^+$ acts as the final electron acceptor and forms NADPH, thereby storing some of the energy acquired by the absorption of light by the pigments.

7. During electron transport, a proton (H^+) gradient is formed across the thylakoid membrane. This proton gradient drives the synthesis of ATP.

8. These two compounds, NADPH and ATP, are the first molecules in which light energy absorbed by chloroplasts is stored as chemical energy.

9. Electrons move from water to replace those lost by chlorophyll in the reaction center P_{680}. H^+ and O_2 are formed.

10. Enzymes catalyzing the carbon cycle are found in the stroma.

11. In the stroma, the H^+ and electrons of NADPH are transferred to organic compounds and are used in a series of reactions that reduce CO_2 and produce molecules of sugar. ATP is used as an energy source.

12. Two major pathways of carbon fixation are the C_3 and C_4 pathways. In C_3 plants, photorespiration may release up to 50% of the previously captured CO_2. This major loss in captured CO_2 does not happen in C_4 plants, which are therefore more productive at high temperatures.

13. In CAM plants, CO_2 is captured at night when stomata are open. The resulting organic acid releases CO_2 to photosynthesizing cells during the day, when stomata are closed.

14. Photosynthesis may be limited by CO_2, light, temperature, water, minerals, and the plant's hereditary efficiency.

15. The atmosphere's concentration of CO_2 has been rising for the past two centuries because of the removal of forests and the burning of fossil fuel. As a result, global temperatures are expected to rise.

Questions

1. Write the general equation for photosynthesis in words.

2. Describe an experiment to demonstrate which colors of light are used in photosynthesis.

3. When light is absorbed by chlorophyll in a leaf, what happens to an electron in the reaction center?

4. What is the significance of the proton gradient that is established across the thylakoid membrane during electron transport in the thylakoid?

5. What are the first two molecules that store energy from light absorbed by a leaf during photosynthesis?

6. Outline the methods used by scientists to determine which intermediate compounds are formed during the synthesis of sugar from CO_2.

7. Name and briefly distinguish among the three major carbon cycles of photosynthesis.

8. Under what environmental conditions would CAM metabolism and C_4 metabolism be advantageous to a plant? Why?

9. Why is photorespiration a wasteful process?

10. A man was growing plants in a greenhouse in the winter. He closed the greenhouse and heated it to keep it warm. His plants grew very slowly. Offer some reasonable explanation for this effect.

Further Readings

Salisbury, F. B., and C. W. Ross. 1992. *Plant Physiology*. 4th ed. Belmont, Calif.: Wadsworth. Advanced plant physiology text with an up-to-date treatment of photosynthesis.

Tolbert, N. E., and J. Preiss, eds. 1994. *Photosynthetic Carbon Metabolism and Regulation of Atmospheric Carbon Dioxide and Oxygen*. London: Oxford University Press.

Wolfe, S. L. 1993. *Molecular and Cellular Biology*. Belmont, Calif.: Wadsworth. A well-illustrated text containing a modern treatment of photosynthesis.

11 ABSORPTION AND TRANSPORT SYSTEMS

1. Water flows into, through, and out of a plant in response to a gradient of forces. The various forces that move water into, through, and out of the plant are combined in the concept of *water potential*. As water flows through the plant, it is moving "downhill" from regions of higher to lower water potential.

2. To replace water lost from leaves by transpiration, water is pulled through the xylem. Plants control the rate of water loss by opening and closing their stomata.

3. Mineral elements in solution in the soil are taken up by root cells through an active, energy-requiring process. A fraction of the minerals are secreted into the xylem and rise to the shoot in the transpirational stream.

4. The products of photosynthesis are transported in the phloem from the leaves (where they are produced) to other parts of the plant (where they are used or stored). An osmotic pump provides the pressure that pushes sap through the phloem sieve tubes.

11.1 TRANSPORT AND LIFE

Is there just a heartbeat between life and death? Often we characterize life in animals by movement, and in higher animals it is specifically the movement of a beating heart that circulates oxygen, nutrients, and hormones around the body in the blood. Plants clearly lack a heart and blood, yet they have the same general need as animals for transporting substances from one organ to another. Although the mechanisms by which plant transport systems work differ from those of animals, the processes are just as effective and impressive.

The need for transport in plants derives from their complex anatomy and their photosynthetic lifestyle. Plants obtain their energy and carbon from photosynthesis. Photosynthesis requires a constant source of carbon dioxide (CO_2), which comes from the air into the leaf (through open stomata) and then spreads into the air spaces between the mesophyll cells. However, the existence of such open pathways between the photosynthetic cells inside the leaves and the air outside means that water vapor can move *out* of the leaf, into the drier air, at the same time as CO_2 is moving *in*. That water must be replaced. Plant cells need a supply of water for maintaining structures, photosynthesis, and growth, and they die if they become dehydrated. The replacement water comes from the soil through the roots. Thus there must be an effective transport system to get the water from the soil into the roots and on up to the leaves.

Growth of the plant requires mineral nutrients, such as nitrogen, phosphorus, potassium, and iron, as well as carbon. Because emerging leaves and growing stems may be a considerable distance from the soil (the source of minerals), plants must also have a system for transporting minerals to meristematic regions.

Photosynthesis produces carbohydrates that provide energy as well as carbon skeletons (covalently linked carbons) for the synthesis of other organic molecules. Energy and carbon are needed in all parts of the plant, especially in meristematic regions of stems and roots, and also in flowers, seeds, and fruits. Consequently, there must be a means of transporting carbohydrates from photosynthetic organs to living cells throughout the plant.

The three sections of this chapter describe the mechanisms by which water, mineral nutrients, and carbohydrates are transported from one part of a plant to another (Fig. 11.1).

11.2 TRANSPIRATION AND WATER FLOW

Water is the most abundant compound in a living cell. Without water solutes cannot move from place to place, and enzymes cannot acquire the three-dimensional shape that they need for catalytic activity. Water is a substrate or reactant for many biochemical reactions, and it provides strength and structure to herbaceous plant organs through the turgor pressure it exerts (see Chapter 3). As much as 85 to 95% of the weight of a growing herbaceous plant is water. Because water molecules are connected to each other by hydrogen bonds (see Chapter 2), the water in a plant forms a continuous network of molecules. This network extends into every apical bud, leaf, and root cell; it permeates, with a few exceptions, every cell wall and much of the intercellular space. Because the network is continuous, a loss of water from one area affects the entire system.

Although the cuticle that covers the stems and leaves of terrestrial plants is relatively hydrophobic (and thus mostly impermeable to the diffusion of water), stomata, lenticels, and cracks in the cuticle allow a loss of water vapor from the interior of the plant. This loss is called **transpiration**. The amount of water transpired by plants is considerable. In one study, a single corn (*Zea mays*) plant in Kansas transpired 196 liters of water between May 5 and September 8. One hectare (2.5 acres) of such plants (14,800 plants) would transpire over 3 million liters (800,000 gallons) of water during the season, an amount equivalent to a sheet of water 28 cm (11 inches) deep over the entire hectare.

A Variety of Factors Affect the Flow of Water in Air, Cells, and Soil

To understand how water moves within a plant and why it is transpired, we must first find out why water moves at all. There are five major forces that move water from

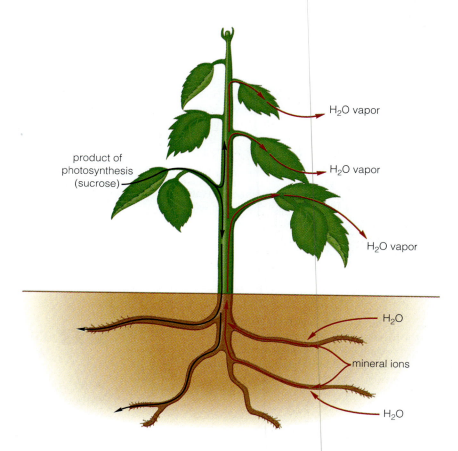

product of
photosynthesis
(sucrose)

H₂O vapor

H₂O vapor

H₂O vapor

H₂O

mineral ions

H₂O

Figure 11.1 The conduction of
water, mineral elements, and
organic nutrients through a plant.

place to place: diffusion, osmosis, capillary forces, hydrostatic pressure, and gravity.

The description of these forces begins with molecules, which, far from being static, are in constant motion. In a gas, the individual molecules move independently in random directions. Overall, however, the net flow is from regions of higher concentration to regions of lower concentration. This principle applies to every element and compound in air, and in particular it applies to water vapor, which flows from areas of higher humidity to areas of lower humidity. The net flow of molecules from regions of higher to lower concentration is called **diffusion**. Diffusion is a major force directing the flow of water in the gas phase (Fig. 11.2a).

Liquids are quite different from gases, but liquid water and solute molecules are also in constant motion and also diffuse from regions of higher to lower concentration. To witness diffusion, place a drop of dye in a glass of water. You will see the dye disperse throughout the water. What you may not see is that the water also diffuses into the region of dye. This process becomes very important when the dye (or any other solute) is confined by a differentially permeable membrane (such as the plasma membrane) that allows water but not solutes to flow across it. Solutes displace water, so liquid with a higher concentration of solute has a lower concentration of water. The diffusion of water across a differentially permeable membrane from a dilute solution (less solute, more water) to a more concentrated solution (more solute, less water) is called **osmosis**.

A device that uses osmosis to power a flow of water out of a chamber is called an **osmotic pump** (Fig. 11.2b). An osmotic pump is one that works by pressure generated through osmosis. It is easy to set up an example in the laboratory. One ties a bag formed from a dialysis membrane around a tube. A dialysis membrane is a thin sheet made of modified cellulose fibers. It has pores large enough to pass water molecules but small enough to retain large solute molecules, such as proteins or starch. Inside the bag is a solution of large solute molecules; outside the bag is pure water. Osmosis pulls water into the bag. As the volume of water in the bag increases, pressure builds up inside the bag. Because there is an outlet through the capillary tubing, the pressure forces solution through this outlet. Later in this chapter, we will see how osmotic pumps function in a plant.

Osmosis represents a very potent attractive force pulling water into cells. As explained in Chapter 3, so long as the apoplast solution—the solution outside of the plasma membrane—is more dilute than cytoplasm, water will tend to flow into cells from the apoplast solution. As the cell wall expands, it exerts a **hydrostatic pressure** to oppose the flow (see Fig. 3.7). Hydrostatic pressure in cells is called turgor pressure and is important because it stiffens the cells and the tissue they constitute.

Outside the cells, water is pulled into the small spaces between the hydrophilic cellulose microfibrils of the cell wall and is tightly held there. This is because the water molecules form hydrogen bonds with each other and also with the carbohydrates of the walls (Chapter 2). Water

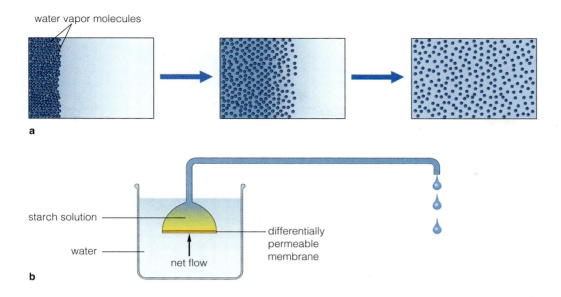

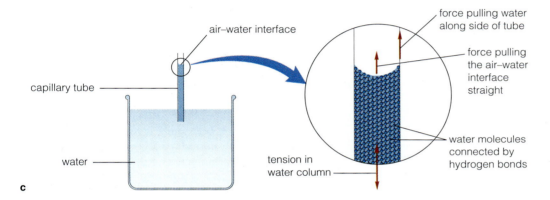

Figure 11.2 Some of the major forces that move water. (**a**) Diffusion of water vapor molecules gives a net movement from high to low concentration until the concentration at all points is equal. (**b**) An osmotic pump pulls water from a dilute solution to a more concentrated solution. The pressure generated will push solution out of an opening. (**c**) Capillary forces pull water into a narrow tube or any other narrow spaces.

molecules are **cohesive**: They stick together. They are also **adhesive**: They stick to hydrophilic molecules such as carbohydrates. The hydrogen bonds between water molecules and the cellulose surface tend to drag the water along so that it covers as much surface as possible; the hydrogen bonds between the water molecules at the interface between water and air tend to minimize the area of the interface by pulling it level. The liquid is pulled along behind the interface by the hydrogen bonds that connect the individual molecules. Together these forces can generate a great tension that pulls water into the smallest spaces. We can visualize the forces by putting a glass capillary tube into water: The water is pulled up the tube until enough water has risen that its weight balances the pull (Fig. 11.2c). Because such a tube is called a *capillary tube*, the forces pulling water into it are called **capillary forces**. Capillary forces produce a tension in the water like that in a stretched rubber band. The strength of the bonds holding water molecules together is surprising. In one type of experiment, investigators were able to exert a tension of 1000 atmospheres (1000 times the pressure of air at sea level—about 15,000 pounds per square inch) on a column of water before it broke. The

amount of tension depends on the cross-sectional area of the space, with the smallest spaces producing the greatest tensions. Also, once the capillary spaces are filled, there is no more tension; but if water is removed (for instance, by evaporation), tension will be reestablished.

Soil particles are also hydrophilic, and they have small spaces. For this reason, water is pulled into the soil and held there by capillary forces, just as it is pulled into the cell walls. The strength of these forces depends on how much water is present. If the soil is very wet, the forces holding the water will be weak; if the soil is dry, the forces will be stronger. In order for a plant to pull water from the soil, the root cells must generate an attractive force greater than the force holding the water in the soil.

Water also moves in response to gravity. It takes pressure to move water upward. With small herbs, plant physiologists need not be concerned with this effect because from the bottom of the root to the top of the shoot there is little change in height. However, gravity can be a significant factor in tall trees. To move water up a 100-meter (325-foot) tree takes a pressure of about 2 MPa (20 atmospheres).

Water Potential Defines the Direction of Water Flow

To describe how a combination of forces determines the direction of water flow, plant physiologists use the concept of **water potential**. Water potential takes into account the many forces that move water and combines them to determine when and where water will move through a plant. If we know the relative water potentials in two regions, we know the direction that water will flow: *Water always tends to flow from a region of high water potential to a region of low water potential.* This is true even if we are thinking of quite different phases, such as liquid water in a solution and water vapor in air. For instance, if the water potential of a solution is greater than the water potential of the water vapor immediately above it, the solution will evaporate. If the water potential of the soil around a root is less than the water potential of the root cells, water will flow out of the root into the soil—and vice versa. It is possible to calculate the water potential of a particular plant tissue (or air or soil) from physical measurements. This is particularly useful to agriculturalists, who must estimate the water needs of their plants and the availability of water in their soil.

Transpiration Pulls Water Through the Plant

Under most conditions, the flow of water through a plant is powered by the loss of water from the leaves. Water is *pulled* up the plant by transpiration, not pushed by pumping from the roots. In transpiration, the primary event is the diffusion of water vapor from the humid air inside the leaf to the dry air outside the leaf. The loss of water from the leaf generates a very strong attractive force that pulls water into the leaf from the vascular system, up the vascular system from the roots to the shoot, and, eventually, into the roots from the soil. There are many steps in the pathway by which the water moves and many factors that influence and control the rate of movement. The following sections discuss these.

DIFFUSION OF WATER VAPOR THROUGH THE STOMATA
The intercellular air spaces in the leaves are close to being *in equilibrium* with the solution in the cellulose fibrils of the cell walls. This means that they are nearly saturated with water vapor, whereas the bulk air outside the leaves is generally quite dry. The difference means that there is strong pressure for diffusion of water vapor out of the leaf. This diffusion can occur if there is a pathway with reasonably low resistance. Most of the leaf is covered with the epidermal cuticle, which has a high resistance to water diffusion. However, stomata have a low resistance when open, and water vapor diffuses out through them. This is the route by which most water is lost from a plant.

Water molecules that leave the leaf first pass through the boundary layer, an unstirred layer of air close to the

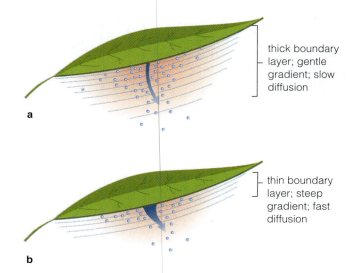

Figure 11.3 The diffusion of water vapor out of stomata through the boundary layer. Note that the thinner the boundary layer, the steeper the water vapor gradient.

leaf, and then enter the bulk air. The rate of diffusion out of the stomata depends in part on the steepness of the gradient of water vapor concentration (Fig. 11.3). All else being equal, a thick boundary layer has a more gentle gradient, and a thin boundary layer has a steeper gradient. Thus, transpiration across a thick boundary layer is slower than that across a thin one. Wind stirs up the air close to the leaf and makes the boundary layer thinner. That is why plants transpire much faster on a windy day than on a still one. Anatomical features of a leaf may slow the rate of diffusion by stabilizing a relatively thick boundary layer. For instance, a dense layer of trichomes on the surface of a leaf tends to preserve a boundary layer of relatively motionless air. **Stomatal crypts** (Fig. 11.4), depressions of the leaf surface into which the stomata open, form an effective boundary layer because the air in the crypts is quite still.

Temperature has a major effect on the saturation value of air. Warm air holds much more water than cool air. Thus, a volume of air with a given amount of water vapor has a much lower relative humidity when it is warm than when it is cool. For this reason, plants tend to lose water to warm air faster than to cool air.

FLOW OF WATER WITHIN LEAVES The loss of water vapor from the intercellular spaces of a leaf lowers the relative humidity of those spaces. Water evaporates from the surrounding cell walls to replace it. (Note that this evaporation occurs inside the leaf, not outside.) The removal of water from the cell walls partially dries them, producing capillary forces that attract water from adjacent areas in the leaf (Fig. 11.5).

Some of the replacement water will come from the inside of the leaf cells across the plasma membrane. As water leaves the cells, they become smaller. This will decrease turgor pressure (as the cell wall springs back from a more extended to a less extended state) and concentrate the solutes, increasing the net osmotic effect. If much water is removed, the turgor pressure falls to zero. That is, the cells become flaccid and lose their ability to support the leaf. We recognize this when the plant wilts.

If the plant is well watered, the water lost from cell walls and from inside the cells will be replaced by water from the xylem. This water flows out of tracheids through the pits in the lignified secondary cell walls and into the fibrous primary cell walls of the mesophyll cells. The spaces between the fibrils of the primary cell walls are very small, but the distances are short—most cells in a leaf are within two to six cell lengths of a small vein—so the resistance to flow is fairly low. This means that the replacement of the cell-wall water is rapid, so long as water can be pulled out of the xylem.

FLOW OF WATER THROUGH THE XYLEM The flow of water out of xylem tracheids has an interesting effect: It pulls on the rest of the water in the tracheid, and it pulls on the walls of the tracheid (Fig. 11.6). Pull one water molecule out of the central space of the tracheid, and the force is transferred by hydrogen bonds to a network of molecules. The water that is removed from the tracheid cannot be replaced by air because the tiny pores and pits leading into the central space are too small for air bubbles to pass through. The water that is removed cannot collapse the walls, because they are strong and rigid. Thus the loss of water results in a hydrostatic tension on the rest of the water in the tracheid. The tension pulls water from adjacent tracheids and vessels, which in turn induces a tension in them. If water continues to flow from the leaf tracheid into the leaf cell walls, there will be a constant stream of water flowing up the xylem, powered by a gradient of tension.

The smaller tracheids, which are separated by pits, provide fairly high resistance to the water flow and thus require a relatively steep gradient of tension to maintain an adequate flow rate. Vessels, with their larger diameters and longer lengths uninterrupted by pits, pose a much lower resistance to water flow. On the other hand, vessels are more easily inactivated by air bubbles. Air bubbles may form in xylem when gases dissolved in the xylem fluid come out of solution under the influence of

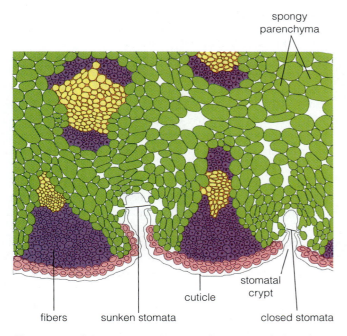

Figure 11.4 Sunken stomata in crypts of a yucca leaf. Note also the thick epidermal cuticle, which limits evaporation from the leaf surface, and the large bundles of fibers, which keep the leaf from drooping if the parenchymal cells wilt.

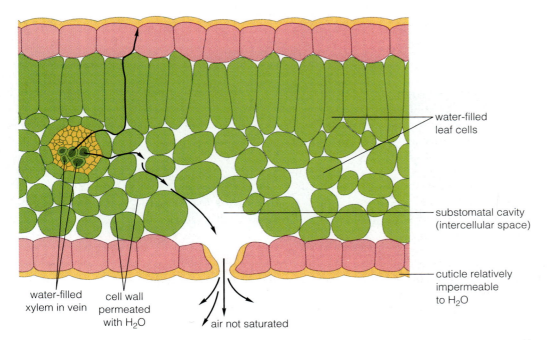

Figure 11.5 Diagram of the flow of water from minor veins along cell walls to the leaf cells, evaporation into the leaf intercellular spaces, and diffusion of water vapor out of a stoma into the surrounding air.

high tension or freezing temperatures. A bubble in xylem under tension acts like a cut in a stretched rubber band: It breaks the tension. That is because once formed, the bubble expands almost instantly to fill the tracheid or vessel. However, it stops at the end of the tracheid or vessel because it cannot pass through the pores in the pits. Because there are many small tracheids, a bubble in one tracheid may make that tracheid useless but will have no effect on the overall flow. Because there are fewer vessels, a bubble in a vessel may block a substantial fraction of the water flow. Conifers have only tracheids, no vessels. This is thought to provide an advantage in dry, cold climates (where these trees are often the dominant vegetation) because these are the conditions most likely to produce bubbles in the xylem.

SYMPLASTIC AND APOPLASTIC FLOW THROUGH ROOTS

Eventually, the loss of water throughout the xylem lowers the water potential in the xylem of a growing primary root. This pulls in water from the apoplast of the **stele** (central region inside the endodermis—see Chapter 5) of that root, as the xylem is directly connected to (and is really part of) the apoplast. In turn, this water is replaced by water flowing into the stele from the root cortex and into the cortex from the soil.

The path of water flow through the cortex of the root and into the stele involves both the apoplast and the symplast (the interconnected cytoplasms of adjacent cells—Fig. 11.7). Because there is no cuticle over the epidermis of a primary root, water may flow in between the cells of the epidermis directly into the apoplast of the cortex and all the way to the endodermis. However, it cannot cross the endodermis in the apoplast because of the Casparian strip (the suberized cell walls that separate the apoplast of the cortex from the apoplast of the stele). To move further into the root, the water must enter the symplast by crossing the plasma membrane of an endodermal cell. Water may also cross the plasma membrane of cells at the root hairs or in the cortex. Once it does this, it can flow from cell to cell through the symplast via the plasmodesmata. It can cross the endodermis in the symplast, then enter the apoplast, and flow into the xylem.

Note that water must pass through at least two plasma membranes to reach the root xylem from the soil. There is a significant resistance to the flow of water

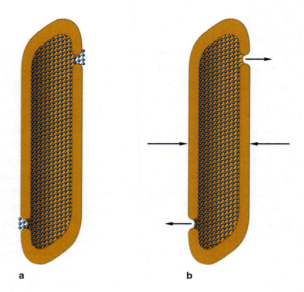

Figure 11.6 How capillary forces can convert the loss of water (for instance, by evaporation) into a tension within a tracheid. The diagram shows pits in the wall of a tracheid (exaggerated size); dots are water molecules; forces of adhesion and cohesion are shown as short lines. (**a**) Before evaporation, little tension. (**b**) After evaporation, high tension.

Figure 11.7 Apoplastic and symplastic pathways of water transport through the epidermis and cortex of the root. Note that water must flow through the symplast of the endodermis in order to enter the stele.

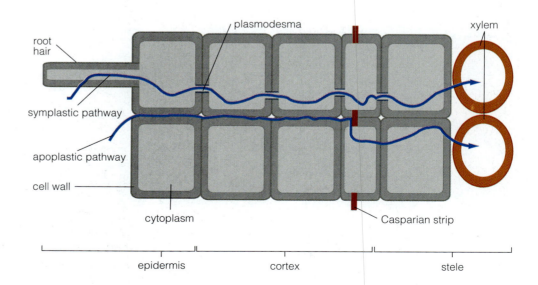

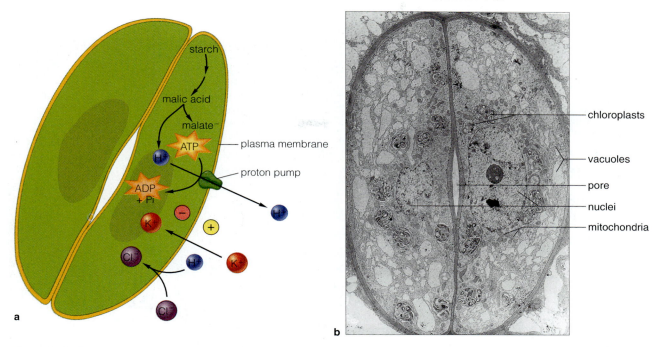

Figure 11.8 Events leading to opening of a stoma. The production of malic acid and the influx of K+ and Cl− pull water into the cells by osmosis.

through these membranes, although it is not easy to measure. One indication of this resistance is that transpiration—and therefore water flow—increases if a plant's roots are placed in boiling water long enough to kill the root cells and destroy the cell membranes.

FLOW THROUGH SOIL The flow of water from the soil into the roots reduces the amount of water in the region of soil around the roots. Most of the water in soil is in small capillary spaces between the soil particles and is bound to the hydrophilic particles by hydrogen bonds. The removal of water near the roots increases the capillary forces that hold water in the soil particles. This results in a gradient of capillary forces between the soil particles near the roots and those farther away. Water flows along the surfaces of the particles and through the capillary spaces in response to this gradient, finally reaching the roots.

Because the capillary spaces are small and the distances may be long, there can be considerable resistance to the flow of water through the soil. This resistance, together with the resistance to flow across the root membranes and through the xylem, is important because it limits the rate at which water can reach the leaves. Even a plant in well-watered soil will wilt in a sudden hot, dry breeze because water cannot move into the roots and up the xylem quickly enough to replace the water lost from the leaves. This is called *temporary wilt*, because such a plant will recover in a few minutes if the loss of water can be stopped. *Permanent wilt* occurs if the osmotic forces pulling water into cells are not as great as the attractive forces holding water to the soil particles.

CONTROL OF WATER FLOW The daily cycle of transpiration is very striking. For most plants, transpiration is very low at night; it increases starting some minutes after the sun comes up; it peaks at noon; and it decreases to its night level over the afternoon. By testing plants in experimental growth chambers, researchers have shown that the rate of transpiration is directly related to the intensity of light impinging on the leaves. Other important environmental factors are temperature, relative humidity of the bulk air, and wind speed.

Microscopic examination of the leaf surfaces shows that light affects the opening of the stomata. In dim or no light, the stomata of most plants are closed; as the light intensity increases, the stomata open up to some maximum value. The mechanism by which light controls stomatal aperture has been the subject of investigation for many years, and we can give a detailed (although not complete) description of many steps.

The primary sensing organs of the stomata are the guard cells. Under illumination, the concentration of solutes in the vacuoles of the guard cells increases. How does the solute concentration increase? First, starch, a storage carbohydrate in the chloroplasts of the guard cells, is converted into malic acid (Fig. 11.8). Second, the proton pump in the guard cell plasma membrane is activated (see Chapter 3). The proton pump moves H+, some of which comes from malic acid, across the plasma membrane. (After malic acid loses an H+, it is called *malate* ion.) This increases the electrical gradient and the pH gradient across the plasma membrane. Potassium ions (K+) flow into the cell through a channel in response to the charge difference, and chloride ions (Cl−), in association with H+ ions, flow into the cell through another channel in response to the H+ concentration difference.

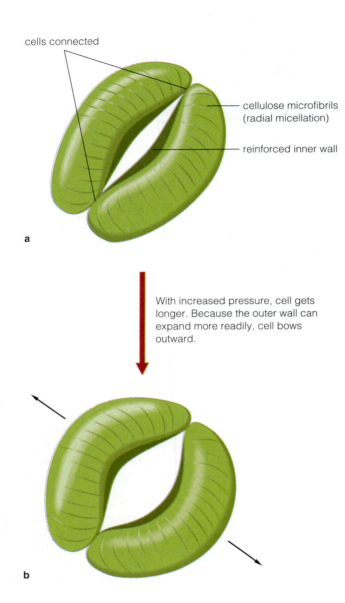

cells connected

cellulose microfibrils
(radial micellation)

reinforced inner wall

a

With increased pressure, cell gets
longer. Because the outer wall can
expand more readily, cell bows
outward.

b

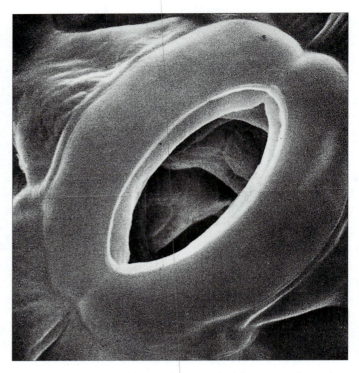

Figure 11.9 How radial micellation and reinforcement of guard cell walls force the expanding cells to bow outward.

The accumulation of malate, K^+, and Cl^- increases the osmotic force drawing water into the guard cells. The signals that turn on the enzymes that form malate and activate the proton pump in the plasma membrane include both red light and blue light, but these seem to act in different ways.

As just stated, the increased solute concentration increases the force drawing water into the guard cells. As the cells expand to accommodate the extra volume, turgor pressure rises until it balances the force drawing water into the cells. Most cells expand in one or two or sometime three dimensions as their internal pressure rises; in contrast, guard cells bend away from each other to open the stoma between them (Fig. 11.9). This is because they have specialized cell walls: first, an arrangement of cellulose microfibrils wrapped around the long axis of the cells (radial micellation); second, a heavier and

less extensible wall adjacent to the stoma. The radial micellation is like a girdle, limiting the direction of cell expansion. The cells get longer, not thicker, when they expand. The less extensible central wall means that the cells bow outward as they get longer.

Darkness reverses the process. As light decreases, there is a reduction in the charge difference and the pH difference across the plasma membrane; there is also an opening of channels to conduct K^+ and Cl^- out of the cells. Other stimuli that have the same effect are an increase in the CO_2 concentration inside the leaf and a loss of water to the point where the leaf wilts. The wilting stimulates mesophyll cells to produce abscisic acid (ABA; see Chapter 15), a plant hormone, which diffuses to the guard cells and specifically stimulates these cells to release K^+ and Cl^-. As these ions leave the guard cells, the osmotic forces cease to balance turgor pressure, and water leaves the cells. The guard cells contract, and the elastic properties of the cell walls pull the cells together, effectively closing the stoma.

While the opening and closing of stomata is the primary mechanism by which plants regulate the loss of water from their leaves (and thus also the flow of water through their roots and stems), there are other ways that they can moderate the rate of water loss when the stomata are open. The unstirred boundary layer of air close to the leaf is the key to such control. The larger the unstirred layer, the slower the loss of water. As already

FERTILIZERS

Good soils provide most of the inorganic nutrients needed by a plant for its growth, but these nutrients can eventually be depleted—absorbed by plants and removed in the harvest or leached (washed) away by water percolating through the soil. Adding fertilizers replenishes the soil. *Complete fertilizers* contain the three elements whose lack is most likely to limit plant growth: nitrogen (N), phosphorus (P), and potassium (K). Containers of commercial fertilizers have a three-number code that tells how much of each nutrient is found in the preparation. One "all-purpose plant food," for instance, is labeled 12-10-12, which means that it contains 12% nitrogen, 10% phosphorus (measured as P_2O_5), and 12% potassium (measured as soluble potash, K_2O). The label generally also tells the form(s) of nitrogen in the preparation. These may include nitrate (NO_3^-), which is soluble and absorbed quickly by plant roots; ammonium ion (NH_4^+, often called *ammonical nitrogen*), which may bind to the soil particles and be released slowly to the plant as nitrate through bacterial action; and organic nitrogen (urea or methylenediurea), which is available to the plant even more slowly as it is broken down by bacteria. Various fertilizers have different proportions of these nutrients, which are sometimes recommended (by the manufacturers) for specific kinds of plants.

Some fertilizers also contain trace elements such as boron, copper, manganese, molybdenum, zinc, and especially iron. These additions may be important for houseplants, growing in containers with a limited amount of soil, and for plants growing in nutrient-poor soils, such as sandy soils and overly leached acidic soils in high rainfall areas. Some fertilizers are formulated to acidify soils, which releases bound trace elements such as zinc and cobalt. Plants such as camellias, liquidambars, and cranberries (*Vaccinium* sp.) need acid soils in order to take up iron and other trace elements.

Organic fertilizers provide N, P, and K from digested animal or plant matter—one liquid fish emulsion is labeled 5-1-1. These are generally good fertilizers, with slow-release forms of nitrogen and possibly with trace elements (generally unspecified on the label); but they do not provide anything beyond what inorganic fertilizers do. For legumes (peas, *Pisum sativum*, and beans, *Phaseolus vulgaris*), you can purchase an inoculum of *Rhizobium*, which will promote the formation of nitrogen-fixing nodules. Incomplete fertilizers are often cheap and effective. Potassium nitrate (KNO_3), for example, gives a quick shot of two of the major nutrients, as does ammonium phosphate (($NH_4)_3PO_4$). Gypsum (calcium sulfate, $CaSO_4$) is a good soil acidifier; so is sulfur, which soil bacteria oxidize to acidic sulfur oxides, including sulfuric acid (sulfate). Ferric sulfate ($Fe_2(SO_4)_3$) provides iron in an acidic environment. If you use these or any fertilizers, be careful to read the labels or consult a gardening handbook, as it is possible to overdose plants.

What should you feed your plants? There are a few common signs of nutrient stress that can help you decide. Yellowing of the lower leaves (Fig. 11.10d, e, p. 175) suggests that the plant is deficient in nitrogen or possibly sulfur; a high-nitrogen complete fertilizer or some ammonium sulfate (($NH_4)_2SO_4$) would help. A dark purplish color (Fig. 11.10b) suggests a lack of phosphorus; the plants could use a high-phosphorus complete fertilizer or bone meal (calcium phosphate, $Ca_3(PO_4)$). Light yellow or white leaves at the shoot tip (Fig. 11.10g) indicate a lack of iron; iron sulfide (FeS) or an acidic iron chelate (for example, Fe-EDTA) is indicated. If the tips of the leaves are dying, it may be a sign that the plants are taking up and accumulating too many salts. This may come from overfertilizing or from minerals in the water. Give potted plants distilled water or enough tap water to leach high concentrations of minerals out through the bottom of the pot.

---

mentioned, anatomical adaptations can make the boundary layer thicker and stabilize it. A large concentration of leaf hairs (trichomes) tends to stabilize the air around them. Stomatal crypts, indentations of the leaf surface into which the stomata open, are especially effective. These adaptations are most often found in plants that live in deserts and other dry or windy areas, where the loss of water could be extreme. Some leaves tend to curl up as they dry out, effectively forming a crypt containing unstirred air.

11.3 MINERAL UPTAKE AND TRANSPORT

One of the roles of water in plants is to dissolve mineral elements and then transport them. These elements occur in natural and synthetic fertilizers (see sidebar, "Fertilizers," above). Although the need of plants for fertilizer has been known since the development of agriculture, it has been only in this century that the important components in fertilizer have been identified. Although organic fertilizers (for instance, fish extracts) are often effective, plants do not need to take up organic compounds (such as proteins, vitamins, or carbohydrates) to grow. They synthesize all these compounds themselves. What they need are elements that are substrates or catalysts for the synthetic reactions. The specific elements that are required have been determined by growing the plants with their roots in aerated solutions (*hydroponically*) containing different combinations of elements (Fig. 11.10).

Certain elements are required in fairly high amounts: potassium (K), calcium (Ca), nitrogen (N), phosphorus (P), magnesium (Mg), iron (Fe), and sulfur (S). Also needed are carbon (C), hydrogen (H), and oxygen (O), but carbon and oxygen come from the air; oxygen also comes from water, as does hydrogen, and we have already considered the transport of this compound. (The classical mnemonic for remembering these elements is: C. HOPKiNS CaFe—Mighty good.) These elements form the common compounds synthesized by plant cells. For instance, carbon, hydrogen, and oxygen are essential parts of all protein, carbohydrate, and nucleic acid molecules. Nitrogen is found in both proteins and nucleic acids, sulfur in proteins, and phosphorus in nucleic acids. Calcium is found in the cell wall and helps hold the other components of the wall together. Magnesium is a component of chlorophyll, and iron is part of some photosynthetic and respiratory enzymes; both atoms help activate certain other enzymes.

Other elements are required in smaller amounts: manganese (Mn), boron (B), molybdenum (Mo), copper (Cu), zinc (Zn), and chlorine (Cl). These compounds are needed in the active sites of certain enzymes with important synthetic functions.

Mineral Nutrients Are Solutes in the Soil Solution

SOIL TYPES Most of the elements that plants need exist in the soil. **Soil** is the part of the earth's crust that has been changed by contact with the biotic and abiotic parts of the environment. Soil is a weathered, superficial layer typically only 1 to 3 meters (3 to 10 feet) thick, made up of physically and chemically modified mineral material associated with organic matter in various stages of decomposition (see sidebar, "Soils," p. 176).

Each environment creates its own unique soil. Soils differ in depth, texture, chemistry, and sequence of layers. Soils are formally named, described, and classified based on these differences. The basic soil classification unit is the *soil type*. Soil types are grouped into soil series, families, and orders. In the entire world there are only 11 soil orders; representatives of the orders, including more than 14,000 soil series and probably more than 50,000 soil types, occur within the United States. The distribution of particular types of plants is often correlated with the presence of particular soil types. Rhododendrons, for instance, grow only in acid (low-pH) soils.

Plant cells, like cells of other organisms, take up mineral elements only when the elements are in solution. In general, that means that the elements are taken up as simple or complex ions—for instance, potassium as K^+, calcium as Ca^{2+}, phosphorus as $H_2PO_4^-$, nitrogen as NH_4^+ or NO_3^-. These ions are found in the soil solution. They enter the soil solution from the dissolution of crystals in rock and soil particles or from the decomposition of organic matter in the soil.

SOIL FORMATION The process of dissolving elements from rock starts with acidic rain (Fig. 11.11). Rain becomes acidic by dissolving CO_2 and, to a lesser extent, sulfur and nitrogen oxide gases like SO_2 and N_2O_5. In solution, these combine with the water to form acids: H_2CO_3, H_2SO_3, and HNO_3. These acids dissociate (lose H^+), raising the concentration of H^+ and making the raindrop slightly acidic. When this rain falls on rock, it can rapidly or gradually dissolve many of the rock's crystals. Limestone ($CaCO_3$) dissolves relatively rapidly through the reaction $CaCO_3 + H^+ \rightarrow Ca^{2+} + HCO_3^-$. Other minerals—such as hematite (Fe_2O_3) and feldspar ($K(AlSi_3O_8)$), sulfides such as chalcopyrite ($CuFeS_2$), and magnesium and calcium phosphate rocks—dissolve more slowly.

The rate at which crystals dissolve depends on the amount of crystal surface area in contact with water. Many processes increase the surface area. The dissolving of crystals itself forms cracks; other crystals on the sides of these cracks come into contact with water. Freezing and thawing of water in a crack breaks off pieces of rock and forms new fissures. This starts the process of soil formation. Water and wind erosion pulverize rock particles; and the smaller the particles, the greater the total surface area. Once there is a small amount of soil solution, lichens and small plants can start to grow. These, too, accelerate the processes of soil formation and dissolution of minerals. Their rhizoids and roots enlarge the cracks through turgor pressure and emit respiratory CO_2, which forms H_2CO_3 (and, by dissociation, H^+ and HCO_3^-).

Young soils, such as those formed as described above, can be a good source of minerals, depending on the composition of the parent rock and the size of the soil particles (with smaller particles providing a richer source of minerals). However, the best soils are not those with the highest concentrations of minerals in their soil solutions. A high concentration of ions raises the osmotic effect of the soil solution and thus limits the movement of water into plants. More importantly, high concentrations of certain ions, such as aluminum (Al^{3+}) and sodium (Na^+), are toxic to plants. Even the essential ion Mg^{2+} is toxic at high concentrations, and the nutrient $H_2BO_4^-$ is toxic at a concentration no more than twice the optimal concentration. It is better to have a lower concentration of the nutrients, with a source that releases ions into the solution as they are taken up by plants. Mature soils contain clay particles and decomposed organic matter that have fixed

Figure 11.10 Tomato plants grown in nutrient culture solutions to show visual symptoms of mineral deficiencies. (**a**) Comparison of a whole plant in complete solution and one in solution lacking nitrogen. (**b**) Solution lacking phosphorus. (**c**) Solution lacking potassium. (**d**) Solution lacking nitrogen. (**e**) Solution lacking sulfur. (**f**) Solution lacking calcium. (**g**) Solution lacking iron. (**h**) Solution lacking magnesium.

SOILS

Soils vary greatly in their physical properties and chemical compositions, as well as in the organisms that live in them. The constituents of soils depend on their parent material and the processes by which they were formed. Different soil constituents include: (1) sand and silt; (granular particles of quartz, feldspar, or basalt, with sand particles being much larger than silt); (2) clay (microscopic particles of complex minerals such as mica); and (3) humus (decayed organic matter). Soils occur in layers or *horizons*. The topsoil (A horizon) generally has the smallest particles and the most organic material. Subsoil constitutes the B horizon, and parent material (C horizon) consists of underlying rock fragments that extend to bedrock (Figure 1).

In judging the capacity of a soil to support plant growth, several properties are important:

Water-holding capacity is the amount of water that stays after the soil is allowed to drain by gravity, less the amount that is held to soil particles by forces stronger than those the plant can exert. This is the maximum amount of water available to plants. Well-balanced soils with considerable humus have the best water-holding capacity. Sandy soils tend to allow water to drain away; high-clay soils may become compacted, with little space for water beyond that held very tightly to the clay particles.

Aeration is the amount of air in the spaces between soil particles. Root growth and function require respiration and thus oxygen. The best soils have about equal amounts of air and water in their interparticulate spaces. Sand, which promotes drainage, also promotes aeration. Swampy soils often become anaerobic.

Cation-exchange capacity describes the quantity of cations that can bind to the soil material but be released in the presence of acid generated by the respiration of roots and microorganisms. Clays and humus are the soil constituents that contribute cation-exchange capacity.

The *pH* of a soil indicates where the soil solution stands on the acid–base scale. The amount of acidic and basic parent material is also important. Neutral soils are good at supporting root function. In basic soils, iron and other elements may be bound into compounds that make them unavailable to plants; in acidic soils, these elements are free, but they may leach from the soil and thus be depleted. In high-rainfall areas such as tropical rain forests, the soils are acidic and generally nutrient poor because of leaching. The growth of plants depends on the nutrients that are held in decaying organic material and that are delivered by wind and water from other regions.

Salinity and *toxicity* measure the presence of high concentrations of elements that interfere with root growth or

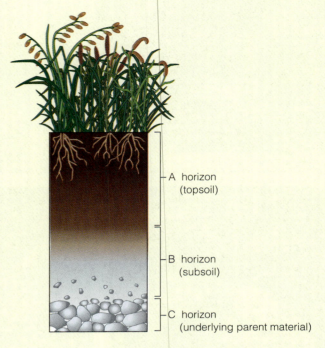

Figure 1 The soil of a grassland. The A horizon is alkaline, dark, and rich in humus. The B horizon consists of clay and calcium compounds. The composition of the C horizon depends on the underlying rock.

A horizon (topsoil)

B horizon (subsoil)

C horizon (underlying parent material)

function. High concentrations of sodium (Na^+) and other elements in saline soils are generally caused by poor drainage (for example, in a "dry lake" with no outflowing river) or by the presence of seawater. Toxic concentrations of some ions may reflect the composition of the parent rock. For example, serpentine soils have toxic concentrations of Mg^{2+} and nickel and low concentrations of Ca^{2+} because they are formed from a magnesium silicate rock containing crystals of heavy metals.

Biota, the organisms living in the soil, may be extremely important in determining the nutrient value of the soil. Nitrogen-fixing bacteria and other bacteria contribute to the nitrogen cycle. Mycorrhizal fungi promote the uptake of phosphate and perhaps other nutrients. Worms, insects, and small mammals break apart compacted soil and improve its texture. All these organisms contribute humus to the soil composition. On the negative side, herbivores and pathogens may kill roots and thus limit nutrient uptake into the plant.

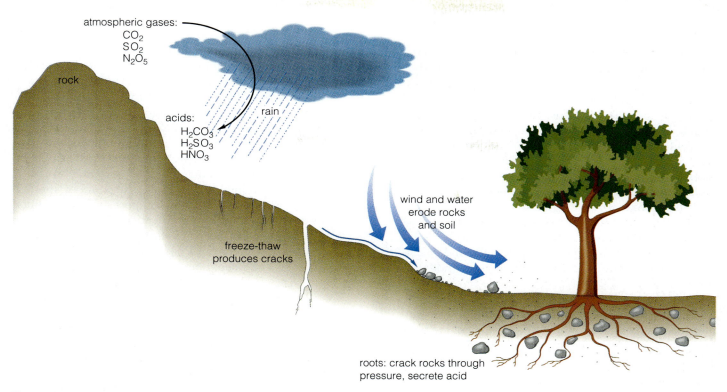

atmospheric gases:
CO_2
SO_2
N_2O_5

rock

acids:
H_2CO_3
H_2SO_3
HNO_3

rain

wind and water
erode rocks
and soil

freeze-thaw
produces cracks

roots: crack rocks through
pressure, secrete acid

Figure 11.11 Factors involved in soil formation.

negative charges. These negative charges bind electrostatically to cations (positively charged ions), lowering their concentrations in the solution but releasing them as they are needed. These soils are said to have a high **cation exchange capacity**. As roots grow, they respire CO_2, which acidifies the soil, releasing H^+. The H^+ binds to the fixed negative charges of the soil in exchange for mineral cations.

NITROGEN FIXATION Nitrogen is a special case. This element is needed in relatively large amounts by plants, but few soils contain much of it. From a global viewpoint, the major storehouse of nitrogen is nitrogen gas (N_2) in the atmosphere, but plants cannot use that form of nitrogen. N_2 must first be converted to NH_4^+ or NO_3^- by a reaction known as **nitrogen fixation**. Lightning and meteors can oxidize N_2 to NO_3^-, but most nitrogen fixation by far is catalyzed by enzymes in certain bacteria, which reduce N_2 to NH_4^+. Some of these are free-living bacteria in the soil. Some have developed mutualistic associations with specific plants, so that they donate a part of the nitrogen they fix and receive carbohydrate and other favorable living conditions in the plant. The best-known example is the bacterium *Rhizobium*, which lives in root cells of legumes (beans, clover, and so forth; see Chapter 19). However, more examples have recently been discovered. Alder trees (*Alnus* sp.), for instance, form associations with bacteria of the actinomycete type.

Although nitrogen can be taken up in either of two forms, NH_4^+ or NO_3^-, it is unusual to find soils that have large quantities of either ion. NH_4^+ (in equilibrium with NH_3) is volatile. Although it is converted to NO_3^- by some soil bacteria during a process known as **nitrification**, NO_3^- is very soluble and is easily leached from the soil. This is why the application of nitrogenous fertilizers is so effective in increasing crop yield. The nitrification of NH_4^+ and fixation of N_2 are just two examples of the chemical reactions that interconvert nitrogen-containing molecules. NO_3^-, once it is taken up by plants, can be converted to NH_4^+, a process known as *nitrate reduction*. NO_3^- in the soil can also be converted by certain bacteria to N_2; this is **denitrification**. Collectively, these reactions are known as the **nitrogen cycle**, in recognition that nitrogen moves back and forth between the abiotic and biotic components of the environment (see Fig. 27.4).

Minerals Are Actively Accumulated by Root Cells

All plant cells, particularly those in meristematic regions, require a source of minerals. Young root cells can absorb minerals directly from the soil solution, but the minerals must be transported through the plant to shoot cells, often over long distances. Because the minerals are in water solution, they are in part transported passively in the stream of water that is pulled through the plant by transpiration. However, there are some active processes that contribute to the uptake and transport of mineral ions. Active processes are ones coupled to strongly downhill reactions, like the hydrolysis of ATP or the oxidation of NADPH (see Chapter 8).

MAINTAINING A SUPPLY OF MINERALS If a plant is actively taking up minerals, it will deplete the supply in the area immediately around the roots. There are three processes that replenish the supply of minerals.

First is the bulk flow of water in response to transpiration. Water in the soil is pulled toward the roots by the gradient of attractive forces created by the removal of water around the roots. Ions in solution are swept along with the water, although local concentrations of charged organic molecules might withdraw ions from the stream (or release ions to it).

Second is diffusion, the tendency of materials in solution to move down their concentration gradients. Even if the bulk flow of water were to stop, there would be a net movement of ions toward the roots to replace those taken up by the roots.

The third process depends on the plant, not the solution. This process is growth: Roots continue to grow, more or less rapidly, throughout the lifetime of the plant. The rate of growth can be surprisingly rapid. The uptake of ions occurs just behind the root tip, so that as a root grows it moves to new regions of soil, where it comes in contact with a new supply of ions.

UPTAKE OF MINERALS INTO ROOT CELLS The next step in the travels of a mineral ion into a plant is its transport across the plasma membrane into a root cell. As mentioned above, this occurs just behind the growing root tip, in the region where primary tissues (for instance, epidermis, endodermis, xylem) have differentiated but where secondary growth has not begun. Upon entering the epidermis, the ions can move along the symplast—that is, through the plasmodesmata toward parenchymal cells in the center (stele) of the root. Ions may travel as far as the endodermis through the apoplastic pathway. Just as water molecules must cross a plasma membrane in order to cross the endodermis, so must mineral ions. In fact, this is probably the primary importance of the endodermis. Forcing ions to cross a plasma membrane before they enter the vascular system allows a plant to exclude toxic ions and to concentrate needed nutrients that are present at low concentrations in the soil solution.

A great deal of research has been done in the last few years to understand how ions cross the plasma membrane. In general, this involves pumps and channels (see Chapter 3). Root cells have the ability to accumulate ions against their concentration gradient—that is, to pull them into the cell, even though their concentration in the soil solution is less than their concentration in the cytoplasm. This requires energy (ATP) and thus active metabolism. It is one of the most important reasons why the growth of plants requires live, healthy roots.

ATP-generated energy is used to accumulate ions in at least two ways (Fig. 11.12). Both start with the plasma membrane proton pump, which pumps H^+ (protons) from the cytoplasm to the apoplast. The pumping of the protons generates a membrane potential difference, with

the inside of the membrane negative (by over 100 millivolts) relative to the outside of the membrane. Such a potential difference tends to pull cations, such as K^+, into the cell. A potential difference of approximately 60 millivolts will support a concentration difference of 10-fold; that is, K^+ will tend to be pulled into the cell until its concentration in the cytoplasm is 10 times higher than that immediately outside the plasma membrane. The relation between membrane potential and concentration difference is logarithmic, so a potential difference of approximately 120 millivolts will support a concentration difference of 100-fold. The potential difference does not ensure that the cations will in fact enter the cell, however. To enter, the cations must pass through specific channels in the plasma membrane, and these channels must be open. The need for specific channels is the molecular mechanism by which the root excludes toxic ions such as Al^{3+}.

The plasma membrane proton pump also generates a proton gradient across the plasma membrane, with a higher concentration of H^+ at the outside. At certain channels, protons form complexes with such anions (negatively charged ions) as NO_3^- and HSO_4^- (Fig. 11.12). The protons neutralize the negative charges of the anions (which would tend to keep the anions out of the cell) and in fact tend to pull the anions into the cell through those channels by diffusion (because the concentration of H^+ and thus of the complexes is higher outside the cell).

MYCORRHIZAE It is a remarkable fact that some roots in soil do not live alone but are closely associated with filaments of fungi, as mycorrhizae (see Fig. 5.22). There are different types of mycorrhizae, depending on the plant and fungus. In some cases, the fungi form a sheath around the root and grow into the spaces between the cortical cells; in others, the fungi penetrate the cortical cell walls. One might be tempted to think of the fungus as a parasite on the plant, obtaining food without providing anything to the plant in exchange; however, plants with mycorrhizae often grow better than plants without. Pine trees (*Pinus* sp.) in many infertile soils grow only in the presence of mycorrhizal fungi. It turns out that mycorrhizal fungi have high-affinity systems for taking up phosphate, systems lacking in the root cells. The fungi provide phosphate to the root, and the root provides carbon- and energy-rich nutrients to the fungi. This is a classic example of a mutualistic association, one that operates to the mutual advantage of both participants.

Ions Are Transported from the Root to the Shoot

Once mineral ions have entered the *root* cells, they can function in metabolic reactions in those cells. However, if they are to promote growth in the *shoot*, they must first be transported there. This requires, first, that the ions move from the stelar cells into the apoplast of the stele. Little is

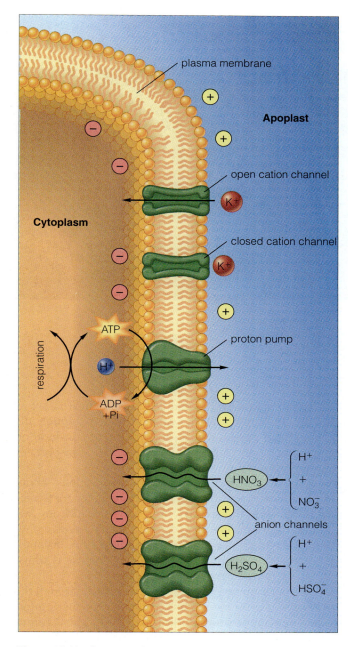

Figure 11.12 Pumps and channels in the plasma membrane. The ATP-utilizing proton pump, which forces H$^+$ out of the cell, provides the major energy charge across the plasma membrane. Channels allow the accumulation of ions in the cytoplasm.

than in the cytoplasm. An alternate hypothesis suggests that ions are accumulated in the vacuoles of developing tracheary elements, probably to a high concentration. Then the ions are released into the apoplast and xylem stream when the xylem elements become mature and their cytoplasm breaks down (see Chapter 4). So far, there is not enough evidence to reject either of these hypotheses.

Ions secreted into the apoplast can be swept into and through the xylem in the transpiration stream. This process takes the ions to whatever region of the plant has stomata open and transpiration occurring.

Ions that have been transported to the shoot may be taken up into the shoot cells. The uptake requires processes like those in root cells, except that the concentrations of ions are probably higher (because they accumulate and the solvent water evaporates). If the salt concentrations in the xylem stream are high, and if the rate of evaporation is high, and if the production of new leaves is slow, then salts may be secreted from the leaves and appear on the surface as crystals. This secretion may occur from specialized trichomes with salt gland cells that actively accumulate and secrete the salts. Even if the salt concentrations in the soil solution are low, salts will tend to build up in the leaves. Eventually, they may reach toxic concentrations. The dead tips of the older leaves of slow-growing houseplants are a sign that salt has accumulated to a toxic level in those leaves. To slow this process, one should water these plants infrequently, but thoroughly. Allow excess water to drain through the pot, carrying away ions that have been accumulated in the soil. Fertilize the plants infrequently and only as long as they show signs of growth.

Root Pressure Is the Result of an Osmotic Pump

As described above, mineral ions are taken up into root cells, passed through the symplast into cells of the stele, and then secreted into the apoplast of the stele. The concentration of ions in the stelar apoplast and in the flowing xylem sap depends in part on the rate of water uptake. If transpiration is occurring, the concentration of ions in the xylem sap may be quite low. But if no transpiration is occurring, ions can accumulate in the apoplast of the stele.

The accumulation of ions in the stele has an osmotic effect. If the soil is saturated with water, the water concentration of the soil solution will be higher than that in the xylem. In this case, water tends to enter the root and stele, building up pressure (*root pressure*) in the xylem and forcing the xylem sap up into the shoot (Fig. 11.13a). The flow of xylem sap, caused by the accumulation of osmotically active salts in the stele, is an example of an osmotic pump. In some grasses and small herbs, water is forced out special openings in the leaves called *hydathodes*. We may see droplets of *guttation water* on the tips of grass leaves on a cool, moist, still morning (Fig. 11.13b). This is

known about the mechanisms involved in this process. One hypothesis suggests that there are special mechanisms by which ions are transported across the plasma membrane of stelar cells out into the apoplast. If this is true, these mechanisms are regulated in a quite different way from those in cortical cells (which take up the ions). The mechanisms are probably selective (requiring special channels) and possibly active. If they are active they require ATP, and they can pump ions into the apoplast even if those ions are more concentrated in the apoplast

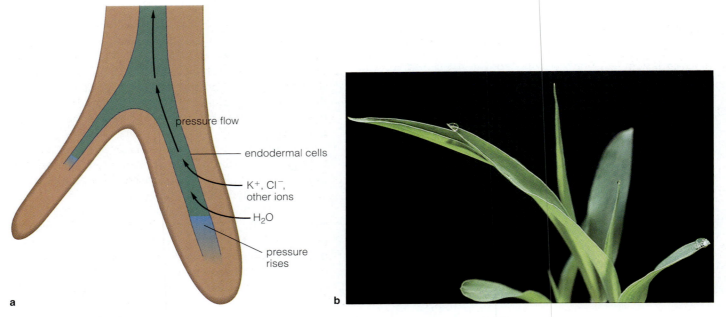

a

b

Figure 11.13 Root pressure. (**a**) Root pressure is generated by an osmotic pump. Solutes accumulate in the stele. Water is pulled in and forced up the xylem. Note that the endodermal cells provide the differentially permeable membrane needed for osmosis. (**b**) Guttation from the tips of barley (*Hordeum vulgare*) leaves caused by root pressure.

water forced out of hydathodes by root pressure and should not be confused with dew (water vapor condensed on a cold surface).

11.4 PHLOEM TRANSPORT

Osmosis Can Pump Solutions

Of all the nutrients that must be transported through the plant, the most important is carbohydrate, the product of photosynthesis. Carbohydrate is a source of carbon for the synthesis of all other organic molecules. It is also a source of energy, which is converted into a more useful form (ATP) when the carbohydrate is metabolized in respiration (Chapter 9). Transport of carbohydrate is sometimes called **translocation**.

Carbohydrate is synthesized by photosynthesis in the chloroplasts of mature leaf cells. It may be stored temporarily as starch in the chloroplasts. Later it may be exported from the leaf in the form of sucrose (common sugar) or, less commonly, another sugar.

The carbohydrate is transported through the phloem. That this is so can be seen by *labeling* the carbohydrate with radioactive CO_2. (Such CO_2 contains the carbon isotope ^{14}C, whose nucleus decays spontaneously, producing high-energy radiation that can be tracked). When the petioles are sampled shortly after the labeling, sectioned, and tested, the radioactivity is found in the sieve tubes. Using this same labeling technique, it is possible to measure the rate of carbohydrate transport. This rate (1 to 2 cm or 0.4 to 0.8 inches per minute) is faster than diffusion

or transport from individual cell to cell, but it is not as fast as the rate at which water is pulled through the xylem. This suggests that the mechanism of translocation is different from diffusion, or cell-to-cell transport, or transpiration.

One of the most interesting observations about phloem transport is its ability to change direction. Generally, sucrose is transported from leaves, which are net producers of carbohydrate, to roots, which use carbohydrate for growth or storage. Sometimes (for instance, when roots are sprouting new shoots) sucrose is transported up from the roots to the shoots.

The mechanism by which sucrose moves through the phloem was the subject of much research and many arguments by scientists in the 1960s and 1970s. The accumulated evidence, however, now supports the idea that the sucrose flows through sieve tubes as one component in a bulk flow of solution. The flow is directed by a gradient of hydrostatic pressure and powered by an osmotic pump. This is variously called the Münch, pressure-flow, or **mass-flow hypothesis**.

Plants Use Osmotic Pumps to Transport Sucrose Through Sieve Tubes

The phloem can be seen as a dynamic osmotic pump, with a **source** of solute at one end and a **sink** at the other. We have already seen one example of an osmotic pump in the section on root pressure. The difference in concentration of ions between the apoplast of the stele (where it was higher) and the apoplast of the cortex and the soil solution (lower) forced water into the stele. The resulting

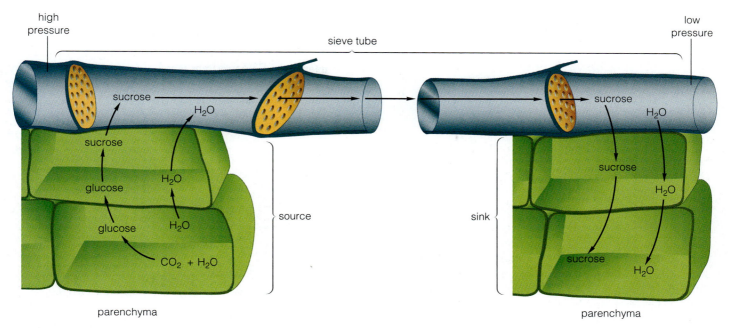

Figure 11.14 The mass-flow mechanism of phloem transport. Sucrose is actively transported into the sieve tubes at the food source region of the plant (leaves or storage organs) and removed at the sink regions (regions of growth or storage). Water follows by osmosis, raising the pressure in the sieve tubes at the source region and lowering the pressure at the sink region. The sieve tube contents flow *en masse* from high- to low-pressure regions.

buildup of hydrostatic pressure pushed the xylem sap up the xylem and into the shoot.

In the phloem, sucrose is the main osmotically active solute. It is pumped from photosynthetically active leaf parenchymal cells into sieve tubes of the minor veins (Fig. 11.14). The mechanism of pumping probably involves a carrier (channel) that transports sucrose together with a proton across a plasma membrane. The pH and electric charge differences across the membrane provide the pumping energy. The exact pathway from parenchymal cell to sieve tube is still uncertain.

The accumulation of sucrose in a sieve tube pulls water into the sieve tube from the apoplast by osmosis. This increases hydrostatic pressure inside the sieve tube at the source (for example, on photosynthetic tissue or on tissue releasing stored food). The pressure (which can reach more than 20 times atmospheric pressure) starts a flow of solution that will travel to any attached sieve tube in which the pressure is lower.

The pressure gradient along the sieve tube depends on the gradient established by differences in sucrose concentration. Because the flow of solution along the sieve tube carries sucrose along with it, you would expect the concentration of sucrose at the source sieve-tube member to drop eventually and the concentration at the sink sieve tube element to rise. This would eliminate the concentration gradient, and the flow would stop. The loss of the concentration gradient is prevented by two mechanisms: the continual pumping in of sucrose at the source and the removal of sucrose at the sink.

These two processes explain how the direction of phloem transport can change. If an organ is a source, it pumps sucrose into the phloem, the hydrostatic pressure rises, and phloem sap flows out of it. If an organ is a sink, it removes sucrose from the phloem and lowers the hydrostatic pressure, so that more phloem sap containing sucrose flows toward it. Growing tissues, such as shoot and root meristems and expanding fruits, are always sinks. The parenchyma cells of stems and roots may, however, be sinks or sources at different times. A storage root, such as a carrot root, is a sink during the first growing season. It removes sucrose from the phloem and stores starch and sugar in its parenchyma cells. After a winter (and once the shoot starts to *bolt* or flower), the carbohydrate in the root is converted to sucrose and pumped into the sieve tubes. At that point the root becomes a source.

Research indicates that the osmotic pump is actively regulated. The mesophyll cells of a young dicot leaf start to photosynthesize early, producing carbohydrate, but the leaf is also growing and using carbohydrate. On balance, a growing leaf is a net user of carbohydrate, so it imports sucrose from the phloem system. Once mature, though, the leaf exports sucrose to the phloem system. One might imagine that the transition is gradual as the leaf's rate of growth tapers off. Instead, measurements show that it is abrupt. This suggests that an off-on switch regulates the direction of sucrose transport at the sieve-tube plasma membrane, but the nature of that switch is still a mystery.

SUMMARY

1. Plants lose water to the atmosphere by transpiration, which involves the diffusion of water vapor from air spaces inside the plant through stomata, lenticels, or cracks in the cuticle to the outside air. The amount of water transpired by plants is very great.

2. The movement of water depends on diffusion, osmosis, capillary forces, hydrostatic pressure, and gravity. The concept of water potential allows one to predict how water will flow in response to a combination of these forces.

3. Under most conditions, water is pulled through a plant. As a first step, water vapor diffuses outward through a relatively unstirred boundary layer of air into the bulk atmosphere. Water evaporates from the cell walls to replace what was lost. Water is pulled by tension up the xylem and into the cell walls. Finally, water moves from the soil to the xylem of young, growing roots through symplastic and apoplastic pathways.

4. Transpiration is controlled by the size of the stomatal opening, which depends on the guard cells. Under certain conditions (in the light, with low CO_2 concentration), guard cells accumulate solutes. A resulting influx of water raises their turgor pressure and stretches them in a way that opens the stoma. In the dark, in high CO_2, or under drought conditions, the process is reversed.

5. The essential minerals for plants are C, H, O, K, N, P, S, Ca, Mg, and Fe, and, in smaller amounts, Mn, B, Co, Cu, Zn, and Cl.

6. Mineral elements are taken up by roots from the soil solution. The minerals reach the soil solution by dissolving from rock crystals into acidified rainwater or through the decomposition of organic matter.

7. Minerals are taken up into root cells actively and specifically. This means that the processes transporting minerals use ATP and channels in the membrane that pass only the desired mineral ions. The hydrolysis of ATP is used to pump protons out of the cell. Part of the energy released is saved in the proton and electrical gradients across the membrane. The proton and electrical gradients are used to force ions into the cell, where they can be concentrated to 100 times or more their concentration in the outside solution.

8. In the absence of transpiration, the accumulation of ions in the root stele pulls water into the stele by osmosis and increases the hydrostatic pressure. This so-called *root pressure* forces water up to and out of leaves, where it appears as guttation.

9. Sucrose, a carbohydrate formed from the products of photosynthesis, is transported from sources (photosynthetic or source tissue) to sinks (growing or expanding tissue) in the sieve tubes of the phloem.

10. The pressure that forces sucrose and other compounds through the sieve tubes comes from a gradient of hydrostatic (turgor) pressure, high near sources and low near sinks. The gradient is produced and maintained by the pumping of sucrose into the sieve tube at the source and its removal at the sink.

Questions

1. Match the forces with their effects:

osmosis	creates a tension in the water within tracheids
gravity	tends to push water out of a cell
capillary forces	moves water vapor molecules out of a leaf
turgor	opposes the transpirational flow of water up a tree
diffusion	tends to pull water into a cell

2. Predict how each of the following changes will affect the tendency of water to move into or out of a leaf cell.
   a. placing the leaf in a closed, humid chamber
   b. doubling the concentration of solutes inside the cell
   c. soaking a piece of leaf containing the cell in a concentrated solution of sucrose

3. Explain why the rate of transpiration rises when:
   a. dawn breaks and the sun comes up
   b. the weather becomes very windy

4. Is the circumference of a tree's trunk greater, less, or the same during the day (when the tree is transpiring rapidly) as during the night (when there is little transpiration)? Hint: Consider what happens to a single xylem vessel during transpiration.

5. List the seven mineral elements needed in the highest amounts for optimal growth of a plant. How does each contribute to the structure or function of the plant?

6. What role does acid rain (in moderate amounts) play in making minerals available to a plant?

7. Can a K^+ ion travel all the way from the soil to the xylem of a root by either the symplastic or the apoplastic pathway? Explain your answer.

8. Why is respiration an important process in roots? Explain how biochemical energy is used to power the uptake of mineral ions into the cytoplasm of a plant cell.

9. Explain the difference between a *Rhizobium* and a mycorrhiza.

10. What causes the guttation water seen on the tips of grass leaves early in the morning? Why does one not see guttation water in the afternoon?

11. Does sucrose exported from photosynthesizing leaves in the phloem always travel down the stem to the roots? Explain your answer.

12. Which is greater: the hydrostatic pressure in a sieve tube of a photosynthesizing leaf or the hydrostatic pressure in a sieve tube of a developing flower?

Further Readings

Salisbury, F. B., and Ross, C. W. 1992. *Plant Physiology.* 4th ed. Belmont, Calif.: Wadsworth.

12 LIFE CYCLES: MEIOSIS AND ALTERNATION OF GENERATIONS

1. Life perpetuates itself through reproduction. Reproduction is the transfer of genetic information from one generation to the next, and this transfer is our definition of *life cycle*. Reproduction can be asexual or sexual.

2. Asexual reproduction requires a cell division known as *mitosis*. Asexual reproduction offers many advantages over sexual reproduction, one of which is that it requires only a single parent. A significant disadvantage of asexual reproduction is the loss of genetic diversity and the likelihood of extinction when the environment changes.

3. Sexual reproduction involves the union of two cells, called *gametes*, which are usually produced by two different individuals. Another kind of cell division, known as *meiosis*, is ultimately necessary to produce gametes.

4. Every species in the kingdom Plantae has both diploid and haploid phases—that is, plants whose cells are all diploid or all haploid. These phases are called *generations*, and they alternate with each other over time.

5. The fossil record reveals that the most recent groups to evolve—those plants that are most complex and adapted to dry land—have sporic life cycles in which the gametophyte generation is smaller and the sporophyte generation is more dominant.

12.1 LIFE CYCLES AND THE PASSAGE OF GENETIC INFORMATION

A basic characteristic of life is that it perpetuates itself. The process can be sexual or asexual. In **asexual reproduction**, each generation is genetically identical to the last. Asexual reproduction occurs in both unicellular and multicellular organisms. For example, a single-celled alga floating near the surface of a lake can divide asexually to produce two single-celled offspring (Fig. 12.1a). This individual cell divides to produce two new cells; then each of those will divide to produce others, and so on for generations of cells. Similarly, a cell within the root tip of a corn plant can reproduce asexually, first to generate two identical offspring cells and then eventually a whole tissue, layer, or region of the root, made up of thousands of identical cells (Fig. 12.1b). The succulent air-plant (*Kalanchoe pinnata*) is able to asexually produce miniature plantlets along leaf edges, each of which can fall off, take root, and become a new plant identical to the parent (see Fig. 7.18c).

Asexual reproduction requires only a single parent cell or parent organism, and all of the progeny will be identical to that parent. The collection of identical individuals is called a **clone**. Many plants in nature produce clones; for example, strawberries, aspens, and coast redwoods (Fig. 12.2). Strawberry plants (*Fragaria* sp.) produce horizontal aboveground stems (stolons or runners) that period-

a

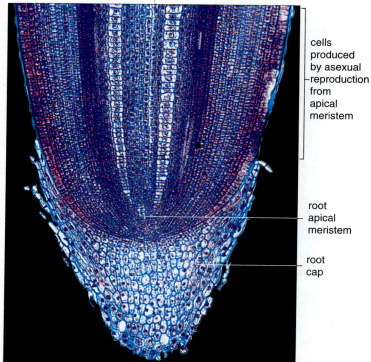

b

Figure 12.1 Asexual reproduction occurs in single-celled organisms and in the tissues of multicellular organisms. (**a**) *Micrasterias*, a single-celled freshwater green alga, has just undergone cell division to produce two daughter cells. (**b**) Longitudinal section of the root tip of corn (*Zea mays*). Cell division in the apical meristem produces several different tissues behind the tip.

cells produced by asexual reproduction from apical meristem

root apical meristem

root cap

ically take root at the nodes and produce new leaves, flowers, and fruits there. If the runners are severed, the new plants remain alive and capable of a fully independent life. When a redwood (*Sequoia sempervirens*) trunk is burned by fire or removed by timber harvest, buds buried under the bark at the base of the stump are stimulated to begin growth. Within decades, a cluster of young redwood trees, all sharing the same root system but having separate trunks, will exist. In time, the parental stump will decompose and disappear, leaving a circular clone of equal-aged offspring. Trembling aspen (*Populus tremuloides*) trees produce special roots that grow laterally under the soil instead of down. These roots periodically give rise to stems—and in time to mature trees—some distance from the parent tree. An entire grove of aspen, occupying several acres of ground, can be a single clone of hundreds of genetically identical trees, seemingly independent but actually all connected to each other below ground.

Figure 12.2 Asexual reproduction also produces clones of many genetically identical individuals. (**a**) Strawberry (*Fragaria*) plants produce horizontal, aboveground stems (stolons or runners), which periodically develop roots and shoots. Here the clone consists of three plants. If the stolon or runner is cut, the members of the clone continue to function independently. (**b**) Looking up at a circle of coast redwood trees (*Sequoia sempervirens*) that all began life at the same time as stump sprouts attached to the base of a burned parent tree. (**c**) Clone of trembling aspen (*Populus tremuloides*) in the Sierra Nevada of California. It is likely that all individuals in this photograph are connected to each other underground.

Sexual reproduction, in contrast, causes each generation to be genetically different. For example, when a bean plant produces flowers, bean pods, and seeds, each seed germinates to produce a plant that is slightly different from the parent plant: perhaps taller, maybe a bit more frost tolerant, possibly with slightly different petal colors, or capable of flowering a few days sooner than the parent.

Both kinds of reproduction result in the transfer of genetic information from parent to offspring, which is the definition of **life cycle** as used in this book. Previously, from other contexts, you might have defined *life cycle* as the process of growth, maturity, and senescence that occurs in any organism between the moments of birth and of death. But in this book we equate *life cycle* with the passage of genetic information from one generation to the next, like the transfer of a baton of DNA between relay runners.

Asexual Reproduction Passes On Unchanged Genetic Information Through Mitosis

Asexual reproduction requires a particular kind of cell division called **mitosis**, which was described in Chapter 3 (Fig. 12.3). Briefly, mitosis proceeds first by a doubling or copying phase in which every chromosome duplicates itself. By the end of this first phase (prophase), each chromosome consists of two identical parts called *sister chromatids*, which are joined together at one point. The chromosomes/chromatids line up on the cell plate during metaphase. In anaphase, the duplicate chromatids pull apart from each other, and each chromatid can now be called a *chromosome*. In telophase, the two groups of chromosomes become separated by a cell plate, thus

forming two offspring cells. Each of the offspring cells retains the same number of chromosomes as the parent. There is no mixing of chromosomes from different parents; as a consequence, each offspring cell retains the exact same complement of chromosomes (and genes) as that of the parent.

There are several advantages to asexual reproduction. First, only a single parent is required; therefore, any isolated individual can still produce offspring and populate a new part of the species' range. Second, asexual reproduction produces offspring that are going to be just as successful in the habitat as the parent was—and because the parent lived to reproduce, so too should the next generation (providing the environment stays unchanged). Third, asexual reproduction generates more offspring, faster than sexual reproduction, so an invading species can come to dominate the landscape quickly. Fourth, asexual reproduction costs less in terms of metabolic energy than sexual reproduction. This is because sexual reproduction requires a plant to invest in reproductive tissue (flowers) even though successful seed formation, dispersal, and germination might not occur in any given year. Asexual reproduction, in contrast, always works.

The only disadvantage to asexual reproduction is that genetic diversity remains fixed. New plants are genetically identical to parents. If the environment changes, all the organisms are equally susceptible to any new stress, such as a new disease, a drought, or a migratory invasion of herbivores. Sexual reproduction results in genetic diversity, and this gives the maximum probability for continued existence of a species over long time periods. The geologic history of earth has consistent themes of environmental change and of extinction for those groups of organisms that possessed the least genetic diversity.

Mitosis

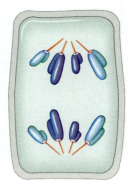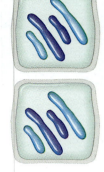

Prophase

Each duplicated chromosome (consisting of two sister chromatids) condenses from threadlike form to rodlike form.

Metaphase

All chromosomes are now positioned at the cell's equator.

Anaphase

Sister chromatids of each chromosome are separated from each other. These new, daughter chromosomes are moved to opposite poles of the cell.

Telophase

When the cytoplasm divides, there are two cells. Each is diploid (2n)—*it has the same chromosome number as the parent cell.*

Figure 12.3 A review of mitosis. A cell with four chromosomes (each chromosome consists of two sister chromatids) is shown in prophase and metaphase. During anaphase the chromatids separate. In telophase, the two daughter cells still have four chromosomes each.

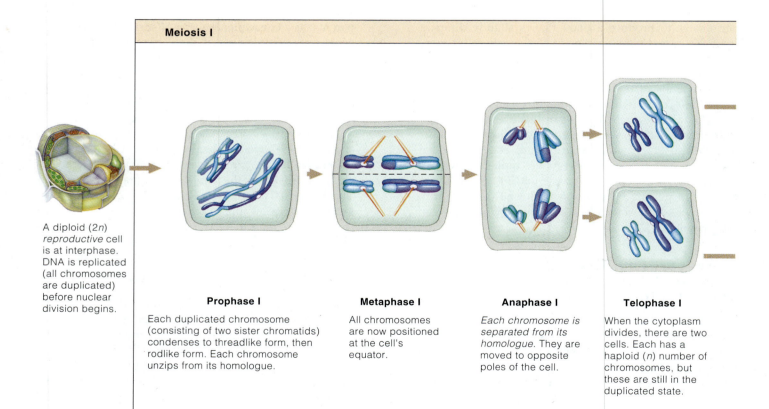

Meiosis I

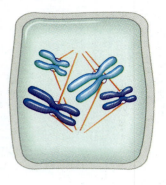

A diploid (2n) *reproductive* cell is at interphase. DNA is replicated (all chromosomes are duplicated) before nuclear division begins.

Prophase I

Each duplicated chromosome (consisting of two sister chromatids) condenses to threadlike form, then rodlike form. Each chromosome unzips from its homologue.

Metaphase I

All chromosomes are now positioned at the cell's equator.

Anaphase I

Each chromosome is separated from its homologue. They are moved to opposite poles of the cell.

Telophase I

When the cytoplasm divides, there are two cells. Each has a haploid (n) number of chromosomes, but these are still in the duplicated state.

Figure 12.4 The two divisions of meiosis. A cell with four chromosomes (each chromosome consists of two sister chromatids) is shown. Meiosis I is a reduction division, during which homologous chromosomes are separated and the number of chromosomes in each cell is halved. Each progeny cell has two chromosomes, each chromosome with two sister chromatids. Meiosis II is a normal mitotic division during which sister chromatids are separated and the number of chromosomes remains constant. Each progeny cell has two chromosomes in this example.

Sexual Reproduction Passes On New Information Through Meiosis

You might think that sexual reproduction requires the union of two organisms, but technically it is the union of two cells, called **gametes**. Unlike asexual reproduction, sexual reproduction requires two parental cells—two gametes—not one. The two gametes must find each other, physically join, and mix their chromosomes in order to create a single offspring cell. Sexual reproduction thus poses two problems. One is to find a way for the gametes to be brought together; the other is to reduce the number of chromosomes. If two gametes with the normal number of chromosomes fuse, then the resulting offspring cell will have twice the normal number. A repetition of this over several generations would create cells with so many chromosomes that normal cell division would be impossible. The solution to this latter problem is a type of cell division—**meiosis**—that reduces the number of chromosomes by half. *Meiosis* comes from a Greek word meaning "to diminish."

Let us examine the steps in meiosis. The terminology of mitosis applies to meiosis as well. The preparation for meiosis begins with replication of new DNA. By the time meiosis begins, each chromosome consists of two identical "sister" chromatids joined together at one point. Meiosis opens with prophase I (Fig. 12.4), in which the chromosomes coil and thicken and become visible when seen through a light microscope.

An event unique to meiosis now occurs. We see that every pair of chromosomes has a slightly different shape. For example, in every dandelion plant (*Taraxacum officinale*) nucleus we can count 24 chromosomes, and we can see that there are 12 distinct pairs of chromosomes; in every cotton plant (*Gossypium hirsutum*) nucleus there are 52 chromosomes organized in 26 pairs; and in every human nucleus there are 46 chromosomes in 23 pairs. The two organisms with the fewest and largest number of chromosomes are both plants: A little California desert annual called *Machaeranthera gracilis* has 4 chromosomes (2 pairs), whereas the adder's tongue fern (*Ophioglossum reticulatum*) has 1260 chromosomes (630 pairs). Because chromosomes are paired the chromosome number is said to be **2n** or **diploid**. Thus, the 2n or diploid number of chromosomes in adder's tongue fern is 1260.

The members of each pair are called **homologous chromosomes**, or **homologues**. (Do not confuse homologous *chromosomes* with *chromatids*. At the beginning of

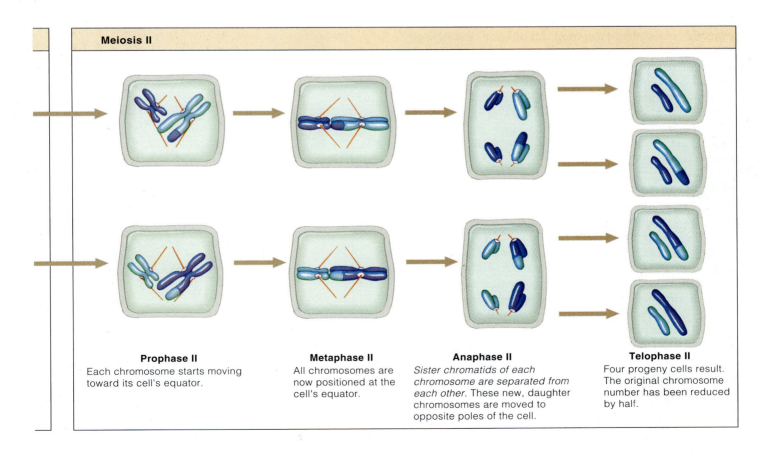

Meiosis II

Prophase II
Each chromosome starts moving toward its cell's equator.

Metaphase II
All chromosomes are now positioned at the cell's equator.

Anaphase II
Sister chromatids of each chromosome are separated from each other. These new, daughter chromosomes are moved to opposite poles of the cell.

Telophase II
Four progeny cells result. The original chromosome number has been reduced by half.

Figure 12.5 Life cycle of cherry (*Prunus*). Starting with the flower, which is part of the sporophyte tree, first read to the top left to see how meiosis occurs in pollen sacs and leads to the production of the pollen tube gametophyte. Then read from the flower to the top right to see how meiosis occurs in ovules and leads to the production of another gametophyte (called the *embryo sac*). Inside the pollen tube is a gamete called a *sperm*, and inside the embryo sac is a gamete called an *egg*. When the two gametes meet and fuse they produce a diploid zygote cell (not shown). As the zygote grows and divides by mitosis, it produces an embryo. The embryo germinates into a seedling, and the seedling grows back into a cherry tree.

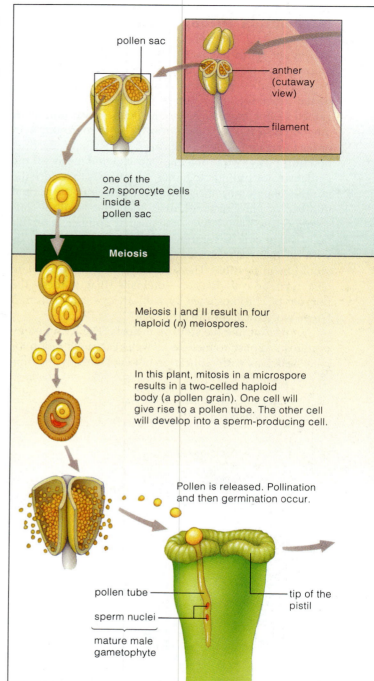

pollen sac

anther (cutaway view)

filament

one of the 2*n* sporocyte cells inside a pollen sac

Meiosis

Meiosis I and II result in four haploid (*n*) meiospores.

In this plant, mitosis in a microspore results in a two-celled haploid body (a pollen grain). One cell will give rise to a pollen tube. The other cell will develop into a sperm-producing cell.

Pollen is released. Pollination and then germination occur.

pollen tube

tip of the pistil

sperm nuclei

mature male gametophyte

prophase I every chromosome is composed of two attached chromatids, and each chromosome has a physically separate homologous chromosome.)

In metaphase I, all the pairs of homologous chromosomes line up on the equator of the cell, one on each side. In anaphase I, the chromosome pairs separate. Recall that in mitosis the chromatids separate during anaphase, but understand that in meiosis homologous chromosomes separate, one homologue moving to each pole. Because the chromosomes at the two poles will belong to different cells once the new cell plate forms in telophase I, this cell division reduces the number of chromosomes by half. The daughter cells each are said to have *n, 1n*, or a **haploid** number of chromosomes (Fig. 12.4). This first division in meiosis is called a **reduction division** because the chromosome number in each cell has gone from 2*n* to *n*.

There is a pause (**interkinesis**), and then the two progeny cells undergo a second cell division. In this second round of prophase II, metaphase II, anaphase II, and telophase II, chromatids are separated in the same manner as occurs in mitosis (Fig. 12.4). This second division is, then, a normal mitotic division. The result of the two divisions is a set of four haploid cells. If the original diploid parent cell had 20 chromosomes (2*n* = 20; 10 pairs), each of the four resulting cells would have 10 chromosomes (*n* = 10; no pairs).

12.2 ALTERNATION BETWEEN DIPLOID AND HAPLOID GENERATIONS

Every organism in the plant kingdom has diploid and haploid generations. A **diploid generation** is represented by a plant whose every cell contains the diploid complement of chromosomes. A **haploid generation** is represented by a different plant whose every cell contains the haploid number of chromosomes. Both plants are different phases of the same species. They alternate in time with each other: The diploid plant undergoes meiosis to produce a haploid plant, which later develops gametes that fuse and produce a diploid plant. This alternation of phases is called the **alternation of generations**.

Except for mosses, plants of the diploid generation in the kingdom Plantae are large and visible, whereas haploid plants are small and hidden. Every cell of a grass leaf, a tree branch, a potato tuber, or a carrot root has this diploid condition. Thus, a grass plant, a tree, a potato plant, and a carrot plant are all examples of the diploid

generation. A synonym for a diploid plant is **sporophyte** (spore plant); each grass plant, tree, potato plant, and carrot plant is a sporophyte. Sporophytes in the plant kingdom are relatively large, long-lived, and complex.

The haploid generation of a grass plant or a carrot plant is microscopic in size and is found only in certain parts of the flower. These haploid plants are called **gametophytes** (gamete plants). Gametophytes in the plant kingdom are relatively small, short-lived, and simple in structure.

The life cycle of a cherry tree (*Prunus*) offers a good

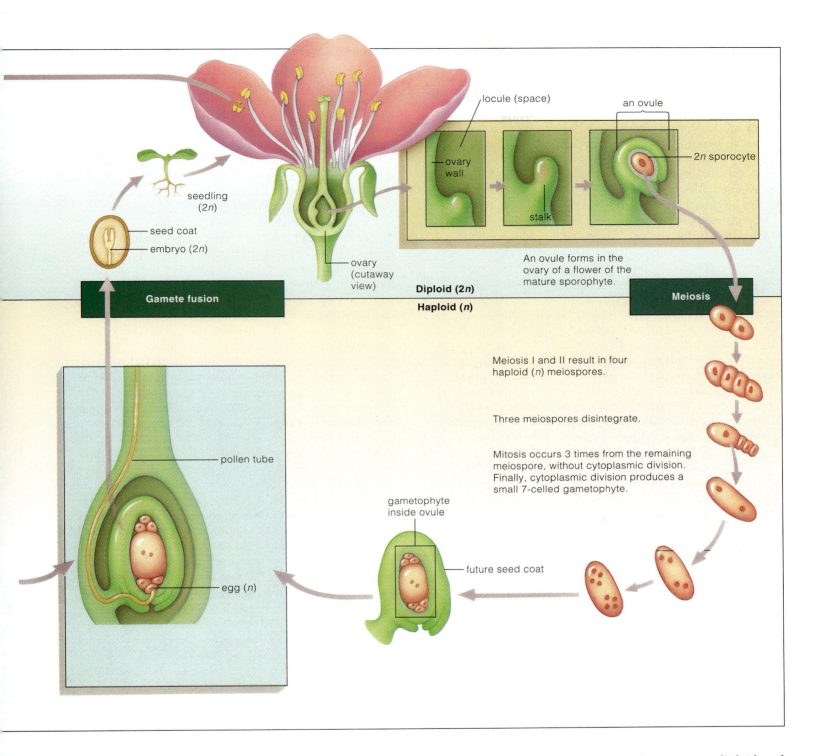

locule (space)

an ovule

ovary wall

stalk

2n sporocyte

seedling (2n)

seed coat

embryo (2n)

ovary (cutaway view)

An ovule forms in the ovary of a flower of the mature sporophyte.

Diploid (2n)

Haploid (n)

Gamete fusion

Meiosis

Meiosis I and II result in four haploid (n) meiospores.

Three meiospores disintegrate.

Mitosis occurs 3 times from the remaining meiospore, without cytoplasmic division. Finally, cytoplasmic division produces a small 7-celled gametophyte.

pollen tube

gametophyte inside ovule

future seed coat

egg (n)

example of sporophyte and gametophyte generations in the life cycle of a flowering plant (Fig. 12.5). The sporophyte of cherry is a tree, with the normal complement of root, stem, leaf, and flower organs. Only in certain parts of the flower are any haploid cells to be found. A detailed discussion of flower structure and function will come later, in Chapter 13; for now just focus on the innermost parts of the the flower. In cherry, there is a whorl of slender filaments, and at the tip of each is a swollen region called the **anther**. When an anther is sectioned, one can see that it is composed of four **pollen sacs**. Initially, all the cells that make up the pollen sacs are diploid and divide by mitosis, but eventually an innermost group of cells divide once by meiosis, each producing four haploid cells. These haploid cells are called spores. A **spore** is a reproductive cell capable of producing a new plant by itself. In this particular case, the spores are produced by meiosis, and they are therefore more precisely called **meiospores**. (In other plants, spores may be produced by mitosis, and these are called **mitospores**.)

Each meiospore develops a thick outer wall, undergoes mitosis once or twice, and becomes a microscopic pollen grain. A **pollen grain** is a young, not completely developed gametophyte. The sacs split open, releasing the pollen grains to passing air currents or animals.

In the very center of the flower is a vase-shaped **pistil** whose swollen base is called an **ovary**. Chambers in the ovary are lined with microscopic bulges of tissue, each called an **ovule**. At first, all the cells of each ovule are diploid, but eventually an innermost single cell undergoes meiosis and produces four haploid cells—four meiospores. The meiospores are not liberated, as they are in the pollen sac; instead they remain inside the ovule. Three meiospores degenerate, but the fourth divides by mitosis to produce a tiny, several-celled gametophyte. One of the cells acts as a gamete; it is called the **egg**.

In the process of **pollination**, pollen grains leave the pollen sac and reach the tip of a pistil. Pollination is the transfer of pollen from the anther to the pistil. If the transfer is within a single flower or between flowers on the same individual plant, it is called **self-pollination**. If the transfer is between flowers on different individuals, it is called **cross-pollination**. Each pollen grain that is genetically compatible with the tip of the pistil ruptures (that is, it *germinates*), and it sends a slender **pollen tube** growing down through the pistil toward an egg. Inside the tube a final mitosis occurs, producing two haploid gametes called **sperm**. This pollen tube, with its sperm, is a complete, mature gametophyte.

When the pollen tube reaches an egg, one of the sperm will escape from the tube and fuse with the egg. At first the two gamete cells fuse, but their nuclei do not. This phase of cytoplasmic fusion is called **plasmogamy**. The chromosome condition is neither n nor $2n$, but $n+n$, a **dikaryotic** state. Usually the nuclei fuse soon after, in a process called **karyogamy**, producing a $2n$ nucleus. Technically, sexual reproduction is the production, liberation, and fusion of gametes. A synonym for gamete fusion is the term **fertilization**.

The new diploid cell that results from the fusion of two gametes is a **zygote**. The zygote begins to divide by mitosis immediately, and in time it forms an **embryo** within a seed coat. The embryo is simply a dormant, multicellular, very small, juvenile sporophyte. When conditions are appropriate, the seed germinates into a seedling, and the seedling grows by mitosis into a mature sporophyte cherry tree, bringing the life cycle back to where we began.

Here Is the Basic Vocabulary of a Mythical *Superplant* Life Cycle

Not all plants reproduce like the cherry tree. Later in this book, we shall discuss how conifers, ferns, mosses, algae, and fungi have significantly different details in their life cycles. All of their life cycles, however, progress through the same general stages. If we study a diagram of the most general life cycle—one that has all possible stages—then the concepts of such a life cycle will apply to all the

thousands of variations. By mastering the modest vocabulary of this general life cycle, we will be able to simplify and more easily understand these variations.

Let's call this generic life cycle the **superplant life cycle** (Fig. 12.6). It is a mythical life cycle. No single species exhibits all these components, but collectively they all do appear in one species or another.

The superplant life cycle begins with a multicellular diploid plant, a sporophyte (upper right of Fig. 12.6). We have already identified a cherry tree, a grass plant, and a potato plant as sporophytes, but so too are conifer trees, ferns, parts of mosses, and certain algae.

Many sporophytes can reproduce asexually with rhizomes, stump sprouts, and bulbs. Some simple plants such as algae can reproduce asexually by *fragmentation*—the splitting off of a part of the sporophyte body. Each part can regenerate the missing part, resulting in two complete organisms. Any of these means of asexual reproduction is represented by the small loop on the sporophyte in Fig. 12.6 called the *asexual cycle*.

At particular times of the year, or under particular stresses, some cells of the sporophyte enlarge and differentiate to produce a structure that contains spores. The structure is called a **sporangium** (plural, *sporangia*). Sometimes a sporangium is simply one large cell; sometimes it is an elaborate multicellular structure. In the cherry tree, the pollen sacs and the ovules are essentially sporangia. Two kinds of sporangia can be produced on a sporophyte: one in which mitosis occurs and one in which meiosis occurs. If mitosis occurs, the spores that form are diploid. The sporangium will eventually burst, releasing mitospores. Mitospores may have flagella and move on their own power, or they may have a thick wall and simply be moved passively by wind or water. In either case, when they land on a suitable substrate they germinate and develop by mitosis back into another sporophyte. This kind of spore is simply another version of the asexual cycle loop already referred to.

If meiosis occurs, meiospores are released. Meiospores may have flagella or not; they may look like mitospores or appear different. The only way that meiospores and mitospores always differ is in their chromosome number. When a meiospore lands on a suitable substrate, it germinates and develops by mitosis into a gametophyte. In the cherry tree, not all meiospores are liberated to move around; those in the ovule stay in place, germinate in place, and develop into gametophytes in place.

A gametophyte may reproduce asexually just like a sporophyte. It can fragment, it can produce rhizomelike structures, or it can produce mitospores—all examples of the asexual cycle loop shown in Fig. 12.6, lower right. Eventually, when environmental conditions are appropriate, sexual reproduction occurs. Structures that contain gametes appear. These structures are called **gametangia** (singular, *gametangium*). They may be single enlarged cells or more elaborate multicellular structures. Gametan-

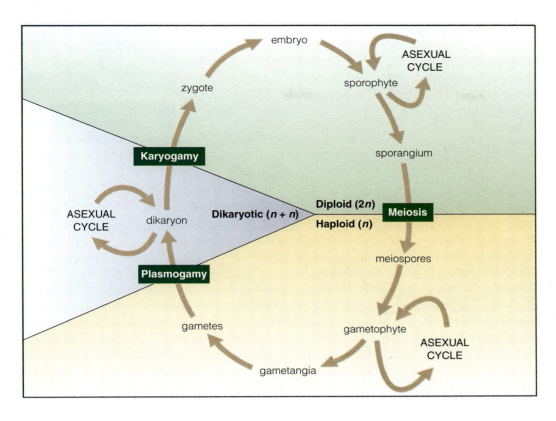

Figure 12.6 The superplant life cycle, a general life cycle within which any individual variation can be placed. The terms used in this cycle are very basic, and many synonyms exist for particular species or variations. Mastery of these basic terms makes an understanding of all life cycles much easier.

gia may look like sporangia, and the enclosed gametes may look like mitospores; but there is always one consistent difference: Gametes cannot produce new individuals by themselves. Gametes must fuse in pairs in order to produce a new organism.

In the most complex life cycles, the gametes that pair and fuse look different and function differently. For example, in the cherry tree one gamete (the egg) is large and immobile; the other gamete (the sperm) is small and mobile. In such cases, the two gametes can be given descriptive names and called *male* and *female*. In certain algae and fungi, however, the two gametes may be equally mobile and equal in size. Even though we cannot visually distinguish among such gametes, there must be profound, invisible metabolic and genetic differences because only certain ones will pair. In such a case, we call them *mating types*, one strain of gamete *plus* and the other *minus*. Just as two male sperm gametes do not fuse, neither will two plus or two minus gametes.

When two appropriate gametes meet, their cytoplasmic contents fuse, but their nuclei remain separate for a time. In the cherry tree, the time between plasmogamy and karyogamy is measured in seconds, but in some fungi the two events can be separated by weeks. In such a case, the dikaryotic *n+n* cell may divide by mitosis and produce a new multicellular *n+n* plant (called a **dikaryon**) which is neither a sporophyte nor a gametophyte. This dikaryon is capable of asexual reproduction by fragmentation or by the production of mitospores; so it has its own asexual loop in Fig. 12.6, middle left.

Eventually karyogamy will occur in certain cells of the dikaryon, turning those cells into zygotes. If plasmogamy and karyogamy follow each other closely there is no dikaryotic phase, and the gametes immediately form a zygote. The zygote may undergo a resting stage, or it may begin dividing by mitosis immediately and develop into a multicellular sporophyte. In seed plants, the development of the zygote into the sporophyte is interrupted by an embryo phase, in which the young sporophyte is protected and packaged within a seed coat and held in an arrested state of metabolism. Embryos of some species may remain dormant but alive within their seeds for decades or even centuries, until the appropriate conditions for germination occur.

Only One Generation Is Multicellular in Zygotic or Gametic Life Cycles

The superplant life cycle is a **sporic** life cycle. In a sporic life cycle, both the sporophyte and the gametophyte organisms are multicellular. Cherry has a sporic life cycle. Recall that there are two different gametophytes in the cherry tree: One produces male gametes, and the other produces female gametes. Cherry has a **heterosporic** variant of a sporic life cycle. That is, meiosis will yield some meiospores that germinate to produce female gametophytes and other meiospores that germinate to produce male gametophytes. In contrast, mosses and ferns generally have a single bisexual gametophyte that

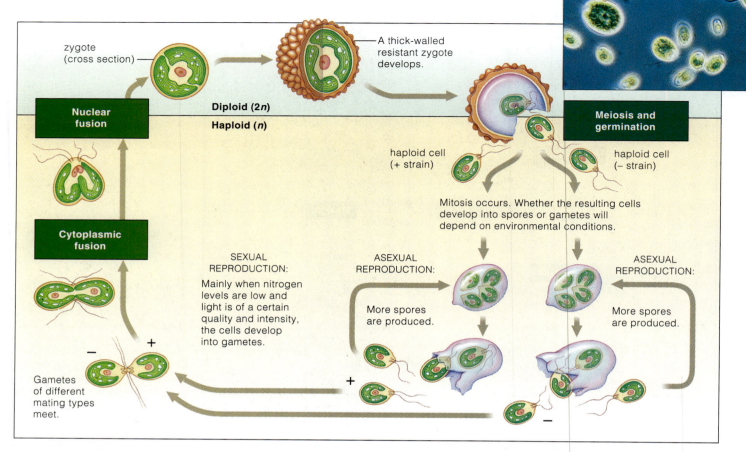

Figure 12.7 Zygotic life cycle of a single-celled species of *Chlamydomonas*, one of the most common green algae of freshwater habitats. *Chlamydomonas* reproduces asexually most of the time. It also reproduces sexually under certain environmental conditions.

Within the figure:

zygote (cross section)

A thick-walled resistant zygote develops.

Nuclear fusion

Diploid (2*n*)
Haploid (*n*)

Meiosis and germination

haploid cell (+ strain)

haploid cell (– strain)

Mitosis occurs. Whether the resulting cells develop into spores or gametes will depend on environmental conditions.

Cytoplasmic fusion

SEXUAL REPRODUCTION: Mainly when nitrogen levels are low and light is of a certain quality and intensity, the cells develop into gametes.

ASEXUAL REPRODUCTION: More spores are produced.

ASEXUAL REPRODUCTION: More spores are produced.

Gametes of different mating types meet.

produces both male and female gametes; there is only one gametophyte phase. This is a **homosporic** variant of a sporic life cycle.

All members of the kingdom Plantae have sporic life cycles, but many algae and fungi do not. A **zygotic** life cycle, such as that of the green alga *Chlamydomonas* (Fig. 12.7), begins with a gametophyte generation. As shown in the lower right corner of Figure 12.7, *Chlamydomonas* gametophytes are single, motile cells commonly found in freshwater habitats. Each cell has a single haploid nucleus. Cells appear to be similar, but genetically they exist as either plus or minus mating types. Occasionally, a gametophyte nucleus will undergo mitosis, producing several haploid spores. The parent cell bursts, releasing these spores, and each spore matures into a new gametophyte-generation cell. Thus, in this case, the gametophyte cells have acted as spores in an asexual part of the life cycle. Under other conditions, one plus and one minus gametophyte cell are attracted to each other in pairs (lower left of the figure). Plasmogamy and karyogamy occur, resulting in a 2*n* zygote cell. In this case, the gametophyte cells have acted as gametes in a sexual part of the life cycle. The zygote may rest in a dormant stage

for some time, but it does not develop into a multicellular sporophyte-generation organism. The zygote will eventually undergo meiosis (upper right of figure) and release many haploid cells. Each cell will mature into either a plus or minus gametophyte-generation cell. *Chlamydomonas* has a zygotic life cycle because there is no multicellular 2*n* phase.

A **gametic** life cycle such as that of the brown alga rockweed (*Fucus*, Fig. 12.8) begins with a multicellular sporophyte. The sporophyte is relatively large and complex, with rootlike, stemlike, and leaflike regions. Within body concavities, special cells enlarge and become sporangia, and their nuclei undergo meiosis. One type of sporangium produces large meiospores, and another type produces small ones. The meiospores are not released from their sporangia at this time. Each large meiospore first differentiates into a female gamete (an egg), and each small meiospore differentiates into a male gamete (a sperm). Then the gametes are released into the surf in such huge numbers that many eggs and sperms are brought together. Only eggs from one parent and sperm from another can fuse; eggs and sperm produced by the same plant will not be attracted to each other. Plas-

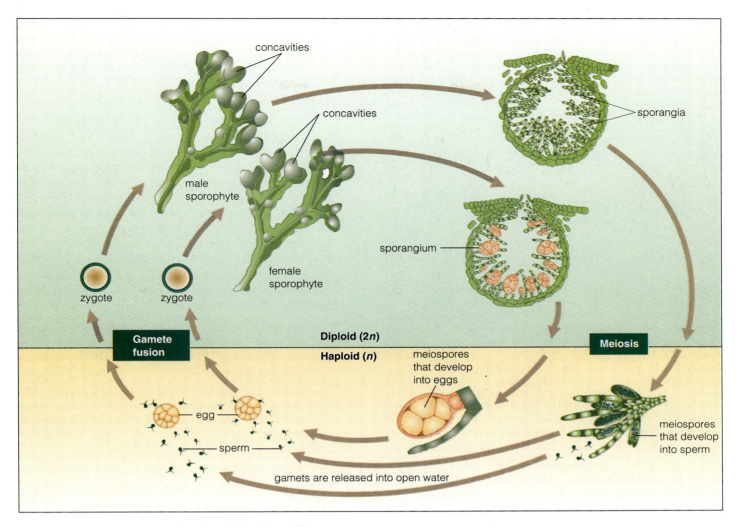

Figure 12.8 Gametic life cycle of the intertidal rockweed *Fucus*. In this life cycle, only the unicellular gametes are haploid.

mogamy and karyogamy occur, and the zygote immediately begins to divide and grow into a sporophyte. As it enlarges, it sinks to the bottom of the intertidal zone, becomes attached to a rock, and then grows into maturity. *Fucus* has a gametic life cycle because the only haploid phase is a single-celled gamete. There is no multicellular gametophyte generation in a gametic life cycle.

The Diploid Generation Has Become Dominant over Evolutionary Time

The fossil record of algae, fungi, and plants is described in some detail later in this book. For our purposes here, it is sufficient to know that the chronological appearance of photosynthetic organisms on earth began with algae, followed by mosses and lower vascular plants like ferns, then by conifers, and last by flowering plants. There are parallel relationships between this evolutionary path and

life cycles. Gametic and zygotic life cycles are common among algae but absent from any of the more advanced plants. In contrast, sporic life cycles are the rule among the complex, more recently evolved terrestrial plants. (Sporic life cycles do exist among some algae such as kelps and seaweeds, but they are the exception.)

The generations of sporic life cycles are not equal among all terrestrial plants. There is a clear trend of increasing dominance by the sporophyte in groups that are more recent in the fossil record (Fig. 12.9). *Dominance* here means that the sporophyte becomes longer-lived, larger, more structurally complex, and more independent than the gametophyte. There must be advantages, in our modern environments, to protecting and limiting the haploid phase of the life cycle. What could these advantages be?

The diploid condition permits many recessive genes to be carried along from generation to generation, each one masked by the dominant gene on the other homolo-

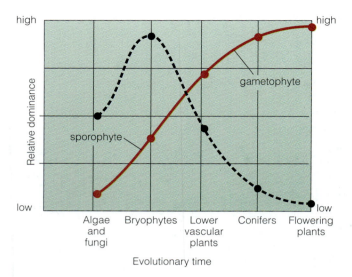

Figure 12.9 Life cycle trends, from the algae and fungi through the flowering plants. The more recently evolved the group and the more adapted it is to dry land, the more dominant is its sporophyte generation. *Dominant* means larger, more complex, and longer-lived than the gametophyte generation.

gous chromosome. Such recessive genes may have no value in the current environment (indeed, they may even be harmful), but they could be valuable in some different, future environment. The recessive genes carried now could contribute to the species' future success, especially as the genes already exist and do not have to be created (with some lag time) by future mutations. There is, however, a potentially unsafe phase in the life cycle for carrying recessive genes, and that is in the gametophyte generation. In the haploid cells of gametophytes there are no recessive genes because there is only one set of chromosomes; every gene's expression shows through in this phase. If, however, the gametophyte is small, short-lived, and protected by the sporophyte, then the expression of potentially deleterious genes might be tolerable. Perhaps the ability to be genetically diverse resulting from the diploid chromosome condition is the explanation for the modern dominance of sporophytes on land.

SUMMARY

1. Life perpetuates itself through reproduction. Reproduction is the transfer of genetic information from one generation to the next, and this transfer is our definition of *life cycle*. Reproduction can be asexual or sexual.

2. Asexual reproduction requires a cell division known as *mitosis*. Each chromosome duplicates itself into a pair of attached chromatids. During anaphase and telophase these chromatids separate, creating two offspring cells that each contain the same number of chromosomes as

the parent cell. Asexual reproduction offers many advantages over sexual reproduction, one of which is the need for only a single parent. A significant disadvantage of asexual reproduction is the loss of genetic diversity and the likelihood of extinction when the environment changes.

3. Sexual reproduction involves the union of two cells, called *gametes*, which are usually produced by two different individuals. Another kind of cell division, known as *meiosis*, is ultimately necessary to produce gametes. Meiosis contains a reduction division step, which separates homologous chromosomes rather than sister chromatids. As a consequence, the number of chromosomes declines from $2n$ (the diploid number) to n (the haploid number). When the gametes later join and combine their chromosomes, the result is a cell with the original diploid number of chromosomes. Were it not for meiosis, sexual reproduction would produce cells with ever-increasing numbers of chromosomes.

4. Every species in the kingdom Plantae has both diploid and haploid phases—that is, plants whose cells are all diploid or all haploid. These phases are called *generations*, and they alternate with each other over time. Except for the mosses, diploid individuals are relatively large, complex, and long-lived compared to haploid individuals. The cherry life cycle, for example, has diploid plants in the form of large trees, but gametophyte plants are microscopic entities containing only a few cells.

5. Basic superplant life cycle terms, which apply to all the thousands of variations of life cycles, are relatively few: *sporophyte, sporangium, mito-* and *meiospores, male* and *female gametophytes, gametangium, plus* and *minus* or *male* and *female gametes, plasmogamy, dikaryon, karyogamy,* and *zygote*. Many synonyms exist for these basic terms; hence, a mastery of the basic terms makes it much easier to understand any single species' variations.

6. Three major types of life cycle exist: sporic, zygotic, and gametic. A sporic life cycle has multicellular gametophyte and multicellular sporophyte phases. Cherry, for example, has a sporic life cycle, and so do all other members of the kingdom Plantae. In contrast, algae and fungi typically have zygotic or gametic life cycles. A zygotic life cycle has only a single-celled zygote to represent the sporophyte phase; the rest of the life cycle is haploid. The green alga *Chlamydomonas* has a zygotic life cycle. A gametic life cycle has only single-celled gametes to represent the gametophyte phase; the rest of the life cycle is diploid. The seaweed *Fucus* has a gametic life cycle.

7. The fossil record reveals that the most recent groups to evolve—those plants that are most complex and adapted to dry land—have sporic life cycles in which the gametophyte generation is smallest and the sporophyte generation is most dominant. We may imagine, then, that there are advantages to protecting and limiting the haploid phase of the life cycle.

Questions

1. Why do plants reproduce?

2. What are some advantages and disadvantages of asexual and sexual reproduction?

3. What is the key difference between mitosis and the reduction division step of meiosis?

4. What is meant by the expression *alternation of generations*? How do sporophyte and gametophyte generations—in species of the kingdom Plantae—typically compare in terms of size, complexity, and life span?

5. Summarize the life cycle of a cherry tree. Describe where meiosis occurs, what the gametophyte plants look like, how the two gametes are brought together to form a diploid zygote, and how the zygote develops back into a cherry tree.

6. The cherry life cycle is sporic. Certain algae and fungi have nonsporic life cycles. Fundamentally, how do they differ from plants with a sporic life cycle?

7. Why do you think that the most advanced terrestrial plants have life cycles dominated by the diploid (sporophyte) generation, whereas less advanced plants, algae, and fungi have life cycles dominated by the haploid (gametophyte) generation?

Further Readings

Normally, we suggest short, semitechnical articles as suitable further reading. For this chapter, however, the most appropriate readings are two standard, classic textbooks on the plant kingdom. Each is well illustrated and offers in-depth information about the life cycles of representative organisms for each division of plants. The first book is a smaller and less expensive paperback.

Bold, H. C., and J. W. La Claire II. 1987. *The Plant Kingdom*. 5th ed. Englewood Cliffs, N.J.: Prentice-Hall.

Gifford, E. M., and A. S. Foster. 1987. *Morphology and Evolution of Vascular Plants*. 3d ed. New York: Freeman.

13 THE FLOWER AND SEXUAL REPRODUCTION

1. Flowers are the specialized structures where sexual reproduction occurs in flowering plants. They are composed of four whorls of leaflike structures: the calyx (composed of sepals) serves a protective function; the corolla (composed of petals) usually functions to attract insects; the androecium (composed of stamens) is the male part of the flower; and the gynoecium (carpels or pistil) is the female part.

2. Stamens produce pollen. The stamen's female counterpart is the pistil, which is composed of the stigma (collects pollen), the style, and the ovary. The ovary produces eggs.

3. Sexual reproduction in plants involves pollination, which is the transfer of pollen from the anther of one flower to the stigma of the same or a different flower. The pollen lands on the stigma and grows into the style (as the pollen tube) until it penetrates the ovule and releases two sperm.

4. Double fertilization is unique to flowering plants. One sperm fuses to the egg to form the zygote; the other fuses to the polar nuclei to form the primary endosperm nucleus. The zygote forms the embryo, and the primary endosperm nucleus forms the endosperm (a nutritive tissue). The fertilized ovule develops into the seed.

5. There are basically two types of pollination: self-pollination, in which the pollen comes from within the same flower or plant, and cross-pollination, in which the pollen comes from different plants. Pollination strategies involve abiotic factors such as wind and water and interesting relationships between plants and animals.

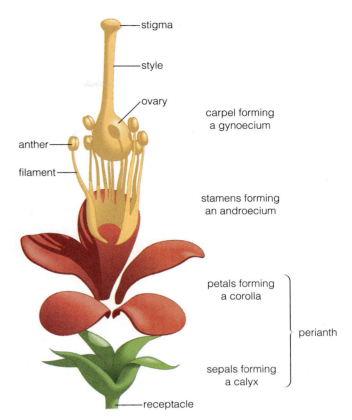

Figure 13.1 A flower, showing the whorls of parts. The perianth consists of two whorls of sepals and petals (calyx and corolla). There is one whorl of stamens (collectively called the *androecium*). A single carpel (a pistil) forms the central whorl of floral parts, the *gynoecium*.

13.1 THE FLOWER: SITE OF SEXUAL REPRODUCTION

Poets and dreamers and people in love often express themselves through flowers. The vivid colors and intricate fantasy of patterns of flower parts have always attracted the notice of humans and captured their imagination. The symmetry and brightness of petals and the interesting scents from flowers are evolutionary designs not for people, but to attract pollinators (mostly insects and birds) to visit flowers. The result of such visits is often a reward of nectar (sugary water); in exchange, the insect will fly away carrying pollen to a different flower. This fine-tuned relationship between the flower and the animal guarantees that sexual reproduction continues between plants, which are stuck in place and are not able to roam about to select a mate. People will always love flowers, but the most important function of the flower is as the site where sexual reproduction occurs, leading to the formation of seeds. Imagine a world without wheat seeds to make bread, corn for products we use every day, or barley to make beer; we'd all live very different lives without seeds.

Of all the characteristics of flowering plants, or **angiosperms**, the flower and fruit are the least affected by changes in the environment such as age, light, water, and nutrition. Consequently, the appearance of flowers and fruits is important to understanding evolutionary relationships among angiosperms.

The development of the flower begins the sexual reproductive cycle in all flowering plants. The function of the flower is to facilitate the important events of gamete formation and fusion. The essential steps of sexual reproduction, meiosis, and fertilization, described in Chapter 12, take place in the flower. The complete sexual cycle involves (a) the production of special reproductive cells following meiosis, (b) pollination, (c) fertilization, (d) seed and fruit development, (e) seed and fruit dissemination, and (f) seed germination.

13.2 THE PARTS OF A COMPLETE FLOWER AND THEIR FUNCTIONS

A typical flower is composed of four whorls of modified leaves—(1) sepals, (2) petals, (3) stamens, and (4) carpels—which are all attached to the terminal end of a modified stem, the **receptacle** (Fig. 13.1).

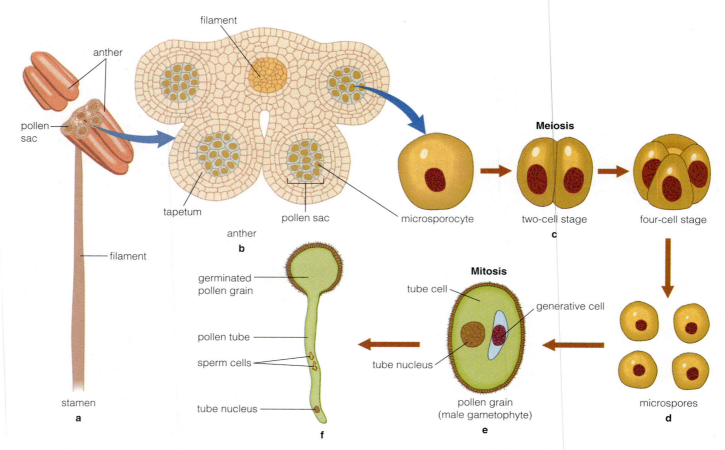

Figure 13.2 Development of pollen from microsporocyte to pollen grain. (**a**) Stamen. (**b**) Cross section of anther. The tapetum immediately surrounds the pollen sac. (**c**) Development of four-cell stage from the microsporocyte by meiosis. (**d**) Four microspores. (**e**) Pollen grain. (**f**) Germination of pollen grain.

The **sepals** are generally green and enclose the other flower parts in the bud. All the sepals collectively constitute the **calyx**, and they function to protect the reproductive parts inside the flower. The **petals** are usually the conspicuous, colored, attractive flower parts, which together constitute the **corolla**. Their most important function is to catch the attention of pollinators.

The **stamens** form the whorl just inside the corolla. Each stamen has a slender stalk, or **filament**, at the top of which is an **anther**, the pollen-bearing organ. The whorl or grouping of stamens is called the **androecium**, and this constitutes the male part of the flower (Fig. 13.1).

The female part of the flower is called the **gynoecium**. It is composed of one or more **carpels**, which are modified leaves folded over and fused to protect the attached ovules. Sometimes an individual carpel or group of fused carpels is called a **pistil**. The term is used simply because its shape is reminiscent of the pistil (or pestle) used to grind objects in a mortar. There may be more than one pistil in the gynoecium. The carpels are usually located at the center of the flower (Fig. 13.1).

Carpels generally have three distinct parts: (1) an expanded basal portion, the **ovary**, which contains the **ovules**; (2) the **style**, a slender supporting stalk; and (3) the **stigma**, at the very tip (Fig. 13.1).

The term **perianth** is applied to the calyx and corolla collectively. The flower organs necessary for sexual reproduction are the stamens (androecium) and carpels (gynoecium). The perianth, composed of calyx and corolla, protects the stamens and pistil(s). It also attracts and guides the movements of some pollinators.

The Male Organs of the Flower Are the Androecium

As mentioned earlier, the androecium is the whorl of stamens, with each stamen consisting of an anther at the end of a filament. The anther usually is made up of four elongated lobes called **pollen sacs** (Fig. 13.2a). Early in the development of the anther, each pollen sac contains a mass of dividing cells called **microsporocytes** (Fig. 13.2b). Each microsporocyte divides by meiosis to form four haploid (*n*) microspores (Fig. 13.2d). The nucleus of each microspore then divides by mitosis to form a two-celled **pollen grain**, which contains a **tube cell** and a smaller **generative cell** (Fig. 13.2e). The role of this two-celled, haploid, **male gametophyte** (gamete-producing plant) is to produce **sperm cells** for fertilization.

a   b

Figure 13.3   Pollen grains as viewed with the scanning electron microscope. (**a**) Iris, ×174. (**b**) Ragweed (*Ambrosia*), ×700.

The pollen grain is surrounded by an elaborate cell wall. The pattern on this wall is genetically fixed, and it varies widely among major groups of plants (Fig. 13.3). The walls contain *sporopollenin*, a very hard material that resists decay. As a result, pollen grains make good fossils, and botanists have found them very useful in studying the evolutionary history of seed plants.

After the pollen grains are mature, the anther wall splits open and the pollen is shed. In various ways discussed later in this chapter, pollen grains are transported to the stigmas of adjacent or distant flowers. This process is called **pollination**. The dry pollen grain then absorbs water from the stigma and becomes hydrated. It also secretes proteins, including some that are involved in pollen recognition and compatibility reactions with the cells of the stigma. This recognition mechanism is important because it inhibits pollen from the wrong species from germinating and growing. These proteins are the same ones that cause allergies and hay fever.

The pollen grain then **germinates** to form a **pollen tube**, which grows toward the ovary. The pollen tube grows between cells in the center of the style, a region called the *transmitting tissue* (Fig. 13.4). In the lily (*Lilium* sp.), the style is hollow, and the transmitting tissue is reduced to a layer of secretory cells on its inner surface; the cells secrete a special material in which the pollen tube grows. In solid transmitting tissues, the pollen tube secretes digestive enzymes that partially break down the transmitting tissue cells, making it easier for the pollen tube to grow.

The pollen tube grows by elongation very near its tip. It is filled with many small vesicles containing cell-wall building materials needed for tube growth. Endoplasmic reticulum and other organelles are also present. Very rapid cytoplasmic streaming (see Chapter 3) occurs in pollen tubes.

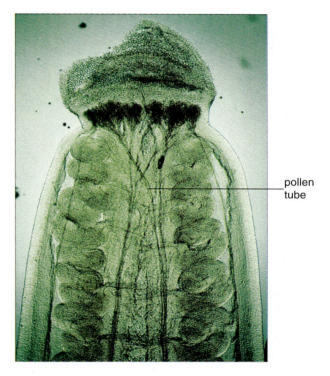

pollen tube

Figure 13.4   Style of jewel plant (*Streptanthus*) flower with pollen tubes that have grown toward the ovary, ×62.

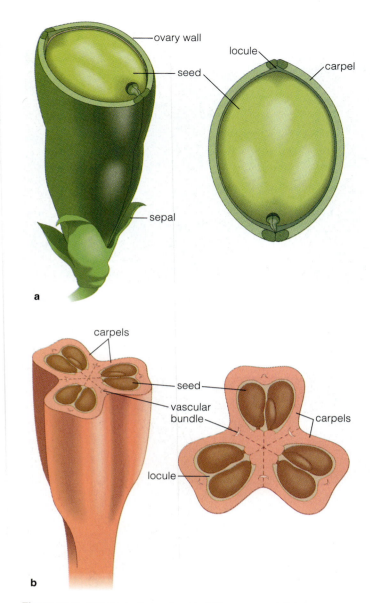

Figure 13.5 Simple and compound pistils compared. (**a**) A section through the simple pistil of pea (*Pisum*), showing an ovary composed of a single carpel. (**b**) A section through the compound pistil of tulip (*Tulipa*), showing an ovary with three fused carpels.

The Female Organs of the Flower Are the Gynoecium

In its simplest form, the gynoecium consists of a single folded carpel or **simple pistil** (Fig. 13.5a) in which the ovary resides. In more complex flowers, the gynoecium may consist of several separate carpels or several groups of fused carpels. A group of fused carpels is called a **compound pistil** (Fig. 13.5b).

The ovary is a hollow structure having from one to several chambers, or **locules** (Fig. 13.5b). The number of carpels in a compound pistil is generally related to the number of stigmas, the number of locules, and the number of sides on the ovary. The pea has a single stigma, and its ovary has one locule (Fig. 13.5a). The tulip pistil has three stigmas, its ovary has three sides, and it contains three locules and three carpels (Fig. 13.5b).

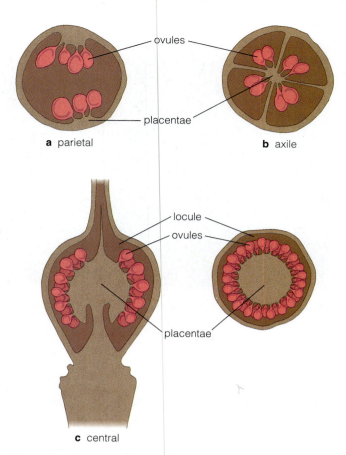

Figure 13.6 Three types of placentation. (**a**) Parietal (*Dicentra*). (**b**) Axile (*Fuchsia*). (**c**) Free central (*Primula*). (Redrawn from Priestley and Scott, *An Introduction to Botany*. © Longman's, Green, 1949. Reprinted with permission of the Longman's Group.)

The tissue within the ovary to which an ovule is attached is called a **placenta** (Fig. 13.6). The manner in which placentae are distributed in the ovary is termed **placentation**. When the placentae are on the ovary wall, as in bleeding heart (*Dicentra*; Fig. 13.6a), the placentation is **parietal**. When they arise on the axis of an ovary that has several locules, as in *Fuchsia*, the placentation is **axile** (Fig. 13.6b). Less frequently, the ovules form on a central column, which is **central** placentation (*Primula*; Fig. 13.6c).

Emanating from the ovary is the style—a stalk whose length and shape vary by species—with a stigma at its upper end (see Fig. 13.1). It is through stylar tissue that the pollen tube grows. In general, the style withers after pollination. Often, the surface of the stigma is covered with short hairs that aid in holding the pollen grains, and sometimes they secrete a sticky fluid that stimulates pollen tube growth. In many wind-pollinated plants, such as the grasses, the stigma are elaborately branched or featherlike.

Within the ovary is the ovule, the structure that (if fertilized) will eventually become the seed. As the ovule matures (Fig. 13.7), it forms one or two outer protective layers, the **integuments**. The integuments do not fuse, leaving a small opening called the **micropyle** (Fig. 13.7b). At the same time, one of the internal dividing cells of the ovule, the **megasporocyte**, is enlarging in preparation for

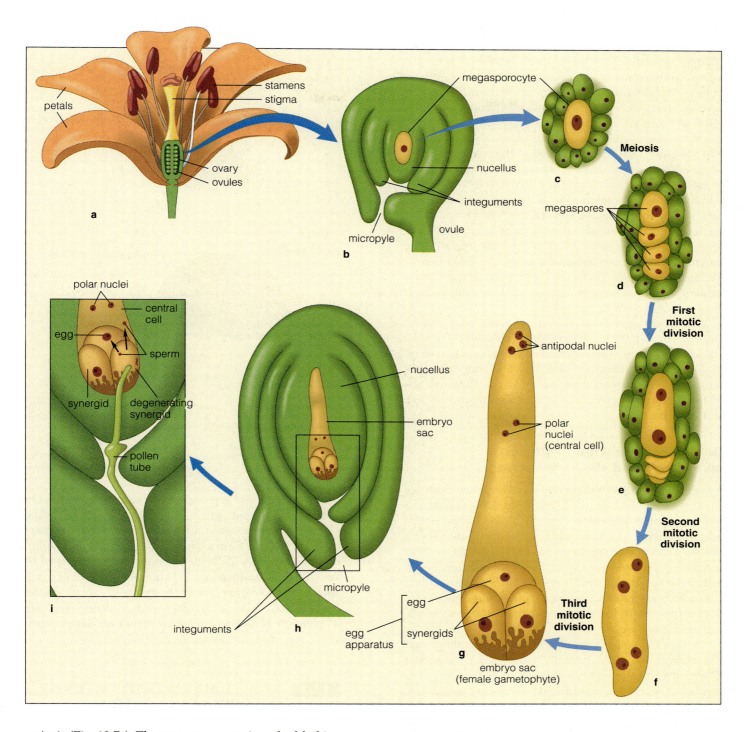

Figure 13.7 Embryo sac development in cotton (*Gossypium hirsutum*). (**a**) Mature flower. (**b**) Magnified longitudinal section of an ovule. (**c**) Megasporocyte before meiosis. (**d**) Four *n* megaspores after meiotic division of the megasporocyte; the three megaspores closest to the micropyle disintegrate. (**e**) First mitotic division of the megaspore. (**f**) Second mitotic division. (**g**) Third mitotic division and the resulting three antipodal nuclei, the egg, two synergids, and the two polar nuclei. All have an *n* chromosome number. (**h**) Mature embryo sac; the antipodals disintegrate at this stage in cotton. (**i**) Pollen tube penetration through the nucellus into a partially degenerated synergid. (**b-h** redrawn from U. R. Gore, *Amer. J. Bot.* 19 (1932): 795–802; **i** modified and redrawn from W. A. Jensen, *Amer. J. Bot* 52 (1965): 781–97.)

meiosis (Fig. 13.7c). The megasporocyte is embedded in a tissue called the **nucellus**. The ovule thus is composed of one or two outer protecting integuments, along with the micropyle, megasporoctye, and nucellus.

The Embryo Sac Is the Female Gametophyte Plant

As a result of meiosis of the megasporocyte, a row of four cells called **megaspores** is produced in the nucellus (Fig. 13.7d). Each megaspore is haploid (*n*). As a rule, the three cells nearest the micropyle disintegrate and disappear, whereas the one farthest from the micropyle enlarges greatly. This megaspore develops into the mature **embryo sac** in several stages: (a) A series of three mitotic divisions occurs to form an eight-nucleate embryo sac (Fig. 13.7e–g), (b) the nuclei migrate, and (c) a cell wall forms around the nuclei.

At the end of this process (Fig. 13.7g), an **egg cell** and two **synergid cells** are positioned at the micropylar end of the embryo sac. Because it is frequently difficult to differentiate the egg cell from the other two cells, all three cells are sometimes referred to as the **egg apparatus**. The two nuclei that migrate toward the center are **polar nuclei**; they lie in the center of the large **central cell**. The three nuclei at the end of the embryo sac opposite the micropyle form three **antipodal cells**. As described in Chapter 12 (see Fig. 12.5), the embryo sac is the **female gametophyte** phase of the flowering plant's life cycle; it is a haploid or *n* plant.

Not every species of flowering plant produces an embryo sac with the same number of cells, nor do they all follow the same developmental sequence. In cotton (*Gossypium hirsutum*), for example, the antipodal cells disintegrate before fertilization (Fig. 13.7h). However, the mature embryo sac must contain, at a minimum, an egg and polar nuclei.

Double Fertilization Produces an Embryo and the Endosperm

As we have seen, germination of a pollen grain produces a pollen tube, which grows down through the stigma and style and enters the ovary (Fig. 13.7i). While the pollen tube is growing, the generative cell within it divides by mitosis to form two sperm cells. In some plants, such as sunflower (*Helianthus annuus*), the sperm cells form even before the pollen are shed from the anther. Many pollen grains may germinate, and their pollen tubes may grow through the pistil, but only one usually enters an ovule and its embryo sac.

How is the pollen tube directed into the embryo sac? The answer has been partially worked out for cotton (*Gossypium*). In the embryo sac, one or both of the synergid cells begins to shrivel and die before the pollen tube enters. It is conjectured that, in the process, the synergids release chemicals that may influence the direction in which the pollen tube grows. Two observations favor this idea. First, the cell wall at the base of the synergids is highly convoluted (see Fig. 13.7g); in other parts of the plant, cell walls like these are associated with active chemical secretion. The second bit of circumstantial evidence is that sometimes the pollen tube actually penetrates one of the synergids and, upon rupturing, empties its contents into the synergid. The two sperm then pass through the incomplete upper cell wall of the synergid, one moving to fuse with the egg and the other with the central cell.

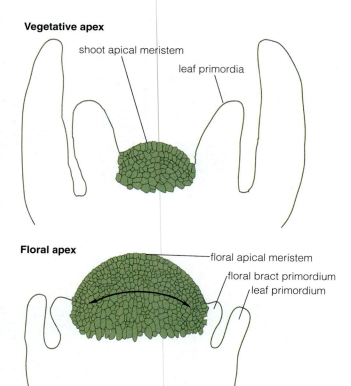

Figure 13.8 Comparison of vegetative and floral apices.

The sperm and egg nuclei fuse to form a diploid (2*n*) **zygote**, which will grow into the embryo. The other sperm nucleus fuses with the polar nuclei of the central cell (Fig. 13.7i) to form a triploid (3*n*) **primary endosperm nucleus**, which will divide to become a food reserve tissue called the **endosperm**. This double fusion of egg with sperm and polar nuclei with sperm is called **double fertilization**. The antipodals and synergids usually degenerate. Now the conditions are set for development of the seed and fruit (discussed in Chapter 14).

13.3 APICAL MERISTEMS: SITES OF FLOWER DEVELOPMENT

During vegetative growth, the shoot apex produces stems and leaves. At some time during the growing season, a signal—for example, day length or temperature—will trigger a change in the metabolism of the shoot apex, thereby starting its transformation into a *floral* apex. The first step in the transition of the apex is a broadening of the apical dome (Fig. 13.8a, b), accompanied by a general increase in RNA and protein synthesis and in the rate of cell division in the apical dome. The first organs to form from the floral apex are **bracts**. These modified leaves develop at the lower periphery of the floral apex. Floral

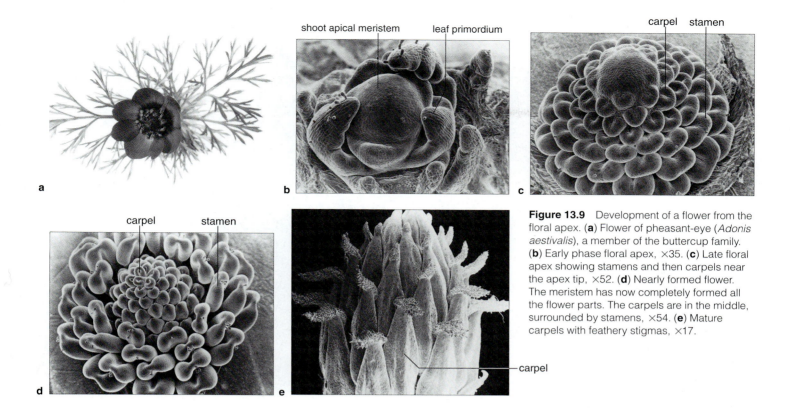

Figure 13.9 Development of a flower from the floral apex. (**a**) Flower of pheasant-eye (*Adonis aestivalis*), a member of the buttercup family. (**b**) Early phase floral apex, ×35. (**c**) Late floral apex showing stamens and then carpels near the apex tip, ×52. (**d**) Nearly formed flower. The meristem has now completely formed all the flower parts. The carpels are in the middle, surrounded by stamens, ×54. (**e**) Mature carpels with feathery stigmas, ×17.

organs usually form in whorls or in spirals, and the internodes between successive sets of floral organs are usually very short. The floral organs are modified leaves, so the flower itself is really a much shortened and very modified branch.

In pheasant-eye (*Adonis*), the sepals and petals form after the bracts, followed by several spirals of stamens and carpels (Fig. 13.9b–d). Each new floral organ forms closer to the floral apex. Eventually, carpels develop at the tip, terminating any further growth of the floral apex (Fig. 13.9d, e).

The development of an **inflorescence**, a group of flowers from the same apex, follows the same spatial sequence: Bracts form first, followed by flowers starting at the periphery of the inflorescence apex and progressing toward the center and the tip. The parts of the flowers (sepals, petals, stamens, and carpels) on the inflorescence develop in the same sequence as they do in individual flowers.

Flowers Vary in Their Architecture

Flowers may be complete or incomplete, perfect or imperfect. A flower that develops all four sets of floral leaves—sepals, petals, stamens, and carpels—is said to be a **complete flower**. An **incomplete flower** lacks one or more of these four sets.

Unisexual flowers are either **staminate** (stamen bearing) or **pistillate** (pistil bearing) and are said to be imper-

fect. Bisexual flowers are **perfect**. When staminate and pistillate flowers occur on the same individual plant, as they do in corn (*Zea mays*), English walnut (*Juglans regia*), and many other species, the plant is called **monoecious** (Fig. 13.10). When staminate and pistillate flowers are borne on separate individual plants, as in *Asparagus* or willow (*Salix*), the plant is said to be **dioecious**.

Flowers may also be classified according to their symmetry, which may be *regular* or *irregular*. In many flowers, such as stonecrop (*Sedum* sp.) (Fig. 13.11a), the corolla is made up of petals of similar shape that radiate from the center of the flower and are equidistant from each other. Such flowers are said to have **regular** symmetry. In these cases, even though there may be an uneven number of parts in the perianth, any line drawn through the center of the flower will divide the flower into two similar halves. The halves are either exact duplicates or mirror images of each other.

Flowers with **irregular** symmetry, such as mint (*Salvia*) (Fig. 13.11b), have parts arranged in such a way that only one line can divide the flower into equal halves; the halves are usually mirror images of each other.

The parts of the flower may be free or united. In the flower of stonecrop (*Sedum* sp.; Fig. 13.11a), the parts of the flower are separate and distinct. Each sepal, petal, stamen, and carpel is attached at its base to the receptacle. In many flowers, however, members of one or more whorls are to some degree united with one another. The union of parts of the same whorl is termed **connation**.

pistillate flower

stigma

ovary

a

pistillate catkin

b

staminate (male) catkin

one stamen

anther

staminate flowers

c

Figure 13.10 Flowers of English walnut (*Juglans regia*), a monoecious plant. (**a**) Branch with staminate and pistillate inflorescences (*catkins*). (**b**) Pistillate flower. (**c**) Staminate flowers.

a

b

Figure 13.11 Floral symmetry. (**a**) Regular flower of stonecrop (*Sedum* sp.). (**b**) Irregular flower of mint (*Salvia* sp.).

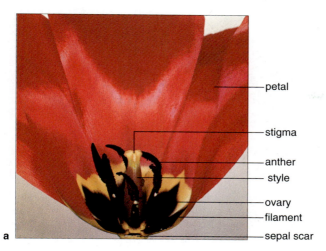

petal

stigma

anther

style

ovary

filament

sepal scar

a

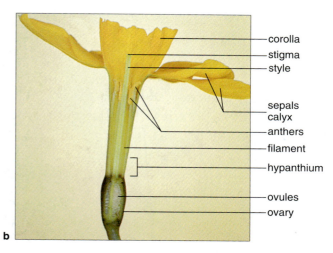

corolla

stigma

style

sepals
calyx

anthers

filament

hypanthium

ovules

ovary

b

Figure 13.12 Variations in the elevation of floral parts relative to the ovary. (**a**) Superior ovary in the tulip flower (*Tulipa*). (**b**) Inferior ovary in daffodil flower (*Narcissus pseudonarcissus*). (**c**) In almond flower (*Prunus* sp.) the hypanthium is not fused to the ovary.

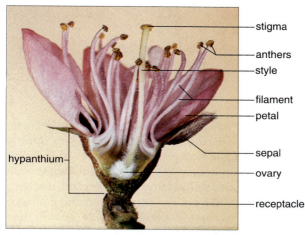

stigma

anthers

style

filament

petal

sepal

ovary

receptacle

hypanthium

c

Union of flower parts from two different whorls is known as **adnation**.

The position of the ovary differs among flowers. In tulip flowers (Fig. 13.12a) the receptacle is convex or conical, and the different flower parts are arranged one on top of another. The gynoecium (whorl of pistils) is thus situated on the receptacle above the points of origin of the perianth (whorls of sepals and petals) and androecium (whorl of stamens). An ovary in this position is said to be **superior**. In the daffodil (Fig. 13.12b), the ovary appears to be below the apparent points of attachment of the perianth parts and the stamens. This is an **inferior ovary**. In a flower with an inferior ovary, the lower portions of the three outer whorls—calyx, corolla, and androecium—have fused to form a tube, the **hypanthium**, which is also fused to the ovary. In some flowers, such as those of cherry, peach, and almond (*Prunus* sp.), the hypanthium does not become fused to the ovary (Fig. 13.12c).

Often Flowers Occur in Clusters, or Inflorescences

In many flowering plants, flowers are borne in clusters or groups known as inflorescences. An *inflorescence* is a flowering branch, and some of the most common ones are illustrated in Figure 13.13.

A very simple type of inflorescence, called a **raceme**, is found in such plants as currant (*Ribes*) and radish (*Raphanus*). The main axis has short branches, each of which terminates in a flower; the short branches are **pedicels**. The oldest flowers are at the base of the inflorescence and the youngest at the apex. A branched raceme is called a **panicle** (Fig. 13.13).

In a **spike**, the main axis of the inflorescence is elongated, as in a raceme or panicle, but the flowers have no pedicels. A **catkin** is a spike that usually bears only pistillate or staminate flowers. Examples are walnut (see Fig. 13.10) and willow (see Fig. 13.20).

An **umbel** inflorescence has a short floral axis, and the flowers arise umbrella-like from approximately the same level (Fig. 13.13). Onion (*Allium*) is a good example.

A **head** is an inflorescence in which the flowers lack pedicels and are crowded together on a very short axis. Members of the sunflower family have this type of inflorescence (Fig. 13.13).

In a **cyme**, the apex of the main axis produces a flower that involves the entire apical meristem, so that the axis itself does not elongate (Fig. 13.13). Other flowers arise on lateral branches farther down the axis. The youngest flowers in any cluster occur farthest from the tip of the main stalk. Chickweed (*Cerastium*) is an example.

Figure 13.13 Types of inflorescences.

13.4 REPRODUCTIVE STRATEGIES: SELF-POLLINATION AND CROSS-POLLINATION

The importance of pollen to plant reproduction was first demonstrated in the 1760s by Joseph Koelreuter, who created artificial hybrids by dusting the stigma on one species with the pollen of another. Three decades later, Christian Sprengel correctly distinguished between self-pollinating and cross-pollinating species and described the role of wind and insects as pollen vectors. Koelreuter and Sprengel are the founders of a field of study called *pollination ecology.*

There are essentially two different kinds of pollination. Transfer of pollen from the anther to the stigma in the same flower and transfer of pollen from one flower to another on the same plant are types of **self-pollination** or **selfing**. Genetic recombination (see Chapter 16) does not result from such crosses because only one plant is involved. The other kind of pollination, **cross-pollination** or **outcrossing**, involves transfer of pollen from one genetically distinct plant to the stigma of another. Cross-pollination results in genetic recombination, leading to genetically diverse offspring.

Various plant traits encourage the success of either selfing or outcrossing. Separation of sexes onto different individual plants ensures outcrossing. Plants that have both sexes on the same individual can prevent selfing by inhibiting pollen tube growth through the style or by inhibiting the formation of a zygote. Selfing is also prevented when the anthers mature or release their pollen some time before (or after) the stigmas on the same plant mature and are receptive to pollen.

Selfing does have advantages, even though genetic diversity is sacrificed. Selfing is a necessary means of reproduction for scattered populations in extreme habitats such as arctic tundra, where pollinating insects are few and the odds of pollen from one plant reaching another are low. Selfing is also common among plants of disturbed habitats, and it permits weedy species (for example, many grasses) to multiply and spread even if only one parent reaches a site. Selfing also saves pollen (and the metabolic energy required to produce it) because selfers generally produce fewer pollen grains per flower or per ovule than outcrossers. Selfing also increases the probability that pollen will reach a stigma because the distance travelled—and the time needed for travel—are so short. The time element is important because pollen is quite sensitive to low humidity and to ultraviolet rays of sunlight; consequently pollen ordinarily has a short life span, measured in hours or days for most species.

Extreme examples of self-pollinating species have flowers that never open. They remain small and budlike, and the pollen falls from the anther sacs onto the stigma while the flower parts are still close together.

Production of Some Seeds Does Not Require Fertilization

Apomixis refers to a form of asexual reproduction in which no fusion of sperm and egg occurs. Normally, egg nuclei will not embark on the series of changes that leads to an embryo unless fertilization has occurred. Occasionally, however, an embryo does develop from an unfertilized egg. This type of apomixis is called **parthenogenesis**. Another type occurs when the embryo plant

Table 13.1 Pollination Syndromes: Traits of Flowers Pollinated by Different Vectors

Trait	Vector							
	Beetle	Fly	Bee	Butterfly	Moth	Bird	Bat	Wind
Color	Dull white or green	Pale and dull to dark brown or purple; sometimes flecked with translucent patches	Bright white, yellow, blue, or UV	Bright, including red and purple	Pale and dull red, purple, pink, or white	Scarlet, orange, red, or white	Dull white, green, or purple	Dull green, brown, or colorless; petals may be absent or reduced
Nectar guides	Absent	Absent	Present	Present	Absent	Absent	Absent	Absent
Odor	None to strongly fruity or fetid	Putrid	Fresh, mild, pleasant	Faint but fresh	Strong and sweet; emitted at night	None	Strong and musty; emitted at night	None
Nectar	Sometimes present; not hidden	Usually absent	Usually present; somewhat hidden	Ample; deeply hidden	Abundant; deeply hidden	Abundant; deeply hidden	Abundant; somewhat hidden	None
Pollen	Ample	Modest in amount	Limited; often sticky and scented	Limited	Limited	Modest	Ample	Abundant; small, smooth, and not sticky
Flower shape	Large, regular dishlike; erect	Funnel-like or a complex trap	Regular or irregular; often tubular with a lip; erect	Regular; tubular with a lip; erect	Regular; tubular without a lip; closed by day; pendant or horizontal	Regular or irregular; tubular without a lip; pendant or horizontal	Regular; trumpet-like; closed by day; pendant or borne on trunk	Regular; small; anthers and stigmas exserted
Examples	Tulip tree, magnolia, dogwood	Skunk cabbage, philodendron	Larkspur, snapdragon, violet	Phlox	Tobacco, Easter lily, some cacti	Fuchsia, hibiscus	Banana, agave, sausage tree	Walnut, grasses

arises from diploid tissue surrounding the embryo sac. These *adventitious* embryos occur in *Citrus* and other plants.

Apomixis also has other variations, but all forms of this phenomenon involve the origin of new individuals without nuclear or cellular fusion of sperm and egg. In some plants, deposition of pollen on the stigma is a prerequisite to apomictic embryo development, even though a pollen tube does not grow down the style and nuclear fusion does not take place. In such instances, there is evidence that hormones formed in the stigma, or furnished by the pollen, move to the unfertilized, but diploid, egg cell (in this case meiosis is omitted during embryo sac formation), initiating embryo development.

Pollination Is Effected by Vectors

Pollen may be moved by either biotic (animal) or abiotic (wind and water) vectors. The set of unique flower and pollen traits that adapt a plant for pollination by a particular vector is its **pollination syndrome** (Table 13.1).

Animal pollinators are not altruistic. They visit flowers for some reward, and only incidentally do they transfer pollen. The most common rewards are pollen and nectar, but sometimes they are waxes or oils.

Because of the great cost and consequences of wasting rewards, plants have devised various strategies to prevent rewards from being stolen by nonpollinating animals. For example, the rewards are offered in such a pre-

Figure 13.14 A beetle-pollinated plant. Tulip tree (*Liriodendron tulipifera*) has the characteristic syndrome of beetle-pollinated flowers, though it may be visited by other insects as well. Some of the numerous stamens and carpels will be destroyed by the chewing behavior of pollinating beetles.

Figure 13.15 A fly-pollinated plant. Skunk cabbage (*Lysichitum americanum*) is a member of the arum family. The many small flowers give off a putrid odor. The funnel-shaped "corolla" is actually a modified leaf called a *spathe*.

cise way that only pollinators are attracted or able to reach them. However, plants are not above chicanery. In more than a few cases, instead of offering any real reward, plants mimic insects or food to trick the pollinator into visiting its flower. In effect, such plants are parasitizing animals for their energy as carriers of pollen. When a reward is actually offered, the interaction is mutualistic—of benefit to both partners.

Pollen is an excellent food for animals. Its tough outer wall resists digestion, but once it is pierced or chewed, the cytoplasm within typically provides protein (15 to 30% by weight), sugar (around 15%), fat (3 to 13%), starch (1 to 7%), plus trace amounts of vitamins, essential elements, and secondary substances. Most pollen is yellow to orange, and highly noticeable. Many pollen grains have a distinctive odor. The timing of pollen maturation can be very precise, coinciding with the seasonal or daily activity of pollinators. The anthers of corn (*Zea mays*), for example, split open in the morning; those of crocus (*Crocus* sp.) split in midday, and those of apple in the afternoon. Certain bat-pollinated flowers release pollen only at night.

Nectar is sugary water transported by the phloem into specialized secretory structures called **nectaries**. The sugar solution from nectaries may run down into elongated petal spurs, or it may form pools at the base of the floral tube. Like pollen, it may have a very limited, precise time of availability. Nectar usually contains 15 to 75% sugar (glucose, fructose, and sucrose). Amino acids are present only in minor amounts, but these are significant to pollinators such as butterflies that are completely

dependent on nectar for all their nutrition for a period of several months. All 13 essential amino acids for insects are present. Lipid is also often a minor component.

BIOTIC POLLEN VECTORS Beetles, flies, bees, butterflies, moths, birds, and bats are all pollinators. Beetles are among the oldest insect groups, already in existence when flowering plants first evolved. The flowers pollinated by beetles today have many primitive traits: regular symmetry, large single flowers, bowl-shaped architecture, and many floral parts that are not fused. Beetles are clumsy fliers. The bowl-shaped flowers attract beetles by odor (Table 13.1), and they provide a large target. Beetles chew a path through the sexual organs. By possessing many stamens and pistils, flowers are sure to have some that not only survive the attack but are pollinated. Many beetle-pollinated species are tropical. The temperate-zone tulip tree (*Liriodendron*) flower (Fig. 13.14) is a good example of this syndrome.

Flies are very diverse, and some have evolved to mimic bees, bumblebees, wasps, and hawk moths. For this reason, there is no single syndrome of floral traits for fly pollination. However, one family of fly-pollinated flowers, the arum lilies (Fig. 13.15), is unique. Their trumpet-shaped petals or bracts often have a checkered mosaic of opaque patches. In a breeze, this moving pattern, along with the epidermal hairs flicking in the wind, attracts flies. The effect may be a mimicry of many moving flies, which will attract other flies. These flowers emit the odor of decaying protein in carrion or dung. Some fly-pollinated flowers trap the insects for several hours.

Figure 13.16 *Phlox diffusa*, a butterfly-pollinated flower.

a

b

Figure 13.17 Bird-pollinated flowers. (**a**) Columbine (*Aquilegia formosa*), pollinated by hummingbirds, is a tubular type. (**b**) *Dryandra hewardiana*, pollinated by honey eaters, is a brush type.

Bees and butterflies are active by day, and they must have a landing platform, usually the flower petals. Moths can hover, so they don't require a corolla lip; and they are active by night or at dawn and dusk. Bees, butterflies, and moths harvest nectar as their reward, but the flowers they pollinate are strikingly different (see Table 13.1). Butterfly flowers are vividly colored, emit faint odors, have a broad blossom rim, are erect, and exhibit prominent **nectar guides** (Fig. 13.16). These are various markings that direct the pollinator to the flower's sexual organs and source of nectar. Moth flowers are white or faintly colored, emit heavy odors that penetrate the night air, have a fringed blossom rim, are pendant or horizontal, and have no nectar guides. Moth flowers are often closed during the day. Both butterfly and moth flowers have long, narrow tubes (impossible for bees and beetles to enter) with pools of nectar at their base. Bee pollination is discussed in the sidebar, "Bee-Pollinated Flowers," (p. 210).

Birds were not recognized by botanists as pollinators until the 20th century. Now we know that thousands of plant species are pollinated by birds in many parts of the world—by hummingbirds in North and South America, sunbirds and sugar birds in Africa and Asia, honeyeaters and honey creepers on Pacific islands, and honey parrots or lorikeets in Australia. Flowers that birds visit for nectar have strikingly similar syndromes. The flowers range from scarlet to red to orange and generally lack nectar guides. They have very deep tubes without a landing platform, are pendant or horizontal, and have abundant nectar but emit no odor (see Table 13.1). Columbine flowers have this syndrome (Fig. 13.17a). Other bird-pollinated flowers have brush-type flowers, with elongate stamens that stick out in all directions (Fig. 13.17b).

Most bats eat insects, but some are vegetarian; these have longer snouts and tongues, smaller teeth, larger eyes, and a better sense of smell. Bat flowers open at night, just as moth flowers do. They are positioned below the foliage of the parent tree—hanging pendant on a long pedicel or attached to the trunk or low limbs. Bats are color-blind, so bat flowers are drab white, green, or purple. The flowers exude a strong musty odor at night, reminiscent of fermenting fluid, cabbage, or bats themselves. Bat flowers are large and tough, with lots of pollen (some have more than 1300 anthers) and nectar (7 to 15 ml). Bat flowers include *Musa* (banana; Fig. 13.18), *Adansonia* (baobab), *Kigelia* (sausage tree), *Ceiba* (kapok), and *Agave*.

ABIOTIC POLLEN VECTORS Wind and water both carry pollen. Typical wind-pollinated flowers are small, colorless, odorless, and lacking in nectar. Petals are often lacking or are reduced to small scales. The flowers or inflorescences are positioned to dangle or wave in the open. Trees such as walnut (*Juglans* sp.), hazelnut (*Corylus* sp.), and oak (*Quercus* sp.) produce flowers before new leaves emerge in the spring. Grasses and sedges (for example, *Cyperus* sp.) position their flowers well above the leaves so they are exposed to wind currents.

The pollen grains of such plants are generally smaller, smoother, and drier than those of animal-pollinated species. Wind-carried pollen grains are 20 to 60 μm in diameter, while insect-carried pollen grains are 13 to 300 μm. The pollen often changes shape from spherical to

BEE-POLLINATED FLOWERS

Bees are among the most "intelligent" insects. Their sense of color is well developed, in contrast to beetles and flies. They are receptive to blue, yellow, white, and ultraviolet wavelengths. They cannot see red, but they will visit red flowers that also reflect ultraviolet light (Fig. 1). These colors are produced by carotenoid and flavonoid pigments in the chromoplasts (Chapter 3) or vacuoles of petal cells. Splotches, dots, or lines of contrasting color, called nectar guides, often lead toward the sexual organs (Fig. 2).

Bees are sensitive to sweet odors and to sugary nectar, and both are often present in bee-pollinated flowers, especially during the day (Fig. 2). Odors come from organic molecules in the pollen or petals. Bees can discern the intensity of sweetness, and they favor the sweetest nectars. Unless the sugar concentration is greater than 18%, honeybees operate at a loss; that is, their expenditure of metabolic energy exceeds that of the reward.

Pollen is a major reward, and bees carry the pollen in their crop, on abdominal hairs, or in hairy "baskets" on their hind legs (Fig. 3). One insect can carry as many as a million pollen grains. A bee's proboscis (tubelike snout) is very sensitive, and petal hairs often force the insect to enter the flower by a certain path that leads it past both anthers and stigmatic surfaces.

Only recently has the importance of oil as a reward for pollinating bees been recognized. More than 2000 species in 13 plant families produce oil instead of nectar. Their flowers are visited by specialized bees with long, densely hairy forelegs; these forelegs slice open *elaiophores* (the oil-containing equivalent of nectaries) and soak up oil like a wick. The bees also wag their abdomens over the petals to mop up oil onto body hairs. The energy content of oil—twice that of sugar—makes it a suitable food for bee larvae.

A few genera of orchids (including *Calypso* and *Ophrys*) mimic the odor, color, and shape of female bees (Fig. 4). Males are attracted, land on the flower, and try to copulate with it. In the process, wads of sticky pollen become attached to their bodies, and these are transferred to another orchid when copulation is again attempted. Some South American species of the orchid *Oncidium* vibrate in the wind and mimic male bees; these are attacked by territorial bees. The diving attack often results in pollen becoming attached to the bee and then transferred to another flower. Other orchids mimic insect prey of female wasps, which land on and sting the petals. They pick up sticky pollen in the process and take it to the next mimic. In these cases, there is no reward for the insect.

Figure 1 A flower that reflects ultraviolet (UV) light. Evening primrose (*Oenothera fruticosa*) is photographed with normal light-sensitive film on the left and with UV-sensitive film on the right. Unlike humans, bees are able to see the reflected UV light.

a   b

Figure 2 Nectar guides. (**a**) In *Viola*, splashes of color and lines lead to the throat of the corolla. (**b**) *Nasturtium* flower, sectioned to show guide lines and hairs leading to the entrance of the tubular spur where the nectar accumulates.

Figure 3 The scotch broom landing platform corresponds to bee size and shape. A bee's weight forces its petals apart, releasing the stamens. These dust the bee with pollen. A bee grooms itself and packs pollen (the *orange mass*) inside "baskets" of leg hairs.

Figure 4 Some orchids imitate the odor, shape, and color of insects, and they attract pollinators without offering any rewards. Species of *Ophrys* attract male bees that attempt to copulate with the flower.

Figure 13.18   Banana (*Musa velutina*), a bat-pollinated flower.

Figure 13.19   Wind-pollinated flower. The architecture of jojoba (*Simondsia chinensis*) flowers and leaves creates vortices of air currents that bring pollen grains to the stigmatic surfaces. (Redrawn and used with permission, K. J. Niklas and S. L. Buchmann, *Amer. J. Bot.* 72 (1985): 530–39.)

Frisbee-like upon release to dry air, improving its aerodynamic form. Wind-pollinated flowers also produce more pollen grains per ovule (500 to 2.5 million) than animal-pollinated flowers (5 to 500). The volume of pollen released is tremendous: A single rye (*Secale cereale*) or corn (*Zea mays*) plant releases 20 million grains, and one plant of dock (*Rumex acetosa*) can release 400 million.

Stigmatic surfaces are enlarged and elaborate, often extending outside the flower. Architecture of the flower and the inflorescence creates vortices that trap pollen and permit the grains to settle onto stigmas at a rate greater than predicted by chance (Fig. 13.19).

To their advantage, many wind-pollinated species are also visited by insects. For example, arroyo willow (*Salix lasiolepis*; Fig. 13.20) flowers have nectaries, and bees are responsible for more than 90% of seed set. Pollen of corn and of some sedge (*Carex*) species is scented and attractive to insects. Bees are known to collect pollen from grass flowers, even though they lack nectar, color, and odor. So wind pollination and insect pollination syndromes are not mutually exclusive.

Most aquatic plants produce flowers that project above the water surface. Some of these—such as pondweed (*Potamogeton*)—are pollinated by wind; others—including water lily (*Nymphaea*)—by insects. Other plants produce flowers just at the water surface, and the pollen floats from anther to stigma; ditch-grass (*Ruppia*) is an example. Only a few plants actually transfer pollen in the water, below the surface, and these employ a variety of syndromes. The pollen grains of water-nymph (*Najas*) are heavy and spherical and sink from the anthers into trumpet-shaped stigmas below them. In eelgrass (*Zostera*), the pollen grains are elongate and threadlike. They actively coil around any narrow object, such as a stigma, that they encounter.

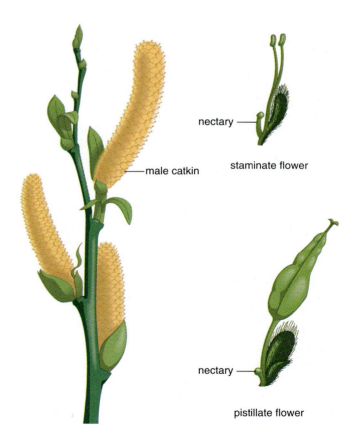

Figure 13.20   Wind- and insect-pollinated flowers. The staminate and pistillate flowers of willow (*Salix*), a dioecious plant, have typical features of a wind-pollinated plant: They lack a corolla and are grouped into catkins that appear before the leaves emerge in spring. Yet the flowers have nectaries, and the pollen grains are rather large and intensely yellow. Some species of willow rely predominantly on bees as pollen vectors (wind being relatively unimportant).

SUMMARY

1. The flower, the distinguishing structure of the flowering plants (angiosperms), is formed of four whorls of parts specialized to carry out sexual reproduction, including pollination, fertilization, and seed production.

2. The four whorls of floral leaves are (a) the calyx, composed of sepals; (b) the corolla, composed of petals; (c) the androecium, composed of stamens; and (d) the gynoecium, composed of carpels.

3. The stamen has two parts: the pollen-producing anther and the filament.

4. The gynoecium consists of one or more pistils. A simple pistil is made of only one carpel; if it consists of two or more fused carpels, it is a compound pistil.

5. A pistil consists of three parts: the stigma, the style, and the ovary. The stigma is receptive to pollen, and the ovary encloses the ovules. At the time of fertilization, the ovules consist of integuments, the nucellus, and the embryo sac.

6. Meiosis takes place in both the anthers and the ovule.

7. Pollen is a two-celled haploid plant or male gametophyte, protected by a cell wall, that is frequently elaborately sculptured. Two sperm cells are produced by the generative cell.

8. The embryo sac is a seven-celled haploid plant, or female gametophyte. There are two synergids, one egg, three antipodals, and a central cell with the two polar nuclei (seven cells and eight nuclei).

9. Pollination is the transfer of pollen from anthers to stigmas. A cellular recognition system regulates the germination of compatible pollen. A pollen tube that grows down the style to the ovule carries the two sperm to the embryo sac.

10. Double fertilization involves the union of one sperm with the egg (to form the zygote) and the union of a second sperm with the polar nuclei (to form the primary endosperm nucleus). This double process is unique to the flowering plants.

11. The embryo plant develops from the zygote. The endosperm develops from the primary endosperm nucleus, and it becomes the nutrient source for the developing embryo.

12. A floral apex has lost its ability to elongate and its potential for vegetative growth.

13. Species that have male and female sexes on different plants are dioecious. Species with staminate and pistillate flowers on the same plant are monoecious.

14. Variation in floral architecture arises from (a) variation in number of parts in a given whorl; (b) symmetry in floral parts; (c) connation of floral parts; (d) adnation of floral parts; (e) superior or inferior ovary; and (f) the presence or absence of certain whorls.

15. Flowers are either solitary on flower stalks or grouped in various inflorescences such as heads, spikes, catkins, umbels, panicles, racemes, and cymes.

16. There are essentially two types of pollination: cross-pollination, which involves more than one plant and results in much variation in progeny, and self-pollination, which involves a single plant and results in progeny having great similarities.

17. Apomixis is a type of asexual reproduction in which an unfertilized egg develops into an embryo.

18. Pollination occurs by means of biotic and abiotic vectors. Biotic vectors are animals that transfer pollen accidentally and secondarily; their primary objective is to obtain rewards for visiting flowers. The rewards commonly include pollen, nectar, oil, wax, and odors. The rewards cost the plant metabolic energy to produce, but this cost is balanced against the surety of cross-pollination. The plant–animal interaction is mutualistic because there are benefits to both partners. Some flowers attract pollinators without providing any rewards; in such cases the relationship is parasitic rather than mutualistic.

19. Animal vectors include beetles, flies, bees, butterflies, moths, birds, and bats. Each animal visits flowers with a particular syndrome of traits. The traits—shape, color, odor, timing of opening, nature of rewards offered—attract the animal and permit it to function as a pollinator.

20. Abiotic vectors include wind and water. Wind-dispersed pollen is distinctly different from animal-dispersed pollen; it is smaller, the outer wall is smoother, and the grains are not sticky. The amount of pollen produced is also much greater, and the ratio of pollen grains to ovules is orders of magnitude larger.

Questions

1. Define each of the following terms:

sepal, petal

pistil, carpel, stigma, style, ovary

stamen, filament, anther

calyx, corolla, gynoecium, androecium

2. Make a detailed diagram showing the steps that occur in the anther leading to the formation of the pollen grains (the male gametophyte).

3. Make a detailed diagram showing the steps that occur in the ovule leading to the formation of the embryo sac (the female gametophyte).

4. Make labeled diagrams of a mature pollen grain and a mature embryo sac. Show all of the cells and nuclei found in each of them.

5. Describe the process of double fertilization. What are the products of each sperm fusion?

6. Explain why a flower is actually a shortened shoot and the flower parts are modified leaves.

7. Explain the following terms:

regular or irregular flowers

complete or incomplete flowers

united flower parts

8. Distinguish between a simple flower and an inflorescence.

9. Explain the difference between self-pollination and cross-pollination. Is one of these strategies better than the other? Why?

10. Pollination syndromes involve pollination strategies that have evolved between particular flowers and particular vectors—for example, bees. Describe an example of a pollination syndrome involving a biotic vector and another involving an abiotic vector.

Further Readings

Barth, F. G. 1985. *Insects and Flowers*. Princeton, N. J.: Princeton University Press. This is a text for advanced readers on the relationships between plants and insects.

Bowman, J. L., D. R. Smyth, and E. M. Meyerowitz. 1989. "Genes Directing Flower Development in *Arabidopsis*." *The Plant Cell* 1: 37–52. This article shows how specific genes are responsible for the different whorls of the flower. It's an advanced article, however, and not for the timid reader.

Fahn, A. 1990. Chapter 19 of *Plant Anatomy*. 4th ed. Oxford: Pergamon Press. Here is one of the most current plant anatomy textbooks. This chapter on flowers concentrates on anatomical features of flower parts.

Gifford, E. M., and A. S. Foster. 1989. Chapters 19 and 20 of *Morphology and Evolution of Vascular Plants*, 3d ed. New York: Freeman. This is a classic textbook and is relatively easy to read and understand. Serious students of flowers will enjoy these chapters.

Greyson, R. I. 1994. *The Development of Flowers*. Oxford: Oxford University Press. This is the newest text on all aspects of flower development.

14 SEEDS AND FRUITS

1. In flowering plants, seeds are the structures containing the embryo plant for the next generation. Seeds are surrounded by a seed coat and contain the embryo axis and the cotyledons. They contain either one cotyledon (monocotyledonous plants) or two (dicotyledonous plants). Cotyledons contain stored food.

2. Germination of seeds involves the activation of processes in the embryo, such as mobilization of food reserves and starting cell division and elongation. The embryo radicle becomes the root system of the seedling plant, and the epicotyl becomes the shoot system.

3. There are several different types of seeds. Seeds have differing mechanisms and specialized structures for dispersal.

4. A fruit is a ripened ovary. There are several different types of fruits.

5. The function of the fruit is to aid in dispersal of the seeds. Several different vectors—wind, water, and animals—are involved in fruit/seed dispersal.

14.1 SEEDS: DORMANT EMBRYOS

Seeds and fruits are without doubt the most important source of food for people and other animals, and they always have been. Seeds and fruits are filled with stored foods intended to help the embryo germinate and grow—or to attract an animal to eat the fruit and inadvertently carry the seeds away to spread them elsewhere. Early people recognized the nutritional value of seeds and fruits, and they harvested them from wild plants. Other, more advanced peoples managed to figure out how to grow them for food.

Rice (*Oryza* sp.), corn (*Zea mays*), and barley (*Hordeum* sp.) seeds (which are actually fruits) feed the majority of the people in the world. Because seeds and fruits are so important, stories, myths, and legends abound. Even the Bible describes "forbidden fruit" in the Garden of Eden.

In the 1500s the Doctrine of Signatures was widely believed. The idea of the Doctrine was that the appearance of a plant or plant part would reveal its inner secrets and possible uses to people. Others expanded this idea to include shape, color, and smell as indicators of use. The scales of a pinecone, for example, look like teeth, so medieval people made a concoction of pinecones mixed with vinegar to gargle for teeth and gum problems. (Actually, the vinegar probably did the trick by itself.) Seeds of viper's bugloss (*Echium* sp.) resemble a snake's head, so the belief arose that a mixture made from these seeds could be used as a remedy for snakebite. Seeds of snapdragon (*Antirrhinum majus*) worn in a linen bag around the neck were supposed to prevent one from being bewitched. Some of these stories may have held a little truth simply by happenstance; but their real importance lies in the fact that they arose in the first place because people found plants, and their seeds and fruits, important to their very survival.

Seeds are mature ovules that contain the embryonic plants of the next generation. The tremendous production of seeds ensures the renewal of plant populations. Each seed is constructed and packaged to ensure its dispersal to a favorable site for successful germination and growth. The fruit is the packaging structure for the seeds of flowering plants. In this chapter we will discuss the structure and development of seeds and fruits and their adaptations for dispersal.

The Seed Is a Mature Ovule

The seed completes the process of reproduction initiated in the flower. Following fertilization, the zygote develops into an embryo, the primary endosperm nucleus develops into the endosperm, and the integuments of the ovule develop into the seed coat.

For a short time after fertilization (Fig. 14.1b, d), the zygote nucleus divides frequently while the primary endosperm nucleus divides rapidly to form the endosperm, the nutrient-rich storage tissue that will feed the seed when it germinates. After the endosperm has developed, the zygote nucleus divides to form a filament of several cells (Fig. 14.1c). The cell farthest from the micropyle begins a series of divisions that produces the early stage embryo or proembryo. At about the same time, the cell closest to the micropyle elongates and divides, becoming the **suspensor**, which supports the embryo in the endosperm (Fig. 14.1d, e). Further divisions result in a globular stage (Fig. 14.1e) and, finally, a heart-shaped stage, after the two cotyledons have developed (Fig. 14.1f). In addition, the embryo develops a radicle at one end and a shoot tip at the other (Fig. 14.1g).

The specific steps just described apply to cotton (*Gossypium hirsutum*); there are many variations in the details of embryo development for other plants. For example, the major distinction between embryos of dicotyledonous and monocotyledonous plants is the number of cotyledons (two or one, respectively).

While the embryo is developing, the nucellus, endosperm, and integuments are also undergoing changes that are characteristic of the group of plants to which the seed belongs. In the great majority of plants, the nucellus and endosperm are required only for the initial stages of embryo development. This is particularly true of the nucellus, which is generally used as a nutritive source in early embryo stages. It persists as a food storage tissue, the *perisperm*, in seeds of sugar beet and a few other species. The endosperm persists as a food reserve in seeds of many monocot plants, such as onion (*Allium cepa*) (Fig. 14.2); these include grasses of such major economic importance as rice (*Oryza sativa*) and serious weed pests such as yellow foxtail (*Setaria lutescens*) (Fig. 14.3). Endosperm persists as a food storage tissue in relatively few dicot seeds, castor bean being an exception (Fig. 14.4).

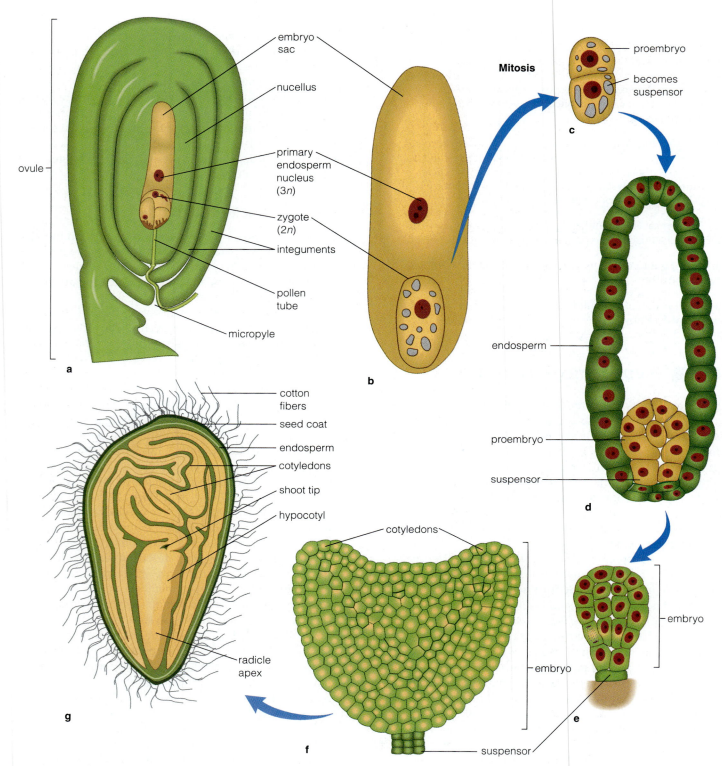

Figure 14.1 Embryo and seed development in cotton (*Gossypium hirsutum*). (**a**) Ovule after double fertilization. (**b**) Embryo sac after fertilization. (**c**) The zygote divides by mitosis; one of the two cells is destined to become the embryo, and the other the suspensor. (**d**) Early stage of embryo (proembryo) and endosperm development. (**e**) Early embryo as a small globe of cells. (**f**) Heart-shaped stage of the embryo, with newly formed cotyledons. (**g**) The cotton seed is the mature embryo, with highly folded cotyledons, surrounded by a seed coat. Note the seed coat fibers (really epidermal hairs), for which cotton is harvested. (a, c, d, and e, redrawn after U. R. Gore, *Amer. J. Bot.* 10 (1932): 795–807.)

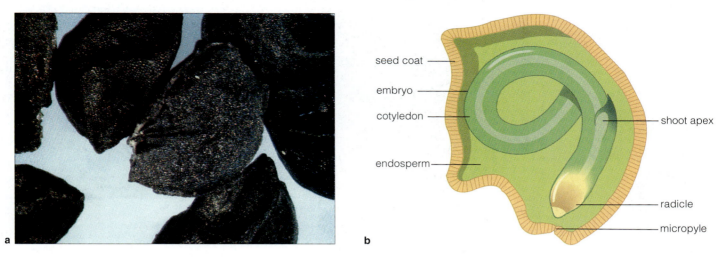

Figure 14.2 Onion seeds (*Allium cepa*). (**a**) External view, ×11. (**b**) Longitudinal section showing the embryo coiled within the endosperm.

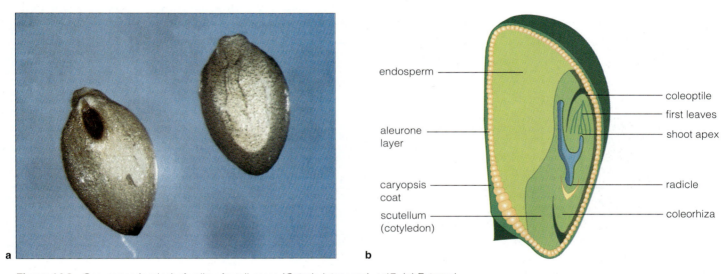

Figure 14.3 Caryopses (grains) of yellow foxtail grass (*Setaria lutescens*), ×17. (**a**) External views. (**b**) Median section.

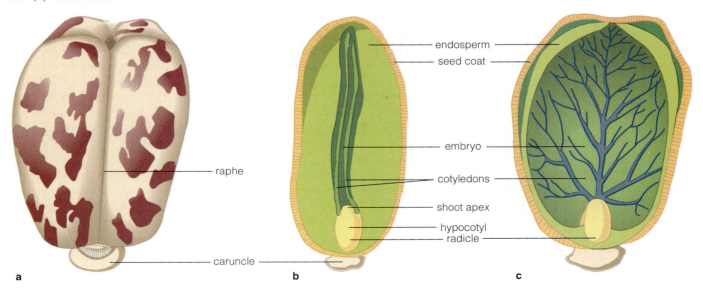

Figure 14.4 Castor bean (*Ricinus communis*) seed, a dicot with endosperm in the mature seed. (**a**) External view. (**b**) Section showing edge view of embryo. (**c**) Section showing flat view of embryo.

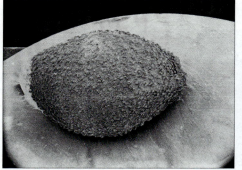

Figure 14.5 Sculptured seed coats. (**a**) Field bindweed (*Convolvulus arvensis*), ×19. (**b**) California poppy (*Eschscholtzia californica*), ×118.

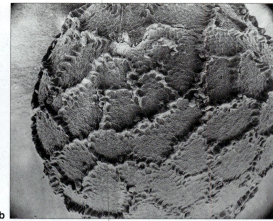

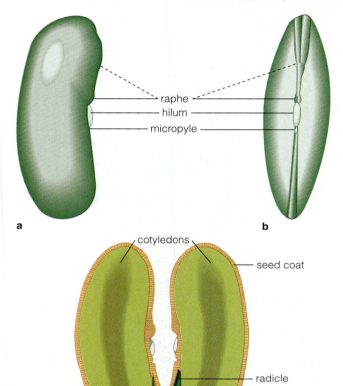

a   b

c

Figure 14.6 Bean (*Phaseolus vulgaris*), a typical dicot seed lacking an endosperm when mature. (**a**) External side view. (**b**) External face view. (**c**) Opened to expose embryo.

When food storage occurs within the embryo, the normal vascular tissues of the embryo convey the solubilized food to the meristems of the emerging plant, where it is required for growth. Food stored in the endosperm, outside the embryo, is absorbed through epidermal cells.

The integuments become hard **seed coats** in the most mature seeds. Scanning electron microscopy shows the seed coats to be variously and sometimes beautifully sculptured (Fig. 14.5). The seed coat acts as a protective shell around the embryo and sometimes contains chemical substances that inhibit the seed from germinating until the temperature, light, or moisture conditions are exactly right for germination.

Seed Structures Vary

Seed structure varies widely. This means that plants have evolved many solutions to propagating themselves successfully. We will briefly describe, as examples of variations in seed structure, the seeds of two dicot plants—bean and castor bean—and two monocot plants—a grass and onion.

COMMON BEAN The bean (*Phaseolus vulgaris*) seed is shaped like a kidney. External structures on the seed are the hilum, micropyle, and raphe (Fig. 14.6). The *hilum* is a large oval scar left when the seed breaks away from its placental connection, the *funiculus*. The **micropyle** is a small opening in the seed coat at one end of the hilum; it is the opening through which the pollen tube enters the ovule. The *raphe* is a ridge at the end of the hilum opposite the micropyle and is the base of the funiculus.

When the seed coat of a soaked bean is removed, what remains is the embryo; no endosperm is present. The bean embryo consists of two fleshy cotyledons and the *embryo axis*. The embryo axis is composed of the embryonic root or radicle at one end and the embryonic shoot or **epicotyl** at the other end. The **hypocotyl** is just below the cotyledons.

CASTOR BEAN The castor bean (*Ricinus communis*) seed has an external structure called the **caruncle**, which is a spongy outgrowth of the outer seed coat. The hilum and micropyle of the castor bean are covered by the caruncle, and the raphe runs the full length of the seed (see Fig. 14.4). The caruncle functions in absorbing water, which is needed during germination.

The castor bean embryo is embedded in a massive endosperm. The embryo consists of two thin cotyledons, a very short hypocotyl, a small epicotyl, and a small radicle. Castor bean oil was used by the Egyptians as a laxative; a paste made from the seeds was also used as a treatment for toothache. These seeds should be handled with care, however, because they contain a very toxic substance, called *ricin*. If ingested, ricin causes nausea, muscle spasms, and convulsions; as few as eight seeds can cause death.

GRASSES The so-called seed of grasses is really a fruit, the **caryopsis** or **grain** (see Fig. 14.3). It is a one-seeded, dry fruit in which the pericarp or ovary wall is firmly attached to the seed coat and seed. The starchy endosperm constitutes the bulk of the caryopsis and is surrounded by a layer of cells, the **aleurone** layer, that contain proteins and fats but little or no starch.

The grass embryo has an axis with a shoot apex and a root apex. The shoot apex, together with several rudimentary leaves, is ensheathed by a **coleoptile**. The radicle is surrounded by the **coleorhiza**. A relatively large part of the grass embryo is a very specialized, shield-shaped cotyledon called the **scutellum**. The outer cells of the scutellum secrete enzymes that digest the adjacent stored foods in the endosperm. These digested foods move from the endosperm through the scutellum to the growing parts of the embryo.

The process of milling to make polished white rice removes the caryopsis fruit coat, the entire protein-rich aleurone layer, and the outer layers of the endosperm. This means that most of the nutritious protein of the grain is removed before the rice is packaged for human consumption. The milled material, called *bran*, is now a popular food itself.

Popcorn is also the fruit of a grass plant. Americans love popcorn. We eat billions of gallons of it every year. Have you ever wondered why popcorn pops? Have you ever wondered why so many kernels are left unpopped at the bottom of the kettle? Scientists at the University of Illinois conducted experiments to answer these questions. Popcorn pops when heated because steam builds up inside the grain. The resistance of the fruit coat holds the steam back until the pressure becomes so high that the grain bursts. All the starch inside becomes fluffy. The "duds" don't pop because their fruit coat has cracks in it that allow the steam to escape.

ONION Like grasses, the onion (*Allium cepa*) is a monocot; but unlike grasses, its seed coat encloses only a small amount of endosperm. The embryo is very simple, the radicle and single cotyledon being quite prominent. The shoot apex is located close to the midpoint of the axis and appears as a notch. The embryo is coiled, with the radicle end usually pointing toward the micropyle (see Fig. 14.2).

14.2 GERMINATION

Germination, the first step in the growth of the embryo, begins with the uptake or **imbibition**, of water. This is a critical step because seeds are quite dry, containing only 5 to 10% water. The cells of dry seeds are tightly packed with stored proteins, starch, and lipids (Fig. 14.7). This stored food is packaged into cytoplasmic organelles called *protein bodies*, *lipid bodies*, and *amyloplasts*, which

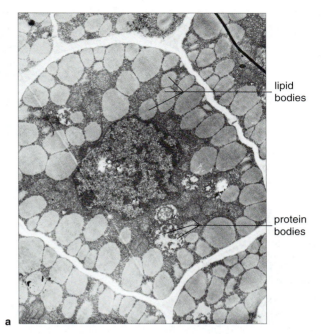

lipid bodies

protein bodies

a

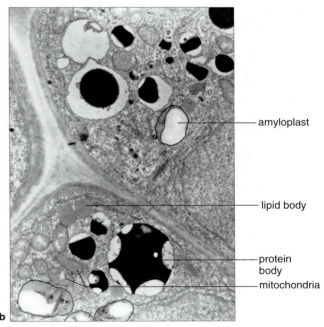

amyloplast

lipid body

protein body

mitochondria

b

Figure 14.7 Cellular changes brought about by germination, demonstrated in transmission electron micrographs of yellow foxtail grass (*Setaria lutescens*). (**a**) Scutellum cell in dry caryopsis before germination; note the abundance of storage organelles such as lipid bodies and protein bodies, ×8,600. (**b**) Root cell after 65 hours of germination; the cell now has fewer storage organelles and more organelles involved in metabolism, such as mitochondria and amyloplasts, ×6,500.

store starch (see Chapter 3). After imbibition, enzymes are activated and rapidly released to digest the stored food into smaller molecules which can then be trans-

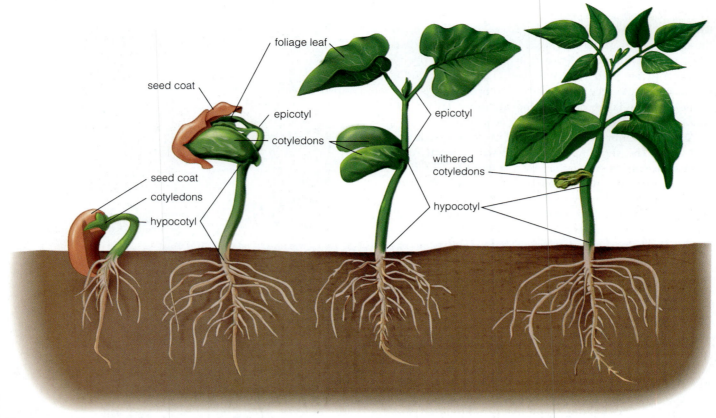

Figure 14.8 Stages in the epigeal germination of a bean (*Phaseolus vulgaris*) seed.

ported and converted into energy needed for growth. Consequently, the cells of imbibed embryos contain fewer storage organelles and more mitochondria, ribosomes, and endoplasmic reticulum (Fig. 14.7b)—all organelles involved in metabolism.

The first indication that germination has begun is generally the swelling of the radicle. It imbibes water rapidly and, bursting the seed coat and other coverings that may be present, starts to grow downward into the soil.

The Germination Process Differs Among Plants

Although the succeeding steps of germination are essentially similar in all plants, there are variations. In the germination of beans, peas, castor beans, and onions, a structure with a sharp hook is first forced upward through the soil. The structure forming the hook is different in each case. In bean (*Phaseolus vulgaris*), the hypocotyl elongates (Fig. 14.8). In pea (*Pisum sativum*), the epicotyl elongates (Fig. 14.9). In both cases, cotyledons and shoot apex remain below ground at first. The hook straightens once it is above ground and exposed to light. In the case of bean, the straightening of the hypocotyl raises the cotyledons and shoot apex toward the light. This is **epigeal germination**. When the pea epicotyl

straightens, the cotyledons remain below ground, and only the apex and first leaf are raised upward. This is called **hypogeal germination** (Fig. 14.9).

Castor bean (*Ricinus communis*) has epigeal germination (Fig. 14.10). It is different from common beans in that its cotyledons first function as absorbing organs, facilitating the transfer of food from the endosperm to the rest of the seedling. When the reserve food supply in the endosperm is exhausted, the cotyledons of the castor bean embryo emerge from the seed coat. They enlarge, become green, carry out photosynthesis for a time, and then eventually wither and die.

In onion (*Allium* sp.), a sharply bent cotyledon breaks the soil surface and slowly straightens out. The cotyledon of the onion is tubular, and its base encloses the shoot apex (Fig. 14.11). The first leaf finally emerges through a small opening at the base of the cotyledon.

In grasses such as corn (*Zea mays*), the situation is more complex. The shoot and root are enveloped by tubular sheaths of the coleoptile (Fig. 14.12) and coleorhiza. The primary root rapidly pushes through the coleorhiza. Adventitious roots then arise from the lower nodes of the stem. The coleoptile elongates and emerges above ground, becoming 2 to 4 cm (about 1 to 2 in) long. At this time, the uppermost leaf pushes its way through the coleoptile and, growing rapidly, becomes part of the photosynthesizing shoot.

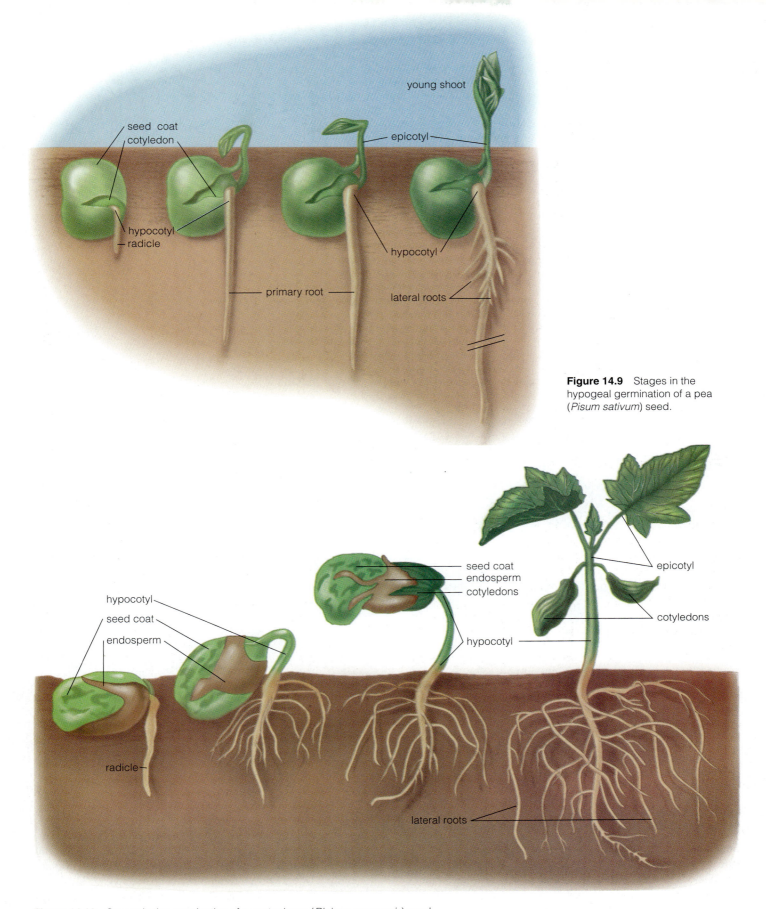

Figure 14.9 Stages in the hypogeal germination of a pea (*Pisum sativum*) seed.

Figure 14.10 Stages in the germination of a castor bean (*Ricinus communis*) seed.

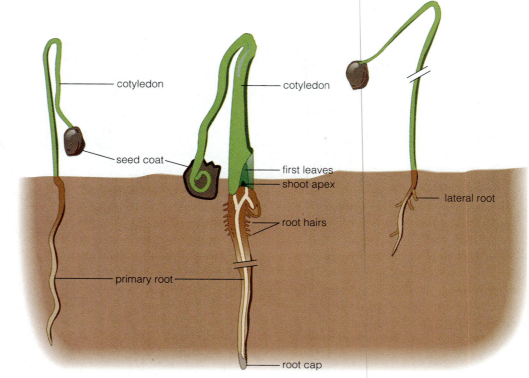

Figure 14.11 Stages in the germination of an onion (*Allium cepa*) seed.

Labels: cotyledon, cotyledon, seed coat, first leaves, shoot apex, root hairs, lateral root, primary root, root cap

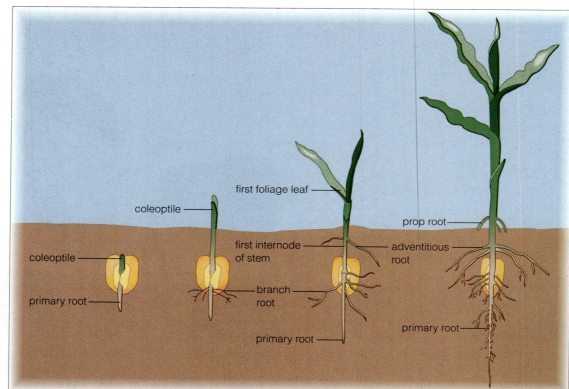

Figure 14.12 Stages in the germination of corn (*Zea mays*). After the primary root emerges, it branches to form the root system. Adventitious roots emerge from the lower stem, and prop roots form to hold the stem upright. The emerging young leaves are protected by a sheathlike coleoptile.

Labels: first foliage leaf, coleoptile, prop root, coleoptile, first internode of stem, adventitious root, primary root, branch root, primary root, primary root

Germination May Be Delayed by Dormancy

Seeds can remain viable for remarkably long periods. In one study, jars containing seeds from several different plant species were buried. At 5- and 10-year intervals, the jars were opened, and the seeds were tested for germination. Most species remained viable for at least 10 years, and one species, the moth mullein (*Verbascum blattaria*),

germinated after more than 90 years. This is not a record for seed longevity, however. Viable seeds from the Oriental lotus (*Nelumbo nucifera*) have been removed from archaeological sites known to be more than 1000 years old.

Many viable seeds will not germinate even when supplied with water, oxygen, and a favorable temperature because they are in a state of **dormancy** (inability to ger-

KEY TO FRUITS

A key is a tool to help students identify things when there are several possible choices. This key can be used to identify the different types of fruits. It is a dichotomous key, in which there are two choices at each level. For example, the first level asks if the fruit is formed from a single ovary or from several. Progress through each level in the key to determine the fruit type.

I. Fruit formed from a single ovary of one flower. *Simple fruits*.

  A. Pericarp fleshy.

    1. The ovary wall fleshy and containing one or more carpels and seeds. *Berry*. (tomato, *Lycopersicon* sp.)

      a. Ovary wall with a hard rind. *Pepo*. (watermelon, *Cucumis melo*)

      b. Ovary wall with a leathery rind. *Hesperidium*. (orange, *Citrus* sp.)

    2. Only a portion of the pericarp fleshy.

      a. Exocarp thin; mesocarp fleshy; endocarp stony; single seed and carpel. *Drupe*. (cherry, *Prunus* sp.)

      b. Outer portion of pericarp fleshy, inner portion papery, floral tube fleshy; several seeds and carpels. *Pome*. (apple, *Malus* sp.)

  B. Pericarp Dry.

    1. Dehiscent fruits.

      a. Composed of one carpel.

        i. Splitting along two margins. *Legume* or *Pod*. (pea, *Pisum* sp.)

        ii. Splitting along one margin. *Follicle*. (individual fruits in *Magnolia* multiple fruit)

      b. Composed of two or more carpels.

        i. Dehiscing in one of four different ways. *Capsule*. (poppy)

        ii. Separating at maturity, leaving a persistent partition wall. *Silique*. (mustard)

    2. Indehiscent fruits.

      a. Pericarp bearing a winglike growth. *Samara*. (maple)

      b. Pericarp not bearing a winglike growth.

        i. Two or more carpels, united when immature, splitting apart at maturity. *Schizocarp*. (carrot)

        ii. One carpel; if more, not splitting apart at maturity; one-seeded fruits.

          a) Seed united to the pericarp all around. *Caryopsis* or *Grain*. (rice)

          b) Seed not united to the pericarp all around.

            i) Fruit large, with thick, stony wall. *Nut*. (walnut)

            ii) Fruit small, with thin wall. *Achene*. (sunflower)

II. Fruits formed from several ovaries.

  A. Fruits developing from one flower. *Aggregate fruit* (classify the individual fruits in key for simple fruits). (strawberry)

  B. Fruits formed from several flowers. *Multiple fruit* (classify individual fruits in key for simple fruits). (pineapple)

---

minate because of reduced physiological activity). Various factors can break dormancy. For instance, light is necessary for the germination of some lettuce (*Lactuca*) species. Scarring or breaking through the seed coat is required before the seeds of some plants (including many kinds of legumes) will germinate; their hard, dense seed coats restrict the movement of water and gases. In orchids, seeds are dispersed while the embryos are immature, and they must develop further before germinating. Seeds of some cool-temperate zone plants, (for instance, gooseberry, *Ribes speciosum*), will not germinate unless they are first subjected for a time to temperatures close to freezing, while moist. At the other extreme, the seeds of some pines (*Pinus* sp.) will not germinate unless they have been subjected to the rather high heat of a fire. Another type of dormancy is produced by natural chemical inhibitors, which occur in many fruits or in seed coats Yellow foxtail grass (*Setaria lutescens*) is an example.

14.3   FRUITS: RIPENED OVARIES

A **fruit**, the ripened ovary of a flower, is an important auxiliary structure in the sexual life cycle of angiosperms. Fruits protect seeds, aid in their dispersal, and may be a factor in timing their germination. Because fruits are highly constant in structure, even when grown in different environments, they play an important role in the classification of angiosperms. The sidebar ("Key to Fruits," above) is a key to help you identify the different types of fruits described in this section.

In everyday usage, the term *fruit* usually refers to a juicy and edible structure such as an apple (*Malus* sp.), plum (*Prunus* sp.), peach (*Prunus persica*), or grape (*Vitis vinifera*). Structures that are commonly called *vegetables*, such as string beans (*Phaseolus vulgaris*), eggplant (*Solanum melongena*), okra (*Hibiscus esculentus*), squash (*Cucurbita* sp.), and cucumbers (*Cucumis* sp.), are all fruits in a

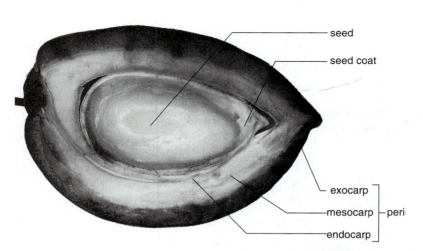

Figure 14.13 Mature almond (*Prunus amygdalus*) fruit showing split outer portion of ovary walls (pericarp): exocarp, mesocarp, and endocarp.

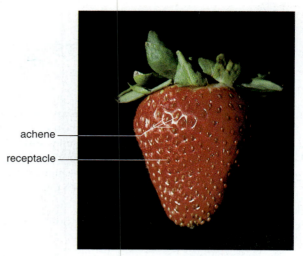

Figure 14.14 Strawberry (*Fragaria*), an aggregate fruit. The tiny individual fruits, called *achenes*, are simple, dry fruits. They are embedded in an enlarged fleshy receptacle.

botanical sense—as are grains of corn (*Zea mays*), and oats (*Avena sativa*), and other cereals.

The Nature of the Ovary Determines the Structure of the Fruit

In spite of considerable variation along family lines, fruits share basic developmental and anatomical characteristics. For instance, all fruit development is initiated by fertilization which stimulates the ovary wall to undergo development and differentiation into three layers. The fruit wall (which develops from the ovary wall) is called the **pericarp**; its three more or less distinct layers (in order, beginning with the outermost) are **exocarp**, **mesocarp**, and **endocarp** (Fig. 14.13). When the fruit is mature, floral structures—such as pedicel, calyx, withered stamens, style and stigma, and even remnants of the corolla—may also be present.

Tissues other than the ovary wall that form part of a fruit, are referred to as *accessory*. Much of the fruit of pineapple (*Ananas* sp.), apple (*Malus* sp.), and strawberry (*Fragaria* sp.) can be called accessory. In the strawberry, the edible portion is a thickened, pulpy central receptacle in which achenes (dry, one-seeded fruits) are embedded.

Fruits May Be Simple or Compound

There are three main categories of fruits. **Simple fruits** are derived from a single ovary. They may be dry or fleshy; the ovary may be composed of one or more carpels, and the fruit may be **dehiscent** (splits open when mature) or **indehiscent** (does not split open). **Compound fruits** are composed of more than one fruit. There are two types: aggregate fruits and multiple fruits. **Aggregate fruits** are derived from many separate ovaries of a single flower, all attached to a single receptacle (for example, strawberry, Fig. 14.14). **Multiple fruits**, such as pineapple

(Fig. 14.15), are the enlarged ovaries of several flowers grown more or less together into a single mass. The receptacle of some multiple fruits, such as fig (*Ficus* sp.), enlarges and is actually the edible part (Fig. 14.16).

In classifying the different kinds of fruits, the following criteria are taken into account:

1. the structure of the flower from which the fruit develops

2. the number of ovaries involved in fruit formation

3. the number of carpels in each ovary

4. the nature of the mature pericarp (whether the fruit wall is dry or fleshy)

5. whether or not the pericarp splits (dehisces) at maturity

6. if the pericarp dehisces, the manner of its splitting

7. the role, if any, that accessory tissues may play in formation of the mature fruit

These criteria were used in creating the fruit key shown in the sidebar.

Simple Fruits Are from Single Ovaries

Simple fruits come in several forms. Their pericarp (fruit wall) may be dry or fleshy. If dry, the pericarp may dehisce or not.

PERICARP DRY AND DEHISCENT   The **legume** or **pod** is the type of fruit found in nearly all members of the pea family (Fabaceae). A pod arises from a single carpel, which at maturity generally dehisces along two sides (Fig. 14.17). In the pea (*Pisum sativa*) pod, the shell is the pericarp, and the pea is the seed. Pods may be spirally twisted or curved, as in Scotch broom (*Cytissus scoparius*). However, a number of legumes such as alfalfa (*Medicago sativa*) have pods that do not dehisce.

berry

a

central axis

floral parts
from several
separate flowers

b

Figure 14.15 Pineapple (*Ananas comosus*), a multiple fruit. The individual fruits, called *berries*, are embedded in the swollen edible receptacle. Jagged-edged bracts extend out over the fruits. (**a**) Exterior view. (**b**) Section through center showing fruits.

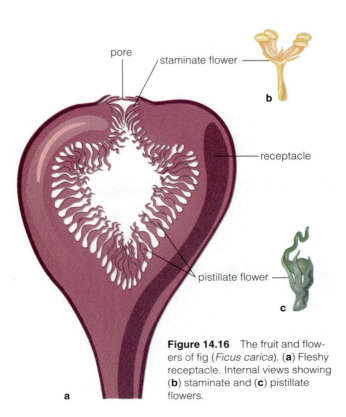

pore

staminate flower

b

receptacle

pistillate flower

c

a

Figure 14.16 The fruit and flowers of fig (*Ficus carica*). (**a**) Fleshy receptacle. Internal views showing (**b**) staminate and (**c**) pistillate flowers.

midrib

seed

funiculus

pericarp

style

stigma

Figure 14.17 Opened pea pod (*Pisum sativum*), showing developing seeds attached to carpel margins.

Figure 14.18 Dehiscing follicle of a *Magnolia* sp.

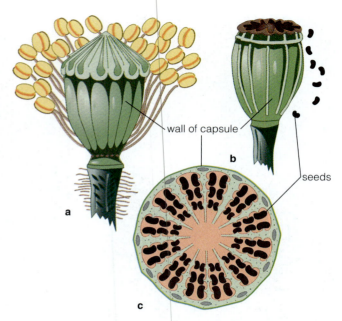

Figure 14.19 Poppy (*Papaver* sp.) capsule. (**a**) Side view, before dehiscence. (**b**) Mature capsule dehiscing by pores at the top. (**c**) Cross section of capsule, showing the position of seeds. (Redrawn from E. Krosmo, *Unkraut in Ackerbau der Neuzeit*. © 1930 by Springer-Verlag.)

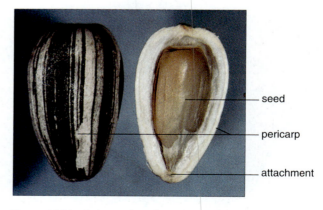

Figure 14.20 Silique of *Mathiola* sp.

Figure 14.21 Achene of sunflower (*Helianthus annuus*), unopened and opened to show attachment of seed.

An example of a **follicle** fruit is the magnolia (*Magnolia grandiflora*) (Fig. 14.18). The follicle develops from a single carpel and opens along only one side.

Capsules are simple fruits derived from compound ovaries (an ovary composed of more than one carpel). Each carpel produces a few to many seeds. Capsules dehisce in various ways along the top surface. Poppy (*Papaver* sp.) is an example (Fig. 14.19).

The **silique** is the characteristic fruit of members of the mustard family (Brassicaceae). The silique (Fig. 14.20) is a dry fruit derived from a superior ovary consisting of two locules. At maturity, the dry pericarp separates into three portions; the seeds are attached to the central, persistent portion.

PERICARP DRY AND INDEHISCENT The **achene** is a dry, one-seeded fruit. Sunflower (*Helianthus annuus*) achenes are usually called seeds, but carefully breaking open the pericarp reveals that the seed is inside. The pericarp is easily separated from the seed coat, which is a thin, filmy layer surrounding the sunflower embryo (Fig. 14.21).

The caryopsis or grain is the fruit of the grass family (Poaceae), which includes rice (*Oryza sativa*) and wheat (*Triticum aestivum*). The grain is a dry, one-seeded, inde-

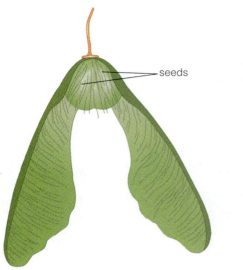

Figure 14.22 Two-seeded samara of maple (*Acer*).

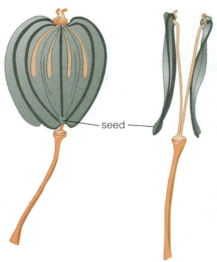

Figure 14.23 Schizocarp of cow-parsnip (*Heracleum hirsutum*). Left: face view; right: side view.

a

hiscent fruit (see Fig. 14.3). It differs from the achene in that pericarp and seed coat are firmly united all the way around the embryo.

The **samara** may be a one-seeded simple fruit, as in elm (*Ulmus* sp.), or a two-seeded one, as in maple (*Acer* sp.) (Fig. 14.22). These fruits are typified by an outgrowth of the ovary wall, which forms a winglike structure that aids in seed dispersal.

The **schizocarp** is a fruit characteristic of the carrot family (Apiaceae), which includes celery (*Apium graveolens*). The schizocarp consists of two carpels that split, when mature, along the midline into two one-seeded, indehiscent halves (Fig. 14.23).

The term **nut** is popularly applied to a number of hard-shelled fruits and seeds. Botanically speaking, a typical **nut** is a one-seeded, indehiscent dry fruit with a hard or stony pericarp (shell). Examples are chestnut (*Castanea* sp.) and walnut (*Juglans* sp.). An acorn, the fruit of the oak (*Quercus* sp.) (Fig. 14.24), is partially enclosed by a hardened cup. The outer husk of the walnut, which is removed during processing, is composed of bracts, perianth, and the outer layer of the pericarp. The hard shell is the remainder of the pericarp. Note that unshelled almonds (*Prunus* sp.), are really not nuts but fleshy fruits known as *drupes*, from which the hulls—exocarp and mesocarp—have been removed. Brazil nuts (*Bertholletia excelsa*) are seeds, not nuts, and the unshelled peanut (*Arachis hypogaea*) is really a pod.

PERICARP FLESHY The fruits in this category are popular for eating. They feature a fleshy fruit wall (pericarp). The fleshy part is usually attractive to animals, who eat the fruit and in turn carry away the seeds. The seeds found in these fruits tend to have a hard seed coat that is not broken down as the seed passes through the animal and is deposited in its feces.

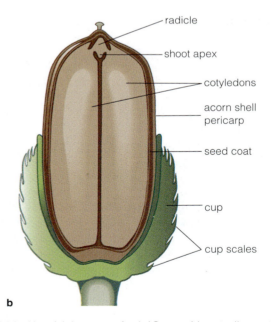

b

Figure 14.24 Nut. (**a**) An acorn of oak (*Quercus*) is actually a nut; the cup is made of fused bracts or cup scales. (**b**) Internal view of acorn, showing the embryo.

Figure 14.25 Berry of tomato (*Lycopersicon esculentum*). (**a**) External view. (**b**) Cross section.

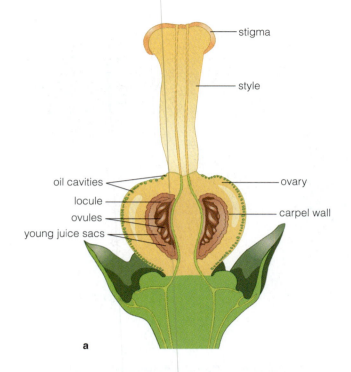

Figure 14.26 A citrus fruit is a hesperidium orange (*Citrus sinensis*). (**a**) Flower, showing a lengthwise section of maturing ovary. (**b**) Cross section and external view of mature fruit.

Cherry, almond, peach, and apricot (all are *Prunus* sp.), in the rose family (Rosaceae), are examples of **drupes**. The olive (*Olea* sp.) (family, Oleaceae) fruit is also a drupe. Derived from a single carpel, the drupe is usually one-seeded. It has a hard endocarp consisting of thick-walled sclereids (see Chapter 4), and a thin exocarp forms the skin. The mesocarp is the edible fleshy portion. The pit of a cherry (*Prunus* sp.) is a seed, with a thin seed coat, plus the stony inner layer (endocarp) of the ovary wall. In almond (*Prunus* sp.) fruit (see Fig. 14.13), the mesocarp is fleshy like a typical drupe when the fruit is young. As it develops, however, the mesocarp becomes hard and dry and forms the hull. The shell of the almond is endocarp. This is an instance in which the seed, not the outer part, is the edible part of a drupe.

A **berry** is a fleshy type of fruit that is derived from a compound ovary. Usually, many seeds are embedded in the flesh (Fig. 14.25), which is pericarp, although the line of demarcation may be difficult to see. It comes as a surprise to many that tomatoes, lemons (*Citrus limon*), and cucumbers (*Cucurbita* sp.) are berries but strawberries (*Fragaria* sp.) and blackberries (*Rubus* sp.) (in spite of their common names), are not. These are good examples of botanical names and common names that don't match.

Lemons, oranges, limes, and grapefruits (all *Citrus* sps.) are a type of berry called a **hesperidium**. Its thick, leathery rind (peel) with numerous oil cavities is exocarp and mesocarp; the thick, juicy pulp segments (endocarp) are composed of several wedge-shaped locules (Fig. 14.26). The juice forms in juice sacs or vesicles that are outgrowths from the endocarp walls. Each mature juice sac is composed of many living cells filled with juice.

The fruits of watermelon (*Citrullus vulgaris*), cucumber, and squash—all members of the cucumber family (Cucurbitaceae)—are a kind of berry called a **pepo** (Fig. 14.27). The outer wall (rind) of the fruit consists of receptacle tissue that surrounds and is fused with the exocarp. The flesh of the fruit is principally mesocarp and endocarp.

Apples (*Malus* sp.) and pears (*Pyrus* sp.) (in the Rosaceae family) are examples of **pomes**. This fruit is derived from a flower with an inferior ovary (Fig. 14.28). The flesh is enlarged hypanthium (a fleshy floral tube), and the core is from the ovary.

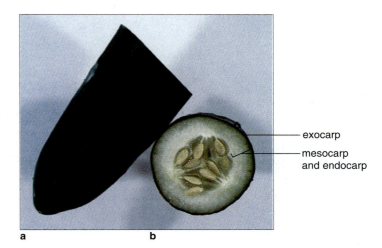

Figure 14.27 Cucumber (*Cucumis sativus*) fruit, a type of berry (pepo). (**a**) Whole fruit. (**b**) Transverse section.

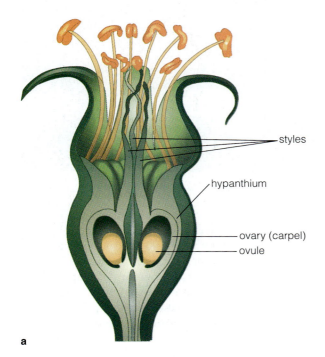

Figure 14.28 Pome fruit of apple (*Malus sylvestris*). (**a**) Median longitudinal section of flower, showing maturing fruit. (**b**) Median section of fruit. (**c**) Cross section of fruit.

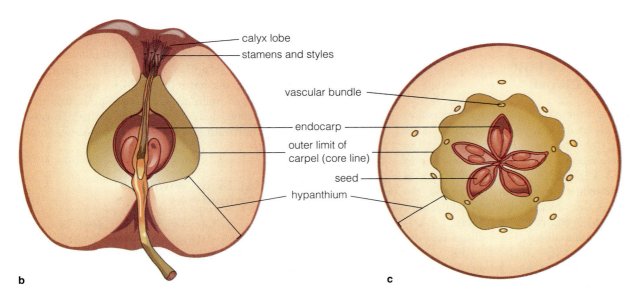

Compound Fruits Develop from Several Ovaries

An aggregate fruit is formed from numerous carpels of one individual flower. These fruits are made up of many simple fruits attached to a fleshy receptacle. The strawberry flower has numerous separate carpels on a single receptacle. Each carpel develops into an achene (see Fig. 14.14). Flowers of raspberry, blackberry, and other species of *Rubus* have essentially the same structure as strawberry, except that the attached fruits are small drupes (Fig. 14.29).

A multiple fruit is formed from individual ovaries of several flowers, all clumped together. The fig (*Ficus* sp.) (see Fig. 14.16) and the pineapple (*Ananas comosus*) (see Fig. 14.15) are examples of multiple fruits; the individual fruits composing them are drupes in fig and berries in pineapple. The fig fruit we eat is an enlarged, fleshy receptacle. Its flowers are small and attached to the inner wall of the receptacle.

Not All Fruits Have Seeds

In some plants, normal fruit may develop without seeds being enclosed. Fruits that develop without fertilization are called **parthenocarpic**; consequently, such fruits are seedless. Thompson seedless grapes (*Vitis* sp.) were thought to be parthenocarpic until it was shown that fertilization does take place but that the ovules fail to mature into seeds. Such situations have led to a broader use of the word *parthenocarpy* to mean simply, seedless fruits.

Parthenocarpic (seedless) fruits are quite regularly produced in such cultivated plants as eggplant, navel orange, banana, pineapple, and some varieties of apple and pear. In certain plants, seedless fruits may be induced by pollen that is incapable of fertilizing the ovules. For example, in some orchids, placing dead pollen or a water extract of pollen upon the stigma may start fruit development. Parthenocarpy is commercially induced in some plants by spraying the blossoms with dilute aqueous solutions of growth substances like auxin.

14.4 ADAPTATIONS FOR SEED DISPERSAL

The role of ripe fruit is threefold: to aid in the dispersal of the seeds inside; to deter inappropriate seed-dispersing animals from taking the fruit or seed; and to protect the seeds from herbivores who merely consume seeds but do not disperse them. It is important to realize that there is no nutritional relationship between the fruit and the seeds within it. That is, the stored food in the fruit cannot be utilized by dormant seeds or by germinating seedlings. The only stored food available to seedlings is in the endosperm and cotyledons within the seed itself.

Both seeds and fruits are rich in a variety of chemical resources: sugar, starch, protein, lipid, amino acids, and a

Figure 14.29 Aggregate fruit of the blackberry (*Rubus ursinus*). The individual fruits are drupes.

variety of secondary compounds. The average caloric value of this material is about 5100 kilocalories (kcal) per gram dry weight, a value approaching that of healthy animal tissue (about 6000 kcal per gram). In contrast, leaf, root, stem, and other vegetative tissues average only 4000 kcal per gram. This means that it costs the plant more to make seeds and fruits than it does to make vegetative organs. This expense is necessary to ensure that the materials are present inside seeds and fruits to guarantee the successful dissemination of the seeds. Several different mechanisms have developed to aid their dispersal. Fruits and seeds may catch the wind and fly (Fig. 14.30a–c), float in seawater, attract the eye and the stomach of a bird (Fig. 14.30g), entice an ant to carry a seed, or permit the fruit to hook onto the hairs of a passing mammal (Fig. 14.30d–f). Just as in pollination, the animal vectors used are sometimes rewarded, sometimes exploited; in other words, the relationship is sometimes mutualistic, sometimes parasitic.

Dispersal May Be by Wind, Water, and Animals

Common abiotic vectors for fruit and seed dispersal are wind and water. Winged and plumed fruits (Fig. 14.30a–c) are common adaptations for wind dispersal. In some cases, the seeds are ballistically exploded by a violent dehiscence of the pericarp (Fig. 14.30h, i). Some

Figure 14.30 Fruits and seeds, showing various devices aiding in dispersal. Wind: (**a**) *Clematis*, ×4. (**b**) Dandelion (*Taraxacum vulgare*), ×5. (**c**) Seed of Coulter's big-cone pine (*Pinus coulteri*), ×1. Attachment: (**d**) Cranesbill (*Geranium*), ×3. (**e**) Foxtail (*Hordeum hispida*), ×3. (**f**) Bur clover (*Medicago denticulata*), ×4. (**g**) Fleshy edible fruit of *Cotoneaster*, ×2. Violent dehiscence of pericarp: (**h**) Vetch (*Vicia sativa*), ×2. (**i**) California poppy (*Eschscholtzia californica*), ×2.

a

b

Figure 14.31 Plants with water-dispersed fruits or seeds. (**a**) Coconut (*Cocos* sp.) germinating on a tropical beach. The coconut fruit (or husk) is fibrous, and the seed within is large and buoyant, capable of floating hundreds or even thousands of miles in seawater. (**b**) The seeds of the desert smoke tree (*Dalea*) are carried away from the parent by flash floodwaters in the adjacent arroyo.

sedges (*Carex* sp.) have a fruit with a membranous envelope containing air, and these are spread by floating on water. The coconut (*Cocos* sp.) is a tropical group of plants famous for growing on midoceanic islands thousands of miles from other land. The coconut fruit is capable of floating for many days and then germinating when washed onto a sandy beach and leached of salt by rainwater (Fig. 14.31a). Many other tropical beach plants have similar (although much smaller) floating fruits. A number of weed species of irrigated farmland are dispersed by water along irrigation ditches. In deserts, for example, the smoke tree (*Dalea*) and desert willow (*Chilopsis*) growing along arroyos (dry stream beds) have hard seeds that are carried away from the parent by flash floods (Fig. 14.31b). The rushing water also pushes the seeds against rocks, scraping the seed coats and scarifying them. Without that scarification, the seeds would remain dormant. Water in this case is more than a dispersal agent; it pretreats the seed and makes it receptive to germination cues.

Common animal vectors include ants, birds, bats, rodents, fish, ruminants, and primates. They are attracted to fruit by its color, position, seasonal availability, odor, and taste. Sometimes, the vector eats only the fruit and discards the seeds; this is true of some primates. In other cases, the vector swallows the seeds unchewed; after passing unharmed through the gut, the seeds are ex-

creted some distance from where they were consumed. This is often the case with birds. Birds are attracted to fleshy, colored berries and, after dining on them, may fly long distances before regurgitating or excreting the hard seeds. Cattle eat legume pods of mesquite (*Prosopis fuliflora*) and *Acacia* in southwestern grasslands and later pass many undamaged seeds out in their excrement. These germinate, and the seedlings grow well in their fertilized microenvironment. In yet other cases, the animals eat many seeds but cache others. Squirrels and jays, for example, may carry walnuts, hickory nuts, acorns, and pine seeds from parent trees to distant hiding places. Apparently, they forget some of the caches and never revisit them; in effect, they have planted these seeds, and clusters of seedlings will emerge later.

Ants are responsible for dispersing many seeds of herbs in temperate-zone forests and grasslands. They harvest the hard, small seeds and deposit them in granaries below ground. Some seeds escape consumption and germinate. Other plant species have **elaiosomes**, or food bodies, at one end of their seeds (Fig. 14.32). Ants harvest the seeds only for that reward and then toss them out. Discarded outside the ant nest (several meters from the parent plant), the unharmed seed may then germinate.

The relationships above are mutualistic because there is some reward for the animal. Plants can also use animals in a more parasitic fashion; in such cases, there is no reward.

Figure 14.32 Elaisomes. (**a**) Flannel bush (*Fremontodendron californicum*), a chaparral shrub of California. (**b**) Each seed of the flannel bush has an elaiosome at one end. The seeds are harvested by ants, taken to their nests, stripped of their elaisomes, and discarded—whereupon the seeds are free to germinate without competing with the parent plant.

For instance, seeds of some aquatic and marsh plants stick to the feet of birds in mud and are carried for long distances. Mistletoe (*Phoradendron* sp.), a parasite of other plants, has naturally sticky seeds, and birds can carry them on their feet to new host trees. Seeds with beards, spines, hooks, or barbs catch a ride to a new site by adhering to animal hair and human clothing (see Fig. 14.30e, f).

Some Plants Have Evolved Antiherbivore Techniques

At the same time that the fruit attracts dispersal vectors, it must repel herbivores. Techniques of discouraging herbivores include reducing the time of fruit availability, making the fruit or seed coat physically hard, and making the fruit or endosperm chemically repellent.

Many perennial plant species do not reproduce every year, or at least the magnitude of their reproduction varies from year to year. Such species produce fruit and seed abundantly only during what are called **mast** years. Because food supply is a limiting factor in population size, the relatively low amount of seed produced in off years keeps the number of seed eaters in check. As a result, seed-eating mammal, bird, and insect populations are not large enough to consume all the seeds available during a mast year, so some seeds escape consumption and germinate.

Those species of plants that do reproduce every year often limit the time when ripe fruit is available. Large fruits that require a long time to develop are kept green, hard, and relatively small until just before the final maturation stages. Then color, texture, size, and sweetness change suddenly, and the fruit is available to herbivores for only a brief time. This limits the number of fruits (and seeds) they can consume before dispersal.

Our common notion of *fruit* is a juicy, soft organ, but many plants produce fruits that are partly or completely dry and hard. Examples include drupes, which have a hard endocarp, and such completely hard fruits as nuts. Biting or boring insects are prevented from invading the fruit or seed by the sclerenchyma tissue (usually sclereids), which also prevents the seed from being damaged by the grinding action in the crops of birds or the mouths of chewing mammals. Legume seed coats are notoriously hard and often pass through animal guts unharmed.

Chemical protection mechanisms are widespread and diverse. Many fruits are rich in **secondary compounds**, chemicals produced by a plant partly or entirely for the effect they have on other organisms. In the case of herbivore defense, the effect is negative and often toxic. Secondary compounds in fruits and seeds include lectins (which cause red blood cells to clump), enzyme inhibitors, cyanogens (which release cyanide, a potent nerve toxin), saponins (a detergent), alkaloids (like opium), and unusual amino acids. When present in the endosperm, these secondary compounds may later have a primary function as well; that is, they may be metabolized by the seedling into valuable nontoxic resources.

During the course of evolution, it appears that the metabolic quirks of at least one species of animal has successfully defused each chemical defense originated by seeds and fruits. This is referred to as *coevolution*. Thus, there are specialized insects capable of eating plant tissue toxic to nearly every other animal. In some cases—as in the larvae of the monarch butterfly, which is able to feed on milkweeds rich in metabolic by-products called cardiac glycosides—the animal not only eats the plant but uses the toxin for its own benefit, to deter a predator.

An excellent example of plant–herbivore coevolution is the series of plant defenses and herbivore responses

Table 14.1 Legume Defenses and Seed Weevil Adaptations

Plant Defense	Weevil Adaptation
Gum production by seed pods following penetration of first larva from egg mass; this may push off remaining eggs or drown or otherwise obstruct young larvae.	A period of quiescence in embryonic development in the egg until seed maturation is completed. Eggs laid singly instead of in clusters.
Dehiscence, fragmentation, or explosion of pods, scattering the seeds to escape from larvae coming through the pod walls.	Eggs laid on seeds only after they have been scattered. Attachment of seeds to one another or to the pod valve.
Production of a pod free of surface cracks, which prevents eggs from adhering. Also, indehiscent pods, excluding weevils, which lay eggs only on exposed seeds.	Eggs laid into the soft, fleshy exocarp. Chewing holes in the pod walls, in which eggs are then laid. Gluing the eggs to the pod surface.
A layer of material on the seed surface that swells when the pod opens and detaches the attached eggs.	Attachment of the eggs by anchoring strands to allow for substrate expansion. Entry of the larva through the pod into the seed before the pod opens.
Production of allelochemic substances.	Mechanisms for avoidance and detoxification.
Flaking of the seed pod surface, which may remove eggs laid on that surface.	Eggs laid beneath the flaking exocarp. Rapid embryonic development or feeding on immature seeds so that entry occurs before flaking begins.
Immature seeds remaining very small throughout the year and then abruptly growing to maturity just before being dispersed.	Entry and feeding upon immature seeds. Delay of embryonic maturation until seeds are mature.
Seeds so small or thin that weevils cannot mature in them.	Utilization of seed contents more completely. Feeding on several seeds.

From T. D. Center and C. D. Johnson, "Coevolution of Some Seed Beetles (*Coleoptera: Bruchidae*) and Their Hosts." *Ecology* 55: 1096–1103. Copyright © 1974 by the Ecological Society of America. Reprinted by permission.

summarized in Table 14.1 for a group of closely related tropical legumes. The plants exhibit a range of herbivore defenses including chemistry, texture, size, and timing. For each defense, however, some species of weevil has evolved a solution.

Distant Dispersal of Seeds Is Not a Universal Aim

The benefit of fruit and seed dispersal is the spread of a species far from its parent. There is a cost as well, however. Many fruits and seeds are wasted because they are eaten and deposited in places inappropriate for successful germination and seedling establishment. In some stressful habitats, only very few safe sites exist, and these are scattered within a large, hostile area. Because parent plants ordinarily already reside within one of the safe sites, it is advantageous to prevent or limit dispersal away from the parents.

One method of limiting dispersal is self-planting. The morphology of the fruit lends itself to lodging near the parent and drilling into the ground. Many grasses have long, bent *awns* (slender bristles) that twist as air humidity fluctuates. The awns function as levers that drive the grain into the soil. Stiff hairs at the base of the grain prevent it from pulling backwards, out of the soil. A few

nongrass herbs, such as cranesbill (see Fig. 14.30d), employ a similar technique.

The peanut (*Arachis hypogaea*) inclines its fertilized flowers down to the ground, and the fruits become buried as they mature. Seeds never leave the immediate proximity of the parent.

Sea rocket, a common annual beach plant of temperate-zone shores, has a bipartite fruit. The top half is easily dislodged when mature, and its corky texture allows the enclosed seed to be carried to distant beaches via ocean currents. The bottom half is firmly attached to the parent, and its seed is buried with the dead parent by shifting sand at the end of the growing season (Fig. 14.33). The following year, hundreds of seedlings mark the place where the parent plant grew the year before. This two-pronged dispersal strategy serves both to spread the species and to maintain it in safe sites year after year, even though it is an annual plant.

Plant species of isolated islands, when compared to close relatives that grow on distant continents, often exhibit a loss of dispersal mechanisms. It is possible that they evolved on continents first and then by rare chance were spread to islands, where the process of evolution modified their seeds and fruits so that dispersal became very limited—in keeping with the limited size of the islands.

Figure 14.33 Two-pronged seed-dispersal strategy. (**a**) Sea rocket (*Cakile maritima*) is a common succulent annual plant along many coastlines. Note the immature, green fruits along some stems. (**b**) Diagram of the fruit. Each part has one seed. The top half is carried away by ocean currents, the bottom half stays with the parent.

SUMMARY

1. A seed is a plant embryo surrounded by a seed coat. Seeds may store food within or outside the embryo. In most dicotyledonous plants such as bean, food is stored in the two cotyledons. Cotyledons may also serve as absorbing and, later, as photosynthesizing organs. Food may be stored in an endosperm rather than in the cotyledons.

2. In seeds of monocotyledonous plants, food is usually stored in an endosperm. In grasses, such as corn, the single cotyledon-like structure (scutellum) is a specialized organ that absorbs the nutrients from the endosperm. In seeds such as those of onion, the cotyledon emerges from the seed coat and becomes green, but its tip continues to absorb food from the endosperm.

3. The first step in germination is the imbibition of water. This water facilitates the activation of enzymes involved in digesting stored food, which is converted to energy for growth.

4. In germination, the cotyledons may be elevated above the ground (epigeal), sometimes becoming photosynthetically active, or they may remain below the ground (hypogeal).

5. Mature seeds may be dormant and, depending on the species and the immediate environment, may remain viable and dormant from a few months to many years.

6. Dormancy is usually broken by providing the seed with moisture, oxygen, and a favorable temperature. Other factors, such as light, the removal of chemical inhibitors, or the destruction of the seed coat, may be required in some instances.

7. A fruit is a ripened ovary plus other closely associated floral parts.

8. There are three different kinds of fruits, classified on the basis of the number of ovaries and flowers involved in their formation:
a. simple fruits, derived from a single ovary
b. aggregate fruits, derived from a number of ovaries belonging to a single flower and on a single receptacle
c. multiple fruits, derived from a number of ovaries of several flowers more or less grown together into one mass

9. Simple fruits may have either a dry or a fleshy pericarp. If the pericarp is dry, it may be either dehiscent (splitting at maturity to allow seeds to escape) or indehiscent (not splitting).

10. The role of the fruit is to aid in the dispersal of the seeds within and to deter herbivores from eating the seeds without dispersing them. There is no nutritional link between the fruit and the seed of the germinating seedling.

11. Vectors of fruit and seed dispersal include wind, water, and animals. Common animal vectors include

ants, birds, bats, rodents, ruminants, and primates. The animals are sometimes rewarded with food for their dispersal activities, but at other times they simply carry seeds and fruits to other sites.

12. Fruits protect seeds from herbivores by the timing of their ripening, their hardness, and their chemical composition. Fruits may contain secondary compounds that deter feeding by all but the most metabolically specialized herbivores.

13. Some fruits prevent or limit dispersal, thus ensuring that the site occupied by the parent plant will be occupied by its offspring well into future growing seasons.

Questions

1. A seed is actually a mature ovule. Define each of the following terms:

integuments

seed coat

suspensor

embryo

cotyledon

hilum

raphe

micropyle

radicle

epicotyl

2. Describe the processes that occur during germination.

3. What is seed dormancy? Why is it an important process?

4. A fruit is a ripened ovary. What are the functions of fruits?

5. What are the differences between simple, aggregate, and multiple fruits?

6. Buy several different fruits at the grocery store and use the "Key to Fruits" in this chapter to identify their fruit types.

7. One role of fruits is to aid in seed dispersal. Describe two different ways that fruits aid in dispersal by abiotic factors. Describe two different ways that fruits aid dispersal by biotic factors.

Further Readings

Esau, K. 1977. Chapters 22–24 of *Anatomy of Seed Plants*. 2d ed. New York: John Wiley. This is the classic textbook on plant anatomy. These are very readable chapters on the basic anatomy of fruits and seeds.

Fahn, A. 1990. Chapters 20 and 21 of *Plant Anatomy*. 4th ed. Oxford: Pergamon Press. Another plant anatomy textbook. This one is more encyclopedic and a bit more rigorous reading.

Stiles, E. W. 1984. "Fruit for All Seasons." *Natural History* 43–53. This is an interesting paper on fruits and their uses in nature.

15 CONTROL OF GROWTH AND DEVELOPMENT

1. The development of a plant includes morphogenesis, the production of new organs of defined shapes, and differentiation, the creation of specific characteristics in a cell or tissue. Morphogenesis involves cell division and expansion; differentiation involves the controlled expression of genes at particular times and places.

2. Hormones are chemical signals that coordinate the development of an organ with what is happening in other parts of the plant and with environmental conditions. The well-characterized plant hormones include auxins, gibberellins, cytokinins, ethylene, and abscisic acid.

3. Light influences plant development through different receptors that detect different colors of light. Phytochromes, which detect red and far-red light, influence the timing of seed germination, the greening of seedlings, and internode elongation in adult plants. They also are involved in the plant's mechanism for measuring day length.

4. Because changes in the relative lengths of day and night forecast changes in seasons, plants can control developmental processes in a way that adapts them to coming seasons by measuring the length of day and night.

15.1 PRINCIPLES OF PLANT DEVELOPMENT

A plant is a whole organism, not a collection of independent cells. As it grows and matures, tissues and organs develop in predictable patterns. Some depend only on internal conditions (governed by heredity, their position in the plant, or the age or size of the plant), and some respond to the external environment. The shapes of leaves and their arrangement on the mature stem, for instance, form a fairly constant set of inherited traits, very similar among plants of a species no matter where they grow. The length of a stem and the direction in which it grows, on the other hand, may depend more on the light available in its environment than on heredity. For either hereditary or environmentally influenced traits, it is difficult to imagine that the patterns of growth and development could occur without signals—signals that somehow communicate to the constituent cells what is happening throughout the plant and outside it. This chapter describes some of these signals and their effects, and it outlines some of the recent research on how these signals control growth and developmental processes.

What Is Development, and Why Does It Matter?

The term **development** includes many events that occur during the life cycle of a plant. It includes growth—cell division and enlargement—and differentiation in both shoot and root. The formation of storage tissues and the events that lead to the mobilization of storage materials when they are needed—say, when a seed germinates or when a tree's buds start to grow in the spring—are also considered to be developmental processes. Development also includes the induced changes that allow cells to resist herbivory by insects or infection by pathogenic microorganisms, and it includes the distinctive growth patterns that distinguish juvenile, adult, and flowering shoots (Fig. 15.1).

Some of these events reflect **morphogenesis:** developmental changes that lead to the formation of specific shapes such as the cylindrical shape of a shoot or root, the flat shape of a leaf (perhaps with a sculpted outline), or the specialized shape of flower petals. Because plant cells are attached to their neighbors by common cell walls, they cannot move around within the plant. This means that the shapes of new organs are determined by the directions in which cells divide and elongate (see Fig. 5.6).

Other events represent **differentiation:** any process that makes cells functionally specialized and different from one another. Differentiation often occurs through the expression of genes. Gene expression includes all those steps by which genes direct the production of proteins. Many proteins in one way or another make a cell distinctive—for instance, the chlorophyll-binding proteins of photosynthetic cells, enzymes that make pigments in the cells of flower petals, or carriers in root-cell plasma membranes that take up mineral nutrients.

Plants Have Indeterminate as Well as Determinate Growth Patterns

The very pattern of growth in plants differs in a fundamental way from that in animals. In most animals, the pattern of growth is **determinate**, which means "having defined limits." In the context of developmental biology, it means that the cells formed from the zygote proceed through a predictable series of cell divisions, movements, and differentiation processes. At the end of the series, cell division and differentiation stop, and the result is the final, adult organism.

In plants, the growth of shoots and roots is **indeterminate**, that is, the shoots and roots will continue to grow until stopped by an environmental or internal signal. The ability to divide is maintained among the cells in the meristem (see Chapter 4) and is not lost at any predictable point. The cells of the meristem divide continuously, producing new cells. Half of these stay with the meristem, and half become part of the plant body—dividing a few times, enlarging, and then differentiating into the various tissues. The meristem cells do eventually lose the ability to divide, but they do not have an inherent, limited number of divisions after which they must stop. Not all plant growth is indeterminate: Dicot leaves and organs that are formed from modified leaves—such as bud scales, bracts, petals, sepals, stamens, and

Figure 15.1 Events in the growth and development of plants. (**a**) An embryo dissected from a seed of mouse-eared cress (*Arabidopsis thaliana*). (**b**) Spring bud-break in a London plane-tree (*Platanus acerifolia*). (**c**) Juvenile and mature Eucalyptus shoots. The blunt, oval leaves (upper left of frame) mark a juvenile shoot; the long, sharp leaves (right of frame) denote a mature shoot. (**d**) Induction of flowering shoots in an orchid (*Masdevallia veitchiana*). (**e**) The brilliant color of a senescent *Liquidambar styraciflua* leaf in the fall. Each picture represents a new or changed phase of growth and development.

carpels—all show a limited, determinate growth pattern, which varies from species to species.

One important implication of the indeterminate growth pattern of plants is that much of development occurs by the repetition of a small number of "programs." A module of the shoot—which includes the leaf, lateral bud, and internode (see Chapter 6)—might result from one of these programs. The instructions for forming leaves, starting branch shoots, and growing secondary tissues might be thought of as subroutines within the module program.

Determination and Competence Reveal Stages in Differentiation

Developmental biologists have demonstrated that cells go through a series of stages on their way to becoming mature, differentiated components of an adult organism. This is especially clear in the embryonic development of an animal. The first cell in the developmental pathway, the zygote, is **totipotent**. This means that it has the capability of making, through cell division and the other processes of development, all the cells in the future organism. The two cells resulting from the first division may also be totipotent, meaning that each one could make a complete organism if it were separated from the other and incubated under the appropriate conditions. However, after three or four divisions, the cells formed are no longer totipotent; separated from the others, they might form only a few tissues, or they might die. These cells are said to be **determined**, which means that their potential to differentiate is limited. (The term *determined* should not be confused with *determinate* or *indeterminate*, as previously defined.)

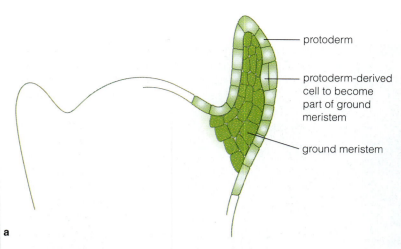

protoderm

protoderm-derived cell to become part of ground meristem

ground meristem

a

b

Figure 15.2 An example of reversible determination in plant cells. The protoderm and epidermal cells of English ivy (*Hedera helix*) lack the ability to make chlorophyll, but the ground meristem and its derivatives in a leaf produce chlorophyll normally. When a protoderm cell divides so that one of its resulting cells becomes part of the ground meristem (**a**), mesophyll in the section of leaf produced from that cell will lack chlorophyll and be light green or white (**b**). The cell that became part of the ground meristem lost its determination as protoderm.

To a degree, plant cells, like animal cells, may also be determined. During the formation of the embryo from the zygote, certain cells become shoot cells, and others become root cells. The formation of a shoot or root meristem is a first step in determining the course of further development. It is unusual for cells in a shoot meristem to make rootlike structures, and vice versa; thus, plant cells are at least partially determined. Then, during the primary growth of the shoot, the cells just below the apical meristem become parts of the three primary meristems: the protoderm, ground meristem, and procambium. Cells in these three primary meristems are even more determined, in that they have started to show characteristics that distinguish each from the other and that suggest they are not interconvertible.

The determination of plant cells is often reversible, however. For instance, adventitious roots form from shoot tissue (see Fig. 5.5), and vice versa, an indication that the determination of cells as shoot or root cells may be changed. Sometimes, protoderm cells divide so that the new cell wall is parallel to the surface of the apex (forming inner and outer cells). Then, the inner cells become part of the ground meristem and produce leaves (Fig. 15.2), an indication that the protoderm cells lose their determination as presumptive epidermal cells and form normal mesophyll cells. Also, removing pieces of tissue from shoots or roots and placing them in culture conditions can lead to the formation of whole plants (the conditions needed for this are discussed later). This means that at least some cells within the mature shoot or root tissue—cells that have differentiated—can regain their totipotency.

Just as determination in a developmental program channels cells toward certain possible fates and away from others, other processes prepare cells for differentia-

tion. This idea comes from the observation that some differentiation processes occur after the cell has received a stimulus. For example, mesophyll cells produce chlorophyll only after being illuminated, and procambial cells produce secondary cell walls after being stimulated by sucrose. However, not all cells respond to stimuli in the same way. Procambial cells do not turn green when exposed to light, and palisade parenchyma cells do not make secondary cell walls in the presence of sucrose. The cells that *do* respond appropriately must have gone through preparatory steps that make them **competent** to respond to the stimulus.

Gene Expression Controls the Development of Traits

The processes of differentiation, by which cells gain properties that allow them to play specialized roles in the life of the organism, depend on the expression of genes. This fact has been known for many years because it has been possible to identify mutations that block certain types of differentiation and to show that the mutant genes are transmitted from parent to progeny like any other gene.

An example of gene expression is the production of red and purple pigments in the petals of petunia (*Petunia* sp.) flowers. These pigments, called *anthocyanins*, are made in petal cells by a sequence of chemical reactions, each of which is catalyzed by an enzyme. The structure of each enzyme is specified by its own gene. Because all the enzymes in the sequence are needed to make the pigments, all the genes must be expressed. If one of the genes is expressed only in certain regions, just those regions will be colored (Fig. 15.3).

One particularly important enzyme in the sequence is called *chalcone synthase*. Chalcone synthase is coded by

Figure 15.3 Differential effects of gene expression on the color of petunia flowers. The dark purple flower (second from left) expresses all the genes needed to make the most complex anthocyanin pigment. The flowers third, fourth, and fifth from the left, by losing genes in the anthocyanin biosynthetic pathway, have progressively simpler, less effective pigments. The pink flower on the left has all the genes in the anthocyanin biosynthetic pathway, but it also has a gene that changes the pH of the vacuole—and thus the color of the pigment.

the *Chs* gene, which is in the nucleus of the cells. The expression of the *Chs* gene involves a long sequence of steps, the first of which is activating the DNA in the portion of the chromosome containing its gene. Inactive parts of chromosomes are wound around proteins to form nucleosomes and more complex structures that fit into the nucleus. Activation may involve the removal of nucleosome proteins and unwinding of the DNA. The exact steps are still somewhat mysterious to us, but they may occur during the early development of the flower (rather than just before the synthesis of the enzyme). It is possible that the activation step is one of the processes leading to determination of the petal cells.

The next step is transcription of the *Chs* gene. As noted in Chapter 2, this involves the enzyme RNA polymerase and other proteins, which form a transcription complex, a structure that separates the template and nontemplate strands of the DNA and starts the process of RNA synthesis. Some of these proteins are always present in the cell; some may be synthesized specifically to start the transcription of a particular gene. For instance, certain genes must be expressed before the *Chs* gene itself can be expressed. It could be that they produce proteins needed for the transcription complex at the *Chs* gene. It is possible that the production of transcription complex proteins is one of the steps in a developmental program that makes a cell competent to respond to a stimulus.

The RNA molecule transcribed from the *Chs* DNA is called a nuclear RNA (nRNA). It is processed into messenger RNA (mRNA) and transported to the cytoplasm, where it is translated by ribosomes to make the protein. The protein may require further processing—for instance, the removal of a short sequence of amino acids—before being transported to the area in the cell where it functions. The *Chs* gene product, chalcone synthase, is thought to catalyze a biochemical reaction in the cytoplasm. This reaction is the production of chalcone, which serves as a substrate (reactant) for the production of the red or purple pigments in the flower.

Each step in a gene's expression must occur in order for the gene to be expressed and the enzyme to be functional. The lack or inhibition of any one step would inhibit the synthesis of the enzyme and prevent the production of chalcone. Thus, turning off just one step has a major effect (no color—that is, a white segment of petal); turning it on reverses that effect. To learn how a gene's expression relates to development, we need to learn what steps are turned off during some stages of development and what signals turn them on at the appropriate stages.

The Coordination of Development Requires a Series of Signals

As stated at the beginning of this chapter, cells, tissues, and organs form in predictable patterns. The existence of patterns implies that the cells within a tissue or organ "sense" their position and develop accordingly. This in turn means that there must be signals, presumably from adjacent cells, that identify the position within a tissue or organ. Furthermore, organs grow in a balanced fashion. A plant generally has a characteristic shoot branching pattern, a pattern of accumulation and mobilization of storage materials, and a pattern of flowering—all of which suggest coordination among the different organs. For such coordination to occur, signals must exist that inform one part of the plant about conditions in another part. By the same reasoning, for plants to respond to the environment in appropriate ways they must perceive signals from the environment. These signals include light, temperature, length of the day, water, chemicals such as calcium ion, and mechanical disturbances such as wind and wounding by herbivores.

Some signals may stimulate new patterns of gene expression; others may activate preexisting proteins or other cell components. In either case, the signal must be perceived by the target cells. That means that the cells must have a **receptor**, a protein that is activated when it binds to or otherwise detects the presence of the signal. Although it is possible that the perception process may end there, recent studies of developmental control suggest that reception of the original signal generally produces a *second messenger*, which may start a signal *cascade*: a series of events in which one signal leads to another and another—(Fig. 15.4). Thus, the original signal might be perceived by receptors on the plasma membrane (many are found there) yet, through a cascade system, trigger events at a distance—such as transcription in the nucleus. In addition, a single signal and receptor can, through a branched cascade, affect several different developmental processes.

The original signals that coordinate the growth and

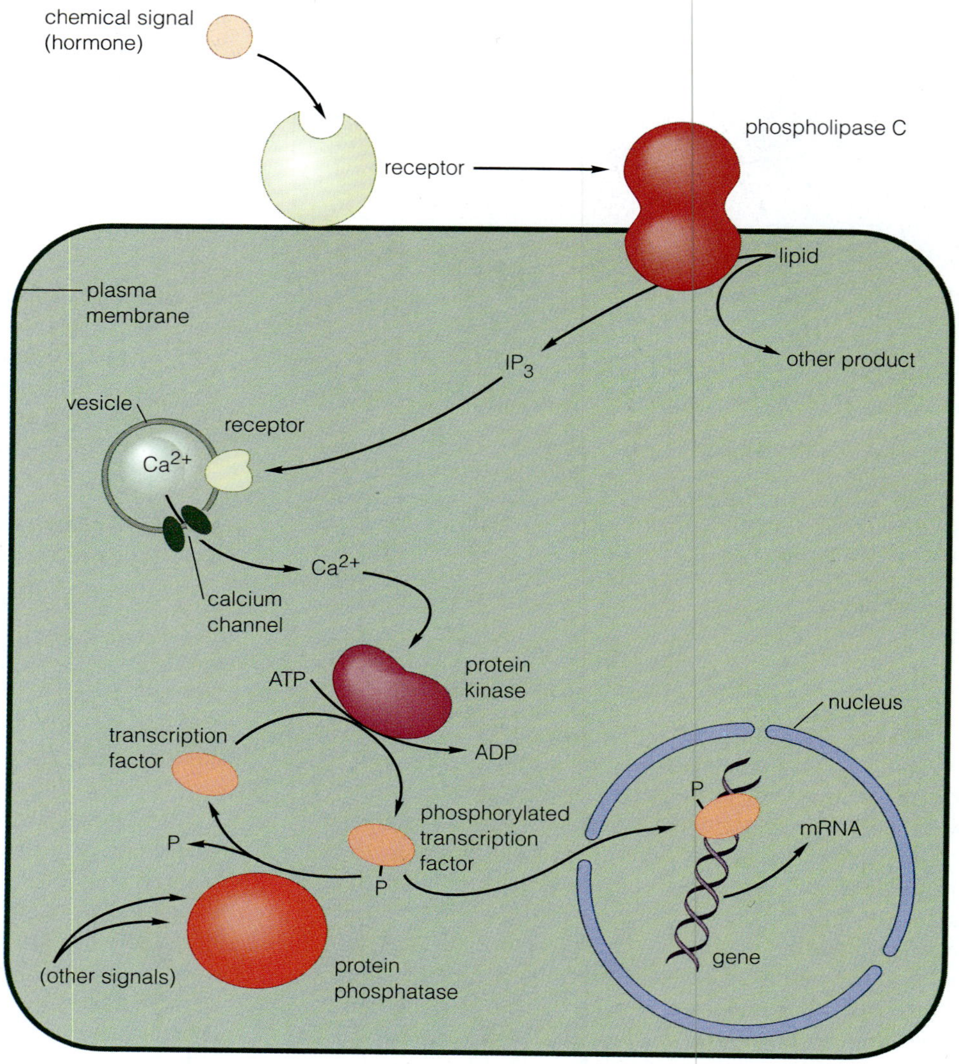

Figure 15.4 Proposed steps of a signal cascade in a plant cell. The arrows represent a sequence of activations or enzymatic reactions. The receptor is a protein in the plasma membrane that binds the hormone and is stimulated to activate phospholipase C. Phospholipase C is an enzyme in the plasma membrane that breaks down certain membrane lipids. IP_3, a product of the reaction, stimulates a receptor on a vesicle, which releases calcium ions (Ca^{2+}) into the cytoplasm. The Ca^{2+} stimulates an enzyme, protein kinase, which adds a phosphate group (from ATP) onto various proteins. One of the proteins may be a transcription factor, which stimulates the expression of a particular gene. A separate signal cascade could stimulate a protein phosphatase, which would inactivate the transcription factor.

development among cells are generally **hormones**, chemicals that diffuse or are transported from one part of the plant to another. Other possible signals are transient electrical impulses (that travel like nerve impulses from cell to cell along a stem or root) and hydraulic impulses (changes in the tension of the water column in the xylem). Environmental stimuli, such as light or temperature, can also serve as initial signals.

The receptors are proteins. Hormone receptors have **binding sites** with a three-dimensional shape into which the hormone fits. One common effect of activating a receptor is a transient increase of calcium in the cytoplasm. The normal calcium concentration in plant cell cytoplasm is very low, but there is much calcium both outside the cell and inside, in the vacuole and in vesicles. Some stimulated receptors allow calcium to enter the cytoplasm, raising the concentration by as much as 50 times. Because several enzymes are sensitive to calcium

concentration, this change can have several effects. One possible effect is the activation of an enzyme that attaches a phosphate group onto other proteins, thereby either activating or inhibiting them. These proteins may lead to another link in the cascade of events. The cascade finally ends either in the activation of a transcription factor—a protein that stimulates the reading of a particular gene—or in the activation (or inactivation) of an enzyme that produces the required differentiation of the cell.

15.2 HORMONES

Intercellular chemical signals were among the first to be discovered in plants. Very low concentrations of certain chemicals produced by a plant can promote or inhibit the growth or differentiation of various plant cells and coordinate development in different parts of the plant

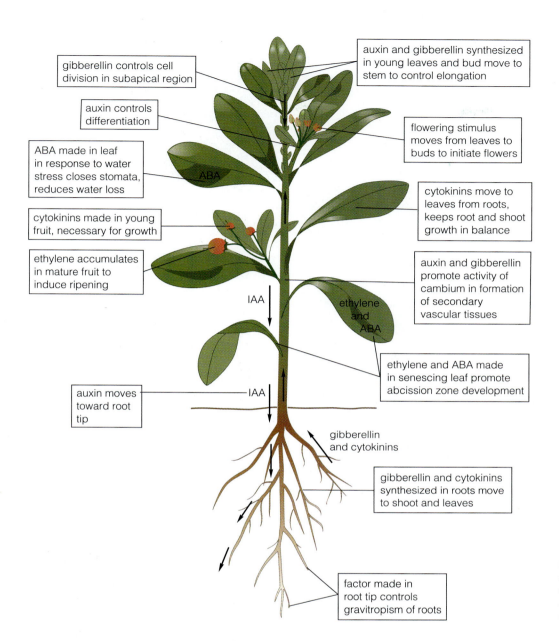

gibberellin controls cell division in subapical region

auxin controls differentiation

ABA made in leaf in response to water stress closes stomata, reduces water loss

cytokinins made in young fruit, necessary for growth

ethylene accumulates in mature fruit to induce ripening

auxin and gibberellin synthesized in young leaves and bud move to stem to control elongation

flowering stimulus moves from leaves to buds to initiate flowers

cytokinins move to leaves from roots, keeps root and shoot growth in balance

auxin and gibberellin promote activity of cambium in formation of secondary vascular tissues

ethylene and ABA made in senescing leaf promote abcission zone development

auxin moves toward root tip

gibberellin and cytokinins

gibberellin and cytokinins synthesized in roots move to shoot and leaves

factor made in root tip controls gravitropism of roots

ABA

IAA

ethylene and ABA

IAA

Figure 15.5 The hormonal coordination of growth in different parts of a plant. *IAA* stands for indole-acetic acid (an auxin), and *ABA* stands for abscisic acid.

(Fig. 15.5). These chemicals have found many uses in agriculture (see sidebar, "Using Plant Hormones," p. 245).

By analogy with chemicals in animals that are secreted by glands into the bloodstream and influence development in different parts of the body, the chemical signals in plants have been called *hormones*. However, some scientists apply a rather strict definition to the term *hormone*: It must be secreted from one organ or tissue and influence another organ or tissue in specific ways. While the chemicals in plants are diffusable, they often influence the same cells that produce them (as well as others); so they are sometimes called **growth regulators** to distinguish them from the hormones that carry systemic or long-distance signals. A variety of chemical compounds serve as hormones or growth regulators (Fig. 15.6). The following sections describe some commonly recognized

plant hormones. Not presented are the many compounds that affect plant development in low concentrations but whose function as hormones under natural conditions is still under investigation.

Auxin Promotes Cell Wall Expansion

Auxin, the first known plant hormone, was discovered during studies of the growth of oat coleoptiles, a model system that has been used for investigations of growth since Charles Darwin and his son Francis worked on "the power of movement in plants" in the 1880s. A coleoptile is a modified tubular leaf that surrounds the first true leaves and the stem of a germinating seedling of a member of the grass family (Poaceae). Moistened and incubated in the dark, an oat seed germinates, and its

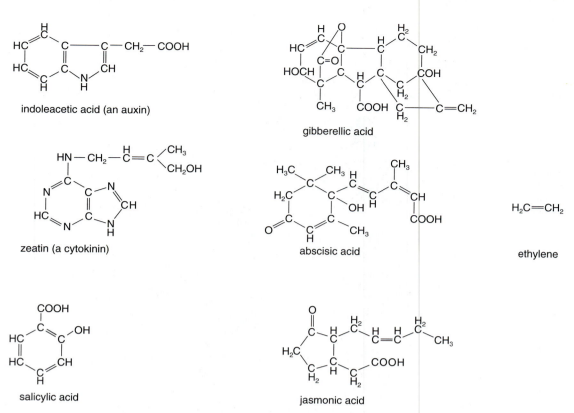

Figure 15.6 The chemical structures of some plant hormones: indoleacetic acid (IAA, the most common natural auxin), gibberellic acid (GA₃, one of more than 50 compounds with gibberellin activity), zeatin (a cytokinin), abscisic acid, ethylene, and two hormones involved in the stimulation of resistance to stress and pathogens: salicylic acid and jasmonic acid.

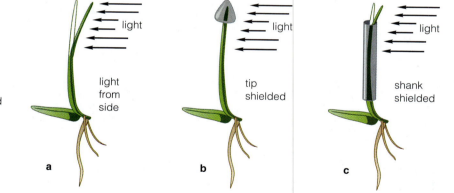

Figure 15.7 The Darwins' experiments on the dependence of oat (*Avena sativa*) coleoptile growth on light. The coleoptile is a modified hollow leaf. It grows upward in the dark but bends toward a source of light (**a**). If the tip is shielded from the light, there is no bending (**b**); if the shank is shielded but the tip is exposed, there is bending (**c**), showing that the perception of the light occurs at the tip.

coleoptile grows rapidly straight up. By the time it is 1 to 2 cm (0.4 to 0.8 in) long all cell divisions have ceased, and further elongation reflects only the enlargement of the cells' walls.

The Darwins' experiments demonstrated that the coleoptile tip was necessary for the elongation of the shank (Fig. 15.7). A hypothesis to explain this suggests that a substance produced at the tip is transported down the shank, where it stimulates growth. The coleoptile could also be induced to bend toward a light, a movement called **phototropism** (blue light was most effective). If the coleoptile was placed on its side, it would bend

upward, a movement called **gravitropism** (see sidebar, "Gravitropism," p. 246). A reasonable explanation is that the light and gravity cause a movement of the growth-promoting substance to one side of the shank. The faster growth at that side causes the bending.

Experiments to test the idea of a diffusable growth-promoting substance were devised by N. Cholodny and Frits Went in the 1920s (Fig. 15.8). They cut off a coleoptile tip and placed it on a small block of agar (a gelatinlike substance), leaving it there long enough for diffusable substances to move into the agar. When they placed the agar on a decapitated coleoptile, the shank resumed

USING PLANT HORMONES

Plant hormones are used extensively to manage plant growth and development in home gardens and in agriculture. The first hormone to be discovered, auxin, is also the one most used by home gardeners. However, it is generally not easy to tell when an auxin is present by inspecting the labels of products in a garden shop. The word *auxin* is seldom mentioned on labels, and the chemicals are generally synthetic analogs of the natural auxin (indoleacetic acid). Concentrations of auxins above the optimal inhibit plant growth and kill meristems. Dicots are particularly sensitive, but monocots are much less so. Therefore auxins such as 2,4-dichlorophenoxyacetic acid (2,4-D) are used as weed killers to remove dicot weeds from monocot grass lawns. Auxins stimulate fruit growth; therefore, a preparation of chlorophenoxyacetic acid is available for spraying on tomato blooms to increase fruit set. Because auxins stimulate the formation of roots, indole-3-butyric acid is sold as a "root stimulator and starter."

Here are just a few of the uses of hormones in commercial horticulture and agriculture:

Auxins suppress dicot weeds in fields of corn and other monocots; stimulate root growth in cuttings of roses and other nursery plants; delay fruit and leaf drop.

Gibberellins stimulate germination of seeds (many species); promote the growth of grape inflorescence internodes, leading to more open berry clusters that are less susceptible to fungal infections; promote berry growth in grapes; increase malt production for brewers; stimulate sugarcane elongation in winter, leading to an increase in yield of 0.7 metric tons per hectare (2 tons per acre); stimulate early cone formation in conifers, leading to faster breeding programs.

Cytokinins, in a mixture with gibberellins, promote the growth and change the shape of apple fruits. Together with auxin, cytokinins are used in tissue culture propagation of, for instance, orchids.

Ethylene (applied as a solution of 2-chloroethane phosphonic acid, known as *ethephon* or by the tradename Ethrel, which releases ethylene when taken up by plant cells) stimulates and synchronizes flowering of pineapples; thins (causes a partial drop of) the fruit of cotton, cherries, and walnuts; induces the ripening of apples, tomatoes, and other fruits. Reducing the amount of ethylene or inhibiting ethylene action is an important method of delaying fruit ripening, allowing fruits to be stored longer and shipped to markets farther away. Apples, for example, can be stored for long periods under high CO_2 concentration and low atmospheric pressure, which inhibits ethylene action.

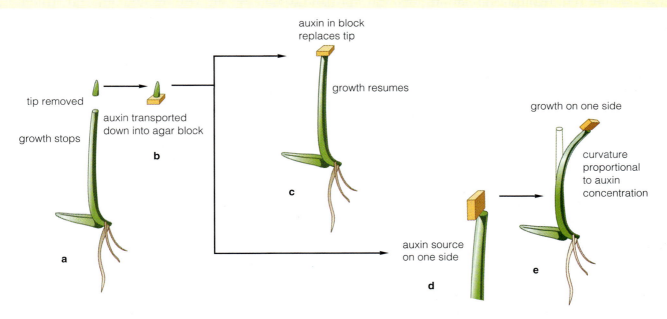

Figure 15.8 The dependence of oat coleoptile growth on auxin. If the tip of the coleoptile is removed, growth ceases (**a**); the tip produces some factor needed for growth. If the tip is placed on an agar block (**b**), and later the block is placed on the decapitated shank, growth resumes (**c**). This shows that the growth factor is a diffusable chemical (auxin). If the agar block is placed on one side of the shank (**d**), the shank bends as it grows (**e**), suggesting that the light-induced curvature might be caused by a movement of auxin to the side away from the light.

GRAVITROPISM

The ability of plant organs to orient their growth up or down used to be called *geotropism* (*geo-*, "earth"; *trope*, "turn"). The phenomenon was renamed *gravitropism* following the recognition that gravitational acceleration was the determining factor and that the process would occur in such exotic environments as other planets and rotating space stations. Most shoots tend to grow "up," away from the pull of gravity (negative gravitropism), and many roots tend to grow "down," toward the pull of gravity (positive gravitropism); but many organs specifically grow "sideways," neither up nor down (diagravitropism).

In any of these cases, it is clear that sensing the direction of the gravitational field is a critical first step in the growth orientation process. The tissues that sense the field are at the tips of the organs—near the apical meristem of shoots and in the root cap of roots. The sensors are small, dense organelles called **statoliths**. These are actually amyloplasts, a type of plastid generally containing starch grains. The starch grains make the statoliths denser than the rest of the cytoplasm, causing them to sink to the bottoms of their cells. When a shoot, coleoptile, or root is oriented straight up or straight down, the statoliths lie on a transverse wall. They do not accumulate to one side or another. The expansion of cells, being the same on both sides, keeps the organ straight. When a shoot or root is turned on its side, the statoliths fall down onto a side wall. This produces—in a way that is still unclear—a movement of the growth regulator auxin toward the down side.

In shoots and coleoptiles, the cells on the down side respond to the increased amount of auxin with faster and greater expansion; cells on the up side correspondingly show slower and more limited expansion. The asymmetric distribution of larger and smaller cells bends the organ so that the tip turns upward. Bending continues as the cells of new regions continue to expand, so long as the distribution of auxin is uneven. When the tip reaches the vertical, the distribution of auxin once more becomes symmetrical, and the organ continues to grow in that direction.

In roots, the cells are more sensitive to auxin. Very small concentrations stimulate growth, but higher concentrations (such as those that stimulate shoots) inhibit growth. The cells on the down side of a root respond to the increased amount of auxin by expanding more slowly and less; the cells on the up side expand more rapidly and attain larger sizes. Thus, the roots begin to curve downward and continue until the root tip is pointing straight down.

How do the sinking statoliths influence the flow of auxin? Here are some suggestive findings: Within a few minutes after a root is placed horizontally, the ionic currents and electric fields around the root tip change. It is thought that the statoliths influence ion pumps in the plasma membranes of their cells, perhaps by pulling on microfilaments. The altered electric fields may pull the negatively charged form of auxin toward the bottom of the root. Calcium ion (Ca^{2+}) moves toward the down side, and Ca^{2+} may regulate carrier proteins. All this is speculation, however. Much about gravitropism remains a mystery. For instance, can you imagine how the explanation above can apply to diagravitropic roots?

growth—showing that the agar had received chemicals from the tip that stimulated the coleoptile's growth (Fig 15.8c). If the agar was placed on one side of the coleoptile the growth was uneven, and the coleoptile bent away from the side that received the agar (Fig. 15.8d, e).

The growth of decapitated coleoptiles demonstrated the presence of a growth-promoting substance, which was named *auxin*. When auxin was purified from urine and later from plant extracts, it was identified as indoleacetic acid, a compound formed by metabolic conversion of the common amino acid tryptophan. (In our bodies, proteins are continually being broken down to their constituent amino acids, including tryptophan. A portion of the tryptophan is metabolized to indoleacetic acid and excreted in urine; it is simply a waste product.)

In plants, auxin is normally produced by the young leaf primordia. It is actively transported down the stem toward the roots. It stimulates the primary growth of the stem and the root, and it has other effects. It stimulates cell division in the vascular cambium (in concert with hormones called *gibberellins*) and promotes the formation of secondary xylem. It stimulates the formation of new root apical meristems—lateral meristems from the roots and sometimes adventitious root meristems from the basal parts of shoots (Fig. 15.9). Auxin also tends to inhibit the activity of lateral meristems near the apical meristem, restricting the formation of shoot branches, a phenomenon called **apical dominance**. The auxin signals the presence and activity of the shoot apical meristem. In pruning a plant one removes apical segments, including the apical meristem and its auxin. With the inhibitory influence of the auxin gone, the lateral buds grow, producing a bushier plant.

The mechanism by which auxin stimulates the expansion of cells has inspired active research since the hormone was discovered. Cells enlarge when the internal pressure (turgor pressure) causes an irreversible stretching of the cell wall. It is easy to show (by removing turgor pressure and exerting a known force on an auxin-treated coleoptile) that the major effect of auxin is to increase the plasticity of the cell wall. This means that the cell wall deforms more easily when it is pulled out of shape by the experimental force and fails to snap back to its original size after the force is removed. We can assume that the same thing happens when the cell walls of auxin-treated cells are stretched normally by internal turgor pressure, and that this increased plasticity represents the reason for the increased growth rate induced by the auxin.

How does auxin stimulate the increase in plasticity? There are two hypotheses, and both may be correct. One, called the *acid–growth hypothesis*, suggests that the main effect of auxin is to cause cells to secrete acid (H$^+$ ions) and that the acid stimulates the changes in plasticity. Experimental evidence to support this hypothesis comes from an increase in the rate of acid secretion by coleoptiles or stem tissue when they are treated with auxin. Auxin stimulates the activity of the plasma membrane proton pump, which results in the active secretion of protons. Also, coleoptile or stem sections that have been frozen and thawed several times (a treatment that kills their cells but leaves the components of their cell walls intact) become more plastic when they are soaked in acidic solutions. Notice that this hypothesis does not require an induction (stimulation of the expression) of special genes by the auxin treatment.

The second hypothesis suggests that auxin works by inducing the expression of genes that make growth-promoting proteins. There is evidence to support this *induced gene expression hypothesis*, too. The application of chemicals that stop ribosomes from translating mRNA also prevents cells from increasing their growth rate in response to auxin; translating mRNA is an essential step in gene expression. In addition, it has been possible to identify mRNAs that increase in concentration rapidly within 10 to 20 minutes after stem sections have been treated with auxin, although we still do not know what proteins all these mRNAs encode. Soybean hypocotyls that bend under a gravitropic stimulus (lying on their side) show an asymmetric distribution of auxin-stimulated mRNAs (with more appearing on the lower side of the hypocotyl, before the bending starts). This suggests that auxin-stimulated mRNAs, and the proteins for which they code, might be involved in the bending process, which is related to growth.

There is some evidence indicating the existence of special proteins that affect the cell wall's plasticity. Their activities may promote acidic conditions; for instance, one of the proteins induced by auxin stimulates the activity of the plasma membrane proton pump, which acidifies the cell wall. This is one way to explain how the two hypotheses could both be true. Other proteins may depend on acidic conditions. A recently discovered protein called *extendin* has been extracted from apical re-

Figure 15.9 Holly (*Ilex opaca*) shoots form roots at their base faster when the bases are treated with an auxin. The ends of these shoots were dipped for 5 seconds in solutions containing 50% ethanol and (from left to right) 0, 0.1, and 0.5% naphthalene acetic acid, a synthetic auxin. They were then rooted in moist vermiculite for two weeks.

gions of cucumber seedlings; when added to stretched, heated stem segments, it increases their plasticity, but only at pH values less than 6.0 (acidic conditions). Another protein has an enzymatic activity (*transglycosylation*) that fits what we might expect for an enzyme that promotes plasticity and growth. It breaks carbohydrate chains and re-forms them in new configurations that can result in a more extended wall (Fig. 15.10). It seems that discoveries regarding the exact mechanism of auxin-induced expansion may be just around the corner.

Gibberellins Stimulate Internodal Growth

Gibberellins are a group of similar plant hormones that stimulate the elongation of stem internodes. They were discovered in Japan by plant physiologists studying a disease of rice (*bakanae*—"foolish seedling disease") caused by a fungus named *Gibberella fujikuroi*. The infected rice seedlings grew much faster than uninfected ones, but they fell over and died before they could produce grain. The physiologists showed that chemicals produced by the fungus stimulated the early growth. Later it was found that the same chemicals are normally made in low amounts in young leaves and transported throughout the plant in the phloem.

A spectacular demonstration of the effect of gibberellins on internode growth is found in the **bolting** of plants with a rosette morphology (Fig. 15.11). Rosettes are plants with internodes so short that the leaves look almost as if they spring from a single node. Iceberg lettuce (*Lactuca sativa*), cabbage (*Brassica oleracea*), carrots (*Daucus carota*), and beets (*Beta vulgaris*) are all rosette plants. When these plants flower, the stem elongates quickly. Flowers form on the new parts of the stem. Also, the internodes between the older nodes lengthen, so it

Figure 15.10 Model for the mechanism of cell wall elongation. A transglycosylation reaction breaks bonds between sugar molecules in cell wall polysaccharides and re-forms longer polysaccharides in a manner that allows the cell wall to elongate.

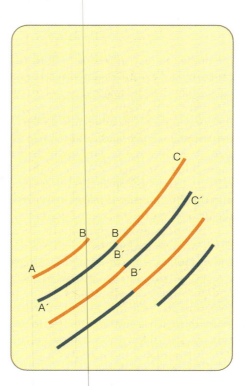

ABC breaks at B; A´B´C´ breaks at B´ A´B´reconnects to BC

Figure 15.11 Gibberellin substitutes for a cold requirement in the bolting and flowering of a carrot plant (*Daucus carota*). (**a**) The rosette form of the carrot, maintained without a cold or gibberellin treatment. (**b**) A carrot plant that flowered in response to a treatment with gibberellin (no cold). (**c**) A carrot plant that flowered in response to a treatment in the cold (no gibberellin).

becomes clear that the leaves of the rosette form at different nodes. Under natural conditions bolting is triggered by environmental signals, such as a lengthening of the day or warming after a period of cold temperature; but it can also be stimulated by spraying the plant with gibberellin. Furthermore, it can be shown that the natural signal that stimulates bolting first induces the plant to synthesize gibberellin. Together, these two observations suggest that rosette plants have that original shape because they are deficient in gibberellins and that gibberellins are a natural mediator in the environmentally stimulated growth of their internodes.

Caulescent plants, those that have naturally long internodes, presumably make more gibberellins while they are growing. Evidence for that suggestion comes from experiments in which plants are sprayed with a chemical that inhibits gibberellin synthesis; such plants develop shorter, stumpier stems with short internodes.

Gibberellins also promote the germination of many types of seeds, and they promote a resumption of growth in stems that have gone dormant for the winter. In all these systems, both cell division and cell elongation are stimulated. The mechanism of the stimulation remains obscure, although pea stems treated with gibberellin have an increased concentration of one of the cell wall-loosening enzymes described in the section on auxin. In addition, gibberellin apparently causes a reorientation of microtubules and (as a result) cellulose microfibrils, so that fewer microfibrils oppose the elongation of cells.

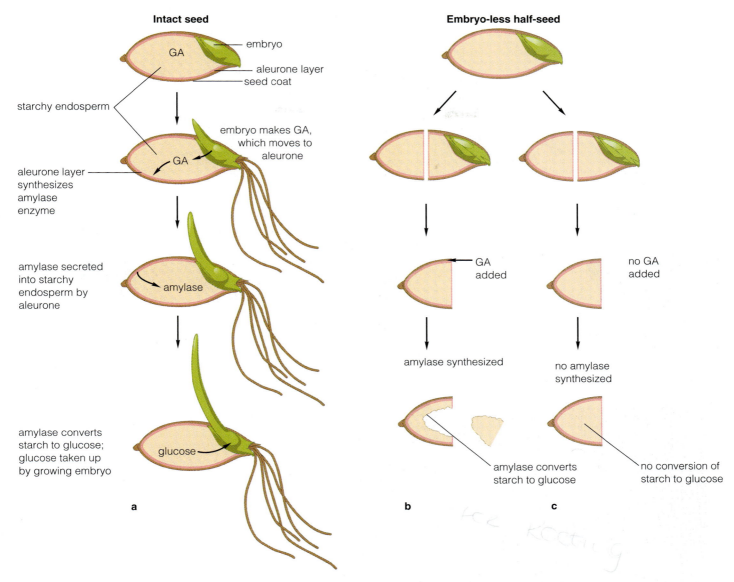

Intact seed

GA
embryo
aleurone layer
seed coat
starchy endosperm

embryo makes GA, which moves to aleurone
GA

aleurone layer synthesizes amylase enzyme

amylase secreted into starchy endosperm by aleurone
amylase

amylase converts starch to glucose; glucose taken up by growing embryo
glucose

a

Embryo-less half-seed

GA added

no GA added

amylase synthesized

no amylase synthesized

amylase converts starch to glucose

no conversion of starch to glucose

b

c

Figure 15.12 The effect of gibberellin on the germination of a barley (*Hordeum vulgare*) seed. (**a**) In an intact seed, gibberellin (GA) from the embryo stimulates the synthesis of amylase, which breaks down starch to form the glucose that nourishes the embryo. Removing the embryo removes the source of gibberellin. Unless gibberellin is added (**b**), no amylase will be synthesized, and the endosperm starch will not break down into glucose (**c**).

In at least one very well-studied monocot system, barley (*Hordeum vulgare*) seeds, gibberellins promote the metabolic breakdown of storage materials (Fig. 15.12). Barley seeds have been studied for many years because the conversion of their starch to sugar is the key event in the malting process, the initial process in the brewing of beer. In malting, barley seeds are moistened with water, which starts the germination. As the embryos start to grow, they form enzymes (α-amylase, maltase) that convert the starch in the endosperm of the seed to the simple sugar glucose. The glucose would be used to nourish the growing seedling, except at this point the brewmaster bakes the *malted* seeds at high temperature to kill the seedlings and inactivate the enzymes. Caramelization of the sugars in the baking process gives the beer some of its flavor. The dry malt is ground up and then dissolved in water, ready for fermentation.

Without the embryos, moistened barley seeds soften, but they do not produce the enzymes that break down the starch. However, if the softened seeds are treated with low concentrations of gibberellins, a special layer of cells on the outer edge of the endosperm (the aleurone layer) synthesizes the enzymes and releases them into the starchy tissue. The gibberellins induce the transcription of the genes for α-amylase in that layer of cells. Newly made mRNA serves as a template for the synthesis of the enzyme. Under normal conditions, the gibberellins are made by the germinating embryo. Thus, they represent a signal from the embryo to the endosperm, announcing the need for nutrients.

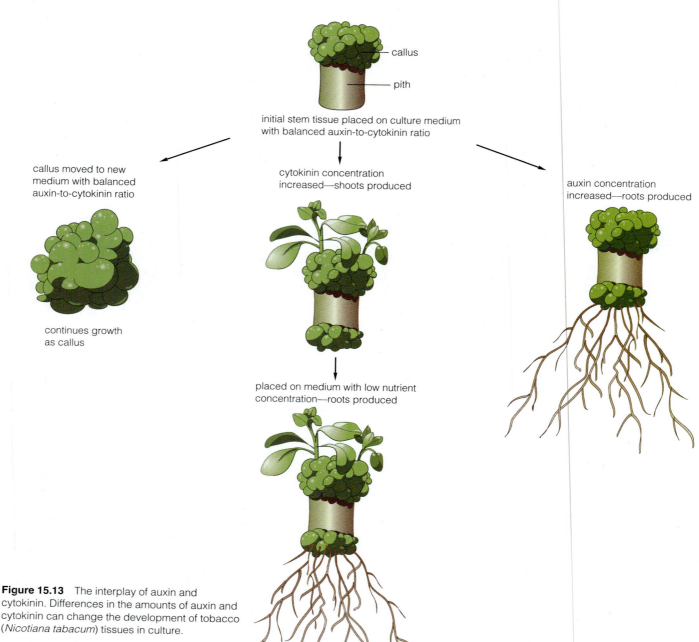

callus

pith

initial stem tissue placed on culture medium
with balanced auxin-to-cytokinin ratio

callus moved to new
medium with balanced
auxin-to-cytokinin ratio

cytokinin concentration
increased—shoots produced

auxin concentration
increased—roots produced

continues growth
as callus

placed on medium with low nutrient
concentration—roots produced

Figure 15.13 The interplay of auxin and
cytokinin. Differences in the amounts of auxin and
cytokinin can change the development of tobacco
(*Nicotiana tabacum*) tissues in culture.

Cytokinins Coordinate Shoot and Root Growth

A third type of growth-stimulating hormone, **cytokinin**, is found in embryos and endosperm. In mature plants, it is produced in the roots and transported to the shoots in the xylem sap. It seems to be an important signal in coordinating shoot growth with root growth, communicating to the shoot the presence of an active, healthy root.

Cytokinin was discovered during experiments designed to define the conditions needed for culturing plant tissues. In the tissue culture procedure, pieces of stem, leaf, or root are removed from a plant. They are briefly sterilized on their surface to remove any microbes and then placed in an enclosed flask containing a nutrient medium. The medium contains a carbon source such as sucrose or glucose, nutrient minerals (see Chapter 11), and certain vitamins. To induce the cells to divide and grow, hormones must also be added. Auxin is very important, but most plant cells placed in a medium containing only auxin as a hormone enlarge without dividing. Early studies showed that a second hormone, found in solutions of boiled DNA and in coconut milk (which is a liquid endosperm), is also necessary. The active ingredients turned out to be cytokinins, modified forms of adenine (a component of DNA and RNA). In a medium containing all these components, including auxin and cytokinin, plant cells could divide and grow rapidly.

Further experiments showed that auxin and cytokinin influence the development, as well as the growth, of plant tissue cultures (Fig. 15.13). When the relative concentrations of the two hormones are balanced, with about 10 times more auxin than cytokinin being added to the cell culture, growth is undifferentiated. The amorphous mass of cells, often loose and fluffy, is called a **callus**. When the auxin concentration is further increased, or when the nutrient concentration in the medium is reduced, the callus produces roots. But when the cytokinin concentration is increased, the callus becomes green and compact and produces shoots. These are general observations; in practice, the precise response of plant tissues to different auxin and cytokinin concentrations depends on the plant species and other growth conditions.

Cytokinins promote shoot growth in intact plants, too. A small amount of cytokinin applied to lateral buds of pea (*Pisum sativum*) shoots stimulates them to grow, even though their growth was previously inhibited by the apical dominance effect of auxin. Also, cytokinin delays **senescence**, the well-controlled process of deterioration that leads to the death of cells. The senescence of leaves, seen as they turn yellow, involves the breakdown of chlorophyll and proteins and the export of the products through the phloem to the meristems or other parts of the plant where they can be used. Leaves detached from the stem and placed in water undergo this process, but adding cytokinin delays it by several days (Fig. 15.14).

Abscisic Acid Promotes Dormancy

The hormone **abscisic acid** is more associated with the suspension of growth than with its stimulation. It was discovered at nearly the same time by two different groups of researchers. While at the University of California at Los Angeles, F. Addicott discovered a compound that promoted the abscission (breaking off—see Chapter 7) of cotton cotyledon petioles from their stem. In Wales, P. F. Wareing and his coworkers found a compound that was associated with the dormancy of woody shoots in the winter. These compounds turned out to be the same: abscisic acid.

Abscisic acid accumulates in the shoots of perennial plants and, as noted above, stimulates dormancy. This involves the formation of protective bud scales (modified leaves that are small, nonphotosynthetic, dry, and tough) around the apical meristem before cell division stops; so dormancy is more complex than a simple inhibition of cell activity.

Abscisic acid also is associated with dormancy of some seeds. It is accumulated in the seed coat during the seed's development; in the presence of this hormone, the embryo does not germinate, even if it is hydrated. Some of these seeds (apple—*Malus sylvestris*—or cherry—*Prunus* sp.—seeds, for instance) require a long period under cool, wet conditions before they can germinate, conditions that stimulate the breakdown of the abscisic acid. Abscisic acid's role in seed development is more than a simple inhibitor, however. The presence of the hor-

Figure 15.14 The effect of cytokinins on senescence. Cytokinins applied to the right-hand primary leaves of this bean (*Phaseolus vulgaris*) seedling inhibited its normal senescence and promoted the senescence of the left-hand primary leaf.

mone late in seed development induces the formation of large amounts of certain proteins by stimulating the transcription of their genes. These proteins are thought to store nitrogen, other elements, and energy for use by the embryo when it germinates. Thus this hormone may have multiple roles in the coordination of seed development.

Abscisic acid has another role, which is not concerned with development but with the control of the photosynthetic system under stress. When water becomes so scarce that leaves wilt, it is important that the stomata close to prevent any further water loss. This is true even though closing stomata will cut off the supply of CO_2 and shut down the photosynthetic system. Abscisic acid is a signal of this emergency situation. Under drought conditions, wilted mesophyll cells of a leaf rapidly synthesize and excrete abscisic acid. This abscisic acid diffuses to the guard cells, where an abscisic acid receptor recognizes the presence of the hormone and acts to release potassium ion (K^+), chloride ion (Cl^-), and H_2O, closing the stomata (see Chapter 11).

Initial experiments suggest that the abscisic acid receptor works through a signal cascade that includes phosphorylated proteins (see Fig. 15.4). A mutant of mouse-eared cress (*Arabidopsis thaliana*) that lacks the ability to respond to abscisic acid in any way lacks a protein phosphatase (an enzyme that removes the phosphate groups from various proteins). This suggests that removing phosphates is one step in the signal chain by which, in various tissues, abscisic acid stimulates storage protein synthesis, maintains dormancy, or closes stomata.

Ethylene Inhibits Growth and Stimulates Senescence

Another hormone associated with the inhibition and modification of growth is **ethylene**. Ethylene is a simple

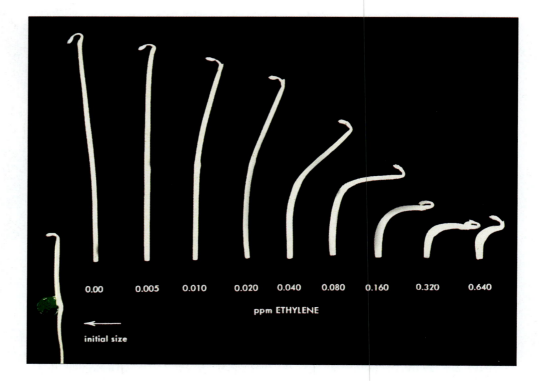

Figure 15.15 The effect of ethylene on growth. An increase in the concentration of ethylene makes dark-grown pea (*Pisum sativum*) seedlings shorter and thicker. Very low concentrations of ethylene are effective. The abbreviation *ppm* means "parts per million." In this case, even the most strongly affected seedling has been treated with less than one part of ethylene per million parts of air.

compound (Fig. 15.6) but an unusual hormone because it is a gas at normal temperatures and pressures. Ethylene is produced by almost any wounded plant tissue and by unwounded tissues whose growth has been restricted, and it can move by diffusion to other nearby organs.

Ethylene has several effects on plants. It causes the petioles to bend downward (away from the stem apex). The orientation of a normal petiole depends on the relative sizes of the cells on opposite sides of the organ, so a bending must signify a change in the sizes of these sets of cells.

Ethylene also slows the growth of stems and roots (Fig. 15.15), and a hypothesis has been advanced to explain this. In very small concentrations the gas disrupts the organization of microtubules close to the cell wall. These microtubules normally lie in parallel rows, and they are thought to control the direction of synthesis of the cellulose microfibrils so that the microfibrils lie parallel to the microtubules (see Chapter 3). Growth then occurs perpendicular to that direction. In the presence of ethylene, the microtubules are absent or disorganized, and the cellulose microfibrils lie in all directions. Growth, too, then occurs in all directions. Cells that develop that way are short and round, rather than long and thin. A stem or root growing under the influence of high concentrations of ethylene is short and stumpy because it is formed from short, round cells.

By far the most characteristic effect of ethylene is its stimulation of senescence. It takes special enzymes—

chlorophyllases and proteases, by which chlorophyll and proteins are broken down—to start this process. Ethylene triggers the expression of genes leading to the synthesis of these enzymes. In many plants, senescence is associated with abscission, which is caused by enzymes that digest the cell walls in a localized region at the base of the petiole (the abscission zone). The petiole breaks from the stem at that point. The activity of the enzymes involved in both senescence and abscission is increased by ethylene. The mechanism by which ethylene induces the synthesis of these enzymes is not well understood, but a specific protein receptor that recognizes and binds to ethylene has been found. Many other pieces of the puzzle will probably be discovered in the next few years.

The ripening of fruit is a variation on the process of senescence. It may involve the conversion of starch or organic acids to sugars, the softening of cell walls, or the rupturing of the cell membrane, with the resulting loss of cell fluid to form a dry tissue. In each case, ripening is stimulated by ethylene (Fig. 15.16). There is an autocatalytic aspect to this effect. Like wounded plant tissues, senescent plant tissues (including ripe fruit) form ethylene. An overripe banana or apple is a potent source of ethylene. The ethylene that these fruits emit can stimulate senescence (ripening) in other, adjacent fruits. This is the physiological truth behind the statement, "One bad apple spoils the barrel." By the same token, the ripening of fruit can be controlled by reducing the concentration of ethylene in the atmosphere. Picked apples, for instance, are often stored for months in an atmosphere containing low concentrations of ethylene and high concentrations of CO_2 (which inhibits the effect of ethylene, possibly by binding to its receptor).

A Variety of Compounds Serve as Stress Signals

A very few years ago, a plant physiologist might have told you that auxin, gibberellin, cytokinin, abscisic acid, and ethylene were the only major hormones used by plants to coordinate their growth and development. As the attention of plant scientists has shifted to include developmental responses to biotic stress and infection, however, new signaling compounds have been discovered.

One type of stress experienced by plants is attack by fungi and fungus-like parasitic protists. Once they detect a fungal infection or the threat of an infection, plants may develop various resistance mechanisms near the site of infection. They may, for instance, produce hydrogen peroxide (H_2O_2), which is thought to act as an antibiotic, killing the infectious agent. They may produce enzymes that break down fungal cell walls. The plant cells surrounding the infection site may die, producing tannins (organic polymers related to lignin) in the process. This is effective because food and water are cut off from the site—and therefore from the fungus—and the tannins are hard for the fungus to digest. Thus, the fungus has difficulty spreading through the dead area to new, live cells. Plants recognize the presence of a fungus from oligosaccharides, short chains of sugar molecules (see Chapter 2) released from the fungal cell wall or from the plant cell wall under fungal attack. There are many kinds of oligosaccharides, but only the few kinds released from fungal cell walls signal the presence of fungus. It is possible that other oligosaccharides may influence developmental processes of healthy plants. For instance, certain oligosaccharides stimulate the formation of flowers from sections of leaves grown in tissue culture.

The resistance induced by the perception of an infection may spread throughout a plant. That is, a plant that has been infected by a virus, bacterium, or fungus and survived may become less susceptible to a later invasion into other parts of its body by the same pathogen (and some other pathogens, too). This phenomenon is known as **systemic acquired resistance** (Fig. 15.17). The mechanism is not fully understood, but it involves the transmission of a signal from the infected organ to new leaves, probably through the phloem, and the induction of several types of antibiotic proteins in the new leaves. (It does not involve antibodies of the type that appear in an immunized animal.)

Recent research has focused on identifying the signals that induce resistance in leaves distant from the primary site of infection. Several compounds have been tentatively identified as being involved. Salicylic acid, a com-

Figure 15.16 The effect of ethylene on the ripening of fruit. The box of tomatoes on the right was kept for three days in a room with an atmosphere containing 100 parts per million of ethylene. The box on the left had not yet been treated with ethylene.

Figure 15.17 Systemic acquired resistance. Both mouse-eared cress (*Arabidopsis thaliana*) plants were inoculated with the pathogenic bacterium *Pseudomonas syringae*. To stimulate its resistance mechanisms, the plant on the left had been treated earlier with a synthetic compound thought to mimic one of the natural signals that induces systemic acquired resistance.

pound related to aspirin, is formed in infected plants, and it induces systemic resistance when it is applied to a plant. Both jasmonic acid, which is formed by the oxidation of a membrane lipid molecule during infections, and hydrogen peroxide can stimulate resistance responses when applied to leaves. Perhaps the most interesting compound is a protein dubbed *systemin*, which is produced in response to infections and is required for establishing systemic resistance. If systemin is the signal that moves from the infected leaf to induce resistance in new leaves, it is the first protein hormone to be discovered in plants.

a

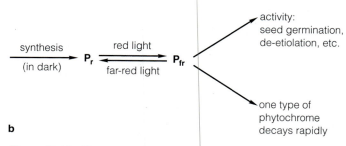

b

Figure 15.18 The red–far-red response. (**a**) Pea (*Pisum sativum*) seedlings were grown in the light (left) and in the dark (right). (**b**) Light changes the form of phytochromes.

15.3 LIGHT AND PLANT DEVELOPMENT

Plant development is strongly influenced by environmental factors as well as by internally produced chemicals. Light is the most important of these factors. The form, growth rate, and reproduction of plants are often influenced by the light in the environment. Recognizing and responding to light is a major way in which plants adapt to their surroundings and match their activities to the time of day and season of the year. These responses result from mechanisms in addition to photosynthesis, which is itself a major light-influenced process. Plants "sense" three colors of light, which correspond to three (or more) distinct light receptors. Red light activates receptors known as **phytochromes**; blue and near-ultraviolet (black) light stimulate different receptors, sometimes called *cryptochromes*; intermediate-wavelength ultraviolet light (the UV-B radiation from the sun that causes sunburn) is detected by a third, so far unnamed, receptor.

The Red–Far-Red Response Acts like an On-Off Switch

Some small seeds do not germinate in darkness. But after a brief soaking, a 1-minute exposure to red light will induce them to germinate. Surprisingly, the effect of the red light can be canceled if, immediately after the red light, the seeds are exposed to 5 to 10 minutes of far-red light. Far-red light includes those wavelengths of the spectrum that are longer than visible red light but shorter than the infrared radiation commonly thought of as heat radiation. A second exposure to red light, given after the far-red light, will stimulate germination again, showing that the far-red light just reverses the first red signal, but does not inhibit the seeds in any irreversible way. It is as though an on-off switch were controlled by the two colors of light.

The red–far-red response governs many aspects of plant development, including the growth of seedlings. A pea seedling grown in the dark is long and light yellow, with unexpanded leaves and a tight hook just below the apical meristem (Fig. 15.18). This syndrome is called **etiolation**. Its purpose is to allow the shoot apex of these seeds, which may germinate several centimeters underground, to reach the surface as rapidly as possible. Exposure of the seedlings to red light starts a process of de-etiolation: The hook opens, the leaves expand, and the growth of the stem slows. However, the de-etiolation process is retarded if the red signal is followed immediately by far-red light.

The red–far-red reversibility reflects the structure and function of phytochromes. Phytochromes are a type of protein containing a pigment molecule related to heme (the oxygen-carrying molecule in animal blood cells). When phytochrome is synthesized in the dark, its pigment can initially absorb red light. This form of phytochrome is called P_r, for red-absorbing; it is thought to be inactive. If this phytochrome is irradiated with red light, it changes its form, and its pigment becomes capable of absorbing far-red light. This is the active form of the phytochrome, called P_{fr}, the form that stimulates seed germination and de-etiolation. When P_{fr} is exposed to far-red light, most of the phytochrome returns to its original form. Because P_r is inactive, while P_{fr} is active, irradiating phytochrome with red and then far-red light is equivalent to turning a switch on and then off. Scientists have not identified with certainty what a phytochrome actually does, but one possibility is that it attaches phosphates to other proteins.

In at least some plants, there are actually three or more types of phytochromes, each coded by a different gene. All types are present in young, dark-grown seedlings. One type, present in higher concentrations than the others, is degraded by light. That is, in the light its P_{fr} form is rapidly broken down, and the gene that codes the protein is turned off. In adult plants growing in the light, only the more stable types of phytochrome are present. These, however, are adequate to make sure that adult plants show some red–far-red responses. For instance, the internodes of many plants are longer in the shade of a leafy canopy and shorter in bright light, a response that helps shoots grow out of darker areas and into the sun. This response occurs because chlorophyll in the canopy leaves absorbs red light more than far-red

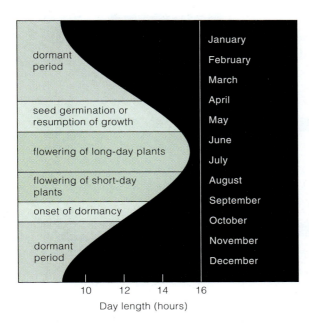

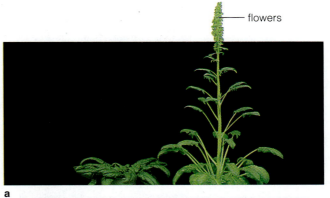

Figure 15.19 The effect of day length on flowering and other plant activities in temperate regions of the northern hemisphere.

light. Thus, under the canopy the ratio of far-red light to red light is higher than in unshaded areas. This means there will be more of the P_r form of phytochrome, which allows (does not inhibit) internode growth. Out from under the canopy, there is a higher ratio of red to far-red light, and so more P_{fr} is present, which inhibits internode growth.

Photoperiodic Responses Are Controlled by a Biological Clock

Over most of the Earth's surface, there are pronounced seasonal differences in temperature, water availability, and illumination. Many plants show developmental changes that prepare them for the coming of both harsh and mild seasons. Before the winter, dormant buds are formed at the shoot apical meristems. The leaves of deciduous plants become senescent, turning colors as their chlorophyll and proteins are broken down. As spring comes, buds break open and start to grow, and the plant may flower. Of all the Earth's seasonal variables, the most reliable and the most useful for anticipating changes in climate is the annual cycling of day and night lengths (Fig. 15.19). Plants have a system that measures the lengths of the days and nights. The control of development by this system is called **photoperiodism**.

Many plants use photoperiodism to time their flowering. These plants fall into two major groups, which traditionally are called **long-day plants** and **short-day plants** (Fig. 15.20). As you will see, these names are somewhat misleading, but they are too firmly embedded in the language of physiologists to be changed. (Many other plants time their flowering without reference to the length of the

Figure 15.20 Flowering in response to day (night) lengths. (**a**) Spinach (*Spinacia oleracia*), a long-day plant, initiates flowers and elongates its stem when the days are 14 hours or longer (nights are 10 hours or less). (**b**) *Chrysanthemum*, a short-day plant, flowers in the autumn, when the nights are long. The plant at the right received one hour of light near the middle of each night and did not flower.

day, and these are called **day-neutral plants**.) Long-day plants begin flowering sometime between January and June, when the days are getting longer. Each variety has a characteristic day length and begins flowering when the days are longer than that day length. Similarly, short-day plants begin their flowering between July and December, when the days are getting shorter. Each variety of short-

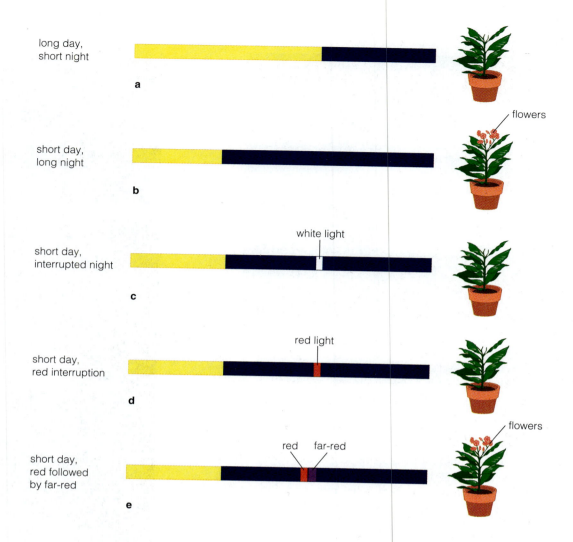

long day,
short night

a

flowers

short day,
long night

b

white light

short day,
interrupted night

c

red light

short day,
red interruption

d

flowers

red   far-red

short day,
red followed
by far-red

e

Figure 15.21 The effect of
night length on short-day
plants. Experiments (**a**), (**b**),
and (**c**) demonstrate that the
length of night, not day, is the
critical signal that stimulates
flowering. (**d**) and (**e**) show that
phytochrome is the receptor by
which the plant perceives an
interruption of night.

day plant begins flowering when the days become
shorter than its characteristic day length.

A simple experiment demonstrates that the plants
actually measure the length of the night, rather than the
day (Fig. 15.21a–c). Short-day plants are grown to maturity under conditions of long days (light periods) and
short nights (dark periods); then they are placed in a
growth room in which the days are shorter than the characteristic length and the nights are correspondingly
longer. Under these conditions, the plants will flower.
However, if a plant receives a pulse of light in the middle
of the night, it does not flower. A pulse of darkness during the day has no effect. Because a light pulse at night
blocks the signal to start flowering, we can conclude that
it is the length of the uninterrupted night that is the
important part of the signal.

This same type of experiment can be used to show
that phytochrome is the receptor by which the short-day
plant perceives light (Fig. 15.21d–e). If different wavelengths (colors) of light are used to interrupt the long
night period, it is red light that inhibits flowering, and
only a short pulse of red light is needed. However, if
a short pulse of red light is followed by 10 minutes of far-
red light, flowering proceeds as though the interruption

never happened. The red–far-red reversibility is a characteristic of phytochrome.

Finally, this experiment can be modified to show that
a **biological clock** is involved in measuring the length of
the night. Biological clocks are known in all eukaryotic
organisms. They are seen, for example, as daily (*diurnal*)
rhythms in the internal temperature of animals and in the
pattern of spore formation in fungi. The leaflets of some
plants, such as *Mimosa*, open during the day and close at
night. These **nyctinastic movements**, which are caused
by the transport of ions and the resulting changes in the
turgor pressures of cells on opposites sides of the petiole,
provide clear evidence for the operation of a clock. Biological clocks are generally reset every day by exposure
to light; but because they continue to operate even in continuous darkness, they are called **endogenous**—that is,
coming from within. They probably represent a basic biochemical or physical oscillation within cells, although the
nature of this oscillation remains a mystery.

In the experiment, mature short-day plants are given
their short day and then placed in the dark for 48 hours.
Different plants are given pulses of light at different
times after the start of the dark period. The effect of the
light depends on the time it is given, with a periodicity

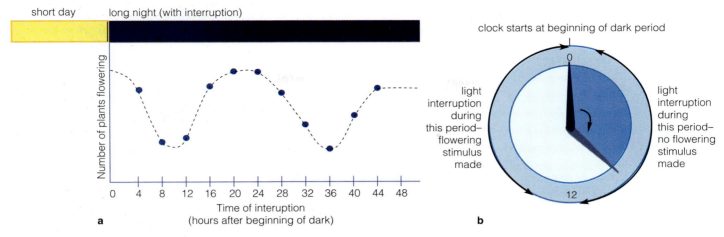

Figure 15.22 The evidence that a biological clock regulates flowering in short-day plants. (**a**) Each point on the graph represents a set of plants that was placed in the dark but was given a light pulse (interruption of the dark) at the time shown on the *x*-axis. Plants interrupted at different times showed different degrees of flowering, revealing a sensitivity to the light pulse that varied with a 24-hour period. (**b**) An interpretation of the dark interruption experiment. An internal biological clock starts running when the lights are turned off. When the lights are turned on, the clock is reset. The effect of light depends on the position of the clock when it is reset.

that cycles about every 24 hours (Fig. 15.22a). Because the effect of light on the flowering responses of the experimental plants changes periodically, the responses are thought to be controlled by interaction between light and the endogenous biological clock. In this experiment, the action of the biological clock is made visible by how the plants respond to light, as if the clock turned on and off some aspect of the light-reception process (Fig. 15.22b).

The signal pathway that initiates flowering may be complex. The perception of day or night length occurs in the leaves, but flower development occurs at the apical or lateral meristem. Stimulating a single leaf by exposing it to a critical night length can be enough to channel development of the meristem toward flowering; in fact, grafting a stimulated leaf onto an unstimulated plant can induce the plant to flower. These observations have led to the hypothesis that a chemical compound, nicknamed *florigen*, is produced in the leaf and transported to the meristems. However, the chemical nature of florigen has not been determined, and even its existence is in doubt. Gibberellins stimulate flowering in some plants, especially long-day plants and plants that require a period of cold temperatures (see Fig. 15.11), but gibberellins do not work on many of the plants thought to respond to florigen. Auxin transport is needed for flower formation, but auxins are present in the meristems even when they are not flowering. Neither gibberellins nor auxins appear to be florigen. It is possible that florigen is a compound (like an oligosaccharide or a protein) that is difficult to preserve once it has been extracted from a plant because enzymes destroy it in the extract; or perhaps it is difficult to detect because it cannot normally be taken up into plant tissues once it has been extracted.

Other responses that are controlled by photoperiodism, probably through the interaction between light and the endogenous clock, include dormancy and senescence in the fall and the resumption of growth in the spring.

Plants Respond to Light in Many Ways

Many responses of plants to light involve the blue part of the spectrum. These include such phototropic responses as bending toward light, described in the section on auxin. They also include the induction of enzymes that synthesize the red pigments in the skins of such fruits as apple. Actually, the latter example is a good way of demonstrating the complexity of light responses. The red pigments are anthocyanins, water-soluble compounds often seen in the lower epidermis of leaves and the petals of flowers. The induction of one of the enzymes involved, chalcone synthase, was described earlier. In some seedlings, these pigments are formed after irradiation with red light, with phytochrome as the receptor. In seedlings of other species, these pigments are formed only after blue light irradiation. Still other plants require both red and blue light for a full response, although either red or blue light will stimulate the formation of some pigment. It seems that induction of the enzymes synthesizing these compounds may be connected to one or both light-signaling pathways.

Although bean seedlings that are grown in the dark will start the process of de-etiolation after a very short red-light exposure, their leaves will not turn green if that is the only light they receive. The green color is chlorophyll, and the synthesis of chlorophyll requires a longer exposure to red or to blue light. The receptor, protochlorophyll, is a precursor to chlorophyll, and it must be excited by light photons before it can be converted to chlorophyll.

Thus the response of plants to light is complex, both at the molecular level and at the level of the whole plant. This should not be surprising, given the importance of light to the survival of a photosynthetic organism.

SUMMARY

1. The development of plants depends on both morphogenesis and differentiation. Morphogenesis is the creation of shape, based on the frequency and direction of cell division and cell elongation. Differentiation is the acquisition of properties (based on differential gene expression), that distinguish a cell, tissue, or organ from other cells, tissues, or organs.

2. Shoots and roots have indeterminate growth patterns: They may continue to grow without an obvious stopping point so long as conditions are favorable. An indeterminate growth pattern depends on the existence of a meristem. Some plant organs, such as flowers and dicot leaves, have a determinate growth pattern, which means that their growth stops at a predetermined size or age.

3. A *totipotent* cell can give rise to any cell type in a plant. Once a cell has become *determined*, however, it and its progeny have a limited number of possible differentiated states. A *competent* cell is primed to respond to a signal by differentiating in a predetermined manner.

4. The expression of a gene includes all of the following steps: activation of chromosomal DNA, transcription, processing of the transcript (nRNA) to form mRNA, translation, processing of the translation product to form an enzyme, transport of the enzyme to its site of action, activation of the enzyme, and formation of the product of the enzymatic reaction.

5. Chemical signals (hormones) coordinate development among different parts of a plant. Signals are perceived by receptors in a cell; perception often starts a chain of events (a cascade) that results in the activation of specific genes, resulting in growth or differentiation.

6. The hormone auxin stimulates growth through cell enlargement in leaves, stems, and roots. It also stimulates cell division and differentiation in the vascular cambium, promotes the formation of root meristems, and inhibits stem branching (through apical dominance).

7. The stimulation of cell enlargement by auxin reflects an increase in cell wall plasticity. Two hypotheses have been advanced to explain this: The acid–growth hypothesis states that auxin works by stimulating the excretion of protons from the cell; the induced gene expression hypothesis states that auxin works by inducing the synthesis of cell wall proteins.

8. The hormone gibberellin stimulates the growth of stems, the germination of seeds and the resumption of growth in winter-dormant buds, and the metabolism of stored starch in germinating monocot (barley) seeds.

9. The hormone cytokinin stimulates cell division and inhibits senescence. Different ratios of cytokinin to auxin concentrations lead to the induction of different types of organs (callus, shoots, or roots) in plant tissue cultures.

10. The hormone abscisic acid is responsible for inducing dormancy in the shoot buds of perennial plants and for maintaining dormancy in seeds. It stimulates abscission of petioles from cotton seedlings and induces the synthesis of certain storage proteins during seed development. It also is a signal of drought stress in leaves, stimulating the closure of stomata.

11. Ethylene, a gaseous plant hormone produced by senescent plant tissues, stimulates senescence of leaves and fruits. It also inhibits stem and root growth by disrupting the organization of cellulose microfibrils in the cell wall.

12. Several chemicals with hormone–like signaling activity are involved in the induction of resistance to disease. These include oligosaccharides (parts of fungal cell walls), salicylic acid, hydrogen peroxide, jasmonic acid, and a protein called *systemin*.

13. Plants have several light-receptive systems that influence their development: phytochromes respond to red and far-red light; cryptochrome(s) respond to blue and near-ultraviolet light; and a separate system responds to intermediate ultraviolet (UV-B) radiation.

14. Phytochromes are proteins that have two interconvertible forms: P_r, which absorbs red light, and P_{fr}, which absorbs far-red light. P_r is converted to P_{fr} in the presence of red light; P_{fr} is converted to P_r in the presence of far-red light. Seedlings that develop in darkness have only the P_r form. They have long stems, poorly developed leaves, and undeveloped chloroplasts—and are said to be *etiolated*. The conversion of P_r to P_{fr} by red light de-etiolates the seedlings by inhibiting stem growth, stimulating leaf growth, and starting the maturation of the chloroplasts.

15. Plants use night length to control developmental processes in a way that adapts them to coming seasons. Plants that flower in response to short nights are called long-day plants, and those that flower in response to long nights are called short-day plants. In perennials, night length also influences the establishment of winter dormancy and the regrowth of buds in the spring.

16. The perception of day length by plants involves phytochrome and an endogenous oscillator (the biological clock). Plants seemingly measure the period of darkness between two light events, so a light pulse (which forms P_{fr}) at a critical time during a dark period will stimulate flowering in a long-day plant and inhibit flowering in a short-day plant.

Questions

1. Which of the following plant organs show indeterminate growth: leaf, vegetative shoot, flowering shoot, carpel, root, stolon?

2. Describe how each of the following concepts applies to plant development: the formation of determined cells; the appearance of competent cells; the preservation of totipotent cells; indeterminate growth and modules.

3. If pieces of stem tissue are placed in the appropriate tissue culture medium, they will regenerate a whole plant. Discuss possible reasons why this can be done with plants and not with humans.

4. The expression of genetic information in the production of a protein involves several sequential steps. Place the following steps in the correct order: translation; activation of chromosomal DNA; transcription; activation of the enzyme; processing of nuclear RNA; processing and transport of the translation product; enzyme activity.

5. Discuss the following statement: Each hormone affects plant development by turning on the expression of a specific gene.

6. Using the concept of second messengers, describe how it is possible for both auxin and gibberellin to stimulate the elongation of young internode cells.

7. Name the hormone that is probably involved in the following responses:
   a. In early summer, your cabbage plants bolt and flower.
   b. In the fall, your ash tree stops growing and develops protective bud scales over its shoot apexes.
   c. The ficus tree in your apartment is leaning toward the window.
   d. The last of your peaches is getting soft.
   e. You spray your ficus tree with a new chemical, which causes lateral buds on the tree to start growing branch shoots.

8. Four oat coleoptiles have been cut from their seed. The first (a) has been slit vertically; the second (b) has its lower section covered by a light-tight shield; the third (c) has its upper section covered by a light-tight shield; and the fourth (d) has had its tip removed. The coleoptiles have been placed in a support with light coming from one side. Which coleoptiles will bend toward the light? Assume that no coleoptile shades another coleoptile.

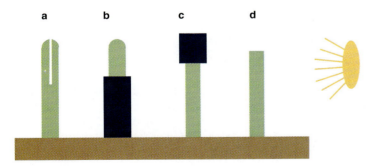

9. Mature plants become *leggy*—have longer-than-normal internodes—when they grow under a thick canopy of leaves. Chlorophyll in the canopy absorbs more red than far-red light from sunlight. It is hypothesized that one form of phytochrome inhibits stem growth more than the other. In this hypothesis, which form of phytochrome, P_r or P_{fr}, inhibits stem growth? Explain your choice.

10. A scientist has suggested that the synthesis of phytochrome is controlled by phytochrome itself because the enzyme is synthesized less rapidly in the light than in the dark. Which of the observations below would most convincingly support her suggestion? Discuss your answer.

   The mRNA for phytochrome is present only in the light.

   High intensities of light are necessary for the plant to make normal amounts of phytochrome.

   The rate of synthesis of phytochrome is greater in plants treated with a pulse of red light followed by a pulse of far-red light than in plants treated only with red light.

   Phytochrome is synthesized only in red light.

Further Readings

Meyerowitz, E. M. 1994. "Genetics of Flower Development." *Scientific American* 271 (November): 56–65.

Rennie, J. 1993. "DNA's New Twists." *Scientific American* 268 (March): 122–32.

Salisbury, F. B., and Ross, C. W. 1992. *Plant Physiology*. 4th ed. Belmont, Calif.: Wadsworth.

Sunset Western Garden Book, 6th ed. 1995. Menlo Park, Calif.: Sunset.

Wayne, R. 1993. "Excitability in Plant Cells." *American Scientist* 81: 140–51.

16 GENETICS

1. The hereditary traits of an organism are determined by units of information called *genes* and are encoded in the base sequence of the organism's DNA. A mutation in a DNA base sequence produces a new allele of a gene, which can be identified by a detectable change in a trait.

2. In plants, DNA is divided among several chromosomes, with each gene occurring at a specific place along the length of one of the chromosomes.

3. Meiosis and fertilization work together to assure that different combinations of the alleles of different genes appear in new generations.

4. The transmission of chromosomes, genes, and alleles from parents to progeny follows clear rules, so that the traits of progeny can be predicted from the traits of their parents and more distant ancestors. However, the processes of meiosis and fertilization include chance events, so the predictions must describe probabilities rather than certainties.

5. A knowledge of genetics allows plant breeders to develop crops with improved qualities and yields.

16.1 TRAITS, GENES, AND ALLELES

Like other organisms, the plants of a particular species can be described by a series of **characters**, such as flower shape and color, the length of the stem, the shape and arrangement of the leaves, the type of fruit, and the shape of the seed. All these characters are specified to some degree by internal factors called **genes**. Genes provide the instructions to the plant cells on how to grow and develop and how to respond to environmental cues. Genetic information is inherited—that is, passed from parents to progeny, from generation to generation in the process of sexual reproduction. *Genetics* is the study of how genes work and the rules by which they are inherited.

Different Alleles Produce Different Traits

Some basic principles of genetics were discovered by Gregor Mendel, an Austrian monk, botanist, and teacher who used garden peas (*Pisum sativum*) as his experimental organism. His report, published in 1866, was generally unappreciated until his principles were rediscovered around 1900. One of Mendel's principles was the idea that some plant characters are determined by heritable factors: genes. When Mendel looked at his pea plants, he noticed that certain characters occurred in more than one form. For instance, for the character of stem length, pea plants could be short or tall; the character of flower color could be red or white; seed shape could be round or wrinkled. He reasoned that the genes for those characters came in more than one form. Today, each variant is called

a **trait**; the alternate forms of a gene are called **alleles**. One allele of a gene for stem length makes a pea plant tall; another makes it short.

Traits Reflect the DNA Code

Genes are sequences of nucleotides in DNA, or **base sequences** (referring to the parts of the nucleotides that give them their identities—see Chapter 2). What we know about genes has come from the work of many scientists, but no doubt the greatest breakthrough in knowledge was made by James Watson and Francis Crick, the discoverers of the double helical structure of DNA. They recognized how the sequence of bases along a DNA molecule could be thought of as a code that specifies the sequence of amino acids in a protein. Proteins have numerous important functions in a cell, including their role as enzymes. As we saw in Chapter 2, the functions of proteins depend on their three-dimensional structures, which in turn depend on their amino acid sequences. An example is sucrose-phosphate synthase, an enzyme needed for the synthesis of sucrose. This protein works because it has a three-dimensional structure that binds to the reactants (uridine-diphosphoglucose—an active form of glucose—and fructose-phosphate) and catalyzes the reaction that combines the glucose and fructose-phosphate to form sucrose-phosphate. That three-dimensional structure is a consequence of the order of the amino acids in the protein molecule. The sequence of the amino acids is determined by the base sequence of its gene.

Sometimes a plant with a different form, a **mutant**, may appear spontaneously in a collection of similar plants from the same parental stock. Figure 16.1 shows several mutant corn (*Zea mays*) plants compared to their "normal," or **wild-type**, relatives. Different forms of a trait arise through **mutation**, changes in the base sequence of DNA. Sometimes, the mutation is as simple as a change in one base. At other times the mutation is more drastic: a complete disruption of the base sequence by the addition or deletion of a line of bases. When a mutation changes the base sequence of a gene that previously coded for an enzyme, the gene no longer contains the information to code for the correct amino acid sequence. Therefore, the enzyme either is not made or does not function properly. For instance, a mutation in the gene encoding sucrose-phosphate synthase could result in a plant's not being able to make the functional enzyme. Such a plant would lack the ability to form sucrose-phosphate in some organ where the enzyme normally functioned, and this might result in a change of a visible trait.

Not all mutations are bad. Some may make the plant more suited to its environment; others may simply make the plant different. Table 16.1 describes a variety of mutants in peas, maize, and *Brassica campestris* (a relative of mustard and its cousins). Once a mutation occurs, the altered DNA may be passed from the mutant organism to its progeny according to the rules that Mendel discovered.

Figure 16.1 Mutations in corn plants. (**a**) An albino mutant seedling among wild-type green seedlings. (**b**) An attached-leaf mutant in which the cuticle has changed, causing the leaves to stick together as they develop. (**c**) A stripe mutant, in which chlorophyll accumulation decreases periodically as the leaf grows from the base. (**d**) Mature ears, in which different kernels are expressing different combinations of alleles of genes controlling the production of anthocyanins (red and purple pigments).

16.2 TRANSMISSION OF GENES

Chromosomes Carry Genes

In plants, as in all other eukaryotic organisms, most of the DNA is found in the nucleus (Chapter 3) and is combined with proteins to form **chromatin**. The chromatin is divided into **chromosomes**, each of which contains a single, linear strand of DNA. Each gene is a portion of the DNA between 300 and 3000 bases long (or longer). Every gene has a particular position—its **locus**—on one of the chromosomes (Fig. 16.2). It is separated from the adjacent genes by long stretches of DNA believed to be nonfunctional. Because the DNA in a chromosome may be a hundred million bases long, there is plenty of room for several hundred genes on one chromosome. The collection of all the genes in an organism is called its **genome**. The smallest number of chromosomes in a plant is four, found in the desert plant *Haplopappus gracilis* (a relative of the sunflower) and some other species. The coast redwood (*Sequoia sempervirens*) has nearly a hundred chromosomes; some ferns have several hundred.

All the cells that make up the root and shoot of a plant are formed from an original single cell by a sequence of mitotic cell divisions. As described in Chapter 3, mitotic cell divisions follow the S phase of the cell cycle, during which the DNA in each chromosome is faithfully replicated. Mitosis assures that each of two progeny cells has one of every chromosome. This means that every allele (mutant or wild-type) found in the original cell will also

be present in all the cells of the plant. When plants reproduce by vegetative reproduction, which occurs solely through mitotic cell divisions, the progeny plants have the same alleles as the parent plants.

Each different type of chromosome has a different set of genes. However, as explained in Chapter 12, all diploid organisms, including flowering plants, have two copies of each type of chromosome, one from the sperm and one from the egg. The two copies are said to be **homologous** because they have the same set of genes. That does not mean, however, that the genes at the same locus on two homologous chromosomes are necessarily identical; they may have different alleles.

Meiosis Segregates Alleles, and Fertilization Combines Them

Diploid organisms—including plants—reproduce sexually, a mechanism that offers greater genetic variety than mutation alone. As described in Chapter 12, the gametes of a plant—sperm and egg—are haploid, with each having only one chromosome of each type. Their union forms a new single cell, the zygote, which has two sets of chromosomes. Thus the zygote, and the plant that develops from it, is diploid. To reproduce, the diploid plant must form new haploid cells by meiosis.

Table 16.1 A Small Selection of Mutations in Three Plant Species

Wild-Type Form	Alternate (Mutant) Form	Name of Gene
Pisum sativum (garden peas)		
Yellow cotyledons	Green cotyledons	*I*
Red flower petals	White flower petals	*A₁*
Smooth seed surface	Wrinkled seed surface	*R*
Tall (more than 20 internodes)	Short (10–20 internodes)	*T*
Green foliage	Yellow-green foliage	*O*
Axillary flowers	Terminal flowers	*Fa*
Straight pod	Curved pod	*Cp*
Tendrils	No tendrils	*N*
Zea mays (maize, corn)		
Filled endosperm	Shrunken endosperm (lacks sucrose synthase)	*sh*
Yellow endosperm	White endosperm	*y*
Colored (red) endosperm	Yellow endosperm	*R*
Normal endosperm	Waxy endosperm (altered starch-synthesizing enzyme)	*wx*
Dormant seed	Viviparous (germinates on cob)	*Vp*
Has isocitrate dehydrogenase	Lacks isocitrate dehydrogenase	*idh*
Brassica campestris (rapid-cycling Brassicas, fast plants)		
Normal internode length	Short internode (gibberellin-insensitive)	*dwf*
Normal internode length	Short internode (rosette—gibberellin-sensitive)	*ros*
Red pigment on hypocotyl	Lacks pigment	*anl*
Waxy covering on leaves, stem	Lacks waxy covering (glossy)	*glo*
Green leaves	Yellow-green (chlorophyll deficient)	*ygr*
Normal internodes	Elongated internodes (lacks phytochrome B)	*ein*

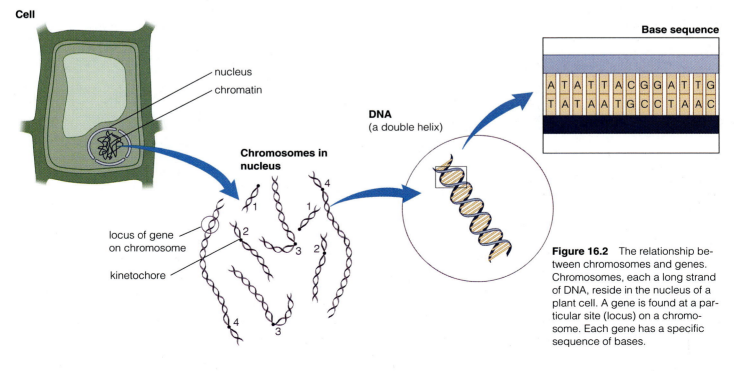

Figure 16.2 The relationship between chromosomes and genes. Chromosomes, each a long strand of DNA, reside in the nucleus of a plant cell. A gene is found at a particular site (locus) on a chromosome. Each gene has a specific sequence of bases.

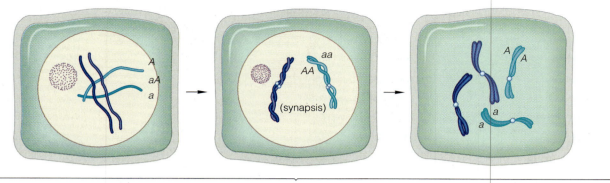

Prophase I

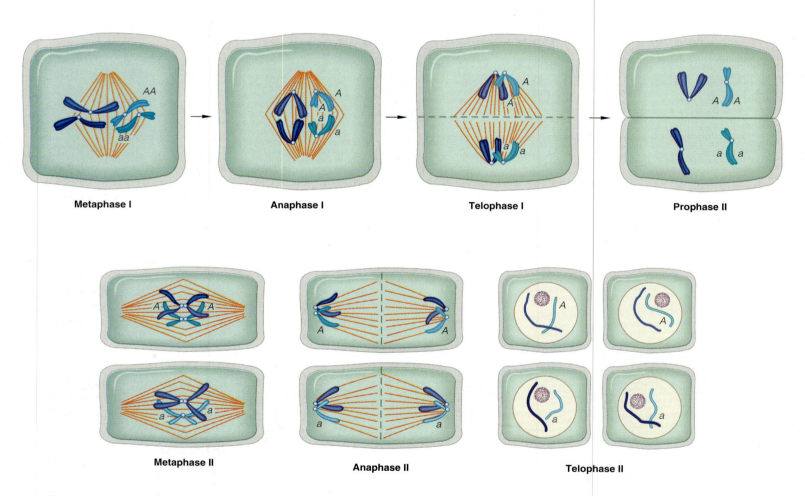

Metaphase I   **Anaphase I**   **Telophase I**   **Prophase II**

Metaphase II   **Anaphase II**   **Telophase II**

Figure 16.3   The distribution of alleles of one gene (locus on a chromosome) to the products of meiosis.

If we assume that a plant (and thus every cell in that plant) has two different alleles (which we designate *A* and *a*) at some locus on a chromosome, we can trace the allocation of those alleles to gametes during the process of meiosis (Fig. 16.3). The preparation for meiosis begins in the preceding S phase with the synthesis of new DNA, after which each chromosome has two identical sister **chromatids** (Fig. 16.4a). Meiosis opens with prophase I, in which the chromosomes coil, shorten, and thicken, becoming more visible. During this period, homologous chromosomes come together to form pairs (synapsis). The chromatids of homologous chromosomes may exchange some corresponding pieces with each other in a process called **crossing over**. The cross formed by the chromatids during the exchange (and visible with a microscope) is known as a **chiasma** (from the Greek word for "cross," plural *chiasmata*). The process results in the formation of rearranged chromatids possessing fragments from both of the homologous chromosomes. If the homologous chromosomes of the parent cells have different alleles, then crossing over produces chromatids with new combinations of alleles (Fig. 16.4b), an effect known as **recombination**. (In contrast to mutation, which actually changes alleles, recombination simply forms new combinations of alleles already present in the genome.) Toward the end of meiosis, the chromatids are separated, so each daughter cell gets one chromatid (now referred to as a *chromosome*) of each type and only one copy of each gene (see anaphase II and telophase II in Fig. 16.3). If the original diploid cell had different alleles for a particular gene, each resulting haploid gamete would have only one of the two alleles. In Mendel's language, the two alleles would have segregated (separated) during meiosis. (This is sometimes known as Mendel's Law of Segregation.) Amazingly, even though Mendel did not know about chromosomes, genes, alleles, and meiosis, he deduced from his experiments how traits are passed from parents to progeny.

Simple Crosses Yield Predictable Results

Mendel was not the first person to study heredity, but he was the first to identify clear rules that predicted and explained the inheritance of traits. His conceptual breakthrough was the identification of several individual characters in pea plants, each of which occurred in one of two forms. Mendel grew plants that were identical except for differences in one, two, or three of these characters, so he could focus on the inheritance of the alleles of their genes.

SINGLE GENES The most basic type of genetic experiment involves a mating between two plants that are genetically identical except for one character, in which they express different traits. If the character is controlled by one gene, then the two plants have different alleles for that gene.

An example comes from one of Mendel's experiments. He carefully took pollen from the anthers of a dwarf variety of peas and dusted it on the stigma of a tall variety. The resulting seeds were planted the next season. All the plants that grew from these seeds—called the F1 (first filial) generation—were tall. The same was true when pollen from tall plants was dusted on stigmas of dwarf plants. Then in a second mating, flowers of the tall progeny were self-pollinated; that is, pollen from each flower was dusted on the stigma of the same flower. The seeds resulting from these self-pollinated flowers—the F2 (second filial) generation—produced 787 tall plants and 277 dwarf plants. In other words, about three-fourths of these progeny were tall, one-fourth dwarf.

Why were all the plants tall in the offspring of the first mating if one of the parents was dwarf? And how did some of the plants from the second mating come to be dwarf when none of their parents were? These results can be explained by the model shown in Figure 16.5. Since there are two traits for this character, tall and dwarf, we can assume that there are two alleles involved and call them *T* (for tall) and *t* (for dwarf). Every adult diploid plant has two copies of each gene because it has two sets of chromosomes. Mendel took pains to make sure that his varieties were genetically homogeneous; so the tall variety had two tall alleles, and the dwarf variety had two dwarf alleles. Geneticists call the visible traits of an organism its **phenotype** and its collection of alleles its **genotype**. In this case, the plant with the tall phenotype had the genotype *TT*; the plant with the dwarf phenotype had the genotype *tt*. Plants with two copies of the same allele are called **homozygous** for that allele because the zygotes from which all of the plants' cells originated had two copies of the same allele. When plants form gametes, every gamete gets one allele from the parent (in this case, *T* or *t*, depending on which parent). Gametes from a homozygous parent all carry the same allele. The genotype of a zygote is the combination of the alleles from the two gametes that fused to form that zygote.

In the first mating, the progeny all received one *T* allele and one *t* allele, and so they all had the genotype *Tt*. Plants that have different alleles of a gene are said to be **heterozygous**. The model suggests that the progeny all had the same genotype; this fits with the observation that they all had the same phenotype.

But why were they tall, if they had copies of both alleles? To explain why the progeny were tall, we have to assume that the expression of the tall allele in some way overshadowed the expression of the dwarf allele. We say that the tall allele was **dominant**; the dwarf allele was **recessive**. This means that the presence of one *T* allele is enough to produce the tall phenotype. Only a *tt* genotype will produce a dwarf plant.

The relationship between alleles is not always dominant-recessive; for some genes, heterozygous plants show traits intermediate between those of the parental homozygotes. In these cases, the alleles are said to be **codominant** or **incompletely dominant**. For instance, in snapdragons (*Antirrhinum majus*) the combination of a

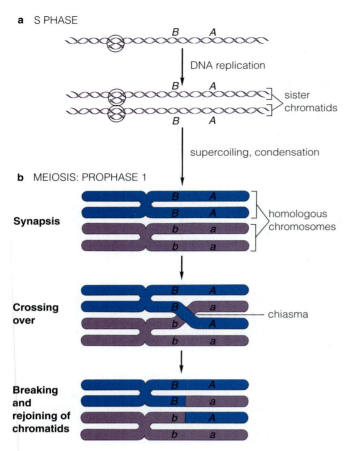

a S PHASE

B   *A*

DNA replication

B   *A*

} sister chromatids

B   *A*

supercoiling, condensation

b MEIOSIS: PROPHASE 1

Synapsis

B   *A*
B   *A*
b   *a*
b   *a*

} homologous chromosomes

Crossing over

B   *A*
B   *a*
b   *A*
b   *a*

— chiasma

Breaking and rejoining of chromatids

B   *A*
B   *a*
b   *A*
b   *a*

Figure 16.4 Steps leading to the recombination of alleles. (**a**) The replication of chromosomes in the S phase of the cell cycle results in two sister chromatids. (**b**) Crossing over between two homologous chromosomes during prophase I of meiosis can recombine alleles for the genes on that chromosome. *A* and *a* represent different alleles of one gene at a particular locus on the chromosome; *B* and *b* represent different alleles of another gene. In the crossing over shown, nonsister chromatids of a homologous pair of chromosomes exchange *A* and *a* alleles.

red allele and a white allele for the flower color gene produces pink flowers.

To explain the results of the second mating, we need to make additional assumptions. The first is that the two alleles in the *Tt* parents are distributed randomly to their gametes; so half the gametes receive a *T* allele, and half receive a *t* allele. The second assumption is that the gametes fertilize randomly, so that combinations of alleles are formed in proportion to their frequency in the gametes.

According to the model, the *Tt* parent that donated pollen produced half *T* and half *t* pollen. The *Tt* parent that was pollinated produced half *T* eggs and half *t* eggs. At fertilization, there were four combinations, and they occurred in equal proportions. One-fourth had the *TT* genotype, one-half had the *Tt* genotype, and one-fourth had the *tt* genotype. Because the *TT* and *Tt* genotypes

yield tall plants, three-fourths of the progeny were tall. The remaining one-fourth were dwarf.

A Punnett square (named for the genetics professor who popularized it) is useful for keeping track of the combinations of alleles formed during fertilization. It is constructed by drawing a square and listing alleles of the male gametes available for the mating along one side of the square and those of the female gametes along the perpendicular side (Fig. 16.5). The combinations of alleles at the intersections of the columns and rows show the expected genotypes of the progeny. We can use the mathematical principles of probability to make quantitative predictions about the phenotypes and genotypes of the progeny (see sidebar, "Probability and Mendelian Genetics," p. 268).

Performing this experiment with two plants that differ in one trait constitutes a test of the assumptions we have made. If the progeny of the first cross were not all identical, we would have to abandon the assumption that the parents were homozygous. If the progeny of the first cross were identical, but the proportions of the second cross did not come out 3:1, we would need to question the assumption that the trait was controlled by a single gene.

TWO GENES Plants, like other organisms, are guided by thousands of genes acting in concert. The overall phenotype—for instance, the branching pattern of a tree together with the shape of its flowers—is determined by combinations of alleles of many genes, rather than by the alleles of a single gene. How are combinations of alleles inherited? Mendel simplified this question by studying the inheritance of pairs of genes.

What will be the result of mating two pea varieties that differ in two characters—for instance, height (tall or dwarf) and form of seeds (round or wrinkled)? Let us assume that we have available a strain of plants that breeds true for tall plants and round seeds and one that breeds true for dwarf plants and wrinkled seeds. "Breeds true" means that when we mate these plants, we will find—as Mendel did—that the resulting progeny are all tall and have round seeds. This fits a model in which one strain has a genotype of *TTRR*—it is homozygous for both the dominant (tall) allele for height and the dominant (round) allele for seed form—and the other strain has the genotype *ttrr*—it is homozygous for the recessive (dwarf) allele for height and the recessive (wrinkled) allele for seed form. The progeny would all have the genotype *TtRr*, representing the combination of the genes from *TR* and *tr* gametes. These are **dihybrid** plants, meaning that these offspring are the combination (hybrid) of two plants that differ in their alleles for two genes. Because both the tall (*T*) allele of the height gene and the round (*R*) allele of the seed gene are dominant, all the F1 plants will be tall and will come from round seeds.

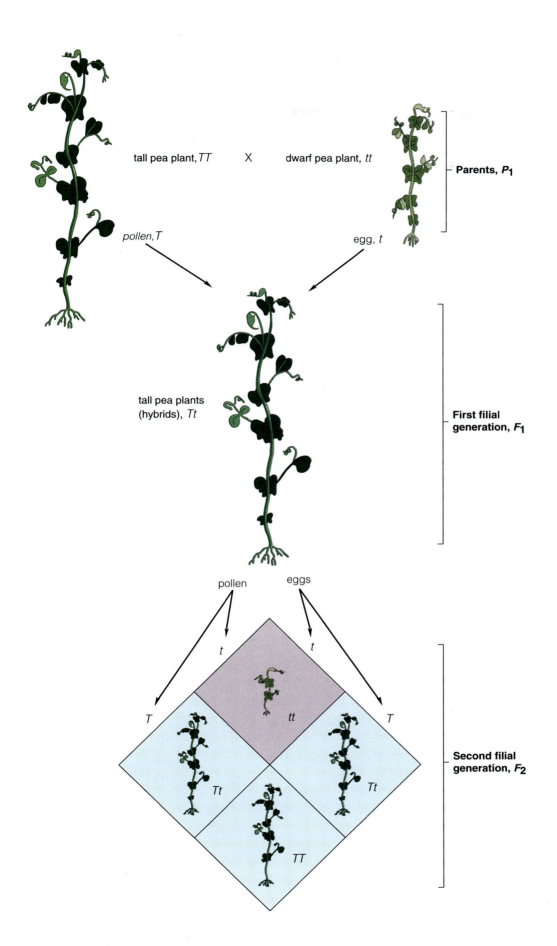

tall pea plant, *TT*   X   dwarf pea plant, *tt*

Parents, *P*₁

pollen, T

egg, t

tall pea plants
(hybrids), *Tt*

**First filial
generation, *F*₁**

pollen   eggs

t   *t*

T   *tt*   *T*

Tt   *Tt*

TT

**Second filial
generation, *F*₂**

Figure 16.5 Diagram of a cross
between tall (*TT*) and dwarf (*tt*)
pea plants. All the plants in the
first filial generation (F1) are
heterozygous (*Tt*), meaning that
the zygote—and therefore the
cells—have both alleles for the
trait, one from each parent. The
dwarf allele (*t*) is not outwardly
visible in the F1 plants, which all
exhibit a tall phenotype. When the
F2 plants are self-pollinated, the
second filial generation (F2) produces (on average) three plants
with the tall phenotype (*TT*, *Tt*, *tT*)
to every one with the dwarf (*tt*)
phenotype.

PROBABILITY AND MENDELIAN GENETICS

One of Mendel's most important innovations was his use of the concepts of mathematical probability and statistics in the study of heredity. These concepts, first developed by Blaise Pascal and Pierre de Fermat to analyze games of chance, can be applied to any event or experiment in which the results are subject both to random fluctuations and to well-ordered processes. The purposes of such application are to separate the well-ordered features of the experiment from the random features, to identify the patterns that the experiment follows, and to use these patterns to predict future results.

A critical concept in probability is the idea of the *outcome* of an event or experiment. The outcomes are unit results, only one of which can occur at any one trial of the experiment and all of which together constitute all possible results of the experiment. If we toss a coin to see which side lands up, the outcomes are "heads" and "tails." If we mate two parents and observe the gender of a resulting offspring, the outcomes are male and female.

A basic operation in the analysis of probability is to assign numbers to each of the various outcomes, numbers that indicate our feeling about the likelihood that each outcome will occur. If we are certain that a particular outcome will occur, we assign it a probability of 1.0 (denoted P = 1.0); if we are certain that an outcome will not occur, we assign it a probability of 0.0 (P = 0.0); if we feel an outcome will occur half the time, we assign it a probability of 0.5 (P = 0.5).

Some of the principles useful in assigning and interpreting numerical probabilities are described below, together with examples taken from problems in Mendelian genetics.

1. In an experiment with *n* equally likely outcomes, the probability of the occurrence of any one outcome is $1/n$.

Example: In the mating of two hybrid plants with genotypes *Gg*, a Punnett square diagram indicates that the progeny may have any of four different genotypes (that is, outcomes): *GG*, *Gg*, *gG*, and *gg*. The probability of any one occurring is 1/4.

2. If the probability of a particular outcome is $1/n$, then in a *large* number of trials, that outcome will tend to represent $1/n$th of the total.

Example: In the experiment described above, the probability of finding the *gg* genotype in any one seed is 1/4. If we looked at 1000 progeny seeds, we would expect to find $1/4(1000) = 250$ seeds with the *gg* genotype. We would also expect to find close to 250 seeds each with the *GG*, *Gg*, and

gG genotypes. Notice, however, that if we look only at *small* numbers of seeds, we should not be surprised if the outcomes do not reflect their respective probabilities. If we look at 10 seeds, we do not expect 2½ seeds with each genotype; if we look at 100 seeds, we do not expect *exactly* 25 seeds with each genotype. But the more seeds we consider, the closer the relative frequency of appearance of each genotype should be to its probability.

3. The probability of an occurrence represented by two or more different (*mutually exclusive*) outcomes is the sum of the probabilities of the separate outcomes.

Example: Assume that the experiment above refers to seed color. The dominant allele, *G*, gives yellow seeds; the recessive allele, *g*, gives green seeds when present in a homozygote. The occurrence, "A seed is yellow," is represented by three different outcomes (the genotypes *GG*, *Gg*, and *gG*). The three outcomes are mutually exclusive because only one can occur in any one seed. The probability that a seed is yellow therefore is the sum of the probabilities for these three genotypes.

$$P(\text{seed is yellow}) = P(GG) + P(Gg) + P(gG)$$
$$= 1/4 + 1/4 + 1/4 = 3/4$$

Likewise, the probability that the seed is heterozygous is

$$P(\text{seed is heterozygous}) = P(Gg) + P(gG)$$
$$= 1/4 + 1/4 = 1/2$$

4. If the probability of a certain outcome in one experiment is $1/n_1$, and the probability of another outcome in a second (independent) experiment is $1/n_2$, the probability that *both* outcomes will occur is $(1/n_1)(1/n_2)$.

Examples: In the experiment described above, the probability of finding a green seed, genotype *gg*, is 1/4; the probability of finding a yellow seed, genotype *GG* or *Gg* or *gG*, is 3/4. Each seed represents a separate mating event, so the probability that the first seed investigated is green and the second seed investigated is yellow is

$$P(\text{1st seed green } and \text{ 2nd seed yellow})$$
$$= P(gg) \cdot P(GG \text{ or } Gg \text{ or } gG) = (1/4)(3/4) = 3/16$$

The probability that the first seed, the second seed, and the third seed are all green is

$$P(\text{1st seed green } and \text{ 2nd seed green } and \text{ 3rd seed green})$$
$$= P(gg) \cdot P(gg) \cdot P(gg) = (1/4)(1/4)(1/4) = 1/64$$

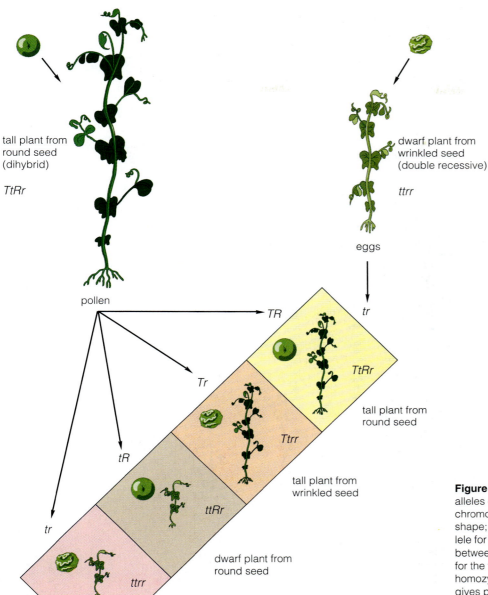

tall plant from
round seed
(dihybrid)

TtRr

dwarf plant from
wrinkled seed
(double recessive)

ttrr

pollen

eggs

TR

Tr

tR

tr

tr

TtRr

tall plant from
round seed

Ttrr

tall plant from
wrinkled seed

ttRr

dwarf plant from
round seed

ttrr

dwarf plant from
wrinkled seed

Figure 16.6 Inheritance of combinations of alleles of unlinked genes (genes on different chromosomes). *R* is the round allele for seed shape; *r* is the wrinkled allele. *T* is the tall allele for height; *t* is the dwarf allele. A cross between a dihybrid pea plant, heterozygous for the two genes (*RrTt*), with a plant that is homozygous recessive for both genes (*rrtt*) gives progeny with four different genotypes and phenotypes. The phenotypes of the progeny reflect the genotypes of the gametes produced by the dihybrid parent.

What happens when we use these dihybrid progeny plants as parents? When cells undergo meiosis, the alleles from these two genes (*TtRr*) are distributed so that each gamete receives one allele of each gene. There are four possible combinations in the gametes: *TR, Tr, tR,* and *tr.* If the two genes are on different chromosomes (which act independently in meiosis), the alleles will be distributed randomly and the four combinations of alleles will occur in equal numbers. If we mate such a plant with the homozygous recessive parent (*ttrr*), the phenotypes of the progeny will reflect the genotypes of the gametes from the dihybrid (Fig. 16.6)—that is, four different phenotypes will occur in equal numbers. Mating an experimen-

tal plant, such as the dihybrid described above, with a plant known to be homozygous recessive in all the genes of interest is called a **test cross** because the phenotypes of the progeny indicate the genotypes of the gametes from the plant to be tested. It is an easy technique for determining whether combinations of alleles are produced in equal numbers in the gametes of a dihybrid plant.

A more complicated case occurs if we use dihybrid plants for both parents. In this case there are 4 possible genotypes for both the male and the female gametes. Combining these genotypes at fertilization gives 16 possible genotypes in the zygote (Fig. 16.7). The Punnett square now becomes a very useful tool for arranging the

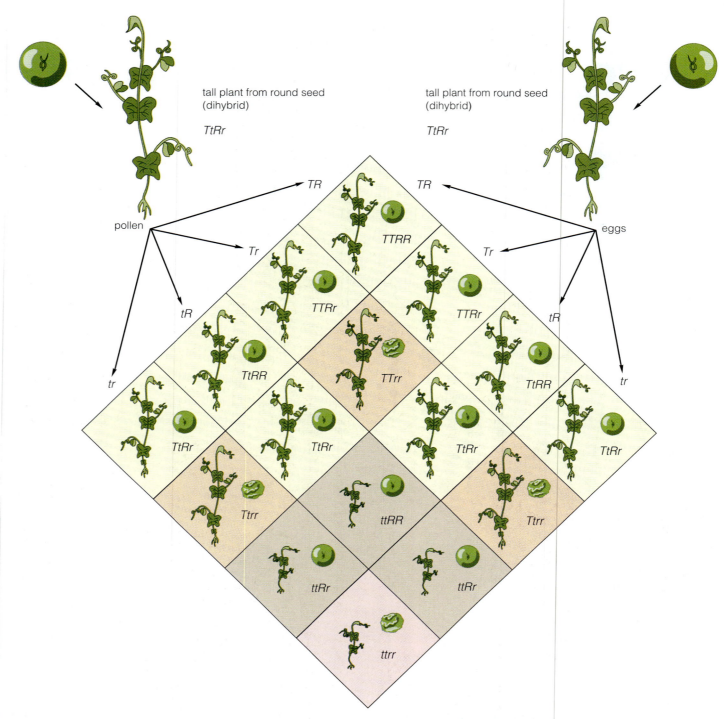

tall plant from round seed
(dihybrid)

TtRr

tall plant from round seed
(dihybrid)

TtRr

pollen

eggs

TR

Tr

tR

tr

TR

Tr

tR

tr

TTRR

TTRr

TTRr

TtRR

TTrr

TtRR

TtRr

TtRr

TtRr

TtRr

Ttrr

ttRR

Ttrr

ttRr

ttRr

ttrr

Figure 16.7 Inheritance of combinations of alleles of unlinked genes, part 2. A cross between two dihybrid pea plants, both heterozygous for the genes for height and seed shape (*RrTt*), gives progeny with 16 different possible combinations of alleles and 4 phenotypes. Because each of the possible combinations (squares on the Punnett square diagram) is equally likely, the phenotypes are expected to occur in the ratio of 9:3:3:1.

zygote genotypes and grouping the phenotypes for the different genotypes. We find that the mating of two dihybrids gives tall plants with round seeds, tall plants with wrinkled seeds, dwarf plants with round seeds, and dwarf plants with wrinkled seeds—in the ratio 9:3:3:1. The 9:3:3:1 ratio is typical for all crosses in which both parents are dihybrid for the same genes.

Mendel, who knew nothing of meiosis or chromosomes, correctly interpreted data from crosses such as these to mean that pairs of heritable factors assort independently in gamete formation. (This is sometimes called Mendel's Law of Independent Assortment.) Today we say that two genes that are independently assorted are **unlinked**, meaning that they are on different chromosomes or that they are so far apart on the same chromosome that they need not be transferred together during meiosis.

When Genes Are Linked, Traits Are Inherited Together

What happens if, in a test cross of a dihybrid plant, we do not find equal numbers of the four possible genotypes in the progeny? This result may mean that the two genes are on the *same* chromosome. In this model (Fig. 16.8), one chromosome of the dihybrid parent carries the dominant alleles of two genes—for example, genes for tendrils and seed shape—and the other chromosome carries the recessive alleles. We say the genes are **linked** because during the formation of gametes, these alleles migrate as a unit rather than independently. So, in an uncomplicated course of events, when the chromatids separate in meiosis, half the gametes receive the two dominant alleles, and half, the two recessive alleles (Fig. 16.8a). The plants that come from round seeds have tendrils; the plants that come from wrinkled seeds do not.

But it is not quite that simple. In prophase I, chromatids may exchange alleles via crossing over (see Fig. 16.4b). This can result in the formation of chromatids with **recombinant** combinations of alleles (Fig. 16.8b). In the situation we postulated, in which the two dominant and two recessive alleles are the original (**parental**) combinations, a recombinant combination of alleles would be the dominant allele of one gene and the recessive allele of the other. Some plants from round seeds will lack tendrils, and some from wrinkled seeds will have them. Notice, however, that the situation could be reversed if we had chosen different parents: The parental combinations could be the dominant allele of one gene and the recessive allele of the other. The probability of crossover events occurring depends on how close together the genes are on the chromosome. Assuming that crossing over occurs randomly along the chromosome, the farther apart two genes are, the more likely crossing over will occur between them, breaking their linkage and forming recombinant combinations of alleles. However, no matter how far apart the two genes are, the fraction of gametes with recombinant combinations is never greater than the fraction of gametes with the parental combinations of alleles (that is, never greater than 50%).

Maternal Inheritance Involves Organellar Chromosomes

Some characters do not follow the rules of inheritance described in the preceding sections. One of these is the response to the herbicide atrazine. Atrazine is a chemical that blocks electron flow in photosynthesis. Most plants are sensitive to it and will die after an atrazine treatment and exposure to light; a few mutant plants, however, show resistance. Atrazine sensitivity (atr^s) and atrazine resistance (atr^r) are inherited traits; they represent alleles of one gene.

If we pollinate an atrazine-sensitive stigma with pollen from an atrazine-resistant plant, all the progeny plants will be sensitive (Fig. 16.9). If, on the other hand, we pollinate an atrazine-resistant stigma with pollen from an atrazine-sensitive plant, all the progeny plants will be resistant. This is not at all what we would expect under the Mendelian model described above. Regardless of whether the sensitive allele or the resistant allele was dominant, or both alleles were codominant, the results of these two matings should have been the same. Instead, it seems that the progeny inherited their sensitivity (or resistance) allele only from the maternal parent.

To explain this unusual type of inheritance, we must recognize that although most of the genes in a plant cell are located on the nuclear chromosomes, not all of them are. There are chromosomes of DNA in the plastids and in the mitochondria as well. These organellar chromosomes are smaller than those in the nucleus, and they are different in other ways; but they do contain genes, and these genes can mutate. During fertilization, only the chloroplasts and mitochondria from the egg are incorporated into the zygote. Chloroplasts and mitochondria from the sperm cells either do not enter the egg or degenerate during the fertilization process. Thus, the chloroplast and mitochondria genes in the zygote all come from the egg, and all the alleles of these genes show maternal inheritance.

16.3 PLANT BREEDING

In the past 40 years, the population of the world has doubled, but the number of acres of land under cultivation has dropped. Despite the increase in population and the decrease in farmed land, the amount of food produced per person is at least as great as it was in the 1950s. The ability of farmers to increase their output results in great part from the breeding of new, more productive plants. New varieties have been formed—with characteristics

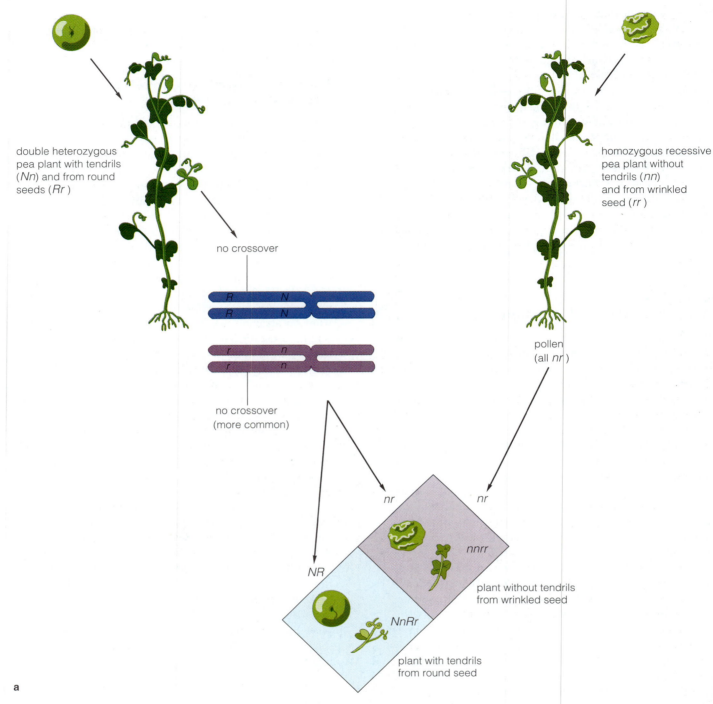

double heterozygous pea plant with tendrils (*Nn*) and from round seeds (*Rr*)

homozygous recessive pea plant without tendrils (*nn*) and from wrinkled seed (*rr*)

no crossover

| *R* | *N* |
| *R* | *N* |

| *r* | *n* |
| *r* | *n* |

no crossover
(more common)

pollen
(all *nr*)

nr

nr

nnrr

plant without tendrils
from wrinkled seed

NR

NnRr

plant with tendrils
from round seed

a

Figure 16.8   Inheritance of linked genes. *R* is the round allele for seed shape; *r* is the wrinkled allele. *N* is the allele for possession of tendrils; *n* is the tendrilless allele. In the mating of a dihybrid pea plant that is heterozygous for the seed shape and tendril genes (*RrNn*) with a plant that is homozygous recessive (*rrnn*), the parental combinations of alleles (*RN* and *rn*) are passed to gametes together unless crossing over occurs during meiosis. (**a**) Without crossing over between the *R* and *N* genes in the formation of the pollen. (**b**) With crossing over.

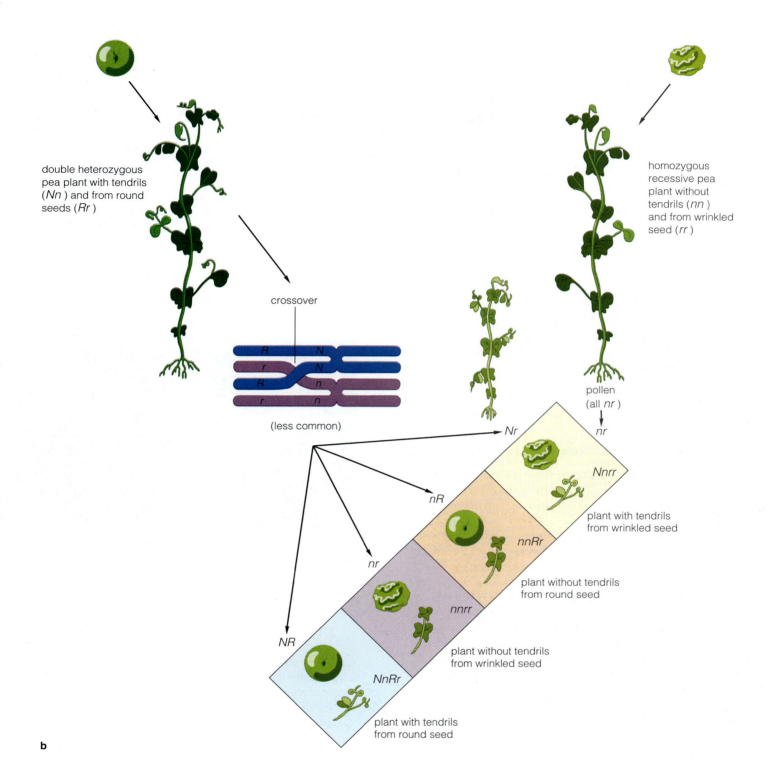

double heterozygous pea plant with tendrils (*Nn*) and from round seeds (*Rr*)

homozygous recessive pea plant without tendrils (*nn*) and from wrinkled seed (*rr*)

crossover

(less common)

pollen (all *nr*)

nr

Nr

nR

nr

NR

Nnrr

plant with tendrils from wrinkled seed

nnRr

plant without tendrils from round seed

nnrr

plant without tendrils from wrinkled seed

NnRr

plant with tendrils from round seed

b

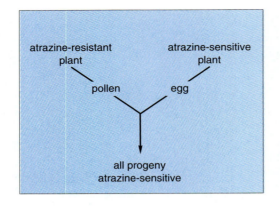
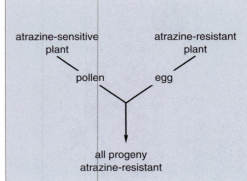

Figure 16.9 Maternal inheritance of a gene on chloroplast DNA.

that make them easier to grow or harvest, with resistance to disease or stress, or with edible parts that are more attractive or nutritious. For instance, modern rice (*Oryza sativa*) varieties are shorter and less likely to fall over when the grain is mature. (Plants that fall over are difficult to harvest, and their grain is more likely to mold.) Modern tomato (*Lycopersicon esculentum*) varieties are bred with resistance to *Verticillium* and *Fusarium* fungi.

Mating Plants Combines Useful Traits

The principles of genetics are used extensively to develop plants with new and useful traits (Fig. 16.10). However, there are some tricks to plant breeding that go beyond the basic genetic principles.

Suppose we want to breed a tomato variety that will resist infection by a fungus that has recently appeared. We may mate a successful but fungus-susceptible commercial variety with a wild variety that shows resistance but has inedible fruit. If the resistance allele is dominant (let us assume it is), the progeny will be resistant, but they will probably have inedible fruit. We then would mate the progeny with the commercial variety (a **back cross**) and test the progeny of that mating for resistance. By chance recombination, some of the resistant progeny will also have acquired some of the genes needed for edible fruit. The most resistant progeny would then again be mated with the commercial variety, and the most resistant progeny again selected. After several cycles, we would have a strain that has both commercial fruit and resistance to the fungus.

Multiple Genes Explain Continuous Variation

The traits that we have discussed in the previous paragraphs occur in one of two (or three or four) forms; they are said to have **discrete variation**. Many important traits, however, vary *continuously* over a certain range; these are said to have **continuous variation**. Examples in

Figure 16.10 Breeding trials for maize varieties at a research farm of Pioneer Hi-Bred International, Inc. Each plot represents a different stage in the development of a new variety, or a different variety, or a different condition (for example, soil type or amount of fertilization or irrigation).

crop plants are size of the harvested organ (Fig. 16.11), sugar content, or firmness of the fruit.

There are several factors that can lead to a continuous variation: (a) The most important is the involvement of multiple genes, each of which influences the trait of interest. Individually, alleles of the different genes may have rather small effects on the phenotype. Together, they can combine to provide a wide range of variation. If there are many genes, there are many combinations of alleles, and this situation leads to a distribution of phenotypes in which the differences among many individual forms are small, relative to the total range of possibilities. (b) Sometimes a gene has multiple alleles, each with a different degree of activity. This increases the number of possible phenotypic forms. It may occur because different changes of the base sequence of a gene produce different

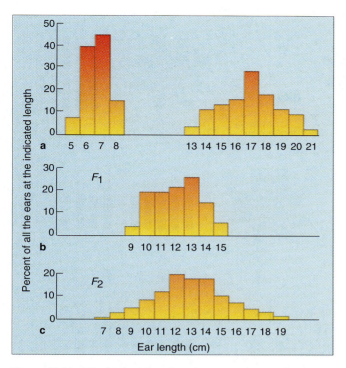

Figure 16.11 Distribution of ear lengths among two maize varieties and their progeny. Both parental varieties (**a**) and the F1 (**b**) and F2 (**c**) progeny show wide and continuous variation in ear lengths, probably because several genes contribute to the trait (1 cm = 0.4 in).

amino acid substitutions in an enzyme, leading to different amounts of enzyme activity. (c) Environmental effects may alter the form of the phenotype. For instance, crowded conditions produce shading and tend to make bean plants grow longer internodes; hot, dry conditions will give grape berries a higher sugar concentration. The randomness of environmental effects tends to blur the distinction among genotypes.

The inheritance of multiple genes involved in a continuously variable trait is no different from that of other genes. We can combine alleles by mating plants with different phenotypes, and we can select progeny on the basis of their phenotypes for further matings with one another. We must recognize, however, that if we do not know the number of genes, we do not know how many are heterozygous and how many are homozygous. Obtaining a strain in which all the relevant genes are homozygous may be important (such a strain will breed true). But with a continuously variable trait, it may be difficult to tell when this has been achieved.

Heterosis Can Give Vigorous Progeny

It is sometimes found that the progeny from the mating of two inbred (highly homozygous) strains are much larger and healthier than either of the parents. This effect is called **hybrid vigor** or **heterosis**. Most corn (*Zea mays*)

planted in the United States comes from hybrid seed. It poses a special problem for breeders because hybrids (heterozygotes) do not breed true—that is, when mated with themselves, they produce both heterozygous and homozygous progeny.

One solution is to produce new hybrid plants solely through vegetative reproduction. Potato (*Solanum tuberosum*) varieties that are reproduced by germinating buds cut from potato tubers all have the same genotype as the plants that produced the tubers.

Another solution is to produce hybrid seed by mating two homozygous strains. To do this, seed companies have developed strains that are male-sterile: They do not produce anthers or viable pollen. When a homozygous male-sterile strain of corn is planted close to another homozygous strain, it gets pollen only from the other strain, and all its seed will be heterozygous.

Plants Are Often Polyploid

In contrast to animals, plants become polyploid relatively easily. **Polyploidy** means having more than two sets of chromosomes. This may occur spontaneously, when a cell reduplicates its DNA and separates the resulting chromatids—but then fails to complete cell division. This commonly occurs in the last stages of development of tracheary elements and storage tissues, although it is less common in the meristem. If a meristematic cell becomes polyploid, the new nucleus, with twice the original number of chromosomes (tetraploid, if the parent cell was diploid), can undergo new rounds of DNA synthesis and mitosis. It may take over the meristem and produce a shoot whose cells are all polyploid. The cells are still capable of undergoing meiosis; so they can form spores, and the spores can form gametes. If a polyploid plant fertilizes itself, its progeny will also be polyploid.

Plant breeders have found that polyploid plants are often larger and more vigorous than their parental types. They also have found that certain chemicals that interfere with the formation of the spindle in mitosis (colchicine, for example) can increase the probability of forming polyploid cells, and these chemicals have been used to produce new horticultural varieties. In nature, polyploids seem more tolerant of such environmental stresses as short, cool growing seasons, aridity, or high temperatures.

In plants, it is sometimes possible for different species to interbreed—that is, to form viable progeny. Often the original progeny are sterile, but they can become fertile if their cells become polyploid (Fig. 16.12). The most important requirement for fertility is successful meiosis, and this requires that each chromosome be present in two copies. This is automatically true in a polyploid cell. The common bread wheat (*Triticum aestivum*) is hexaploid—having six copies of each chromosome—and is thought to have arisen in the Middle East by hybridization of three species from two related genera.

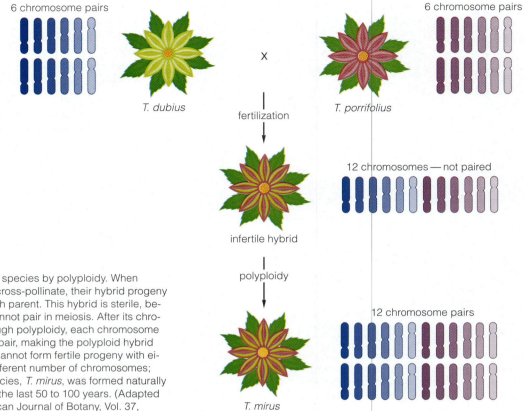

6 chromosome pairs

T. dubius

×

6 chromosome pairs

T. porrifolius

fertilization

infertile hybrid

12 chromosomes — not paired

polyploidy

T. mirus
(fertile polyploid hybrid)

12 chromosome pairs

Figure 16.12 The formation of a new species by polyploidy. When *Tragopogon dubius* and *T. porrifolius* cross-pollinate, their hybrid progeny has one set of chromosomes from each parent. This hybrid is sterile, because its two sets of chromosomes cannot pair in meiosis. After its chromosome complement is doubled through polyploidy, each chromosome has an identical one with which it can pair, making the polyploid hybrid fertile. However, the polyploid hybrid cannot form fertile progeny with either of its parents, because it has a different number of chromosomes; thus, it is a new species. This new species, *T. mirus*, was formed naturally in the American west, probably within the last 50 to 100 years. (Adapted from information in M. Ownbey, American Journal of Botany, Vol. 37, pp. 407–486, 1950.)

SUMMARY

1. The characters of plants are controlled by genes. Different alleles of a gene (mutant or wild-type) produce different traits, alternate forms of the character. The alleles, and thus the traits, are inherited.

2. Chemically, genes are made of DNA, and they control the production of proteins, which serve many functions in the cell.

3. Mutations are changes in the base sequence of the DNA. The modified base sequence may then change the amino acid sequence of an enzyme, or it may affect the rate or timing of expression of another gene.

4. Each gene is found at a particular locus on a chromosome. Different plants have different numbers of chromosomes, but each diploid cell in a plant has two copies of each chromosome.

5. Mitosis preserves the number of chromosomes and the alleles that are found on each chromosome. Vegetative reproduction occurs through mitosis and yields progeny that are genetically identical to the parent plant.

6. The sexual life cycle is dependent on meiosis. In meiosis, the two alleles of a gene in the diploid cell *segregate* (separate), so that half the haploid gametes receive one allele from the parent, and half receive the other allele.

7. If two genes are on different chromosomes, the segregation of the alleles of one gene does not affect the segregation of the alleles of the other. This means that if there are four possible combinations of alleles from the two genes, each combination is equally likely, and one-fourth of the gametes will receive each combination.

8. If two genes are linked (on the same chromosome), they tend to move into a gamete as a unit, rather than separately. The result is that the two parental combinations of alleles are more likely to occur in gametes than the two recombinant combinations of alleles. Recombinant combinations of alleles can be formed by crossing over of chromatids during meiosis.

9. Genetic techniques can be used to combine useful alleles of different genes in agricultural plants, even if the relationship between genes and traits is not simple.

10. Continuously variable traits can be explained by the influence of multiple genes, by the possibility of several different alleles, and by environmental effects. Even if a desired trait is controlled by several genes, a series of matings (with careful selection of progeny) can usually isolate plants with the correct combination of alleles.

11. Hybrids (heterozygotes) are often particularly vigorous. Special techniques are needed to produce hybrid seed for agricultural use.

12. Polyploidy (formation of plants with more than two chromosomes of each type) sometimes produces particularly vigorous plants. Polyploidy can also allow hybrids of distantly related plants to be fertile and form new plant species.

Questions

1. Provide the appropriate terms:

  a. *Gene* is to *character* as _____ is to *trait*.

  b. *Phenotype* is to *character* as *genotype* is to _____.

  c. A *base sequence* is to a *gene* as _____ is to a *protein*.

  d. _____ chromosomes have similar sets of genes.

  e. _____ organisms have two identical alleles for one gene;

  _____ organisms have two different alleles for one gene.

  f. A new allele can be produced by _____ ; a new combination of alleles can be produced by _____ during meiosis.

2. Trace the pathway by which information in the genes becomes a visible trait, by listing (in order) the processes that must occur.

3. Some wild radish plants have white flowers, and some have yellow flowers. Explain what you would do to determine whether the flower color is controlled by one gene with two alleles.

4. Explain why it takes plant breeders many growing seasons to produce a new crop plant (for instance, a tomato plant) with both good fruit and a new useful trait, such as disease resistance.

Genetics Problems

The abstract but regular nature of heredity has led biology teachers to adopt a tradition of assigning word problems in genetics. Here are some problems, all concerning plants, that will test your understanding of the concepts described in this chapter.

1. In peas, the dominant allele of the A_1 gene for flower color gives red flowers; the recessive allele a_1 (when homozygous) gives white flowers. Which is more likely: two plants with red flowers producing progeny plants with white flowers, or two plants with white flowers producing progeny plants with red flowers?

2. The wild-type, dominant allele of the *ein* gene in a mustard plant produces phytochrome, and the phytochrome regulates growth so that the plants have normal-length internodes. The recessive allele is associated with elongated internodes. In a greenhouse full of normal-sized mustard plants, one plant stands higher than the rest. If you assume that the increased height of this plant is because of a mutant allele of *ein*, then do you also assume that the plant is homozygous or heterozygous for this allele?

3. The color of pea cotyledons is controlled by several genes. One gene (probably the one that Mendel studied) has a dominant allele that suppresses the green color and gives yellow cotyledons. The recessive allele gives green cotyledons when it is homozygous. Assume that this is the gene involved in the following problem: A plant that grew from a green seed is pollinated with pollen from a yellow-seeded plant. A yellow seed from this mating is grown, and the resulting plant is pollinated with pollen from a plant that grew from a green seed. When the pods from this second mating mature, what proportion of the seeds will be yellow? What proportion will be green?

4. A geneticist collected seeds from a pea plant whose genetic constitution she did not know. All the seeds had yellow cotyledons. One pea plant, which grew from one of those yellow seeds, self-pollinated; the first pod she opened had 1 green and 9 yellow seeds. She then collected 100 more seeds. Of these 100, how many did she expect would be green? Would she expect the same proportion of green seeds in a second self-pollinated plant grown from another of the original yellow seeds?

5. The flower color of sweet peas is controlled by two genes (*C* and *P*). To have red flowers, a plant must have a dominant allele of each gene (for example, *CCPP, CCPp, CcPp, CcPP*—note that the dominant alleles are capitalized). If a plant has two recessive alleles for either gene (or both) (for example, *ccpp, ccPP, CCpp, ccpP, Ccpp*), its flower will be white. Two white plants are mated, and they produce 100% red-flowered progeny. What were the genotypes of the parents?

6. In corn, the genes for red endosperm and shrunken endosperm are unlinked. The dominant allele of the red gene gives red seeds (*RR, Rr*); homozygous recessive seeds (*rr*) are yellow. The dominant allele of the shrunken gene gives normal-shaped seeds (*ShSh, Shsh*); homozygous recessive seeds (*shsh*) are shrunken. A plant that is heterozygous for the red gene and homozygous recessive for the shrunken gene is used to pollinate the silks (stamens) on a plant that is homozygous recessive for the red gene and heterozygous for the shrunken gene. What will be the genotypes and phenotypes of the seeds in the resulting cob, and in what proportions will they be expected to appear?

7. A corn plant that is heterozygous for three genes (*AaBbCc*) is crossed with a plant that is homozygous recessive for all three genes (*aabbcc*). Four phenotypes of progeny seedlings are found, in equal numbers. The four phenotypes correspond to the following combinations of alleles: *ABC, aBC, Abc, abc*. Which genes are linked, and which are unlinked?

8. In corn, the gene for red endosperm, *R*, is located on the same chromosome as the gene for white seedling, *W2*. A test cross of a plant that was heterozygous for the two genes with a homozygous recessive plant produced progeny with the following genotypes and in these numbers: *RrW2w2*, 20; *Rrw2w2*, 172; *rrW2w2*, 180; *rrw2w2*, 28. (*R* and *W2* are the dominant alleles; r and *w2* are recessive.) Draw a diagram of the homologous chromosomes in the heterozygous parent, showing which alleles were on which chromosomes (that is, show the parental combinations of alleles). What fraction of the pollen produced by the heterozygous parent was recombinant?

9. As a breeder of tomatoes, you have discovered a strain of wild tomato that is resistant to a destructive virus. The resistance is controlled by the dominant allele of a single gene (*Rv*). Unfortunately, the wild tomato has bad-tasting fruit and other undesirable traits, so you want to breed a domestic tomato with only the resistant gene from the wild strain. You cross the wild tomato with a domestic tomato (cross 1) and select resistant progeny; you cross the progeny from the first cross with the domestic tomato (cross 2) and select resistant progeny; and so on. What fraction of the progeny of cross 1 will be resistant? of cross 2? of cross 10? What fraction of the progeny of cross 1 will have alleles from the wild tomato of genes unlinked to *Rv*? of cross 2? of cross 10?

10. Several years ago, corn geneticists isolated a mutant in which anthers did not develop. The mutant was called *male-sterile*. They pollinated this mutant with wild-type pollen; all the resulting

plants were also male-sterile. They pollinated these plants with wild-type pollen; all these progeny plants were also male-sterile. At this point, the geneticists decided that the gene for male sterility was maternally inherited. Explain why they drew this conclusion.

Further Readings

Corcos, A. F. and F. V. Monaghan. 1993. *Gregor Mendel's "Experiments on Plant Hybrids": A Guided Study*. New Brunswick, N.J.: Rutgers University Press.

Mendel, G. 1959. "Experiments in Plant Hybridization." Translation in *Classic Papers in Genetics*. Edited by J. Peters. Englewood Cliffs, N.J.: Prentice-Hall.

Strickberger, M. 1985. *Genetics*. 3d ed. New York: Macmillan.

17 BIOTECHNOLOGY

1. DNA can be purified. Purified DNA can be broken into pieces that contain individual genes, and different pieces can be recombined to form new combinations of genes.

2. Recombined DNA can be inserted into plant cells. The plant cells may express the genes contained by the inserted DNA.

3. The genetic manipulation of plants can produce useful agricultural and horticultural varieties with traits that would have been extremely difficult or impossible to select by traditional breeding methods.

17.1 USEFUL MATERIALS FROM LIVING ORGANISMS

Biotechnology, a word that has been in the news only since about 1985, could mean many things. Because *bio* refers to life, and *technology* is the application of science to the creation of products for human use, biotechnology could refer to almost any branch of agriculture, to the fermentation industries, and to some food processing industries. However, new and spectacular discoveries about the structure and function of genes and enzymes have tended to overshadow more traditional fields, and the word *biotechnology* now often implies the genetic manipulation of organisms to give them new capabilities or improved characteristics. In this chapter, we will describe some of the techniques that allow genetic engineers to modify the genes of plants and some of the advantages we can expect from the new organisms.

17.2 STEPS TOWARD GENETIC ENGINEERING

DNA Can Be Purified, Recombined, and Cloned

Years of basic research and many technological breakthroughs were needed before scientists learned to manipulate genes in ways that were truly useful.

PLASMIDS The technology for the chemical manipulation of DNA began in the 1960s with the discovery of **episomes** in bacteria. Episomes are small pieces of DNA, generally not required for the survival of the bacterial cell. Sometimes episomes are incorporated into the main bacterial chromosome, but often they are separate, autonomous, circular units—minichromosomes. Some episomes are the dormant chromosomes of viruses. Other episomes carry bacterial genes that help the cell survive in unusual environments; these episomes are called **plasmids**. Some plasmids carry genes for resistance to antibiotics or to certain viruses; some carry genes for enzymes that metabolize

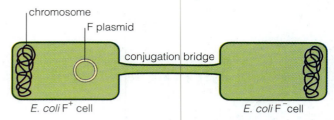

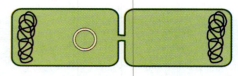

1 Conjugation begins when an F⁺ and an F⁻ cell become attached by a conjugation bridge.

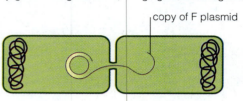

2 The conjugation bridge retracts, bringing the cells together.

3 A copy of the F plasmid is transferred to the F⁻ cell.

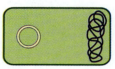

4 The DNA is duplicated and formed into a circular plasmid. Both cells are now F⁺.

Figure 17.1 The F plasmid (or fertility factor) in the bacterium *Escherichia coli*. The F plasmid has genes for constructing a conjugation bridge, through which copies of the F plasmid can be transferred to an *E. coli* cell lacking the plasmid. F⁻ denotes absence and F⁺ presence of the plasmid.

rare compounds; and some provide unusual biochemical pathways. An important characteristic of an episome is its ability to be replicated in a cell, just like the main chromosome. An example is a plasmid called pBR322, which comes from the gut bacterium *Escherichia coli*. It contains two genes—one that provides resistance to ampicillin and one that provides resistance to streptomycin—as well as a site where DNA replication can start.

Episomes are useful because they are easy to purify and work with. The main chromosome of the bacterium *E. coli* has 4.7 million base pairs and thousands of genes, but a plasmid might have only 2000 base pairs and two genes. This makes the plasmid more stable in a test tube and easier to analyze. Furthermore, circular plasmids can be transferred between bacterial cells. One mechanism for such a transfer that has been extensively analyzed involves the F plasmid, which contains genes that make a **conjugation** bridge (Fig. 17.1, see also Fig. 19.3). The conjugation bridge, a tube made from protein, connects a

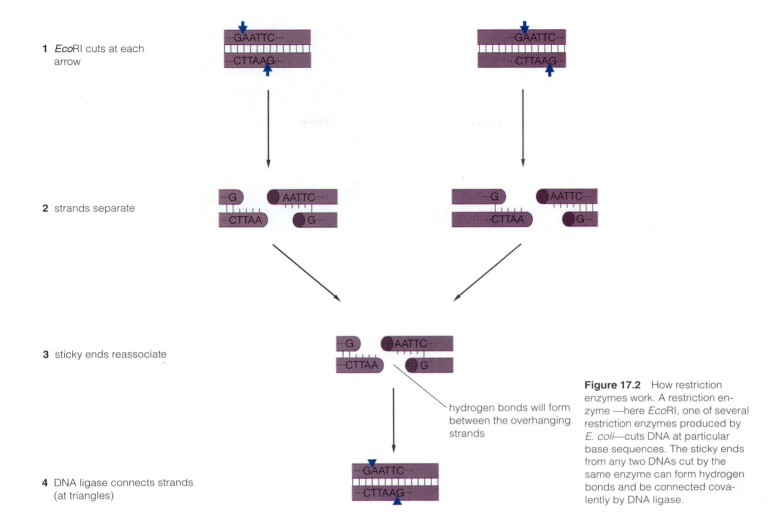

1 *Eco*RI cuts at each arrow

2 strands separate

3 sticky ends reassociate

hydrogen bonds will form between the overhanging strands

4 DNA ligase connects strands (at triangles)

Figure 17.2 How restriction enzymes work. A restriction enzyme —here *Eco*RI, one of several restriction enzymes produced by *E. coli*—cuts DNA at particular base sequences. The sticky ends from any two DNAs cut by the same enzyme can form hydrogen bonds and be connected covalently by DNA ligase.

bacterium that has the F plasmid to one that does not. A copy of the F plasmid is synthesized and transported across the conjugation bridge into the cell that originally lacked it. This means that new genetic information is provided to the recipient cell.

RECOMBINANT DNA Also in the 1960s, microbiologists discovered that bacteria contain enzymes capable of cutting DNA at specific base sequences. These enzymes are called **restriction endonucleases** or **restriction enzymes** because their function is to protect the cell by restricting invasion by foreign DNA. Cells develop various ways, chemically modifying their own DNA to make sure it is not cleaved by the restriction enzymes. Different restriction enzymes (from different species or even strains of bacteria) recognize different sequences of bases in DNA.

When isolated and purified, restriction enzymes were found to be extremely useful because they allow scientists to cut purified plasmid DNA in very specific, reproducible places (Fig. 17.2). Furthermore, the cuts can be reversed. Many restriction enzymes make cuts with *sticky ends*, at which there are overlapping regions of complementary DNA strands. At lower temperatures these ends stick together, and the DNA can be covalently connected (*ligated*) using another enzyme, DNA ligase.

Pieces of DNA from different sources can be combined because the sticky ends formed by a particular restriction enzyme all have the same base sequence, no matter what their source. These pieces will stick together and can be ligated, forming a **recombinant DNA** molecule (Fig. 17.3). If the recombination process inserts a new gene into a plasmid, and if the DNA becomes circular (and is not too large), the new gene can be taken up with the plasmid by a receptive bacterium. In this case, the plasmid is called a **vector** for the transfer of the new gene.

If a researcher adds to a test tube a mixture of pieces of DNA (plasmids and isolated copies of a gene, for instance), all cut with the same restriction enzyme, and allows them to recombine, the result will be a random mixture of DNA molecules of various sizes. Some pieces of DNA will stick their own ends together to form very small circles; two plasmids (or two genes) may combine together just as well as one plasmid with one gene; finally, three, four, or more pieces may join end to end to form really large circles. Cutting the DNAs with *two* restriction enzymes simplifies the situation. Because each molecule has sticky ends with different base sequences, single pieces cannot re-form circles. Still, much of the art in genetic engineering is in selecting the desired combination of genes—through a procedure known as *cloning*.

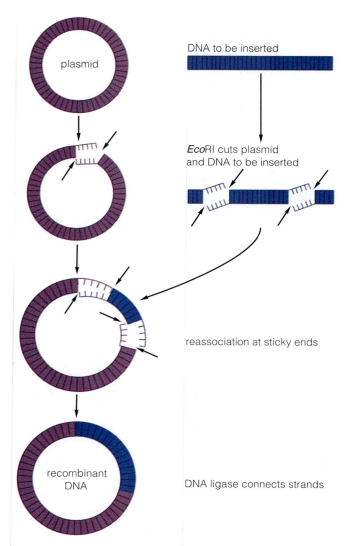

Figure 17.3 The formation of recombinant DNA. Restriction enzymes (such as *Eco*RI) can be used to insert a piece of DNA into a plasmid. If the plasmid has only one site recognized by the restriction enzyme, the foreign DNA can be inserted only there.

In the diagram:
- plasmid
- DNA to be inserted
- *Eco*RI cuts plasmid and DNA to be inserted
- reassociation at sticky ends
- recombinant DNA
- DNA ligase connects strands

CLONING A **clone** is a colony or group of cells or organisms, all of which have the same genes. **Cloning** is replication of the cells in the colony. In the context of DNA biotechnology, cloning is a simple method for separating and eventually characterizing individual molecules of DNA. It works because an individual molecule inserted into a single bacterial cell can be replicated many times as the cell divides, until the colony contains hundreds of thousands of copies of the same molecule.

To see how it works, assume that we have prepared a mixture of recombinant DNA molecules formed from a plasmid and a gene of particular interest to us. We use a special plasmid, pUC19, which has the gene for resistance to the antibiotic ampicillin and a gene that makes the enzyme β-galactosidase. We treat the plasmid with a restriction enzyme that makes one cut in the middle of the β-galactosidase gene, then add the new gene, and ligate. We combine the mixture of DNA molecules with a

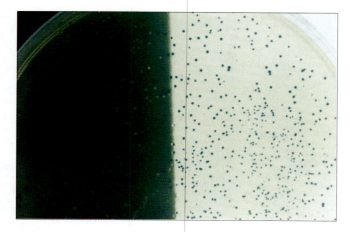

Figure 17.4 Colonies (or clones) of *E. coli* on an agar plate. The cells in the blue colonies (most visible at the right) are making β-galactosidase, which liberates the blue dye from its colorless complex with the sugar galactose. The cells in the white colonies (most visible against the black background at the left) do not make β-galactosidase because their gene for that enzyme has been interrupted by the insertion of a foreign piece of DNA.

suspension of bacterial cells in such a way that each cell will on the average take up only one DNA molecule. Only circular molecules will be taken up. Next the bacteria are spread out on a petri dish containing nutrient agar and ampicillin (Fig. 17.4). The bacteria are diluted enough so that each cell is separated from every other. Those bacteria that do not contain plasmids will be killed by the ampicillin. Those bacteria that do contain plasmids will survive and grow into individual colonies (clones). We know that each clone must contain a plasmid.

In the nutrient agar is a chemical that turns blue in the presence of β-galactosidase. Those colonies of bacteria that contain intact β-galactosidase genes will synthesize β-galactosidase and turn blue. Colonies that contain plasmids in which the gene of interest has been inserted into the middle of the β-galactosidase gene will be unable to make β-galactosidase and will remain white. We can select the bacteria containing those plasmids because it is likely that they contain the gene we want. By growing them in nutrient medium, we can have an inexhaustible supply of that particular gene.

We can also determine the exact base sequence of the DNA we are interested in, although this step is not necessary for us to reproduce the DNA, transfer it to other bacteria, or recombine it with other pieces of DNA. The basic strategy is to break the DNA at different bases, so that the size of the fragments relates to the positions of the bases (see sidebar, "Determining the Base Sequence of DNA"). It is possible to separate DNA fragments of different sizes by electrophoresis (a technique that moves chemicals through an electric field at a rate proportional to their charge and/or molecular weight), resolving fragments that differ by just one base pair. This allows us to see directly the relative positions of different bases—the base sequence.

DETERMINING THE BASE SEQUENCE OF DNA

All the information in DNA is encoded in its sequence of bases. Determining the sequence is useful because it allows researchers to tell whether the DNA could code for a protein and, if so, what the amino acid sequence of the protein would be. It allows them to compare the DNA being studied with other known DNAs to detect mutations or potential evolutionary relationships. Sequencing DNA is now a major activity of biotechnologists, and many thousands of DNA sequences are stored in computer databases.

There are two main procedures for determining DNA sequences. Both work by making DNA fragments of sizes that correspond to the positions of particular bases. We illustrate the procedure devised by F. Sanger and his colleagues, which uses DNA synthesis to make the different-sized fragments.

Assume that you have cloned a piece of DNA of unknown sequence in a vector. The cloned DNA is double-stranded; you heat the strands to separate them so that you can use one strand as a template for DNA synthesis. Then you add a short piece of DNA with a sequence complementary to the vector DNA that is next to the sequence to be determined; this short piece, the *primer*, binds to its complementary sequence in the vector DNA and serves as the starting point for DNA synthesis.

You now add this template–primer complex to four separate DNA-synthesis reaction mixtures. Each mixture contains the DNA-synthesizing enzyme DNA polymerase and activated nucleotides, which are substrates (reactants) for making the new DNA. Each mixture also contains a small amount of a special substrate that stops DNA synthesis at particular bases. One mixture stops synthesis at A bases, one at G bases, one at C bases, and one at T bases (Fig. 1a).

When you run the DNA-synthesis reaction, you will be making four sets of DNA fragments. Each set is itself a mixture of DNAs of different lengths. The mixture with the A-terminating reagent, for instance, has a mixture of DNAs. All of them started at the primer, but some of them stopped at the first A of the synthesized DNA, some at the second, some at the third, and so forth. The relative sizes of these fragments indicate the relative positions of the A bases (and their complementary T bases in the template); see Figure 1b.

The newly synthesized DNA is separated by electrophoresis. The four mixtures are placed in an artificial plastic gel and subjected to an electric field. The DNA fragments all have negative ionic charges, so they move away from the negative and toward the positive pole. The holes between the gel molecules are small, so fragments move at different speeds, with short fragments moving through the gel more quickly than long ones. Even fragments that differ by only one base in length can be distinguished. At the end of a period of electrophoresis, the positions of the different fragments are visualized. The presence of a band of a certain length in the mixture with the A-terminating reagent means that there was an A at that position. The locations of the bands that are one base longer and shorter show the bases adjacent to that A. In practice, the sequence of the DNA can be read directly off the electropherogram, as shown in Figure 1c.

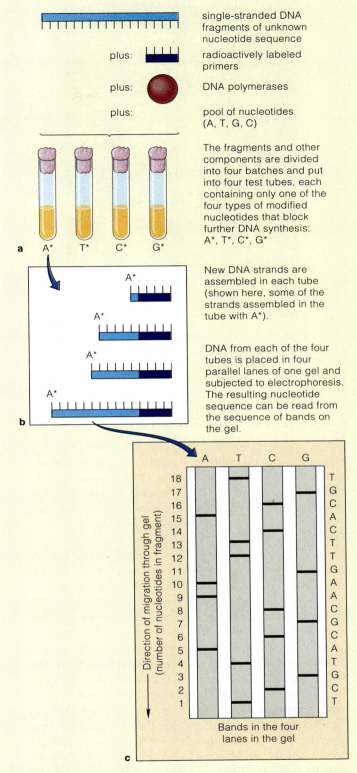

single-stranded DNA fragments of unknown nucleotide sequence

plus: radioactively labeled primers

plus: DNA polymerases

plus: pool of nucleotides (A, T, G, C)

The fragments and other components are divided into four batches and put into four test tubes, each containing only one of the four types of modified nucleotides that block further DNA synthesis: A*, T*, C*, G*

New DNA strands are assembled in each tube (shown here, some of the strands assembled in the tube with A*).

DNA from each of the four tubes is placed in four parallel lanes of one gel and subjected to electrophoresis. The resulting nucleotide sequence can be read from the sequence of bands on the gel.

Figure 1 The sequence of a section of DNA can be determined by synthesizing pieces of DNA (**a**) Each horizontal band in an electrophoretic gel, diagrammed in (**c**), represents a piece of a particular length (**b**): The closer the band is to the top of the gel, the longer the piece of DNA. The DNA in all bands starts at the same base, determined by the base sequence of the primer; the DNA in all bands in a given vertical lane ends at the same type of base (G, A, T, or C), determined by the base on the modified nucleotide. The relative positions of the bands then give the relative positions (sequence) of the different bases.

Figure 17.5 Crown galls are formed when *Agrobacterium tumefaciens* infects wounded plant tissue. The wounds often occur around the crown (area between stem and root) but can also be higher on the stem, like the gall on this walnut (*Juglans* sp.) tree. The gall tissue grows actively in the laboratory.

Scientists Use a Special Bacterium to Insert DNA into Plants

The techniques described in the previous section were developed using DNA from bacteria. DNA in plants has the same chemical structure as bacterial DNA, and it can be extracted from plant cells and manipulated in a test tube using many of the same techniques, including cloning. This means that we may be able to select and purify plant genes. It does not, however, automatically mean that we can put the genes back into plant cells in a way that allows the cells to survive and use the new genetic information.

In searching for a way to insert genes into plant cells, scientists focused on a condition called **crown gall disease**. This disease, caused by the soil bacterium *Agrobacterium tumefaciens*, results in tumors (galls) growing in places where a plant has been wounded (Fig. 17.5). The disease is called *crown gall* because plants are often wounded at the *crown*, the junction between shoot and root. *A. tumefaciens* cells attach to the walls of plant cells exposed in the wound and cause the plant cells to start dividing. The fact that the plant cells continue to divide even when the bacteria are killed with antibiotics shows that the bacteria *transform* the plant cells, apparently by turning off their normal mechanisms for limiting cell division. The result is rather like an animal cancer, and in fact much of the early research on crown gall disease was financed by the National Cancer Institute.

What mechanism do the bacteria use to transform the plant cells? For many years, biologists suspected that a transfer of genes was responsible, and in the middle 1970s the DNA that contained those genes was identified. Infectious strains of *A. tumefaciens* have a large plasmid, called the **Ti** (tumor inducing) **plasmid** (Fig. 17.6), part of which the bacteria inject into plant cells (Fig. 17.7). The region that is injected (the T-DNA) contains three main genes that cause the cells to divide and grow. Two of the genes code for enzymes that make the plant hormone auxin (indoleacetic acid); one of the genes codes for an enzyme that makes another plant hormone, a cytokinin (isopentenyl adenine). It has been known for years that an excess of auxin and cytokinin will cause plant cells, especially the parenchyma cells at the surface of a wound, to divide and grow.

Another T-DNA gene is for an enzyme that synthesizes an unusual type of amino acid called an *opine*. Opines are generally combinations of two normal amino acids (there can be different combinations, depending on the enzyme). Once synthesized in the plant cells, opines leak out into the intercellular spaces. The bacteria growing in the intercellular spaces of the tumor also make enzymes that allow them to take up and metabolize opines, using them as a source of carbon, nitrogen, and energy. Thus, the plant cells transformed by the bacteria—that is, injected with new bacterial genes—nourish their bacterial parasites. In short, *A. tumefaciens* has evolved a mechanism for conducting genetic engineering.

For human genetic engineers, this system is valuable because the T-DNA of the Ti plasmid can serve as a vector to carry genes into plant cells (Fig. 17.8). In practice, researchers start with T-DNA that has lost the genes for auxin and cytokinin synthesis, so that it does not form tumors. They insert the gene of interest, together with a gene for antibiotic resistance that allows them to select for cells that have stably incorporated the T-DNA (by treating the cells with the antibiotic). This recombinant T-DNA, generally in the form of a miniplasmid, is transferred to an *A. tumefaciens* cell that has a Ti plasmid lacking its own T-DNA. Spread on the cut surface of a piece of leaf, the bacteria transfer the recombinant T-DNA to the plant cells. The leaf is transferred to a medium containing antibiotics that kill the bacterial cells and also select for those plant cells that have incorporated the resistance gene in the T-DNA. After these steps, the plant cells can be coaxed to regenerate new plants by using standard techniques of tissue culture (Chapter 15).

There Are Other Methods for Injecting DNA into Plants

While the *A. tumefaciens* system is the favorite method of transforming plant cells, there are a few other ways to add new genetic material to these cells. One is called **biolistics**, a play on the word *ballistics*. The apparatus is also called a gene gun. DNA containing the gene of interest is adsorbed onto the surface of small (subcellular-sized) particles of gold or tungsten. These particles are pressed onto the front of a bullet. The bullet is then

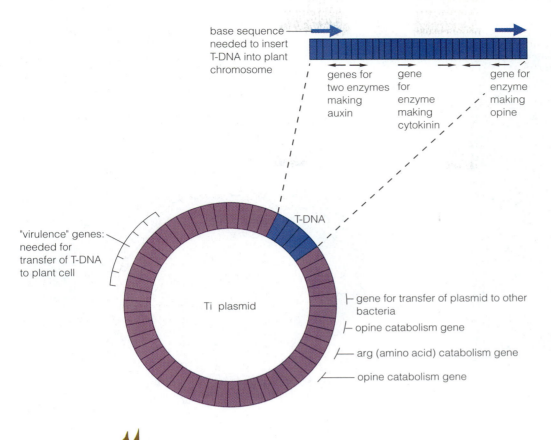

base sequence needed to insert T-DNA into plant chromosome

genes for two enzymes making auxin

gene for enzyme making cytokinin

gene for enzyme making opine

T-DNA

"virulence" genes: needed for transfer of T-DNA to plant cell

Ti plasmid

gene for transfer of plasmid to other bacteria

opine catabolism gene

arg (amino acid) catabolism gene

opine catabolism gene

Figure 17.6 The Ti plasmid from a strain of *Agrobacterium tumefaciens* showing the T-DNA that is transferred to plant cells.

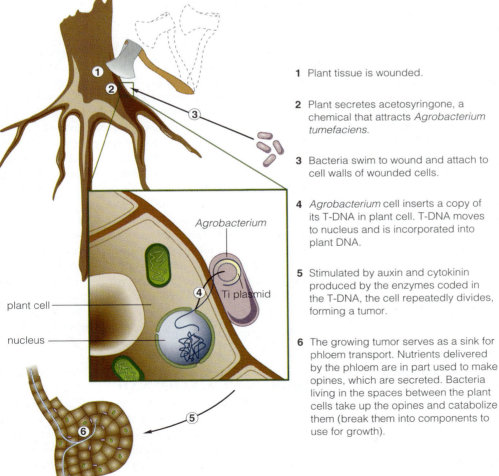

1 Plant tissue is wounded.

2 Plant secretes acetosyringone, a chemical that attracts *Agrobacterium tumefaciens*.

3 Bacteria swim to wound and attach to cell walls of wounded cells.

4 *Agrobacterium* cell inserts a copy of its T-DNA in plant cell. T-DNA moves to nucleus and is incorporated into plant DNA.

5 Stimulated by auxin and cytokinin produced by the enzymes coded in the T-DNA, the cell repeatedly divides, forming a tumor.

6 The growing tumor serves as a sink for phloem transport. Nutrients delivered by the phloem are in part used to make opines, which are secreted. Bacteria living in the spaces between the plant cells take up the opines and catabolize them (break them into components to use for growth).

Agrobacterium

Ti plasmid

plant cell

nucleus

Figure 17.7 The sequence of events that leads to a crown gall tumor.

Figure 17.8 Plants can be transformed when infection by a genetically engineered *Agrobacterium tumefaciens* Ti plasmid introduces new genetic information. This tobacco (*Nicotiana tabacum*) plant glows in the dark because the new gene that was inserted (which came from a firefly), produces the enzyme luciferase.

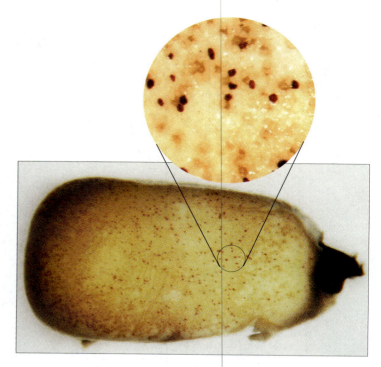

Figure 17.9 A maize (*Zea mays*) seed treated with biolistic pellets that had DNA containing an *R* gene, which turns on the formation of red anthocyanin dyes. The dark (red) spots represent cells in the aleurone layer that express the gene.

loaded into a gun, and the gun is fired at the plant tissue. Originally, a common .22 caliber gunpowder-powered bullet was used, but a pellet for a high-pressure air (or helium) pistol is also effective and avoids the toxic gases produced by the gunpowder. A special metal plate with a hole that is smaller than the cross section of the bullet abruptly stops the flight of the bullet before it reaches the tissue, but the gold or tungsten particles keep going with enough energy to penetrate the cells. They are so small that they appear not to leave holes in the cell membranes after they pass through; they also appear not to harm the cells in any other way. After the particles come to rest inside the cells, the adsorbed DNA dissolves into the cytoplasm. In at least some cells, the DNA reaches a place where it can be used as a template for RNA synthesis, and its genetic information is expressed (Fig. 17.9).

Another method for getting DNA into a plant cell is **electroporation**. This is based on the discovery that a short, high-voltage charge of electricity can produce temporary holes in a plasma membrane without permanently harming the cell. For this technique, it is necessary to make **protoplasts**—cells without walls—from the recipient plant cells. The cell walls can be removed by treatment with polysaccharide-hydrolyzing enzymes (cellulase and polygalacturonase) while the cells float in an osmotically balanced medium to prevent water from entering and breaking them open. The protoplasts are then placed between two electrodes in an ice-cold solution that contains the DNA. A few pulses of electricity, each less than 0.1 sec-

ond long, produce the membrane holes. These close up in seconds or minutes, but during this time some DNA can enter the cells. Cultured under the proper conditions, protoplasts regenerate their cell walls, begin dividing, and even regenerate whole plants—plants that may express the genes of the DNA that entered the protoplasts.

Viruses can also be used to inject genes into plants. Plant viruses (subcellular particles of nucleic acid and protein), infect plant cells, multiply, and spread throughout the plant (Chapter 19). The genes of most plant viruses are usually made of RNA rather than DNA, but genetic engineers can produce DNA copies of the viral genes and connect them to genes of interest. In the plant, the introduced genes are expressed wherever the virus spreads. This method does not produce a permanently transformed plant because the viral and introduced genes are not incorporated into the plant's own DNA, and they are not passed to seed formed by the infected plant. However, proteins made by the infected plant in response to the introduced genes can be extremely useful.

17.3 APPLICATIONS OF BIOTECHNOLOGY

Because bacteria were the first organisms to be transformed with new genes, they have been the first to be exploited as a source of new products. The most valuable have been proteins. For instance, diabetics who require

Figure 17.10 Petunia plants that are transformed with additional flower color genes produce unusual patterns, including sectors that have no pigment at all.

daily injections of insulin now get human insulin produced by bacterial cells transformed with the human gene for insulin. Previously, they received pig insulin, purified from the pancreases of slaughtered pigs; this procedure was costly, and the insulin was not as satisfactory. Other proteins having medical uses that are now produced by genetic engineering include somatotropin, erythropoietin, clotting factors, and interferon. Genetically engineered bacteria (or yeasts) also produce the enzymes added to laundry detergents. And of course, many of the restriction enzymes on which genetic engineering depends are produced in large quantities by genetically engineered bacteria.

Plants Can Serve as a Source of Biochemicals

Though bacteria can grow very rapidly, nothing can beat plants for the economical production of large quantities of protein. The most prevalent single protein in the world is thought to be ribulose bisphosphate carboxylase, the first enzyme in the C_3 cycle of photosynthesis. It might be possible to harness the metabolic capacity of genetically engineered plants to produce large quantities of a protein valuable for industrial uses.

In addition to proteins, plants synthesize oils as storage materials for their seeds. By changing the enzymes of the synthetic pathway, it may be possible to design plants that produce specialized oils or waxes. These materials could be useful in detergents, cosmetics, pharmaceuticals, and lubricants. Most biological oils are biodegradable. In the past, this has been seen as undesirable in an industrial product such as a motor oil because it means that the product has an uncertain (and possibly short) lifetime. The risk of polluting the environment and the cost of disposing of nonbiodegradable wastes, however, have emphasized the advantages of easy degradability.

Plants may be engineered to produce vaccines that are cheaper and more acceptable to people in developing countries. A vaccine contains an inactivated or weakened bacterial or viral pathogen (or a part of the pathogen). Injected into a person's body, it stimulates the immune system to produce antibodies or immune cells that will kill an active pathogen if it infects the person at a later time. Some vaccines (for instance, the Sabin polio vaccine), can be taken orally. With this in mind, scientists are designing and testing food plants that contain genes for proteins from important pathogens. Two intriguing examples are a banana (*Musa sapientum*) that makes a protein from the hepatitis B virus and an alfalfa (*Medicago sativa*) sprout that contains a part of the cholera toxin. In the future, for the cost of a few seeds and some careful farming, even an impoverished region may be able to protect its population from serious lethal diseases through such vaccination.

Gene Manipulations Can Produce Many Types of Useful New Plants

NEW HORTICULTURAL VARIETIES Transforming plants with new or altered genes may yield plants with new traits that are useful or attractive. Much is known about the genes for the enzymes that make anthocyanins—the red, blue, and purple pigments found in plant flowers as well as other organs. Plant molecular biologists have produced plants that have additional genes for enzymes in the anthocyanin pathway and that, as a result, have flowers with unusual colors or patterns (Fig. 17.10). There is hope of producing a blue rose, something not found in searches among wild varieties and random mutants.

MORE PEST RESISTANCE Predation by insects and damage caused by viral, bacterial, and fungal diseases are among the major factors that limit the productive harvest from food crops. Classical plant breeding has traditionally selected varieties that resist these pests, and with ge-

Figure 17.11 Plants can be transformed with genes that provide resistance to insects and viruses, as well as bacteria and fungi. (**a**) Caterpillars dine on a cotton (*Gossypium hirsutum*) plant, but (**b**) not if it is genetically modified to resist attack. (**c**) A normal potato (*Solanum tuberosum*) plant is susceptible to viral attack, but (**d**) a genetically modified strain is resistant.

netic mating techniques it has been possible to combine resistance genes with genes for high-quality vegetables, cereal grains, and fruit. But classical genetic techniques are inefficient, requiring many cycles of back crossing and selection. It seems likely that, as we identify the genes for resistance, we will be able to transfer them quickly and efficiently using modern molecular techniques (Fig. 17.11). For instance, seeds of common beans produce a protein that blocks the digestion of starch by two insect pests, cowpea weevil and Azuki-bean weevil. The gene for this protein has now been transferred to the garden pea. In tests, it protected stored pea seeds from infestation by these insects, killing or slowing the development of the larvae.

Furthermore, we can identify methods of resistance that plants never evolved. *Bacillus thuringiensis* is a bacterium that makes a protein toxin that kills insects. Scientists have already inserted the gene for *B. thuringiensis* toxin into some crop plants. These plants synthesize the toxin and kill many of the insects that graze on them. Tobacco mosaic virus infects crops in the plant family Solanaceae (tomato, potato, eggplant, green pepper). In-

serting the gene for the viral coat protein into the plant genome seems, for unknown reasons, to make the plant resistant to infection by the virus. Scientists in Mexico have used the same technique to produce potato varieties immune to potato viruses X and Y, and in Hawaii a genetically engineered papaya resistant to a ringspot virus is being tested. Protecting agricultural plants against viral infection increases the yields of fruits or other useful crops that can be harvested from those fruits.

Protecting plants against competition from weeds takes a different strategy. Some of the first to be engineered have been plants resistant to herbicides. Tomatoes and soybeans that are resistant to glyphosate (trade name Roundup) have already been produced and tested, and similar cotton and canola plants are in the process of development. A resistant crop allows farmers to use herbicides to kill weeds in the middle of a field of crop plants. The hope is that this technique will allow a more flexible use of the safer herbicides.

IMPROVED QUALITY OF FRUIT AFTER HARVEST In many countries, a large proportion of harvested crops never reaches consumers because it spoils in route to the market. In America, a great deal of the energy and cost associated with food production results from the complications of getting food to market in edible and attractive form. Controlling the atmosphere (limiting the amounts of oxygen and ethylene) and reducing the temperature are two expensive techniques for preventing certain fruits from spoiling. Many fruits are picked green, stored, and then artificially ripened with an ethylene treatment. How much more effective and efficient it would be if we could insert genes that slowed the rate of senescence (aging) and thus slowed spoilage. One of the enzymes needed for the softening of tomatoes as they rot is polygalacturonase. Scientists at Calgene (Davis, California) have inserted into tomato plants a gene that blocks the synthesis of polygalacturonase and thus delays the senescence of the fruit. This FlavrSavr tomato (Fig. 17.12), which can be shipped ripe, is considered to taste better than fruits that must be shipped green.

IMPROVED TOLERANCE TO ENVIRONMENTAL STRESS One of the most difficult challenges is to produce plants that grow better under stressful conditions. We know that certain plant varieties have genetic information that allows them to survive saline conditions, droughts, high light intensities, or chilling temperatures. In some cases, we even know how the resistance works, in a general way. But very seldom do we understand what genes or proteins are involved. Often, stress resistance depends on several genes, which complicates the analysis. Much current research is directed toward identifying the genes that differ between stress-tolerant and stress-sensitive varieties. There is hope that it will be possible to breed varieties of plants for a wider range of climate and soil conditions.

Figure 17.12 The FlavrSavr tomato (*Lycopersicon esculentum*), the first genetically engineered fruit to be marketed. Because it lacks the enzyme polygalacturonase, it does not ripen (soften) as rapidly.

17.4 IS BIOTECHNOLOGY DANGEROUS?

In 1992, a group of 1000 chefs signed a pact promising not to use any genetically engineered foods in their cooking. What were they or their customers worried about? It is true that new technologies may cause problems. If the technologies are widely and carelessly applied, these problems can become serious. The problem of disposing of radioactive and toxic chemical wastes is one example. In part, the worry reflects general antitechnology feelings. Much of the controversy about genetically engineered plants reflects the public perception that little is known about genes and especially about the genes that are "unnaturally" introduced into organisms.

One counterargument points out that crops have been genetically manipulated for centuries—by breeding and selection. Familiarity and our positive experiences have made the use of new plant varieties acceptable. Few people realize how little is known about the genetic composi-

tion of new crop varieties that have been developed through random mutagenesis (production of mutants) or crossbreeding with wild types. We may be able to say that we know more about changes in the genes of a crop modified by genetic engineering than about changes in the genes of one developed through classical breeding.

In the early days of genetic engineering, when little was known about engineered organisms, scientists themselves placed regulations and limits on different types of experiments involving recombinant DNA. Different types of experiments required different levels of care. For instance, experiments with cancer viruses and pathogens that infect mammals required (and still require) stringent containment procedures, whereas simple experiments cloning genes in a genetically weakened and well-understood strain of bacteria could be performed in a regular laboratory with simple aseptic procedures. The manipulation of plant genomes is considered a very low-risk experiment. The marketing of engineered plants has more risks; but these have been examined, and ways for eliminating uncertainties have been identified. The scientific issues to be evaluated in the approval of a genetically engineered food are: (a) Does the product contain any new allergic material that might affect especially sensitive groups? (b) Are new toxic compounds introduced into the food supply, or are existing toxins raised to unacceptable levels? (c) Are nutrient levels adversely affected? (d) Will the use of genes for antibiotic resistance (used to indicate when a plant has been stably transformed) compromise the use of important therapeutic drugs? If the answers to these questions are all negative, the food is considered to be safe.

A second type of problem involves the possibility that new genes from the desired recipient crop species could be transferred to a related wild, weedy species. This is a concern when the new gene confers protection against natural pests or chemical herbicides. Research is being conducted to determine the probability of genes being transferred between crop and wild species and the degree of danger inherent in the escape of pest-resistance genes.

The problems involved with plant biotechnology should be balanced against the advantages. Biotechnology provides an excellent example of information-intensive solutions replacing energy-intensive solutions to problems. Planting crops that resist pests takes much less energy (in terms of fuel, chemicals, and labor) than applying insecticides. Designing crops to avoid spoilage reduces the waste of energy (fertilizers, labor, and fuel for both farm operations and distribution) that occurs when spoiled crops are discarded. For this reason, it is not surprising that some developing nations are actively conducting biotechnological research and aggressively moving to adopt the results to increase and stabilize their food supplies. In this growing field, research is the key. It is certain that the more we understand about plant and animal physiology and ecology, the more safely and effectively we can use biotechnology to improve our lives.

SUMMARY

1. Biotechnology is the application of biology to the creation of products for human use, including particularly the genetic manipulation of organisms to give them new capabilities or improved characteristics.

2. DNA can be manipulated through chemical and biochemical techniques. This allows a scientist to isolate, identify, and reproduce (through cloning) a gene.

3. Genes can be inserted into plant cells by using the Ti plasmid from *Agrobacterium tumefaciens*, a parasitic bacterium that naturally transfers DNA to plant cells, or by using physical techniques, including biolistics and electroporation. In either case, if the inserted genes include one that is selectable (for instance, that provides resistance to a toxic antibiotic), it is possible to find cells that have stably incorporated the genes into their own chromosomes.

4. The techniques of molecular biology and genetic engineering are used to produce bacteria and fungi that synthesize new pharmaceutical drugs and industrial compounds.

5. New varieties of plants produced by genetic engineering can serve as a source of biochemicals—for instance, proteins or lipids useful in industrial processes.

6. Changing the genes of plants can produce new horticultural varieties with attractive and unusual flowers.

7. Genetic engineering may produce pest-resistant crops more efficiently than classical genetic methods. And by creating plants resistant to herbicides, genetic engineering may allow chemical methods of weed control to be used more safely.

8. Certain genes may be altered by genetic engineering to produce fruit that maintains its quality longer after harvest.

9. Genetic engineers are working to identify genes that will give plants improved tolerance to environmental stresses such as salt, heat, cold, or drought.

10. Although genetic engineering, and especially the impending appearance of genetically engineered foodstuffs on the market, have engendered concern among members of the public, genetic engineering of plants involves little more hazard than does classical plant breeding.

Questions

1. A mutant *E. coli* bacterium without a plasmid cannot grow on a medium containing the antibiotic ampicillin or streptomycin; it dies. When the same strain of *E. coli* is infected with the pBR322 plasmid, it can grow on ampicillin or on streptomycin. Inserting a new gene into the middle of the ampicillin-resistance or the streptomycin-resistance gene (in the pBR322 plasmid) inactivates that gene. Assume that Strain A of *E. coli* contains a pBR322 plasmid that has a piece of DNA inserted into the ampicillin gene and that Strain B contains a different pBR322 plasmid that has a piece of DNA inserted into the streptomycin gene. Complete the table below.

	Grows on ampicillin	Grows on streptomycin
Bacterium without pBR322 plasmid	No	No
Bacterium with normal pBR322 plasmid	Yes	Yes
Strain A	?	?
Strain B	?	?

If you had a suspension that contained a mixture of Strain A and Strain B, how could you select and grow a culture of Strain A cells?

2. Some investigators have found DNA associated with ancient plant fossils. They suggest that this DNA comes from the cells of the original plants. (Alternatively, the DNA might come from ancient bacteria or fungi or from a modern contaminant.) In order to analyze the DNA, they must have large quantities of individual pieces, which they can obtain by cloning the pieces. Describe the process by which they might obtain these quantities.

3. Match the following terms and their definitions:

polygalacturonase	piece of bacterial DNA independent of the chromosome
anthocyanin	plasmid used to carry a gene into a cell
Agrobacterium tumefaciens	enzyme that digests plant cell walls
cloning	enzyme that cuts DNA at specific sites
restriction enzyme	technique for making plasma membrane permeable to DNA
plasmid	technique for isolating a single strain of bacterium or sequence of DNA
electroporation	colored compound in a plant cell
vector	bacterial strain used to transfer plant cells

4. Outline the arguments for and against the following projects:

a. Producing a bacterium that could live in the human gut and break down cellulose to the sugar glucose. Such a bacterium might allow humans to derive more nutrition from what is now indigestible plant material.

b. Producing a strain of bacterium that could fix nitrogen symbiotically with *any* species of plant (as *Rhizobium* does with legumes).

c. Producing a strain of wheat, the kernels of which produce a toxic compound that kills mildew fungi.

d. Producing a banana that makes a protein from the hepatitis B virus.

Further Readings

Gasser, C. S., and R. T. Fraley. 1992. "Transgenic Crops." *Scientific American* 266 (June): 62–69.

Joyce, G. F. 1992. "Directed Molecular Evolution." *Scientific American* 267 (December): 90–97.

"Planted Out." 1993. *Economist* 327 (June 12): 93–94.

Walden, R. 1988. *Genetic Transformation in Plants*. Milton Keynes, U.K.: Open University Press.

18 EVOLUTION, SYSTEMATICS, AND PHYLOGENY

1. Taxonomists classify organisms into a hierarchy of taxa (species, genera, families, and so on) based on combinations of traits. A major goal is to group organisms on the basis of evolutionary relationships.

2. Life evolves because the DNA that passes from one generation to the next is altered by mutation and recombination —and because environmental interactions allow some organisms to reproduce more effectively than others (the principle of natural selection). Species can vanish by extinction or can multiply by dividing into populations that evolve separately.

3. Cladistic methods use computers to deduce evolutionary relations among organisms. Scores are assigned to many traits, and computer programs seek paths of evolution that could lead to the modern array of species with the least number of changes. As traits, these studies use both body form (morphology) and molecular structures.

4. Most biologists accept five kingdoms of life: the Monera (prokaryotes), three phylogenetic kingdoms of eukaryotes (Animalia, Plantae, and Fungi), and a fourth artificial kingdom of eukaryotes (the Protista). Molecular studies suggest that we need a taxon higher than the kingdom—the domain. There are two prokaryotic domains (the Bacteria and the Archaea) and one eukaryotic domain (the Eucarya).

18.1 THE DIVERSITY OF LIFE

Ranging from oak trees to bacteria and from whales to mushrooms, life takes on a colossal number of forms. As many as 10 million species may live on Earth, and fossils suggest that a hundred times that many species have arisen, flourished, and died out in the 3.5 billion years that life has occupied this planet.

How did such great diversity in life-forms come into being? How can we sort the many forms of life so that we can study them? Where did all the species come from? Have they changed, and will they change again? Did they descend from common ancestors? If they did, how are modern species related?

How long humans have been grappling with these problems is not known, but written records show that it's been at least since the ancient Greeks, around 400–300 B.C. The fields of study dealing with these issues are taxonomy, evolution, systematics, and phylogeny. **Taxonomy**, the science of classification, has given us an excellent system of naming different groups and the species they contain. Every biologist must be acquainted with it in order to understand the scientific literature and communicate with colleagues. **Evolution** is the process by which life-forms change and new forms come into being. **Systematics** is the effort to find how modern organisms are related. Systematists try to deduce the history of change that led to modern forms of life. The evolutionary steps that led from ancient

Figure 18.1 This California field contains many populations, including grasses, wildflowers, and insects. Each population is a separate species.

to modern life-forms are called **phylogeny**, a term that means "the origin of groups." This chapter introduces all four fields of study, beginning with a concept that is fundamental to all the fields: the *population*.

18.2 THE PROBLEM OF THE SPECIES

A **population** consists of all the organisms in a given geographic area that belong to the same species. Most areas contain populations of many species. For example, the field shown in Figure 18.1 contains populations of wildflowers, insects, and so on. To define a population, one must distinguish among different species. But what is a species? No definition is entirely satisfactory, yet biologists use the term every day. Roughly speaking, a **species** is a group of organisms that have a combination of traits not found in other organisms. For example, oranges are the fruit of a species called *Citrus sinensis*; lemons are fruits of the species *Citrus limon*; and grapefruits come from the species *Citrus paradisi*. Few people would mistake one of these fruits for the other; they differ in characteristic ways, as do the trees that bear them.

With the *Citrus* species as examples, we can discuss the difficulty in defining the term *species*. Let's begin with a practical problem: Given an organism, how can we tell whether it belongs to a particular species? The surest method is to compare the organism with a type specimen at a museum, botanical garden, or university. A **type specimen** may be a sample that was placed on file when the species was first named, or it may be another specimen that was established by comparing it with the original type specimen. But when you make comparisons, you can expect your specimen to differ from the type specimen in some ways. How much can organisms differ without being viewed as separate species? The answer depends on how much natural variation occurs within the species. Differences occur with age, and the course of development can be altered by environmental influences and differences in genes. For example, immature and mature oranges differ in color, size, and taste. Consequently, a single type specimen is not enough to define a species. Many specimens are needed to show the natural range of variation.

The *Citrus* example illustrates yet another problem in defining species: The boundaries we draw will depend on the traits we use for comparison. Oranges, lemons, and grapefruits have many traits in common: a leathery rind that secretes pungent oils; segmented fruit in which the flesh consists of swollen hairs; and so on. The trees, too, are similar in many ways. Why, then, don't we view all of these plants as members of the same species? The answer is that biologists take many other traits into account—traits that differ among orange, lemon, and grapefruit trees. For example, lemons and oranges and grapefruits have characteristic differences in flavor. To acknowledge the differences, biologists assign these plants to three different species. At the same time, biologists acknowledge the similarities by grouping all these species into a higher category called a *genus* (*Citrus*). We'll say more about the *genus* concept later. For now, the point is that when we define a species by a combination of traits, the definition will depend on which traits are chosen.

For sexual species, a **mating test** resolves most of the difficulty in defining species. To make the test, two populations are brought together under natural conditions. If they mate and produce fertile progeny, they belong to the same species. If they coexist in the same area without interbreeding, they are different species. If all organisms reproduced sexually, the species concept could be defined in terms of the mating test. However, many species reproduce only by asexual means. For example, biologists recognize some 14,000 species of asexual fungi. These species can be defined only by combinations of traits that do not occur in other organisms. Furthermore, even in sexual species the mating test does not always give clear results. Two populations may interbreed to a limited extent, producing offspring that are weakly fertile. This is true, for example, of the California wildflowers called *tidy tips* (genus *Layia*), in which matings between populations yield offspring of which only 0.5 to 30% are fertile. How fertile must they be to pass the mating test? The issue is debatable.

Because of such problems, taxonomists sometimes argue about boundaries between species. Nevertheless, most species are clearly distinct entities that pass on their unique traits from one generation to the next. They make a solid foundation for taxonomy and systematics.

18.3 TAXONOMY: GROUPING SPECIES IN A HIERARCHY

People have always recognized that certain species are variations on a common theme. *Citrus* is one example, and there are thousands of others (oaks, pines, and so on). Since ancient times, people have had common names for the most familiar groups. Unfortunately, common names vary so that a given name may refer to different species in different locations. For scientific communication, we need a set of names that everyone accepts, as well as a formal system for assigning new names. This is the province of taxonomy, the science of classification.

Taxonomy began with ancient Greek naturalists, but the era of modern taxonomy began two and a half centuries ago with a Swedish botanist named Carl von Linné (1707–1778), who is better known as Carolus Linnaeus because he chose to apply the strategy he devised for naming organisms to his own name. He worked at a time when oceangoing ships were bringing many exotic species to Europe, and scientists were fascinated with the diversity of life. Many exotic plants were established in botanical gardens, such as Kew Gardens (which still exists; see Chapter 25 sidebar) in London. Working in the great gardens, Linnaeus invented a simple, reliable way to name organisms—a method that became the foundation for all taxonomy. He paid special attention to the structure of flowers, which have many well-defined traits that vary from one species to another. Vegetative parts such as leaves have fewer traits and are more subject to environmental change.

In 1753 Linnaeus published a book, *Species Plantarum*, in which he named about 6000 species and assigned them to 1000 groups called *genera* (singular, *genus*). A **genus** is a group of species that are similar enough to be obviously related, as in the *Citrus* example (oranges, lemons, and so on). Many of the genera and species that Linnaeus defined are still accepted today.

For each species, Linnaeus wrote a short descriptive phrase in Latin. He regarded this phrase as the formal name, but for convenience he wrote a single word in the margin that could be combined with the genus name to provide an abbreviated name.

Not surprisingly, taxonomists favored the two-word abbreviated name, or **binomial**. Today every species is given a binomial, or **species name**, which is always underlined or printed in italics. The first word (always capitalized) is the genus. The second word is the **specific epithet**. More details of the naming process are given in the sidebar, "How to Name a New Species," on page 294.

The binomial system of Linnaeus has only these two levels of classification. But genera also share important traits and can be grouped into sets. Genera with similar traits make up a **family**. For instance, all the plants that have roselike flowers constitute the family Rosaceae. This family includes garden roses, cherry trees, almond trees, and many others. In turn, families can be grouped into **orders**, orders into **classes**, classes into **divisions**, and divisions into **kingdoms**. Some biologists take a further step and group kingdoms into **domains**. Table 18.1 illustrates these main categories, using the sweet pea (*Lathyrus odoratus*) as an example. Note that plant taxonomists have agreed on a standard set of letters to end the

HOW TO NAME
A NEW SPECIES

Few thrills in science can match the discovery and naming of a new species. On a trip to a tropical rain forest, you could easily stumble over many new species or even new genera (Fig. 1). Millions of hectares of dense forest have never been visited by any scientist, and the species diversity is higher than anywhere else on Earth. It is harder to find new species in areas such as the United States, which have been more thoroughly explored. But even here, new species are named every year.

Holding a strange organism in your hand, how can you tell if it's a known species? The first step is to consult a handbook called a **key**. You can see the general nature of a key by examining the key to fruit types in Chapter 14. A key gives a series of choices that narrows down the possibilities. Each choice lists two opposing traits, which you judge by examining the specimen. For example, one trait might be, "Has flower parts in groups of three," and the alternative choice might be, "Flower parts not in groups of three." Each answer directs you to another choice, and so on, until you arrive at a single species. A comprehensive key for all plants would occupy several volumes. Field keys are smaller and less complete. They may stop at the family level, for example; or they may cover just one division, such as the conifers.

It's easy to make errors in consulting a key because some traits are hard to judge. So if the key suggests that your specimen might be a known species, the next step is to compare it with a type specimen. Institutions called **herbaria** keep preserved type specimens of plant species. In addition, living specimens of many species are kept in **botanical gardens**. People often bring or send plants to herbaria for identification. (Preserved specimens of animals are kept in bestiaries; living animals are kept in zoological gardens.)

If your work with keys fails to match the specimen with a known species, you need to consult an expert taxonomist to

Figure 1   Some species in this view of the Amazon rain forest are probably unknown to science.

be sure that you have a new species. You have probably been able to identify the specimen's family, and you could contact a university to find a taxonomist who specializes in that group.

If it really is a new species, you have the privilege of naming it, subject to international rules. For plants, the rules were set up at the International Botanical Congress of 1930 and are periodically revised. You must give the plant a binomial name, in which you assign the species to the genus that fits it best. You can choose the specific epithet at will. You must publish a descriptive paragraph in Latin, and you must present specimens to a herbarium for storage.

The formal, complete name of the species consists of the genus name, the specific epithet, and the name of the person who first described the species. When scientists publish research articles on a given species, the discoverer's name is usually indicated by an initial or abbreviation after the binomial. For example, in the name *Lathyrus odoratus* L., the *L* at the end signifies that Linnaeus first named the species.

names at most levels. The kingdoms are described in Chapter 1, and the divisions discussed in this book are listed in Table 18.2.

Although the levels named above may seem sufficient to an unpracticed eye, taxonomists need extra levels to fit the multitudes of living species. For example, two or more families may be grouped into a superfamily, or a single family may be divided into several subfamilies. A species can be divided into subspecies, varieties, subvarieties, cultivars, races, and forms. Subdivisions below the species level are especially important in cultivated plants, where breeding programs have led to varied forms within a species. For example, *Prunus persica* variety (var.) *persica* is the peach, while *Prunus persica* var. *nectarina* is the nectarine.

In discussing taxonomy and systematics, there is often a need to speak of groups in general. Here the term **taxon** (plural, **taxa**) is useful. For instance, a species is a taxon, a kingdom is another taxon, and so on.

18.4   EVOLUTION: THE REASON FOR LIFE'S DIVERSITY

As taxonomists named millions of species, they confirmed what we all know from everyday experience: Life is immensely diverse. How did such diversity arise? Through most of recorded history, people attributed life's diversity to a divine plan of creation. Supernatural powers were thought to have established an ideal, unchang-

Table 18.1	Classification of the Sweet Pea	
Classification Level	Name	Ending
Specific epithet	*odoratus*	*
Genus	*Lathyrus*	*
Family	Fabaceae	-aceae
Order	Rosales	-ales
Class	Magnoliopsida	-ida
Division (phylum)	Magnoliophyta	-phyta
Kingdom	Plantae	*
Domain	Eucarya	-a

*Names at this level do not have a consistent ending.

ing pattern or body type for each species. The belief was that growing organisms strive to match the pattern and they differ because the physical world is imperfect. The taxonomist's goal was to deduce the perfect pattern by examining the common features of individuals.

This way of thinking began to fall out of favor in the 19th century as scientists explored **fossils** (preserved forms or impressions of leaves, shells, bones, and other evidence of organisms that are embedded in rocks). Observing the way rocks form today, scientists concluded that the more deeply buried fossils represented earlier forms of life. Those forms differed greatly from modern forms, challenging the view that each species is unchanging. The difference between dinosaurs and modern reptiles and birds is a familiar example, but equally great changes have taken place in the world of plants. For in-

Table 18.2	Kingdoms and Divisions Covered in This Book
Taxon	Example or Common Name
Kingdom Monera	Bacteria
Kingdom Fungi	
Division Zygomycota	Bread molds
Division Ascomycota	Sac fungi
Division Basidiomycota	Club fungi
Division Deuteromycota	Imperfect fungi
Kingdom Protista	
Division Chytridiomycota	Chytrids
Division Myxomycota	Plasmodial slime molds
Division Acrasiomycota	Cellular slime molds
Division Oomycota	Egg fungi; water molds
Division Bacillariophyta	Diatoms
Division Pyrrhophyta	Dinoflagellates; dancing algae
Division Chlorophyta	Green algae
Division Rhodophyta	Red algae
Division Phaeophyta	Brown algae
Kingdom Plantae	
Division Bryophyta	Liverworts, hornworts, mosses
Division Psilophyta	Whisk ferns, genus *Psilotum*
Division Lycophyta	Club mosses
Division Sphenophyta	Horsetails
Division Pterophyta	Ferns
Division Pinophyta	Conifers
Division Cycadophyta	Cycads
Division Ginkgophyta	Maidenhair tree, genus *Ginkgo*
Division Gnetophyta	Mormon tea and others
Division Magnoliophyta	Flowering plants

stance, all the modern marsh plants called horsetails (genus *Equisetum*; see Chapter 23) are a meter (3 ft) or less in height. But 300 million years ago, there were tree-sized horsetails. Coal is partly made of their fossilized remains. Still farther back, 600 million years ago, there were no land plants at all. The only photosynthetic organisms were bacteria (Chapter 19) and algae (Chapter 21).

By the 1850s, fossil studies had built up a convincing impression that life's history emphasized change rather than constancy. In consequence, many scientists were ready to believe that the hereditary characteristics of species could change, or evolve, over the course of many generations. All that scientists needed was a clear idea of how such evolution might occur. As chance would have it, two English naturalists (Charles Darwin and Alfred Wallace) came up with the required idea at about the same time. Darwin's ideas began to take shape during a trip around the world but took many years to mature. Meanwhile, Wallace built his ideas in the rain forests of Southeast Asia. A letter from Wallace prompted Darwin to publish his work, which came out in 1859 and was called *The Origin of Species*. In it, Darwin proposed a mechanism based on four tenets:

1. Variation exists among the organisms in a population.

2. Some of the variations are hereditary.

3. Populations produce more offspring than the environment can support.

4. The members of the population best adapted to the environment will leave the most offspring.

Working together for many generations, Darwin's four factors could gradually alter the genetic makeup of a population and could ultimately lead to new forms of life.

Darwin's work showed that there is no ideal type for each species. Members of a species resemble one another because they inherit body-making genes from a common ancestor. They differ because heredity is subject to change. The quest for perfect patterns was replaced by a search for the paths of evolution that led to modern forms.

Darwin's book was immensely influential because it presented a detailed, careful argument with much supporting evidence. In a few decades, most of the scientific community adopted his view. Today, no principle in biology is more important than evolution.

Variations in Heredity Arise from Mutations and Recombination of Genes

Darwin guessed that changes in heredity occur at random; they do not arise to fit the needs of an organism. He also presumed that new traits are faithfully passed on to succeeding generations. Darwin's ideas were bold speculations, for little was known of heredity in his day. Noth-

Normal   Mutant

Figure 18.2 Normal and mutant flowers of the plant *Arabidopsis thaliana*, which is widely used in research on plant gene control. The mutation was produced by treating seeds with EMS.

ing was known of DNA or meiosis, and Mendel's studies were still unpublished. But Darwin was right. We know today that DNA carries the information guiding heredity. DNA is copied and passed on to offspring, but it incurs changes often and randomly. When such changes occur in the DNA of an organism's reproductive cells, and the organism reproduces, the altered DNA may produce new characteristics in the mutant offspring.

Many chemicals and environmental radiations are **mutagens**; that is, they cause changes in the sequence of bases in DNA. Some chemicals, such as nitrous acid, can alter the bases in DNA molecules. Others interfere with the enzymes that copy DNA, causing errors when cells replicate their DNA. Still others resemble normal bases enough to be incorporated into DNA, but their presence causes copying errors. Chemical mutagens occur in polluted air and soil, as well as in natural foods. Many cancer-producing compounds act by altering DNA, so that cells lose the hereditary control of growth and division. Today, many plant biologists use the chemical mutagen ethyl methane sulfonate (EMS) to induce mutations for scientific study. The mutant plants may have abnormal responses to the environment—or abnormal flowers, stems, and other parts (Fig. 18.2). Such artificial mutants are very useful in exploring the way genes normally control plant development. They may ultimately lead to improved crop plants.

Among radiations, ultraviolet (UV) light is especially important because cells at the body surface are exposed to sunlight and absorb its UV wavelengths. UV light causes bonds to form between DNA bases, and the organism's effort to repair the damage may lead to changes in base sequence. Such errors are more likely in people who are exposed to much sunlight, and they can lead to skin cancer. Other mutagenic radiations include medical

X rays, cosmic rays from outer space, and high-energy particles that are released when radioactive atoms decay in the body. These radiations can alter DNA bases, and they pack enough energy to break DNA molecules. The broken pieces can rejoin in wrong combinations.

Besides mutations, recombination is another important means by which DNA changes naturally, but it requires DNA from a second organism. There are several ways in which DNA from one organism can get into another organism, creating new combinations of hereditary information. Viruses sometimes carry DNA from one host organism to another. However, sexual reproduction is the most common source of recombination.

Sexual reproduction usually brings together DNA from two parents with many small differences. When the offspring make reproductive cells by meiosis (Chapter 12), parts of chromosomes may be exchanged by crossovers (see Fig. 16.4). Further, during meiosis the pairs of chromosomes line up and move to the poles without regard to parental origin. Therefore, each gamete has genes from both of the organism's parents (Fig. 18.3). The combinations vary widely among gametes, so offspring differ from one another and from both parents.

Endosymbiosis May Join DNA from Very Different Organisms

In theory, a species could also evolve by assimilating the DNA from a very distantly related species. Strong barriers usually prevent organisms of very different kinds from merging their DNA; however, exceptions can occur when cells of one species become permanent residents inside cells of another species. Such residency is called **endosymbiosis** and can be witnessed among modern organisms, such as the unicellular green algae that live inside cells of the giant clam and certain bacteria that become permanent residents inside host cells. In these cases, each organism retains its genetic identity. But a prolonged association of this kind might lead to a transfer of genetic information between the organisms, as is believed to have happened with mitochondria and chloroplasts.

Much evidence suggests that mitochondria and chloroplasts arose through endosymbiotic associations between bacteria and primitive eukaryotic cells, between 1.5 and 2 billion years ago. In the case of mitochondria, a respiring bacterium either invaded or was engulfed by a eukaryotic host cell. In the case of chloroplasts, a photosynthetic cyanobacterium was engulfed by a eukaryotic host. The resident bacteria are believed to have multiplied so that all the descendants of the host cell also contained resident bacteria. Eventually, most of the DNA from the resident bacteria was transferred to the host cell's nucleus, and the genes were gradually modified so that the bacteria evolved into the forms we now recognize as mitochondria and chloroplasts.

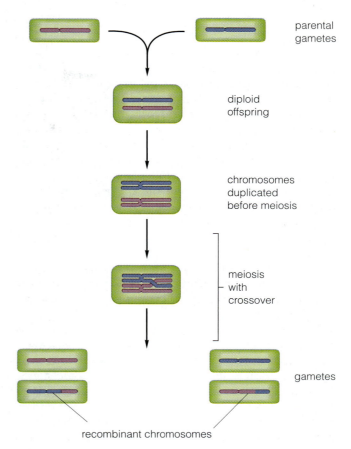

Figure 18.3 Recombination through sexual reproduction. Color is used to distinguish the chromosomes of the two parents. For simplicity, only one kind of chromosome in a single cell is shown. Genes from separate individuals come together in the offspring. When an offspring makes gametes, meiosis gives each gamete a combination of genes from the two parents.

Molecular evidence supports this contention. Both organelles contain their own DNA and use its information to synthesize their own ribosomes. Other ribosomes in the cell are made in the cytoplasm, using information from nuclear DNA. The ribosomes in mitochondria and chloroplasts resemble those of bacteria more than they resemble cytoplasmic ribosomes. The easiest way to explain these findings is to assume that mitochondria and chloroplasts once were independent bacteria.

By merging DNA from different sources, endosymbiotic associations can open the way to major innovations and great increases in the diversity of life. Mitochondria and chloroplasts demonstrate this point. Mitochondria gave eukaryotes the respiratory energy to become large and complex, and chloroplasts allowed some eukaryotes to evolve into the plant kingdom and giant algae.

The Environment Guides Evolution by Natural Selection

Although changes in DNA provide new information for evolution, the environment also plays an important role by determining which information will pass to future generations. Every organism has traits that make it easier to survive and reproduce in a particular environment. Those traits are **adaptations** to that environment. For example, winter-deciduous trees drop their leaves in autumn. This avoids the damage that would occur if leaves were to freeze while still attached. In a warmer climate, the ability to drop all leaves at a certain time offers no advantage. In fact, the tree in its leafless stage would be unable to compete with surrounding plants.

Biologists began to accept the concept of evolution when Darwin showed how a species might acquire adaptive traits. Darwin proposed that the environment plays a key role, which he compared to the artificial selection that a farmer practices when improving livestock or crop plants: Individuals vary, and those with desirable traits are allowed to reproduce while others are not. Darwin imagined that the environment imposes a similar **natural selection** on wild species.

To understand natural selection, suppose that some plants in a population have flowers the color of leaves, while other plants have lighter-green flowers as a result of differences in alleles. Pollinating insects would notice the lighter flowers and pollinate them more often, so the lighter flowers would make more seeds than their darker relatives. Each year, the population regrows from seeds of the past year. The plants with lighter flowers contribute more than their share of seeds, so they represent a higher proportion of the plants with every passing year. In time, alleles for green flowers are eliminated, replaced by alleles for lighter flowers. If new mutations provide material for further selection, the changes may accumulate until the population consists mostly of plants with bright, showy flowers.

As the example shows, natural selection alters the organism's reproductive success. Thus, biologists often describe natural selection as **differential reproduction**. The environment affects evolution only when it allows some organisms to leave more reproducing offspring than others.

The effect of natural selection depends on how well the species is adapted to local conditions. When a population is exposed to a new environment, the average individual may have many traits that are ill-suited to the environment, and the best-suited individuals may be in the minority. However, those that are best-suited will leave more offspring, until they are the most common type. In other words, the average has changed. Natural selection that causes such a change in the average is called **directional selection**.

Figure 18.4 illustrates directional selection as it might have occurred when cacti were adapting to life in deserts. There is good reason to believe that the ancestors of cacti had broad leaves. This is suggested partly by the presence of modern broad-leaved plants that are closely related to cacti and partly by the fossil record. In contrast, most modern cacti have replaced leaves with spines. The change may have involved a gradual narrowing of the leaves. In Fig. 18.4, the upper graph shows how leaf

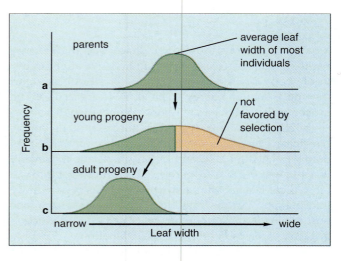

Figure 18.4 Directional selection. (**a**) Distribution of leaf widths among parental plants. (**b**) Distribution of widths among young offspring. Mutations and recombination have expanded the range of variation, but natural selection acts against the offspring with the broader leaves (shown by shading). (**c**) Distribution of leaf widths among offspring that reach reproductive age. The average width is less than in the preceding generation.

width may have varied in the original population. Mutation and recombination may have led to offspring with greater variation, but natural selection acted against those with broad leaves (shading in the middle graph). This would happen in desert habitats where evaporation of water from leaves is a hazard. After selection, the adults that reached reproductive age had a narrower average leaf width than their parents (bottom graph). Repeated over many generations, this process could have narrowed the leaves to the size of spines.

When a population is well adapted to its habitat, almost all the changes that occur in genes will lessen the degree of adaptation. In this instance, natural selection favors the most common types within the population, with the environment weeding out the less common types such as new mutants. This is called **stabilizing selection**. The plants called horsetails illustrate stabilizing selection; they closely resemble horsetail fossils found in rocks that are 300 million years old.

Stabilizing selection tends to prevent a well-adapted species from changing its characteristics. This is illustrated in Figure 18.5. Mutation and recombination lead to young offspring with greater variation than their parents, but stabilizing selection acts equally against offspring at both ends of the distribution curve because these are the plants that deviate most from the average traits. As a consequence, the plants that reach reproductive age have the same traits, distributed over the same range, as their parents.

Both directional and stabilizing selection tend to reduce the variety within a population. However, too much loss of variation can be dangerous. If all the individuals in a population are highly adapted to a specific environment, a change in the environment can extinguish the population.

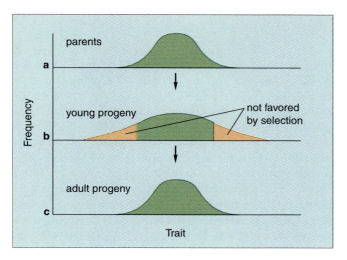

Figure 18.5 Stabilizing selection. (**a**) Variation among parental plants. (**b**) Variation among young offspring. Mutation and recombination have expanded the range of variation. However, natural selection acts against those that are least like the parents (shading). (**c**) Variation among offspring that reach reproductive age. The final variation in traits is the same as in the preceding generation.

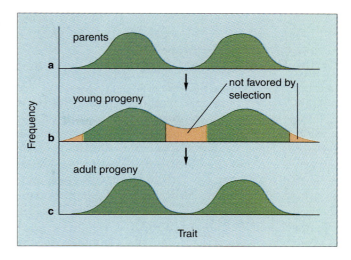

Figure 18.6 Diversifying selection. (**a**) Variation among parental plants. (**b**) Variation among young offspring. Mutation and recombination have increased the frequency of intermediate individuals. Natural selection acts against the intermediate types as well as types that are extremely different from either parental group (shading). (**c**) Variation among progeny that reach reproductive age. Because of selection, two quite different types of plants of the same species remain common.

For example, suppose that a species evolved flowers that could be pollinated only by a certain insect species. If those insects were to leave the area, the plants would never be pollinated and would ultimately die out. But there is one kind of natural selection that preserves (or even increases) variety. This is called **diversifying selection** (or **disruptive selection**). It can promote survival of a population, for a varied population may include some individuals that persist when the environment changes. Diversifying selection (Fig. 18.6) occurs when the environment favors two distinct types in the population—not a single type as in the other forms of selection.

A species of grass (*Agrostis tenuis*) that grows on mine tailings in Wales provides an example. Mine tailings are piles of discarded rock that often contain high amounts of exotic minerals. Some of the grass plants thrive on the tailings but do poorly on normal soils, whereas other plants of the same species grow poorly on tailings but well on normal soils. Because both habitats occur side by side, the two kinds of individuals often exchange pollen and produce progeny with intermediate mineral requirements. The intermediate types do not compete well with the more specialized types on either mine tailings or normal soil, and their numbers are kept small by the fact that intermediate habitats are limited in area. In this case, the occurrence of two divergent habitats side by side selects for two distinct types in the population.

Population Genetics Shows Whether Natural Selection Is Acting on a Gene

In the early decades of the 20th century, geneticists developed an elaborate mathematical system called **popula-**

tion genetics to determine whether natural selection is altering the genetic makeup of a population. These quantitative methods focus on changes in the ratio of alleles for a specific gene. For example, a gene for flower color might have alleles B and b, which call for blue and white flowers, respectively. The question is whether natural selection acts on the gene for flower color. To decide, population geneticists will study a population that makes both blue and white flowers, and they will look for changes in the ratio of B and b alleles from one generation to the next. Breeding programs will show the ratio. For example, suppose B is dominant over b. If so, then white individuals have genotype bb. The genotypes of blue individuals can be discovered by breeding them with white-flowered plants; those with genotype BB will have all blue progeny, but those with genotype Bb will have both white and blue progeny.

Suppose the ratio of B to b changes from one generation to another. Does this mean that natural selection is acting on the gene? If the population is large, the answer is yes. Proof resides in a mathematical relationship called the **Hardy-Weinberg Law**, which says that the ratio of alleles in a large population will not change if the two alleles do not differ in their effect on reproductive success (you can find a more detailed treatment of this law in any genetics text). If the ratio does change, one allele must be favored over the other: It leads to more matings, or more offspring per mating, or better survival of offspring, and so on. In other words, natural selection is taking place.

Many genes in various populations have been tested in this way. Some have stable ratios and are said to be in *Hardy-Weinberg equilibrium*; that is, they are not subject to selection. In other cases, the ratios show considerable change and are proof that natural selection occurs in wild populations.

Chance Affects Evolution in Small Populations

Because of accidents, the best-adapted individuals are not always the ones that reproduce most effectively. Lightning may strike a vigorous tree while leaving a less healthy tree undamaged. These chance effects are not important in large populations, but they can have a major impact on evolution in small populations. If a population is so small that only a few individuals have a given trait, they may all be eliminated by chance despite the adaptive value of the trait. Such chance elimination of alleles in a population is called **genetic drift**.

Another effect of chance on small populations, the **founder effect**, occurs when a few individuals from a large population establish a small, isolated population. Chance determines which of the main population's alleles are present in the founders. As a result, the founders may have a combination of traits that is uncommon in the old population. This may start the new population on an evolutionary path quite unlike that of the old. The founder effect has often been cited in studies of offshore islands, where wind, water, and birds occasionally bring seeds from mainland plants. Island plants may be related to mainland species, but their traits may differ in many ways.

18.5 SPECIATION

Natural selection can alter a population in many ways, as in the conversion from broad-leaved shrubs to modern, spiny cacti. But this process by itself does not increase the number of living species. Rather, it merely changes an existing species. Nevertheless, the fossil record shows many instances in which the number of species has increased with time. For example, no fossils of flowering plants occur in rocks older than about 130 million years, whereas today there may be half a million species of flowering plants. How did all the flowering species arise? The answer requires something more than mutation, recombination, and natural selection. The process that converts one species into two is called **speciation**.

Speciation Requires Isolation and Divergence

For one species to become two, an original population must split into two populations that are **reproductively isolated** from one another, so they cannot exchange genes. Then, if the populations are exposed to different environments, they will accumulate different adaptive traits. The accumulation of different traits is called **divergence**.

Reproductive isolation usually begins with **geographic isolation**: Some members of a population migrate into a new region that is too remote to allow reproduction with the parent population. For example, a storm may blow a few seeds to an offshore island.

Polyploidy can lead to even faster reproductive isolation. As we saw in Chapter 16, most plants have two chromosome sets and are diploid. As a plant grows, its meristematic cells repeatedly duplicate the chromosomes and divide. But sometimes a cell fails to divide after duplicating the chromosomes (Fig. 18.7). This cell and all of its progeny have four sets of chromosomes instead of the normal two sets. Plants with more than two sets of chromosomes are polyploid. If the polyploid cells grow into a branch that forms adventitious roots, an independent polyploid plant may be formed. Comparisons of chromosomes suggest that at least one-fourth of flowering plant species are polyploid. Examples include wheat, potatoes, and cotton. By the mating test, a new polyploid plant is an instant new species because matings with its diploid antecedents will not produce fertile offspring. Figure 18.7 shows why. A polyploid plant makes gametes with two sets of chromosomes, whereas a diploid plant makes gametes with one set of chromosomes. If the two kinds of gametes fuse, they produce a triploid plant, with three sets of chromosomes. The triploid may be a vigorous

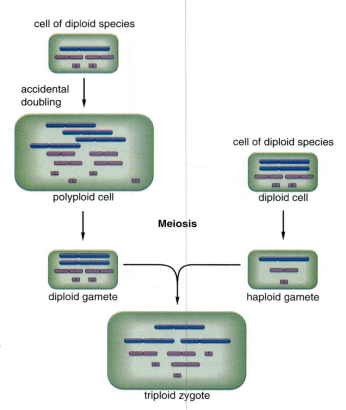

Figure 18.7 Why polyploidy leads to reproductive isolation and thereby to a new species. The original diploid species has three pairs of chromosomes. If a cell doubles the chromosomes and then fails to divide, a polyploid cell results. When a plant with those cells goes through meiosis, it makes gametes with two sets of chromosomes, whereas the parent species makes gametes with one set. If gametes from diploid and polyploid plants fuse during fertilization, the result is a triploid plant. Having an odd number of chromosome sets, the triploid cannot perform meiosis and is sterile.

plant, but it cannot make gametes. Meiosis requires a pairing between chromosomes, and pairing is impossible with an odd number of chromosome sets.

Hybridization is another common source of new plant species. A hybrid arises by fusing gametes of two different species. The mule is a familiar example; it results from a cross between a horse and a donkey. Hybrids are common among plants because wind and animals transfer pollen between species. If the species are close relatives, the hybrid may be vigorous and perhaps fertile. An example is the common sycamore tree (*Platanus X hybrida*), which arose naturally when European and American species hybridized. But hybrids are often sterile, usually because the two parent species have different numbers or kinds of chromosomes. This is true of hybrids between radish and cabbage. In such cases, meiosis fails because the chromosomes do not pair properly (Fig. 18.8). In that case the hybrid is sterile. Fertility can be restored if a cell of the hybrid accidentally becomes polyploid, allowing the chromosomes to pair. The polyploid plant is reproductively isolated from the parent species because its gametes will have a different number of chromosomes from the parent species. An important example is *Triticosecale*, a human-made hybrid between wheat (*Triticum*) and rye (*Secale*) that combines the high productivity of wheat with the disease-resistance of rye. The sterile hybrids became fertile when plant breeders doubled the chromosome number.

Adaptive Radiation Is the Formation of Many Species from One Ancestral Species

When speciation occurs many times, it can build a large group of divergent taxa from an unspecialized ancestral species. This is called **adaptive radiation**. For example, in a period of 20 to 40 million years, the flowering plants expanded from an unknown founding species to the largest division in the plant kingdom. Adaptive radiation occurs when the ancestral species has the opportunity to exploit a variety of environmental resources. An opportunity can arise in three ways: (1) through extinction of competing groups, (2) through the evolution of advantageous new traits, or (3) through a major change in climatic conditions.

The adaptive radiation of mammals became possible with the extinction of competitors. Dinosaurs dominated most land habitats for 160 million years, while mammals were limited to small, ratlike forms. About 65 million years ago, the dominant dinosaurs died out. Mammals promptly multiplied into many diverse forms, occupying habitats where dinosaurs once ruled.

Flowering plants had an even greater adaptive radiation, probably triggered by evolution of advantageous new traits. In evolving the flower, they established a pollinating relationship with insects that greatly improved reproduction. This happened some 130 million years ago,

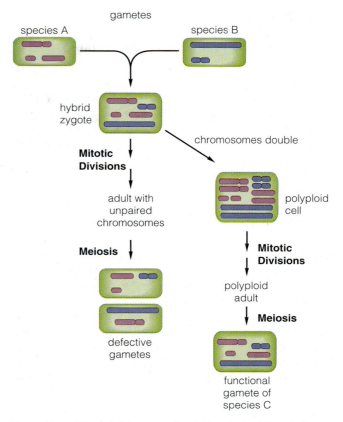

Figure 18.8 Why hybrids are sterile and how a doubling of chromosomes (polyploidy) restores fertility. Top: A hybrid forms by fusing the gametes of two species with differing chromosome sets. Left branch: The hybrid's chromosomes do not form matched pairs; hence, they are distributed abnormally in meiosis, yielding nonfunctional gametes. Right branch: Chromosome doubling permits normal pairing in meiosis; hence, the gametes receive one chromosome of each kind and are fully functional.

well before the dinosaurs died out. Thereafter, the number of flowering plant species grew explosively, with different types of angiosperms taking over many habitats that other plants had previously occupied.

The fossil record suggests that a process called *coevolution* played a large part in the adaptive radiation of flowering plants. **Coevolution** occurs when two groups of organisms interact so closely that evolution in one group causes evolution in the other. In many instances today plant and insect species have matching traits, suggesting that they evolved together. Competition for pollinators might favor plants with distinctive flowers, leading to increasingly varied plant species. Flower traits would be especially valuable if they gave the plant an exclusive pollinating relationship because that would lead to efficient pollen transfer. The advantage of an exclusive food supply would lead to matching evolution in the insects. If these surmises are correct, the adaptive radiation of flowering plants would have been matched by an adaptive radiation of insects. The fossil record shows exactly that.

18.6 CLASSIFYING ORGANISMS ACCORDING TO THEIR GENETIC RELATIONSHIPS

Having seen how taxa evolve, we can examine the way systematists group species into meaningful taxa. Their task is not easy because they have only indirect evidence about evolutionary links among species. Thus, systematists often disagree about the way to group species. This section explains some of the pitfalls and shows how systematists work around them.

Classification Can Be Phenetic or Phylogenetic

Of the two methods for organizing the diversity of life into groups, the oldest approach is to group organisms purely on the basis of appearance. This is called **phenetic classification** (from the Greek word meaning "to show"). Theophrastus, the first taxonomist, used this method 2300–2400 years ago. He divided plants into trees, shrubs, and herbs.

Phenetic classification is still important today. But since the acceptance of evolution, most systematists prefer to define groups based on common ancestry. This is called **phylogenetic classification**. The presumed paths of evolution are often shown in diagrams called **phylogenetic trees**, in which the branch points represent speciations that lead to new taxa, and modern species are at the ends of twigs (Fig. 18.9). The benefit of classifying on the basis of ancestry is that all the genera in a family, for example, would be related. The same would be true of families within an order, and so on. This is not necessarily the case when organisms are classified based on their appearance.

Most systematists compromise between phenetic and phylogenetic approaches. Organisms are grouped phylogenetically when good information is available about their evolutionary connections. They are grouped phenetically when the ancestry is less clear.

Fossils Form an Incomplete Record of Plant Origins

To make a phylogenetic classification, we must deduce the ancestral connections among modern groups. At first glance, fossils might seem to offer the most promising source of information. Biologists speak of the fossil record as if it were a book, showing how life has changed through time. In a sense, it *is* a book. Fossils form when parts of organisms are buried in mud or sand that gradually turns to stone. Sediments pile up layer upon layer, and the embedded fossils are almost like plants pressed between pages of a book.

If fossils were abundant enough, they could show all the branch points in evolution. If two plant species differ

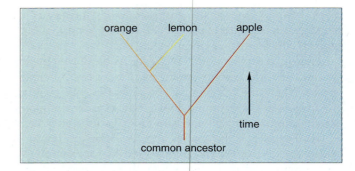

Figure 18.9 A phylogenetic tree, constructed on the assumption that differences accumulate with time after two taxa separate from their common ancestor. Multiple similarities imply recent common ancestry. The orange and lemon trees are more closely related to each other than to an apple tree, but all three originated from a common ancestor.

today, a series of fossils would show how they gradually changed from a common ancestral species. Unfortunately, the fossil record is both fragmentary and biased. Millions of species exist at each moment, but only a small fraction of the most abundant organisms leave fossils. Because fossils form in sediments, most fossilized plants originally lived near swamps or lakes. Pollen grains, fragments of wood, and imprints of leaves make far more abundant fossils than delicate parts such as flowers. As a result, the fossil record shows few of the branch points in evolution.

Comparing Characteristics Among Plants Provides Clues About Their Relatedness

Because fossils are inadequate, systematists try to reconstruct the paths of evolution by comparing modern organisms. The basic idea is that taxa with many differences probably separated from a common ancestor long ago, whereas taxa with few differences separated more recently. Orange, lemon, and apple trees illustrate the principle. As flowering plants, they share a common ancestor. But lemons resemble oranges more than they resemble apples. This suggests the phylogenetic tree shown in Figure 18.9: An early branch separated the ancestors of apple trees from the ancestors of modern citrus; later, the citrus line branched to give orange and lemon trees.

Modern phylogenetic analysis is a sophisticated science, with strong computer support and a set of standards for judging the results. The principles are expressed most clearly in a discipline called **cladistics**, in which organisms are grouped according to similarities that are derived from a common ancestor. A central concept is the **clade**, a branch point and all the species that descended from it.

To determine whether a group of taxa form a clade, the systematist begins by listing observable traits, called **characters**, that differ among the taxa. Characters include such things as xylem vessels, oil glands, flower pigments, or a particular base at a certain location in the DNA. Each character can have two or more *character states*, such as "present" versus "absent."

Until recently, systematists relied mostly on *morphological characters*—aspects of plant form, such as the presence of tracheids or the number of flower parts. Such characters will always be important, if only because they relate directly to the environmental factors that govern natural selection and evolution. But *molecular characters* have become increasingly important in recent decades. They include the detailed structure of proteins and nucleic acids, as well as the ability to make particular molecules, such as pigments. Recently, molecular biologists have invented easy ways to read the base sequence of DNA. The DNA work is particularly important because evolution is the accumulation of changes in DNA. Organisms duplicate their DNA and pass it on with minor changes in each generation. Thus, by looking at the differences in DNA beween two species, we can estimate how much time has passed since the two species parted from their common ancestor. Furthermore, DNA studies provide a multitude of characters for cladistic analysis. Each nucleotide along a DNA molecule is a character, with four possible states (bases A, C, G, or T). Because there are millions of bases in a DNA molecule, the base sequence includes millions of characters.

Further analysis depends on judgments about the sequence in which character states evolved. A character state that evolved early is **primitive** or **ancestral**; a state that evolved later is **derived**. For example, suppose we can establish that early plants lacked oil glands in the leaves but that some plants later evolved the ability to make oil glands. Then the absence of glands is a primitive state, and the presence of glands is a derived state.

Biologists judge a trait to be primitive if it is present in many taxa that are otherwise quite different. For example, consider the conducting cells in xylem. Tracheids occur in the great majority of plants, including ferns and conifers as well as flowering plants. In contrast, vessels occur almost exclusively in flowering plants and their close relatives. The simplest explanation is that tracheids arose in an ancestor that gave rise to all these groups. Vessels, by contrast, arose in organisms that led to flowering plants after they had branched off from the other groups. With this reasoning, the primitive condition is to have tracheids but not vessels. Accordingly, the presence of vessels is a derived trait.

The central task of cladistic analysis is to build a phylogenetic tree based on shared derived characters, which is called a **cladogram**. The goal is to deduce the path of evolution that led to the modern array of taxa from a presumed common ancestor. This is done by drawing cladograms that show the sequence of character changes needed in various paths of evolution. Computers are essential for this task because the number of possible cladograms increases very rapidly with the number of taxa. If 15 taxa are included, over a trillion cladograms are possi-

ble! A computer can sort through the multitude of possible cladograms in a short time and find the pattern that is best supported by the data.

Many studies focus on the genes that code for ribosomal RNA (rRNA). The genes for rRNA make an excellent evolutionary clock because they have regions that are known to change rapidly, over evolutionary time, and other regions that are known to change more slowly—a molecular analogy to the minute hand and hour hand of a clock. To judge the relations among genera and among species, researchers focus on the highly variable regions of rRNA, which accumulate differences rapidly. To judge the relations among higher taxa such as kingdoms or divisions, researchers focus on the more stable regions of rRNA. Higher taxa departed from their common ancestors hundreds of millions of years ago. It takes that kind of time for the stable regions of rRNA to accumulate significant changes.

Cladistic methods are revolutionizing systematics. They can help reveal which character states are primitive and which are derived, and they can reveal cases in which the same state was derived independently in two or more clades. This can be useful in exploring the way characters evolve. More broadly, cladistic analysis throws into question the entire system of higher taxa (families, orders, and so on) that has been built up by less objective methods. Cladistics confirms many established boundaries between taxa. However, it also shows many cases in which an established taxon combines species that probably came from different lines of evolution. For example, cladistic analysis suggests that the category of reptiles is artificial and that crocodiles are related more closely to birds than to turtles. Among flowering plants, perhaps the most striking discovery is that monocotyledons diverged from dicotyledons much earlier than had previously been thought. Such observations may one day revolutionize the treatment of higher taxa.

Convergence Can Lead to Mistaken Conclusions About Plant Ancestry

The greatest problem with phylogenetic classification comes from the assumption that shared derived characters imply common ancestry. The assumption is usually sound, but sometimes it fails and leads to wrong conclusions.

Failures occur because similar traits sometimes arise independently in two groups of organisms. This is called **convergence**. It occurs when populations with different origins evolve under similar environmental pressures. For example, some plants in the deserts of Africa resemble the cacti of American deserts (Fig. 18.10). They have fleshy green stems, spines, and reduced leaves. This might prompt us to put the African and American species in the same family. But a much larger set of anatomical traits shows that the African and American species belong to different families, the Euphorbiaceae and the Cac-

a

b

Figure 18.10 Desert plants of the southwest United States (**a**) and Africa (**b**). The similarities are attributed to convergence. On the basis of floral and anatomical features, these plants are placed in separate families: the Cactaceae (**a**) and the Euphorbiaceae (**b**). The two families also include plants that are very different.

taceae, respectively. Less specialized members of these families do not have the shared cactuslike traits. Evidently, the desert members of the two families evolved similar traits when they were subjected to similar selective pressures for a very long time.

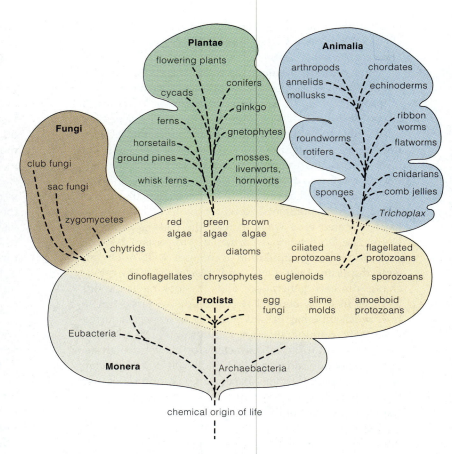

Figure 18.11 The five kingdoms of life.

18.7 DOMAINS AND KINGDOMS

In striving for phylogenetic taxa, biologists are rethinking the way they draw boundaries between the kingdoms of life. The plant ecologist Robert Whittaker first defined five kingdoms on the basis of structure and nutrition. He placed all prokaryotes (bacteria) in kingdom Monera, and he placed unicellular eukaryotes in kingdom Protista. Among multicellular eukaryotes, kingdom Plantae included all photosynthetic species, kingdom Animalia included all species that have muscles and ingest solid food, and kingdom Fungi included all species that lack muscles and absorb dissolved food. With these criteria for defining kingdoms, Whittaker's scheme is a phenetic classification.

To define kingdoms on a phylogenetic basis, most taxonomists favor a modified version of the Whittaker scheme (Fig. 18.11) in which the animal, plant, and fungal kingdoms are defined by a combination of traits that is unlikely to have arisen more than once. In this scheme, all the prokaryotes (bacteria and other organisms without nuclear membranes) make up kingdom Monera. Kingdom Animalia is limited to nonphotosynthetic organisms that lack cell walls but form embryos. Kingdom Plantae

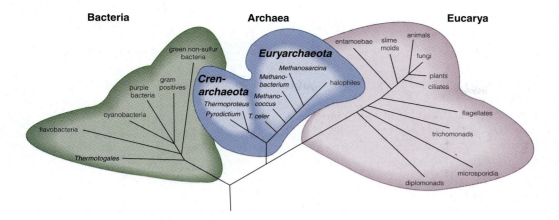

Figure 18.12 Phylogenetic tree based on comparisons of rRNA, showing the three domains of life (Bacteria, Archaea, and Eucarya).

is limited to organisms (usually photosynthetic) that have cellulose walls and form embryos. Kingdom Fungi consists of chitin-walled nonphotosynthetic organisms that lack a motile stage and reproduce by spores. Kingdom Protista includes all the eukaryotic algae and other organisms that do not fit into any of the preceding kingdoms. With this definition, the Protista include both unicellular and multicellular organisms.

No system is perfect, and this one has its limitations. The kingdom Protista is artificial; its members make up several lines of evolution that are not closely related to one another. To be consistent, the Protista should be divided into at least a dozen tiny kingdoms. Such a splitting has been avoided because the idea of a kingdom implies a group of considerable size.

Comparisons of rRNA show that prokaryotes form two quite distinct groups; these have been named the archaebacteria or **Archaea** and the eubacteria or **Bacteria**. Besides the differences in rRNA, they differ in cell wall and membrane molecules, transfer RNAs, sensitivity to antibiotics, and habitats. In terms of those traits, the two groups of prokaryotes differ from each other as much as they differ from eukaryotes or **Eucarya** (Fig. 18.12). This suggests that the three groups arose about the same time from a common ancestor. Thus, we should not group all the prokaryotes into a single kingdom. This poses a problem for taxonomists. Because we are not likely to stop thinking of animals and plants as kingdoms, some biologists have proposed that the term *domain* be used for a taxon higher than a kingdom. In this view, there are two prokaryotic domains (Bacteria and Archaea), each including just one kingdom, and one eukaryotic domain (Eucarya), which includes at least four kingdoms.

SUMMARY

1. Organisms are assigned to species according to their combinations of traits or by subjecting them to a mating test.

2. Taxonomists have defined a hierarchy of taxa, consisting of species, genera, families, orders, classes, divisions (or phyla), kingdoms, and domains. Every species is given a binomial name, consisting of the genus name and a specific epithet.

3. Early philosophers subscribed to the notion of an ideal pattern for each species, but current evolutionary thinking conceives of a species as a group of organisms that can change.

4. Evolution requires spontaneous changes in DNA. The changes can result from mutations (which alter DNA) or from recombination (which brings together DNA from two organisms). Mutations are caused by heat energy or by mutagens (radiations and chemicals) that lead to errors in copying DNA or chemical changes in DNA molecules. Recombination can result from such events as sexual reproduction and viral infections. Endosymbiotic associations can also bring together DNA from two organisms, as was probably the case in the origin of mitochondria and chloroplasts.

5. Through natural selection, the environment allows some organisms to reproduce more abundantly than others. Over many generations, this can lead to major changes in genetic constitution and body forms. The altered body forms are better adapted to a specific environment. Directional selection leads to changes when a population is adapting to a new environment. Stabilizing selection maintains the adapted state of a population. Diversifying selection can maintain diversity in a population.

6. The mathematical system of analysis called population genetics can show whether a population is undergoing natural selection.

7. In small populations, chance affects the direction of evolution by eliminating some types and allowing others to prosper without regard to natural selection. Such changes make up genetic drift and lead to a founder effect when small colonies are established.

8. Speciation (multiplication of species) occurs when part of a population becomes reproductively isolated

from the rest, followed by divergence as the isolated population adapts to a different environment. Isolation can be geographic, or it can be genetic—as when accidental hybridization or polyploidy occur.

9. Adaptive radiation (multiple speciations in a short time) occurs when a founding species gains a new ecological opportunity through elimination of competitors, climatic changes, or evolution of a valuable new trait. Adaptive radiation creates higher taxa, as the founding species multiplies into a genus that may expand into a family, and so on.

10. Systematists prefer to define higher taxa based on common ancestry (phylogenetic classification) rather than common appearance (phenetic classification). Both approaches are used because the paths of evolution are often obscure.

11. Fossils give clues about phylogeny, but gaps in the fossil record force systematists to rely mostly on comparisons of modern organisms that reveal shared characters.

12. In comparing suggested paths of evolution, systematists favor the paths with the fewest changes. Cladistic methods use computers to deduce the phylogenetic path with the least number of changes. The results are expressed as cladograms. Comparisons of nucleic acids, particularly rRNA, are increasingly useful in deducing phylogenetic relations among taxa. It is possible, however, to infer relatedness where none exists when studying groups that have experienced convergent evolution.

13. Most biologists accept five kingdoms of life: the prokaryotic kingdom Monera and four kingdoms of eukaryotes (Protista, Animalia, Plantae, and Fungi). Protista is considered an artificial kingdom because its members are not closely related.

14. Molecular studies suggest that we need a taxon called the *domain*, which is higher than the kingdom. There are two prokaryotic domains (the Bacteria and the Archaea) and one eukaryotic domain (the Eucarya), which includes all four eukaryotic kingdoms.

15. To identify an organism, taxonomists use keys that call for serial choices between alternative traits. Final identification requires comparison with type specimens.

Questions

1. How do biologists determine whether two organisms belong to different species?

2. How does a taxonomist differ from a systematist?

3. What are the main taxonomic levels between the domain and the species?

4. Describe the roles in evolution of mutagens, sexual reproduction, viral infections, endosymbiosis, and interactions between organism and environment (natural selection).

5. What causes natural selection to be stabilizing, directional, or disruptive?

6. Describe a research program that would show whether natural selection is occurring in a wild population.

7. What causes the founder effect and genetic drift, and how do these processes affect evolution?

8. Why do geographic isolation, polyploidy, and hybridization lead to more species?

9. In the view of most biologists, what caused the adaptive radiation of flowering plants?

10. Given a table of character states, determine which states are primitive and which are derived.

11. Why does convergent evolution lead to errors in phylogenetic analysis?

12. Describe the five kingdoms and three domains of life.

13. You have found an organism unlike anything you have seen before. How can you determine whether it is a new species?

Further Readings

Cronquist, Arthur. 1988. *The Evolution and Classification of Flowering Plants*. 2d ed. New York: New York Botanical Garden. A good reference on the title topic.

Eldridge, Niles. 1995. *Reinventing Darwin*. New York: John Wiley. New perspectives on evolution, written by a leader in the field.

Hancock, James F. 1992. *Plant Evolution and the Origin of Crop Species*. Englewood Cliffs, N.J.: Prentice-Hall. An interesting study of how fiber, spice, and food plants reached their present forms.

Maddison, Wayne P., and David R. Maddison. 1992. *MacClade Version 3*. Cambridge, Mass.: Sinauer Associates. Introduces cladistic theory and methods, and includes a popular computer program for cladistic analysis.

Margulis, Lynn, and Rene Fester, eds. 1991. *Symbiosis as a Source of Evolutionary Innovation*. Cambridge, Mass.: MIT Press. Persuasive arguments for the importance of symbiosis.

Slatkin, Montgomery, ed. *Exploring Evolutionary Biology: Readings from* American Scientist. 1995. Cambridge, Mass.: Sinauer Associates. Introduces major themes and concepts in the science of evolution.

19 MONERA AND VIRUSES

1. The cellular organization of the simplest monerans (bacteria) seems much less complex than that of eukaryotic cells. However, some moneran cells develop complex, specialized structures such as flagella, thylakoid membranes, and spores.

2. All moneran organisms can be divided into two groups: the Archaebacteria and the Eubacteria. Members of the Archaebacteria, thought to be the most primitive organisms alive, inhabit harsh environments such as hot springs, salt flats, and anaerobic mud flats. Members of the Eubacteria are more specialized and thus are considered more highly evolved.

3. Monerans obtain their energy for movement, metabolism, growth, and reproduction from one of three classes of chemical reactions: Heterotrophs oxidize organic compounds from other organisms (living or dead); chemolithotrophs oxidize inorganic compounds; and phototrophs oxidize and reduce compounds using the energy of light (photosynthesis).

4. Many monerans have close, symbiotic associations with other organisms. Some associations are mutualistic: They benefit the moneran and its host. The relationship between the nitrogen-fixing *Rhizobium* and legumes is one example. Some associations are parasitic: The moneran harms the host. *Erwinia amylovora*, which causes fireblight in apples, is an example of a plant parasite.

5. Viruses are genes, wrapped in a coat of protein, that infect cells. They appropriate the biochemical machinery of the cells, using it to reproduce themselves and sometimes to kill the cells. Viral diseases of plants cause serious reductions in crop yield and quality.

19.1 MONERA, VIRUSES, AND THE STUDY OF PLANTS

Monera is one name for the kingdom that includes all the organisms with cells that lack a nucleus. Another name is **Prokaryota** (the prokaryotes); a third is **bacteria**. The organisms in this kingdom generally have a simpler cell structure than the members of other kingdoms (collectively, the eukaryotes). Most often, the monerans are single-celled organisms. However, some form colonies, and some form structures with a degree of morphological differentiation. As small and apparently simple as they are, they pervade the world and represent a large fraction of the Earth's biomass. Although their structures are simple, their biochemical abilities are often complex and sometimes unique—that is, they possess enzymes and metabolic pathways not found in any eukaryote. Some of these metabolic capabilities are absolutely essential for maintaining the physical and chemical characteristics of the Earth in a state suitable for life. Monera is thus an important kingdom, intensively studied by microbiologists and bacteriologists. The kingdom is much too large a subject to be covered in any detail by a textbook on plants. Yet, there are three important reasons why a person who studies plants should also study monerans:

1. Many of the biochemical compounds, enzymes, and metabolic pathways of plants are also found in monerans. Photosynthesis, the uptake of mineral nutrients, and response to environmental stresses (such as drought) occur in both monerans and plants. The discoveries made in studying monerans can be used to guide plant research.

2. The evolutionary ancestors of plants were monerans. Not only did the first eukaryotic cell probably evolve from some unknown moneran, but the large organelles in plant cells—the mitochondria and plastids—are probably related to two different, more advanced types of monerans. Studying monerans is necessary for understanding the origin of plants.

3. Plants form ecological associations with monerans. Some of these associations are mutualistic symbioses (for instance, the nitrogen-fixing association between the bacterium *Rhizobium* and legumes); in some associations, the moneran is a parasite on the plant. In either case, it is necessary to study the monerans to understand how the plant copes with its environment.

Viruses are particles constructed of a nucleic acid genome (either RNA or DNA) and a protein coat. They are not monerans. They are not even cells and cannot live independently. They must infect the cells of some organism in order to reproduce, and they are parasites on the cells they infect. Viruses should be studied by plant biologists for many of the same reasons that monerans should be studied. Many of the basic properties of genes and proteins can be investigated using viruses. The discoveries obtained by studying viruses can be used to guide plant research. Some viruses are plant parasites and are thus important in the lives of many plants. The diseases that viruses cause, the ways in which they are transmitted from plant to plant, and the methods plants use to combat viral diseases are all part of the life histories of plants.

19.2 MONERAN CELL STRUCTURE

Many Moneran Cells Have Simple Structures

The basic structure of a moneran cell is very simple relative to a eukaryotic cell (compare Figs. 19.1 and 3.3). The moneran cell is surrounded by a plasma membrane. This plasma membrane fulfills the same roles that it does in plants: It accumulates nutrients, keeping them and other

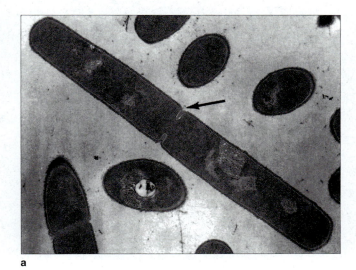

a

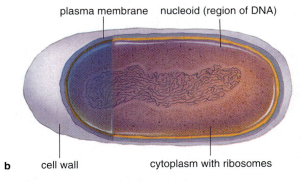

plasma membrane   nucleoid (region of DNA)

b   cell wall   cytoplasm with ribosomes

Figure 19.1 The moneran cell. (**a**) Transmission electron micrograph of *Bacillus cereus*, showing the light-colored nucleoid and darker cytoplasm. At the time the cell was prepared for microscopy, it had recently divided, and a new crosswall (arrow) was forming. (**b**) Organization of the major components of the moneran cell.

cell components (such as enzymes) in the cytoplasm; it excludes toxic compounds; and it is sensitive to aspects of the environment such as water potential or the presence of nutrients. Outside the plasma membrane is (usually) a cell wall. Like the cell wall of plant cells, the cell wall of monerans prevents the cell from bursting when it is in hypotonic solutions (solutions more dilute than the cytoplasm). The cell wall of monerans is generally formed from a **peptidoglycan**, a combination of amino acids and sugars that surrounds the cell like a net. It does not have the cellulose found in plant cell walls. One of the best-known families of antibiotics (which includes penicillin) acts by inhibiting the formation of the bacterial cell wall; bacteria treated with penicillin thus become osmotically sensitive. They will burst when in fluids—including body fluids—that are more dilute than their cytoplasm. Traditionally, bacteria were classified in part by their shapes: **cocci** (small, round cells), **bacilli** (rods), **vibrios** (bent or hooked rods), **spirilli** (helical forms), and

stalked forms. These shapes represent the different patterns of growth of the cell walls of monerans.

In the cytoplasm is a full set of genes and a complete apparatus for expressing them to make proteins. The genes are encoded in DNA, just as in eukaryotic cells. All the housekeeping genes (genes needed for the basic functions of life) are on a single circular chromosome. This chromosome may be very long (in circumference). In the best-studied bacterium, *Escherichia coli*, the chromosome is 1.4 mm long and contains 4.2 million nucleotide pairs (Fig. 19.2)—a very large amount of DNA, considering that the *E. coli* cell is only about 1 μm long and 0.5 μm wide. The chromosome is wound in the cell and is localized in an area called the **nucleoid**. Unlike eukaryotes, the moneran chromosome is not surrounded by a nuclear envelope; so there is no defined nucleus. It used to be thought that the chromosome had no structure and was packed randomly into the cell; however, recent investigations have found that it is complexed with specific structural proteins that organize it into loops.

Many monerans have sets of accessory genes called **plasmids**. Plasmids are relatively small circles of DNA, on the order of 2000 to 200,000 nucleotide pairs long. They often contain functionally related sets of genes. Some plasmids, for instance, have genes that confer a resistance to antibiotics. Some plasmids in parasitic bacteria have genes that make them pathogenic (disease-causing) in their hosts. The plasmids can be replicated (reproduced) independently of the chromosome. Sometimes they are replicated faster than the chromosome and the cell, so that there are many copies in a cell. Sometimes, for instance at high temperatures, the plasmids are replicated more slowly, so that daughter cells are formed that do not have a plasmid.

Some plasmids have genes for enzymes and structural proteins that transfer copies of the plasmid to bacteria that do not have any (Fig. 19.3). This is one method of forming new combinations of genes in bacteria. Although most plasmids are transferred only among cells of one species, some can be transferred to many species of bacteria. This is how it has been possible for resistance to antibiotics to spread so quickly to many pathogenic bacteria in the past few years, a situation that threatens to neutralize many of our modern defenses against disease.

A special type of plasmid, the F plasmid, has the ability to incorporate itself into the main chromosome. When this occurs and the DNA is then replicated, a copy of the new chromosome with the incorporated F plasmid can be transferred to a recipient cell through a tube, called an F pilus, produced by the donor cell. Once the new chromosome is in the recipient cell, there is a possibility of an exchange of pieces of the new chromosome with the original chromosome. This process, called **conjugation** (see Fig. 17.1), allows for a genetic recombination of chromosomal genes. Although it is entirely unlike meiosis, it may play a similar role (promoting diversity) in the pop-

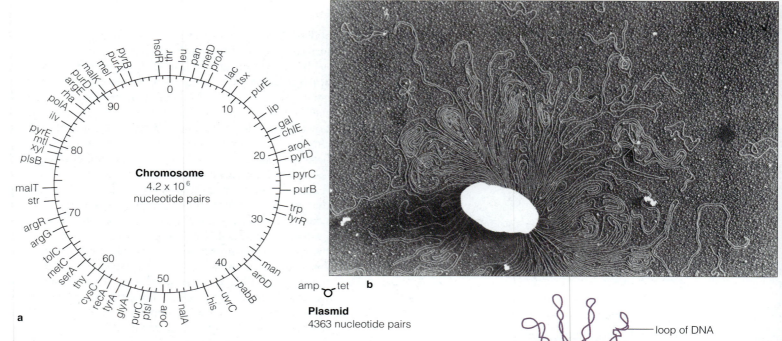

Figure 19.2 Moneran chromosome. (**a**) This diagram of an *Escherichia coli* chromosome shows a fraction of the identified genes. The order of the genes was determined by gene-transfer experiments using F plasmids. The sequence of all 4.2×10^6 nucleotide pairs is known. There are 4288 genes identified from the sequence. Next to the chromosome is a diagram of a genetically engineered plasmid, showing the positions of two genes that confer resistance to two antibiotics. Note that the plasmid has only 1/1000 as many nucleotide pairs as the chromosome. (**b**) Picture and diagram showing an *E. coli* nucleoid spread out for electron microscopy. The loops are held in place by a central matrix of proteins.

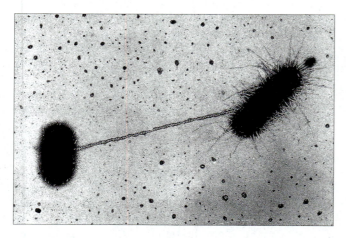

Figure 19.3 Conjugation between two cells of *Escherichia coli*. The donor (F⁺) cell is on the right; the recipient (F⁻) cell is on the left. The long connecting arm is an F pilus. Note that the donor cell also has many smaller pili.

ulation genetics of monerans. (The use of plasmids in biotechnology was discussed in Chapter 17.)

The moneran cell contains ribosomes with a general composition and structure similar to those of eukaryotes. The ribosomes have two subunits, each made of RNA and protein. Moneran ribosomes are smaller than eukaryotic ribosomes, however, and their RNAs have different nucleotide sequences.

Moneran cells reproduce by **binary fission**, meaning that the cell splits in two. Preceding this step, the DNA in the main chromosome replicates, so that there are two full copies of the chromosome. It is thought that each copy of the chromosome may be attached to the plasma membrane and that the separation of the points of attachment (as the cell grows and the membrane enlarges) may pull the chromosome copies apart (Fig. 19.4). This process is very different from, but analogous to, the separation of chromosomes in mitosis. A new crosswall (see Fig. 19.1) forms between the two chromosomes, so that each progeny cell receives a copy. The other organelles, ribosomes, enzymes, and plasmids are probably divided randomly as the crosswall forms. Although cell division in monerans seems much simpler than it is in eukaryotes, this may be because of the small size of the cells and our inability to visualize the intricate details of the division process, as we can with eukaryotes.

Notably lacking in the above description is any component of the eukaryotic endomembrane system: endoplasmic reticulum, Golgi apparatus, vesicles, or vacuole. Monerans have no mitochondria or plastids and no known cytoskeleton (although genes related to the eukaryotic tubulin gene have recently been identified). It would be wrong, however, to think that there are no complex adaptations among the monerans. The next sections describe some of the more common modifications to the basic moneran cell structure.

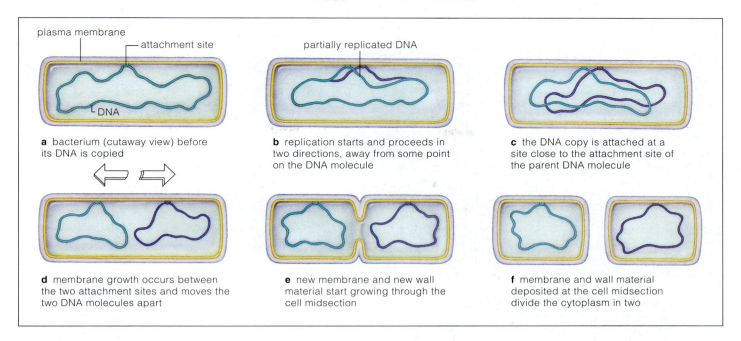

a bacterium (cutaway view) before its DNA is copied

b replication starts and proceeds in two directions, away from some point on the DNA molecule

c the DNA copy is attached at a site close to the attachment site of the parent DNA molecule

d membrane growth occurs between the two attachment sites and moves the two DNA molecules apart

e new membrane and new wall material start growing through the cell midsection

f membrane and wall material deposited at the cell midsection divide the cytoplasm in two

Figure 19.4 Bacterial reproduction by binary fission.

Some Moneran Cells Have Modified Extracellular and Intracellular Structures

Moneran cells are distinguished from each other, in part, by the thickness of their cell wall. One of the earliest ways of classifying bacterial cells, *Gram's stain* (named after the Danish physician who devised it), depends on the chemical nature of the cell wall. In Gram's staining procedure, cells fixed to a microscope slide are stained with a purple dye called crystal violet and a dilute solution of iodine. The slide is washed with water and then briefly with ethanol. Finally, the cells are stained with a red dye called safranin. Some cells, called **Gram-positive** cells, retain the dye during the ethanol wash and look purple; others, called **Gram-negative**, lose it and look pink (Fig. 19.5a). Gram-positive cells have a thick cell wall that restricts the dissolution of the crystal violet–iodine dye complex in the ethanol. Gram-negative cells have a thinner, more permeable wall that permits the ethanol to penetrate the cell and plasma membrane and wash away the crystal violet–iodine complex.

Gram-negative and Gram-positive bacteria also have different extracellular structures. Gram-negative cells have a second membrane, called the **lipopolysaccharide** layer, surrounding the cell outside the cell wall (Fig. 19.5b). This membrane forms a separate compartment of the cell, called the **periplasmic space**. Certain enzymes and other proteins, including proteins that sense the osmotic potential of the medium, are localized in that space. Gram-positive cells sometimes have a slimy polysaccharide sheath outside the cell wall. This sheath protects against being taken up by bacteria-eating cells, including the macrophages (cells in our body that ingest and kill pathogens).

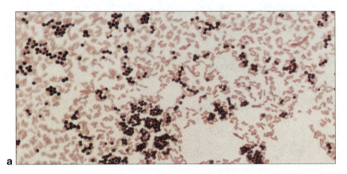

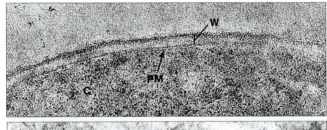

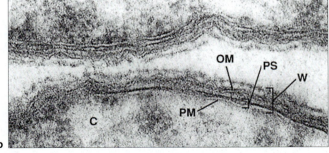

Figure 19.5 Some distinguishing differences between Gram-positive and Gram-negative bacteria. (**a**) Light micrograph of stained Gram-positive cocci (purple spheres) and Gram-negative bacilli (pink rods). (**b**) An electron microscopic comparison of the plasma membrane and cell wall regions of Gram-positive (top) and Gram-negative (bottom) bacteria. W, cell wall; PM, plasma membrane; C, cytoplasm; PS, periplasmic space; OM, outer membrane (lipopolysaccharide layer).

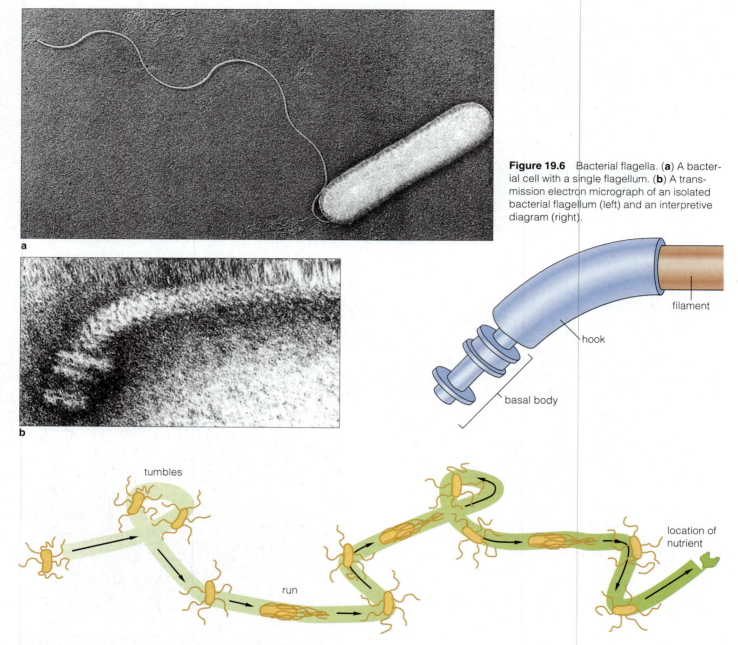

Figure 19.6 Bacterial flagella. (**a**) A bacterial cell with a single flagellum. (**b**) A transmission electron micrograph of an isolated bacterial flagellum (left) and an interpretive diagram (right).

filament

hook

basal body

a

b

tumbles

run

location of nutrient

Figure 19.7 Bacteria control their flagella so as to seek nutrients and avoid stress or toxic compounds. When the flagella turn one way, the bacteria move in a straight line (run); when the flagella turn the other way, the bacteria tumble. When conditions are improving, the runs are longer.

Many moneran cells swim using **flagella**, projections that propel the cell (Fig. 19.6). The moneran flagellum is not related in any way to the eukaryotic flagellum. It is formed from many subunits of the protein flagellin, which line up to form a helical (corkscrew-shaped) filament. The filament rotates, propelled by a motor apparatus—the basal body—at its base, where it penetrates the plasma membrane. The rotation of the basal body is powered by a proton and electrical gradient produced across the plasma membrane. The rotating flagellum pulls the cell through the liquid medium. Under certain circumstances—generally when the systems of the cell sense an unsatisfactory change in the environment—the motor reverses direction. Because the helical flagellum is not as rigid under the reverse stress, it bends (flops), and the cell tumbles. After a short period, the motor reverses again, the flagellum regains its original conformation, and the cell starts swimming in a straight line. If once again the environment is unsatisfactory, the procedure is repeated, and this continues until the cell finds an improvement in its conditions (Fig. 19.7).

Another extracellular organelle is the **pilus** (Latin for "hair"), a thin, hollow, nonmotile projection from the cell. At the end are proteins that recognize and bind to receptors on other cells. A pilus is a conduit for the transfer of plasmid or chromosomal DNA from the donor (the cell with the pilus) to the receiver (Fig. 19.3).

Although monerans are usually described as not having internal membranes, this is not true in all cases. Some cells have cisternae or thylakoid membranes, flattened bladders that enclose separate compartments within the cytoplasm (Fig. 19.8). These structures are related to the cells' energy metabolism. As in chloroplasts, the thylakoid membranes of monerans function in the light reactions of photosynthesis; they separate two compartments with different pHs and electrical potentials and thereby permit the cell to store the free energy generated by electron transport reactions across the membranes (see Chapter 10). Many moneran cells are chemolithotrophic, meaning that they derive energy from inorganic oxidation–reduction reactions, rather than from photosynthesis (see p. 316), and some of these also have cisternae for energy storage. The oxidation–reduction reactions presumably transport electrical charge and protons across the cisternal membranes; as in photosynthesis, the charge and proton gradients can be used to synthesize the high-energy compound ATP. (Not all photosynthetic and chemolithotrophic monerans have internal membranes. Some accomplish the same process by establishing their pH and electrical gradients across the plasma membrane.)

Some Moneran Cells Form Spores

Under harsh environmental conditions, moneran cells will die, but some species have developed ways of surviving by forming tough **spores**. Typically, bacterial spores are small, desiccated cells in a condition of suspended animation. Covered with a specialized, hardened cell wall, they contain a complete genome and sufficient enzymes and metabolites to germinate and reestablish growth when conditions improve. The combination of the hardened wall and the dry, inanimate state make these spores resistant to many environmental insults, including boiling and the action of oxidizing agents (such as sodium hypochlorite, the active ingredient of household bleach) and antibiotic compounds.

The formation of spores is a complex developmental process, involving the activation of specialized genes. In *Clostridium tetani*, an endospore (Fig. 19.9) forms within the vegetative (that is, growing) cell. The nucleoid and some ribosomes are surrounded by the spore wall, which in turn is surrounded by the spore coat. Because the nucleoid is isolated from it, the rest of the cell degenerates. In other monerans, spores develop on specialized organs. In Actinobacteria, which are abundant in soil, a group of spores forms on a vertical stalk that raises the spores above the substrate so that they can be blown to new sites by air currents. Myxobacteria, another group of soil bacteria, form sacs full of spores that are released when the sac is hydrated (Fig. 19.10).

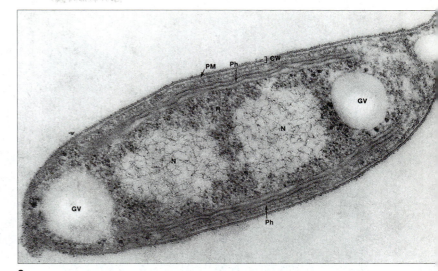
a

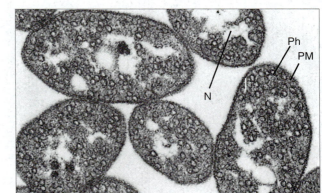
b

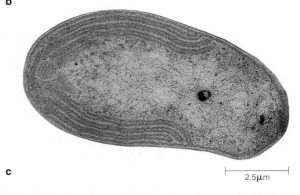
c

2.5μm

Figure 19.8 Monerans with internal membranes. (**a**, **b**) Photosynthetic monerans, showing two different arrangements of thylakoid membranes. (**c**) A chemolithotroph with internal cisternae. PM, plasma membrane; CW, cell wall; Ph, photosynthetic thylakoids; N, nucleoid; GV, gas vesicle (for flotation).

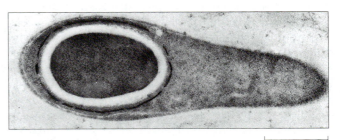

Figure 19.9 Endospores (light area) inside *Clostridium*, an anaerobic bacterium. 2.2μm

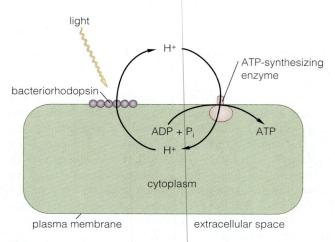

Figure 19.11 The mechanism of photosynthetic energy metabolism in *Halobacterium halobium*.

Figure 19.10 A fruiting form of the myxobacterium *Chondromyces crocatus*. The organism is a large (0.2-mm) colony with some specialized cells. The raised appendages are filled with spores.

Each of the structures of moneran cells plays some role in their survival, growth, or spread. The next section describes the processes by which various monerans find the energy to live, grow, and reproduce.

19.3 LIFESTYLES OF SELECTED GROUPS OF MONERANS

In the late 1970s, when enough ribosomal RNA sequences had been analyzed to start comparing the sequences from diverse organisms, a remarkable division of the moneran kingdom became clear. The majority of well-studied monerans had ribosomal sequences that varied among one another but were clearly based on one model; another set of monerans had sequences that definitely did not fit that model. Several lines of evidence led scientists to suggest that the second set represented the descendants of very ancient organisms. The former group was named the **Eubacteria** (*eu-* from the Greek for "true"); the latter group was named the **Archaebacteria** (*archae-* from the Greek for "ancient"). Interestingly, from their ribosomal RNA sequences, the Archaebacteria seem to be more closely related to eukaryotes than the Eubacteria are. Table 19.1 lists the major types of bacteria in these two groups and some of their characteristics.

Archaebacteria Inhabit Harsh Environments

All the known Archaebacteria genera have the basic moneran structure, with few or no specializations. Their metabolic pathways are likewise comparatively simple. For example, **methanogens** (methane-generating bacteria) derive energy from the reaction

$$CO_2 + 4H_2 \rightarrow CH_4 + 2H_2O$$

CH_4 is methane, a component of natural gas. This reaction is a downhill reaction and so can produce usable free energy, enabling the methanogens to live in reducing (oxygen-poor) environments, such as swamp mud and the rumens of cows. In our normal, oxygen-rich atmosphere, methane tends to be oxidized to CO_2; but without oxygen, and in the presence of H_2, the formation of methane is favored. Therefore, methanogenic bacteria are actually *required* to live without oxygen in order to obtain the free energy they need. All methane on the Earth, including natural gas, comes from this reaction, catalyzed by methanogenic bacteria. The rate of methane production throughout the world, 2×10^9 tons per year, is equivalent to several percent of the rate of photosynthesis.

Another archaebacterium, *Halobacterium halobium*, has a unique and simplified type of photosynthesis. A photoreceptor, bacteriorhodopsin, which is more similar to the mammalian eye photoreceptor than to any type of chlorophyll, is embedded in its plasma membrane. There is no associated electron transport system. When this photoreceptor absorbs light, it forces a proton from the cytoplasm to the outside of the cell, forming a pH and electrical gradient (Fig. 19.11). This gradient can be used to synthesize ATP by the same mechanism that occurs in mitochondria. Thus the bacterium derives useful chemical energy from light, but because there is no photosynthetic electron transport chain, it cannot make carbohydrates by reducing CO_2 (see Chapter 10).

Table 19.1   Selected Types of Monera

Group	Example (Genus)	Comment	Energy Source
Archaebacteria			
Methanogenic bacteria	*Methanobacillus*	$CO_2 + 4H_2 \rightarrow CH_4 + 2H_2O$	Lithotrophic
Thermoacidophilic bacteria	*Thermoplasma*	Grows at pH 1, 60–80°C (140–176°F)	Lithotrophic
Halophilic bacteria	*Halobacterium*	Grows in saturated salt solutions	Phototrophic, heterotrophic
Eubacteria			
Aphragmabacteria	*Mycoplasma*	Causes pneumonia	Heterotrophic
Spirochaetae	*Treponema*	Causes syphilis	Heterotrophic
Sulfur-reducing bacteria	*Desulfovibrio*	SO_4^{2-} + lactate $\rightarrow H_2S$ + pyruvate	Heterotrophic
Nitrogen-fixing aerobic bacteria	*Rhizobium*	Symbiotic N_2 fixer	Heterotrophic
Omnibacteria	*Escherichia, Erwinia, Agrobacterium*	Includes enterics	Heterotrophic
Fermenting bacteria	*Lactobacillus, Bacteroides, Clostridium*	Anaerobic	Heterotrophic
Aerobic endospore-forming bacteria	*Bacillus*	Endospore formers	Heterotrophic
Micrococci	*Staphylococcus*	Gram-positive aerobes	Heterotrophic
Actinobacteria	*Streptomyces*	Antibiotics producers	Heterotrophic
Myxobacteria	*Stigmatella, Chondromyces*	Colonial spore formers	Heterotrophic, lithotrophic
Pseudomonads	*Pseudomonas*	Oxidizer of various organic compounds	Heterotrophic, lithotrophic
Chemoautotrophic bacteria	*Nitrosomonas, Nitrobacter, Methylomonas*	Nitrogen, sulfur oxidizers	Lithotrophic
Anaerobic phototrophs	*Chlorobium, Rhodomicrobium, Rhodopseudomonas*	Green and purple sulfur bacteria, purple nonsulfur bacteria	Phototrophic
Cyanobacteria	*Anabaena*	Oxygen producer	Phototrophic
Chloroxybacteria	*Prochloron*	Oxygen producer	Phototrophic

Adapted from Margulis and Schwartz, *The Five Kingdoms*. New York: Freeman, 1988.

Archaebacteria live in what we would consider very harsh environments. As has been mentioned, methanogens live in oxygen-poor environments. Another group of Archaebacteria, the *thermoacidophiles*, live in hot, acid environments such as volcanic hot springs. For them the optimum temperature is 70–75°C (158–167°F), with a maximum of 88°C (190°F); the optimum pH is 2 to 3 (minimum pH 0.9). *Halophiles*, a third group, are found in saturated salt solutions, for instance in hot springs and drying salt flats (Fig. 19.12). Halophiles have adapted so well to these conditions that they have (and need) little or

Figure 19.12 Evaporation ponds near Blenheim on the north coast of the south island of New Zealand. The red color is caused by *Halobacteria*, among other microorganisms, which thrive in salt-saturated waters. Similar colored ponds can be seen at the south end of San Francisco Bay, where companies evaporate bay water and collect the salt crystals for commercial use.

no cell wall. If a halophile is moved from its normal environment to distilled water, it will burst. One of the reasons that Archaebacteria are thought to be ancient is that these environments are thought to have prevailed on the primitive Earth several billion years ago.

Eubacteria Include the More Specialized Monerans

The Eubacteria are much more diverse, especially in the various ways that they derive energy from their environment. Much of the classification of Eubacteria is based on their energy metabolisms because they reflect basic genetic capabilities. **Heterotrophs** (*hetero-*, "other"; *-troph*, "related to feeding") derive their energy from the breakdown of organic compounds. **Chemolithotrophs** (*chemo-*, "chemical"; *litho-*, "rock") derive their energy from catalyzing inorganic reactions. **Phototrophs** (*photo-*, "light") derive energy by absorbing light photons. Energy metabolism also has ecological implications because it may determine where a species lives and how the species affects its environment.

HETEROTROPHS Heterotrophs live on the organic compounds of living or dead tissue or on the excretions of other organisms. In living tissues, these organisms are parasites and are generally harmful, but in dead tissue and on excretions they play the valuable ecological role of recycling carbon, nitrogen, and other elements that are locked in unused and otherwise unusable materials. Usually, energy is released for the use of the heterotrophs through oxidation of the organic compounds. Various species of bacteria use different electron acceptors in these reactions, including oxygen, nitrate, nitrogen, and sulfate.

$$\text{organics} + O_2 \rightarrow CO_2 + H_2O$$
$$\text{organics} + NO_3^- \rightarrow CO_2 + N_2$$
$$\text{organics} + N_2 \rightarrow CO_2 + N_2O$$
$$\text{organics} + SO_4^{3-} \rightarrow CO_2 + S^0 + H_2S$$

(These reactions are not balanced, but they show the primary substrates and products.)

In the absence of oxygen or another inorganic electron acceptor, some bacteria (for instance, *Lactobacillus*, a milk spoilage bacterium) derive energy from organic molecules by converting them to more stable states—generally by splitting them up (which increases disorder) and by catalyzing the transfer of electrons in internal oxidation–reduction reactions:

carbohydrates → lactic, acetic, formic, and carbonic acids

The most intensively studied moneran, *Escherichia coli* (see Figs 19.2, 19.3), is a heterotroph. It belongs to the group Omnibacteria and the family Enterobacteriaceae (often called the enteric or coliform bacteria, referring to the gut or the colon), whose members live in the soil and in the intestines of animals. In intestines, they get their energy by metabolizing undigested foods; in the soil,

Figure 19.13 Steam vent in Lassen Volcanic National Park in northern California. Some thermophilic bacteria derive energy from reduced sulfur compounds, easily detected by their smell, which are dissolved in the water.

they get it by metabolizing litter, excretions, dead microbes, and other sources of organic material. The presence of coliform bacteria in water supplies is used as an indicator that the water is contaminated with sewage. This indicates a problem because human sewage carries pathogenic bacteria and viruses. The particular coliform bacteria that are detected in a water test may be perfectly safe—some are, after all, natural inhabitants of healthy animals—but some strains of *E. coli* produce toxins that cause severe infections.

CHEMOLITHOTROPHS The accumulation of any compound, organic or inorganic, in a relatively reduced (electron-saturated) state provides an organism in an oxygen atmosphere with the opportunity to obtain energy by catalyzing the transfer of electrons from the reduced compound to the oxygen (or a relatively oxidized substitute). Chemolithotrophs (Fig. 19.13) specialize in the oxidation of inorganic compounds. Like heterotrophs, they are especially important in recycling elements such as nitrogen and sulfur, so that these elements can be used by other organisms. The oxidation of sulfur by bacteria also releases the elements that were chemically bound to it, which can help in the smelting of metals (see sidebar, "Bacteria and Biomining," p. 317). Following are a few of the many possible reactions and the monerans that use them to obtain energy.

Nitrosomonas, found in soil and deep water:

$$2NH_3 + 3O_2 \rightarrow 2H^+ + 2NO_2^- + 2H_2O$$

Nitrobacter, found in soil:

$$NO_2^- + 2H^+ + O_2 \rightarrow NO_3^- + H_2O$$

Thiobacillus, found in hot springs:

$$S^{2-} + 2O_2 \rightarrow SO_4^{2-}$$

Thiobacillus denitrificans:

$$5S + 6NO_3^- + 2H_2O \rightarrow 5SO_4^{2-} + 3N_2 + 4H^+$$

This is one of many denitrifying metabolic reactions, producing the relatively inert nitrogen gas. Denitrification can also be accomplished by heterotrophs.

Methylomonas:

$$CH_4 + 2O_2 \rightarrow CO_2 + 2H_2O$$

Methylomonas does not extract energy from complex organic molecules such as glucose, as heterotrophs do—only from the above reaction. So, in spite of the organic compound methane being a substrate, this reaction is much more like those of the other chemolithotrophs.

PHOTOTROPHS Like plants, moneran phototrophs (Figs. 19.8a, b; 19.14) derive their energy from sunlight. They have light-absorbing pigments (bacteriochlorophyll or chlorophyll) that collect light energy, which excites electrons and promotes their transfer from one molecule to another. In other words, these organisms store light energy by converting it to the potential energy of relatively unstable chemical compounds. There are several types of photosynthetic monerans. They differ not only in their cell structure and ecological habitats, but in the substrates they use in their photosynthetic reactions.

The main types of photosynthetic monerans are the green sulfur bacteria, purple sulfur bacteria, purple nonsulfur bacteria, and the cyanobacteria and chloroxybacteria. Following are the photosynthetic reactions that each performs.

Green sulfur bacteria:

$$2H_2 + CO_2 + light \rightarrow H_2O + CH_2O$$

Green and purple sulfur bacteria:

$$2H_2S + CO_2 + light \rightarrow H_2O + CH_2O + 2S$$

Purple nonsulfur bacteria:

$$2H_3CCH_2OH + CO_2 + light \rightarrow H_2O + CH_2O + 2H_3CCHO$$

Cyanobacteria and chloroxybacteria:

$$2H_2O + CO_2 + light \rightarrow H_2O + CH_2O + O_2$$

Although at first glance these reactions may seem unrelated, they actually follow a standard pattern. In each case, carbon is reduced from an oxidized state—CO_2—to the more reduced state of carbohydrate—CH_2O. CH_2O per se is not really formed in photosynthesis; but its chemical formula, multiplied by six, corresponds to the elemental composition of a simple sugar (such as glucose) that is formed in photosynthetic cells (see Chapter 10). The carbon in CO_2 receives electrons, but what you see in the formulas is that it gets hydrogens. Getting electrons and getting hydrogen atoms are equivalent events because the hydrogens represent protons that follow the electrons. CO_2 also loses an oxygen, which is reduced and appears on the right-hand side of the equation as H_2O. Also, in each case an electron-rich element or compound loses electrons (and protons) and appears on the right-

BACTERIA AND BIOMINING

Monerans are impressive, both in the variety of chemical reactions they catalyze to obtain energy and in the scale at which they operate. Much of the Earth's surface has been molded by the chemical activity of bacteria. It is thought that before the appearance of bacteria, virtually all the iron on Earth was present in a chemically reduced form; now, most beds of iron ore are oxidized. Iron- and manganese-oxidizing bacteria have been isolated and studied.

The abilities of some species of bacteria to chemically modify various elements are now being used in the smelting of ores to extract copper, gold, and phosphate. Copper (Cu) is often found in rocks as an insoluble sulfide, CuS. Traditional energy-intensive smelting methods require that the rock be crushed and heated to extract the metal. A much easier and cheaper method is to soak the ore in an acidic solution containing *Thiobacillus ferroxidans*, a bacterium that obtains energy by oxidizing sulfide to sulfate. This bacterium functions happily under acidic conditions, which also promote the oxidation of reduced copper. The resulting copper sulfate is soluble and can be leached from the remaining insoluble rocks. The metal is then prepared from the copper sulfate solution. At the present time, 25% of all copper produced in the world comes from this *biomining* technique.

The same bacterium can help extract gold from sulfide-containing ores. In low-quality ores, small particles of gold are trapped in an insoluble matrix of copper and iron sulfides. Oxidation of the sulfur and leaching of the remaining sulfates release the gold, which can be recovered by extraction with cyanide. Treating the ore with suspensions of *T. ferroxidans* is cheaper than the traditional high-pressure, high-temperature method of oxidation, and it extracts a higher percentage of the gold.

Phosphate, needed as fertilizer in agriculture and in some industrial processes, is traditionally extracted from ore by using high temperatures or sulfuric acid. A new process dissolves the phosphates in organic acids, metabolic by-products of the bacteria *Pseudomonas cepacia E-37* and *Erwinia herbicola*. This method is more environmentally friendly because the organic acids are themselves biodegradable through the actions of other bacteria.

Current research is aimed at finding, selecting, or perhaps genetically engineering new strains of bacteria that work quickly and effectively and under the harsh physical and chemical conditions that may exist in piles of rock sitting outside mines.

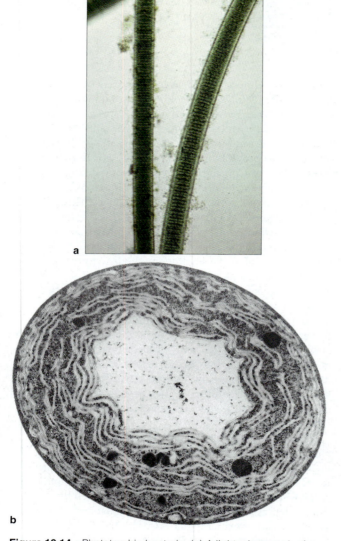

Figure 19.14 Phototrophic bacteria. (**a**) A light micrograph of a species of *Lyngbya*, a filamentous cyanobacterium; (**b**) an electron micrograph of *Prochloron*, a member of the Chloroxybacteria, which has been found only as an endosymbiont in cells of a tunicate (a marine invertebrate).

hand side of the equation in its oxidized state. A classical way of writing these reactions to show the similarities is

$$2nH_2A + nCO_2 + \text{light} \rightarrow nH_2O + nCH_2O + 2nA$$

In this equation, A may be either nothing (H_2A may just be H_2), or S, or an organic compound such as H_3CCHO (acetaldehyde), or O. The number of carbohydrate units produced is shown by n. Note that only cyanobacteria and chloroxybacteria are able to oxidize water and produce O_2. The green and purple sulfur bacteria and the purple nonsulfur bacteria are poisoned by oxygen and must grow under anaerobic conditions. The evolution of the capability to produce O_2 was a major innovation in photosynthetic metabolism. It is thought to be an impor-

tant stage in the development of the Earth as we now know it because all the O_2 in the atmosphere was formed by the ancestors of cyanobacteria and chloroxybacteria.

There are also major differences among the moneran phototrophs in their photosynthetic pigments. The anaerobic phototrophs have bacteriochlorophyll, a pigment related to but different from chlorophyll. Cyanobacteria have chlorophyll *a* in their photosystems, but their major light-harvesting complexes contain proteins called *phycobilins* in a package (called a *phycobilisome*) attached to the surface of the thylakoid membranes. The phycobilin in cyanobacteria—phycocyanin—absorbs yellow-green light; that is why the cyanobacteria look relatively blue-green. Chloroxybacteria have chlorophyll *a* and chlorophyll *b* in their light-harvesting complexes.

Based on the photosynthetic pigments of their present-day members, some primitive cyanobacteria and chloroxybacteria are thought to be evolutionary precursors of the plastids in two major groups of photosynthetic eukaryotes: Rhodophyta (red algae) and Chlorophyta (green algae), respectively. Researchers propose that monerans originally formed an endosymbiotic relationship with a eukaryotic precursor cell (see Chapter 18). Over time, a substantial portion of the genetic material of the moneran was somehow transferred to the nucleus of the eukaryote, and this endosymbiont became an organelle, an actual part of the eukaryotic cell. Cyanobacteria and the eukaryotic Rhodophyta (red algae) both have chlorophyll *a* and phycobilisomes; both also have phycocyanin, although members of the Rhodophyta also have a second phycobilin, phycoerythrin. Chloroxybacteria and Chlorophyta (green algae) both have chlorophylls *a* and *b*. Although the endosymbiotic mechanism is only hypothesized, the existence of two independent moneran/eukaryote pairs with such similar light-harvesting complexes provides strong evidence for its validity.

This section has described the great variety of biochemical pathways by which various monerans obtain their energy for growth and reproduction. While obtaining energy is not the only important activity of a cell, it is one of the more important ones. The chemical formulas just described especially demonstrate that monerans play a major role in the interconversion of inorganic elements and compounds, many of which are nutrients on which plant metabolism depends. Thus, plants depend on monerans for their existence on Earth.

19.4 MONERANS THAT FORM SYMBIOTIC RELATIONSHIPS WITH PLANTS

Monerans and plants also form complex *symbiotic* relationships (arising from living together). Some of these are *mutualistic* (benefit both the moneran and the plant), but others are damaging to the plant.

Rhizobium Forms a Mutualistic Association with Legumes

As described in Chapter 11, nitrogen represents a special nutritional problem for a plant. The world's major store of nitrogen is N_2 in the atmosphere. Plants cannot use this N_2, but some monerans can. The infection of plants with such monerans can be beneficial and even sometimes essential for the plant.

Rhizobium is a heterotrophic bacterium that lives in soil. Along with several other monerans, it synthesizes an enzyme called *nitrogenase*, which gives it the ability to *fix* nitrogen—that is, to convert N_2 to NH_4^+. NH_4^+ can be incorporated by the bacteria and plants into the structures of amino acids and other nitrogen-containing organic compounds. Nitrogen fixation is an energy-using reaction; each N_2 requires the addition of 8 high-energy electrons and the hydrolysis of 16 ATP to form 2 NH_4^+ ions. The bacteria get this energy by metabolizing other organic compounds, especially carbohydrates. Nitrogen fixation works best in a low-oxygen atmosphere because oxygen inactivates the nitrogenase enzyme.

Rhizobium has developed an ability to form a close mutualistic association with legumes, plants of the family Fabaceae. Both partners in the association contribute, and both partners derive benefits. The plant contributes high-energy carbohydrate and a protected environment; the bacterium contributes nitrogenase and other enzymes. The benefit to both partners is the supply of fixed nitrogen. The association occurs in special organs called **root nodules** (see Fig. 5.20).

The establishment of the symbiotic association is a very complex process in which each partner triggers processes in the other, including the induction of genes to make proteins that are used only in the active nodules. The sequence of events, as far as we now know it, is:

1. The root secretes an attractive chemical.

2. The chemical induces *Rhizobium* bacteria in the vicinity to swim toward the root. It also begins the induction of nitrogen fixation genes in the bacteria.

3. The bacteria enter at a root hair and move inward through an infection thread (a tube of plasma membrane), losing their cell wall and synthesizing nitrogen-fixing enzymes as they do so.

4. When the bacteria reach the root cortex, they are released from the infection thread into several cells. The bacteria without their cell walls are called *bacteroids*; they become surrounded by a special membrane, the peribacteroid membrane.

5. Chemicals secreted by the bacteria (or bacteroids) during the formation of the infection thread later induce cell division in the root cortex and pericycle, forming the nodule. These chemicals also induce the synthesis of specialized nodule proteins, including a type of hemoglobin (leghemoglobin) that buffers the oxygen concentration in the central part of the nodule, where the nitrogen is fixed.

A typical nodule in soybean has a core of bacteroid-infected cells surrounded by a cortex of parenchymal cells that contain extensive vascular bundles. The photosynthetically produced carbohydrate sucrose is transported to the nodule via the phloem (Fig. 19.15). The sucrose then diffuses or is actively transported to *Rhizobium* cells in the core of the nodule. There, oxidative respiration of the carbohydrate provides the high-energy electrons and ATP necessary to reduce N_2 to NH_4^+. The oxidative reactions not only provide energy but also reduce the concentration of O_2, a necessary step to preserve the nitrogenase enzyme. Leghemoglobin, present in the plant cells that host the bacteroids, also helps regulate or buffer the concentration of O_2 (Fig. 19.15). The NH_4^+ produced by the bacteroids is used in the synthesis of a key amino acid, glutamine (gln), which can donate nitrogen to other compounds. A substantial fraction of this amino acid is exported to the plant cells in the cortex of the nodule, which convert it into an easily transportable nitrogenous compound. This is secreted to the apoplast and transported up the xylem to the growing parts of the shoot.

Bacteria Can Be Plant Parasites

A symbiotic association is not always mutualistic; those in which one organism benefits at the expense of the other are parasitic. *Agrobacterium tumefaciens*, described in Chapter 17, is a bacterial plant parasite; there are many others. Two important examples are *Pseudomonas syringae* and *Erwinia amylovora*. Bacterial plant pathogens are divided into subgroups called *pathovars*, according to the plants that they infect. Pathovars of *P. syringae* cause the wildfire disease of tobacco; blights of beans, peas, and soybeans; and diseases of several other crops. Pathovars of *E. amylovora* cause fireblight of apple and pear. The names *wildfire* and *fireblight* suggest the speed at which a bacterial infection can spread through a field of tobacco or a large apple tree.

Bacteria are carried to uninfected plants by water, insects, humans, or other animals. *E. amylovora*, for example, is carried from flower to flower in apple or pear orchards by bees, as they make their pollination rounds. The bacteria can enter the plant tissues through natural openings: stomata, lenticels, hydathodes (the openings at the tips of leaves through which guttation water is extruded), and nectarthodes (the openings in flowers from which nectar comes). Once inside the plant, these heterotrophs multiply quickly, absorbing nutrients from the plant cells. Some bacteria secrete enzymes that break down the plant cell walls; some produce chemical toxins; and some apparently cause their damage simply by absorbing nutrients and multiplying. The infected plant

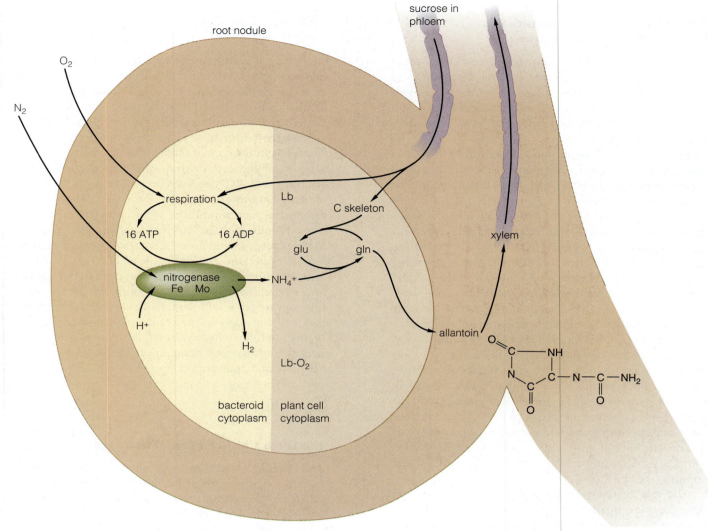

Figure 19.15 The synthesis of fixed nitrogen in a *Rhizobium*–legume nodule. Carbohydrate (sucrose) arrives via the phloem. Some of the carbohydrate is donated to the bacteroids for use in respiration; some is used to make carbon skeletons to which NH_4^+ can be attached. Glutamic acid (glu) and glutamine (gln) are nitrogen-rich amino acids; allantoin is another nitrogen-rich compound. These carry nitrogen to the plant shoot in the xylem stream. N_2 and O_2 arrive at the nodule via air spaces in the soil; they diffuse into the core. O_2 is used quickly in bacterial respiration; this keeps the oxygen concentration low and protects the nitrogenase enzyme with its iron (Fe) and molybdenum (Mo) cofactors from inactivation by oxygen. Changes in the oxygen concentration are buffered by leghemoglobin (Lb), an oxygen-binding protein made by the plant.

tissues often turn brown and die. A large portion of the plant may be involved. Some bacterial pathogens can overwinter in the dead tissue, returning to infect new plant tissue at the next growing season.

Plants have some defenses against the infecting bacteria. Sometimes, depending on the plant and the infecting agent, a plant will react by producing antibiotic compounds, including phytoalexins and hydrogen peroxide. This is called the **hypersensitive response**. The antibiotics may directly kill some pathogen cells, but hydrogen peroxide may also restrict the spread of infection by killing adjacent plant cells. In such a case the infection

may produce a spot of necrosis in a leaf, but it will not kill the whole leaf or the plant.

These two descriptions of symbiotic associations between a plant and a moneran demonstrate the complexity of the relationships between members of the two kingdoms. They also show how important it is to consider the monerans in the environment when describing the biology and ecology of plants.

Viruses—although they are not free-living and do not influence the physical environment—are parasites of plants and also form complex relationships with them. The next section describes some of these relationships.

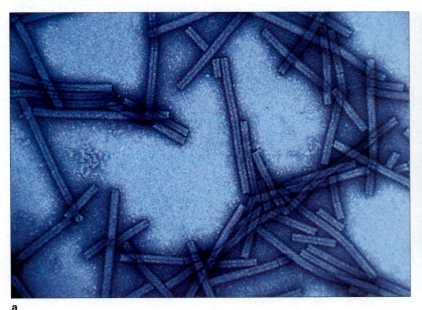

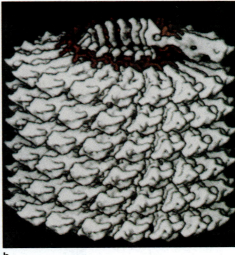

b

a

Figure 19.16 An electron micrograph of tobacco mosaic virus (**a**), together with an interpretive diagram of the virus's arrangement of nucleic acid (in red) and protein (**b**).

19.5 VIRUSES

Viruses are subcellular parasites. They are said to be subcellular because they have neither the internal structures found in moneran or eukaryotic cells nor the capacity to reproduce on their own. They are considered parasites because they invade cells and use their hosts' metabolism to produce more of their kind. Many biologists have debated the question of whether viruses are alive. Their inability to reproduce suggests that they cannot be considered truly alive; yet what functions they do possess are all based on physical and chemical principles basic to normal living organisms.

Viruses Are Infectious Genes

All types of living organisms have viruses that infect them. You are undoubtedly acquainted with the effects of various human viruses: chicken pox, measles, mumps, and influenza are all caused by specific viruses, as is acquired immune deficiency syndrome (AIDS). Insects and other animals, fungi, and bacteria all have their own types of viruses; and so do plants.

Viruses have a simple structure, compared to cells. They contain a nucleic acid genome surrounded by a coat of protein. (The coat sometimes includes lipids.) Surprisingly, the genome may be either RNA or DNA, depending on the type of virus, and it may be single-stranded or double-stranded. Whatever its structure in the virus, the genome contains the information for making virus-specific proteins by using the same genetic code that operates in all cells. The coat may take different forms: It can be a rod, a polyhedron, or a more complex shape.

Whatever its shape and composition, it protects the nucleic acid as it is transferred from one organism to the next and is sometimes instrumental in the infection process. Some viruses also contain one or more enzymes that are used in the infection process, but outside a cell no virus has any metabolic activity (respiration or protein synthesis, for example). One might consider a virus to be an infectious gene.

Tobacco mosaic virus is a typical plant virus. It has a single-stranded RNA genome complexed with coat protein subunits, all wound into a helical rod (Fig. 19.16). When RNA and protein are separated chemically and applied separately to test plants, a plant that receives the RNA will develop a viral infection; a plant that receives the protein will not. The experiment demonstrating this was among the early pieces of evidence that nucleic acids are hereditary material, carrying the instructions for reproduction of the entity from which they came.

Virus Infections Stunt Plant Growth

When the tobacco mosaic virus infects a cell of tobacco, tomato, or another related plant, the coat protein is removed, and the viral RNA serves as a messenger RNA—forming complexes with the cell's ribosomes and using the cell's amino acids, tRNAs, and energy source to make viral proteins. The first protein formed is an enzyme (called an RNA-dependent RNA polymerase) that forms a complementary strand of RNA (that is, a single-stranded RNA with a base sequence complementary to that of the viral genome), using the viral genome as a template. This complementary RNA can then serve as a template for the formation of new viral genomes.

Figure 19.17 Symptoms of viral diseases in plants. (**a**) A mosaic in Chinese cabbage (*Brassica pekinensis*). (**b**) Necrotic local lesions on a tobacco (*Nicotiana tabacum*) leaf. (**c**) Variegation in tulips (*Tulipa gesneriana*), also called color breaking.

Later, a second protein, the coat protein, will be synthesized. This protein will bind to and wind up the new viral genomes into complete virus particles. The viral genome codes at least one other protein, which may promote the transmission of the virus.

Plant viruses, including tobacco mosaic virus, cannot infect plant cells without help. They are too large to pass through a normal primary cell wall, not to mention the barriers (epidermis, cork) that protect a plant from invasions. In nature, plant viral infections are almost always spread from plant to plant by insects, which pick up the viral particles as they chew or suck on infected plants and then transmit them when they chew or suck on uninfected plants. Mites (arachnids, related to spiders), nematodes (roundworms), and fungi also can infect plants when they penetrate the cells. In the laboratory, viral infections are induced by rubbing abrasive-treated leaves with a viral suspension. The abrasive (for example, Carborundum™) makes holes in the cell walls, through which the virus enters. Once in the cell and forming new virus particles, the virus may become systemic; that is, it may spread throughout the plant. Recent evidence suggests that viruses move through the plasmodesmata and phloem and that a viral protein is responsible for opening the plasmodesmata so that particles as large as a virus can pass through. A mutant virus has been found that cannot move systemically. There are also mutant tobacco plants that do not allow tobacco mosaic virus to move systemically but instead confine an infection to a local lesion (a small area of dying or dead cells).

Generally, viral infections of plants do not kill the plants. Because further transmission of a virus that killed its host and made the host unappetizing to insects would be unlikely, natural selection may eliminate plant viruses that are excessively virulent. However, infected plants are usually stunted relative to uninfected plants, and the infection often (but not always) causes changes in the color or shape of the foliage (Fig. 19.17). The mosaic of tobacco mosaic virus, the symptom of a systemic infection, is a pattern of dark- and light-green regions on a leaf. The necrotic local lesions of tobacco mosaic virus are small (approximately one millimeter in diameter) yellow or brown spots. Other viruses cause chlorosis (yellows), concentric white rings on a leaf (ringspot), transparent areas next to the leaf veins (vein clearing), or tumors (galls). Virus infections in flowers sometimes cause beautiful, multicolored patterns (color breaking); variegated tulips are a famous example.

SUMMARY

1. The basic structure of a moneran cell is very simple, relative to a eukaryotic cell. A cell wall and plasma membrane surround the cytoplasm and an undifferentiated nucleoid. Many moneran cells have no other organelles recognizable by electron microscopy. Monerans never have a complex endomembrane system, as eukaryotes do, but some monerans contain internal membranes that provide sites for electron transport reactions and store electrochemical energy.

2. The basic genetic structure of a moneran is also simple. Most genes are lined up on a single circular chromosome. Some specialized genes may be found on smaller DNA circles called plasmids. Plasmids can be transferred from cell to cell; the F plasmid, when incorporated into the chromosome, mediates the transfer of chromosomal genes from a donor cell to a recipient.

3. Some monerans have complex modifications to their extracellular structures, such as an outer lipopolysaccharide membrane (in Gram-negative cells), one or more flagella for motility, and a pilus for transfer of genetic material. Some monerans form specialized spores for reproduction, dispersal, and resistance to harsh environmental conditions.

4. Monerans can be divided into two major groups. Archaebacteria are thought to be the most ancient organisms. Their cell structures and metabolic reactions are relatively simple. Eubacteria are more specialized and often have more complex structures and metabolic capabilities.

5. Monerans can be classified according to the ways in which they obtain energy. Heterotrophs live on the organic compounds of living or dead tissue or on the excretions of other organisms. Chemolithotrophs specialize in the oxidation of inorganic compounds. Phototrophs derive their energy from sunlight. Heterotrophs and chemolithotrophs play major ecological roles by recycling the elements in organic and inorganic waste materials. Phototrophs contribute to this role; like plants, they also contribute to the input of carbon and energy into the organic components of an ecosystem.

6. Nitrogen represents a special nutritional problem for plants. Some plants form mutualistic associations with monerans, cooperating in the fixation of N_2 to form NH_4^+. The plants contribute carbohydrate as a source of energy and provide an oxygen-protected environment; the monerans contribute the enzyme nitrogenase. The development of a root nodule in the legume–Rhizobium symbiotic association is a complex process that involves the expression of special genes in both partners.

7. Some monerans are parasitic on plants, invading leaves and other organs through wounds, stomata, and hydathodes. The moneran cells multiply quickly in the intercellular spaces of the plant, secreting enzymes to break down cell walls and metabolizing and absorbing the contents of the cells. The blighted parts of the plant are destroyed.

8. Plant viruses are subcellular parasites, formed from a nucleic acid genome and a protein coat; they appropriate the plant cells' protein-synthesizing machinery, using it to reproduce huge numbers of themselves. Plant viruses are normally transmitted among plants by insects. Viral infection weakens a plant and causes various visible symptoms, but it does not generally kill the plant.

Questions

1. From the list below, indicate the organelles (a) that are found in every moneran cell, (b) that are found only in specialized moneran cells, and (c) that are found only in plant (or other eukaryotic) cells.

cell wall   ribosome   endoplasmic reticulum

flagellum   peptidoglycan   thylakoid

nucleoid   plasma membrane   nuclear envelope

spore   plasmid   lipopolysaccharide

2. Distinguish between bacteria and viruses. Which do you think appeared first in the evolution of life?

3. Criticize the following statements:
   a. The only way a bacterium can acquire new genetic information is by mutation.
   b. Conjugation in bacteria is just like meiosis in plants and other eukaryotic cells.

4. Archaebacteria are thought to represent an ancient line of organisms that may have existed on Earth for billions of years. Give one reason why they have been able to resist competition from more highly evolved organisms such as eubacteria and eukaryotes. (Hint: Where do archaebacteria live?)

5. For each of the reactions below, indicate whether it would be catalyzed by a heterotroph, chemolithotroph, or phototroph. Which of these reactions would also be catalyzed by a plant cell?
   a. $2H_2O + CO_2 + light \rightarrow H_2O + CH_2O + O_2$
   b. $organics + NO_3^- \rightarrow CO_2 + N_2$
   c. $2NH_3 + 3O_2 \rightarrow 2H^+ + 2NO_2^- + 2H_2O$
   d. carbohydrates $\rightarrow$ lactic, acetic, formic, and carbonic acids
   e. $2H_2S + CO_2 + light \rightarrow H_2O + CH_2O + 2S$

6. Describe the advantages to each partner of the mutualistic symbiotic association between Rhizobium bacteria and legume plants.

7. Distinguish between the methods by which the two plant parasites Erwinia amylovora and Agrobacterium tumefaciens (Chapter 17) derive nutrition by infecting plant cells.

8. A sample of tobacco mosaic virus is separated by chemical procedures into RNA and protein fractions. The leaf of one plant is rubbed with the protein solution. The leaf of another plant is rubbed with the solution of RNA. Which, if either, of these plants will be infected? Will the infected plant produce viral protein, viral RNA, or whole viral particles (RNA plus protein)?

9. Explain the following statement: Agriculturalists have found that the most effective way to stop the spread of a plant viral disease is by the use of insecticides.

Further Readings

Ingraham, J. L, and C. A. Ingraham. 1995. *Introduction to Microbiology*. Belmont, Calif: Wadsworth.

Matthews, R. E. F. 1992. *Fundamentals of Plant Virology*. San Diego : Academic Press.

Moffat, A. S. 1994. "Microbial Mining Boosts the Environment, Bottom Line." *Science* 264 (May 6): 778-79.

Service, R. F. 1994. "*E. coli* Scare Spawns Therapy Search." *Science* 264 (July 22): 475.

20 KINGDOM FUNGI

1. Fungi are eukaryotic heterotrophs that reproduce with spores and have chitinous cell walls.

2. Some fungi destroy crops and stored food, but most are valuable decomposers or symbionts that assist plant growth (in mycorrhizae) or cohabit with algae and cyanobacteria (in lichens). Bread-making yeast is a fungus, and penicillin is a fungal product.

3. Most fungi feed by means of a mycelium, a body composed of fine, branching threads that penetrate into foods.

4. To multiply, some fungi make only asexual spores. This is true of green kitchen mold. In contrast, most mushroom fungi make only sexual spores. Most fungi make both kinds of spores, using asexual spores to multiply and sexual spores to diversify. Divisions of fungi are defined by the way they make sexual spores.

Figure 20.1 A mushroom, the reproductive body of a fungus. The feeding body of the fungus consists of underground filaments (a mycelium) that may be living in symbiosis with a forest tree.

20.1 FUNGI: FRIENDS AND FOES

Everyone has met the fungi, for better or worse (Fig. 20.1). We all know about mushrooms and moldy food, but few people recognize the full importance of fungi in human affairs and world ecology. Some fungi help reduce human hunger by aiding the growth of plants. Other fungi break down wastes and release the elements for use by plants. We make bread, beer, and wine with the help of a yeast fungus. Countless lives have been saved by antibiotics that were first discovered in fungi. In scientific research, a fungus gave the first clear evidence of the way genes act.

Fungi also create problems for humanity. Not many fungi cause human disease, but more than 5000 species attack plants. Dutch elm disease, pine blister rust, and wheat rust are costly examples. Still more damaging are fungi that attack stored food, lumber, and clothing. In the humid tropics, where fungi grow especially well, they destroy a large fraction of the harvests; and in every part of the world, dry rot fungi damage houses.

20.2 FUNGAL TRAITS

What are fungi? **Kingdom Fungi** includes at least 100,000 species, ranging in size from microscopic yeasts to mushrooms up to a meter (about 3 feet) in diameter. Biologists think that all fungi evolved from an aquatic ancestor that moved to dry land about half a billion years ago.

Fungi are heterotrophs; they lack the ability to carry out photosynthesis, so they must get energy as animals and most bacteria do, by taking in organic molecules from the environment. However, lack of mobility hampers active fungi in their quest for food.

Fungi lack the mobility of animals because they have cell walls. This is the trait in which fungi most resemble

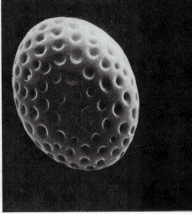

Figure 20.2 Fungal spores, (**a**) *Neurospora* and (**b**) *Gelasinospora cerealis*, as seen with the scanning electron microscopy.

plants. Rapid motion requires flexible cells that can change their shapes. Cell walls prevent such changes. Unlike the walls of plants (which contain cellulose), fungal cell walls contain a modified polysaccharide called **chitin**. Outside of kingdom Fungi, chitin occurs in the walls of one small group of protists, the chytrids (see Chapter 21),* and in the external skeletons of insects, lobsters, and spiders and other arthropods. Chitin synthesis and other similarities suggest that fungi are related more closely to animals than to plants.

Despite their aquatic ancestry, fungi do not have swimming reproductive cells (motile cells that can move about via flagella, as in animal sperm). In this respect they are even less motile than mosses and ferns, which do have swimming sperm cells. Chytrids, the aquatic ancestors of fungi, also have swimming cells. Biologists think fungi lost the ability to make flagella when they moved to dry land.

Lacking mobility, fungi can reach new food sources only by releasing multitudes of tiny reproductive units called **spores** (Fig. 20.2). Spores are not unique to fungi;

---

*Some biologists include the chytrids in the kingdom Fungi.

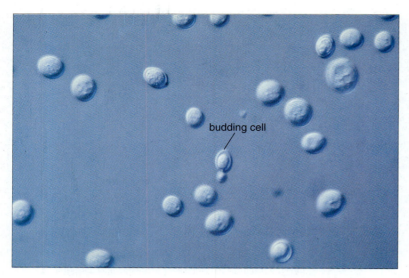

Figure 20.3 Cells of baker's yeast (*Saccharomyces*). One cell is reproducing by budding, ×1700.

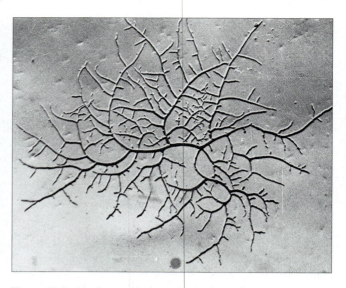

Figure 20.4 Hyphae making up a young mycelium, ×400.

plants and some bacteria also make them. A fungal spore, which is made up of one or a very few cells, can resist dehydration and is small enough to be lifted and carried by a gentle breeze. It contains a food store but remains dormant until it settles into an environment that favors growth. Then the spore quickly grows into an organism. Ever-present spores make it likely that a sandwich, a fruit, or a pile of manure will soon become moldy because it has been attacked by a fungus.

Sometimes common names get in the way of understanding. This is true of the terms *mold* and *fungus*. In nontechnical language, a mold is any visible organism that decomposes organic matter. Thus, fungi are molds. But some molds—including slime molds and water molds—are not part of kingdom Fungi (and are discussed instead in Chapter 21). These molds do not have chitin in their walls.

20.3 THE VARIED ECOLOGICAL ROLES OF FUNGI

Fungi have evolved two solutions to the quest for food. Many fungi are **saprobes**, which digest discarded organic matter. Others take food from living host organisms.

Saprobic fungi compete with bacteria as **decomposers**, the organisms that complete the great ecological cycles of the chemical elements by breaking down complex organic molecules. Without these organisms, such vital elements as nitrogen and carbon would be increasingly tied up in wastes and dead organisms.

Most saprobic fungi are opportunists: They survive by exploiting temporary opportunities such as a fallen fruit or a pile of manure. Many opportunistic fungi are small organisms because they have little time for growth before the food is gone. However, a fungus may grow to cover a large area in a forest or grassland where fallen leaves and dead branches regularly replenish the food supply. Also,

fungal saprobes may grow large by attacking dead trees that contain enough food for prolonged growth.

In **symbiosis**, organisms of separate species have a close relationship in which each organism influences the welfare of the other. Symbiotic relations are classified by the benefit to the partners: If both organisms benefit, the relation is **mutualistic**; If one partner benefits and the other is harmed, the relation is **parasitic**. Fungi engage in both kinds of symbioses.

Mutualistic fungi are widespread and important to plants. For example, most forest mushrooms grow from fungi that have underground connections with tree roots (see Chapter 5). In these associations, called mycorrhizae, the fungus penetrates a plant root, growing between cells and sometimes entering them. The fungus takes nutrients from the root, but there is clear evidence that the plant also benefits; many plants grow poorly without their fungal partners. The full extent of the plant's benefit is still unknown, but research shows that the fungus can pass phosphorus or nitrogen from soil to the plant (see Chapter 11). Fossil studies suggest that mycorrhizal associations may be almost as old as the plant kingdom, dating back at least 450 million years. The majority of present-day plants may have fungal partners.

A few parasitic fungi attack animals, but most of them attack plants. Rust fungi are familiar parasites in the lawn; their spores may color the leaf blades orange. Rust and smut fungi infect crop plants such as corn and wheat, draining energy and reducing the harvest. Wheat rusts cost millions of dollars every year in lost harvests. They evolve rapidly, and continuous efforts are needed to breed new wheat plants that can resist them.

20.4 MYCELIAL GROWTH

The simplest fungi, **yeasts**, are separate, microscopic, rounded cells (Fig. 20.3). However, most fungi develop a multicellular feeding body called a **mycelium** (Fig. 20.4).

The mycelium, which is the vegetative body of the fungus, consists of branching tubular cells called **hyphae** (singular, **hypha**). A few fungi can switch between mycelial and yeastlike growth, depending on the growing conditions. It is easy to overlook the mycelium; its hyphae are extremely slender and are usually buried in a food mass. A mushroom is a reproductive body made of many hyphae tightly pressed together.

A mycelium is well-equipped to penetrate a food mass such as a rotting peach. Hyphae grow at their tip, where the cell wall is soft and extensible. Behind the tip, each hypha periodically forms a branch, a new hypha. In this way, a single hypha quickly branches into a mycelium. Each hyphal tip secretes digestive enzymes that break proteins and other large food molecules into smaller molecules such as amino acids and sugars. These molecules filter through the hyphal cell wall and enter the cytoplasm, sometimes aided by active transport. The accumulated solutes in the hypha cause water to enter the hypha by osmosis. The water builds up pressure, which stretches the cell wall, elongating the hypha and driving it deeper into the food mass. Because these events require water, mycelia become inactive when the substrate dries. Dry rot, so common in houses, is misnamed; the fungus spreads only when wood is wet.

If a mycelium is cut into fragments, the cut hyphae seal their broken ends and continue growing as separate fungi. Even without injury, local regions of a mycelium may compete for food as if they belonged to entirely different mycelia. However, some food sharing does occur within the mycelium. The microscope shows a mass transport of material, like the flow of water in a pipe, from feeding hyphae to sites of reproduction. In addition, higher magnifications show a slower, more controlled movement of particles along tracks in hyphae. This **cytoplasmic streaming** can either oppose or augment the mass transport. These movements can carry particles throughout a whole mycelium.

Most fungi can make nearly all the molecules they need from a few environmental raw materials. They need water, a few minerals, a few vitamins, and an organic compound (a sugar will serve) for carbon and energy. The vitamin most often needed is the same vitamin B_1 that humans need. As a group, fungi can use many materials as organic substrates: wood, paper, glue, leather, manure, fruits, and so on.

How do fungi survive in competition with bacteria, which were highly successful opportunists for two billion years before fungi evolved? To reach new food sources, any opportunist—whether fungus or bacterium —must feed quickly, reproduce abundantly, and disperse its reproductive cells widely. Bacteria excel in fast feeding and reproduction; their tiny cells can double in number every half hour, and each cell is a reproductive unit. But if bacteria live deep within a food source, their cells may be slow to reach the surface for dispersal. The fungal mycelium is a pipeline that quickly moves food from the depths of a food source to the surface, where spores are made and dispersed. These features allow fungi to exploit deeply buried food while bacteria thrive at the surface. In addition, mycelia often secrete acids and antibiotics that slow bacterial growth. Acidic soils, such as those beneath conifers or heaths, are rich in fungal decomposers and poor in soil bacteria.

20.5 FUNGAL REPRODUCTION

Fungi rely on spore production to reach new food sources. Lacking motility, spores cannot actively seek foods; they must drift passively until chance takes them to a resource. The great majority of spores die, so a fungus must build vast numbers of spores as fast as possible. Most fungi meet the challenge by means of asexual reproduction, in which parts of the parental body pinch off, drift for a time, and form new mycelia if they reach a food source in a favorable environment. Sometimes the reproductive cells have no special protective features, but more often they are spores with specialized walls that aid in both survival and dispersal. Spore walls like those in Fig. 20.2 may contain water-resistant compounds that limit dehydration, and other spores may have extensions that catch air currents or hooks that anchor the spores to animals.

Asexual reproduction is the fastest, cheapest way to multiply because it does not require a partner. However, it has one disadvantage: It generates offspring that are genetically identical to the parent. If all the offspring are alike, they lack the variety needed to exploit new kinds of opportunities. They are alike because the nuclei in asexual spores are made by mitosis, in which the chromosomes are copied faithfully and divided between two identical nuclei (see Chapter 12). Asexual spores are called **mitospores** to emphasize their origin.

Most fungi also have sexual reproduction, which can produce spores containing a mixture of genes from two parents. Sexual reproduction is slower and more complex than asexual reproduction because it requires a meeting and fusion of parental gametes, followed by meiotic events that join and sort out chromosomes. In most fungi, the steps in sexual reproduction make up a zygotic life cycle (Fig. 20.5). You may want to contrast this life cycle with the other life cycles shown in Chapter 12.

The cycle begins when a fungal spore germinates and grows into a haploid (n) mycelium. Within the mycelium, the original spore nucleus divides many times by mitosis, making hyphae with many nuclei. Mating occurs when two compatible mycelia meet. Unlike animals, fungi do not make eggs and sperm. Instead, the proximity of a partner causes some hyphae to specialize and play the part of gametes. They fuse in a process called **plasmogamy** and make a special cell that contains nuclei from both parents. The fusion cell may grow into a **dikaryotic** ($n+n$) mycelium, in which every cell contains

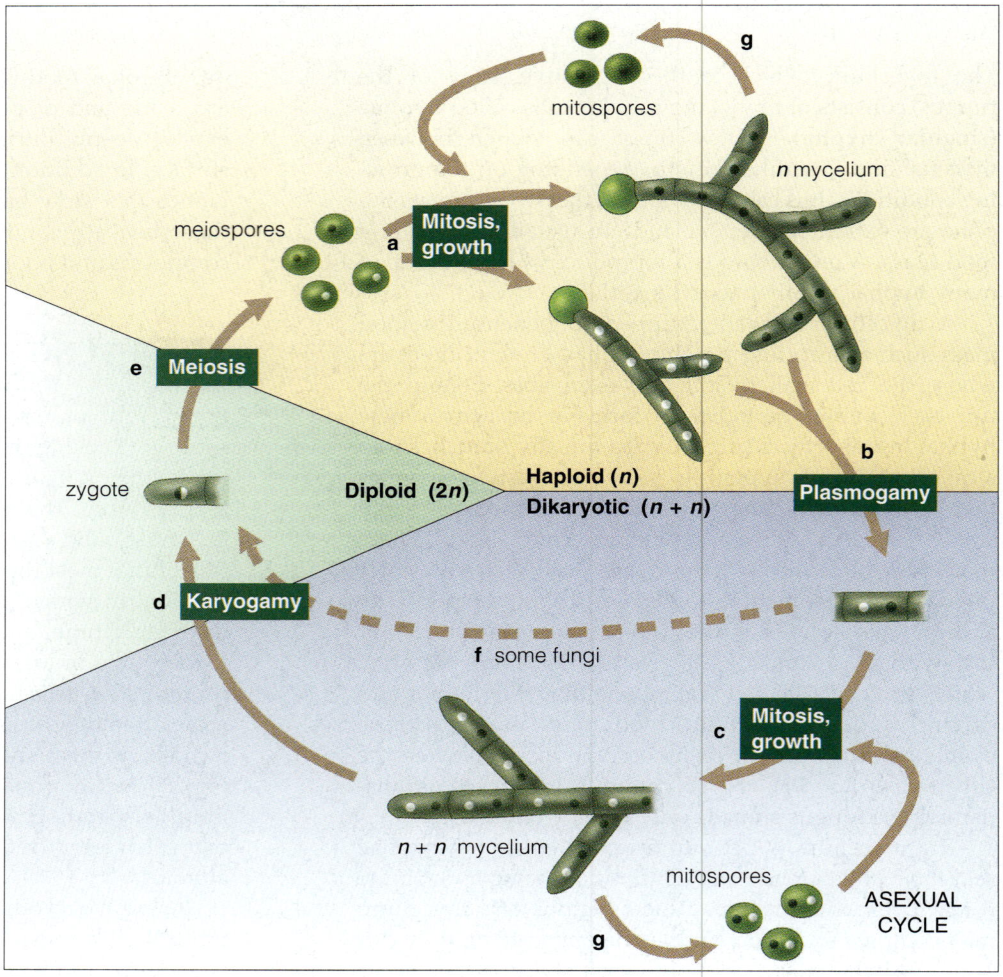

Figure 20.5 Generalized zygotic life cycle of fungi. (**a**) Haploid spores grow into haploid mycelia. (**b**) Hyphae fuse at their tips (plasmogamy), making a dikaryotic (n + n) cell. (**c**) The n + n cell grows into a dikaryotic mycelium with paired nuclei. (**d**) The paired nuclei in some dikaryotic hyphal tips fuse (karyogamy) to make diploid (2n) zygote cells. (**e**) The diploid nucleus goes through meiosis to make haploid meiospores. (**f**) Not all fungi form an n + n mycelium; in some, plasmogamy is followed directly by nuclear fusion (karyogamy) and meiosis. (**g**) Most fungi make asexual mitospores, which may be either dikaryotic or haploid.

one haploid nucleus from each parent. As the hyphae grow, the paired nuclei divide simultaneously by mitosis. Fungi are the only organisms with a dikaryotic stage in the life cycle. Eventually, some of the paired nuclei fuse in a process called **karyogamy**. This creates diploid (2n) zygote cells, which are the only diploid cells in the life cycle of most fungi. The 2n nucleus immediately goes through meiosis (the reason for labeling this a zygotic life cycle), creating four haploid (n) nuclei. During meiosis, the matching parental chromosomes recombine, so the new haploid nuclei contain chromosomes with parts from both parents. Finally, the recombinant nuclei are packaged into new spores. To emphasize their origin, the recombinant spores are called **meiospores**. They can grow into new haploid mycelia, initiating a new life cycle.

20.6 CLASSIFICATION BY SEXUAL REPRODUCTION

Though all fungi reproduce with spores, the details vary enough to provide a basis for classification. Careful studies show that some fungi lack sexual reproduction altogether, whereas the rest make sexual spores in three distinctive ways. Accordingly, mycologists divide king-dom Fungi into four divisions: the Zygomycota, Ascomycota, Basidiomycota, and Deuteromycota.

When fungi in division **Zygomycota** fuse hyphae, they convert the resulting cell into a resting **zygospore** before proceeding with reproduction (Fig. 20.6a). Other fungi do not make zygospores; instead, the cell that results from fusing hyphae immediately proceeds to the next step in sexual reproduction. Fungi in division **Ascomycota** make sexual spores inside a parent cell called an **ascus**, the effect being like marbles in a bag (Fig. 20.6b). (*Ascus* is the Greek word for "sack" or "bag.") Fungi of division **Basidiomycota** make sexual spores on the surface of a parent cell called a **basidium** (Fig. 20.6c). (*Basidium* is the Greek word for "pedestal.")

Besides the fungi in these divisions, there are over 14,000 species in which sexual reproduction has never been observed. They cannot be assigned to other divisions, so how can they be classified? Reluctantly, many mycologists put them into a catchall, artificial division called the **Deuteromycota**, and refer to them as **fungi imperfecti** because their life cycles are incomplete.

The rest of the chapter treats each division in turn, looking at their life histories and the importance of their members. Before we begin, however, a few words on terminology may be helpful. The names of fungal divisions end in **-mycota**, and *-ota* is grammatically plural. Thus,

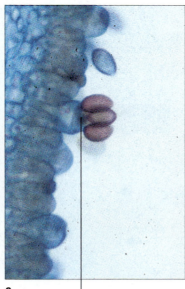

c

basidium

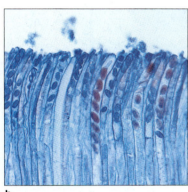

b

Figure 20.6 Characteristic sexual structures of three fungal divisions. (**a**) Zygospore of *Rhizopus*, division Zygomycota. (**b**) Asci of *Peziza*, division Ascomycota. Each elongated bag is an ascus with eight spores. (**c**) Basidium of *Coprinus* with spores; division Basidiomycota.

it's wrong to say "a Zygomycota" when referring to just one member of the division. To speak of one member, mycologists use the term *zygomycete*. Likewise, one member of the Ascomycota is an *ascomycete*, and so on.

20.7 THE ZYGOMYCOTA

Of all fungi, the Zygomycota most resemble the aquatic organisms (chytrids) from which kingdom Fungi probably arose, and they are therefore thought to be the most primitive true fungi. Only 1,056 species have been described, but some of them are very important. The majority of vascular plant species depend on mycorrhizal zygomycetes to collect minerals from poor soil—a great benefit to agriculture. On the negative side, many zygomycetes cause human hunger by spoiling stored foods. You may have seen their coarse, cottony mycelia in the kitchen, where they spoil strawberries and attack other fruits and bread.

Aseptate Mycelia Speed Reproduction

Zygomycetes are unique among fungi in that most of them are **aseptate**, meaning that they lack crosswalls, or **septa** (singular, **septum**), in the growing mycelium (Fig. 20.7). The lack of such barriers enables them to speed food transport from their feeding tips to the surface where spores are made. Fungi in all other divisions have septa at regular intervals and so are said to be **septate**. An aseptate mycelium is a single giant branching cell. As the hyphae grow, mitotic divisions convert the original haploid nucleus into many nuclei. However, the cytoplasm remains a continuous unit in which any nucleus can migrate from one end to the other. Zygomycetes *can* make septa; they do so to abandon parts of the mycelium or to prepare for spore formation. They simply make septa more sparingly than other fungi do.

Zygomycetes Make Spores in Sporangia

As a mycelium grows, it makes asexual spores on aerial hyphae called **sporangiophores** (literally, "sporangium bearers"; the Greek suffix *-phore* means "bearer"). Food flows into the tip of the sporangiophore (Fig. 20.7), causing it to swell into a spherical **sporangium** (spore container). A wall forms at the base of the sporangium, and the contents divide into mitospores.

Zygomycetes disperse spores in varied ways. The sporangiophores may grow several centimeters, using gravitropic and phototropic responses to bring the sporangia into open places where dispersal can occur. In *Rhizopus*, black bread mold, the spores disperse on the wind when the wall of the sporangium breaks open. Many other zygomycetes have sticky spores that are dispersed by animals in the meadows where many zygomycetes dwell. Fungi of the genus *Pilobolus* shoot each ripe sporangium through the air (see sidebar, "Making a Fungal Garden", p. 331).

Zygomycetes can also reproduce sexually. In some species, a single mycelium can mate with itself by fusing the tips of hyphae. However, the vast majority require two mycelia of opposite **mating types**. Because mycelia of both types look alike, the two types are called *plus* and *minus* rather than male and female. Each type releases a sex hormone into the surroundings. When a mycelium detects the proper hormone, it makes special sexual hyphae that grow toward the source of the hormone. When sexual hyphae of opposite types touch, a septum forms behind the tip of each hypha, making two **gametangia**. The tips of the gametangia fuse, and the intervening walls dissolve to make a **zygosporangium**, a large fusion cell with many nuclei from both parents.

Next, the wall of the zygosporangium grows thick and hard, and the cell inside goes into a state of rest that may last for months. The resting cell and the hardened zygosporangium, together, make up a unit that is traditionally called a zygospore (Figs. 20.6 and 20.7). Inside,

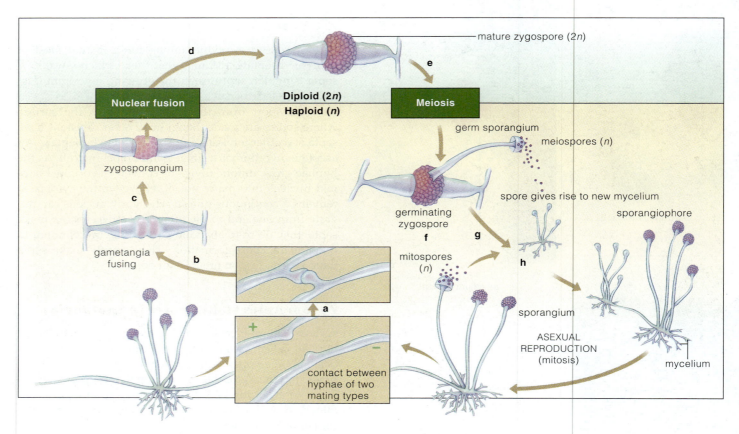

Figure 20.7 Life cycle of a zygomycete (*Rhizopus stolonifer*). (**a**) Tips of aseptate hyphae from two mycelia grow together. (**b**) Septa form behind the two touching tips, making two gametangia with many nuclei. (**c**) The gametangia fuse to make a zygosporangium. Inside, nuclei fuse to make many zygote nuclei. (**d**) A zygospore forms inside the zygosporangium. (**e**) The zygote nuclei go through meiosis. (**f**) The zygospore germinates, making a germ sporangium that releases haploid spores. (**g**) Haploid spores grow into new haploid mycelia. (**h**) Zygomycetes also reproduce asexually from mitotically produced spores.

the two kinds of nuclei fuse in pairs (karyogamy) and produce 2*n* zygote nuclei. Then meiosis converts the diploid nuclei into haploid nuclei with mixtures of genes from the two parents. In other words, recombination has occurred. Eventually an unknown signal causes the zygosporangium to split open and a hypha grows out. In some species, the hypha becomes a mycelium. More often, however, the hypha immediately swells into a germ sporangium. The contents divide into many meiospores, which escape to produce new haploid mycelia. This completes the sexual cycle.

<h2>20.8 ▍ THE ASCOMYCOTA</h2>

The Ascomycota include more species than any division of fungi: Over 32,000 species have been named. Also, the great majority of the 14,000 nonsexual fungi (division Deuteromycota) appear to have descended from ascomycetes.

Most ascomycetes are harmless saprobes, and some are quite useful. For instance, studies of the salmon-colored bread mold *Neurospora crassa* led to the discov-

ery that genes control protein synthesis. Morels (genus *Morchella*) and truffles (genus *Tuber*) are prized by gourmets. Some ascomycetes (including truffles) are mycorrhizal partners of trees. The most valuable fungus of all—*Saccharomyces cerevisiae*, the yeast of baking and brewing—is an ascomycete. When grown without oxygen, *Saccharomyces* gets energy by alcoholic fermentation, a process in which the fungus converts sugar to carbon dioxide and alcohol (see Chapter 9). When this happens in bread dough, the carbon dioxide makes the bread rise. In beer and wine, both alcohol and gas bubbles are valued products. In recombinant DNA technology, *Saccharomyces* has gained a new importance: Its simple life cycle, fast reproduction, and small number of genes (about 1/10 that of animals) make it an ideal vehicle for cloning genes and for exploring the way eukaryotic cells control their genetic information.

Other ascomycetes are costly pests. The brown rot fungi attack stone fruits such as peaches, and powdery mildews attack leaves of grapevines and many other crop plants. The ascomycete *Claviceps purpurea* infects rye plants, replacing some of the grains with purple repro-

MAKING A
FUNGAL GARDEN

It's easy to make a garden of fungi that illustrates fungal ecology. Take a quart jar to a horse or cow barn and put a scoop of fresh dung in the bottom. Cap the jar loosely to allow some air flow without too much drying, and watch it for a week or two. Zygomycetes should appear first. You can distinguish them by their coarse, cobweblike mycelia and sporangiophores, dotted with tiny black sporangia. Deuteromycetes and ascomycetes should come next. Their mycelia are too fine to see, but they have a characteristic dusty coating of green or black conidia. Mushrooms (reproductive bodies of basidiomycetes) are likely to appear last. The sequence in which various fungi appear illustrates ecological succession, the replacement of one species by another in a developing ecosystem. The earliest fungi are the best opportunists.

If you have a hand lens or microscope, look for the zygomycete called *Pilobolus* (Fig. 1). Its tiny golden sporangiophores, a few millimeters long, appear early in succession. Water droplets may give them a jewel-like appearance. With a lens, you can see that they bend toward the strongest light source. In nature, this motion allows the tip to follow the sun across the sky. A swelling just under the sporangium focuses light on a sensitive region, causing growth adjustments (phototropism). Pressure mounts until the sporangiophore bursts, throwing the sporangium several meters through the air. With luck, it sticks to a blade of grass and is swallowed by a grazing animal. The spores survive passage through the animal's intestine and are excreted in manure. Light sensors coordinate these events, so most sporangia are shot away in late morning, when the sun is about 45° above the horizon—the angle at which projectiles travel the farthest. To simulate this effect, cover the jar with foil, leaving a small hole on one side. Look for sporangia on the glass near the hole.

Few fungi shoot sporangia as *Pilobolus* does, but many ascomycetes shoot individual spores into the air. Asci often build up pressure until the tip bursts, expelling spores for a centimeter (about 1/3 inch) or two. If this happens, their spores may form a brown film on the glass.

Other fungi have more subtle ways to disperse spores. If mushrooms grow in your culture, tilt the bottle and leave it in that position for a day or two. The stipes will bend to keep the cap level so that the gills are vertical. This helps spores to fall between the gills, into open air.

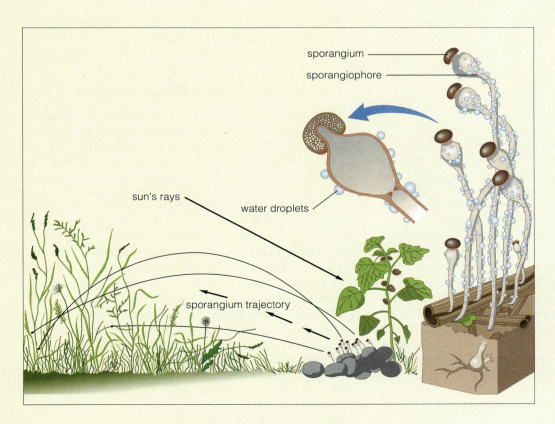

Figure 1 Sporangiophores of *Pilobolus*, the hat-thrower. The swelling just below the sporangium refracts light onto a sensitive region, adjusting growth so that the axis of the sporangiophore points toward the sun. High pressure eventually ruptures the swelling and shoots the sporangium for several meters.

Figure 20.8 Rye plant infected with *Claviceps purpurea*. Dark objects that replace grains are ergots.

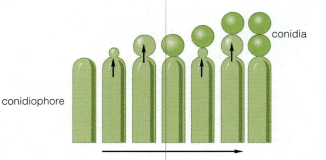

Figure 20.9 One way of making conidia. The conidiophore has a soft wall at the tip; pressure causes the wall to inflate, forming a bud. A crosswall forms at the base of the bud to release the spore and regenerate the conidiophore's inflatable tip. Repetition leads to a chain of spores.

ductive bodies called ergots (Fig. 20.8). The ergots contain several potent alkaloids, including lysergic acid, the parent compound from which LSD is made. When medieval villagers made rye bread from infected grains, the alkaloids caused a disease called St. Anthony's fire, in which victims had delusions and developed gangrene from loss of blood circulation in hands and feet. Today, most ergot-infected grain is eliminated by inspectors, but sometimes it still poisons unwary humans and livestock. In medicine, ergot is often used to control migraine headaches, to limit blood loss, and to stimulate uterine contractions during childbirth.

Septate Mycelia Provide Several Advantages

Those ascomycetes that are yeasts make no mycelia. These microscopic, rounded cells multiply either by **budding** or by dividing in half every time they double in size. To make a bud, the parent cell swells at a soft spot in the wall, much as a bulge forms at a weak place in a tire or balloon. Then a crosswall forms to separate the two cells.

Most ascomycetes, however, do make mycelia. The hyphae are septate, with crosswalls at regular intervals. Septa slow the transport of food but do not stop it, for each septum is **perforate**: It has a hole in the center that allows large molecules and even organelles to pass.

How do perforate septa help the fungus? For one thing, they help control damage to the mycelium. If a hypha is cut open, internal pressure will force cytoplasm out of the broken ends. In zygomycetes, which lack septa, the outflow stops only when the cytoplasm congeals to make a plug (later, a new wall forms over the plug). A damaged ascomycete hypha may lose much less cytoplasm because the small hole in the septum quickly becomes plugged.

Septa can also defend against damage by other mycelia. Some fungi produce toxic materials. If two species

grow into contact and one of them is toxic, the other hypha may die. Somehow, septa prevent the toxin from spreading, so only the touched compartment dies.

Septa may also allow the mycelium to make complex structures in which many hyphae contribute, with a division of labor among parts. Thus, many ascomycetes make intricate reproductive structures that combine many hyphae, whereas aseptate fungi (zygomycetes) do not. In theory, septa may permit such a division of labor because they make it more difficult for nuclei to move through the mycelium. Localization of nuclei can lead to different genes being expressed in different parts of the organism.

Conidia Speed Asexual Reproduction

Like zygomycetes, ascomycetes spread by making asexual mitospores. However, somewhere in their evolution they gave up the sporangia that are so typical of zygomycetes. Instead, most ascomycetes make mitospores called **conidia**, one by one, at the ends of specialized hyphae known as **conidiophores** (Fig. 20.9). Conidia form in a way that resembles blowing bubbles from a pipe: The tip of the conidiophore has a soft wall, which expands to form a spore as pressure drives cytoplasm toward the tip. A crosswall cuts off the spore, and the outer wall of the spore hardens. By repeating these steps, the conidiophore can make a chain of spores. Note how this differs from making spores in a sporangium: There, the wall of the parental hypha remains thin to become the sporangium; spore walls form within the enclosed cytoplasm. With conidia, the hyphal wall becomes the spore wall.

What did ascomycetes gain by switching from sporangia to conidia? The answer is unknown, but one possible advantage is a faster onset of spore formation. In zygomycetes, no spores can form until the sporangium has filled with cytoplasm. By contrast, a conidiophore can start forming a spore as soon as the first food comes in from the feeding hyphae.

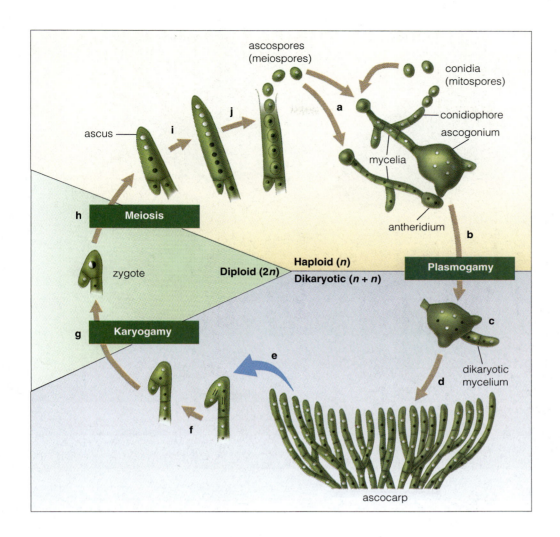

Figure 20.10 Ascomycete life cycle. (**a**) Haploid spores (ascospores) grow into mycelia. (**b**) Two haploid mycelia make specialized hyphae (an ascogonium and an antheridium) that join (plasmogamy) to make a fusion cell with nuclei from two parents. (**c**) Dikaryotic hyphae grow from the ascogonium. (**d**) Haploid and dikaryotic hyphae cooperate to make an ascocarp. Tips of dikaryotic hyphae will develop into asci (**e**, **f**). (**e**) Tip of hypha grows laterally, and nuclei divide mitotically. (**f**) Septa form to create three cells. The middle cell is the ascus. (**g**) Nuclei fuse in the ascus creating a diploid zygote. (**h**) The zygote nucleus goes through meiosis. (**i**) The haploid nuclei divide mitotically. (**j**) Walls form around the nuclei, making ascospores that escape from the tip of the ascus.

Most Ascomycetes Add a Dikaryotic Stage to the Life Cycle

In sexual reproduction, the ascomycetes add several new features. First, one of the two mycelia usually contributes more than the other by expanding its sexual hypha into a structure called an **ascogonium** (Fig. 20.10). The other mycelium contributes an ordinary hypha or a pinched-off cell or else a slightly enlarged hypha called an **antheridium**. Like zygomycetes, most ascomycetes have mating types that determine which mycelia may fuse.

As a second novelty, ascomycetes do not make zygospores. Instead, the fusion cell grows one or more hyphae that form a dikaryotic mycelium (although a few ascomycetes, such as yeasts, omit the $n + n$ stage). Within the new hyphae, the nuclei of the two parents form pairs that divide in unison (Fig. 20.10c). Thus, as the hyphae elongate and make septa, each compartment contains two nuclei (one of each kind).

Eventually, tips of these $n + n$ hyphae make meiospores called **ascospores** (Fig. 20.10j). This begins in a peculiar way (Fig. 20.10e): The tipmost cell grows laterally,

the nuclei divide in parallel, and two septa form. This creates three cells, of which the middle cell has one nucleus of each kind (Fig. 20.10f). That cell is the ascus, in which the ascospores will form. It is not clear why these divisions are needed to initiate the ascus. The paired nuclei in the ascus fuse to make a $2n$ zygote. While the ascus expands, meiosis converts the diploid nucleus into four haploid recombinant nuclei. Those nuclei divide mitotically, making eight nuclei. Finally, walls form around the nuclei and adjacent cytoplasm to make ascospores.

Asci have one feature that may improve on the germ sporangia of zygomycetes: Most asci shoot ascospores into the air, increasing dispersal. This is possible because the ascus builds up pressure as spores mature, until the tip of the ascus ruptures. The ascospores germinate to make new haploid mycelia, completing the life cycle.

A nagging question remains: What did the ascomycetes gain by inserting a dikaryotic stage into the life cycle? The full answer is unknown, but one benefit is easy to see: The $n + n$ stage leads to more meiospores per mating. To see why, compare meiospore formation in zygomycetes and ascomycetes. In both, a mating produces

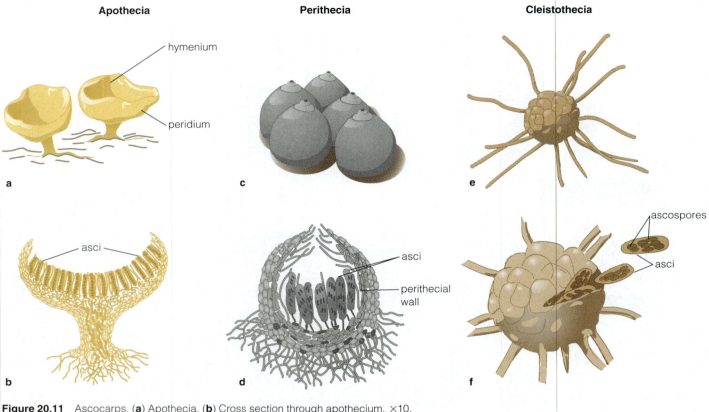

Apothecia

hymenium

peridium

a

asci

b

Perithecia

c

asci

perithecial wall

d

Cleistothecia

e

ascospores

asci

f

Figure 20.11   Ascocarps. (**a**) Apothecia. (**b**) Cross section through apothecium, ×10. (**c**) Perithecia. (**d**) Diagram of perithecium in cross section, ×150. (**e**) Cleistothecium that is still closed. (**f**) Cleistothecium that has opened to release asci. Each ascus contains eight spores, ×500.

a fusion cell with at most a few hundred pairs of nuclei. Zygomycetes fuse the nuclei and put them through meiosis, making four meiospores per pair. In ascomycetes, the $n+n$ mycelium allows the original pairs to multiply by mitosis. This could lead to many more pairs; hence, there are many more meiospores per mating. This could be an advantage when matings are scarce.

Ascocarps Improve Sexual Spore Dispersal

Most ascomycetes make elaborate structures called **ascocarps**,* which shelter ascospores and enhance their dispersal. Haploid hyphae from the parental mycelia grow together with dikaryotic hyphae, making a platform or container for spore formation and dispersal. Ascospores form on one side of the ascocarp, called the fertile layer, or **hymenium** (see Fig. 20.11). The other side, called the **peridium**, is sterile and makes no spores. The most common types of ascocarps are shown in Figure 20.11.

The cup fungi, such as *Peziza*, make ascocarps that are open cups called **apothecia** (Fig. 20.11a, b). They range

*Also called **ascomata**, singular **ascoma**.

from a millimeter to several centimeters in diameter. Asci line the upper surface of the ascocarp. The apothecium may improve reproduction by pointing the asci upright, so that the spores travel a few centimeters into the air. This may lift spores high enough to reach air currents.

Ascocarps can also be flask-shaped **perithecia**, which have a pore that opens to release spores (Fig. 20.11c, d). Dutch elm disease is caused by a fungus (*Ceratocystis ulmi*, sometimes called *Ophiostoma ulmi*) that makes perithecia. The spores are spread with the help of beetles that burrow into tree bark. Once in the burrows, the spores grow into mycelia; these invade the bark tissue and make perithecia that open back into the burrows. The perithecia produce a jelly that carries new spores as it extrudes into the burrow. Beetles accidentally pick up the jelly when they pass and carry the spores to other trees. By this mechanism, Dutch elm disease has all but eliminated the elm forests that once existed all across the eastern United States.

Still other ascocarps are closed spheres, or **cleistothecia**, which dry up and shelter the enclosed asci until the cleistothecium is crushed (Fig. 20.11e, f). The powdery mildews make cleistothecia that remain through winter on dead leaves.

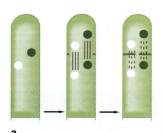
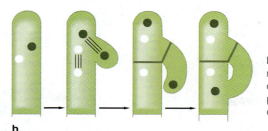

Figure 20.12 Two ways in which basidiomycetes maintain the $n+n$ nuclear condition during growth of the long-lived dikaryotic mycelium. (**a**) overlapping mitotic spindles; (**b**) formation of a clamp connection.

a

b

20.9 THE BASIDIOMYCOTA

The most complex, dangerous, costly, and familiar fungi belong to division Basidiomycota. With more than 22,000 known species, they include the fungi that make edible and poisonous mushrooms, as well as rusts and smuts that cause billions of dollars in crop damage every year. Some basidiomycetes attack lumber and cause dry rot in houses, creating additional millions of dollars in damage. Others are beneficial mycorrhizae, growing as partners of forest trees.

Basidiomycetes Have a Long Dikaryotic Stage

In their haploid mycelia, basidiomycetes look superficially like ascomycetes: Their mycelia have perforate septa, and some of them make conidia. They all have a dikaryotic stage. Thus, basidiomycetes may have evolved from ascomycete ancestors. In sexual reproduction, however, the basidiomycetes differ sharply from other fungi.

One striking difference is that basidiomycetes have a much longer dikaryotic stage. In ascomycetes, the $n+n$ mycelium is short-lived and serves only to make ascospores. The feeding mycelium is haploid. By contrast, basidiomycetes make dikaryotic mycelia that may feed and live for many years. The prolonged $n+n$ stage multiplies the number of paired nuclei, vastly increasing the number of sexual spores per mating. The $n+n$ mycelium may produce reproductive bodies (for example, mushrooms) many times in its long life. A dikaryotic mycelium grows in the soil, feeding on leaf litter or drawing energy from tree roots. When conditions are favorable, mushrooms form around the perimeter of the mycelium, where they give the impression of stools around a dance floor (hence the traditional term *fairy ring*). As the mycelium grows, each year's ring of mushrooms has a wider diameter. Judging from the present diameter and annual growth rate of the ring, mycologists estimate that some dikaryotic mycelia are hundreds of years old.

A long-lived $n+n$ mycelium needs a special mechanism to keep the nuclei in pairs. They must divide at the same time, and the products must be sorted into new pairs. Microscopists have found two mechanisms by which basidiomycetes sort their nuclei. In rusts and smuts, the mitotic spindles overlap one another, and the new septum forms in the overlap zone (Fig. 20.12a). Other basidiomycetes have a more elaborate system for sorting nuclei. As the tipmost cell of a hypha doubles in length, it makes a backward-pointing branch (Fig. 20.12b). The two nuclei divide in parallel, placing one nucleus in the branch. Then a septum walls off the branch, and a second septum walls off the trailing nucleus. The branch grows until it meets and fuses with the cell behind it. This gives each cell a pair of nuclei. The branch remains as a C-shaped appendage. Mycologists call this branch a **clamp connection** because it resembles a carpenter's clamp. Clamp connections are a sure sign that a feeding hypha was made by a basidiomycete.

Basidiocarps Are the Most Complex Fungal Structures

The most familiar fungal structures—mushrooms—are made by certain basidiomycetes. They are reproductive structures, or **basidiocarps**.* A basidiocarp is made entirely of $n+n$ hyphae. (Contrast it with ascocarps, where haploid hyphae take part.) There is an upright stalk, the **stipe**, which grows from a feeding mycelium (Fig. 20.13a). At its upper end, the stipe bears a cap, or **pileus**. The underside of the pileus has a spore-bearing hymenium, which often takes the form of gills. **Basidiospores** (meiospores) develop from the tipmost cells of hyphae in the hymenium and fall down between the gills to reach the open air. To prevent spores from landing on the opposite gill, the stipe senses gravity and grows so that the gills are nearly vertical.

The order of fungi that includes edible mushrooms also contains the most poisonous mushrooms. Those of the genus *Amanita* (Fig. 20.13a) are especially dangerous. Although poisonous mushrooms are much less common than benign ones, mycologists advise enthusiasts to consult an expert before sampling mushrooms they have gathered in nature. Deadly ones can easily be mistaken for edible species.

---

*Also called **basidiomata**, singular **basidioma**.

Figure 20.13 Basidiocarps. (**a**) *Amanita muscaria*, one of the most poisonous mushrooms. (**b**) A bracket fungus growing on a tree. (**c**) Puff balls. (**d**) A coral fungus. Basidia occur all over the surface of the basidiocarp. (**e**) A bird's nest fungus.

cap (pileus)

gills

stalk (stipe)

a

b

c

pore

d

e

rotting wood

"eggs" (spore-bearing structures)

Some mushrooms, such as *Amanita muscaria*, found in the forests of Europe and Asia, contain hallucinogenic compounds. In the Americas, the Aztecs and Mayans used mushrooms of the genus *Psilocybe* to induce visions and prophecy, and some Native Americans still use this mushroom in religious rites. It is not advisable to experiment with mushrooms, however; the dosage that induces hallucinations can be perilously close to the dosage that causes severe illness or death.

Basidiocarps vary widely. Instead of gills, **pore fungi** have tiny vertical pores lined with basidia under the cap. **Bracket** or **shelf fungi**, found on tree trunks and logs, omit the stipe and connect the cap directly to a food source (Fig. 20.13b). Some brackets have pores and grow for many years, becoming woody and adding to the ends of the pores every year.

Puffballs are another prominent kind of basidiocarp

(Fig. 20.13c). These egg-shaped structures are common on the forest floor and in meadows. Their flesh is solid white when immature. Later, the internal cells generate basidiospores. As the basidiocarp matures and dries, the outer covering may flake off, or it may remain as a leathery bag with a hole in the top. When an animal steps on the mature puffball, spores puff into the air from the hole.

Coral fungi, bird's-nest fungi, and jelly fungi are also basidiocarps. The names are descriptive: Coral fungi make a branching basidiocarp that resembles certain marine corals (Fig. 20.13d). In a bird's-nest fungus, the basidiocarp consists of a vase-shaped structure with several mushroomlike units inside (Fig. 20.13e). In jelly fungi, the supporting tissues have a gelatinous texture.

Basidiocarps may produce colossal numbers of spores. The common grocery mushroom, *Agaricus bisporus*, makes about 40 million spores every hour for two

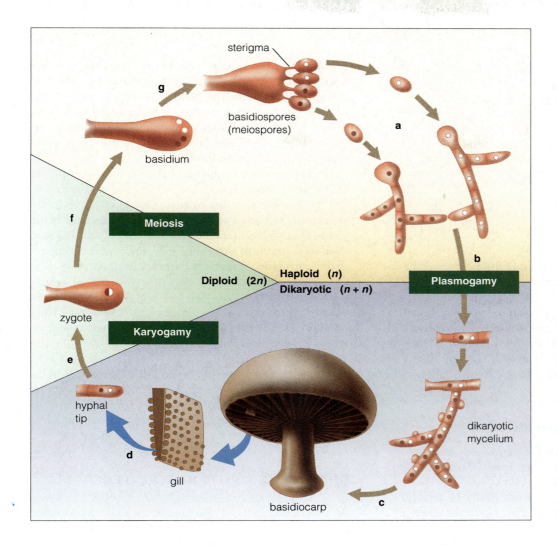

Figure 20.14 Life cycle of a basidiomycete. (**a**) Two haploid spores grow into mycelia. (**b**) Tips of the hyphae fuse (plasmogamy) to make a dikaryotic mycelium. (**c**) Dikaryotic hyphae grow together, forming a basidiocarp. (**d**) Enlargement of a gill. Tips of dikaryotic hyphae at the gill surface will become basidia. (**e**) Nuclei in a hyphal tip fuse (karyogamy), making a diploid zygote. (**f**) Meiosis creates four haploid nuclei in each basidium. (**g**) Haploid nuclei and cytoplasm are forced through sterigmata, making basidiospores.

days. *Ganoderma applanatum*, a shelf fungus that attacks living trees, can make 3 billion spores per day for six months out of each year.

Basidia Extrude Meiospores

The main events in the sexual reproduction of basidiomycetes can be illustrated with a typical mushroom (Fig. 20.14). Events begin when basidiospores germinate to make haploid mycelia. Most species have genes that prevent mating unless the mycelia have different alleles. If they are compatible, two hyphae fuse to initiate a dikaryotic mycelium. Further growth produces a large dikaryotic mycelium. When the time comes to make basidiospores, hyphae grow together and form the mushroom (basidiocarp). The spores form at hyphal tips on the gill surfaces. In the tipmost cell, the two nuclei fuse to make a 2*n* zygote nucleus. Meiosis follows immediately, making four recombinant haploid nuclei. At the same time, the cell broadens to form a pedestal, or basidium (see Fig. 20.6c). Next, conical projections called **sterigmata** (singular, *sterigma*) arise on the surface of the basidium. Pressure in the basidium forces one nucleus and

some cytoplasm through each sterigma, much like blowing a bubble. The wall of each "bubble" hardens to make a basidiospore. Then the spores are pushed away, to fall out of the mushroom.

This method of spore formation resembles the way conidia arise; in both cases, the spores inflate one by one, and the hyphal wall becomes the spore wall. This suggests that basidiomycetes evolved their method of making basidiospores by "borrowing" instructions from the programs that make conidia. By contrast, asci and sporangia make spore walls within the cytoplasm, enclosed by the hyphal wall.

The basidiomycetes that make basidiocarps have two additional features that do not occur in other fungi. First, they spread chiefly by means of meiospores, whereas other fungi spread by mitospores (usually conidia). These basidiomycetes do not make asexual spores, though hyphae sometimes divide into short segments that can generate new mycelia.

Second, basidiomycetes that make basidiocarps have a unique kind of septum in their hyphae. In ascomycetes and other basidiomycetes, the septa are simply crosswalls with a central perforation. But in these basid-

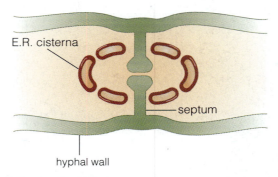

E.R. cisterna

septum

hyphal wall

Figure 20.15 Diagram of a dolipore septum.

iomycetes the septum thickens around the pore, and a caplike structure made of cisternae from the endoplasmic reticulum forms on each side. The cap usually has many openings that permit molecules to pass through the pore. The function of this elaborate **dolipore septum** (Fig. 20.15) is still uncertain. However, it probably restricts the movement of organelles more effectively than simple septa, and it may provide better damage control when hyphae are injured.

Smuts and Rusts Are Costly Pests

Two groups of basidiomycetes—the smuts and rusts—merit special attention because they often infect food crops, reducing harvests, raising the cost of food, and increasing world hunger. These fungi can easily be overlooked because they lack basidiocarps. However, they have the most complex life histories of all fungi.

Smut fungi replace seeds with black masses of fungal spores. The smuts vary in the way they attack the host: Some infect only seedlings, others infect flowers, and still others form localized infections wherever the spores happen to strike the host plant. For example, *Ustilago maydis*, which attacks corn (*Zea mays*), forms pustules or tumors at each infection site. Infected kernels become much enlarged because of the development of a mycelium that produces many black mitospores. The swollen kernels are edible before spores form (indeed, they are considered a delicacy in parts of Mexico), but their appearance can be daunting to the uninitiated. Each smut fungus attacks only one host species and can be controlled by breeding varieties of crop plants that resist the fungus.

Rust fungi are named for the rusty color of some spores that form on the leaves and stems of infected plants. The most important rusts, which attack cereal grains and other grasses, belong to the species *Puccinia graminis*. Varieties of the fungus attack different grasses. The variety *avenae* attacks oats, while the variety *tritici* attacks wheat.

It is difficult to eradicate rust fungi because they have two alternate hosts. In the case of wheat rust, the hosts are wheat and a wild shrub called *barberry*. These two hosts have different roles in the life of the fungus. The wheat plant serves to multiply and spread the existing genotypes of the fungus population. The fungus reproduces asexually on wheat, draining food reserves from the leaves and stem. The plant is usually not killed, but its seed production is reduced. The barberry host is the sexual ground for the rust fungus. Here, matings between different mycelia produce spores with new genotypes that invade grain fields.

Biologists use two approaches to control wheat rust. One approach is to breed new varieties of wheat that resist the common strains of rust. This is effective in the short term and has a side benefit, in that genes for better yields are also built into the new varieties of wheat. However, new strains of rust fungi arise every year because of sexual recombination on the barberry hosts. Among them are strains that can attack the new varieties of wheat. Within a decade, the once-resistant wheat cannot be planted without risk of a major rust epidemic. Thus, plant breeders must continually introduce new wheat varieties to keep ahead of the fungus.

The second method of controlling wheat rust is to eliminate barberry plants. If successful, it would stop the formation of new rust strains. Unfortunately, barberry inhabits mountainous areas as well as wheat-growing regions. This prevents complete eradication, so the battle between humans and wheat rust may never end.

20.10 THE DEUTEROMYCOTA

As mentioned early in this chapter, sexual reproduction has not been found in at least 14,000 species of fungi. Because taxonomists assign fungi to divisions based on their sexual reproduction, these fungi are placed in a division of their own, the Deuteromycota. Mycologists think that most members of the Deuteromycota evolved from ascomycetes that lost the ability to reproduce sexually, because asexual reproduction in the great majority of deuteromycetes closely resembles that in ascomycetes. Most of the remaining deuteromycetes probably derived from basidiomycetes. Occasionally, a scientist discovers sexual reproduction in a fungus that has been classified as a deuteromycete. When this happens, the fungus is reclassified. For instance, the imperfect fungus *Penicillium vermiculatum* is now known to make asci like those of the ascomycete genus *Talaromyces*; hence, this fungus is also called *Talaromyces vermiculatum*. However, the old deuteromycete name is kept, too, because the identification manuals used by field workers are keyed to asexual characteristics. Because the classification of deuteromycetes is doubtful, the groupings within this division are called **form genera** or **form families** to indicate that the members may not have a hereditary relationship.

Is genetic recombination possible in deuteromycetes, even though they lack sexual reproduction? The surprising answer is, yes. Mycelia sometimes fuse and bring together nuclei of different origins. Somehow, the nuclei exchange genes. Meiosis is not involved, so this

21

THE PROTISTA

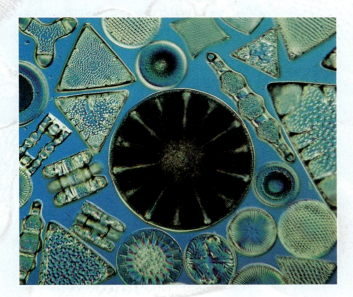

1. The division Protista is an artificial collection of animal-like, fungus-like, and photosynthetic eukaryotic organisms. Superficially, they resemble one another in having simple bodies, but they represent widely different evolutionary lines. This chapter focuses on the photosynthetic organisms, the algae.

2. *Algae* is a common name for a group of many divisions, collectively containing 20,000 to 30,000 species that differ in cell wall construction, types of chlorophylls and pigments, morphologies, and habitat ranges. Algae differ from plants in that their gametangia are single cells, their life cycles lack an embryo stage, and their bodies lack an epidermis with stomata.

3. The most widespread divisions of algae include the Bacillariophyta (diatoms), Pyrrhophyta (dinoflagellates), Rhodophyta (red algae, including many seaweeds), Phaeophyta (brown algae, including large kelps), and Chlorophyta (green algae). Phycologists believe that ancient green algae were ancestral to higher plants.

4. Microscopic, floating forms of algae are ecologically important because they are at the base of aquatic food chains. Larger seaweeds and kelps are economically important because of cell wall materials such as agar, carrageenan, and algin.

5. Algal life cycles include gametic, zygotic, and sporic types. In many cases, however, asexual reproduction—by cell division, fragmentation of filaments, or mitospores—is more common than sexual reproduction.

21.1 THE PROTISTS: SIMPLE AQUATIC ORGANISMS

The past two chapters have described two of the five kingdoms of life: Monera and Fungi. In this chapter we will describe a third kingdom, the **Protista**. The kingdom Protista is artificial, having been devised as a holding place for all eukaryotes that are neither animals, fungi, nor plants. Protists are unicellular or simple multicellular organisms that do not usually exhibit specialized organs such as leaves, roots, or stems. Protist cells are eukaryotic like those of fungi, with an organized nucleus, double-stranded DNA, specialized organelles in the cytoplasm, and the capacity for sexual reproduction and spore production.

This kingdom contains animal-like, fungus-like, and plant-like organisms. Animal-like protists include unicellular protozoans such as *Paramecium* (Fig. 21.1a), microscopic foraminiferans, and large sponges from the ocean. Because of space constraints, these protists will not be discussed in this chapter. Fungus-like protists include slime molds (Fig. 21.1b) and pathogens such as late blight of potato. They differ from fungi in that they usually don't form hyphae nor have chitin in their cell walls, but they do have motile gametes capable of swimming through water. These protists are discussed in the next section. Plant-like protists are photosynthetic organisms informally called **algae** (singular, **alga;** Fig. 21.1c). This chapter will focus on algae because they are probably the evolutionary precursors of plants.

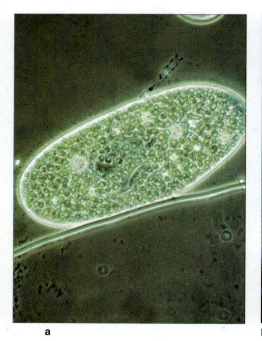
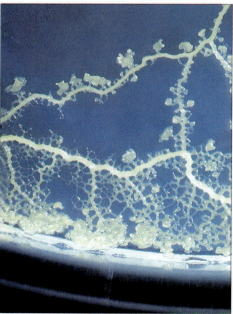

a b c

Figure 21.1 Diversity in the kingdom Protista. (**a**) *Paramecium*, a protozoan. (**b**) *Physarum*, a slime mold. (**c**) The feathery red alga *Polysiphonia*.

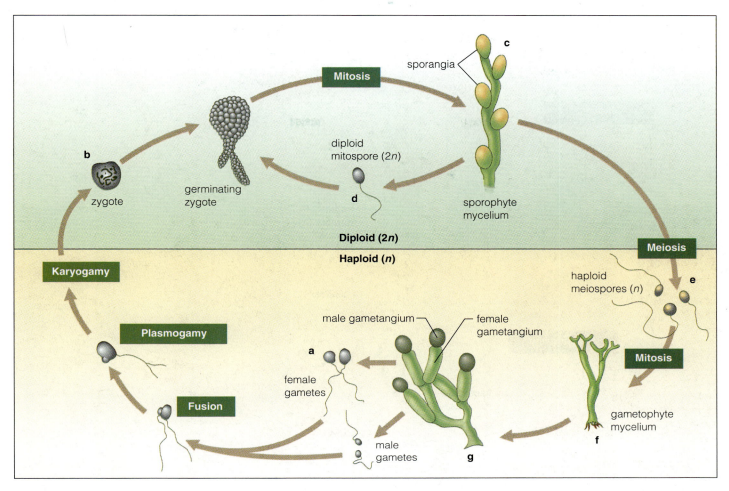

Figure 21.2 Sporic life cycle of the chytrid *Allomyces macrogynus*. (**a**) Gametes are motile, with the female gamete using a hormone to attract a male gamete. Cellular and nuclear fusion of gametes creates a zygote (**b**), which—upon germination and mitosis—results in a sporophyte mycelium (**c**). The sporophyte has two types of sporangia: those that release diploid mitospores (**d**) and those that release haploid meiospores (**e**). The mitospores germinate to form sporophytic mycelia, whereas the meiospores grow into (haploid) gametophytic mycelia (**f**). The gametophyte produces male and female gametangia (**g**), which release male and female gametes, renewing the cycle. (From Alexopolous and Mims, *Introductory Mycology*, 3d ed. New York: John Wiley, 1979.)

21.2 FUNGUS-LIKE PROTISTS

Chytrids May Be the Ancestors of Fungi

Chytrids, slime molds, and oomycetes are fungus-like protists. The kingdom Fungi probably evolved from one of these nonphotosynthetic, aquatic members of the Protista. The best candidates are in the division **Chytridiomycota (chytrids)**. Most of the 450 species live as aquatic saprobes or as parasites on algae and fungi. A few species live in the soil, including some that are important parasites of crop plants. Some are unicellular without any mycelium, but the largest form a delicate fringe of mycelium around their food, such as decaying seeds. Their walls contain chitin, like the fungi, but their spores and gametes are motile, quite different from the fungi; and they do not have a prolonged dikaryotic life cycle stage.

The best-known chytrids are in the genus *Allomyces* (Fig. 21.2). The life cycle begins when swimming gametes fuse to make a zygote, which grows by mitosis into a diploid sporophyte, composed of a mass of branched, aseptate hyphae. As the mycelium grows, some hyphae produce sporangia that contain diploid mitospores. When released, the spores swim to new substrates and produce new diploid mycelia by mitosis. Other hyphae produce sporangia in which meiosis occurs, yielding haploid meiospores. These are also motile, and when released they swim to new substrates and produce haploid mycelia by mitosis. *Allomyces* has multicellular sporophyte and gametophyte life cycle phases; that is, it has a sporic life cycle. Furthermore, the gametophytic and sporophytic mycelia are **isomorphous** (they look alike). True fungi do not have sporic, isomorphous life cycles.

Gametophytic hyphae swell at the tips, producing two kinds of gametangia. One liberates small, motile

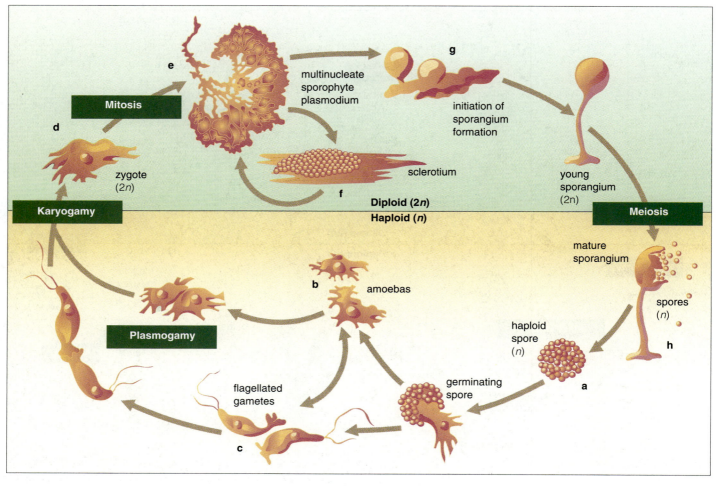

Figure 21.3 Stages in the life cycle of a plasmodial slime mold. (**a**) Haploid spores germinate into either amoeboid (**b**) or flagellated (**c**) cells. Two amoeboid or two flagellated cells join to form a diploid zygote (**d**), which—through mitosis—forms the sporophyte, a multinucleate plasmodium (**e**). The plasmodium can survive dry periods by temporarily hardening into a sclerotium (**f**). The plasmodium forms sporangia (**g**), in which meiosis creates haploid spores. These are released (**h**), to start the cycle over again.

male gametes, and the other liberates somewhat larger female gametes. Female gametes secrete a hormone that attracts the male gametes. Plasmogamy (cell fusion) and karyogamy (nuclear fusion) occur, bringing us back to the diploid zygote. The zygote enters a resting stage and then begins to divide by mitosis into a mycelium.

Not all chytrids have bodies as complex as *Allomyces*. The simplest kinds live as unicellular parasites inside a host cell. Others have a mycelium but lack motile gametes. Most have a zygotic life cycle, like fungi, rather than a sporic life cycle; that is, most do not produce a diploid mycelium generation. Despite these variations, there are many similarities between chytrids and the most primitive true fungi, the Zygomycota (bread molds), supporting the notion that they are the ancestors of the kingdom Fungi.

Slime Molds May Be Related to Amoebas

The slime molds have a unique combination of traits that places them at the boundary between fungi and animals. There are two divisions of slime molds: the **Myxomycota**, or **plasmodial slime molds**, and the **Acrasiomycota**, or **cellular slime molds**. The multicellular life cycle phase of plasmodial slime molds contains thousands of nuclei without cell membranes between them; the multicellular phase of cellular slime molds are smaller and the cells do have membranes between them.

Slime molds resemble animals in that they lack cell walls, engulf food, and have motile cells at some phase of the life cycle. On the other hand, they resemble fungi and plants in that they form sporangia and nonmotile spores with cell walls.

A typical plasmodial slime mold life cycle begins with haploid spores (Fig. 21.3). Depending on the circumstances, a spore germinates into either a flagellated cell or a cell resembling the protozoan *Amoeba*, capable of **amoeboid motion**. Such movement occurs when a local part of the cell membrane softens while other parts stiffen; then the contents of the cell flow into the softened region, moving the cell in that direction. If the spore germinates into a flagellated cell, then the beating motion of the flagella propels it. The amoeboid or flagellated cells move among leaf litter or rotting wood. The two cell forms are continuously interconvertible. Both take in food particles—spores and microbes—by phagocytosis, a process in which the cell forms a pocket around a food object and the pocket pinches off inside the cell to become a food vacuole. The cell then secretes enzymes into the food vacuole to digest the food. Indigestible particles are later excreted by fusing the food vacuole with the outer cell membrane.

Eventually, two amoeboid or two flagellated cells join together, acting as gametes. The diploid zygote undergoes many mitoses, and all of the resulting diploid nuclei lie free in a **plasmodium**, which is the equivalent of a sporophyte made up of a single, large, multinucleate cell. A mature plasmodium can be several centimeters (about one inch) in diameter and weigh several grams (1/10 ounce); it is a network of channels through which the contents flow as various regions are softened or stiffened. The plasmodium is the sporophyte generation (see Fig. 21.3e). It can withstand dry periods by hardening into a brightly colored compact crust (a **sclerotium**) that softens back into a moving plasmodium when humid conditions return. Eventually, the plasmodium develops many stalked sporangia. Meiosis occurs within the sporangia; they split open and release haploid spores, which starts the cycle anew.

Cellular slime molds have a similar amoeboid stage. As food resources become depleted, the haploid amoebas come together to form a plasmodium. In this case, however, the plasmodium remains haploid and cellular. Certain amoebas secrete a chemical signal, cyclic AMP. Other amoebas migrate to these cells along a gradient of cyclic AMP. As they cluster, the cell membranes do not disappear. The plasmodium is septate, not multinucleate as in the Myxomycota. To emphasize this difference, it is called a pseudoplasmodium. The pseudoplasmodium migrates like a small slug, barely visible to the naked eye, moving toward light, warmer temperature, and nutrients. Eventually it stops and forms a single stalked sporangium. Cells within the sporangium take on a rounded shape, secrete a cellulose cell wall about themselves, and become haploid mitospores. When released, each spore can germinate into an amoeba, and the cycle repeats.

The Oomycota May Have Evolved from Algae

A third group of fungus-like organisms in the Protista is the division **Oomycota**, which includes egg fungi, water molds, and downy mildews. They resemble fungi in having hyphae, producing spores, and lacking chlorophyll. However, they share cellulose cell walls, swimming spores, certain cellular details, and unique metabolic pathways with some algal divisions. As a consequence, many biologists think that the Oomycota are algae that have lost their chloroplasts.

Most egg fungi are benign decomposers that live in soil or freshwater habitats, but a few are pathogens of important crops. Downy mildews, for examples, infect

Figure 21.4 Grape leaf infected with downy mildew, *Plasmopora viticola*.

beans, grasses, and melons, among other plants. They are easily identified by the dense web of sporangia-bearing hyphae (sporangiophores) that makes the infected leaf appear to be covered with soft down (Fig. 21.4). Downy mildew of grape, *Plasmopora viticola*, nearly destroyed French vineyards in the 19th century. Before this time, the disease had been restricted to North America, where it infected wild grapes. These wild species were seen to be valuable rootstocks for European grapes, so they were imported to France. Some carried spores of downy mildew that were carried by the wind to European plants, causing disease symptoms that were first noticed in 1878. The disease then spread rapidly. Investors began planting vineyards in countries outside France, believing the French wine industry was doomed. But the mycologist Alexis Millardet saved the day by discovering that a mix of lime and copper sulfate, applied as a dust on leaves and stems, killed downy mildew. The French wine industry recovered, and opportunistic owners who had overplanted vineyards in other countries, such as Italy, went into bankruptcy. Millardet called his mixture *Bordeaux mix*, and it is still used today.

Another 19th-century egg fungus changed the history of Ireland. The potato blight of the 1840s was caused by *Phytophthora infestans* (the name literally means "plant destroyer"). The potato is a New World plant, originally restricted to the cold, rocky soils of Andean farms. Its value as a food crop on marginal farmland elsewhere was obvious, and it soon became the main crop in Ireland. One farm family could subsist on an acre of potatoes and a cow. An unusual series of warm, humid summers permitted the fungus to become epidemic, rotting the potatoes and killing the plants. A quarter of a million people died of starvation, and a million more emigrated to the United States, adding an important new ethnic component to American culture. Decades later, mycologists learned that potato blight could be controlled by spraying infected fields with poisons that kill the blight and by carefully disposing of infected potato tubers.

21.3 ## THE ALGAE: A GROUP CONTAINING MANY DIVISIONS

The algae include 20,000–30,000 species in nine divisions (Table 21.1). The divisions differ in such traits as pigments and color; hence the names of many divisions refer to a color (red algae, green algae, golden algae, brown algae). They range in shape from unicellular to colonial (clusters of cells) to filamentous to sheetlike (Fig. 21.5).

Table 21.1 Characteristics of Major Algal Divisions

Division	Number of Species	Chlorophylls and Accessory Pigments	Cell Wall	Storage Product	Flagella	Description
Chrysophyta (golden algae, class Chrysophyceae only)	300	*a* and *c*; fucoxanthin	Cellulose; some with silica scales	Fats, oils, laminarin	Two, unequal, anterior	Mainly unicellular and found in fresh water
Bacillariophyta (diatoms)	8000+	*a* and *c*; fucoxanthin	Silica and pectin	Fats, oils, laminarin	Generally none	Mainly unicellular; prominent in phytoplankton; sometimes placed in Chrysophyta
Xanthophyta (yellow-green algae)	400	*a* (and *c* in some)	Cellulose	Oil or laminarin	Two, various	Mainly unicellular and found in fresh water; sometimes placed in Chrysophyta
Pyrrhophyta (dinoflagellates)	1000	*a* and *c*; peridinin	None or plates of cellulose	Starch, oil	Two, unequal, lateral	Unicellular; 95% of species are marine; prominant in phytoplankton
Euglenophyta (euglenoids)	450	*a* and *b*	None	Paramylum	Two, equal, anterior	Unicellular, with animal-like properties; found mainly in polluted fresh water
Rhodophyta (red algae)	4000	*a* and *d*; phycobilins	Cellulose and sometimes agar or carrageenan	Fats, oils, laminarin	None	Mostly multicellular seaweeds; 98% of species are marine; often found in warm, deep water; complex life cycles
Phaeophyta (brown algae)	1500	*a* and *c*; fucoxanthin	Cellulose and algin	Laminarin, mannitol	Two, unequal, lateral	Multicellular, including large kelps; 98% of species are marine; often found in cool, shallow water
Chlorophyta (green algae)	7000	*a* and *b*; carotene	Cellulose	Starch	Two, equal, anterior	Diverse morphology; possible precursor to Kingdom Plantae
Charophyta (stoneworts)	250	*a* and *b*	Cellulose	Starch	Two, equal, anterior	Filamentous and differentiated into nodes and internodes; multicellular gametangia; sometimes placed in the Chlorophyta

*A few small groups have not been included—for example, the cyanobacteria from Chapter 19.

Figure 21.5 Algal diversity. (**a**) Unicellular *Micrasterias*. (**b**) Colonial *Gonium*. (**c**) Colonial *Volvox*. (**d**) Filaments of *Spirogyra*. (**e**) *Caulerpa*. (**a–e**) are green algae in the division Chlorophyta. Seaweed forms include (**f**) *Porphyra* (red algae, Rhodophyta) and (**g**) *Fucus* (rockweed, brown algae, Phaeophyta).

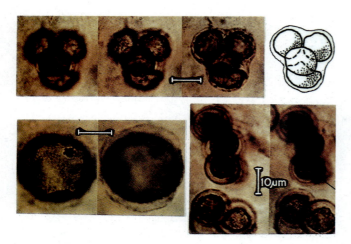

Figure 21.6 Microfossils of green algae in chert approximately 1 billion years old.

Some are complex enough to exhibit marked differentiation into specialized tissues or even organs. Most algae occur in wet habitats, where their simple bodies are supported by buoyancy in water. They produce no water-conducting or strengthened cells. Every cell in most algal bodies can carry on photosynthesis and obtain water and nutrients directly from its surroundings by diffusion. Such a simple body is called a **thallus**. Algae differ from members of the kingdom Plantae in several ways: (1) The gamete-producing structures of algae are much simpler, lacking any protective layer of sterile cells around the gametes; (2) algae do not have an embryo stage in their life cycle; and (3) algae do not have an epidermis with cuticle and stomata.

Although algae are best adapted to wet environments, many species survive in seasonally dry habitats by going dormant between wet seasons. Some algae drift or swim in open water; others attach to the bottoms of streams or shallow seas, to surface soil particles, to tree trunks, to other algae, or to rocky cliffs battered by surf. Still others form symbiotic associations with fungi, higher plants, or animals. We commonly see algae growing on the sides of fish tanks, around leaking faucets, in garden pools, and as a scum on the surface of ponds in summertime.

Algae occur in the most severe habitats on earth: on snow with perpetually freezing temperatures; in hot springs with temperatures of above 70°C (158°F); in extremely saline bodies of water such as the Great Salt Lake in Utah; beneath 150 m (500 ft) of water in lakes or seas; even surviving—within 1 km (0.6 mi) of ground zero—a 20-kiloton atomic bomb test in Nevada.

The fossil record for algae extends back nearly 1 billion years, to siliceous rocks that formed from sediment on the bottom of shallow seas or bays. (Siliceous rock contains silica, which makes up quartz, sand, and glass.) At this early time, the atmosphere is thought to have been so low in oxy-gen that ultraviolet radiation (UV) from the sun was able to reach the earth's surface. (Today, a form of oxygen called *ozone* absorbs most of the UV light.) UV radiation is harmful to living tissue; therefore it would have been impossible for life to exist on the surface of land 1 billion years ago. The only suitable habitat would have been underwater because water is very effective at absorbing UV.

Thin sections of a billion-year-old chert (a type of quartz) examined under a microscope reveal unicellular and filamentous organisms, and even some cellular details, that resemble certain species of living algae (Fig. 21.6). Traces of complex molecules that are breakdown products of chlorophyll are also in the rocks. Botanists believe that these early fossils are ancestral forms of the division Chlorophyta, the green algae. The fossil record for algae shows continued growth in algal diversity through ensuing millennia, in particular an explosive growth of new forms and divisions about 400 to 600 million years ago, when plants first began to inhabit the land.

The various divisions of algae (Table 21.1) differ from each other in biochemistry, cellular characteristics, and habitat. **Phycologists** (those who study algae) consider variations in cell wall construction, number and placement of flagella, and type of chlorophyll to be quite fundamental and significant, revealing evolutionary relationships. We shall briefly review five of the divisions.

The Bacillariophyta Are the Diatoms

Single-celled **Bacillariophyta**, commonly called **diatoms**, are important members of the **phytoplankton**—the collection of floating, photosynthetic, microscopic algae. As seen in face view, their silica walls are exquisitely ornamented with perforations (Fig. 21.7). The silica is embedded in a pectin matrix. The shape of the cell can be circular, triangular, oval, or more elaborate. In cross section, the cell wall is seen to have two parts (valves) that fit over each other like the halves of a petri dish. Diatoms contain chlorophylls *a* and *c* and the accessory pigment **fucoxanthin**. Fucoxanthin gives these cells their characteristic color, ranging from olive brown to golden brown. Diatoms store food reserves as an oil. Some diatoms inexplicably have a gliding motion, even though cilia and flagella are absent.

Diatoms also exist as attached, stalked, single cells or as filaments. Many stalked diatoms grow as **epiphytes** (literally, upon other plants) on seaweeds and kelps. They do not parasitize the host plant but use it as a base to gain access to light near the water's surface. It is estimated that the productivity of the extra layer of epiphytes can be as great as that of their larger hosts. Thus, epiphytes add greatly to the first tier of the food chain (the transfer of energy in an ecosystem) in shallow water.

Diatoms may also become attached to nonliving surfaces. For example, the undersurface of icebergs is coated

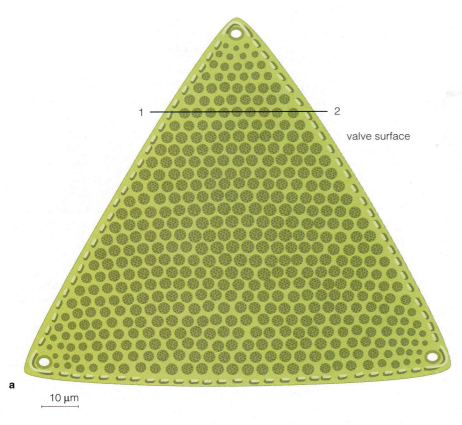

1 ——————— 2

valve surface

a

10 μm

upper valve

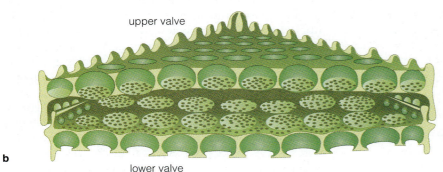

b

lower valve

Figure 21.7 The triangular diatom *Triceratium favus*, showing details of the cell walls. (**a**) Face view. (**b**) Cross-sectional view taken along the line 1–2 in part (**a**).

with diatoms, particularly of the genera *Navicula*, *Nitzchia*, *Podosira*, and about two dozen others. The diatoms are abundant enough to stain the ice a yellow-brown color; their density is several orders of magnitude higher than that in the surrounding water. The base of the arctic food chain, then, is enriched because of these ice-encrusting diatoms. Diatoms also create *algal turfs*, which coat shallow rocks in quiet freshwater or marine habitats. These turfs have as high a daily productivity per square meter of surface as does a tropical rain forest; therefore they play an important part in the food chain of these habitats as well. The surfaces of salt marsh mudflats are also dominated by diatoms, which are grazed on by insects.

The Pyrrhophyta Cause Red Tides

The **Pyrrhophyta** (also known as **dinoflagellates**) are very important members of the phytoplankton. Most species in this division are unicellular, motile, and marine. Some have a platelike exterior made of cellulose (Fig. 21.8), whereas others lack a cell wall entirely and are enclosed only by a thickened membrane. A few are colonial or filamentous. Usually, two flagella are present; both emerge from the same pore, but otherwise they are different. One is flat and ribbonlike and encircles the cell in a groove around the middle; it is responsible for rotational movement. The second flagellum trails behind and provides forward movement, while acting at the same time as a rudder.

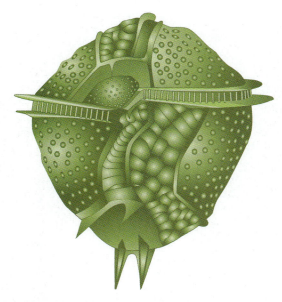

Figure 21.8 The dinoflagellate *Ceratocorys aultii*. Flagella, which ordinarily lie within the grooves or trail behind, are not shown.

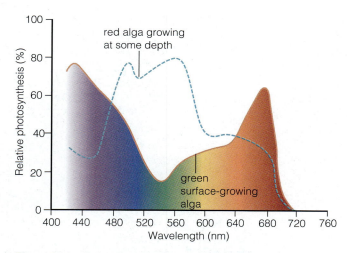

Figure 21.9 Action spectrum for photosynthesis in a green, surface-growing alga (solid red line) and a red alga growing at some depth (dashed blue line). The red alga is able to use light in the blue-green wavelengths (440–580 nm), which is not absorbed by the surface layers of water.

Dinoflagellates contain chlorophylls *a* and *c* and a brown pigment called **peridinin**, which gives the cells a green-brown or orange-brown color. As with plants, starch is stored as a food reserve. Some forms are luminescent and contribute to the glow of water when it is disturbed at night by surf or by the wake of a ship. At certain times of the year, and along some coasts, the density of dinoflagellates in the phytoplankton increases manyfold, turning the water a reddish color (hence the name *red tide*) and releasing toxins into the water.

The Rhodophyta Are the Red Algae

The **Rhodophyta**, or red algae, are almost exclusively marine and are most abundant in warm water, in which they often extend to some depth. Most are multicellular and large enough to be called seaweeds. Morphological diversity is based on an elaboration and aggregation of filaments. The simplest red algae are small, branched, delicate filaments (see Fig. 21.1c). More complex forms have filaments compressed into three dimensional tissue (that resembles parenchyma) and a body with a rootlike **holdfast** (which anchors the body to a surface), a stemlike **stipe**, and a leaflike **blade**. Some forms are lime-encrusted and participate in building tropical reefs. Cell walls contain cellulose, sometimes augmented with agar or carrageenan. Red algae exhibit some of the most complex life cycles of any group, having three generations.

The Rhodophyta have some traits in common with cyanobacteria. Both contain only chlorophyll *a* (some reds have a slightly different form called chlorophyll *d* as well) and similar accessory pigments called **phycobilins**. Neither posesses motile gametes, but both form an animal storage product called amylopectin. Chloroplasts are present in red algae, but they lack grana. Furthermore, carbohydrate accumulates in the cytoplasm, not in the chloroplast.

The accessory pigments of red algae allow them to grow at greater depths than green or brown algae. Both the quality and the quantity of light change with passage through water. Red light is completely absorbed in the upper layers, leaving a blue-green twilight to prevail farther down. Experiments have revealed that aquatic algae have adjusted their metabolism to the light at different depths. Fig. 21.9 shows the action spectrum of photosynthesis for both the green alga sea lettuce (*Ulva taeniata*), which grows at very shallow depths along rocky coasts, and for the red alga *Myriogramme spectabilis*, which grows permanently submerged in deeper water. You can see that the most important wavelength of light for photosynthesis shifts with depth to center around the blue-green region, 440 to 580 nm. The phycobilins in red algae are able to trap light in this region of the spectrum, transferring the energy to chlorophyll.

Eventually, at about 170 m (560 ft) in water of average clarity, the amount of light becomes so depleted that no alga—regardless of pigments—can scavenge enough light to support growth. The record depth for a red alga is 268 m (879 ft) in an unusually clear part of the Caribbean Sea. Algae in such low light grow very slowly and are possibly hundreds of years old.

The Phaeophyta Are the Brown Algae

The **Phaeophyta** (brown algae) are almost exclusively marine, but in contrast to the reds they are most diverse and abundant in cool, shallow waters. There are no unicellular forms. The simplest brown algae are filamentous and sheetlike; the most complex are **kelps**. Kelps are large species with well-differentiated regions that resemble roots, stems, and leaves and with internally distinctive tissues and regions (see sidebar, "The Kelp Forest Ecosystem," p. 352, and also Fig. 21.10).

Chlorophylls *a* and *c* are present, plus the pigment fucoxanthin, which also occurs in the diatoms. Carbohydrates, stored as mannitol or laminaran, accumulate as granules in the cytoplasm. Chloroplasts do not contain grana. Cell walls are made from cellulose and algin. Motile cells have two unequal flagella attached along the side of the cell.

Macrocystis is the largest kelp known, and it has the fastest growth rate of any multicellular plant. In the course of a single growing season, it grows from a single-celled zygote into a young plant attached to a rocky bottom 6 to 30 m (20 to 100 ft) below the surface and then into a mature giant 60 m (200 ft) long (much of it floating on the surface). A mature *Macrocystis* (Fig. 21.10) consists of an anchoring holdfast, stemlike stipes, and numerous leaflike blades that arise all along the stipes. The base of each blade is inflated into a gas-filled **bladder**, which increases buoyancy.

Kelps are complex anatomically as well as morphologically. If the stipe is sectioned and examined under the microscope, several regions are apparent (Fig. 21.11). Cells in the outermost layer are protective and in addition are meristematic and contain chloroplasts. To distinguish this unique tissue from the much simpler epidermis of higher plants, it is given the name **meristoderm**. A broad **cortex** region beneath the meristoderm is composed of parenchyma-like cells. Mucilage-secreting cells line canals through the cortex. Loosely packed filaments of cells fill the innermost part of the stipe, a region called the **medulla**.

Some cells in the transition zone between cortex and medulla function as sieve elements. They have sieve plates (Fig. 21.11c), form callose, adjoin one another to make continuous tubes, and are known to move mannitol at a rate resembling sugar movement in vascular land plants. The value of a photosynthate-conducting system in these large plants is easy to understand: The mass of floating fronds on the surface shades the lower part of the stipes and the holdfast so much that the shaded parts cannot produce enough carbohydrate to maintain themselves and must have additional amounts translocated to them. Unlike plants, kelps contain neither tissue that resembles xylem nor any cells with lignified secondary walls.

The Chlorophyta Are the Green Algae

Members of the **Chlorophyta**, or green algae, occur predominantly in freshwater habitats. They also exist in salt water, on snow, in hot springs, on soil, and on the leaves and branches of terrestrial plants. This division is second only to the diatoms in number of species. The Chlorophyta

Figure 21.10 The giant kelp *Macrocystis pyrifera* growing in water 4 m deep.

and the closely related division **Charophyta** (placed within the Chlorophyta by some phycologists) share many traits with higher plants. They all have chlorophylls *a* and *b*, similar accessory pigments, starch as a storage product, and cellulose cell walls. Because of these similarities, as well as some details of cell division and flagellar attachment, most

THE KELP FOREST ECOSYSTEM

The California coast is one of the richest areas in the world for kelp. Here, giant kelp ("sequoia of the sea," *Macrocystis pyrifera*) dominate nearshore waters on the continental shelf. This is the largest kelp species in the world, and it forms stands so magnificent in size and so dense that they have been called kelp forests or kelp beds (Fig. 1).

The geographic range of this species in the northern hemisphere extends from near San Francisco south to the midpoint of the Baja California peninsula. In the southern hemisphere, it is found along the Chilean coast, off scattered islands in the South Atlantic and Indian Oceans, and into the western Pacific all the way to Tasmania and New Zealand.

The species is well known because of its economic importance as a source of algin. Kelp boats cruise offshore, harvesting the blades and stipes that float on, or just below, the surface. The material is then processed onshore to extract and purify the algin. Giant kelp grows so rapidly that the same bed may be reharvested every six weeks. The U.S. seaweed industry produces more than $120 million worth of algin each year. Kelco Company, the largest California-based firm to harvest giant kelp commercially, has monitored kelp bed acreage for many years, and its data show dynamic fluctuations over time. What has caused these changes? Why haven't kelp beds been stable? The intensity of human harvesting does not seem to have been a factor. Instead, the answer seems to be that the environment has not been stable: There have been fluctuations in temperature, turbidity, sewage, sea urchins, and sea otters.

Giant kelp grows in water between 5 and 20 m (15 to 65 ft) deep; only the uppermost blades lie partly exposed to air when they float on the surface. The water in which giant kelp grows is relatively cold because local upwelling currents bring deep water to the surface. This deep water is also nutrient-rich. Water temperatures off the California coast normally change very little from month to month or year to year, staying in the range of 12 to 20°C (54 to 68°F). Experiments show that young kelp grow very poorly when the water temperature is above 20°C and when dissolved nitrate becomes low.

Giant kelp's sensitivity to high temperatures and low nutrient levels means that it suffers declines during El Niño events: changes in the pattern of wind and water circulation in the Pacific Ocean that occur every three to eight years and last for about a year. El Niños cause surface water temperatures to rise several degrees, which produces phytoplankton blooms that use up nitrate in the water. Phytoplankton blooms also reduce the amount of solar radiation that penetrates into water. Mature kelp requires about 50% of full sun to achieve light saturation and maximum growth. Normally, a depth of 20 m in clear water has light at this level, but even moderate turbidity in the water from phytoplankton or sewage means that optimal light can penetrate to only half the normal depth.

Figure 1 Kelp forest as viewed from underwater, off the California coast.

Another factor that affects the vigor of kelp beds is sewage pollution. Sewage creates turbidity and a muddy ooze on the nearshore ocean bottom. Young *Macrocystis* sporophytes must attach to exposed rock; the ooze both prevents the plants from anchoring themselves and may also cover young plants and kill them. Sewage also increases sea urchin populations, inducing these bottom-dwelling invertebrates to graze more intensively on kelp. Their feeding severs the upper body of the kelp from the holdfast, setting it free to be washed up as beach drift.

Sea urchin populations used to be controlled by a natural predator, the sea otter. By the early 20th century, this animal had been hunted nearly to extinction for its fur. There are now laws that protect sea otters, and their numbers are rising. Where sea otter populations are highest, kelp forests are most vigorous; where sea otter populations are lowest, kelp forests have declined during this century. The resurgence of sea otters, then, pleases kelp harvesters. At the same time, shellfish harvesters are displeased because another favorite prey item of sea otters is abalone, a commercially valuable shelled marine animal used as seafood.

Marine ecosystems show many connections between organisms and their environment. These complex connections make for a robust ecosystem so long as the environment remains constant, but they are capable of rapid unraveling when people or nature change the environment.

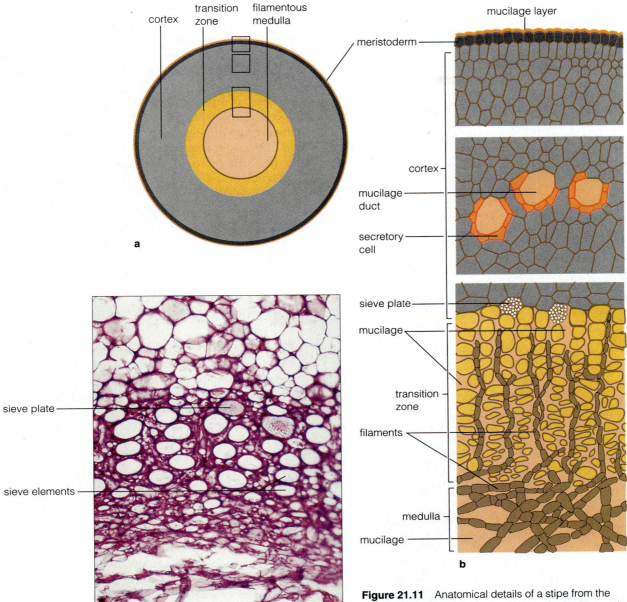

Figure 21.11 Anatomical details of a stipe from the giant kelp *Macrocystis pyrifera*. (**a**) Diagram of the entire stipe cross section. (**b**) Details of portions of the cross section (see dotted parts in **a**). (**c**) Detail of sieve plate at one end of a sieve element.

phycologists hypothesize that ancient green algae are ancestral to higher plants.

Morphologically, they include unicellular forms, colonial groups, filaments, and sheets. *Chlamydomonas* is an example of a unicellular form (Fig. 21.12). Each cell has two flagella at the anterior end and a single, large, cup-shaped chloroplast. Most species have a red-colored carotene body called the *eye spot*, which is capable of sensing light. The two flagella enter the cell at an angle, and they end in **basal bodies**, which are connected by a bridge of many microtubules. Forward movement results from flagellar motion resembling a swimmer's breast-stroke: The flagella extend straight ahead, sweep back-

ward without bending, and then fold and come back to the forward position.

Chlamydomonas-like cells may link together into permanent multicellular colonies; *Gonium* and *Volvox* are examples (see Fig. 21.5b and c). *Volvox* is a hollow sphere of cells connected to each other by plasmodesmatal strands of cytoplasm. One colony may contain as many as 20,000 cells. The cells of a colony are clearly integrated because their flagella are synchronized and able to move the colony in a given direction.

Some filamentous forms of green algae have complex morphologies that imitate higher plants. *Caulerpa*, for example, forms extensive mats on the bottom of shallow

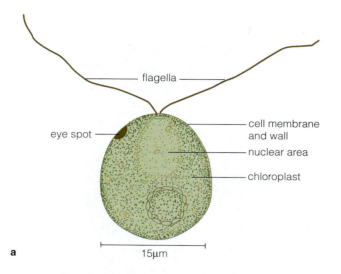

a

15μm

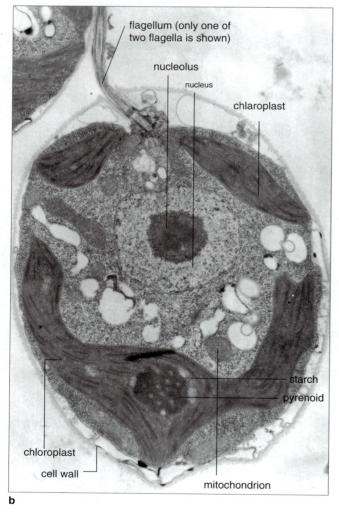

flagellum (only one of two flagella is shown)

nucleolus

nucleus

chlaroplast

starch

pyrenoid

chloroplast

cell wall

mitochondrion

b

Figure 21.12 The unicellular green alga *Chlamydomonas*. (**a**) Drawing. (**b**) Electron micrograph.

bays beneath the Mediterranean Sea. Each individual consists of a horizontal, creeping portion (which sends branching filaments down into the substrate) and featherlike or sheetlike blades that reach upward into the water (see Fig. 21.5e). There are, then, rootlike, stemlike, and leaflike regions.

Another complex filamentous green alga is the stonewort *Chara*, which has an anchoring region and a stemlike region differentiated into nodes and internodes (Fig. 21.13). *Chara* also possesses multicellular gametangia in which the gametes are protected by a shell of sterile cells. No other group of algae has these advanced, plantlike structures. The fossil record indicates that members of the Chlorophyta and Charophyta were present when the first land plants appeared on Earth, and their unique body details give botanists additional reasons—beyond the chemical and cellular factors already mentioned—to suspect they are the ancestors of land plants.

21.4 LICHENS

Anyone who walks in nature has seen crusty or leafy lichens on rocks, tree trunks, and fence posts (Fig. 21.14). A **lichen** is a composite organism consisting of a fungal mycelium that encloses cells of either green algae or cyanobacteria. About 20,000 species of lichens are known. In the great majority of species, the fungal component is an ascomycete (in fact, there are more species of ascomycetes in lichens than there are free-living). Fewer than 50 lichen species involve basidiomycetes or imperfect fungi. In contrast to the enormous diversity of fungi, only about 20 species of algae form the other half of the partnerships.

Lichenologists classify lichens into four general categories. **Crustose** (crustlike) **lichens** are a thin layer of tissue that tightly adheres to a substrate (Fig. 21.14a). These lichens often grow on exposed rock, coloring cliffs brilliant colors of orange or yellow. **Fruticose** (shrublike) **lichens** have twiglike projections that arise from the substrate (Fig. 21.14b). Much of the *reinder moss* so important as a food supply for arctic caribou and reindeer is *Cladonia alpestris*, a fruticose lichen. **Foliose** (leaflike) **lichens** have leafy plates of tissue attached to the substrate along one margin (Fig. 21.14c). Some reach a large size: Species of *Lobaria pulmonaria* may have a diameter of 40 cm. **Twofold** lichens change shape through their life cycle: A crustose or foliose vegetative body forms first, followed by a secondary fruticose body with reproductive structures.

Lichens are epiphytes. They use other organisms or structures merely as places of attachment and do not parasitize the plants they grow upon. Within the lichen, the fungal and algal partners have a mutualistic relationship that benefits both organisms. The fungus gains carbohydrate from the alga, and the alga gains a humid environment and mineral nutrients from the fungus. In some

sex organs

branch

female gametangium

male gametangium

Figure 21.13 The stonewort *Chara*, in the division Charophyta. (**a**) Part of an upright cluster of filaments, showing differentiation into nodes and internodes. (**b**) The female gametangium. Note the shell of of sterile, helical cells on the female gametangium, which protects the gametes within.

Figure 21.14 Lichens. (**a**) The crustose lichen *Caloplaca* paints rock outcrops in the northwest territory of Canada. (**b**) The fruticose lichen *Ramalina* grows near the California coast and experiences summer fog. (**c**) The foliose lichen *Peltigera* grows beneath a boreal forest in Alaska.

cases, it has been possible to experimentally separate the alga and fungus and to grow them separately on culture media; therefore, the association is not absolutely essential. However, when the two grow together they always produce characteristic acidic byproducts, and they arrange themselves into a lichen body with consistent shape, color, and habitat traits.

Lichens reproduce asexually by producing bits of intertwined algal cells and hyphae that erupt through special breaks in the cortex. These are carried by wind; if they land on an appropriate substrate, the cells divide by mitosis and differentiate into another lichen body. Only the fungal partner can reproduce sexually.

21.5 ECOLOGICAL AND ECONOMIC IMPORTANCE OF ALGAE

Algae are important in two basic but quite different ways. They are important to the entire biosphere because they are photosynthetic, creating carbohydrates and releasing oxygen (and in so doing forming the base of aquatic food chains and building tropical reefs). They are also economically important to humans because they serve as food, fodder, and fertilizer and have many industrial and pharmaceutical uses.

Planktonic Algae Are at the Base of Aquatic Food Chains

Microscopic, floating algae called phytoplankton (Fig. 21.15) occur in such great numbers over such a large surface of water that they are said to be "the grasses of the sea." That is, they produce the vegetation that feeds grazing animals that in turn feed predators and, ultimately, humans. Phytoplankton are at the base of aquatic, and especially oceanic, food chains. Phytoplankton are unicellular. They must stay close to the surface and avoid sinking below the photosynthetic zone by swimming, storing oil droplets, altering their ionic balance, and developing fine projections that extend out from the cell wall. The average life span of a given cell is probably measured in hours or days, the organism eventually dying and sinking to the ocean bottom.

The density of phytoplankton is normally low in the open ocean, perhaps a few thousand cells per liter (approximately one quart), but the oceanic expanse is enormous. As a result, on a global scale, phytoplankton produce 3.26×10^{17} kilocalories (3 quintillion kcal) of photosynthate each year, which is four times that produced by all the world's croplands.

This production of food energy would be even greater if such limiting factors as temperature and mineral nutrients were ameliorated. Deep oceans and freshwater lakes tend to be cool and low in nutrients. Continental edges and shallow lakes are warmer, have more mixing of water through all depths, and have more nutrients; there, productivity by plankton is greater. If nitrogen- or phosphorus-rich wastewater is added to such areas—or if water temperature rises for some natural reason—a tremendous increase in the density of phytoplankton—a **bloom** or **red tide**—occurs, in which the cells color the water a rusty red. The Red Sea has been given its name from the frequent recurrence of such episodes. Red tides are often caused by the dinoflagellates *Gymnodinium* and *Gonyaulax*. Both produce a potent, water-soluble toxin that affects vertebrate nervous systems. It is related to curare poison and is 10 times as toxic as cyanide. Massive fish kills result from red tides. Invertebrates such as shellfish can accumulate the toxin without being affected, but humans or other vertebrates that eat such tainted shellfish are poisoned. Red tides are often reported along the warm coastal waters of southwest India, southern China, southwest Africa, the eastern United States, the Gulf of Mexico, southern California, Peru, and Japan.

Other blooms are caused by the golden alga *Prymnesium parvum* and the cyanobacteria *Anabaena*, *Aphanizomenon*, *Gloeotrichia*, *Microcystis*, *Nostoc*, and *Oscillatoria*. Some of these phytoplankton also produce toxins and noxious odors.

Algae Help Build Tropical Reefs

Reefs, which form in shallow tropical seas (most often near islands or on continental shelves), provide an important habitat for an incredible variety of marine life. Partly the stony remains of coralline animals, reefs are also constructed by algae. Certain red and green **coralline algae** create a carbonate exoskeleton resembling coral around their filaments; this exoskeleton becomes a physical part of the reef after the algae die. Coralline algae also help build the reef by connecting the corals' exoskeletons. More than half the bulk of many reefs is from algae, not from coralline animals.

In addition, a few species of algae grow symbiotically within the tissues of coralline animals. The tissue of these invertebrate reef-forming animals is photosynthetic because it contains these symbiotic algal cells. The alga, usually the dinoflagellate *Symbiodinium microadriaticum*, has a photosynthetic rate 10 times higher than that of phytoplankton, probably because it is protected and nourished by the animal cytoplasm around it. The relationship between corals and algae is mutualistic—both organisms benefit. The alga produces sugar and oxygen for the animal, exporting more than 90% of the sugar into the animal's tissue and retaining only a small balance for its own growth and maintenance. In exchange, the cells of the coral contribute carbon dioxide, nitrogen, and mineral nutrients to the alga. Together, both prosper.

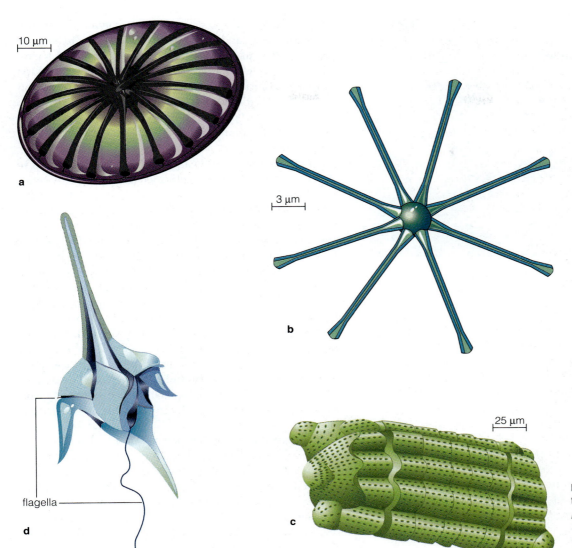

10 µm

3 µm

25 µm

flagella

a

b

c

d

Figure 21.15 Examples of phytoplankton. Diatoms: (**a**) *Asteromphalos elegans*; (**b**) *Asterionella formosa*; (**c**) *Biddulphia biddulphia*. (**d**) The dinoflagellate *Ceratium*.

Algae Serve as Medicine, Food, and Fertilizer

Seaweeds—red, brown, and green marine algae of moderate size—usually grow in the rocky intertidal zone, alternately exposed by low tides and covered by high tides. Seaweeds are an important part of the human diet and medicine chest in several parts of the world. In East Asia, seaweed harvesting has been known for 5000 years.

Shen Nong, the legendary Chinese "father of medicine" prescribed seaweed for a variety of ailments in texts dated 3600 years ago. Later, Confucius praised its curative value. A million metric tons a year of the brown seaweed *Laminaria* are harvested off the China coast as a source of iodine, which is added in trace amounts to the diet to prevent goiter (an enlargement of the thyroid gland in the neck). However, many of the historical claims of the special health-giving value of seaweed have not been substantiated by scientific research.

For centuries, the Japanese have used algae as a tasty supplement to their rice diet. The demand for *nori* (the red alga *Porphyra*) has grown to such an extent that it is cultivated in shallow intertidal bays (Fig. 21.16). The Polynesians in Hawaii utilized and named at least 75 species of *limu* (seaweed) as food sources. Some rare species were cultivated only in marine fish ponds belonging to nobility. Dulce (the red seaweed *Palmaria palmata*) has been used as a food for 12 centuries in the British Isles. It was the Irish who discovered that small quantities of Irish moss (another red alga, *Chondrus crispus*), when boiled with milk, would produce a jelly dessert that the French later called *blancmange*. Nevertheless, with some exceptions, algae do not have much nutritive value; in fact, their major constituents are largely indigestible. Today, algae are used

Figure 21.16 *Nori* (*Porphyra tenera*, Rhodophyta) culture in Sendai Prefecture, Honshu Island, Japan. *Hibi* nets sit about 30 cm (1 ft) above mean low tide in September, at the beginning season of nori culture. Harvest time is January.

more as condiment, garnish, or dessert than as staple—much as we use lettuce, watercress, celery, or herbs.

Seaweeds contain large amounts of potassium, nitrogen, phosphorus, and other minerals characteristic of good fertilizer or cattle feed supplements. In historic times, Native Americans and the Scotch-Irish used Irish moss as a fertilizer to build up poor soils for such crops as corn and potatoes. As a fertilizer, seaweed compares favorably to barnyard manure: It enhances germination, increases the uptake of nutrients in plants, and seems to impart a degree of resistance to frost, pathogens, and insects.

Algal Cell Walls Have Industrial Uses

Physical and chemical characteristics of the cell walls of some algae lead to products with many industrial, pharmaceutical, and dietary applications. It is through these products—primarily diatomite, agar, carrageenan, and algin—that algae have their greatest direct economic value.

Diatomite (or **diatomaceous earth**) is a sedimentary rock composed of fossilized diatom cell walls. Diatoms (Bacillariophyta) are important members of the phytoplankton (see Fig. 21.15); as they flourish and die, their empty walls sink and accumulate in bottom sediments. One of the richest deposits of diatomite—a 300-m- (1000-ft-) thick layer near Lompoc, California—formed this way about 15 million years ago beneath a warm, shallow sea. In more recent geologic time, the Lompoc area was uplifted above sea level, and the diatom deposit was revealed by erosion. Several companies mine the diatomite for industrial and pharmaceutical use.

Diatomite makes a superior filter or clarifying material, both because the microscopic wall pores create a large surface area (230 grams—8 oz—contain the area of a football field) and because the rigid walls are incompressible. It is used in both laboratory filters and swimming pool filters. Diatomite is inert, and it can be added to many materials to provide bulk, improve flow, and increase stability. In these ways diatomite is used in cement, stucco, plaster, grouting, dental impressions, paper, asphalt, paint, and pesticides. Diatomite is also used as an abrasive.

Agar (or **agar-agar**) is a polysaccharide, analogous to starch or cellulose but chemically different. With cellulose, it is a component of the walls of certain red algae, mainly *Gelidium* and *Gracilaria*. Today, agar is a common medium on which bacteria are grown. This use was discovered a century ago by a physician's wife, Frau Fanny Eilshemius Hesse, who used agar to thicken her jam. The noted microbiologist Robert Koch presented her idea to the world in his scientific writings of the late 19th century.

Seaweeds containing agar are commercially harvested by divers who descend 3 to 12 m (10 to 40 ft) into warm nearshore waters off Australia, California, China, Japan, Mexico, South America, and the southeastern United States. In addition to its bacteriological use, agar is important in the bakery trade. When added to icing, it retards drying in open air or melting in cellophane packages. Because agar is virtually indigestible, it is also used as a bulk laxative.

Carrageenan is another polysaccharide that accompanies cellulose in the walls of red algae, mainly Irish moss. The substance takes its name from the town of Carragheen, County Cork, along the south shore of Ireland, where its properties were first documented. Carrageenan reacts with the proteins in milk to make a stable, creamy, thick solution or gel. Consequently, it is used commercially in ice cream, whipped cream, fruit syrups, chocolate milk, custard, evaporated milk, bread, and even macaroni. It is added to dietetic, low-calorie foods to bring back the appropriate mouth "feel" of nondietetic foods. Carrageenan is also used in toothpaste, pharmaceutical jellies, and lotions of many sorts. Irish moss is commercially harvested in the United States off the shore of Maine.

The brown algae (Phaeophyta) have a commercially valuable compound called **algin**. It is a long-chain polymer made up of repeating organic acid units. It is the principal wall component of brown algae, constituting up to 40% of the cell wall by weight. Water is strongly adsorbed to algin, creating a thick solution. One tablespoon of powdered algin added to one liter (about one quart) of water increases the viscosity to that of honey. In nature, algin may be valuable to intertidal algae during low tide due to its ability to retain water. Commercially, it is used as an additive to beer, water-based paints, textile sizing, ceramic glaze, syrup, toothpaste, and hand lotion. Hundreds of algal species contain algin, but only a few are commercially harvested: *Macrocystis pyrifera* along the California coast; species of *Ascophyllum*, *Fucus*, and *Laminaria* off England and the China coast; and *Durvillea* in Australian waters.

21.6 ALGAL REPRODUCTION

The life cycles of many algae are still unknown, but the three basic life cycles—zygotic, gametic, and sporic—have all been found. Most studies have concentrated on red, brown, and green algae.

Each life cycle has sexual and asexual portions. Asexual reproduction occurs more often than sexual reproduction, so it is more commonly seen by researchers. Typical methods of asexual reproduction are cell division (for single-celled algae) and **fragmentation** (the splitting apart of a filament). Vegetative reproduction can also be achieved by the formation, liberation, and germination of motile or nonmotile mitospores produced in sporangia. In many algae, sporangia show little if any difference in appearance from ordinary body cells, except that they may be enlarged. The parent nucleus divides by mitosis several times, and each resulting nucleus accumulates cytoplasm about itself and secretes a bounding wall, forming a mitospore.

Ulothrix Typifies the Zygotic Life Cycle in Algae

The filamentous green alga *Ulothrix* has a zygotic life cycle (Fig. 21.17). The life cycle of *Chlamydomonas*, a common green alga in freshwater habitats, is also zygotic; its cycle was described in Chapter 12 (see Fig. 12.7).

The nucleus of each *Ulothrix* cell is haploid. When sexual reproduction begins, some of the nuclei will divide by

Figure 21.17 The zygotic life cycle of the filamentous green alga *Ulothrix*. Algal filaments are haploid. The nucleus of some cells can divide and produce mitospores (see box in lower left). Each mitospore settles to the bottom of the pond and produces a new haploid filament. Other cells, however, produce + or – gametes, which fuse to produce a resting zygote. The zygote then undergoes meiosis and produces meiospores, which germinate and grow to become haploid algal filaments.

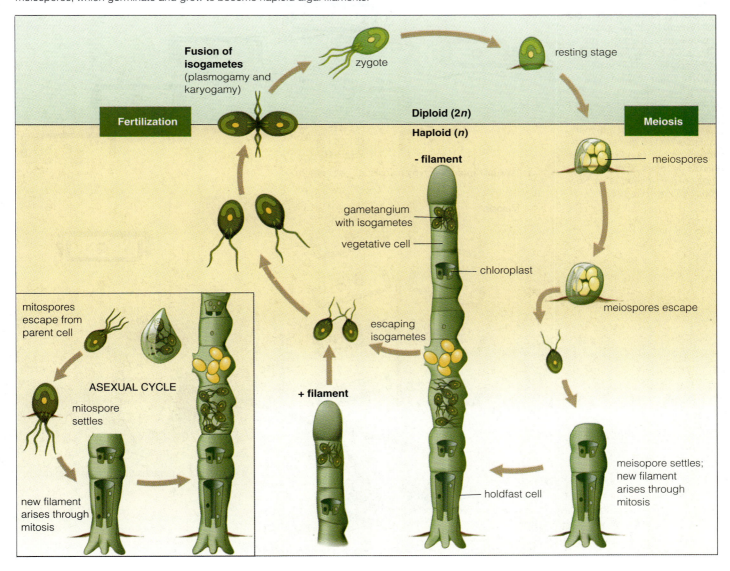

repeated mitosis, producing many motile gametes inside the wall of the parent cell. In such a case, the parent cell is a gametangium. The gametes of *Ulothrix* look very much like individual *Chlamydomonas* cells: They have two anterior flagella, and they swim about in water. If a *Ulothrix* gamete approaches another suitable gamete, the two will pair and fuse, producing a diploid zygote cell with four flagella. Although all *Ulothrix* gametes look alike to us (they are **isogametes**), there must be genetic differences recognizable by *Ulothrix* that make fusion possible between some, but not all, gametes. That is, there appear to be two mating types of gametes, known as plus (+) and minus (−). Since a given *Ulothrix* individual produces only plus or minus gametes, gametes from the same individual will not mate, whereas gametes from different individuals will be able to fuse.

The zygote resulting from gametic fusion becomes spherical, loses its flagella, and enters a resting stage during which it is resistant to such environmental extremes as exposure to dry air. Once growing conditions return, the zygote becomes metabolically active and divides by meiosis to produce plus and minus meiospores, each of which is capable of producing a new individual by itself. The meiospores are dispersed, and each can germinate, divide by mitosis, and produce a plus or minus haploid plant.

When *Ulothrix* begins to reproduce asexually, a vegetative cell becomes a sporangium. Eventually, 16 to 64 pear-shaped mitospores are released; each has four flagella. After a period of activity, these motile mitospores settle to the bottom of a pond, lose their flagella, and begin to produce a new plant by mitosis. Some mitospores may enter a dormant phase and be resistant to environmental stress during that time.

Diatoms Have a Gametic Life Cycle

Diatom cells are diploid. The nucleus undergoes meiosis, producing four haploid nuclei (Fig. 21.18). Only one or

Figure 21.18 Gametic life cycle of a diatom. (**a**) Repeated asexual cell division results in smaller diploid cells because one of the progeny cells is formed within the wall of the smaller half of the parent cell wall. (**b**) When cells become critically small, meiosis occurs. Three of the haploid nuclei degenerate, and the single remaining one becomes the nucleus of a gamete. (**c**) Fusion of two gametes form a diploid zygote. (**a** and **c** redrawn after G. M. Smith, *Freshwater Algae of the United States*. New York: McGraw-Hill, 1933.)

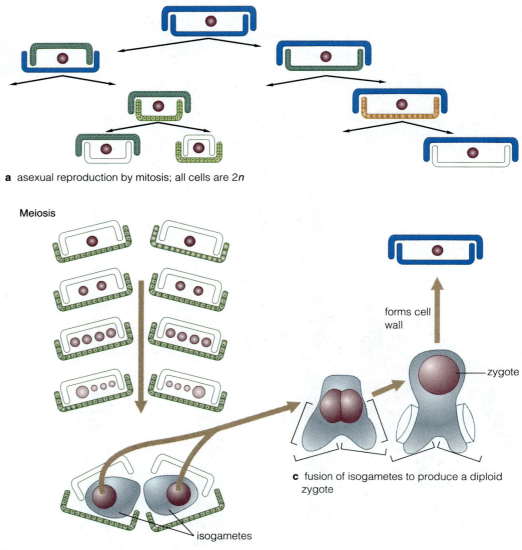

a asexual reproduction by mitosis; all cells are 2*n*

Meiosis

forms cell wall

zygote

c fusion of isogametes to produce a diploid zygote

isogametes

b sexual reproduction by meiosis, leading to isogametes

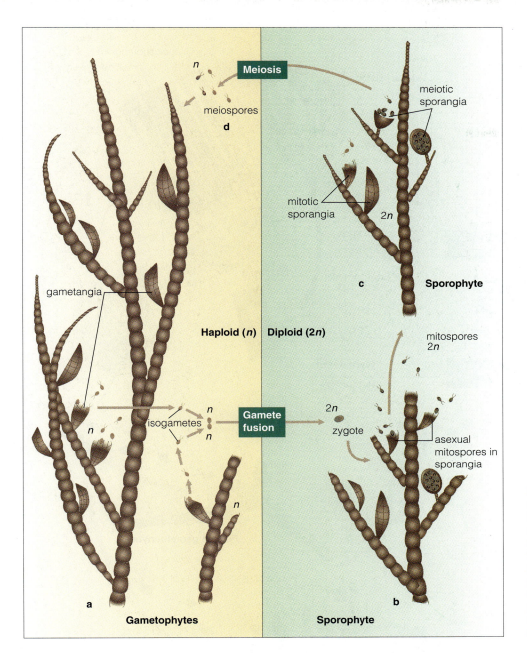

Figure 21.19 The sporic life cycle of filamentous brown alga *Ectocarpus*, which has isomorphic generations. (**a**) The gametophyte generation releases plus and minus gametes. These fuse in pairs to form zygotes. (**b**) and (**c**) Each zygote produces a sporophyte plant. A sporophyte can generate asexual mitospores capable of producing another sporophyte plant (**b**), or meispores that will germinate to form the gametophyte generation once again (**c** and **d**).

two of the nuclei may survive and become gametes. The parental cell wall has served as a gametangium. If two diatom gametangia are near each other, they open; the two gametes emerge and fuse, forming a diploid zygote. The zygote increases greatly in size and may rest for a short time. It then secretes a normal two-part silicate wall around itself and becomes a vegetative diploid diatom.

Diatoms reproduce asexually by cell division. Recall that each cell has two wall segments that fit together like the top and bottom of a petri dish. After division, a new wall forms within the old one. This means that the progeny cell that inherits the smaller wall segment will make a new wall that is smaller yet. As this pattern continues from generation to generation, one line of cells becomes ever smaller. Eventually a critically small size is reached, and division stops. To regain full size, the cell must reproduce sexually, as described earlier.

The common intertidal rockweed *Fucus* (in the Phaeophyta) also has a gametic life cycle; it was described in Chapter 12 (see Fig. 12.8).

Ectocarpus Has a Sporic Life Cycle with Isomorphic Generations

Ectocarpus is a filamentous brown alga commonly found in cool coastal salt water. We begin the life cycle with the haploid phase (Fig. 21.19a), which has curved clusters of gametangia on side branches. Mitosis within each gametangium produces many gametes, which swim when released using two unequal flagella that arise from the middle of the cells. The gametes all look alike—even though they represent plus and minus mating types— hence, they are isogametes. Fusion of a plus isogamete and a minus isogamete in open water yields a diploid zygote.

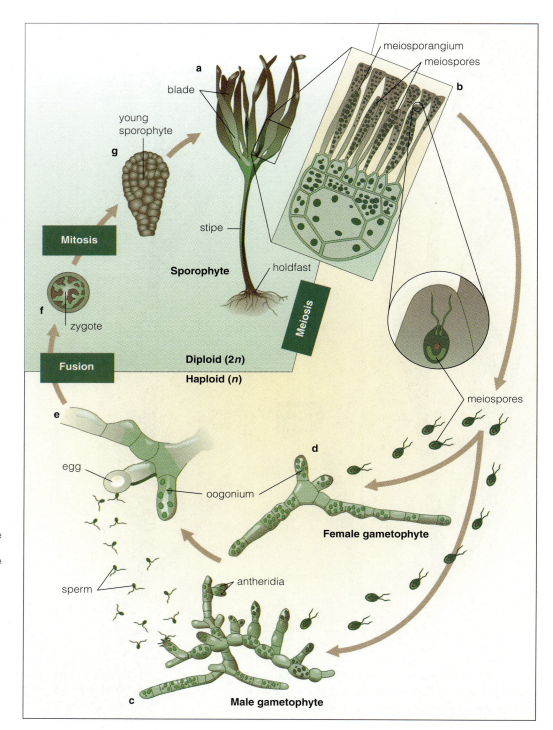

Figure 21.20 The sporic life cycle of the kelp *Laminaria*, which has heteromorphic generations. (**a**) The sporophyte plant. (**b**) Meiosis occurs in sporangia in the blade, producing male and female meiospores, which form male (**c**) and female (**d**) gametophyte individuals, respectively. The male gametophytes produce antheridia and sperm; the females produce oogonia and eggs. Fusion of gametes (**e**) creates a diploid zygote (**f**) which then forms a young sporophyte (**g**).

The zygote settles to the bottom, germinates, and divides by mitosis to produce a diploid plant (Fig. 21.19b and c).

The diploid sporophyte looks identical to the haploid gametophyte. The only difference is the chromosome number within every nucleus. Plants like *Ectocarpus* that have identical-looking gametophytes and sporophytes have **isomorphic generations**.

The sporophyte produces two kinds of reproductive cells. In one, clusters of cells on side branches (looking identical to the gametangia on gametophytes but called *asexual sporangia* because of their function) liberate motile cells. Although the cells look like gametes, they do not pair and produce zygotes. Instead, each is capable of producing a new individual by itself; that is, each is a diploid mitospore. The individual that each mitospore produces is another sporophyte, identical in appearance to the parent sporophyte. The other kind of sporangium is spherical; meiosis occurs within this one, producing many meiospores. The meiospores, when released, look just like gametes and mitospores. Each one can settle down to the bottom, germinate, and produce a gametophyte plant, completing the life cycle.

Laminaria Has a Sporic Life Cycle with Heteromorphic Generations

Plants in which the gametophyte and sporophyte are not identical have **heteromorphic generations**. Many groups of algae exhibit this kind of life cycle, but it is most highly developed in the Phaeophyta, particularly in the kelps. The kelp *Laminaria* has a sporophyte generation with a well-developed holdfast and a long unbranched stipe with a cluster of narrow blades extending from the end (Fig. 21.20a). Sporangia usually occur in groups just below the meristoderm on a blade (Fig. 21.20b). Each sporangium is a single cell within which meiosis occurs (followed by one to several rounds of mitosis), producing 8 to 64 meiospores. The meiospores are released, swim about, settle to the bottom, and produce gametophytes.

The gametophytes are small, branched filaments, quite unlike the sporophytes. Some of the gametophytes will produce male gametes (Fig. 12.20c); others will produce female gametes (Fig. 12.20d). The cells at the tips of some filaments on male gametophytes enlarge and function as **antheridia** (male gametangia); their nuclei undergo mitosis, producing many motile sperm. Other cells on filaments of female gametophytes enlarge and function as **oogonia** (female gametangia); their nuclei undergo mitosis, producing one to several large eggs. The eggs are extruded nearly out of the oogonium, but they remain attached. A sperm cell will fuse with an egg cell, producing first a zygote and then a young sporophyte (Fig. 21.20e, f, and g). At this stage, the sporophyte falls from the gametophyte, is carried passively by currents to the bottom, and begins to develop into a mature *Laminaria* sporophyte. The life cycle is completed.

SUMMARY

1. The kingdom Protista includes animal-like organisms (protozoans, foraminifera, sponges), nonphotosynthetic fungus-like organisms (chytrids, slime molds, egg fungi), and photosynthetic plant-like organisms (algae). All are eukaryotic, many are microscopic, and most have very simple bodies. Ancestors of the kingdoms Animalia, Fungi, and Plantae are all represented in the Protista.

2. Fungi may have evolved from the division Chytridiomycota, a group of 450 largely aquatic decomposer or parasite species.

3. Algae is a common group name for about 20,000 to 30,000 species of photosynthetic organisms classified into nine divisions. Algae differ from plants in that: (1) Their gametangia are single cells, lacking any protective layer of sterile cells around the gametes; (2) they lack an embryo stage; (3) they do not have an epidermis with cuticle and stomata; (4) they are more often aquatic than terrestrial; and (5) their bodies are usually simple and rela-tively undifferentiated. Body shape includes single-celled forms, colonies, filaments, sheets, and three-dimensional packages of cells.

4. The nine divisions of algae differ in cell wall construction, number and placement of flagella, types of chlorophylls, pigments, morphologies, and habitat ranges. The five divisions described in detail are: Bacillariophyta (diatoms), Pyrrhophyta (dinoflagellates), Rhodophyta (red algae), Phaeophyta (brown algae), and Chlorophyta (green algae).

5. The Chlorophyta share several traits with higher plants: chlorophylls *a* and *b*, similar accessory pigments, starch as a storage product, cellulose cell walls, similar morphologies (that include rootlike, stemlike, and leaflike regions—*Caulerpa, Chara*), complex gametangia, a particular pattern of cell division, and unique flagellar architecture. For these reasons, phycologists believe that ancient green algae were ancestral to higher plants. The fossil record of green algae extends back nearly 1 billion years, much longer than that of higher plants.

6. Lichens are symbiotic associations between green algae or cyanobacteria and fungi (usually ascomycetes, rarely basidiomycetes or imperfect fungi). They include crustose, fruticose, and foliose forms. Lichens reproduce asexually. The fungal partner can reproduce sexually and liberate ascospores.

7. Algae are ecologically important in aquatic ecosystems by being at the base of the food chain. Most of their productivity comes from phytoplankton. In polluted or warm, shallow water, phytoplankton can become dense enough to color the water, poison vertebrates, and create noxious odors. Algae are also important as reef builders, both as coralline green and red algae and as symbionts with coral animals.

8. Seaweeds have some economic importance to human societies as medicine, food, and fertilizer. However, their major value comes in the physical and chemical properties of the cell walls of diatoms (diatomite), red algae (agar, carrageenan), and brown algae (agar).

9. Algae can reproduce sexually by zygotic, gametic, and sporic life cycles. *Ulothrix* is an example of an alga with a zygotic life cycle. Diatoms have a gametic life cycle. *Ectocarpus* has a sporic life cycle in which the sporophyte and gametophyte generations are isomorphic. *Laminaria*, in contrast, has a sporic life cycle with heteromorphic generations.

Questions

1. As biologists learn more about organisms that are currently classified in the division Protista, what is likely to happen to the division?

2. How are chytrids, slime molds, and egg fungi different from organisms in the kingdom Fungi?

3. What evidence do biologists have to support their hypothesis that plants evolved from green algae?

4. Make up your own short dichotomous key that separates members of the algal divisions Bacillariophyta, Pyrrhophyta, Rhodophyta, Phaeophyta, and Chlorophyta.

5. How is it possible for microscopic algae to be important enough to fuel oceanic food chains, which culminate in large carnivorous fish and aquatic mammals like whales? Why is it possible to say that reefs are built by algae as much as they are built by coral animals?

6. The divsion Phaeophyta includes complex marine algae called kelps. Describe their specialized body regions, their unusual anatomy, and their sporic life cycle.

Further Readings

Adey, W. H. 1987. "Food production in Low-Nutrient Seas." *Bio-Science* 37: 340–48. Net production in surface waters of deep ocean areas is very low, only 0.2 g carbon per square meter of surface per day. This could be increased by pumping high-nutrient subsurface water to the surface. Algal tufts, however, which grow in shallow seas, have production rates two orders of magnitude higher. If such tufts were regularly harvested in the warm Caribbean Sea, 25 million metric tons of dry algae per year would be obtained—a potential source of human or animal food.

Bold, H. C., and M. J. Wynne. 1985. *Introduction to the Algae: Structure and Reproduction.* 2d ed. Englewood Cliffs, N. J.: Prentice-Hall. A very comprehensive, yet readable review of all algal divisions regardless of habitat.

Brown, M. T. 1987. "Effects of Desiccation on Photosynthesis of Intertidal Algae from a Southern New Zealand Shore." *Botanica Marina* 30: 121–27. The photosynthetic rates of attached seaweeds to alternate submersion and exposure by tidal action correlate with their position in the intertidal zone: The higher in the zone they grow, the greater is their recovery from exposure within three hours.

Dawes, C. J. 1981. *Marine botany.* New York: Wiley Interscience. A narrower review of the biology, ecology, and systematics of marine algae.

Falkowski, P. G., et al. 1984. "Light and the Bioenergetics of a Symbiotic Coral." *BioScience* 34: 705–9. Elegant study of the dinoflagellate *Symbiodinium microadriaticum,* a common symbiont in coral tissue. Even in low light, the alga exports more than 90% of its photosynthate to the surrounding animal cell and provides at least 60% of the animal's growth needs. However, the alga does not provide nitrogen, and this nutrient is obtained from animal prey or detritus.

Foster, M. S., and D. R. Schiel. 1985. *The Ecology of Giant Kelp Forests in California: A Community Profile.* U.S. Department of the Interior. Fish and Wildlife Service. Biological Report 85(7.2). An excellent summary of the kelp bed ecosystem off the California coast, with emphasis on the biology of *Macrocystis pyrifera.*

Garrison, D. L., et al. 1986. "Sea Ice Microbial Communities." *BioScience* 36: 243–50. Diatoms attached to the underside of sea ice in Antarctica are able to grow in temperatures as low as $-2°C$ (28°F) and in light as low as 0.4% of full sun at the ice surface.

Littler, M. M., and D. S. Littler. 1985. "Deepest Known Plant Life Discovered on an Uncharted Seamount." *Science* 277: 57–59. A living, attached red alga found at a depth of 268 m (879 ft).

Radmer, R. J. 1996. "Algal Diversity and Commercial Algal Products." *BioScience* 46: 263–70. Radmer claims that the known number of algae represents a small fraction of probably 200,000 extant species. He reviews their phylogenetic relationships and then describes a variety of products humans utilize for food, paper products, pharmaceuticals, agriculture, and animal feed that come from 20 algal species.

Robinson, D. H., and C. W. Sullivan. 1987. "How Do Diatoms Make Silicon Biominerals?" *Trends in Biochemical Sciences* 12: 151–54. We still don't know much about how diatoms use silicon instead of carbon to fashion cell wall material.

Wharton, R. A., Jr., and W. C. Vinyard. 1981. "Red Snow and Hot Springs." *Fremontia* 8 (4): 11–14. A very readable commentary on various green and red algae that occur in the snow of high California mountains and in hot springs atop Mt. Shasta.

22 PLANTAE: BRYOPHYTES

1. Bryophytes are members of the Kingdom Plantae, but they differ from all other plants in that they lack a vascular tissue (they are nonvascular) and they have a sporic life cycle in which the gametophyte generation is dominant.

2. Modern species within the division Bryophyta may have evolved from more than one common ancestor. Bryophytes are divided into five classes: liverworts (Hepaticopsida), hornworts (Anthocerotopsida), and mosses (Andreaeopsida, Bryopsida, Sphagnopsida). Liverworts may have evolved from brown algal ancestors; mosses and hornworts, from green algae. Most botanists believe that vascular plants did not evolve from bryophytes, but probably evolved from algae in parallel with bryophytes.

3. Unique features, not present in Protista, include an embryo phase in the life cycle and multicellular gametangia that protect and insulate gametes from the environment. The most complex bryophyte sporophytes also contain such novel anatomical specializations as a cuticle, stomata, and primitive water- and sugar-conducting tissue. Bryophyte gametophytes are not able to control their water balance, however, and they dry out rapidly when the habitat becomes dry. Yet, the dry plants are still alive, and they are capable of becoming active within hours of being rewetted.

4. Bryophytes are important elements of many ecosystems. They provide most of the biomass in boreal vegetation such as tundra, they dominate the understory of cool-temperate forests, and they are common in damp microenvironments.

Some species are aquatic, and most require humid conditions; but some are able to colonize dry, exposed habitats such as rock outcrops and desert soil surfaces.

22.1 HALFWAY BETWEEN PLANTS AND PROTISTS

Bryophytes are small, herbaceous (nonwoody) members of the kingdom **Plantae**. They have a combination of traits that place them between algae and vascular plants, capable of bridging both aquatic and terrestrial environments. Bryophytes commonly grow on wet, terrestrial sites: along creek banks or on rocks just above running water; bathed by spray near waterfalls; in the dense shade of damp forests; or on the trunks of trees, where rainwater funnels down from the canopy and frequently runs over them. Some grow as aquatic plants in lakes, but their diversity takes them also onto the leaves of tropical rain forest trees, into the crust of desert and coastal sand dune soil, to the splash zone just above high tide on rocky shores, onto the edge of snowbanks, into rabbit holes and dimly lit caves, within coastal salt marshes and desert salt pans, onto disturbed sites, near hot springs, and onto cold antarctic islands (Fig. 22.1).

The division Bryophyta includes the familiar mosses, with over 14,000 species distributed in a wide variety

Figure 22.1 Morphological diversity and habtats of bryophytes. (**a**) Epiphytic mosses on trees in Olympic National Park, Washington. (**b**) Mosses forming a soil crust beneath the desert shrubs shadscale (*Atriplex confertifolia*) and sagebrush (*Artemisia tridentata*) in Idaho. (**c**) Black-colored *Grimmea* moss growing on bare rock outcrop in the Appalachian mountains. (**d**) A carpet of feathermoss (*Hylocomium*) beneath an Alaskan spruce (*Picea*) forest.

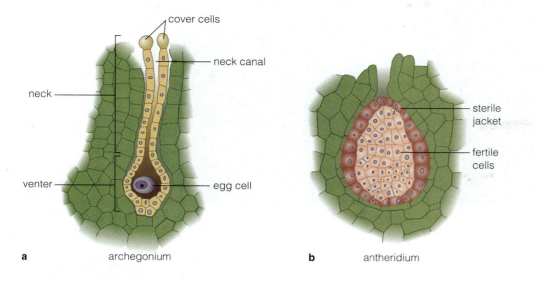

Figure 22.2 A bryophyte archegonium (**a**) and antheridium (**b**). The archegonium has a swollen base (the venter) within which lies one gamete, the egg; a long neck surrounding a canal; and a tier of cover cells at the tip of the neck. The outermost cell layer of the spherical antheridium is sterile (nonreproductive); it protects many fertile cells, which become gametes called sperm.

of habitats; the less-familiar liverworts with 9,000-plus species; and the hornworts, with more than 400 species. Bryophytes are generally small, and their size hides their prevalence and ecological importance. The largest bryophyte is the Australian moss *Dawsonia superba*, which grows to a height of 70 cm (28 in). Some hanging, epiphytic mosses and some aquatic mosses are equally long, but they are not erect. The smallest bryophytes are barely visible without a hand lens.

Bryophytes Aren't Completely Aquatic or Terrestrial

Bryophytes are not exclusively aquatic or terrestrial. They are algalike in that they: (1) produce free-swimming sperm that require water as a medium in which to reach eggs; (2) have no vascular system (although some do have a primitive type of conducting tissue); (3) have no lignified tissue (and hence are nonwoody); and (4) lack differentiation into true roots, stems, and leaves. Because of these traits, bryophytes usually require habitats that are wet for at least some of the time each year. However, some species that live in arid environments can survive dry periods for months or even years. They endure in a shriveled, dormant state and are among the most drought-tolerant land plants that currently exist.

Bryophytes also resemble land plants in a variety of ways. Mosses have organs that superficially resemble leaves, stems, and fibrous roots. Some have a cuticle and stomata that retard dehydration. They also produce asexual, wind-borne spores protected from desiccation by a wax-impregnated wall. Also, certain features of their reproductive cycle mimic those of higher land plants such as ferns and seed plants: (1) They produce gametes within a jacket of protective cells; (2) they form eggs in flask-shaped **archegonia** and sperm in rounded **antheridia** (Fig. 22.2); (3) diploid offspring are sheltered and nourished inside the parental body at an early stage of development (a multicellular embryo); and (4) the life cycle is **sporic**, having both gametophyte and sporophyte generations.

The bryophyte life cycle is unique among land plants in that the haploid gametophyte generation is dominant. The gametophytes are green, independent, relatively long-lived, and morphologically more complex than the sporophytes. The sporophytes are at least partially parasitic on the gametophyte; they are short-lived and comparatively simple. By contrast, in other land plants the diploid sporophyte generation is more conspicuous, and the gametophyte is parasitic on the sporophyte.

The Division Bryophyta Contains Five Classes

At present, we are not at all certain where bryophytes fit in the evolutionary pathway from algae to higher plants. Some biologists speculate that bryophytes represent the ancestral stock for higher plants; some think they are a completely separate side shoot of evolution, off the main track that led to higher plants; and some think they are degenerate forms of primitive vascular plants, evolving after higher plants had already appeared on earth.

The fossil record is of little help in resolving these speculations because the early record for bryophytes is very poor. The earliest fossil bryophytes found so far occur in rocks formed 350 million years ago in the geologic period known as the Devonian. At that time, more advanced plants were already present. Furthermore, these first fossils are remarkably modern in appearance. The oldest fossil bryophyte, *Pallavicinites devonicus*, is very similar to living species in the genus *Pallavicinia*. It is possible that the Devonian was a time of rapid bryophyte evolution. Alternatively, it may be that we simply have not yet discovered fossils that exist from an earlier time.

Based on biochemical, cellular, and morphological traits, most botanists guess that bryophytes evolved from algae as a separate line, in parallel with higher vascular plants. These traits also suggest that there was not a single common algal ancestor for bryophytes, but several: Liverworts may have evolved from brown algae; mosses and hornworts may have evolved from green algae. For

a   b

Figure 22.3 Thalloid (**a**) and leafy (**b**) liverworts.

this reason, some classification schemes put mosses, liverworts, and hornworts into separate divisions. In this book, however, we include all three lines within the single division **Bryophyta** and divide it into five classes: **Hepaticopsida** (liverworts), **Anthocerotopsida** (hornworts), **Andreaeopsida** (granite mosses), **Sphagnopsida** (peat mosses), and **Bryopsida** (true mosses).

22.2   LIVERWORTS

The approximately 9000 species in the class Hepaticopsida are commonly known as liverworts, or hepatics (from the Greek word for "liver"). The name *liverwort* is very old, having been recorded in medieval manuscripts since the 9th century. It was probably applied to these plants because of their fancied resemblance to the liver and the belief that plants resembling human organs would cure diseases of the organs they represented. A prescription for a liver complaint in the 16th century called for "liverworts soaked in wine."

Liverworts are the simplest bryophytes. The great majority grow in moist, shady habitats. The gametophyte is the prominent generation. The gametophyte thallus (plant body) is green and often grows as a ribbonlike, branching plate of tissue that is held to the surface of damp soil by rootlike filaments of tissue called **rhizoids** (Fig. 22.3a). The plant body ranges in size from one to several centimeters (about 1 in) across.

Some liverworts are more complex, with leaflike and stemlike structures called **phyllids** and **caulids**, respectively. The phyllids are blunt-tipped or have lobes and are attached to a caulid in two to three overlapping rows in a two-dimensional plane (Fig. 22.3b). Phyllids and caulids are not true leaves and stems because they lack vascular tissue. The filamentous rhizoids are not true roots because they, too, lack vascular tissue. Liverworts produce sporangia (**capsules**) that split into four valves when they release their spores. The capsules also contain special thickened cells, **elaters,** that aid in spore dispersal.

Our discussion of this class includes *Riccia* and *Marchantia*, in the order Marchantiales, and *Porella*, in the order Jungermanniales. These three examples express most of the morphological variation of hepatic plants.

Figure 22.4 The characteristic appearance of *Riccia*, in the class Hepaticopsida.

Plants in the Marchantiales Have Ribbonlike Bodies

Gametophytes of the order Marchantiales are small, green, thalloid (ribbon-shaped) plants. They grow Y-shaped branches by a simple forking at the growing tip. In some species, a rosette of branches is formed. The thallus is less than 20 cells thick and differentiates into distinct upper and lower layers. The upper layer, just below the epidermis, is composed of chlorenchyma cells adapted for photosynthesis. They form a matrix penetrated by air chambers that connect to the outside through pores in the upper epidermis. Several types of storage parenchyma cells generally make up the lower layer of the thallus. Rhizoids extend down into the substrate and anchor the gametophyte.

Riccia is a widely distributed genus. It is the simplest example of plants in the Marchantiales. Although *Riccia* requires abundant water for active growth, most species can tolerate considerable drought. A few species are aquatic.

The gametophyte (Fig. 22.4) is internally differentiated. The upper layer of tissue in the thallus consists of vertical rows of chlorenchyma penetrated by air chambers (Fig. 22.5a), while the lower layer is composed of

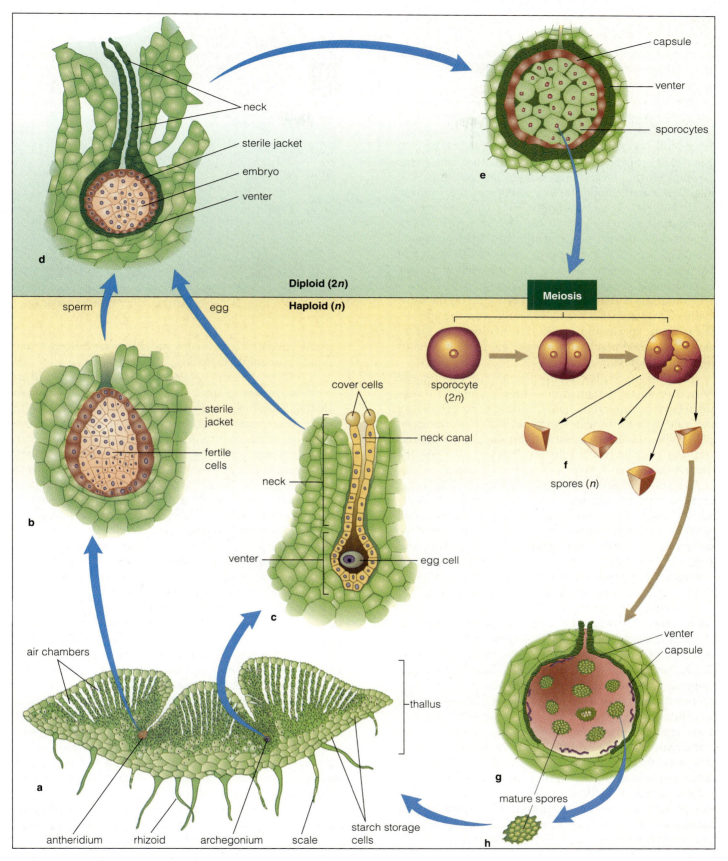

Figure 22.5 *Riccia* life cycle. (**a**) Cross section through a gametophyte thallus. Gametangia (antheridia and archegonia) are in notches on the upper surface. (**b**) An antheridium. (**c**) An archegonium. (**d**) Young sporophyte (embryo) developing in the venter of an archegonium. (**e**) Sporophyte (capsule) with sporocytes about to undergo meiosis, still enclosed in the venter. (**f**) The process of meiosis (followed by mitosis), resulting in (**g**) mature haploid spores that eventually become liberated (**h**). Each spore can germinate to produce a gametophyte thallus (**a**).

colorless parenchyma cells that sometimes contain starch grains. Hepatic gametophytes are independent, free-living organisms.

The gametangia (sperm- and egg-producing structures) are embedded in deep, lengthwise depressions or furrows on the upper surface of the thallus. The male gametangium, or antheridium, is a spherical ball of cells (Fig. 22.5b). Surface cells of this structure are sterile (nonreproductive); they form a layer one cell thick that shields the fertile reproductive cells from the environment. The female gametangium, or archegonium, is a pear-shaped group of cells. It contains two parts: (1) an expanded basal portion, the **venter**, and (2) an elongated **neck** (Fig. 22.5c). Four **cover cells** are located at the top of the neck. A short stalk attaches the archegonium to the gametophyte.

Antheridia and archegonia can be found on the same thallus or on separate thalli, depending on the species. Most hepatic species (but not *Riccia*) are **dioecious**, meaning that a single plant will produce either antheridia or archegonia, not both.

The interior, fertile cells of the antheridium give rise by mitosis and differentiation to sperm. The sperm cells are relatively numerous, small, dense with cytoplasm, and able to move in water with the aid of two long flagella. When sufficient moisture surrounds an antheridium, the exterior sterile cells break apart, allowing the sperm to escape.

The venter of each archegonium contains a single egg. Shortly before the egg cell is mature, the cover cells separate and the cells in the center of the neck dissolve, so that an open canal connects the egg with moisture outside the archegonium. Sperm swim toward the archegonium, and then down the neck, in response to chemical attractants exuded from the egg. Only one sperm fertilizes the egg, forming a 2*n* zygote. The zygote is the first cell of the sporophyte generation.

The zygote undergoes mitosis, producing an embryo, a spherical mass of more than 30 similar, undifferentiated cells within the venter (Fig. 22.5d). The gametophyte thallus provides all nutrients that support this early growth. Further development of the sporophyte involves differentiation of an outer layer of cells; these form a protective jacket of sterile tissue that surrounds internal cells capable of producing spores. These internal cells at first divide by mitosis until many closely packed cells have been formed (Fig. 22.5e). Each of these cells, now called **sporocytes**, undergoes meiosis and produces four haploid spores. Recall that the name for the organ that produces spores is a **sporangium**. The entire mature sporophyte of *Riccia* (the fully developed diploid generation) is simply one sporangium: a jacket of sterile cells enclosing a mass of spores and elaters.

When the spores are mature, the sporangium wall breaks down, but the spores are not released (Fig. 22.5h) until after the death and decay of the gametophyte. As each spore is finally disseminated and lands in an appro-

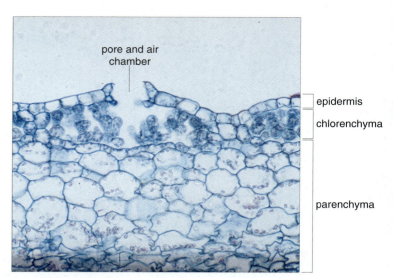

Figure 22.6 *Marchantia* thallus. Cross section showing pore, air chamber, epidermis, chlorenchyma, and parenchyma storage tissue.

priate habitat, it germinates and grows by mitosis into a typical gametophyte thallus.

Plants of the genus *Marchantia*, another thalloid liverwort, may grow in large mats on moist rocks and soil in shaded locations. *Marchantia* is widely distributed and fairly common along the banks of cool streams. It is somewhat better adapted to growing on land than *Riccia*, but abundant moisture is still required for optimal growth and for fertilization. The thallus of *Marchantia* is similar to that of *Riccia*, but there is a prominent midrib, and the tips of the branches are notched. A typical thallus is one to two centimeters (less than an inch) across. Size and degree of branching depend upon growing conditions.

On the upper epidermis are polygonal areas, each with a conspicuous pore in the center (Fig. 22.6). These areas demarcate air chambers below the pore, which bathe chlorenchyma cells in air containing carbon dioxide. The pores function as permanently open stomata. As in *Riccia*, the lower cells of the thallus are colorless parenchyma, modified for carbohydrate storage. Rhizoids and sheets of cells called *scales* project from the lower surface, increasing the surface area in contact with the substrate and anchoring the thallus.

Marchantia reproduces asexually in two ways. Older parts of the thallus die, and younger portions (no longer attached) develop into separate individuals. In addition, small **gemmae cups** form on the upper surface of the thallus (Fig. 22.7), and small disks of green tissue called **gemmae** grow from the bottom of these cups. When the gemmae are mature, raindrops break them free of the cup and scatter them away from the thallus. If the gemmae land on a suitable substrate, each is capable of developing into a gametophyte plant by mitosis.

Marchantia gametophytes are dioecious, producing either antheridia or archegonia. The gametangia are very similar to those of *Riccia*, but they are raised above the thallus on slender stalks called **antheridiophores** or **archegoniophores** (Fig. 22.8).

gametophyte thallus   gemma cups

Figure 22.7 Gemmae cups on upper surface of *Marchantia* thallus.

a   antheridiophore   thallus

Plants in the Jungermanniales Have Leafy Bodies

The largest order of liverworts, the order Jungermanniales, contains two-thirds of all hepatic species. The gametophytes are leafy, in contrast to the ribbonlike thallus of *Riccia* and *Marchantia*. Leaflike phyllids are borne on stemlike caulids in a semierect posture off the ground. Rhizoids anchor the plant to the substrate.

Porella, a very common, widespread genus in this order (Fig. 22.9), forms dense mats on rocks and trees. Young portions of caulids are densely clothed with phyllids arranged in three ranks. The upper two are large and lobed; the lower one is much smaller and unlobed. The leaves are one cell thick and lack a cuticle. Shaded, older, lower portions of caulids lack phyllids, and new rhizoids arise along those naked segments. Some liverwort species, including *Porella*, have very few rhizoids.

Antheridia occur in the axils of caulids and phyllids, and archegonia occur on specialized short, leafy caulids. Sporophytes superficially resemble those of *Marchantia* but differ in having larger elaters and a more complex sporangium (capsule) wall.

b   thallus   archegoniophore   gemma cup

Figure 22.8 *Marchantia* antheridiophores (**a**) and archegoniophores (**b**).

22.3   HORNWORTS

The hornworts belong to the class Anthocerotopsida, and they have the simplest gametophytes of the division. They are small, green plants with a flat thallus like that of

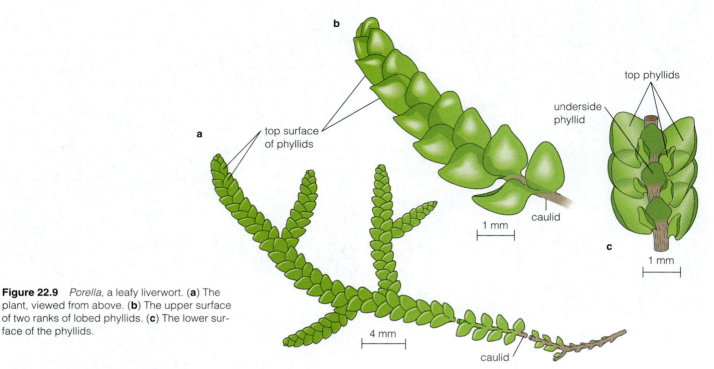

Figure 22.9 *Porella*, a leafy liverwort. (**a**) The plant, viewed from above. (**b**) The upper surface of two ranks of lobed phyllids. (**c**) The lower surface of the phyllids.

some liverworts; but their capsules are uniquely long and pointed, and when they release their spores they split into two valves (Fig. 22.10).

Archegonia and antheridia are similar in appearance to those already described. The antheridia are located in roofed chambers in the upper portion of the thallus; the archegonia are embedded within the thallus. When mature, sperm are released from the antheridia. They swim toward a nearby archegonium, down the neck canal, and fertilize the egg, creating a zygote.

The sporophyte that develops from that zygote is very different from a hepatic sporophyte. It consists of a **foot** (embedded in the gametophyte thallus) and an upright, elongated, pointed capsule (sporangium; Fig. 22.11). The epidermis contains true stomata, and chlorenchyma tissue occurs just below the epidermis. A central cylinder of tissue consists of sporocytes, which undergo meiosis and produce haploid spores; these spores mature first at the tip of the capsule. A meristematic region just above the foot continually adds new cells to the base of the sporangium. The mature tip of the sporangium splits into two valves, releasing spores, and new basal sporangium material replaces it. The sporangium grows upward from the base much like a blade of grass, and its total height may reach several centimeters (1–2 inches; Fig. 22.10).

Under favorable growing conditions, the sporophyte remains alive for many months, outliving the gametophyte. The foot then comes into direct contact with the soil, acting as a root. Conducting tissue capable of transporting water and nutrients may develop at the base of the sporangium. In this unique case, the sporophyte becomes an independent organism, taking on many attributes of higher land plants such as ferns and seed plants.

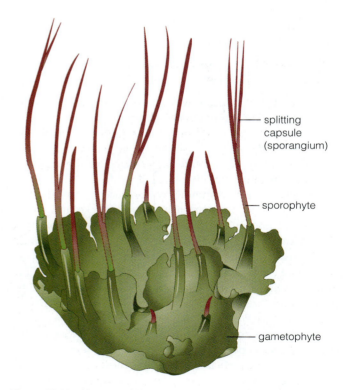

Figure 22.10 Hornwort (*Anthoceros*) gametophyte with mature sporophyte attached. Mature sporangia in this class split into longitudinal strips, from the top down. Rhizoids, which anchor the gametophyte and sporophyte to the ground, are not shown.

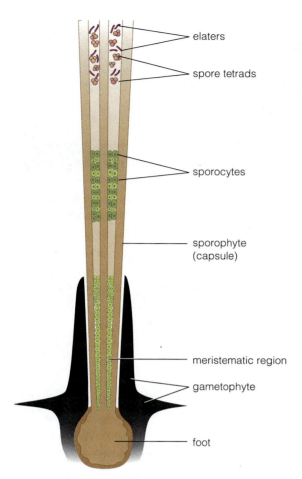

elaters

spore tetrads

sporocytes

sporophyte
(capsule)

meristematic region

gametophyte

foot

Figure 22.11 Longitudinal section through a hornwort sporophyte (*Anthoceros*). Only the lower portion of the sporophyte is shown. Black represents gametophytic tissue in which the sporophyte is embedded. The tubular sporophyte is mainly an elongated sporangium. Sporocyte cells within undergo meiosis to produce spores. Other cells produce elongate elaters, which assist in the ejection of spores from the ripe sporangium. (Redrawn from R. M. Holman and W. W. Robbins, *A Textbook of General Botany*, New York: John Wiley, 1938.)

22.4 MOSSES

The three classes of mosses—Andreaeopsida, Sphagnopsida, and Bryopsida—are the best-known members of the Bryophyta because they are larger, have wider distributions, and have many more species than the liverworts and hornworts. They all have phyllids with sharp tips and a prominent midrib, and the phyllids are spirally arranged around the caulid. Mosses often cluster to form tufts or carpets of vegetation on the surface of rocks, soil, or bark. Filamentous rhizoids anchor the plants to their substrate. The granite moss class (Andreaeopsida) features small, dark-colored plants that grow on rocks in cool climates. Peat mosses (Sphagnopsida) are much larger and are confined mainly to acidic bogs. Of the

three moss classes, the true mosses (Bryopsida) have the widest habitat range and the most species.

The life cycle of the true moss *Polytrichum* (Fig. 22.12) is representative of bryophyte life cycles in general. The life cycle is sporic, with a dominant gametophyte generation. The gametophyte generation is more complex, longer-lived, and more independent than the sporophyte. The sporophyte remains attached to the gametophyte and depends on it for water and nutrients. The sperm are produced in antheridia; the eggs, in archegonia. Water is essential as a medium to carry sperm from antheridium to archegonium. Bryophytes can also reproduce asexually.

Gametophytes Have Protonemal, Bud, and Leafy Phases

Moss gametophytes have three phases: a creeping, filamentous stage called the **protonema**; attached **buds**; and an upright, leafy plant stage bearing small, spirally arranged phyllids around a caulid. The protonema develops from a germinating haploid spore.

The spore is dormant at first, with a low rate of metabolism and low water content. In the presence of liquid water and sunlight, a spore swells and becomes active. Chloroplasts develop in the cytoplasm, chlorophyll is synthesized, starch is broken down, and the rate of photosynthesis increases. An initial filament of cytoplasm protrudes from the swollen spore; then, protonemal strands branch and grow outward in all directions, much like fungal hyphae or algal filaments (Fig. 22.13). The protonemal strands grow at their tips and growth is directed toward red light. Phytochrome is involved (see Chapter 15). The protonema can become dense and extensive enough in some mosses to make the wet surface of soil, rocks, or bark visibly green to the naked eye. Peat mosses have a flat, platelike protonema instead of a filamentous one.

After several weeks, certain protonemal cells divide to create a small mass of cells called a *bud* (Fig. 22.13c). One of the bud cells functions as the tip of a caulid, and it controls the direction of further cell division in such a way that a caulid takes shape with spirally arranged phyllids clothing its surface. Each bud thus develops into a mature gametophyte plant. Colorless rhizoids grow downward from the bud, anchoring the gametophytes to the substrate. Each protonema may produce a dense clone of many genetically identical leafy gametophytes. The formation of buds is under hormonal control, with cytokinins and auxins being important stimulants.

SPECIALIZED TRANSLOCATION TISSUES As the leafy gametophyte grows, caulid cells differentiate and mature into specialized tissues. The simplest mosses have an epidermal layer of small, thick-walled cells called **stereids** surrounding a homogeneous cortex of parenchyma-like tissue. Some species have a thin cuticle over the epider-

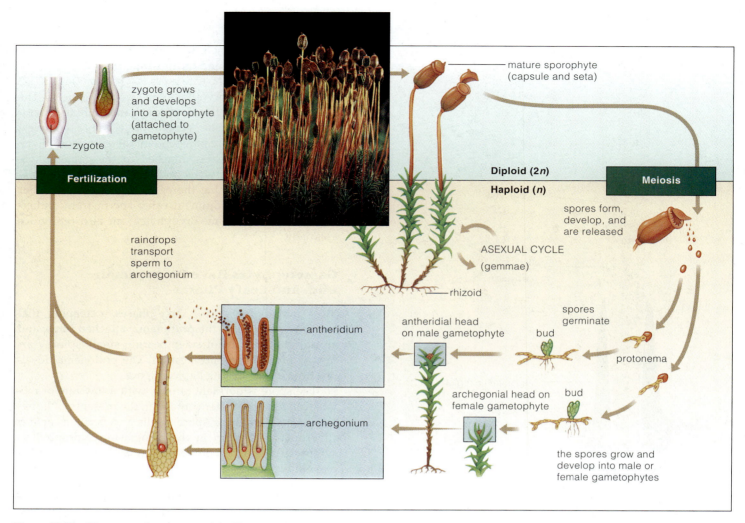

Figure 22.12 Diagrammatic summary of the life cycle of a typical moss.

Figure 22.13 Early gametophyte development in *Funaria*, a moss of the class Bryopsida. (**a**) Germination of spores. (**b**) Protonema. (**c**) Protonema with one bud.

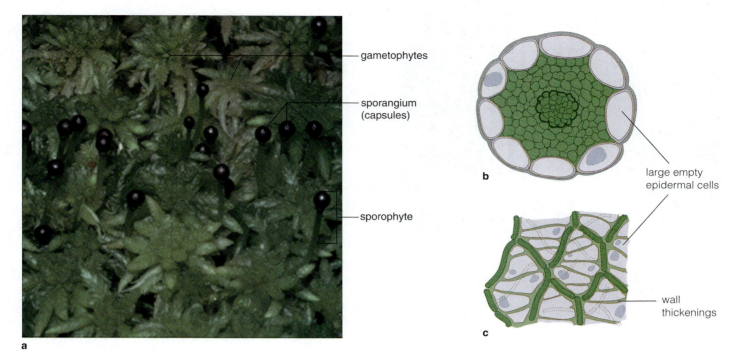

Figure 22.14 *Sphagnum*, a peat moss in the class Sphagnopsida. (**a**) Both gametophytes and sporophytes are visible. (**b**) Cross section of gametophyte caulid, showing the water-absorbing epidermis of *Sphagnum*, composed of large, empty, colorless cells with pores. (**c**) Surface view of phyllid, showing green cells alternating with large, colorless, water-absorbing cells. (Redrawn from R. F. Scagel et al., *An Evolutionary Survey of the Plant Kingdom*. Belmont, Calif: Wadsworth, 1965.)

mis, but others lack a cuticle. Stomata are absent. The epidermis of *Sphagnum* is unique in containing large, empty, clear cells (Fig. 22.14). These cells can fill with water through a pore when wetted and serve as a reservoir of moisture for the moss plant. Because they grow in acidic surroundings, where bacteria are inhibited, peat moss tissue itself is bacteria-free and sterile.

Many mosses have a more complex caulid anatomy, with a central strand of conducting tissue. One kind of conducting tissue is made up of **hydroids**—elongated, thin-walled, dead, empty cells that conduct water (Fig. 22.15). Their end walls are oblique, sometimes very thin, perforated with pores, or partly dissolved. Experiments with dyes show that translocation of water can occur in this tissue. One genus of moss shows continuous hydroid tissue from the caulid to the midribs of the phyllids, imitating the vascular system of higher plants. Hydroids show some resemblance to tracheids but lack their specialized pitting and lignified walls.

Even though many mosses have hydroids and a thin epidermal cuticle, they are not considered vascular plants, equal in complexity to ferns, conifers, and flowering plants. The nature and distribution of their conducting tissue is so limited that mosses are unable to control internal water content very effectively, and they dry as fast as the environment dries. Mosses don't necessarily die upon drying, however; many are exceptionally tolerant of dehydration. In the most drought-tolerant species, the plants become dormant, much like the embryo of a seed; but they resume normal metabolism within 4 to 24 hours of being rewetted—even when they have existed as shriveled tissue for as long as 20 years. In contrast, mosses that live in water or in temperate and tropical rain forests are permanently damaged or killed by aridity.

A few of the most specialized mosses also contain cells resembling the sieve cells of vascular plants. In a zone around the hydroids, elongated cells with oblique end walls may occur. These so-called **leptoids** (Fig. 22.15) are living, but their nuclei are degenerate and inactive. They have many plasmodesmata in their end walls, and callose may be present. Nearby parenchyma cells may function as companion cells. Tracer studies show that sugars may move through these cells at rates of up to 300 mm/hr (1 ft/hr).

PHOTOSYNTHESIS Mosses are very sensitive to light. The light compensation point for photosynthesis (where photosynthesis just equals respiration) is reached at 1% of full sun, and the saturation point (where photosynthesis is maximum) is only at 4–20% of full sun. Some mosses that grow in dim light near the mouths of caves have specialized cells capable of focusing any available light onto chloroplasts. Snowbed mosses tolerate being covered by snow for 10 months of the year because they are able to photosynthesize in the very weak light that filters through 20 to 30 cm (8 to 12 in) of snow.

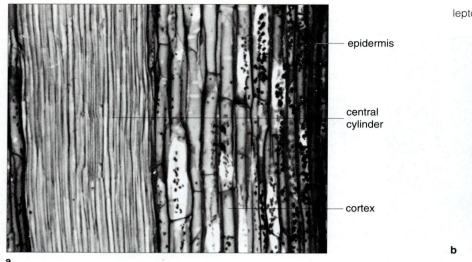
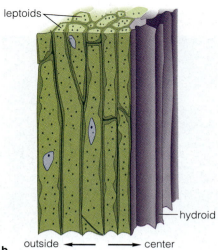

Figure 22.15 Caulid anatomy of mosses in the class Bryopsida. (**a**) *Mnium*, epidermis of stereid cells, cortex of parenchyma cells, and central conducting cylinder. ×111. (**b**) *Polytrichum*, portion of the central region, showing leptoids and hydroids.

Mosses use the C_3 photosynthetic pathway and typically produce the sugars sucrose, glucose, fructose, and mannose. Arctic mosses are unique in storing lipids, rather than starch, as a food reserve. Their high lipid content makes these mosses energy-rich and important in the arctic food chain. Moss gametophytes photosynthesize more slowly than vascular plants, fixing only 2 to 4 mg of CO_2 per gram dry weight of tissue per hour. Photosynthesis is maximum when the plants are well watered and when temperatures are 15 to 25°C (59 to 77°F). Snowbed mosses are exceptional in that they are able to photosynthesize at a temperature of −5°C (23°F), and their optimum temperature for photosynthesis is only 6 to 11°C (43 to 52°F).

ASEXUAL AND SEXUAL REPRODUCTION Asexual reproduction is accomplished in several ways: First, the protonema may continue to produce new buds, so a forest of moss plants may spread outward in all directions, matlike, at the edges of the clone. In this respect, the protonema is analogous to a network of runners or rhizomes. Second, phyllid tissue placed in wet soil may produce protonemal strands that develop buds and new moss individuals. Third, rhizoids can sometimes produce buds. Fourth, lens-shaped tissue called *gemmae* may be produced by rhizoids, on phyllid tips and surfaces, at the ends of special stalks, and even in specialized cups. Gemmae have the same function in mosses as they do in liverworts: If detached from the parent plant and dispersed to a suitable habitat, each can begin mitotic cell division and will differentiate into a gametophyte plant.

The growth rates, causes of mortality, and life spans of moss gametophytes are not well known. Some *Polytrichum* mosses are known to live as long as five years, but asexual reproduction would keep a clone alive much longer than that.

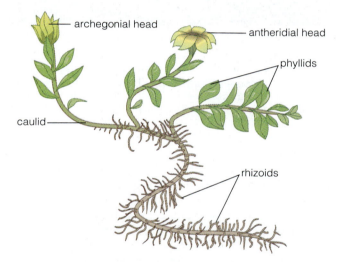

Figure 22.16 Location of gametangia on *Mnium*, a moss that produces archegonia and antheridia on the same individual, but in different locations.

When the gametophyte is mature, it produces gametangia at the apex of the main caulid (Fig. 22.16) or sometimes at the ends of short side caulids. Gametangial formation is stimulated by such factors as substrate pH, air temperature, and tissue concentration of nitrogen and certain carbohydrates in the tissues. Hormones such as auxin, gibberellin, cytokinin, and cyclic AMP may also stimulate sexual development. Reproduction by many moss species seems to be independent of the length of day or night.

Some moss species are dioecious, with separate plants producing either antheridia or archegonia. Other species are **monoecious**, meaning that they have both types of gametangia on a single plant. In the latter case, the two types of gametangia may be produced together or on separate parts of the plant.

a

antheridial heads

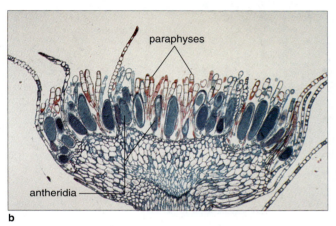

paraphyses

antheridia

b

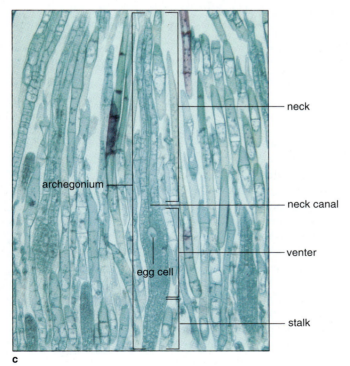

archegonium

egg cell

neck

neck canal

venter

stalk

c

Figure 22.17 Gametangia and gametangial heads on true mosses in the class Bryopsida. (**a**) Antheridial heads of *Polytrichum* seen from above. (**b**) Longitudinal section of an antheridial head of Mnium. (**c**) Archegonium surrounded by paraphyses on an archegonial head of *Mnium*.

Antheridia are elongate, and the outer jacket of sterile cells contains chloroplasts that turn red-orange when the sperm are mature. The antheridium exudes mature sperm when free water is present. Each sperm has an elongated nucleus and is propelled by two flagella similar to those of hepatics. Antheridia are surrounded by club-shaped multicellular hairs called **paraphyses** that have conspicuous chloroplasts (Fig. 22.17). Clusters of antheridia are borne in **antheridial heads** at the tips of certain caulids, or they may be mixed with archegonia.

In *Mnium*, the phyllids that surround antheridial heads are sometimes spread out like the petals of a flower, and the antheridia appear as a single orange spot in the center. The whorl of phyllids functions as a splash cup, using the force of raindrops to eject sperm some distance away. Sperm can remain mobile for as long as six hours, but the distance they can travel on their own power is less than 50 cm (20 in). Sperm of some mosses, such as *Bryum capillare*, hitch a ride on insects that have been attracted to the antheridial heads, either by the red and yellow colors or by secretions from the paraphyses. The insects then transport sperm to **archegonial heads**, clusters of archegonia that form at the tips of certain caulids.

Each archegonium has a neck, a thickened venter containing a single egg, and a long stalk (Fig. 22.17c). When the egg is mature, the neck opens, creating a canal. Attracted by a gradient of sucrose emitted by the egg, sperm swim down the canal toward the egg. Water is necessary to carry sperm from an antheridium to an egg, and the phyllids arranged around archegonial and antheridial heads help retain a film of water over the gametangia. When a sperm reaches an egg, it fertilizes the egg, creating a diploid zygote cell.

Sporophytes Have Complex Capsules

Soon after fertilization, the zygote begins to develop into a spindle-shaped embryo that differentiates into foot, **seta** (stalk), and sporangium regions. The foot penetrates out through the venter and into the caulid. Transfer cells facilitate the movement of water and nutrients from the gametophyte to the parasitic sporophyte for its growth and development. The seta elongates rapidly, raising the yet-to-be-formed sporangium above the top of the leafy gametophyte (Fig. 22.18). The old archegonium increases in size as the sporophyte enlarges and becomes the **calyptra**, which remains for a time as a protective covering for the sporangium. The presence of the calyptra seems necessary for normal growth and differentiation of the sporophyte. The mature sporangium (capsule) and seta can have a complex anatomy, with an epidermis of cuticle, stomata, thick-walled stereids, a cortex region, and a central strand of conducting tissue.

In contrast to a life span of several years for the gametophyte, the sporophyte usually completes development in 6 to 10 months. A few species that live in disturbed

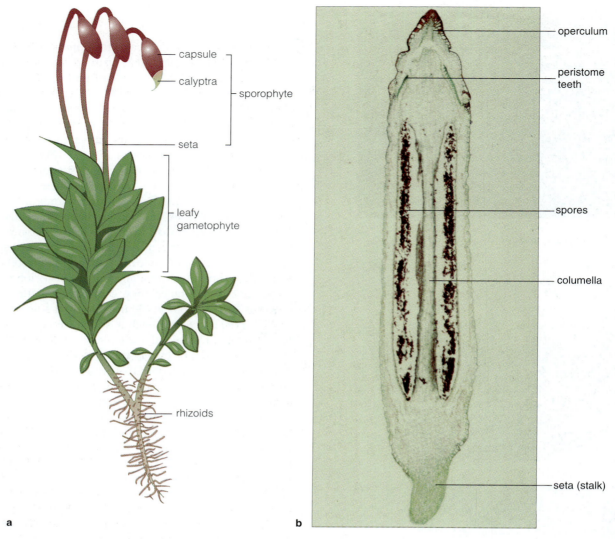

Figure 22.18 Sporophyte of *Mnium*. (**a**) Parasitic sporophytes attach to the leafy gameto-phyte. (**b**) Longitudinal section of a nearly mature capsule. The operculum has not yet fallen off, and the spores have not completed development.

habitats can mature in even shorter periods, whereas arctic moss sporophytes require several years. Most moss sporophytes contain chlorenchyma. The sporophyte can fix 10–50% of the carbohydrate needed for growth and maintenance, the rest coming from the parasitized gametophyte. Mature sporophytes in the class Andreaeopsida, however, lack chlorophyll.

The capsule, in which spores mature, is typically 1 to 3 mm in diameter and 2 to 16 mm long; it is elevated more than 10 mm above the gametophyte by the seta (Fig. 22.18). Granite moss capsules open by slits to disperse the spores. Peat moss capsules have lids that blow off when the capsule dries, violently ejecting all the spores at once. In true mosses, sporocytes fill the capsule except for a central column of sterile tissue called the **columella**. The columella forms a small dome at the top of the capsule. When the capsule is mature, the columella's tissue becomes dry and brittle, forming a lid, or **operculum** (Fig. 22.19). Cells immediately below the operculum form a double row of triangular **peristome teeth**, whose

broad bases are attached to thick-walled cells that form an **annulus** around the upper end of the sporangium. The annulus eventually breaks down, and the operculum falls away, leaving a passageway into the capsule and exposing the peristome teeth. By this time, the columella has collapsed, meioisis has occurred in the sporocytes, and thousands of haploid spores have been formed.

Peristome teeth are rough and very sensitive to the amount of atmospheric humidity. When the air is humid or the teeth are wet, they bend into the capsule's cavity. When the air or teeth are dry, they straighten and lift out some of the spores, which are then disseminated by wind. If a spore comes to rest in a suitable habitat, it germinates and grows into a protonema.

The spores are about 25 μm in diameter. Wind is capable of carrying them for tens to thousands of kilometers (miles), but most spores settle within a few meters (yards) of the capsule. One family of mosses that grows on dung or carrion (the Splachnaceae) has spores dispersed by insects. These cling together in a sticky mass

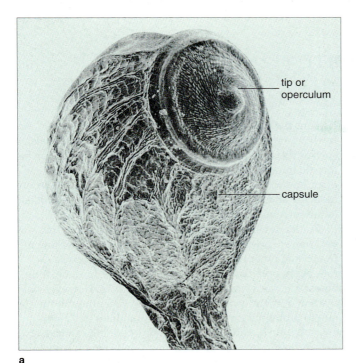

tip or operculum

capsule

a

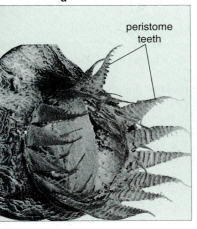

peristome teeth

b

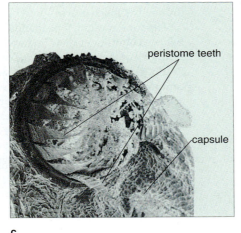

peristome teeth

capsule

c

Figure 22.19 Scanning electron micrograph of a capsule of *Funaria*, a true moss in the class Bryopsida. (**a**) Capsule with operculum in place. (**b**) Peristome teeth are fully open; capsule is empty. (**c**) Operculum has been shed; peristome teeth are partially open, with some spores attached to them.

on the columella, which projects beyond the operculum at maturity. The teeth are bent back and do not respond to humidity. The capsule takes on purple, red, yellow, and white colors and resembles a small flower from a distance. The colors and mushroomlike odors attract flies, which carry the sticky spores considerable distances away. Moss spores have a waxy cuticle and are resistant to aridity. They are capable of remaining dormant but alive for as long as a decade, so long as they are kept in dry conditions.

Mosses Have Significant Economic and Ecological Value

Sphagnum (a large genus of peat mosses) is by far the most economically important bryophyte. Because of their special epidermal cells, *Sphagnum* mosses are frequently added to potting soil to increase its water-holding capacity. Tons of *Sphagnum* are harvested throughout the world and then sold in the nursery trade in many countries. During World War I, *Sphagnum* was used on a large scale as a wound dressing because its acidic, sterile tissue acts as both an antiseptic and an absorbent. In some cold-temperate areas, *Sphagnum* and other mosses accumulate as a thick, compacted, semidecomposed layer atop the mineral soil. This deposit, called *peat*, can be cut out in blocks, dried, and burned as fuel for cooking and heating (see sidebar, "Mining Moss: Peat for Profit," p. 380). On a much larger scale, the Rhode generating station in Ireland burns 2000 metric tons (2200 tons) of peat a day to produce electricity.

Ecologically, bryophytes are important as colonizers of bare rock and sand. Their small bodies trap dust and wind-blown silt, in this way building soil. Mosses create stabilizing soil crusts in such inhospitable places as coastal dunes and inland deserts. Some bryophytes harbor symbiotic nitrogen-fixing cyanobacteria; thus they significantly increase the amount of nitrogen available to the biosphere. In tundra vegetation, bryophytes constitute as much as 50% of the aboveground biomass, and they are an important component of the food chain that supports animals within the ecosystem. Tundra mosses are rich in lipids, but most bryophyte tissue has a caloric value equivalent to that of higher plants. The temperate forests in the South Island of New Zealand and in the Pacific Northwest of North America, as well as the cold-temperate spruce–fir forests of the northern hemisphere, also exhibit a significant amount of moss biomass in the shade of overstory trees (Fig. 22.1).

Bryophytes are also important as tools to improve our basic understanding of fundamental biological processes. Scientists have come to recognize mosses in particular as excellent experimental plants for such studies. They are easy to propagate, to grow in small spaces, to clone into sexually identical replicates, and to observe for growth and developmental changes. Many discoveries about moss genetics, tissue development, and ecology in the past several decades have proved widely applicable to land plants in general. Although bryophytes may not be evolutionary ancestors to higher plants, they have turned out to be very useful models of higher plants.

SUMMARY

1. Bryophytes are small, herbaceous plants with a combination of traits that adapt them to life both on land and in water. Their habitats range from aquatic, to humid or wet terrestrial, to epiphytic, to arid.

2. Bryophytes resemble higher plants in their sporic life cycle, their possession of an embryo stage, and their specialized gametangia (called antheridia and archegonia).

MINING MOSS: PEAT FOR PROFIT

Peat is any kind of plant matter that accumulates under water-soaked conditions and that does not completely decompose because the amount of oxygen is limited. Peat is a mixture of fresh-water marsh plants such as reeds, sedges, and grasses, and it often includes peat mosses belonging to the genus *Sphagnum*. If conditions are right—cold temperatures, abundant moisture, water that is acidic and low in nutrients—peat can accumulate in layers at rates as fast as 25 cm (10 in) per century to a slower 25 cm per millennium. The oldest deposits are tens of meters thick (1 m = 3.2 ft). The thousands of square kilometers of peatlands in the cold temperate regions of the world, multiplied by depth, total more than 200 billion metric tons (220 billion tons) of dry biomass.

Peat has been put to a variety of human uses. It is, first of all, a fuel. Stone Age Europeans found that it could be dug out in blocks, dried, and burned for heat. Today Russia, Ireland, Finland, Sweden, Germany, and Poland extensively harvest peat for fuel. An Irish home may use 15 metric tons of dry peat each year, an amount of peat that can be harvested by one person in a month's time. Two-bladed shovels called *slanes* are used by individuals, but commercial operations employ specialized tractors, millers, harrows, and harvesters. Such costly investment is needed to harvest the large volumes of peat burned in power plants in Russia and Ireland. The use of peat as a fuel in the United States has been very minor, but the energy content of that peat is estimated to exceed all current, combined reserves of coal, petroleum, and natural gas.

Peat is also used in horticulture. When added to potting mixes, peat increases the soil's water-holding capacity and lightens the soil, allowing air and water to move more freely. Peat is used as a mulch for acid-loving ornamentals such as rhododendrons and heaths. Compressed peat can be formed into planting containers for seeds, seedlings, cuttings, and root balls. These containers can take up and retain moisture and also decompose over time, releasing the roots to the surrounding soil.

Peatlands can also be cultivated, if drained. Once the peat becomes drier and aerated, it decomposes, releasing nitrogen and other nutrients. Such cool-climate crops as carrots, beets, potatoes, onions, lettuce, cabbage, broccoli, mint, blueberries, and strawberries do well on peat. However, cultivation over the course of many years causes the soil surface to drop because the peat literally oxidizes and blows away. Within a century, several meters of depth can be lost. Agricultural peatlands in California's delta region have to be protected by levees because their surface elevations have dropped as much as 6 m (20 ft) below sea level.

The conservation of peatlands has become an ecological issue in countries such as England and Ireland, where nearly all of the once-extensive bogs have been stripped. Peat bogs are unique ecosystems, with plant and animal components that have evolved over geologic time. The environmental processes that produce bogs proceed so slowly that it is not possible to restore them in human lifetimes, once they have been modified. We humans certainly have the technology and will to convert entire landscapes and ecosystems, but do we have the right to do so?

Some members of the division have leaflike, stemlike, and rootlike organs. At the same time, they resemble algae in that free water is essential for movement of the sperm to the egg, there is no supportive tissue, and the gametophyte generation is dominant. The sporophyte generation is less well-developed than—and parasitic on—the gametophyte generation.

3. Most plant biologists believe that bryophytes evolved from one or more algal divisions. In this book, the 24,000 living species of bryophytes are classified in a single division, the Bryophyta, which includes five classes: Hepaticopsida (liverworts or hepatics), Anthocerotopsida (hornworts), Andreaeopsida (granite mosses), Sphagnopsida (peat mosses), and Bryopsida (true mosses).

4. Liverworts (class Hepaticopsida) include simple organisms with thalloid gametophytes (*Riccia, Marchantia*) and others with leafy gametophytes (*Porella*). All have rhizoids that anchor the plants to their substrate.

5. Sexual reproduction within the Hepaticopsida involve archegonia, antheridia, embryos, and parasitic sporophytes; but their morphology and placement depend on the species. *Riccia* gametangia are sunken in the thallus, and the sporophyte is simply a sporangium embedded in the gametophyte thallus, which opens to release haploid spores when the gametophyte dies and decays at the end of its life span. *Marchantia* gametangia are located on elevated organs called antheridiophores and archegoniophores.

6. The hornwort *Anthoceros* (class Anthocerotopsida) has *Riccia*-like gametophytes. Its sporophyte has an elongated sporangium with indeterminate growth, and in some cases it may outlive the gametophyte and become essentially free-living.

7. The major points of the moss life cycle are: (1) The life cycle is sporic; (2) the gametophyte generation is dominant; (3) sperm are produced in multicellular antheridia and eggs in multicellular archegonia; (4) water is essen-

tial as a medium to carry sperm from antheridium to archegonium; and (5) an embryo stage is present.

8. Moss gametophytes develop through three morphological phases: (1) a filamentous, algalike protonemal phase that produces (2) buds, which develop by mitosis into (3) plants with leaflike phyllids and stemlike caulids.

9. Specialized tissue in some moss gametophytes includes chlorenchyma, parenchyma, epidermal stereids, water-conducting hydroids, and sugar-conducting leptoids. Moss gametophytes are capable of tolerating extreme dehydration for years and then resuming normal activity within hours of being rewetted. However, despite the presence of conductive tissue in some genera, these organisms are nonvascular plants, and they are not capable of controlling water loss.

10. Moss gametangia are elevated on stalks, rather than being embedded within gametophyte tissue as they are in most hepatics. Sperm are usually carried to eggs by free water, but insects transfer the sperm in some cases. Fertilization ultimately requires free water.

Questions

1. The bryophytes have added several innovations to the algal life cycle, including multicellular gametangia and an embryo stage. How do these innovations contribute to the survival of bryophyte plants?

2. In what ways is it possible to conclude that the gametophyte generation is dominant over the sporophyte generation in the Bryophyta?

3. Describe the difference in morphological and anatomical complexity between the gametophyte of the liverwort *Marchantia* and that of a typical moss; between the sporophyte of *Marchantia* and that of a moss.

4. Describe the conducting tissue of some of the most advanced mosses. Why is it not the same as the vascular tissue of higher plants?

5. Mosses cannot control their water balance, and yet they are important elements of the biotic crust of desert soils. How can mosses persist in a desert environment? When would they reproduce?

Further Readings

Asakawa, Y. 1986. "Chemical Relationships Betwen Algae, Bryophytes, and Pteridophytes." *Journal of Bryology* 14: 59–70. By examining biochemical and metabolic similarities and differences, the author concludes that liverworts evolved from brown algae, that mosses and hornworts evolved from (different) green algae, and that vascular plants evolved in parallel with (separately from) bryophytes.

Chopra, R. N., and P. K. Kumra. 1988. *Biology of Bryophytes*. New York: John Wiley. Probably the most comprehensive, up-to-date global survey of bryophytes that exists in print.

Crum, H. 1988. *A Focus on Peatlands and Peat Mosses*. Ann Arbor: University of Michigan Press. A very readable paperback on the biology, ecology, and economic importance of peat mosses.

During, H. J. and B. F. van Tooren. 1987. "Recent Developments in Bryophyte Population Ecology." *Trends in Ecology and Evolution* 2: 89–93. Bryophytes have an unexpectedly high amount of genetic variability, despite the fact that many populations maintain themselves almost entirely by asexual reproduction. Moss species fill as many different life-history strategies as do higher plant species.

Dyer, A. F., and J. G. Duckett. 1984. *The Experimental Biology of Bryophytes*. New York: Academic Press. Details about moss physiology and development.

Richardson, D. H. S. 1981. *The Biology of Mosses*. New York: Halsted Press/Wiley. The author's enthusiasm and lucid writing make for compelling reading. He discusses water balance, photosynthesis, nutrient requirements, the life cycle, relationships with animals, microbes, and man—including response of mosses to various pollutants.

Smith, A. J. E. 1986. "Bryophyte Phylogeny: Fact or Fiction?" *Journal of Bryology* 14: 83–89. A brief survey of how much (and how little) botanists know about the evolutionary history of bryophytes. His suggestion: Treat the group as a single division even though we don't know for certain that they are monophyletic.

23 PLANTAE: LOWER VASCULAR PLANTS

KEY CONCEPTS

1. Approximately 14,000 species of plants, classified into four divisions, are informally known as the seedless (or lower) vascular plants. Sporophytes dominate the life cycle, and they are large and complex relative to those of bryophytes. The sporophytes are more tolerant of life on dry land than are those of bryophytes because they exercise control of their water balance by possessing vascular tissue, stomata, and a cuticle. The gametophytes, however, still require a seasonally wet habitat, and water outside the plant is essential for movement of sperm from antheridium to archegonium.

2. The division Psilophyta is represented by two living genera, *Psilotum* and *Tmesiptris*. Sporophytes of *Psilotum* are very primitive, consisting of a green, branching stem but lacking both true leaves and roots. *Tmesiptris* is more complex and lives as an epiphyte in tropical trees.

3. The division Lycophyta has five living genera, three of which occur in North America: *Isoetes*, *Lycopodium*, and *Selaginella*. These plants have true leaves, stems, and roots. *Selaginella* has a unique life cycle among lower vascular plants, in that it is heterosporous like the seed plants.

4. The division Sphenophyta has a single living genus, *Equisetum*. Sporophytes are jointed and ribbed, and their stems have a complex anatomy.

5. Most species of lower vascular plants are in the division Pterophyta, the ferns. Ferns are widely distributed throughout the world. Their leaves are complex and varied. The few aquatic ferns are the only heterosporous ferns. Ferns have food value and other uses for a variety of human societies.

23.1 THE STEP FROM WATER TO LAND

Within 1 billion years of the earth's formation, the first forms of life (microscopic bacteria) were living in early seas and pools. An additional 3 billion years were required for land-dwelling plants to evolve from their aquatic ancestors. Surely, then, we can conclude that the step from water to land was a difficult challenge. Once the transition had been made, land plants diversified into many different forms, taxonomic groups, and habitats in the relatively short geological span of 50 million years.

The movement from water to dry land required the evolution of several new structures and chemicals. In water, plants are buoyed and bathed with soluble nutrients, including dissolved gases; their organs are thin, and necessary environmental resources can reach every cell by diffusion. Therefore, plants need no complex structures either for support or for the uptake and transport of nutrients. On land, however, they require special organs in contact with wet soil for the absorption of nutrients; they must have stiff tissues for an erect form; and they require a conducting system for the transport of water throughout the plant body. The reproductive cycle also has to be modified for land. Spores of terrestrial plants will no

Figure 23.1 Representatives of the four divisions of seedless vascular plants. (**a**) Whisk fern *Psilotum nudum*, in the Psilophyta. (**b**) Club moss *Lycopodium obscurum*, in the Lycophyta. (**c**) Horsetail *Equisetum telmateia*, in the Sphenophyta. (**d**) The tree fern *Alisophyta* in a Puerto Rican forest, a member of the Pterophyta.

longer be produced and spread under water, so they must be chemically sealed to avoid desiccation. In fact, all aboveground plant surfaces must be sealed to reduce water loss yet, at the same time, must remain permeable to important gases such as carbon dioxide and oxygen.

Today, the great majority of species in the kingdom Plantae are land plants that have vascular tissue (xylem and phloem). They are informally called **vascular plants**. Some vascular plants reproduce by seeds, but others disperse spores. The latter are called **lower vascular plants** or **seedless vascular plants** because they are more primitive. This group includes four divisions of living plants: the **Psilophyta** (whisk ferns), **Lycophyta** (club mosses), **Sphenophyta** (horsetails), and **Pterophyta** (true ferns). They number fewer than 14,000 species, most of them small herbs (Fig. 23.1). About 200 to 400 million years ago, they were more numerous and diverse.

The lower vascular plants share some traits that adapt them better than the bryophytes to life on dry land. They control their water balance by possessing vascular tissue, stomata, a waxy cuticle, and spores with walls containing

Plantae: Lower Vascular Plants　　**383**

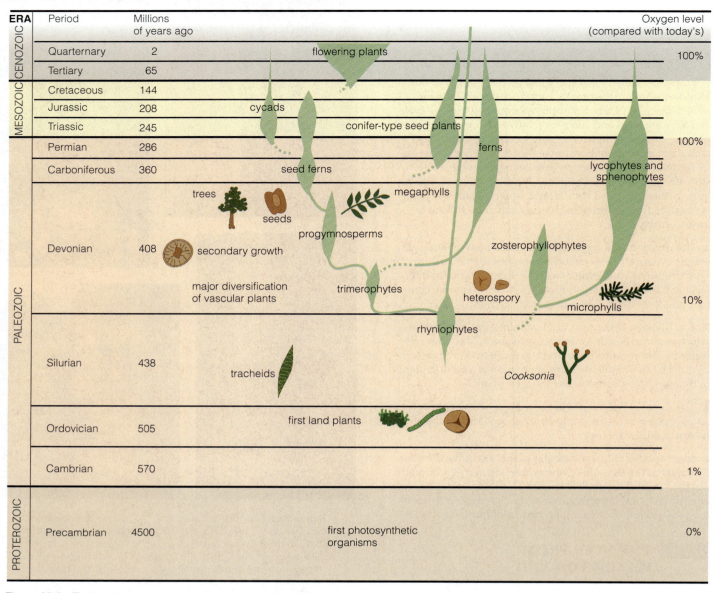

ERA	Period	Millions of years ago		Oxygen level (compared with today's)

The geologic timescale table (Figure 23.2):

CENOZOIC
- Quarternary — 2 — flowering plants — 100%
- Tertiary — 65

MESOZOIC
- Cretaceous — 144
- Jurassic — 208 — cycads
- Triassic — 245 — conifer-type seed plants

PALEOZOIC
- Permian — 286 — ferns — 100%
- Carboniferous — 360 — seed ferns — lycophytes and sphenophytes
- Devonian — 408 — trees, seeds, secondary growth, megaphylls, progymnosperms, major diversification of vascular plants, trimerophytes, zosterophyllophytes, heterospory, microphylls — 10%
- Silurian — 438 — tracheids, rhyniophytes, Cooksonia
- Ordovician — 505 — first land plants
- Cambrian — 570 — 1%

PROTEROZOIC
- Precambrian — 4500 — first photosynthetic organisms — 0%

Figure 23.2 The geologic timescale, showing the fossil record for major groups of plants and the oxygen content of the atmosphere according to the Berkner-Marshall model. The widths of the shaded areas represent the approximate number of species at a particular time. Dashed or broken connections indicate that botanists are unsure about the relationship. (Modified from Gensel and Andrews, 1984.)

sporopollenin. They retain some features of bryophytes—such as free-living gametophytes and swimming sperm—that require water outside the plant for fertilization.

Early Vascular Plant Fossils Date to the Ordovician Period

The fossil record is not very helpful in telling us when the first vascular land plant appeared, for two reasons. First, it is rare to find a fossil of a complete plant. Fossilized leaves, stems, roots, and reproductive parts seldom occur together and linked in such a way that we can be certain they all belong to the same individual. Second, the chemical and structural features that adapt plants to land evolved independently of each other. Thus, a given fossil may have one but not all of the features of a land plant, leaving the botanist to guess whether it was terrestrial or aquatic.

The first evidence of possible land plants is fossils of microscopic plant parts: meiospores of a characteristic shape with thick walls containing sporopollenin, bits of epidermal tissue with a definite cuticle, and cells that resemble tracheids. These elements have been found in several parts of the world in deposits that are 430 to 470 million years old—from the mid-Ordovician to early Silurian periods on the geologic timescale (Fig. 23.2). Unfortunately, we have no idea what the plants that contained these elements looked like.

THE MOMENT FOR COLONIZING LAND Dry land had been available for plant colonization for billions of years. Why did invasion of land begin during the Ordovician? In 1965, two atmospheric scientists, L. Berkner and L. Marshall, proposed that movement onto the land had to wait until enough free oxygen had accumulated in the atmosphere. They reasoned that the primeval atmosphere, lacking free oxygen, was incapable of shielding life on land from ultraviolet (UV) radiation. Sunlight is rich in high-energy UV, which can penetrate living tissue and cause metabolic damage by disrupting chemical bonds. Organisms living in water are safe from such damage, however, because water effectively filters UV radiation. Oxygen, in the form of ozone (O_3), is also capable of absorbing UV rays. Berkner and Marshall theorized that the amount of free oxygen in the atmosphere began to increase slowly after the first photosynthetic organisms evolved, 3 to 4 billion years ago, because oxygen is a waste by-product of photosynthesis.

The Berkner-Marshall model predicts that oxygen in the atmosphere would have reached 10% of the modern level about 400 million years ago. That much oxygen would have been enough to create an ozone shield high in the earth's atmosphere, enabling life to exist on land.

Other botanists have added new layers of speculation to the Berkner-Marshall model. They point out that phenolic compounds (complex alcohols) in plants also absorb UV; therefore, there would have been a selective advantage to individual plants that produced large amounts of phenols. The accumulation of phenols confers another selective advantage on plants, which is that phenols inhibit herbivores from feeding. Furthermore, modern plants polymerize phenols such as phenolic acids into lignin. Lignin adds mechanical strength to xylem elements; in fact, a vascular system isn't possible without it. Lignin is absent from fossil deposits until the invasion of land, approximately 400 million years ago.

SILURIAN AND DEVONIAN PERIODS: RAPID EVOLUTION
The first undisputed whole plant fossils of a vascular plant appeared in mid-Silurian deposits, about 420 million years old, in Czechoslovakia, Ireland, Russia, Scotland, the United States, and Wales. This extinct plant has been given the genus name *Cooksonia* (Fig. 23.3). It was a small herbaceous plant less than 10 cm (4 in) tall, made up of narrow stems that always fork into two branches. Each branch terminated in a sporangium filled with spores in groups of four whose walls contained lignin. The xylem in the stem was a delicate strand of tracheids with lignified secondary walls. There was no pith. We know nothing about the lower part of the plant. It's possible that there was no true root system—just a horizontal rhizome with rhizoids.

Other Silurian vascular plants were *Renalia*, with 30-cm (1-ft) stems that exhibited unequal branching and produced kidney-shaped sporangia; *Baragwanathia*, which had stems as tall as 1 m (3.2 ft), densely clothed with leaves; and *Rhynia* (Fig. 23.4). *Rhynia's* excellent preservation in

sporangia

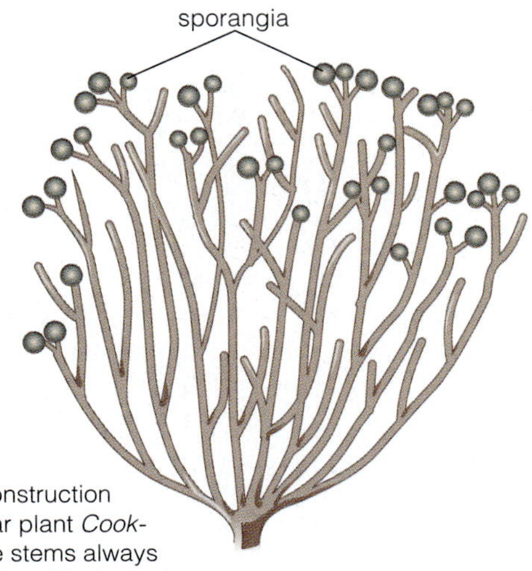

Figure 23.3 Reconstruction of the early vascular plant *Cooksonia*, Note that the stems always fork into two branches. (Redrawn from D. Edwards, *Palaeontology* 13 (1970): 451–61.)

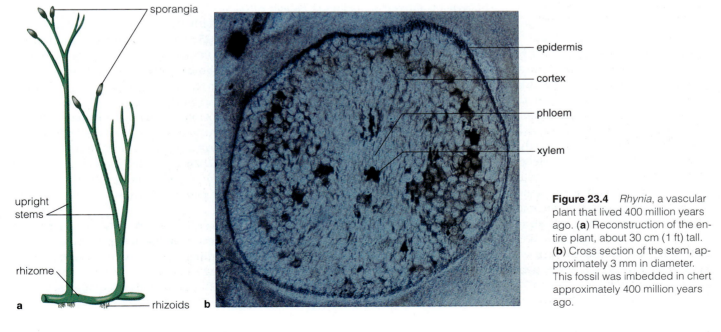

sporangia

upright stems

rhizome

rhizoids

epidermis

cortex

phloem

xylem

a   b

Figure 23.4 *Rhynia*, a vascular plant that lived 400 million years ago. (**a**) Reconstruction of the entire plant, about 30 cm (1 ft) tall. (**b**) Cross section of the stem, approximately 3 mm in diameter. This fossil was imbedded in chert approximately 400 million years ago.

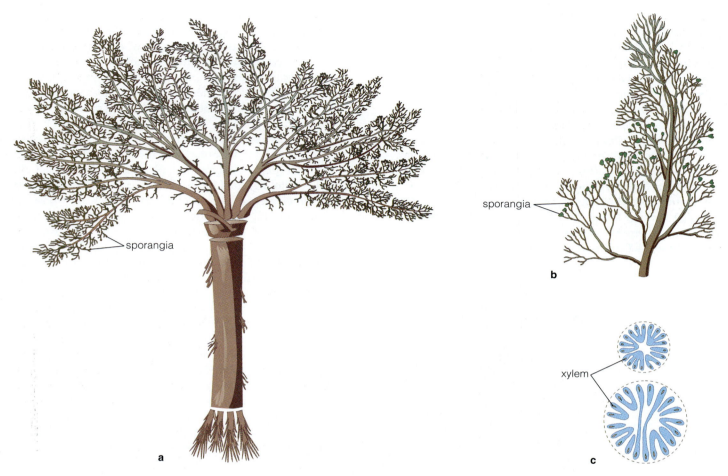

Figure 23.5 Reconstruction of *Pseudosporochnus nodosus*, one of the first trees. (**a**) Entire plant, about 1 m (3 ft) tall. (**b**) Detail of a terminal branch with sporangia. (**c**) Xylem pattern in cross sections of small and large branches. (Redrawn from S. Leclerq and H. B. Banks, *Palaeontographica* 110 (1962): 1.)

the fossil record has made it the best-known primitive vascular plant. It consisted of a slender rhizome from which arose upright stems that branched dichotomously but sparingly. Leaves and roots were absent. The cortex of the stem was probably photosynthetic, taking on the role of leaves. Rhizoids growing from the rhizome absorbed water and nutrients from the soil.

The rate of evolution was uniquely rapid during the 40-million-year Devonian period. In this span of time, ancestral forms of all four modern divisions of seedless vascular plants appeared, and several now-extinct divisions rose and fell. The landscape at the start of the Devonian was probably somewhat monotonous, resembling a modern meadow or marsh. By the end of the Devonian, the vegetation had become complex and layered, with a mosaic of forests, shrublands, and meadows. Herbaceous plants had given rise to shrubs and large trees with woody tissue, bark, secondary growth, roots, leaves, and seeds.

Pseudosporochnus nodosus was one of the first trees, though compared with trees of today it would be a sapling at best (Fig. 23.5). Its main trunk was up to 1 m (3 ft) tall and 8 cm (3 in) in diameter, topped with a crown of finely branched limbs. There were no leaves. Stem xylem was no longer restricted to a central bundle (as in *Rhynia* and *Cooksonia*); instead it was fragmented into many lobes and divisions. Other trees in the Devonian were much larger. *Aneurophyton*, for example, was 13 m (43 ft) tall, and *Archaeopteris* was 25 m (82 ft) tall, and had a trunk 1 m (3 ft) in diameter at the base (Fig. 23.6). Annual growth rings are absent from petrified wood samples of these trees, perhaps because the climate was tropical year-round.

Leaves also evolved, from very small epidermal outgrowths to larger structures having a vascular system of their own. According to a widely accepted theory (Fig. 23.7), true leaves are thought to represent modified branches: First, branching becomes unequal, resulting in

Figure 23.6 Reconstruction of *Archaeopteris*, a Devonian tree about 25 m (82 ft) tall. (Redrawn from C. B. Beck, *American Journal of Botany* 49 (1962): 373.)

short branches and long shoots. Then the short branches turn to lie in the same plane; and finally, epidermal webbing grows out to connect the branches. Each branch has become a vein, and the cluster of branches has become a true leaf, or **megaphyll**. Megaphylls occurred on many vascular plants in the past, but today they are found only on ferns and seed plants. They are fundamentally different in form and evolution from the superficial **microphylls** attached to the first vascular plants (and still found on living members of the Psilophyta, Lycophyta, and Sphenophyta). Microphylls may have evolved by foldings or outgrowths of stem epidermal tissue into which a branch of stem xylem extended. Megaphylls typically have a *network* of veins, whereas microphylls have a *single* vein. Furthermore, the anatomy of the node—the place where leaf and stem join—is much more complex for megaphylls than for microphylls.

The first seeds also developed during the Devonian period. Technically, a seed is an embryo sporophyte embedded in its parent gametophyte and surrounded by a hard seed coat. Fossil seeds of *Archaeosperma* have a cluster of small branches forming a protective layer around several seeds. Unfortunately, we do not know anything about the rest of the plant attached to these fossil seeds. More illuminating fossils of seed plants have not been found in deposits until the Carboniferous period, which followed the Devonian.

Fossil Fuels Began Forming in the Carboniferous Period

Extremely lush swamp forests dominated the landscape of the Carboniferous period (Fig. 23.8). Because the land was low, minor changes in sea level or the water table successively inundated forests, buried them in silt, and then raised new ones on top of the old. The buried

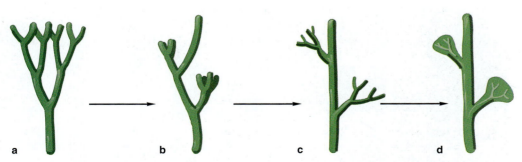

a   b   c   d

Figure 23.7 Theory for the origin of megaphylls (true leaves). (**a**) Stem with equal Y-forked branches, the earliest condition. (**b**) Unequal branching, which yields a main stem axis and short side branches. (**c**) The small branches come to lie in a plane. (**d**) Webbing occurs, producing a leaf blade. (Redrawn from G. M. Smith, *Cryptogramic Botany*, 1955.)

COAL, SMOG, AND FOREST DECLINE

All human societies burn fuel, and in doing so they release by-products into the environment. The particular material that is used as fuel, the reason it is burned, and the type and volume of by-products released differ enormously from place to place and throughout history.

Technological societies today use fuel for heating, cooking, industrial processes such as smelting, producing mechanical energy, and the creating of electricity. Although water, wind, sunlight, and nuclear fission are also used for some of the same objectives, most of our demand for energy is met by the carbon-based fuels of wood, coal, natural gas, and crude oil.

It's ironic that we depend on carbon-based fuels because carbon is a rather uncommon element. It makes up only 0.04% of the earth's mass. Photosynthesis, recall, does not create carbon; it merely rearranges it from part of a gas to part of a sugar. Carbon is finite on earth, therefore, the rate at which it cycles through the biosphere is very critical to life.

Coal is not just a form of carbon, like graphite or diamonds. It is a mixture of organic substances, moisture, and minerals. The major elements that make up the chemical compounds are carbon, hydrogen, oxygen, nitrogen, and sulfur. Coal begins as plant debris deposited in such a way that oxygen is limited. Consequently, the litter is only partially decomposed. Over time, the deposit of litter thickens. The next process in coalification is burial of the organic material in sediment. Over time, if the sediment is deep enough, pressure and heat compress the material further, turning it into a soft rock we call coal.

The heat released by coal when it burns depends upon the pressure and heat it was subjected to over geologic time. Standard names are given to coal of different heat value: lignite (which is the lowest grade), subbituminous coal, bituminous coal, and anthracite. Heat is typically measured in BTUs (British thermal units). (1 BTU = 252 calories, and 100,000 BTUs = 1 therm = 25,200 kcal. On average, 1 metric ton (1.1 ton) of coal contains 278 therms.) Different fuels can best be compared in terms of energy units such as BTUs because otherwise we're comparing gallons, tons, and cubic feet. Coal satisfies about 30% of the world's annual energy demand in BTUs.

Coal reserves are not uniformly distributed throughout the world. Of the earth's estimated recoverable coal reserves of 10 trillion metric tons (11 trillion tons), the United States has nearly 50%. The availability of coal has been an economic boon but an ecological disaster for those countries with large coal reserves. By-products of coal burning include soot, volatile hydrocarbons, and oxides of nitrogen and sulfur. When the English had exhausted their wood reserves for fuel in the 17th century and began to burn coal, London became a notoriously dirty and noxious place to live. Fogs became thicker, because fog droplets could condense around soot particles. Fog became *smog* (smoke + fog). A further health hazard came from the nitrogen and sulfur oxides, which become strong acids in water and can produce respiratory diseases. London became famous for "killer smogs" as its population and its air pollution climbed in the 18th, 19th, and 20th centuries: smogs so unhealthy that people actually died from their effects. (Young children and the elderly are particularly susceptible to smog.) Tall smokestacks alleviated some of the problem locally, but prevailing winds then carried the pollutants eastward, causing complaints from northern Europeans. Burning soft coal, high in sulfur, is no longer allowed in London.

Central and eastern Europe created their own problems. After the Second World War, countries like Poland, Czechoslovakia, and Germany became heavily industrialized, burning sulfur-rich coal without smokestack scrubbers, which remove sulfur oxides before they reach the atmosphere. As a consequence, millions of hectares of conifer forest in central and eastern Europe have suffered a decline, especially noticeable since 1980. Symptoms include premature shedding of needles and cones, discoloration of nee-

Figure 23.8 Diorama reconstruction of a Carboniferous forest.

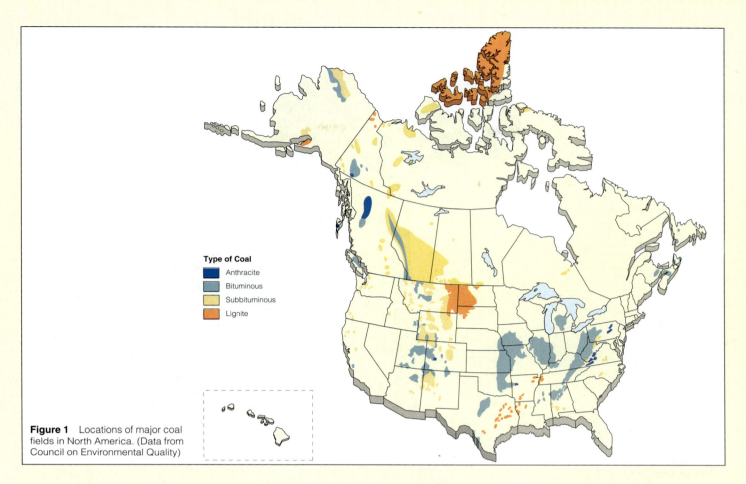

Figure 1 Locations of major coal fields in North America. (Data from Council on Environmental Quality)

Type of Coal
- Anthracite
- Bituminous
- Subbituminous
- Lignite

dles, abnormal branching, loss of mycorrhizal associations, slowed growth, and death. The combination of symptoms and the wide diversity of species affected do not imitate any known disease. Forest ecologists believe that the decline is caused primarily by acidic rain from coal burning and by ozone from automobile exhaust.

In North America, major coal fields exist in the Northwest Territories of Canada, in the Appalachian and Midwest regions of the United States, in North and South Dakota, and in the Four Corners area of Utah, Colorado, New Mexico, and Arizona (Fig. 1). Much of the eastern coal has been mined in underground shafts, whereas western coalfields

are open pits. In both cases, noncoal refuse is left above ground as polluting mine spoil that is very difficult for plants to invade.

Despite the ecological problems that result from the mining and burning of coal, it is likely to remain a prime source of energy for the world for decades to come. Alternative energy sources, such as nuclear fission and the construction of hydroelectric dams, have equally damaging or potentially worse ecological consequences. We can expect continued technological advances that will clean up coal prior to burning, trap and recycle by-products after combustion, and restore the landscape after mining has ceased.

organic remains have since become compressed and changed; today they form the world's coal reserves. For that reason coal is called a *fossil fuel*, and the Carboniferous period is called the *Coal Age* (see sidebar, "Coal, Smog, and Forest Decline," above).

Plants have been buried and transformed into fossil fuel at other times. For example, diatomaceous deposits from the more recent Mesozoic and Cenozoic eras are now mined as oil and gas, and some vascular plant deposits from the same eras are now mined as coal. The Carboniferous deposits are far more extensive than any others, however.

Coal Age forests were dominated by Lycophyta (to which club mosses and quillworts belong). One example is *Lepidodendron* (Fig. 23.9), up to 35 m (115 ft) tall and with a trunk 1 m (3 ft) across. Straplike leaves and sporangia were attached near the ends of the branches. A cross section of the trunk shows pith, primary and secondary xylem, cambium, and an enormous amount of cortex and cork. Very little of the stem area served for conducting or strengthening. The roots were dichotomously branched.

Second in abundance were giant horsetails in the division Sphenophyta. *Calamites*, for example, was smaller

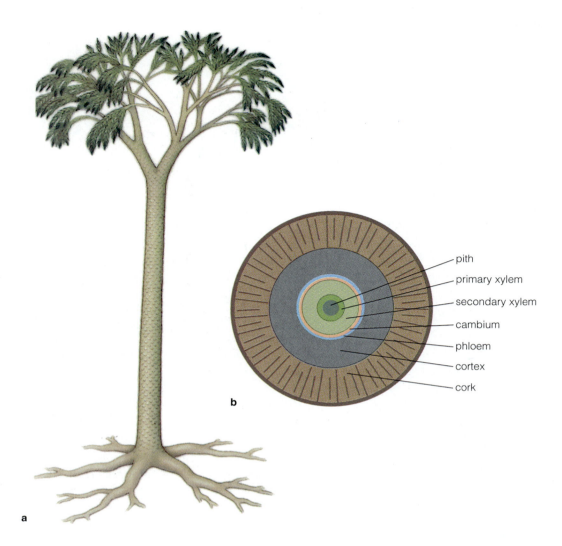

Figure 23.9 *Lepidodendron*, a dominant tree of the Carboniferous forest and a member of the Lycophyta. (**a**) Entire plant, which grew to about 35 m (115 ft) tall. (**b**) Cross section of the trunk, about 1 m (3 ft) in diameter. (Redrawn from M. Hirmer, *Handbuch der Palaobotanik* 1 (1927): 182.)

pith
primary xylem
secondary xylem
cambium
phloem
cortex
cork

than *Lepidodendron* but was still 9 m (30 ft) tall and had a trunk about 30 cm (12 in) in diameter. Whorls of branches arose at nodes, with smaller branches arising from them, and leaves from the smaller branches. The upright stems were attached to a horizontal rhizome system.

Third in abundance—not tall, but dominating the forest floor—were ferns (Pterophyta) and seed ferns (a now-extinct group). Seed ferns were fernlike but possessed seeds rather than sporangia on their fronds. Exactly what ferns evolved from is not clear because their fossil record began late in the Devonian, after all other lines of lower vascular plants had already appeared. Some botanists think that *Archaeopteris* is a likely ancestor of the ferns because of its fernlike branch systems, but other evidence tends to show that *Archaeopteris* is more closely related to gymnosperms (such as modern pine trees). Perhaps ferns and gymnosperms both came from the same stock.

Seedless vascular plants dominated world vegetation during much of the Paleozoic era, to which both the Devonian and Carboniferous periods belong. Their reign was long—about 200 million years—but great extinctions

lay ahead at the end of the Paleozoic, and new forms of plant life replaced them to become dominant.

The rest of the chapter will deal with the four divisions of the lower vascular plants.

23.2 DIVISION PSILOPHYTA: THE MOST PRIMITIVE VASCULAR PLANTS

The modern division Psilophyta is represented by a single family, the Psilotaceae, which has only two living genera, *Psilotum* and *Tmesipteris* (pronounced mes-ip'-tris). These are relatively uncommon plants that grow in tropical and subtropical regions, often as epiphytes. The North American range for *Psilotum*, (whisk fern) includes the states of Arizona, Florida, Hawaii, Louisiana, South Carolina, and Texas; but *Tmesipteris* is restricted to New Caledonia, Australia, and New Zealand in the southern hemisphere.

Psilotum is the simplest, most primitive vascular plant. The sporophyte generation of *Psilotum* is 45 to 60 cm

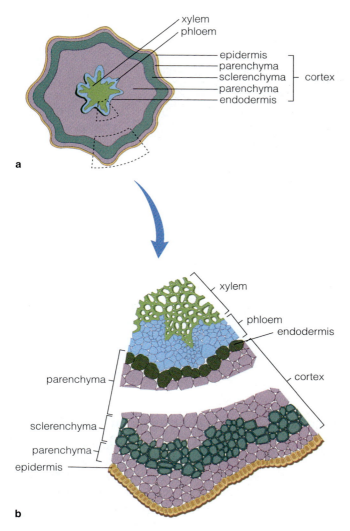

a

xylem
phloem
epidermis
parenchyma
sclerenchyma
parenchyma
endodermis
cortex

xylem
phloem
endodermis
parenchyma
cortex
sclerenchyma
parenchyma
epidermis

b

Figure 23.10 Cross section of a *Psilotum* stem. (**a**) Arrangement of tissues. (**b**) Enlarged section showing cellular detail.

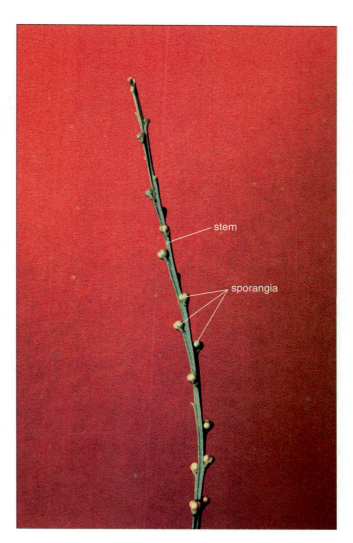

stem
sporangia

Figure 23.11 *Psilotum nudum* branch.

(18 to 24 in) tall and has a very simple architecture reminiscent of extinct *Cooksonia* and *Rhynia* (see Figs. 23.1a, 23.3, and 23.4). The upright, branching stem is slightly ribbed, and it bears **enations** (small, scalelike appendages that lack vascular tissue). The stems arise from a horizontal rhizome system that is clothed with many rhizoids. Cortex cells of the rhizome are infected with mycorrhizal fungi that extend into the soil.

The anatomy of the stem has traits of both shoots and roots (Fig. 23.10). The very center of the stem is rootlike in that it contains primary xylem, and there is no pith. A zone of primary phloem surrounds the xylem. Furthermore, the vascular tissue is surrounded by an endodermis with conspicuous Casparian strips. The cortex, however, is stemlike, with photosynthetic parenchyma that fills most of the stem's volume. All photosynthesis occurs in the stem. The epidermis contains stomata and is bounded by a cuticle.

Sporangia are borne in axils of some of the enations (Fig. 23.11). Meiosis occurs in the sporangia, forming tetrads of meiospores. The meiospores are ultimately released from the sporangia; if they land in a suitable habitat, they germinate and repeatedly divide by mitosis to form a gametophyte.

Psilophyta gametophytes are little more than a nonphotosynthetic bundle of parenchyma tissue several millimeters long and 1 to 2 mm in diameter. Gametophytes may grow on the surface of rock or bark or beneath the surface of the soil. In either case, hyphae of a mycorrhizal fungus spread throughout the gametophyte and extend out into the environment, bringing water and nutrients into the plant. The gametophyte depends upon the fungus for survival. Gametangia are scattered over the surface of the gametophyte: Flask-shaped archegonia contain one egg each, and spherical antheridia contain spirally coiled, multiflagellate sperm.

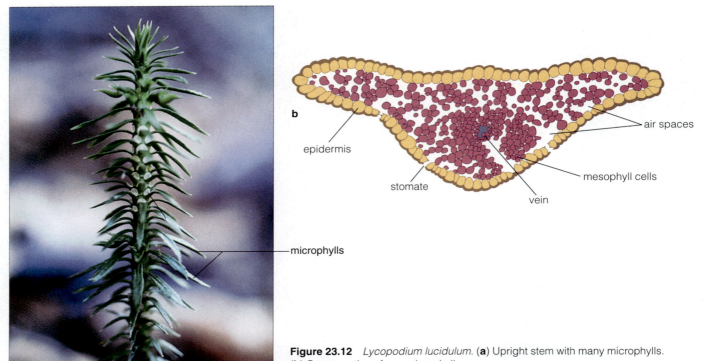

Figure 23.12 *Lycopodium lucidulum.* (**a**) Upright stem with many microphylls. (**b**) Cross section of one microphyll.

epidermis

stomate

vein

mesophyll cells

air spaces

microphylls

b

a

After a sperm is released, it swims through water to fertilize an egg and produce a diploid zygote cell. The zygote begins dividing and forms an embryo sporophyte, which is attached to the gametophyte by a foot. The foot sends haustorial outgrowths into the gametophyte, and the embryo lives as a parasite on the gametophyte until it develops into a larger, independent plant.

23.3 DIVISION LYCOPHYTA: THE CLUB MOSSES AND SPIKE MOSSES

Living members of division Lycophyta are commonly called *club mosses* or *spike mosses*. They are not true mosses, but their small size and green microphylls make them superficially resemble mosses. They are more complex than Psilophyta plants, containing true roots and microphylls with veins. (Coal Age representatives of this division had megaphylls, but living members have only microphylls.) The microphylls are typically arranged in a spiral. The stem has a central core of xylem and phloem, without pith, like the stem of Psilotum.

Three genera of Lycophyta occur in the northern hemisphere: *Lycopodium*, *Selaginella*, and *Isoetes*. Two other genera, *Phylloglossum* and *Stylites*, grow in the southern hemisphere. Most of the species are tropical or subtropical, but some species do extend into temperate, aquatic, and even arid habitats. The remainder of this section deals with these three genera.

Lycopodium Has a Homosporous Life Cycle

The genus *Lycopodium* (club mosses) contains about 400 species. Most are trailing plants, with short upright branches that resemble thick mosses or pine seedlings—hence the common names *club moss* and *ground pine* (see Fig. 23.1). They are widely distributed, occurring in the understory of tropical, subtropical, and temperate forests. Several species grow in humid parts of the United States.

The main stem branches freely and is generally prostrate. Upright stems, approximately 20 cm (8 in) in height, grow from the horizontal stem. Both types of stems are sheathed with numerous small, green microphylls. The microphylls have a well-developed epidermis with stomata, a mesophyll with many air spaces, and a midvein (Fig. 23.12).

The stem has a netlike system of connected strands of xylem with phloem between them (Fig. 23.13). The xylem contains tracheids, and the phloem contains sieve cells and some parenchyma cells. An endodermis encircles the vascular cylinder. Small, well-developed adventitious roots arise irregularly from the underside of the horizontal stem. These are roots, not rhizoids or stems.

Lycopodium has a sporic life cycle (Fig. 23.14). Both generations are independent, capable of living on their own apart from each other. Sporangia are homosporous—that is, all of one type—and occur on special microphylls grouped toward the tips of upright stems on the sporophyte (Fig. 23.14b). These microphylls are called **sporophylls**, meaning "spore-bearing leaves." Sometimes the sporophylls are indistinguishable from microphylls else-

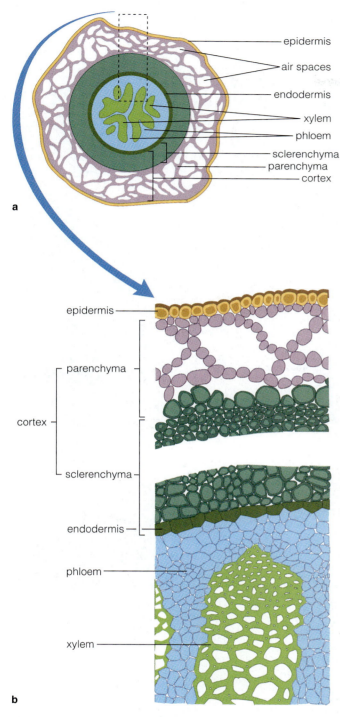

Figure 23.13 Structure of a *Lycopodium* stem. (**a**) Cross section of a stem, showing irregular distribution of xylem and phloem. (**b**) Enlarged sections of cortex and vascular tissue (see dashed lines in **a**), showing cellular detail.

where on the plant, but often they differ in size, shape, position, compaction, and color. Apical clusters of sporophylls form a **cone** or **strobilus** (see Fig. 23.14a).

Meiosis occurs in each sporangium, producing haploid meiospores (Fig. 23.14c). The meiospores are shed, germinate slowly on the ground, and develop into gametophytes. Gametophytes in some species grow aboveground and are lobed and green (Fig. 23.14d); in other species they may be subterranean and lack chlorophyll. They are always associated with a mycorrhizal fungus, just as are *Psilotum* gametophytes.

Both male and female gametangia are borne on the upper surface of the same gametophyte (Fig. 23.14e). When biflagellate sperm are liberated from antheridia, they swim through free water to fertilize eggs in archegonia (Fig. 23.14f–h). The resulting zygote divides into an embryo that possesses a well-developed foot, rudiments of a short primary root, leaf primordia, and a short shoot apex (Fig. 23.14i). The embryo and young sporophyte are parasitic on the gametophyte, but with further development the sporophyte becomes nutritionally independent (Fig. 23.14j).

Selaginella Has a Heterosporous Life Cycle

Selaginella plants (spike mosses) resemble those of *Lycopodium* in general appearance, but they are smaller, more widespread, and have unique life cycle features. The sporophyte generally consists of a branched, prostrate stem with short, upright branches, usually only 10 cm (4 in) high (see Fig. 23.1b). In some species, the stem is upright, slender, and taller. Both horizontal and upright stems are sheathed with microphylls in four rows, or *ranks*.

The 700 species of *Selaginella* occur mainly in tropical, subtropical, and temperate zones. A few are adapted to relatively dry habitats. Several species are commercially grown as ornamental plants, including *Selaginella lepidophylla* (resurrection plant, rose-of-Jericho), *S. willdenovii* (peacock fern), and *S. braunii* (treelet spike moss).

Sexual reproduction in *Selaginella* is heterosporous. In other words, two types of sporangia are produced: large **megasporangia** and small **microsporangia**. The sporangia are clustered in cones or strobili, as they are in *Lycopodium*. A single strobilus usually contains both types of sporangia (Fig. 23.15b, c, f).

A megasporangium is filled with diploid **megasporocytes**. Only one of them will divide by meiosis, however; the rest degenerate. The remaining megasporocyte, nourished by fluid from the degenerated neighbor cells, grows large. Meiosis of the megasporocyte then yields four large **megaspores**, each of which may germinate and grow into a female gametophyte, or **megagametophyte** (Fig. 23.15d).

A microsporangium is filled with about 250 diploid **microsporocytes**. Unlike the megasporocytes, most of the

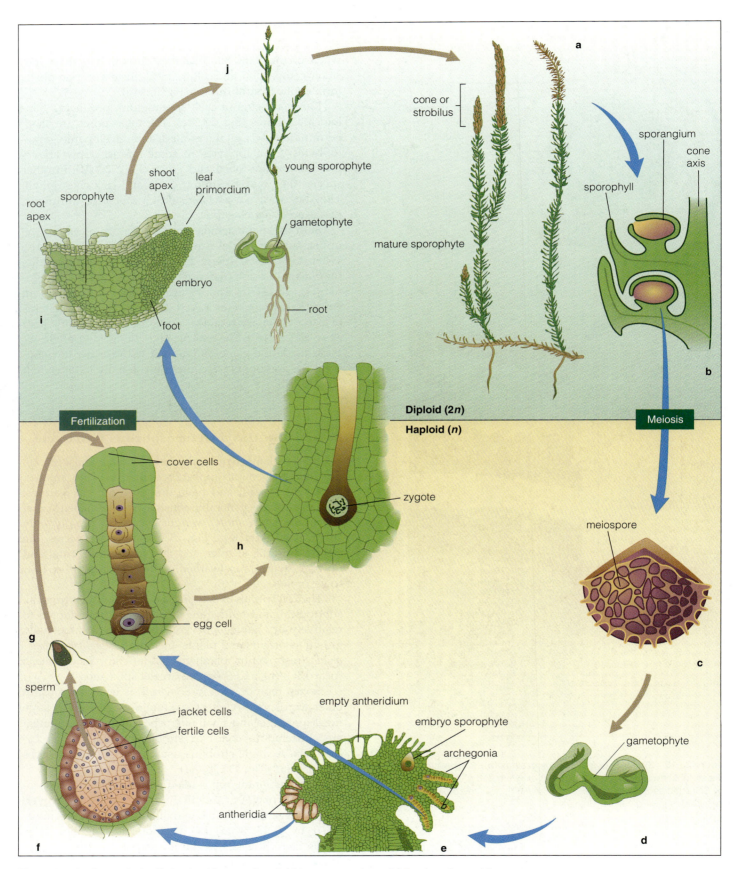

Figure 23.14 Stages in the life cycle of *Lycopodium*. (**a**) Mature sporophyte. (**b**) Section of a strobilus showing location of sporangia on a sporophyll. Meiosis in the sporangium produces meiospores (**c**), which are shed from the strobilus and germinate to form gametophytes (**d**). (**e**) Section of gametophyte, showing location of archegonia and antheridia. Biflagellate sperm produced in antheridia (**f**) swim in a film of water (**g**) to archegonia (**h**), where they fertilize egg cells and form diploid zygotes. Zygotes grow and differentiate into embryo sporophytes (**i**), which remain attached to (and parasitic upon) the gametophyte until further development makes them independent (**j**).

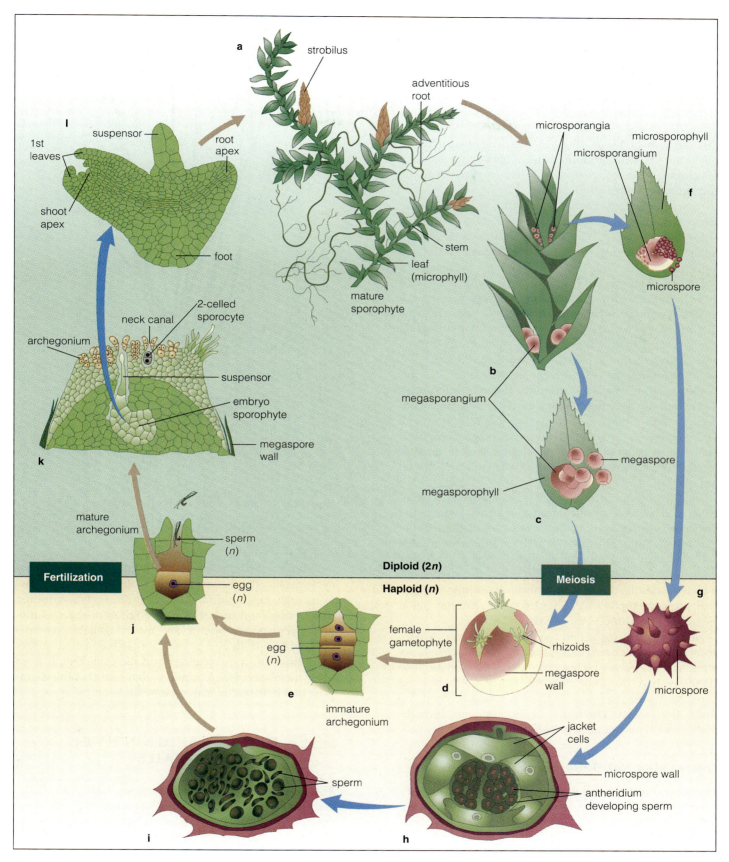

Figure 23.15 Stages in the life cycle of *Selaginella*. (**a**) Shoot of a sporophyte, showing three strobili. Inside each strobilus are two kinds of sporangia: megasporangia (**b**) and microsporangia (**f**). Meiosis within the larger megasporangium yields a few megaspores, one of which divides within the megaspore wall to produce a female gametophyte (**d**). Meiosis within the microsporangium produces many microspores, each of which develops into a male gametophyte while still within the microspore wall (**g**, **h**, **i**). Motile sperm emerge when moisture causes the microspore wall to burst. The sperm swim to a nearby archegonium (**j**), fertilization occurs, and an embryo forms within female gametophyte tissue (**k**, **l**). (Redrawn from R. A. Slagg, *American Journal of Botany* 19 (1932): 106 and from H. Bruchmann, *Flora* 104: 180.)

microsporocytes survive and go through meiosis to produce small **microspores** (Fig. 23.15g).

The megagametophyte is initially retained within the megaspore wall, but at maturity its volume ruptures the wall and a small cushion of colorless gametophyte tissue protrudes. Archegonia and vestigial rhizoids develop on this cushion of tissue (Fig. 23.15e). The megagametophyte may be shed from the cone at almost any stage of development. In some species, it may be retained in the cone until well after fertilization.

Microspores germinate while still within the microsporangium. Each divides into two cells; one of the cells—the prothallial cell—is small and does not divide further. It represents the vegetative portion of the microgametophyte. (The prothallial cell is not shown in Fig. 23.15.) The other cell divides repeatedly and forms an antheridium with a jacket of sterile cells enclosing 128 to 256 sperm, all within the microspore wall. Thus the mature male gametophyte (microgametophyte) is a small plant, basically consisting of a single antheridium imprisoned within the initial microspore wall (Fig. 23.15h)

Microspores are explosively ejected from the microsporangium midway through their development. They continue growing to maturity without a direct connection with either the parent sporophyte or the soil. Those microspores that sift down to megasporophylls below them come to lie near developing female gametophytes. The microspore wall eventually ruptures, releasing sperm.

Fertilization occurs only when enough water (from either rain or dew) is present for the sperm to swim to the archegonia. After fertilization, a zygote immediately begins cell division. The first cell division produces one cell that develops into the embryo and a second cell that grows into an elongated structure called the **suspensor**, which pushes the developing embryo sporophyte deeper into the gametophyte (Fig. 23.15k). The embryo has a foot, a root, two embryonic leaves, and a shoot apex.

In some species, the embryo is held in the megaspore and within the sporangium—a condition that closely resembles that of a seed. However, it is not a seed because (1) the embryo does not enter dormancy, (2) there is no outer hard seed coat, and (3) there is no stored food. Instead, the embryo continues developing into a sporophyte, which may be seen growing out of the cone.

Isoetes May Not Belong to the Lycophyta

The genus *Isoetes*, commonly called *quillwort* and *Merlin's grass*, contains more than 60 species. All are small plants, often growing submerged in shallow water for at least part of the year (which is unusual for plants in this division). Their tufts of microphylls resemble the grasslike leaves of some monocotyledonous plants, and their corms resemble bulbs (Fig. 23.16). Air chambers through the microphylls no doubt aid in the diffusion of oxygen when the plant is submerged.

microphylls

corm

roots

Figure 23.16 *Isoetes*, about 3–5 cm (1–2 in) tall.

Isoetes also differs from other Lycophyta genera in life cycle details. It is heterosporous. Unlike *Selaginella*, however, each megasporangium produces several hundred megaspores, and each microsporangium produces 1 million microspores. Sperm are multiflagellate, whereas those of *Selaginella* and *Lycopodium* are biflagellate. *Isoetes* is so different in morphology and life cycle details from other members of the Lycophyta that some botanists propose that it be placed in a separate division.

23.4 DIVISION SPHENOPHYTA: THE HORSETAILS

This division Sphenophyta contains only one living genus, *Equisetum*, with about 25 species. Many of these species live in cool, moist habitats. A few, such as *Equisetum arvense*, can grow in dry places. Sphenophytes contain silica in their stem epidermis, making them very abrasive. In pioneer days they were used to scour pots and pans—hence one of their common names, *scouring rush*. Their bushy, branching structure gave them another common name, *horsetail* (refer back to Fig. 23.1c).

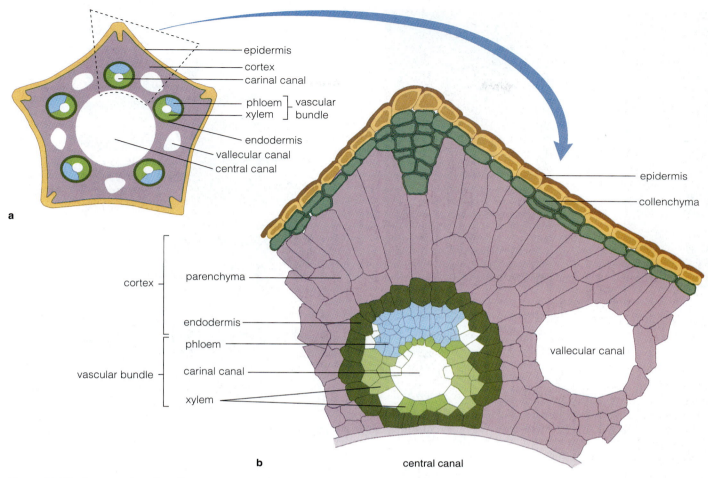

Figure 23.17 Cross section of an *Equisetum* stem. (**a**) Whole stem. (**b**) Enlarged section showing cellular detail within the dashed area of (**a**).

The sporophyte is usually less than 1 m (3 ft) tall, but one tropical species is vinelike and has a stem up to 7 m long. All horsetails are perennial and have a branched rhizome from which upright stems arise. Depending on the species, the upright stems may branch profusely, sparingly, or not at all. They are marked by vertical ridges and distinct nodes. A whorl of branches and/or microphylls attach to each node. The microphylls are relatively small, nongreen, and short-lived. The green stems are the major photosynthetic organs. Roots emerge from rhizome nodes but may also arise from stem nodes if the stem is in contact with moist soil.

Equisetum stems are hollow except at the nodes, where a diaphragm of tissue forms a solid joint. The outer stem ridges are constructed of collenchyma. Strands of collenchyma also project inward, toward vascular bundles (Fig. 23.17). Each bundle has a small air channel called a *carinal canal*. Arms of xylem extend outward from this canal; phloem tissue lies between the radial arms of the xylem. Some species contain vessels in the xylem, but most contain only tracheids. There is an endodermis, but its position varies with the species. It may surround each vascular bundle (as shown in Fig.

23.17), or it may encircle the entire ring of all vascular bundles. Larger air channels, called **vallecular canals**, lie between the vascular bundles. Photosynthetic tissue fills the rest of the cortex. Most of the stem's cross section is a large, hollow **central canal**. The biophysicist Karl Niklas has shown that these hollow, jointed stems are remarkably strong and rigid per unit weight—stronger than solid stems, in fact. *Equisetum* uses its biomass very efficiently.

The **sporangiophores** (sporangium-bearing organs) of *Equisetum* are grouped together in strobili, or cones, at the tips of upright stems and occasionally at the tips of lateral branches. In most species, cones occur on ordinary vegetative shoots, but in a few species they occur only on special fertile shoots that look different from vegetative shoots. The sporangiophores are stalked, shield-shaped structures borne at right angles to the main axis of strobilus (Fig. 23.18). Each sporangiophore resembles an umbrella.

Sporangia are attached to the underside of the umbrella shield, close to its edge, and they extend horizontally inward toward the stem. Each sporangium is filled with sporocytes, which divide by meiosis to produce haploid meiospores. Because all the meiospores are alike,

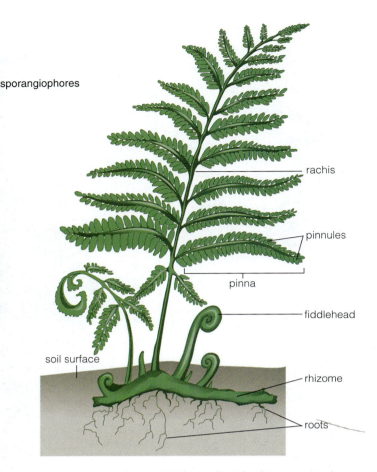

Figure 23.18 (**a**) Cones or strobili of *Equisetum telmateia*. (**b**) One strobilus sectioned to show sporangiophores.

Figure 23.19 *Athyrium felix-femina* (lady fern) has a typical fern body with a perennial underground rhizome that produces compound fronds. The fronds die back annually. Young fronds are tightly coiled, which protects them as they push up through the soil from the rhizome to the surface (dashed line).

Equisetum is homosporous. The meiospores are green and thin-walled, with long, ribbonlike elaters attached to the spore wall. The elaters coil and uncoil in response to humidity. They help discharge the spores when the sporangium splits open at maturity.

The spores germinate readily within days if they fall on a suitable substrate. They then divide by mitosis and produce small gametophytes about the size of a pinhead. The gametophytes have delicate, photosynthetic lobes of tissue and a cushionlike base with rhizoids. Both male and female gametangia are present on the same gametophyte, but they mature at different times, so gametes must cross-fertilize. The sperm is a multiflagellate, spiral shape. Fertilization takes place only when there is enough water to form a film over and among the gametophytes.

The diploid zygote develops into an embryo with no suspensor and a very small foot region, but otherwise it is similar to those already described in this chapter. The embryo and young sporophyte stages are parasitic on the gametophyte, but as the sporophyte matures it becomes fully independent.

23.5 DIVISION PTEROPHYTA: THE FERNS

Excepting only the flowering plants, division Pterophyta contains the largest and most diverse group of species of living vascular plants. Most of the 12,000 species are perennial, herbaceous, shade- and moisture-loving ferns of moderate size with large, typically compound leaves called **fronds**. At one extreme of size are tree ferns more

than 20 m (66 ft) tall (see Fig. 23.1d); at the other extreme are a few small aquatic species 1 to 2 cm (0.6 in) across.

Ferns are unique among lower vascular plants in their possession of macrophylls, in their diversity of growth forms, and in their continuing evolution. Among the kingdom Plantae, only ferns and flowering plants appear to be evolving new species; in contrast, all other groups are pale relicts of past diversity.

The Pterophyta have a rich fossil record extending back to the Carboniferous period, with some ancestral types apparent in the Devonian. Modern ferns, however, do not appear in the geologic record until the mid-Jurassic period, about 160 million years ago. Extant forms comprise four orders, the largest of which is the Filicales, the **true ferns**. We will describe the biology of true ferns in the rest of this chapter.

Fern Sporophytes Typically Have Underground Stems

Fern sporophytes extend themselves through space and time by an underground perennial stem, the rhizome. The rhizome grows from its tip. Roots and fronds arise from nodes along the rhizome (Fig. 23.19). All cell divi-

a

b

c

d

Figure 23.20 Diversity of fern fronds. (**a**) Simple, entire frond of hart's tongue fern, *Phyllitis scolopendrium*. (**b**) Pinnate frond of polypody fern, *Polypodium*. (**c**) Bipinnate frond of cliff brake, *Pellaea*. (**d**) Much-dissected frond of spleenwort, *Asplenium*.

sion is initiated from an apical cell—a pattern common to all seedless vascular plants. The rhizome can remain alive for centuries in some fern species, spreading a clone of fronds over a considerable area. Some bracken fern (*Pteridium aquilinum*) clones are thought to be 500 years old.

Most ferns do not have stems above the ground—only fronds. Young fern fronds are rolled into tight coils, or **fiddleheads** (Fig. 23.19). Coiling protects their still-meristematic tissue as it is pushed up through the soil, from the rhizome to the surface, by the elongating petiole. Above the soil, the frond unwinds from the base upwards, continuing to grow by cell division at its coiled tip. In certain species, the frond tip may remain meristematically active for years, producing leaves 3 to 5 m (10 to 16 ft) long.

A typical fern frond has a well-developed epidermis, with chloroplasts in the epidermal cells and stomata on the lower surface. The leaf mesophyll may be differentiated into palisade and spongy parenchyma layers. The petiole is prolonged to a **rachis** on which secondary and tertiary leaflets (**pinnae** and **pinnules**) are attached (Figs. 23.19 and 23.20).

The vascular tissue of fern stems is organized into one or more bundles, each having a core of xylem, surrounded by phloem and generally encircled by an endodermis. These vascular bundles are interconnecting, so that their arrangement varies from level to level along the stem. However, there are arrangements characteristic of various genera. For example, *Polypodium* (polypody fern) has a ring of small bundles placed well out from the center of the rhizome (Fig. 23.21). The xylem of *Polypodium* contains only tracheids, but the xylem of some genera (the bracken fern *Pteridium* and the aquatic fern *Marsilea*) contain vessels. Sieve cells are present in the phloem.

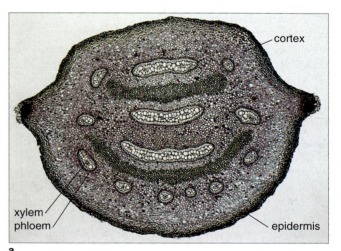

a

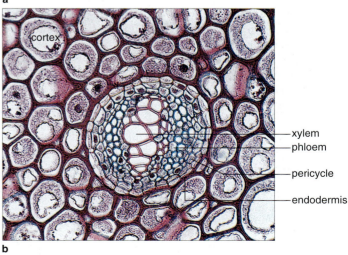

b

Figure 23.21 *Polypodium* rhizome cross section. (**a**) Entire rhizome, showing circular arrangement of vascular strands. (**b**) A single vascular bundle.

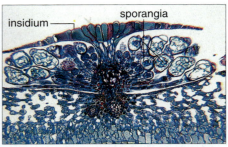

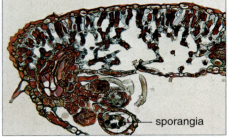

Figure 23.22 Fern sori. (**a**) Underside of holly fern (*Cyrtomium falcatum*) frond, showing many scattered sori. (**b**) Section of a sorus of *Crotonium*, showing the umbrella-like indusium covering the sporangia. (**c**) False marginal indusium and sporangia of cliff brake, *Pellaea rotundifolia*.

Sexual Reproduction Is Usually Homosporous

Fern plants become sexually mature in 1–10 years. The length of the juvenile phase depends on the plant's size and life span—the longer the sporophyte is capable of living and the larger its mature size, the longer the juvenile period. Sexual reproduction begins with meiosis and meiospore production. Meiospores are produced in sporangia that ordinarily develop on the lower surface or margins of fronds. Not all fronds are fertile (capable of producing spores); in some species fertile fronds are different in shape from vegetative fronds. The distribution of sporangia on the surface of a fertile frond varies with the genus and species. Sporangia may uniformly cover most of the lower surface, be grouped in clusters (**sori,** singular *sorus*), occur only along veins, or be restricted to the leaf margin.

Sori may be exposed or be covered by an umbrella-like fold of tissue called an *indusium* (Fig. 23.22a, b). The indusium protects immature sporangia and shrivels away when sporangia are mature and about to shed spores. Marginal sporangia are often protected by the curled edge of the leaf, which functions as a false indusium (Fig. 23.22c).

The sporangium is a delicate, watch-shaped case, consisting of a single layer of epidermal cells, only one row of which possesses heavy walls. This row, which nearly encircles the sporangium in most true ferns, is the annulus. Its function is to open the mature sporangium and to snap back, flinging out the ripe spores. This action throws spores 1 to 2 m (3–7 ft) away; then gravity, wind, water, and even electrostatic factors may carry them much farther. Charles Darwin, sailing hundreds of kilometers from shore on the *Beagle*, recorded in his journal that some of the air samples he collected contained fern spores.

Fern spores are resistant to ultraviolet radiation, low humidity, and very low temperatures. *Dryopteris* (wood fern) spores, for example, have been experimentally held at −254°C (−425°F) for 11 hours without affecting their viability. In the temperate zone, spores tend to be released in the fall, at the end of the growing season; in the tropics, they may be released in any month of the year. Once released, fern spores can remain alive for several months to several years.

Fern spores are about the size of pollen grains, and they are produced in prodigious numbers. Individual fronds typically release 750,000 to 750 million spores per year, and most ferns have 10 to 20 fronds per plant. *Dryopteris psuedomas* (wood fern) probably holds the spore-production record, releasing 1 billion spores per day per plant!

Spore germination can take place under a wide variety of conditions. Ideal conditions include long days, temperatures of 10 to 20°C (50 to 68°F), partial shade, and high humidity. Fern spores contain phytochrome and an unidentified blue-absorbing pigment, which together control germination. Well-watered spores can be inhibited from germinating by exposure to far-red light or blue light; they are stimulated to germinate by red light. Their requirement for red light explains why spores are inhibited from germinating in deep shade. In a few cases, gibberellin (GA) also stimulates spore germination.

After the spore has been triggered to germinate, cell division is very ordered. First, mitoses produce a short filament of cells. In the presence of blue light, and perhaps with the assistance of such hormones as indoleacetic acid (IAA) and gibberellin, the apical cell of the filament begins to divide perpendicular to the filament, in time producing a green, heart-shaped gametophyte 1 to 2 cm (0.6 in) across (Fig. 23.23d). Rhizoids attached to the lower surface anchor the plant to the soil.

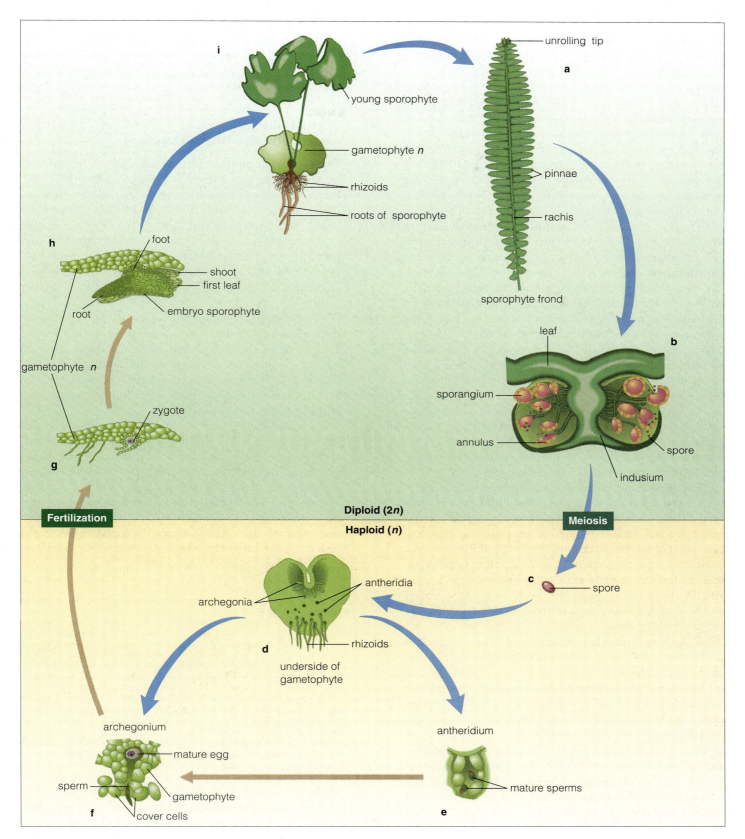

Figure 23.23 Stages in the life cycle of a fern. On the underside of a sporophyte frond (**a**) are many clusters of sporangia (**b**). Meiosis occurs within each sporangium, producing many small spores (**c**), each of which germinates to produce a gametophyte plant (**d**). Antheridia (**e**) and archegonia (**f**) are present on every gametophyte, but usually a sperm must swim from one gametophyte to fertilize the egg on another (**f, g**). The diploid zygote develops into an embryo sporophyte, which at first lives by parasitizing the gametophyte (**h**); but when more mature, it is rooted in the soil and is independent (**i**). Young sporophyte fronds often look different from mature fronds produced the next year (compare **i** and **j**). (Redrawn from M. E. Hartman, *Botanical Gazette* 91 (1931): 252; from D. H. Campbell, *Mosses and Ferns.* New York: Macmillan, 1905; and from R. M. Holman and W. W. Robins, *A Textbook of General Botany.* New York: John Wiley, 1939.)

The lobelike margins are only one cell thick, but the central region is several cells thick. Most fern gametophytes require two to three months to mature, but extremes exist. Some species colonizing disturbed habitats can mature in only four to six weeks, and *Woodwardia radicans* (chain fern), an inhabitant of the understory of cold-temperate conifer forests, requires two years.

Both male and female gametangia are present on the lower surface of the same gametophyte (Fig. 23.23d), but not usually at the same time. Small gametophytes are mainly male; larger ones with more stored carbohydrate are mainly female. The timing of gametangial appearance can be experimentally manipulated. Self-fertilization does occur in a few species, but usually a sperm from one gametophyte does not fertilize an egg on the same gametophyte.

Antheridial formation is stimulated by a hormone chemically similar to gibberellin. Inside an antheridium are helical sperm with as many as 100 flagella. When free water is present, the antheridium bursts (Fig. 23.23e), releasing sperm cells; these swim randomly at first, with a rotating motion. If archegonia are nearby, the sperm swim toward them, drawn by some attractant that exudes from the archegonia.

Archegonial necks spread open in the presence of free water, making a wide path for sperm to navigate through to the egg (Fig. 23.23f). As soon as one sperm has penetrated an egg, the plasma membrane of the egg changes, and no other sperm is able to penetrate.

The diploid zygote cell begins dividing and developing into an embryo, with foot, shoot, and root regions (Fig. 23.23g, h). Although one gametophyte may contain many zygotes, only one or two will mature into embryos. Some sort of selective abortion eliminates the others. In a short time, an embryo develops into a young sporophyte that is larger than the parent gametophyte (Fig. 23.23i). At this stage, the sporophyte has its own green fronds and root system—and so is nutritionally independent of the gametophyte.

A surprising number of fern species show gametophytes with a diploid chromosome number and sporophytes with a haploid chromosome number. The reasons for this are: gametophytes can produce sporophytes without fertilization and sporophytes can produce gametophytes without meiosis. When moisture and sucrose concentrations are high, vegetative tissue of a gametophyte can produce an embryo and a normal-looking—but haploid—sporophyte, without fertilization. This kind of asexual reproduction is called **apogamy**. Young sporophyte fronds, if wounded and placed on wet soil, will produce normal looking—but diploid—gametophytes. This kind of asexual reproduction is called **apospory**. It is clear, then, that the two phases of the fern life cycle are not completely controlled by the chromosome number. Environmental factors also play a role.

Some ferns outside the order Filicales are heterosporous, not homosporous. These are restricted to five genera in the water fern families Marsiliaceae and Salviniaceae. Among ferns, they are also unique for their morphology (Fig. 23.24), their aquatic habitat, and their specialized anatomy.

Ferns Have Ecological and Economic Importance

Ferns are most common in the understory of humid temperate and tropical forests. They are widely distributed, however, and also grow in arctic and alpine tundras, saline mangrove swamps, semiarid deserts, and on coastal rocks swept by salt spray. Some ferns are aquatic or are epiphytes or vines, but most are terrestrial and rooted in soil. Very few species are annuals. Ferns can provide the bulk of biomass in some tropical forests and dominate the understories of some temperate conifer forests.

Throughout their life cycle, ferns are remarkably independent of animals. Animals are not essential vectors for spores, sperm, or embryos; nor are animals attracted to fern tissue for food (because of poisonous secondary compounds synthesized in fern tissue). Ferns rarely possess mechanical defenses such as prickles, spines, thorns, stinging hairs, or hard leaves.

Humans have found uses for certain ferns at certain life history stages. In the fiddlehead stage, fronds of some species are edible, and these have become important greens in the diet of many cultures. The petiole and rachis of other ferns are used as basket-making material. Florists mix fern fronds with flower arrangements, and nurseries propagate ferns as popular indoor houseplants and outdoor landscaping plants. Biologists have found fern gametophytes to be excellent subjects for research on physiology and plant development.

SUMMARY

1. Seedless vascular plants number about 14,000 species in four divisions: Psilophyta, Lycophyta, Sphenophyta, and Pterophyta. Shared traits include a well-developed vascular system (but rarely with vessels), reproduction by thick-walled spores, an alternation of independent generations, biflagellate or multiflagellate sperm that require free water for fertilization, an embryo stage, and a dominant sporophyte stage that has an epidermis with cuticle and stomata.

2. Early vascular plants (*Cooksonia*, *Renalia*, *Rhynia*) were leafless and rootless stems, but evolution was rapid during the 40-million-year Devonian period. Microphylls, megaphylls, true roots, secondary growth, woody plants, and seeds all appeared for the first time in the Devonian.

3. Seedless vascular plants dominated the world's vegetation during much of the Paleozoic era. Diversity of form peaked in the Carboniferous period, 360 to 286 million

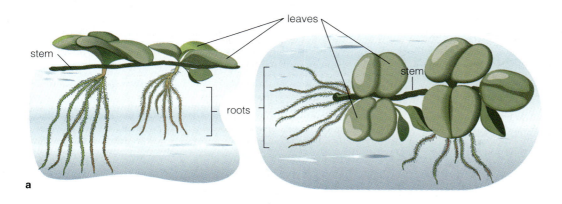

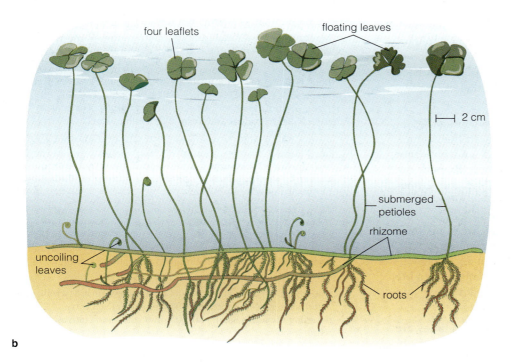

Figure 23.24 Examples of aquatic ferns. (**a**) *Salvinia* (shown in side and surface views) is a floating fern. (**b**) *Marsilea* is rooted in sediment, but the leaves float on the surface. Both are common in calm, warm, shallow freshwater ponds, ditches, or backwater areas.

years ago. *Calamites* and *Lepidodendron*, related to modern horsetails and club mosses, were major forest trees during this period. Their fossil remains are mined today as coal. Other Carboniferous plants included ferns and seed ferns. Many forms of seedless vascular plants became extinct at the end of the Paleozoic.

4. The division Psilophyta is represented by only two living genera, *Psilotum* and *Tmesipteris*. Sporophytes of *Psilotum* (whisk fern) consist of a branched, underground rhizome with rhizoids and an upright, green, branched stem with scalelike enations. Psilotum is homosporous, and the meiospores produce small, nonphotosynthetic gametophytes, which are nourished by symbiotic fungi. Each gametophyte produces both male and female gametes. The embryo sporophyte is at first attached parasiti-

cally to the gametophyte but in time becomes independent.

5. The division Lycophyta contains five living genera, three of which occur in North America: *Lycopodium*, *Selaginella*, and *Isoetes*.

6. Sporophytes of *Lycopodium* (club moss) have true roots, stems, and microphylls with a central vein. Sporangia are borne on special microphylls called sporophylls, which are clustered into cones or strobili near the ends of upright stems. Club mosses are homosporous, and the meiospores produce small gametophytes that are sometimes photosynthetic. The life cycle is much like that of *Psilotum*.

7. Sporophytes of *Selaginella* are similar to those of *Lycopodium*. Vessels are present in the xylem of some species. *Selaginella* is heterosporous, having megasporangia that produce megaspores and microsporangia that produce

microspores. Megaspores grow within the ruptured cell wall into megagametophytes; microspores grow within the spore wall into microgametophytes. Both gametophytes are little more than gametangia with a bit of vegetative, accessory tissue. The developing embryo is pushed deep into the megagametophyte by a suspensor.

8. Sporophytes of *Isoetes* (quillwort) have a small tuft of grasslike microphylls and basal corms, which make them resemble monocotyledonous seed plants. *Isoetes* grows in wet areas, often beneath standing water. It is heterosporous.

9. The division Sphenophyta consists of a single living genus, *Equisetum* (horsetail, scouring rush). Sporophytes have uniquely jointed, ribbed, hollow stems. Whorls of microphylls and/or branches are attached to each node. True roots are present, and vessels are found in the xylem of some species. Sporangia are borne on sporangiophores grouped into complex cones. Horsetails are homosporous, and meiospores produce small, photosynthetic gametophytes. The life cycle is much like that of *Psilotum* and *Lycopodium*.

10. The division Pterophyta contains 12,000 species of ferns. Most are small herbs, but some reach tree stature. In most cases the stem is under ground, and only fern fronds (leaves) appear above ground. Fronds are megaphylls, with complex venation and sometimes mesophyll differentiation into palisade and spongy parenchyma. Vessels are present in some species. Sporangia, located on the underside of fronds, are sometimes clustered into sori and protected by an indusium. Meiospores germinate to produce a small, photosynthetic, heart-shaped gametophyte. The homosporous life cycle is much like that of *Psilotum*, *Lycopodium*, and *Equisetum*.

11. Fern gametophytes and sporophytes sometimes are created by environmental conditions, not solely by chromosome number. Gametophytes can produce haploid sporophytes by a process called *apogamy*, and sporophytes can produce diploid gametophytes by a process called *apospory*.

12. A few ferns are floating aquatics, and these are heterosporous; all other species are homosporous.

Questions

1. The lower vascular plants include several divisions that today are relicts of what they once were. Explain in what ways these groups are relicts.

2. Why did plant invasion of land not occur until the Devonian period?

3. During which periods and eras were lower vascular plants the dominant form of vegetation on land? Describe the composition of a Carboniferous (Coal Age) forest as an example of vegetation dominated by lower vascular plants.

4. In what ways do lower vascular plants exercise more control over their internal water balance, as compared to bryophytes?

5. Why can one say that *Psilotum* is the most primitive vascular plant on earth today?

6. Describe the difference between a homosporous and a heterosporous life cycle.

7. In mosses, the sporophyte is attached to and dependent on the gametophyte. In ferns, however, the two generations live independent lives. How might this separation be important for the development of the larger, more complex fern sporophyte?

Further Readings

Berkner, L. V., and L. C. Marshall. 1965. "On the Origin and Rise of Oxygen Concentration in the Earth's Atmosphere." *Journal of Atmospheric Science* 22: 225–61. The original account of the Berkner-Marshall hypothesis that free oxygen in the atmosphere results entirely from photosynthesis and that the course and timing of evolution were related to increasing levels of oxygen in the atmosphere.

Bold, H. C., and J. W. LaClaire II. 1987. *The Plant Kingdom*. 5th ed. Englewood Cliffs, N.J.: Prentice-Hall. A refreshingly short and reader-friendly textbook. An evolutionary survey of organisms from blue-green algae and fungi to flowering plants.

Gensel, P. G., and H. N. Andrews. 1987. "The Evolution of Early Land Plants." *American Scientist* 75: 478–89. A very readable and well-illustrated summary of what the fossil record tells us about the early diversification of vascular plants during the Silurian and Devonian periods.

Gray, J., and W. Shear. 1992. "Early Life on Land." *American Scientist* 80: 444–56. Why was the evolutionary step from water to land more of an evolutionary challenge than the initial genesis of life? An excellent, readable, and well-illustrated survey of animal and plant evolution during the Silurian to Permian periods of the Paleozoic era.

Mickel, J. 1994. *Ferns for American Gardens*. New York: Macmillan. A colorful reference for more than 400 fern taxa. Each entry describes the fern's habit, plant size and color, and cultivation requirements.

Niklas, K. J. 1989. "The Cellular Mechanics of Plants." *American Scientist* 77: 344–49. An essay on the biophysics of stem architecture, in which the author shows why hollow, jointed horsetail stems are an efficient solution for stability.

Raghavan, V. 1989. *Developmental Biology of Fern Gametophytes*. New York: Cambridge University Press. A detailed summary of what we know about the biology, ecology, and behavior of fern gametophytes.

Rogers, W. P. 1978. *The Coal Primer; A Handbook on Coal for the Non-Coal Person*. Van Buren, Ark.: Valley Press. All you ever wanted to know about the types of coal, methods of mining and processing, and locations of the world's major coal fields.

Silver, C. S., and R. S. Defries. 1990. *One Earth—One Future*. Washington, D.C.: National Academy Press. A superb summary of major environmental issues by scientists from the National Academy of Sciences. Chapters on global warming, ozone depletion, acid deposition, and loss of biodiversity are written elegantly, simply, and for the general reader.

Ulrich, B., and J. Pankrath, eds. 1983. *Effects of Accumulated Air Pollution in Forest Ecosystems*. Dordrecht, Netherlands: Reidel. A thorough account of forest decline in Europe and its possible causes, including acid deposition from the burning of coal.

24 PLANTAE: GYMNOSPERMS

1. Gymnosperms have novel life cycle features that adapt them to life on land more completely than lower vascular plants. The new traits include the ovule, the seed, and pollination.

2. *Gymnosperm* is a general term for four divisions: Cycadophyta, Ginkgophyta, Gnetophyta, and Pinophyta. The latter division is the best known and most economically important; it includes pines, firs, spruces, hemlocks, redwoods, cedars, cypress, yews, and several southern hemisphere genera.

3. The pine life cycle is heterosporous. Male cones are small and seasonal. Each scale has two microsporangia, in which microspores are formed and divide into immature male gametophytes while still retained in the microsporangia. The gametophytes are released to the wind as pollen grains. Female cones are large and require two years to mature. Each scale has two megasporangia, in which megaspores are formed and divide into mature female gametophytes while still retained in the megasporangia.

4. Pollen grains sift down into female cones and germinate, producing a pollen tube, which grows through the megasporangium toward a megagametophyte. The pollen tube is a mature male gametophyte with a very simple body. Both male and female gametophytes are small, simple structures and are parasitic upon the sporophyte generation.

5. Fertilization of one egg is sufficient to trigger the development of a seed. Pine seeds contain an embryo sporophyte, stored food (in the form of a female gametophyte), and a seed coat. The function of the seed is to disperse the next generation away from the parent plant, both in space and in time.

24.1 GYMNOSPERMS: NAKED-SEED PLANTS

Gymnosperm is an informal group name for several divisions of seed plants. The gymnosperms are more truly land plants than seedless vascular plants because they are less dependent on free water. Their life cycles have novel features that make life in dry habitats possible. Seedless vascular plants are terrestrial, but they are tied to a wet environment during part of their life cycle just as thoroughly as amphibian animals are. Their small, delicate, free-living gametophytes, their motile sperm, and their embryos all require high humidity and surrounding films of free water. These links to a wet environment were broken in the course of evolution to seed plants.

Of the two groups of **seed plants**—gymnosperms and angiosperms—gymnosperms have the more ancient lineage. The advancements exhibited by gymnosperms over seedless vascular plants are the ovule, the seed, and pollination. The ovule is a megasporangium that retains and nourishes the female gametophyte within a protective outer integument; thus the gametophyte no longer has to survive a risky, independent existence. The seed is an organ that allows the embryo—in a dormant state—to be moved away from the parent in time and space. Pollination is a means of transferring sperm to the archegonium without free water. Air and plant tissue become the mediums of transport. We'll explain these three features in more detail later in the chapter.

Although both gymnosperms (conifers and their allies) and angiosperms (flowering plants) have seeds and vascular tissue, they differ in life cycle and in anatomical and morphological details. One major difference is how the seed is borne on the plant. In gymnosperms, seeds are formed naked on the surface of modified leaves called **scales**. In angiosperms, seeds are formed within the protection of an ovary; when the seeds mature, the ovary becomes a fruit. *Gymnosperm* means "naked seed." *Angiosperm* means "enclosed seed."

Gymnosperms number about 900 species, all trees, shrubs, or vines. Common examples include cedar (*Cedrus*), cycad (*Cycas, Dioon, Zamia*), fir (*Abies*), juniper (*Juniperus*), maidenhair tree (*Ginkgo*), pine (*Pinus*; Fig. 24.1), redwood (*Sequoia, Metasequoia, Sequoiadendron*), and spruce (*Picea*). Many other species are important in the southern hemisphere but are not well known in the north. The gymnosperms are grouped into four divisions: Pinophyta (pines, firs, spruces), Cycadophyta (cycads), Ginkgophyta (Ginkgo), and Gnetophyta (a mix of three genera that have traits of both angiosperms and gymnosperms).

24.2 THE MESOZOIC: ERA OF GYMNOSPERM DOMINANCE

The gymnosperm line originated in the mid-Devonian period, about 380 million years ago (see Fig. 23.2). *Archaeopteris* (see Fig. 23.6) and *Aneurophyton*, the first trees, are thought to be early ancestors of gymnosperms and so are called **progymnosperms**. Progymnosperms were most abundant in tropical climates. During the Carboniferous and Permian periods, progymnosperms gave rise to seed ferns and true gymnosperms. Seed ferns such as *Medullosa* combined the leaves of ferns with seeds borne naked on the ends of branches. *Cordaites*, a tree up to 30 m (100 ft) tall with 1-m-long (3-ft) strap-shaped leaves and seeds in cones (Fig. 24.2a), was a true gymnosperm present from the Devonian to the Permian.

During the Permian period, which marked the end of the Paleozoic era and the start of the Mesozoic era, climates became cooler and drier. Dominance of the world's vegetation by seedless vascular plants came to an end. New gymnosperm groups evolved and spread into many habitats. Most of the Mesozoic (245 million to 65 million years ago) was the age of gymnosperms, dinosaurs, and

a

b

Figure 24.1 Gymnosperms include the pines. (**a**) Mature pines are typically tall trees, as are these 40-m (130-ft) ponderosa pine (*Pinus ponderosa*) and sugar pine (*P. lambertiana*) in the Sierra Nevada forest of California. (**b**) Pines have needlelike leaves borne singly or in clusters on short shoots; these ponderosa pine needles are 15 cm (6 in) long and in clusters of threes. Pines have seeds borne naked in woody cones. Ponderosa pine cones are about 10 cm (4 in) long, with a sharp prickle at the end of each scale.

a   b

Figure 24.2 Reconstructions of two extinct gymnosperms. (**a**) *Cordaites*, 9 to 30 m (30 to 100 ft) tall, from the Carboniferous period. (**b**) *Williamsonia*, 5 m (17 ft) tall, member of the Bennettitales and a cycad ancestor. (**a** redrawn from D. H. Scott, *Studies in Fossil Botany*. London: Adam and Charles Black, 1920; **b** redrawn from C. A. Arnold, *An Introduction to Paleobotany*. New York: McGraw-Hill, 1947.)

moving continents. Our best geologic reconstruction of the world at the start of the Mesozoic era shows all of the continents joined into a single land mass near the equator. Beginning in the Triassic period, this primeval megacontinent, called **Pangaea**, began to break apart. North and South America split from Europe and Africa, and the Atlantic Ocean formed. Mesozoic gymnosperms included precursors of the cycads (Fig. 24.2b), as well as modern forms of cypress (*Cupressus*), *Ginkgo*, juniper (*Juniperus*), monkey puzzle tree (*Araucarea*), Mormon tea (*Ephedra*), pine (*Pinus*), and redwood (*Sequoia*).

As the Mesozoic era drew to a close, climates and continents continued to change. A southern continent called Gondwanaland split apart, and India became an island. Mountain building created cool weather at high elevations and arid regions in their lee. The climatic gradient from tropics to poles steepened. Many gymnosperms, including all seed ferns, became extinct, and the first flowering plants rose to dominance; dinosaurs declined, while social insects and mammals multiplied and diversified. Today, gymnosperms have fewer species and growth forms, and they occupy fewer habitats, than during the Mesozoic.

24.3 DIVISION PINOPHYTA

The most widely known and economically important gymnosperms belong to the division **Pinophyta**, with seven families and approximately 700 species. Not all of the species have woody cones, but the cone is such a conspicuous feature of many that this division has also been called the Coniferophyta (*conifer* means "cone-bearer"). Conifers that lack woody cones include the junipers, podocarps,

Figure 24.3 *Abies magnifica* variety *shastensis* (shasta red fir) female cone and branch. Note that the cone (12 cm long, or 5 in) is upright in all fir (*Abies*) species.

Figure 24.4 *Picea* (spruce). The cones are about 10 cm (4 in) long and hang down.

yews, and plum yews; in these plants, a berrylike tissue called an **aril** surrounds the seeds, making them resemble the fruits of flowering plants. Botanists, however, interpret the aril as part of a modified cone, or as an elaboration of the integument—not as an ovary.

We will now review the families Pinaceae, Cupressaceae, Taxodiaceae, Taxaceae, Podocarpaceae, and Araucariaceae.

Pinaceae Includes Pines, Firs, and Spruces

The family Pinaceae is important economically for wood, pulp, turpentines, and resins, as well as for ornamentals. Its species make up the bulk of the northern hemisphere's conifer forests. Leaves are needlelike and are borne singly or in clusters called **fascicles**, on special short shoots. Individuals are monoecious, meaning that both male and female gametophytes occur on the same tree. There are 10 genera in the family.

Pinus (pines) is the largest genus in the family, with 93 species. Pines are usually large, long-lived trees with an asymmetrical shape. Bristlecone pines (*P. longaeva*) are the oldest living things, some individuals reaching more than 5000 years of age. Pine needles are clustered, two to five per fascicle (see Fig. 24.1b), except for the single-leaf pinyon (*P. monophylla*). They are oval to triangular in cross section. Cones are pendant (hanging) and vary greatly in size. Usually, cones are shed once the seeds have matured and spilled out. Some **closed-cone** pines, however, keep their scales closed until heated by fire (which may not occur for many years). The ecological advantage of this behavior is that a rain of seeds falls onto the mineral-rich ashes immediately after a fire has removed all competing vegetation. Although the parent generation of pines is killed by the fire, the species retains its dominance because the site is reoccupied by the next generation.

Abies (firs) have a symmetrical, cylindrical, or pyramidal shape. They are neither as long-lived nor as large as pines. Each year's growth is marked by a symmetrical whorl of branches, so that it is possible to age young trees by counting the number of branch whorls from tip to base. Needles are borne singly. Cones are carried erect on the branches (Fig. 24.3), and they shatter at maturity rather than falling as a unit.

Picea (spruces) closely resemble firs, but the needles are angular in cross section and often sharply pointed (Fig. 24.4), rather than flat and blunt as are fir needles. Tree crowns are often narrowly columnar. Cones are pendant and fall as a unit when mature except for those of black spruce (*P. mariana*), an important species all across Canada, which is a closed-cone conifer. Large-scale wildfires periodically sweep through the Canadian forest, and the closed cones of black spruce allow the species to quickly re-establish itself.

Figure 24.5 *Pseudotsuga menziesii* (Douglas fir) cone, about 8 cm (3 in) long, showing bracts sticking out between the scales. The foliage is much softer than that of spruce and fir.

Figure 24.6 *Larix laricina* (tamarack) branch with clusters of deciduous needles.

Figure 24.7 *Juniperus occidentalis* (mountain juniper), a species in the family Cupressaceae. (**a**) Massive trees grow on exposed granite in the Sierra Nevada. (**b**) Branch with fleshy female cones that are called "berries," about 1 cm (0.4 in) in diameter.

Tsuga (hemlocks) are pyramidal with slender, horizontal branches and drooping tops. The needles are somewhat flat, with a short petiole. They resemble the leaves of firs but are much shorter. Cones are small.

Pseudotsuga has only three species. One of these—Douglas fir (*P. menziesii*)—is the most heavily cut timber tree in the United States. It dominates much of the Pacific Northwest, Cascade, and Rocky Mountain regions. It may grow to a height of 60 m (200 ft) and have a trunk diameter of 3 m (10 ft). Its smaller branches hang down, and the singly borne needles resemble those of spruce, but they are much softer. Douglas fir cones are easily recognized by the bracts, which extend out from between the scales (Fig. 24.5).

Larix is the genus of larches. Larches (and some members of the redwood family, Taxodiaceae) are unusual among conifers in being deciduous. They lose all their needles in the fall. Most gymnosperms lose only some of their leaves each year because the life span of a leaf is several years or longer (the record holder, bristlecone pine, retains its needles for 30 years). Larch needles are short and grouped in crowded clusters on short shoots (Fig. 24.6). One-year-old shoots, in contrast, have needles arranged singly and spirally. The American larch, or tamarack (*L. laricina*), is frequently found at the edge of bogs.

Cupressaceae Includes Junipers and Cypress

The Cupressaceae family contains shrubs or trees with small, scalelike or awl-shaped leaves and very open branching patterns. Individuals may be unisexual or bisexual (dioecious or monoecious). Cones may be woody, as in cypress (*Cupressus*), or fleshy, as in juniper (*Juniperus*; Fig. 24.7). The family contains more than 130 species with a worldwide distribution. Some cypress are closed-cone conifers.

Taxodiaceae Includes Redwoods

Members of the Taxodiaceae family are the dawn redwood of China (*Metasequoia glyptostroboides*), the California coast redwood (*Sequoia sempervirens*), the Sierra redwood (*Sequoiadendron giganteum*), and the bald cypress of the eastern United States (*Taxodium distichum*), plus several other genera from eastern Asia. The dawn redwood and the bald cypress are deciduous. Coast redwood is possibly the tallest tree in the world; individuals over 60 m (200 ft) in height are common, and the record height is 112 m (370 ft; see sidebar "The California Coast Redwood Forest," p. 412). Redwood has outstanding

THE CALIFORNIA COAST REDWOOD FOREST

Coast redwood, California redwood, and sequoia are names that bring to mind images of fog, looming trees, silent forests, and a sense of prehistoric time. Like other members of the Taxodiaceae family, *Sequoia sempervirens* grows in a humid climate with moderate winters. Over geologic time, its once-wide range throughout the north temperate zone shrank to a 700-km (435-mi) strip of land hugging the California coast. The redwood belt is less than 35 miles wide, and it occupies coast-facing slopes below an elevation of 800 m (2600 ft). The total area of its modern range amounts to less than 800,000 hectares (2 million acres)—less than 1% of California's area and less than 0.003% of the earth's surface.

Yet, the coast redwood is well known. It dominates a forest that has a remarkable growth rate, and its biomass (weight of organic matter) is greater than any forest in the world. Overstory redwood trees commonly exceed 100 m (330 ft) in height and 3 m (10 ft) across at the base of their trunk, and attain ages of 1000 years. The aboveground biomass is estimated to reach 3600 metric tons per hectare (1600 tons per acre). The tropical rain forests, about which we read so much, have only one-seventh as much biomass.

The redwood forest exists where it does because the nearby ocean keeps air temperatures buffered throughout the year. Winters do not have hard frosts, and summers are mild. In addition, cold, upwelling water just offshore creates fog banks that extend inland far enough to bathe the redwoods during most summer days. Fog reduces temperature and moisture stresses around the needles. Also, the tree canopies comb the fog as it passes through them, forming larger droplets that drip to the ground and add the equivalent of 20 cm (8 in) of rain each year.

Redwood trees are uniquely tolerant of fires, which occur naturally during California's dry fall days when lightning strikes the ground. Redwood bark is thick and usually provides an effective insulation for the vascular cambium deep beneath. If the fire is hot enough to kill the tree's crown, however, the very base of the tree and all the underground roots remain alive (because soil is an even better insulator than bark). At the junction of root and stem lies a band of dormant buds. If the tree trunk is killed, the normal hormone balance is modified, and the buds break dormancy. Many fast-growing shoots emerge through the bark and above the soil. Over time, some of these will mature, producing a circle of trees (Fig. 1), all genetically identical to the parent that used to occupy the space in their center. Few other conifers can vegetatively reproduce in this manner.

Fires often lead to floods during the following wet season because ground cover has been removed, leaving nothing to slow the runoff of rainwater. Redwoods are also tolerant of floods, whereas most other species are killed by the layer of silt left behind by floodwaters, which suffocates the roots. Redwood has the capacity to generate adventitious roots from the buried trunk. New feeder roots then take on the function of the dying deeper roots. Redwood seedlings also survive best on bare mineral soil—either fresh silt or soil burned of its litter. Litter harbors damping-off fungi that otherwise infect and kill virtually every seedling.

Figure 1 A circle of coast redwood (*Sequoia sempervirens*) trees. This is a clone. Each member is genetically identical because all arose as stump sprouts when the parent plant died. The parent has since decomposed, but the offspring are all connected underground to the same root system.

In short, the redwood maintains its hold *because* of episodic catastrophes, not in spite of them. Other species that compete with redwood for dominance are killed by fire and flood. Thus, park managers who try to protect the redwood forest by suppressing fires may instead be enabling the transition over time from a redwood forest to a fir–hardwood forest. Park managers in many other parts of the world are learning a similar lesson: namely, that disturbance is a natural part of the environment and that we cannot preserve some ecosystems without allowing that disturbance to continue.

We do not have many areas of mature redwood forest left to manage. Because of redwood's timber value, it has been intensively harvested for the past 150 years. Only 3% of the original acreage remains, almost all of it in state and federal parks. Although many young redwoods remain in the cutover areas, species composition and forest architecture there are different. It takes more than 200 years for a redwood tree to attain mature size, and only mature trees are capable of supporting certain animals. Marbled murrelets, spotted owls, flying squirrels, and tree voles require large trees for nesting or roosting habitat. Certain insects, fungi, and wood-decaying microbes require a thick layer of litter on the forest floor, which is achieved only in old-growth forests. Shade-loving herbs, mycorrhizal fungi, and small vertebrates do not grow and reproduce successfully until the overstory is deep and dense enough to create the required low-light, humid environment.

A wise management scheme for this small remnant of a once-great forest must be a high priority if future generations of humans are to enjoy it. But how exactly does one manage enormous, long-lived organisms that maintain their dominance only with the help of episodic floods and fires that may recur only once a century? Redwood park management needs a very long-term plan, extending 100 or more years into the future. The human species has a difficult time planning even a few years ahead. Can we create the will, the public agencies, and the policy to achieve century-long plans?

Figure 24.8 Branch of *Taxus* (English yew) with seed enclosed in a fleshy aril. Red arils are attractive to birds, which may be important in dispersing the seeds.

Figure 24.9 *Phyllocladus*, a New Zealand mountain shrub in the podocarp family. Most podocarps are shrubs and trees in the southern hemisphere. They lack woody cones and instead have fleshy cones called "berries," like those of junipers.

lumber qualities, not the least of which is its resistance to decay, conferred by natural by-products that accumulate in the wood. Sierra redwood is the most massive tree in the world; trunks at breast height reach nearly 10 m (33 ft) in diameter. The trunk alone of the largest living tree, the General Sherman tree in Sequoia-Kings Canyon National Park, weighs 625 metric tons (over 680 tons).

Redwoods are currently very restricted in their range, but they were once more numerous and widely distributed. Early in the Cenozoic era, 20 to 60 million years ago, they were part of a very rich forest that covered the cool-temperate zone of the northern hemisphere. This forest had a unique mix of gymnosperm and angiosperm trees—a mix not found anywhere today. Fossil deposits of this forest have been recovered from Asia, North America, Greenland, and Europe. Climatic change and continental drift fragmented the forest and forced redwoods into continually shrinking habitats. Dawn redwood, in fact, is so rare and narrowly distributed that it was known only from the fossil record until living trees were discovered in China's Szechwan Province in the 1940s. Seeds have since been distributed all over the world, and dawn redwoods now grow in many botanic gardens.

Taxaceae Includes Yews, but Plum Yews Belong to Cephalotaxaceae

Members of the Taxaceae are shrubs or trees with dark-colored, broadly linear, sharp-pointed leaves. Individuals are either male or female. The seeds are borne singly and are often covered with an aril (Fig. 24.8). One of the family's more than 20 species is the English yew (*Taxus baccata*), famous for excellent bows made of its wood and its connection to English medieval history and folklore. The quality of the wood results from extra spiral thickenings of secondary wall material in xylem cells. More recently, the Pacific yew of the western United States (*T. brevifolia*) was found to contain an important anti-

cancer compound, taxol. Taxol is present in such small amounts in the bark of Pacific yew (a 100-year-old tree contains only 300 mg), that environmentalists were concerned that deforestation would be justified in the name of medicine. Chemists learned how to synthesize the taxol molecule in the laboratory in 1994, however, making it unnecessary to fell old-growth trees.

The Cephalotaxaceae, a Chinese family, consists of fewer than 10 species of shrubs and trees in the genus *Cephalotaxus*. Leaves are yewlike. Seeds are borne in pairs and are covered with an aril; they are 2 to 3 cm (1 in) long and resemble oval plums. Individual plants are male or female.

Podocarpaceae and Araucariaceae Are Largely Southern Hemisphere Families

Podocarps are shrubs or trees (Fig. 24.9). Depending on the species, leaves may vary from short needles to long and broadly oblong blades. Individuals may be unisexual or bisexual. The seeds are borne singly and enclosed in an aril, similar to juniper berries. As is also true for juniper berries, the flesh of podocarp seeds is attractive to birds, who digest the flesh but defecate the seed. Many excellent lumber trees of Australasia, Africa, and South America are in this family. *Podocarpus dacrydioides* and *P. totara* of New Zealand, for example, reach 60 m (200 ft) in height. It's a large family, with about 140 species. Most species are restricted to the southern hemisphere, but some occur in Japan, Central America, and the West Indies. A number of species are widely planted as ornamentals.

Plants in the family Araucariaceae are relatively large trees in the genera *Araucaria* (monkey puzzle, southern pine) and *Agathis* (kauri). The tallest trees in the tropics, reaching to 89 m (290 ft), are species of *Araucaria*, and *Agathis australis* of New Zealand has trunks rivaling those of redwood in girth. Leaves vary from needlelike to flat and broad. Female cones tend to be large and globose, and—like fir cones—they disintegrate in place when ripe.

Figure 24.10 *Araucaria*, southern pine. This genus includes trees that are among the tallest and most massive in the world. This particular species is the extremely symmetric Norfolk Island pine, *Araucaria heterophylla*.

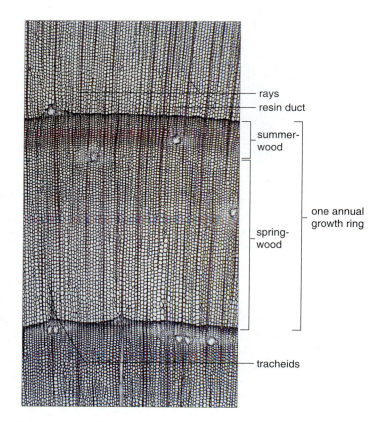

Figure 24.11 Three growth rings of pine, as shown in cross section. Note the narrow, single-cell-wide rays and the absence of vessels. Late wood (summerwood) appears denser than early wood (springwood) because the tracheids are smaller. Several resin ducts are visible here, but some conifers lack resin ducts.

The more than 30 species are exclusively native to the southern hemisphere, but some (such as the strikingly symmetrical Norfolk Island pine, *Araucaria heterophylla*), are widely planted as ornamentals throughout the world (Fig. 24.10).

We now leave this taxonomic section, which has emphasized differences, and move inside the plants to explore common themes of anatomy and reproduction.

24.4 THE VASCULAR SYSTEM OF CONIFERS

Conifers are anatomically and morphologically more complex and longer-lived than any group we have discussed so far in our survey of the kingdom Plantae. Conifer trunks have secondary xylem and phloem, rays, wood with annual rings, and bark. (A three-dimensional reconstruction of a piece of coast redwood is shown in Fig. 6.18.) The architecture of conifer wood is extremely regular. Part of the reason for this regularity is the absence of vessels. Tracheids produced in the spring are largest in diameter; as the season progresses, new tracheids are smaller in diameter and have thicker walls. Some tracheids are so thick that they resemble fibers and are called **fiber-tracheids**. **Xylem rays** are only one

cell wide and are composed of brick-shaped parenchyma cells. Other parenchyma cells may run in vertical columns; these are called **axial parenchyma** (refer back to Chapter 6).

Many gymnosperms produce resin, a mix of complex organic compounds that are the by-products of their metabolism. Typically, resin accumulates and flows in long **resin ducts**, axial chambers bounded by parenchyma cells (Fig. 24.11; see also Fig. 6.19). Resin inhibits wood-boring insects. If insects penetrate the trunk and break through any resin duct, they are encased in a flow of resin and immobilized. Trees that are stressed by drought or disease produce less resin, making them susceptible to insect infestation.

The phloem of coast redwood (*Sequoia sempervirens*) and other gymnosperms contains **sieve cells** (sieve-tube members without companion cells), fibers, ray parenchyma, and axial parenchyma.

Pine needles (leaves) show many anatomical adaptations to aridity. They have a thick cuticle, sunken stomata, a fibrous epidermis, closely packed mesophyll without intercellular air spaces, and veins (vascular bundles) only in the center of the leaf (Fig. 24.12). Leaves are

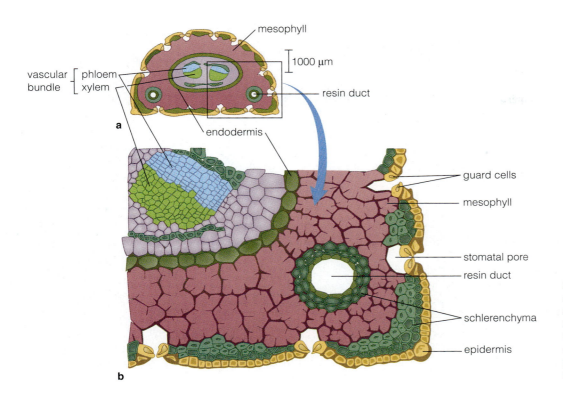

mesophyll

1000 µm

vascular bundle { phloem xylem

resin duct

endodermis

a

guard cells

mesophyll

stomatal pore

resin duct

schlerenchyma

epidermis

b

Figure 24.12 A pine needle in cross section. (**a**) The entire cross section, with major tissues outlined. (**b**) Cellular detail of a wedge, from the epidermis to one of the two vascular bundles. Some pines have only one vascular bundle.

thick, rather than thin, yielding a low surface-to-volume ratio. These traits reduce the loss of water vapor, which is very important in winter (when soil water is frozen and evergreen leaves are subjected to drying winds of low relative humidity). Needles have a life span that ranges from 3 to 30 years, depending on the species. Although the leaves do not continue growing during that life span, their vascular bundles do undergo secondary growth, producing a limited amount of secondary phloem and secondary xylem.

24.5 THE LIFE CYCLE OF *PINUS*, A REPRESENTATIVE CONIFER

We'll use the pine (*Pinus*) life cycle (Fig. 24.13) to represent the typical life cycle of all members of the Pinophyta. Remember, however, that not all species in this division produce cones, and they can differ from pine in many other, more subtle ways. Pine is a convenient plant to use as a model, though, for several reasons: Each tree has both male and female gametophytes, seeds do not have fleshy coverings, and a great deal is known about the life cycle. Pine sporophytes are trees, with one exception: *P. mugo* of Europe is a shrub.

Two Kinds of Cones Are Produced

Pines—like all seed plants—are heterosporous. Male gametophytes are produced in small cones that, for conve-

nience's sake, are called **male cones**. Strictly speaking, this term is not correct because the cone is sporophyte tissue and not a male gametophyte; however, meiosis within it produces microspores that develop into male gametophytes. Male cones are also called **staminate cones** and **pollen cones**. Female gametophytes are produced in larger cones that are commonly called **female cones**. Again, the female cone is sporophyte tissue and not a female gametophyte, but meiosis within it produces megaspores that develop into female gametophytes. Synonyms for female cones include **ovulate cones** and **seed cones**. Male and female cones differ in size, architecture, longevity, and location on the tree.

Male cones average 1 cm (less than 0.5 in) in length and 5 mm (0.2 in) in diameter. They are borne in groups (Fig. 24.14), usually on the lower branches of trees. Each cone is composed of a large number of small sporophylls called *scales*, attached spirally to the cone axis. Two microsporangia develop on the undersurface of each scale. Microspore development proceeds very similarly to that in *Selaginella*, a seedless heterosporous plant described in the previous chapter (see Fig. 23.15). The microsporangium is lined with a layer of nutritive cells called the **tapetum**. Inside are microsporocytes that undergo meiosis and produce haploid microspores (Fig. 24.13f). The nucleus within each microspore divides several times by mitosis within the spore wall. The resulting **pollen grain** contains two viable nuclei and vestiges of several squashed vegetative cells. A pollen grain is an immature male gametophyte.

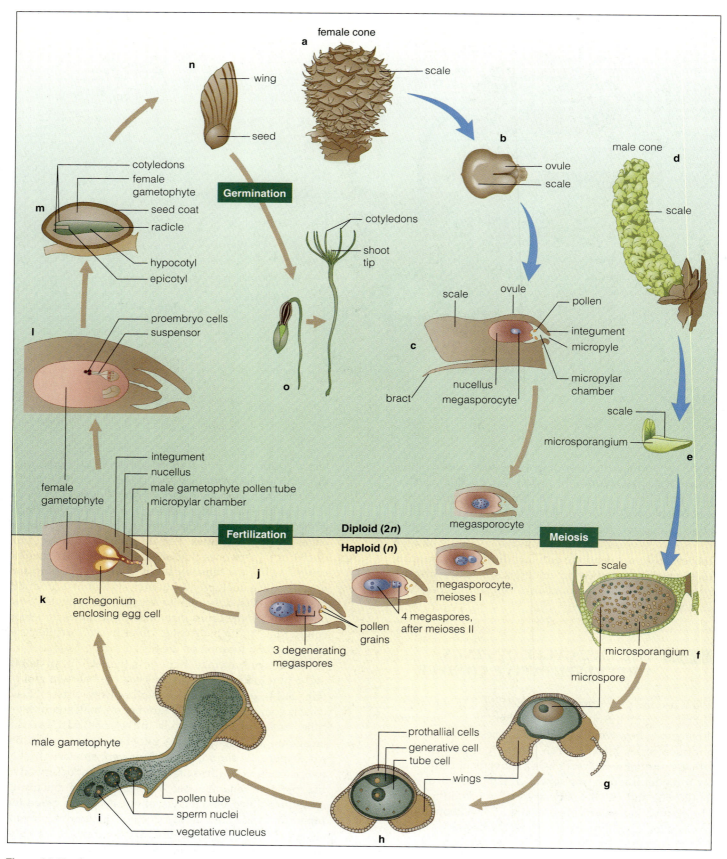

Figure 24.13 Stages in the life cycle of a pine. (**a**) Young female cone. (**b**) Scale from female cone, with two ovules on its upper surface. (**c**) Section of female cone at the time of pollination. (**d**) Male cone. (**e**) One scale from a male cone. (**f**) Section of microsporangium with mature pollen grains inside. (**g–i**) Developing male gametophyte. (**j**) Meiosis in the nucellus of an ovule. (**k**) Mature female gametophyte at the time of fertilization; one egg is being fertilized by a sperm from the ruptured pollen tube. (**l**) and (**m**) Developing embryo and seed. (**n**) Mature seed with wing. (**o**) Stages of germination. (**c**, **g**, and **i** are redrawn from J. J. Coulter and C. J. Chamberlain, *Morphology of the Gymnosperms*. Chicago: University of Chicago Press, 1917.

pollen

male cones

leaf needle

a

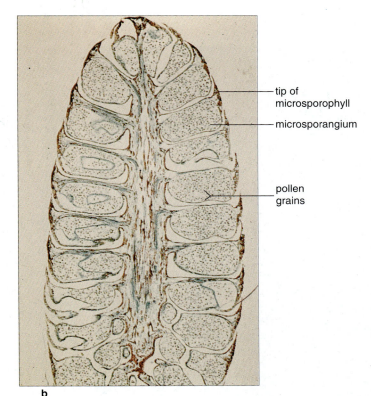

tip of microsporophyll

microsporangium

pollen grains

b

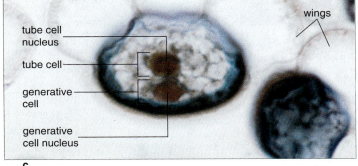

wings

tube cell nucleus

tube cell

generative cell

generative cell nucleus

c

Figure 24.14 Male pine cones. (**a**) Cluster of cones near the end of a branch at the time of pollination. (**b**) Longitudinal section through part of one cone, showing microsporangia attached to the underside of each scale; each microsporangium is filled with pollen grains. (**c**) Pollen grains with wings.

Enormous numbers of pollen grains are eventually shed from the microsporangia of a single tree. The pollen grains are yellow and light in weight (in order to be dispersed by wind). Windrows of pollen are visible on the ground during the period of release, which is generally in spring. Pine pollen grains have two inflated wings that are outgrowths of the wall (see Fig. 24.13g), and these make the pollen grain more buoyant. The pollen grains of many other gymnosperm species lack such wings.

The mature female cone is the cone we commonly associate with pines and other conifers (Fig. 24.15). It is composed of many wooden scales attached to an axis in a spiral arrangement. Fused to the underside of each scale is a bract; this may be relatively small and obscure in some species but is quite large and even longer than the scale in a few (such as Douglas fir, *Pseudotsuga menziesii* and Shasta red fir, *Abies magnifica* var. *shastensis*).

When the female cone is young, it is red and is softer and much smaller than the male cone. Female cones develop singly in early spring at the tips of young branches in the upper part of the tree. Two megasporangia develop on the upper surface of each scale. Remember, a megasporangium and its outer integument layer form the ovule of pine. The ovules first appear as small protuberances,

close to the axis of the cone. The integument, which forms the outermost protective layer of the ovule, is pierced by an open pore facing the axis of the cone. This pore is the **micropyle,** through which pollen grains later enter (Fig. 24.13c). In pine, only one of the many cells filling the young ovule is a megasporocyte. The others form a nutritive tissue called the **nucellus.** When the megasporocyte divides by meiosis, four megaspores arranged in single file are the result. Generally, only one of the megaspores develops into a female gametophyte; the other three degenerate.

The megaspore grows very slowly into a female gametophyte. Most conifers require several months, and pine takes just over a year. The development of the gametophyte takes place entirely within the ovule. Two or more archegonia differentiate at the micropylar end of the growing gametophyte, using energy from digestion of the nearby nucellus. Thus, the gametophyte is completely dependent (parasitic) on sporophyte tissue.

At maturity, the ovule consists of an outer integument, a thin layer of remaining nucellus, and an ovoid fe-

two-month-old female cones

leaves (needles)

fourteen-month-old female cones

scale

ovule

bract

b

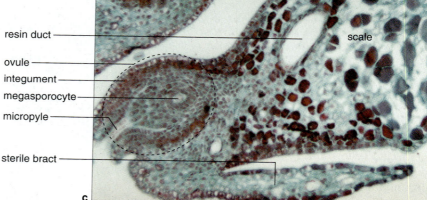

resin duct

ovule

integument

megasporocyte

micropyle

sterile bract

scale

c

Figure 24.15 Young female cones at the time of pollination. (**a**) Two-month-old cones are at the tip of a young branch. Below are larger cones just over a year old. (**b**) Longitudinal section through one two-month-old cone, showing ovules on top of each scale and bracts attached to the bottom of each scale. (**c**) Closer view of one ovule (dashed area). Note the large megasporocyte, inside the nucellus (megasporangium), and the micropyle and micropylar chamber. Meiosis has not yet occurred.

male gametophyte that is undifferentiated except for several archegonia at one end, each with an enclosed egg (Fig. 24.13k). Notice that there is space between the micropyle and the nucellus; this is the **micropylar chamber**.

Pollination Replaces the Need for Free Water

In flowers, pollination is the transfer of pollen from anther to stigma. In conifers, it is the transfer of pollen from male cone to female cone. Wind is the vector that carries conifer pollen from cone to cone. In pine, the placement of female cones above male cones makes it more likely that pollen will be carried from one tree to another, rather than settling on a female cone in the same tree. Thus, cross-pollination is usual. For those conifer families that have dioecious (unisexual) individuals—the Cephalotaxaceae, Cupressaceae, Podocarpaceae, and Taxaceae—self-pollination is impossible; cross-pollination is always the case. Cross-pollination is valuable because it creates more genetic variation in the next generation, increasing the capacity of a species to take advantage of diversity in the environment.

Recent studies of the aerodynamics of pollen near cones have demonstrated that the shape of the young female cone can create unique air currents and eddies that bring pollen grains of the appropriate species close to the open scales, increasing the probability that pollen will land in the cone. Thus, wind pollination is not entirely random.

Pollination occurs when the female cone is about nine months old. Female cone buds are initiated in the late summer preceding the spring of pollination. The cone is preformed in the bud and merely enlarges when the bud breaks in spring. Cone scales at this stage turn slightly away from the cone axis, providing space for pollen grains to drift down to the ovules. A sticky **pollination drop** exudes from the micropyle. The drop is similar in chemistry to the nectar of flowers, containing about 8% sugar (glucose and fructose) and traces of amino acids. It does not seem to attract animals, but it passively traps pollen grains that touch it. A chemical signal diffuses from the trapped pollen to the ovule, triggering rapid absorption of the liquid. This draws pollen grains through the micryopyle and into the micropylar chamber, where they come to lie on the surface of the nucellus (Fig. 24.16).

The pollen grain germinates, slowly developing an elongating pollen tube that grows through the nucellus toward an egg (see Fig. 24.13i). Several nuclear divisions occur in the tube, but no cell walls are formed. Two of the last-formed nuclei are sperm nuclei. This pollen tube, containing two sperm nuclei and several vegetative nuclei, is the mature male gametophyte.

Fertilization Leads to Seed Formation

At the time of pollination, the megasporocyte is undergoing meiosis. As the pollen tube grows, the female gametophyte forms. The small red cone grows larger and turns green, and the scales become tightly closed. Timing of male and female gamete formation is so coordinated that the egg is ready for fertilization by the time the pollen tube with sperm nuclei has reached the archegonium. In pine, this development takes about 12 months.

The sperm nuclei, together with the cytoplasmic contents of the pollen tube, are discharged directly around the egg cell. Sperm nuclei do not possess flagella. Hence, they are not actively motile, but somehow one sperm nucleus comes in contact with the egg and unites with it. This is not the end of its journey, for it must pass through the egg cytoplasm to reach the egg nucleus. Pine egg cells have hundreds of times the volume of a sperm nucleus. How sperm DNA finds its way through that huge space to egg DNA is unknown.

The fertilized egg becomes a diploid zygote, and it begins to divide immediately into a relatively elaborate **proembryo**, the apical cells of which develop into an embryo. The mature embryo consists of several cotyledons, an epicotyl, a hypocotyl, and a radicle (Fig. 24.13m). There is no foot.

Sometimes more than one embryo will form. The multiple embryos can originate in two ways. Each female gametophyte has more than one archegonium, and all can be fertilized if there are enough pollen grains. Furthermore, each proembryo is capable of forming as many as four embryos. In the first case, fraternal siblings would be formed; in the second, identical siblings. Generally, however, only a single embryo survives to maturity.

While the embryo develops, the female gametophyte also continues to grow by digesting the remaining nucel-

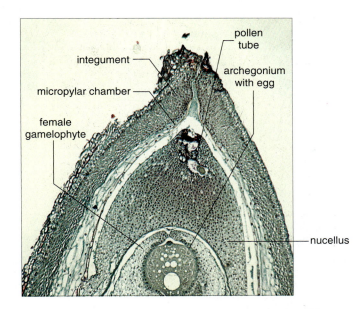

Figure 24.16 Older female cone at the time of fertilization. This section through half of an ovule shows the closed micropyle, a pollen tube growing through the nucellus, and a differentiated archegonium with a large egg cell. Only the micropylar end of the ovule is visible.

lus, and its cells become packed with food. The integument hardens and becomes a **seed coat**. The micropyle closes. The female cone enlarges still more, becomes woody, and loses its green color.

When the seeds are mature, the cone scales open and the seeds fall out. Pine seed coats usually have an outgrowth called a **wing**, which catches the wind and aids in the dispersal of seeds from the parent tree (Fig. 24.13n). Seeds are usually shed immediately upon ripening, in late summer or early fall. The mature, open female cone is two years old when seeds are dispersed (Table 24.1).

Seeds are dispersal packages containing a dormant embryo, stored food, and a protective outer coat. Their richness in carbohydrates has made pine seeds an important food source for human and animal populations. Native Americans ground the seeds into a meal and used the flour, much as other cultures have used grains. Many birds and mammals eat and cache pine seeds.

Seeds that escape predation by herbivores usually lie dormant until the next spring. The dormant embryos have a very low metabolic rate, and their water content is low. They are able to tolerate aridity, anaerobic conditions, and temperature extremes while in this state. Their dormancy is broken by exposure to cold, wet winter conditions. Dormancy can be artificially broken by storing seeds in wet cheesecloth at temperatures a few degrees above freezing for four to six weeks, a process called **stratification**. Dormancy prevents pines from germinating on warm fall days, as that would force seedlings to endure inhospitable winter growing conditions.

When the seed does germinate, the radicle emerges first. Then the hypocotyl elongates, taking the cotyledons and epicotyl above the surface while still enclosed in the

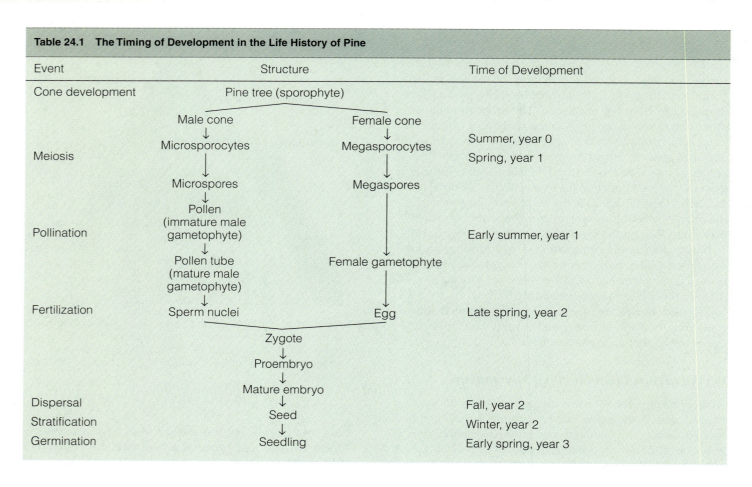

Table 24.1   The Timing of Development in the Life History of Pine

Event	Structure	Time of Development
Cone development	Pine tree (sporophyte)	
	Male cone → Microsporocytes Female cone → Megasporocytes	Summer, year 0
Meiosis		Spring, year 1
	Microspores Megaspores	
	Pollen (immature male gametophyte)	
Pollination		Early summer, year 1
	Pollen tube (mature male gametophyte) Female gametophyte	
Fertilization	Sperm nuclei Egg	Late spring, year 2
	Zygote	
	Proembryo	
	Mature embryo	
Dispersal	Seed	Fall, year 2
Stratification		Winter, year 2
Germination	Seedling	Early spring, year 3

seed coat. The cotyledons absorb nutrients stored in the female gametophyte and then enlarge, pulling themselves out of the seed coat (Fig. 24.13o). When light strikes the cotyledons, they turn green and become photosynthetic. The young pine grows slowly in the first year, and most growth occurs below ground, in the root system. The cotyledons may remain attached into the second year; until then, the true leaves are few in number and are relatively small. In such cases, the cotyledons continue to grow and elongate until they are shed.

24.6   OTHER GYMNOSPERM DIVISIONS

Conifers (Pinophyta) account for almost 80% of all living species of gymnosperms. The remaining divisions are shadows of their former selves, much as are seedless vascular plants. They once were more numerous and widely distributed; now a relatively few species exist in widely separated places. Some of these relicts of the past are so unusual that it is difficult to classify them. Members of the Gnetophyta, for example, are grouped together for convenience, but they probably are not closely related.

Cycadophyta Plants Have Palmlike Traits

Some 200 million years ago, during the early Mesozoic era, members of the **Cycadophyta** made up a large part

of the earth's vegetation and were probably food for at least some herbivorous dinosaurs. Today, this once-diverse group contains only 10 genera and about 100 species, which grow in widely separated areas (largely in the tropics). Only one genus, *Zamia*, occurs naturally in the United States, and it is restricted to southern Florida. The genera *Cycas*, *Dioon*, and *Zamia* are widely propagated as ornamental plants, either outdoors in warmer regions or in greenhouses elsewhere (Fig. 24.17).

These trees are palmlike in appearance, and some of their common names reflect this nature (*Cycas revoluta* is sago palm, for example). *Zamia*'s trunk is largely subterranean, but the trunk of *Cycas* is above ground, though slow-growing (a specimen 2 m (6 ft) tall might be as old as 1000 years). Cycads possess large cones, and in this they resemble the Pinophyta. Pollination in some cycads differs from that in conifers, however; beetles seem to play a role in moving cycad pollen from male cones to female cones (the plants are dioecious). Another difference from conifers is the presence of multiflagellated sperm and enormous egg cells. These eggs are the largest cells in the plant kingdom, reaching 300 μm in diameter and visible to the naked eye.

Ginkgophyta Has Only One Living Species

Only one living representative remains of the very ancient **Ginkgophyta** division: the maidenhair tree (*Ginkgo biloba*, Fig. 24.18). It grows wild today only in warm-

Figure 24.17 Cycads. (**a**) *Cycas revoluta* (sago palm). (**b**) *Zamia* with female cones.

seed cones

Figure 24.18 Branch and leaves of *Ginkgo biloba* (maidenhair tree). The leaves have turned yellow and are about to fall off as winter begins.

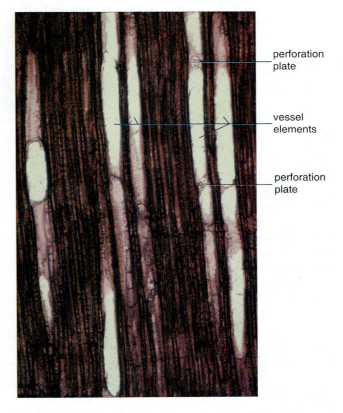

Figure 24.19 Wood of *Ephedra*, a genus in the Gnetophyta, showing xylem vessels. Most gymnosperms lack vessels.

perforation plate

vessel elements

perforation plate

temperate forests of China, but it has been grown for centuries on Chinese and Japanese temple grounds as a traditional decorative tree. It is also widely planted throughout the world as an urban street tree, being very tolerant of pollutants. Chinese herbalists prescribe infusions of the leaves as a medicine for many ailments, and the tree is a symbol of longevity.

Ginkgos are large, diffusely branching trees with characteristic small, fan-shaped leaves divided into two lobes. Some trunks are more than 3 m (10 ft) in diameter. In fall, the leaves turn a brilliant golden color and drop. Like the cycads, *Ginkgo* individuals are unisexual. Male trees are preferred as ornamentals because the seeds on female trees are covered with a foul-smelling, fleshy aril.

Gnetophyta Plants Are Intermediate Between Gymnosperms and Angiosperms

The **Gnetophyta** division consists of only three genera, *Ephedra*, *Gnetum*, and *Welwitschia*. In some respects their

traits place them as intermediate between gymnosperms and angiosperms. For example, they contain vessels in the xylem (Fig. 24.19), their ovules are surrounded by two integuments (not one), and their pollen-producing structures resemble stamens. All these traits are shared by angiosperms and are absent in other gymnosperms. Yet, their seeds are still borne naked, and therefore mem-

Figure 24.20 Examples of plants in the Gnetophyta. (**a**) Branches of the shrub *Ephedra viridis*, about 1 m (3 ft) tall, growing in the White Mountains of California. (**b**) Detail of a stem node and the scalelike leaves; most photosynthesis is conducted in the green stem. (**c**) Seed nearly surrounded by a red aril. (**d**) *Gnetum leyboldii*, a tropical vine.

gnetum

bers of this division are not angiosperms. All have dioecious (unisexual) individuals. Sperm are nonmotile. The three genera look quite different from each other and exhibit important life cycle differences, making their inclusion in a single division artificial.

Only *Ephedra* (also called *joint fir* or *Mormon tea*) is found in North America. Its 40 species are distributed through the warm-temperate Mediterranean rim, India, China, the southwestern deserts of the United States, and mountainous parts of South America. It is a vine or shrub with whorls of small, deciduous leaves at prominent joints (Fig. 24.20a). The leaves are little more than scales, so most photosynthesis is conducted by the green stems. Joint fir is the source of the drug ephedrine, an alkaloid that constricts swollen blood vessels and is also a mild stimulant. An overdose can cause death.

Gnetum is a tropical genus of 30 species, most of which are lianas (climbing vines). The leaves look very much like those of typical dicotyledonous plants (Fig. 24.20d).

Welwitschia is found along the arid southwest coast of Africa. It has two long, leathery, straplike leaves that trail along the soil surface (Fig. 24.21a). When Austrian naturalist Friedrich Welwitsch first saw this plant on a field trip in 1859 near the Angolan coast, he fell to his knees in disbelief. Specimens taken to Europe caused Darwin to describe it as the platypus of the plant kingdom. Local Africans called it *otjitumbo* ("stump"), but the English taxonomist Joseph Hooker named it in honor of its European discoverer plus the Latin word for "wonderful": *Welwitschia mirabilis*.

The plants are slow-growing, most photosynthate going into an exceptionally well-developed tap root. The aboveground portion consists of two leaves that become frayed at the ends. The leaf tissue has sclereids, a considerable concentration of lignin, and numerous crystals of calcium oxalate—all combining to make the leaves very durable and inedible. Stomata are sunken, probably an adaptation to the arid habitat. (Annual rainfall is only 2 cm—less than 1 in—although condensation of fog can add the equivalent of another 5 cm—2 in.) Individuals can attain ages of 1000 to 2000 years.

The basic units of the male cone bear a striking resemblance to flowers (Fig. 24.21b, c). Each unit (Fig. 24.21d) consists of several scales or bracts, six antherlike structures, and a central, sterile structure resembling a pistil. The "pistil" is actually an outgrowth of the integument of an ovule. This outgrowth is drawn out into a stylelike tube, with a stigmalike disc at the apex. A drop of fluid, similar to the pollination drop of pine, may be secreted onto the surface of that disc. The ovule is sterile, however, so the cone is functionally male.

Female cones (Fig. 24.21e, f) are borne on other individuals. One naked ovule is present on each cone scale. Pollen is carried by wind from the anthers of the male cone to the ovules of the female cone. When the seed is mature, its integuments are expanded into four wings (Fig. 24.21f).

This completes our examination of life cycles. We end the chapter with a review of gymnosperm importance to various ecosystems.

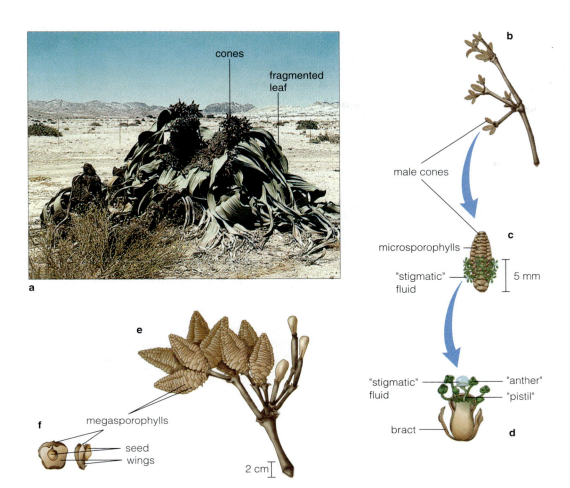

Figure 24.21 *Welwitschia mirabilis*, a plant of the Namib desert of southwestern Africa. (**a**) The two leaves of this plant have become fragmented into many ribbonlike strips. Cones are visible at the juncture of the stem crown and the leaves. A plant of this size—about 1 m (3 ft) tall and 1 m across—is 1000 years old. (**b**) A cluster of male cones. (**c**) A male cone at the time of pollen shedding, when antherlike, pollen-producing structures are fully extended. (**d**) A single male "flower" with several antherlike structures and a nonfunctional pistil-like structure with a stigmatic disc. The pistil-like structure is an outgrowth of the integument of a sterile ovule, and it secretes a drop of stigmatic fluid. (**e**) A cluster of female cones. (**f**) Each bract encloses one ovule; when the ovule has matured, outgrowths of the integument form wings. (Redrawn after Alice Addicot with the assistance of C. H. Bornman.)

24.7 THE ECOLOGICAL AND ECONOMIC IMPORTANCE OF GYMNOSPERMS

Gymnosperms have great ecological importance. Though their habitats range from tropical forests to deserts, their centers of dominance are the cool-temperate zones of the northern and southern hemispheres. In North America, they are part of low-elevation forests across Alaska, Canada, and New England and down the Pacific coast; in addition, they are part of high-elevation forests in the Appalachian, Rocky, and Cascade-Sierran mountain chains. They also occur in more localized habitats within the southeastern coastal plain. The gymnosperm landscape covers more than one-third of the North American land-mass. Within this landscape, the ecosystems are *created* largely by conifers. Conifer foliage is rich in organic acid, so its decomposition in turn makes the soil acidic and relatively low in nutrients. Only those shrub and herb species that can tolerate such soil conditions and the low level of light beneath the dense conifer overstory are able to grow here. The acidity also hinders bacteria but favors fungi, so the decomposer microflora is strongly affected. Conifer foliage and wood are high in secondary compounds that inhibit grazing animals, therefore mammal and insect diversity is low, as is that of insectivorous birds.

Gymnosperms also have great economic importance. They are a major source of lumber, paper pulp, turpentine, and resins, and they are used as fuel for heat. More than one historian has pointed out that humans have always lived in a Wood Age, even if we talk about the Stone Age, Bronze Age, and Iron Age. The fact is that wood—living or tranformed into coal—has provided energy for smelting and heating and cooking, as well as raw building material for habitations and vehicles throughout human history. Minoan, Greek, and Roman civilizations rose and fell depending on their access to forests, often forests made up of gynmosperm trees. In North America, economic growth has depended in particular on lumber of ponderosa pine (*Pinus ponderosa*), eastern white pine (*P. strobus*), and Douglas fir (*Pseudotsuga menziesii*). A significant amount of raw or sawn logs is exported to countries that no longer have this natural resource.

The magnitude of forest exploitation on the North American continent rivals modern deforestation in the tropics. Enormous lowland forests of pine from the Great Lakes states and the Southeast, as well as equally extensive upland conifer forests in the West, have been clear-cut in the past 100 years. An area greater than all of California's 40 million hectares (100 million acres) has been clear-cut in this relatively short time. Clear-cuts in some Canadian provinces are so large that they are one of the few human artifacts visible from orbiting spacecraft. The white pine forests of the Great Lakes area have never recovered from harvesting, and the second-growth forests of the West lack the biotic diversity and ecological stability that the pristine forests had acquired over centuries of slow development.

Gymnosperms are widely used in landscaping because of their evergreen habit and the diversity of their growth forms. They can be pruned into low ground covers, hedges, and dramatically branched trees. Many have rapid growth rates. Some—like the Norfolk Island pine (*Araucaria heterophylla*), cycads (*Cycas*), or the Monterey cypress (*Cupressus macrocarpa*)—have unique canopy architectures; many people find these trees pleasing and so have brought them to countries far outside their original ranges.

SUMMARY

1. Gymnosperms are vascular plants whose life cycles have novel features beyond those of the seedless vascular plants. The new traits include the ovule, the seed, and pollination, all of which are adaptations for survival on land. *Gymnosperm* means "naked seed."

2. Living gymnosperms number approximately 900 species and are grouped into four divisions: Cycadophyta, Ginkgophyta, Gnetophyta, and Pinophyta (sometimes called Coniferophyta, or the conifers). They have great economic and ecological importance, mainly in the temperate zones of the world.

3. The pine life cycle in general represents the Pinophyta. Pine is heterosporous; that is, an individual pine tree will have both male and female cones. Some other families have unisexual individuals.

4. Male cones are small and seasonal. They consist of whorls of scales attached to a central axis. Two microsporangia are attached to the underside of each scale. Microspores will develop into pollen grains within the microsporangium wall. The pollen grains (immature male gametophytes) are liberated from the microsporangium and carried passively on the wind to female cones, generally on another tree.

5. Female cones are the cones we commonly associate with conifers. When young, a female cone is as small as a male cone, but it ultimately becomes larger. It consists of whorls of scales, each scale attached to a bract. Two megasporangia (ovules) are attached to the upper surface of each scale. An immature ovule consists of an outer integument with a micropyle and micropylar chamber at one end, a nucellus inside the integument, and a central megasporocyte. This is the stage of development at the time of pollination.

6. The megasporocyte divides by meiosis into four megaspores, but only one develops into a female gametophyte. At maturity, the female gametophyte is a relatively small bundle of cells with several archegonia at the micropylar end. In the meantime, the pollen grain has been pulled through the micropyle by a shrinking pollination drop until it lies on the surface of the nucellus. It germinates into a pollen tube that grows slowly through the nucellus toward an archegonium and egg. Cell division in the tube produces two nonmotile sperm nuclei, one of which fertilizes an egg, forming a zygote.

7. The zygote divides to produce a proembryo, some of whose cells divide to produce an embryo with cotyledons and epicotyl, hypocotyl, and radicle regions. Surrounding the mature embryo is an enlarged female gametophyte (which grows by parasitizing the nucellus) and a seed coat (which forms from the integument). Embryos are dormant when seeds are shed. Germination is possible after the seeds are exposed to cold, wet conditions for several weeks—a process called *stratification*. Thus, a seed is a dormant embryo with stored food and a protective outer coat; its function is to disperse the embryo away from the parent plant over time and space.

8. The division Cycadophyta contains 100 species of woody tropical plants (the cycads) that combine palmlike and coniferlike traits. Cones are very large. Pollination sometimes involves beetles as vectors. Sperm are multiflagellate.

9. The division Ginkgophyta contains a single species, *Ginkgo biloba*. The species is native to China, but it is

widely used as a street tree and ornamental. Its fan-shaped leaves are deciduous.

10. The division Gnetophyta is artificial because it contains three quite dissimilar genera. All have traits that combine those of gymnosperms and angiosperms. Only the genus *Ephedra* is found in North America. *Gnetum* is a tropical genus, and most species are lianas. *Welwitschia*, native to the arid coast of southwest Africa, has been called the platypus of the plant world because of its bizarre collection of traits. Stamen-like and stigma-like structures are present in its cones, yet it bears naked seeds.

Questions

1. How do the novel life cycle features of ovule, pollen, and seed adapt conifers to life on land?

2. In what ways does conifer wood play an important role in technological economies? Is it going too far to say that the human species has always lived in—and still does live in—a Wood Age?

3. How does a pine seed differ from a *Selaginella* megagametophyte with an embryo?

4. Why is a pollen grain an *immature* male gametophyte? When does it become completely developed?

5. Can you see any pattern of gametophyte-sporophyte relationships in the sequence moss-to-fern-to-conifer?

6. In what ways are conifers different from cycads, ginkgo, and *Welwitschia*? Are these differences big enough to warrant placing them in different divisions?

Further Readings

Bold, H. C., and J. W. LaClaire II. 1987. *The Plant Kingdom*. 5th ed. Englewood Cliffs, N.J.: Prentice-Hall. A refreshingly short and reader-friendly textbook. An evolutionary survey of organisms from blue-green algae and fungi to flowering plants.

Bornman, C. H. 1978. *Welwitschia*. Cape Town, South Africa: Struick. A small book (fewer than 80 pages) that covers the biology, ecology, systematics, and history of a plant Hooker called "the ugliest in the world" and Darwin called "the platypus of the plant kingdom." Written in both German and English. Many color photographs are included.

Burns, R. M., and B. H. Honkala, eds. 1990. Conifers. Vol. 1 of *Silvics of North America*. U.S. Department of Agriculture. Agriculture Handbook 654. Descriptions of 63 species of major conifers of North America. Each description includes aspects of biology, distribution, ecology, economic uses, and growth requirements, with many references to published literature.

Meyen, S. V. 1984. "Basic Features of Gymnosperm Systematics and Phylogeny as Evidenced by the Fossil Record." *Botanical Review* 50: 1–111. Somewhat technical, but recommended because of its thoroughness. Progymnosperms originated in late Devonian and were ubiquitous. Cycads and pines evolved in the Carboniferous, in the tropics, which was the location for most gymnosperm diversification. Ironically, gymnosperms are now most common in the cool-temperate zone.

Niklas, K. J., and S. J. Buchmann. 1987. "The Aerodynamics of Pollen Capture in Two Sympatric Ephedra Species." *Evolution* 41: 104–23. Two species of joint fir have different sizes and densities of pollen grains, and the local eddy patterns set up by each species' leaves and ovules separate the two types of pollen. Wind pollination is not so random after all. Professor Karl Niklas has earlier publications that describe eddy patterns set up by gymnosperm cones ("The Aerodynamics of Wind Pollination." *Botanical Review* 51 (1985): 328–86).

Van Gelderen, D. M., and J. R. P. Van Hoey Smith. 1996. *Conifers: The Illustrated Encyclopedia*. Corvallis, Ore.: Timber Press. A colorful survey of 65 genera of the division Pinophyta, with notes on distribution, classification, biology, and propagation information. Two volumes, and rather expensive; try your local library.

25 PLANTAE: ANGIOSPERMS

1. Angiosperms (a common name for the division Magnoliophyta) have ovules borne inside ovaries, rather than naked on scales. After the ovule is fertilized it becomes a seed, and the ovary becomes a fruit. The new features—ovary and fruit—increase the protection of the gametophyte generation and widen the dispersal of the seed over time and space.

2. A third new structure is the flower, which typically contains both pollen-producing stamens and ovule-producing pistils. Surrounding the stamens and pistils may be variously colored and shaped petals and sepals, which attract and direct animals that transfer pollen from flower to flower.

3. Double fertilization is a unique part of the angiosperm life cycle. When a pollen grain germinates into a pollen tube, the tube contains two sperm nuclei. One nucleus eventually unites with an egg, producing a zygote; the other unites with two other nuclei in the gametophyte, producing an endosperm cell. As an embryo develops from the zygote, the endosperm cell multiplies and surrounds the embryo with stored food. Double fertilization conserves energy because food for the embryo does not accumulate until after fertilization.

4. Angiosperms today dominate most of the world's landscapes. They evolved rapidly in the Cretaceous period, possibly in upland settings where fossils are not easily preserved. Within the kingdom Plantae, only the Magnoliophyta and Pterophyta divisions are currently evolving new species. All other divisions are mere relicts of their past diversity.

5. The Magnoliophyta are typically divided into two classes, Magnoliopsida (dicots) and Liliopsida (monocots), and the two of these are divided into a total of 10 subclasses. Currently, botanists speculate that the most primitive, earliest-to-evolve subclasses contain woody plants with simple and pinnately veined leaves, terminal and few-flowered inflorescences, large and symmetrically open flowers, and seeds with relatively small embryos and large amounts of endosperm.

6. Plant geography is the study of plant distribution throughout the world. Some subclasses, orders, and families are widely distributed, whereas others are narrowly restricted to one part of the world and one type of environment. In the most recent thousand years, human populations have purposely and accidentally carried plants into regions those plants had not yet reached on their own, creating a more homogeneous global landscape. Angiosperms are of exceptional importance to people for food, fiber, pharmaceuticals, building materials, ornaments, and beverages.

Figure 25.1 Angiosperms dominate most of the world's landscapes, including forests, shrublands, and grasslands. This photograph of coastal California was taken in late spring, when the angiosperms were flowering and most obvious.

25.1 PLANTS WITH AN ENCLOSED SEED

Flowering plants today dominate the world's vegetation (Fig. 25.1). Hardwood forests and woodlands, shrublands, grasslands, wetlands, deserts, cold high-elevation tundras, and warm low-elevation rain forests have a biomass (weight of organic matter) composed chiefly of flowering plants. Nearly all of our crop plants, orchard trees, garden plants, and ornamentals are flowering plants. Coffee, tea, and cocoa beverages are made from flowering plants. Cotton and linen are fabrics from flowering plants. Many pharmaceuticals—from aspirin to morphine—come from, or are patterned after, metabolic chemicals in flowering plants. Hardwoods used for fuel and lumber also come from flowering plants.

Angiosperm is a synonym for "flowering plant." It means "seed within a vessel" or "enclosed seed." The defining angiosperm feature is the enclosure of the ovules (and the seeds they become) within surrounding tissue called an *ovary*. The ovary is part of a flower, a structure that occurs only in angiosperms. For some time the technical name for the division has been **Anthophyta** (*anthos* is Greek for "flower"). However, the modern tradition for division names is to root them in an important genus, and there is no genus beginning with the prefix *Antho-*. For that reason, in this book we accept the division name as **Magnoliophyta**, after the genus *Magnolia* of the southeastern United States (magnolia is also the common name; Fig. 25.2).

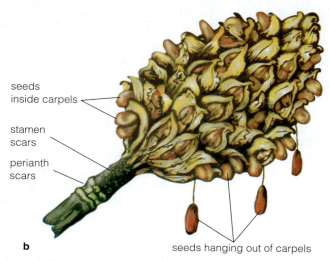

Figure 25.2 Flower (**a**) and fruit (**b**) of southern magnolia (*Magnolia grandiflora*). The division Magnoliophyta (flowering plants, angiosperms, Anthophyta) is named after this genus. Novel structures of plants in the division are shown here: the flower, the pistil (carpel) with enclosed ovules, and the fruit. A flower is a modified leafy branch that attracts animal pollinators. In this case, the flower is pollinated by beetles. Each magnolia flower has tens of pistils, each pistil containing a single ovule. The fruit is a mature, enlarged pistil with an enclosed seed; its function is to help disseminate the seeds. In magnolia, the ripe fruits open while still attached to the tree, and the red seeds dangle in the wind. Birds eat the seeds whole and later excrete them some distance away. Many of the seeds survive the passage and germinate to begin a new generation.

This chapter brings us full circle in this textbook. We began with chapters on the development, structure, physiology, and genetics of flowering plants. Then we examined the key features of fungi, algae, seedless vascular plants, and gymnosperms in a sequence that parallels the evolutionary appearance of these groups in the fossil record. This sequence also parallels the trends of: (1) increasing adaptation to land and (2) increasing complexity

and dominance of the sporophyte generation. The survey, the sequence, and the trends all peak with the division Magnoliophyta. When did this group evolve, and how did it come to dominate the land?

25.2 THE MYSTERIOUS ORIGIN OF THE ANGIOSPERMS

Charles Darwin called the origin of flowering plants "an abominable mystery" because they appeared suddenly in the geologic record, without a clear fossil history showing some transition from other plant groups (see Fig. 23.2). The first angiosperm fossil impressions—from the early Cretaceous period, about 130 million years ago—are numerous, diverse, and complex. Fossil leaves, stems, flowers, fruits, and seeds are abundant. The fossil record does not seem to show organisms that are halfway between angiosperms and gymnosperms.

Bits of wood, pollen, and leaves dating back to the Triassic period are thought by some paleobotanists to be from flowering plants. That evidence is inconclusive, but it has been supported by recent reevaluations of the fossil record and of our ideas about what constitutes an angiosperm trait. Some evolutionists suggest that the Magnoliophyta line theoretically should extend back to the Triassic, linked there to now-extinct gymnosperms in the order Bennettitales or to seed ferns. They argue that such traits as insect pollination, drought tolerance, and early reproduction are not unique to the angiosperms but were inherited from early gymnosperms. Others assert that flowering plants evolved from the Gnetophyta, that anomalous collection of three genera with both gymnosperm and angiosperm traits.

A question still remains: Why did early flowering plants leave no fossils? One hypothesis is that flowering plants evolved in uplands where preservation of plant remains in sediments is a rare event. According to this **upland theory** of angiosperm origins, erosion on mountainous slopes would tend to remove and damage twigs, flowers, and fruits before they could be trapped, covered, and protected in sediments. Upland habitats may have been uniquely suitable for early flowering plants because they were drier and colder than the lowlands and exhibited a great diversity of microenvironments in a small area. Environmental diversity and stresses, such as short growing seasons and aridity, favor evolution.

Why were flowering plants able to supplant gymnosperms? Probably because they had new vegetative and reproductive features that promoted survival and gave them a competitive advantage. One such novel feature that evolved was an improved vascular system. Angiosperm phloem contains sieve-tube members with companion cells, in contrast to the sieve cells of gymnosperms. Sieve-tube members have a larger diameter and larger sieve pores, increasing the efficiency of sugar transport. Similarly, angiosperm xylem typically contains

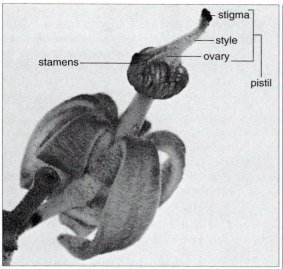

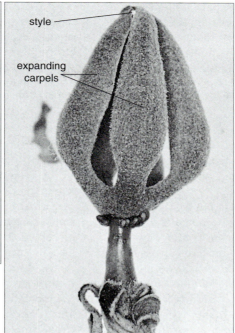

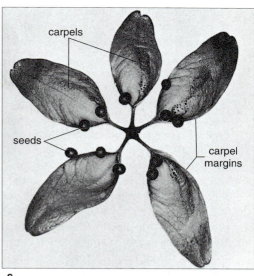

Figure 25.3 *Sterculia plantanifolia.* Notice that the pistils (carpels) resemble leaves. (**a**) A single pistil of the flower. (**b**) After pollination the pistils separate, except along the style and stigma. (**c**) At maturity each ovary splits open, exposing seeds attached to a leaflike surface.

large, relatively thin-walled vessels in addition to tracheids. The movement of water is much more efficient through these vessels.

Another novel feature is the fruit, the ripened ovary with enclosed seeds. Fruits aid in the dispersal of seeds by catching on the wind, adding buoyancy in water, or being moved by animals. Some gymnosperms have fleshy arils or scales that accomplish the same end, but not to the same degree of elaboration as with angiosperms.

William Bond, of Cape Town University, recently combined these features and others to formulate a **seedling hypothesis** as an explanation of angiosperm dominance. His hypothesis compares gymnosperms to a tortoise and angiosperms to a hare. Gymnosperms are woody, slow-growing, and have lengthy reproductive cycles. The juvenile stage is long. Cotyledons and young leaves are thick and evergreen, energetically expensive to manufacture, and not changeable in shape. Gymnosperm tracheids and sieve cells are relatively inefficient. All this leads to a low seedling growth rate. The gymnosperm is a slow, steady tortoise.

In contrast, many angiosperms are herbaceous, fast-growing, and have short reproductive cycles. The juvenile stage can be short. Cotyledons and young leaves are often thin, deciduous, energetically cheap to make, and variable in shape. Vessels and sieve-tube members are highly efficient pipelines. All this leads to a rapid seedling growth rate. The angiosperm is a hare. Bond's theory predicts that gymnosperms will be outcompeted everywhere except where angiosperm *seedling* competition is reduced, as in cold-temperate regions with nutrient-poor soils.

Angiosperms not only produce a novel organ—the ovary around the ovule—but they enclose that organ within a completely new structure, the flower. How could the flower have developed from preexisting organs such as stems and leaves? The flower is thought to be a modified branch whose leaves have become sepals, petals, stamens, and carpels. Two types of indirect evidence support this hypothesis. First, the early developmental stages of floral parts at the apical meristem closely resemble those of leaves (see Chapter 13). Second, the stamens or carpels of some plants—such as *Sterculia plantanifolia* in the cacao family—bear a striking resemblance to leaves. The ovary is made from five simple carpels united only at their stigmatic tips. When the ovary matures, it opens to show five leaflike carpels that bear seeds along their margins (Fig. 25.3). Those plants with the most strikingly leaflike flowers also have other primitive traits, such as xylem with few vessels, woodiness, and a tropical habitat.

25.3 THE RISE OF THE ANGIOSPERMS TO DOMINANCE

Angiosperms diversified and became so abundant in the fossil record of the late Cretaceous period that we can conclude they soon became the dominant plant life on land. Thus, as the Mesozoic era ended, so did dominance by gymnosperms and dinosaurs. The climate cooled rather suddenly at the boundary of the Mesozoic and

Cenozoic eras—so suddenly that some paleobotanists turn to theories of catastrophic events to explain the massive environmental change and extinctions that occurred in such a brief time. One theory proposes that a meteor slammed into what has become the Gulf of Mexico, creating disastrous fires and throwing debris into the atmosphere. During the months or years the debris took to settle back to earth, it reflected solar radiation back to space; consequently, temperatures on earth plummeted, extinguishing entire groups of organisms. One assumption of this theory is that meteoric disturbance has occurred many times—not just at the end of the Mesozoic—and will occur again in the future.

We do not know if such an event ushered in the Cenozoic era. We do know, however, that the Cenozoic's 65-million-year span has been one of continuous continental movement, climatic change (Fig. 25.4), and crustal upheaval, and that through it all, angiosperms held dominance. The Atlantic Ocean widened; India rammed into Asia and created the Himalayan uplands; Madagascar split from southern Africa; Australia and Antarctica became islands; and Saudi Arabia separated from northern Africa, forming the Red Sea. Continual slow crustal movement changed the positions of the continents relative to the poles and equator (and relative to climatic zones). As a consequence of all this movement, vegetation types migrated and shifted throughout the Cenozoic (Fig. 25.5).

Angiosperm Fossils Show Climatic Change During the Tertiary Period

The Cenozoic era is traditionally divided into two periods, the Tertiary and the Quaternary. The Tertiary period extended from approximately 60 million years ago to 2 million years ago. During this time, continents continued to break apart, increasing the diversity of climate and vegetation types. An excellent case-study example of climate and vegetation change during that time comes from the fossil record of the John Day Basin in northeastern Oregon.

In the Eocene epoch, which began about 57 million years ago (see Fig. 25.4), the dominant vegetation of the John Day Basin contained cinnamon (*Cinnamomum*), fig (*Ficus*), cycads, and tropical ferns. Leaves were large, with entire margins. Today, such plants are found in the mountain forests of Central America, which have an annual rainfall of more than 150 cm (60 in) and no winter frost. Paleobotanists often accept the dictum that the present is the key to the past; therefore, they estimate that the climate of the John Day Basin at the time was like that of modern mountain forests in Central America: wet, humid, and mild.

Twenty million years later, in the Oligocene epoch, rainfall was perhaps down to 120 cm per year (48 in), summers were wet and mild, and winters were drier and cold (but with very little snow). The fossil record shows a change to a rich, mixed conifer–hardwood forest. Conifers included coast redwood (*Sequoia*), Sierra redwood (*Sequoiadendron*), dawn redwood (*Metasequoia*), ginkgo, bald cypress (*Taxodium*), cedar (*Cedrus*), true firs (*Abies*), and hemlock (*Tsuga*). Broad-leaved, winter-deciduous angiosperm hardwoods included alder (*Alnus*), beech (*Fagus*), chestnut (*Castanea*), cottonwood (*Populus*), dogwood (*Cornus*), elm (*Ulmus*), hickory (*Carya*), maple (*Acer*), oak (*Quercus*), persimmon (*Diospyros*), Sassafras, sweet gum (*Liquidambar*), tree of heaven (*Ailanthus*), and willow (*Salix*). Some broad-leaved evergreens were also present, such as madrone (*Arbutus*) and tanbark oak (*Lithocarpus*).

In keeping with climatic change from the Eocene, leaves had become smaller, with dentate (toothed) or convoluted margins, indicating a drier habitat (because smaller leaves with lobed or uneven margins are able to be cooled directly by wind instead of by transpiration). Deciduousness implies a colder winter.

We do not find this type of mixed forest anywhere on Earth today because the component species have been fragmented and separated. The closest approximations are along the cool, wet Oregon–California coast, in the Great Smoky Mountains of Tennessee, and in parts of China and Japan.

Twelve million years later, in the Miocene epoch, the shift to deciduous trees was even more pronounced, and conifers became less abundant. Oak, hickory, and maple dominated the forest. Species richness declined. Paleobotanists imagine that the climate was much like that in modern Ohio: 100 cm (40 in) of annual precipitation, with half of that falling as snow in winter, when freezing temperatures were common.

By the Pliocene epoch, fir and spruce—conifers that we associate today with very cold climates—were common, along with winter-deciduous hardwoods. Some leaves were quite small and hard, indicating adaptation to a cold and arid climate.

Today, the John Day Basin is dominated by sagebrush (*Artemisia tridentata*). Trees are absent except along watercourses. Annual precipitation is only 25 cm (10 in) per year, most of that falling as snow in winter. The area is a semiarid cold desert.

The Ice Age Affected the Diversity of Plants in Temperate Zones

The Quaternary period began about 2 million years ago, when the earth's climate cooled by a few degrees—just enough to alter hydrological balances near the poles and at high elevations. Summer melting no longer kept up with winter snowfall, with the result that snow accumulated year after year. Its layers compressed into ice, and great ice sheets called *glaciers* coalesced, moved, and came to cover much of the cool-temperate zone. The balance of water shifted from the oceans toward bodies of fresh water and ice; sea level fell. This Ice Age, which has occupied most of the Quaternary, is known as the Pleistocene epoch.

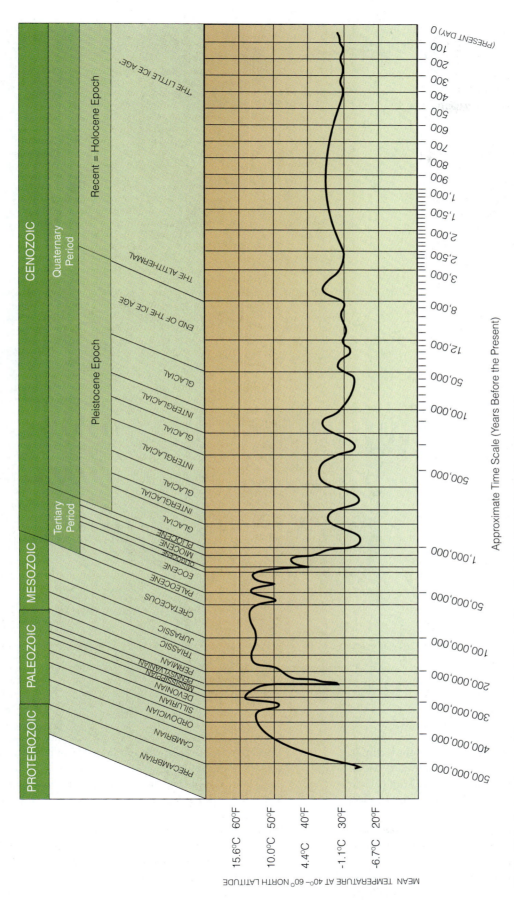

Figure 25.4 Average temperature over geologic time for the earth at 40 to 90° N latitude (the region that is now the north temperate, boreal, and arctic zones). The time scale is distorted to show more detail for the past 1 million years. Today's mean temperature (extreme right, horizontal scale) is approximately 0°C (32°F). Notice that the temperature during glacial advances and retreats (interglacials) in the Quaternary and again 3000 to 8000 years ago (the Altithermal) was only several degrees colder or warmer. In the Cretaceous, when angiosperms first appeared in the fossil record, mean temperature was about 13°C (55°F), much warmer than at present. (Adapted from E. Dorf, *American Scientist* 48 (1960): 341.)

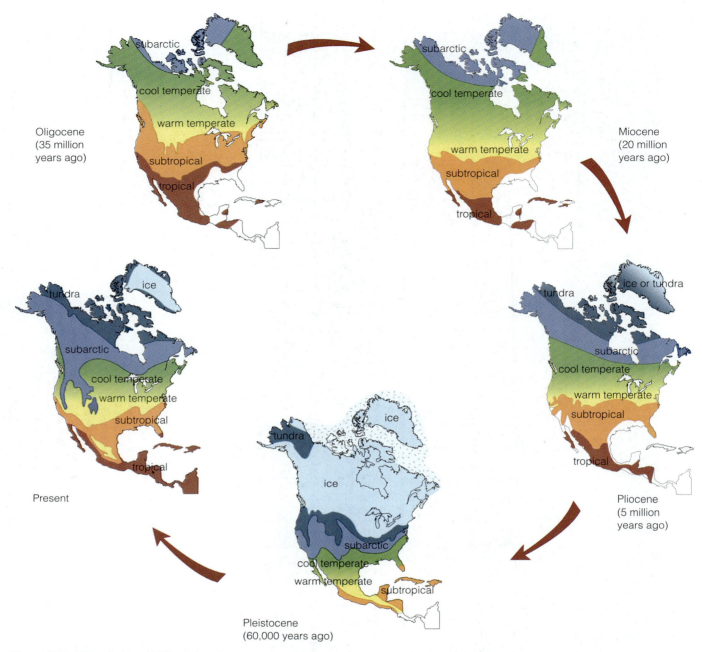

Figure 25.5 Climatic zones of North America during the past 35 million years of the Cenozoic era. (Redrawn from E. Dorf, *American Scientist* 48 (1960): 341.)

Periodically during the Pleistocene, the climate warmed and glaciers retreated; then the climate cooled and glaciers readvanced. Several such glacial-interglacial cycles occurred. During glacial advance, almost all of Canada, the northern third of the United States, most of Europe, all of Scandinavia, and large parts of Siberia were covered by ice. Mountain chains well into the tropics were all capped by glaciers that extended downslope, filling wide canyons and mountain valleys. Ice covered 30% of the world's land surface. Sea level was 60 m (198 ft) lower than at present because so much water was locked in glacial ice.

The cause of the climate change that led to glacial advance is still being debated. We do know that widespread glacial advance has occurred at least twice before in the geologic record: toward the end of the Proterozoic era (about 600 million years ago) and toward the end of the Paleozoic era during the Permian period (about 245 million years ago). It is possible that our third cycle of glacial advance and retreat is not yet over. The last glacial ad-

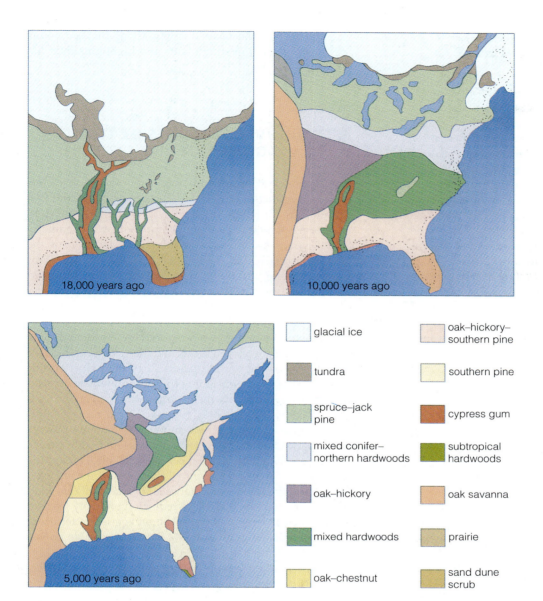

glacial ice

tundra

spruce–jack pine

mixed conifer–northern hardwoods

oak–hickory

mixed hardwoods

oak–chestnut

oak–hickory–southern pine

southern pine

cypress gum

subtropical hardwoods

oak savanna

prairie

sand dune scrub

Figure 25.6 Reconstruction of major vegetation zones in eastern North America since the last maximum glacial advance. (Redrawn from H. R. Delcourt et al., *Quarterly Science Review* 1 (1983): 153–75.)

vance peaked 18,000 years ago, and most of its retreat was completed 10,000 years ago. Some geologists identify 10,000 years ago as the end of the Pleistocene and the beginning of the Recent (or Holocene) epoch. Others contend that we are merely in an interglacial period and another glacial advance will begin in a few thousand years.

When glaciers advanced, vegetation zones were pushed lower in elevation or lower in latitude. For example, at full glacial advance 18,000 years ago (Fig. 25.6), a spruce–pine boreal forest covered the eastern United States as far south as North Carolina. Treeless tundra covered the Appalachian Mountains. A thin strip of conifer–hardwood forest formed a northern cap to oak–hickory deciduous forest, and Florida (much wider then) was covered by sand dune scrub. Only 800 km (500 mi) of north-to-south distance separated the front of an ice sheet from deciduous forest, indicating a sharp gradient in climate.

When glaciers retreated, vegetation zones rebounded, the climatic gradient softened, and landscape diversity increased (Fig. 25.6). Each cycle of advance and retreat caused some extinctions, however, so that the biota of the temperate zone became simpler during the Pleistocene. Many large mammals disappeared from North America during this time: the mastodon, mammoth, giant ground sloth, llama, peccary, giant armadillo, sabertooth tiger, ox, yak, and horse. Some of these survived elsewhere in the world, but others became globally extinct. Some anthropologists believe that hunters migrating across the Bering Land Bridge (from Siberia to Alaska) caused the extinction of many of these species. According to this Pleistocene overkill hypothesis, humans with sophisticated group-hunting methods overexploited prey species that had no evolutionary past experience with such predators.

Angiosperm Evolution Was Affected by Humans in the Quaternary Period

Humans also played a role in plant distribution and evolution during the Quaternary. Humans in hunter–gatherer cultures may at first merely have harvested wild plants as they found them. However, they gradually began to cultivate and select some of those species for convenience and higher yield. They did not sow seeds in geometric patterns and till the land; but they did use fire, pruning, selective harvesting, and sowing (of rhizomes, bulbs, or seeds) *without* cultivation to favor the abundance of certain food plants. Anthropologists call this stage **protoagriculture**. Protoagriculture may have gone on for thousands of years before agriculture and the domestication of crops were fully established.

The "root" crops cassava (*Manihot esculenta*), sweet potato (*Ipomoea batatas*), and taro (*Colocasia esculenta*) were cultivated in southeastern Asia as long as 15,000 years ago. Human-induced selection in these species has gone on for so long that some have lost, or nearly lost, their capacity for sexual reproduction. It is doubtful that they could survive in nature. The earliest archaeological evidence for **seed agriculture** (cultivation of annuals such as rice, wheat, beans, or squash) takes us back 11,000 years. Wild annuals that demonstrated exceptional productivity were valued and propagated. Over time, such artificial selection pressure on the genetic makeup of plants resulted in so much change from their wild relatives that it is now difficult to determine where, and from what wild stock, they were first domesticated.

Humans have also accidentally domesticated and favored the evolution of **weeds**. Weeds are plants that grow well in disturbed or trampled soil, in waste areas rich in nitrogen, or interspersed with crop plants. Some weeds have evolved seeds that are the same size as crop seeds, so they are not easily separated during threshing or sieving. Thus, when the next season's crop is sown, the weeds are sown inadvertently right along with it.

As humans explored new lands, they brought along not only their culture but also their domesticated plants and companion weeds. Sometimes ornamental plants were brought into new regions. The weeds and ornamentals often turned out to be aggressive competitors in their new homes, able to displace native species from the landscape. As much as a third of the flora of some parts of the United States is composed of weeds imported from various parts of the world, including Europe, Australia, Asia, and Africa. Introduced plants have become common in forests, grasslands, woodlands, wetlands, coastal strands, and deserts. Only the most stressful habitats, such as alpine tundra, salt marshes, and rocky outcrops, seem free of introduced species. The accelerated pace of land disturbance, single-species cultivation, and travel are making the flora of the world increasingly homogeneous and lower in biological diversity.

The rest of the chapter summarizes the taxonomic diversity of the flowering plants and highlights their life cycle.

25.4 NOVEL FEATURES OF THE ANGIOSPERM LIFE CYCLE

Angiosperms share with gymnosperms the characteristic of producing a seed, but the life cycles of the two groups are different in several important features. In flowering plants, the size and complexity of the gametophyte generation become reduced, the location of the ovule becomes hidden, fertilization involves two steps instead of one, and the dispersal of the seed is improved by its enclosure within a fruit. These novel life cycle features adapt flowering plants to life on land and conserve their food reserves.

Our survey of the plant kingdom has revealed a trend of decreasing gametophyte size and increasing dependence on the sporophyte. Angiosperms carry the reduction further than gymnosperms, for both male and female gametophytes. In angiosperms, the male gametophyte (in the form of a pollen grain) has only two nuclei and no vegetative cells at all; the body has been eliminated. The female gametophyte has been reduced to seven cells and eight nuclei, and an archegonium is no longer recognizable. Perhaps the three synergid cells represent all that's left of the archegonium, and the three antipodal cells represent all that's left of the gametophyte body. This reduction probably leads to more efficient reproduction: When an egg is not fertilized or when a pollen grain does not land in a suitable flower, less energy is wasted.

Additional savings in energy result when animal vectors transfer pollen grains to pistils (Fig. 25.7). Animals move selectively, not randomly, visiting sequential flowers of the same species. Thus, fewer pollen grains are needed to make seeds. Mutualistic relationships also develop, creating a potential for rapid evolution and the movement of both plant and animal species into new environments.

The enclosure of an ovule within a pistil shelters the ovule against drying and attack by herbivores or pathogens. Enclosure also permits selectivity for appropriate pollen because the pollen has to germinate on a receptive stigma and then grow through an accepting style. Finally, the ovary wall later matures into an organ that protects the seed, enhances seed dispersal, and controls seed germination. The consequences are that seedlings will emerge when and where conditions are most favorable.

Double fertilization (Fig. 25.7) also conserves energy because food storage tissue does not accumulate until *after* fertilization. Energy-rich endosperm tissue is not part of a preexisting female gametophyte; it is a new tissue that is created only by the act of fertilization. If unfertil-

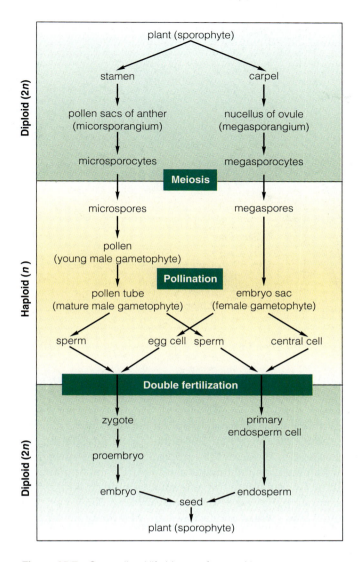

Figure 25.7 Generalized life history of an angiosperm.

ized ovules contained as much energy-rich tissue around them as fertilized ovules, a significant amount of food reserves would be wasted.

The angiosperms and the ferns are the only groups of terrestrial plants forming new species and becoming more diverse at this point in geologic time. Of the two, the angiosperms are far more abundant. About 215,000 species of flowering plants have been described, and taxonomists imagine many thousands more scientifically unknown species exist in poorly explored regions. Certainly the number of described species rises each year. When Linnaeus published *Species Plantarum*, his compendium of the plant world in the mid-18th century, he listed fewer than 8000 species. A conservative estimate is that 250,000 extant species in the Magnoliophyta exist on Earth. Possibly a major reason for the diversity, success, and dominance of this group is their possession of the novel life cycle traits we have just described.

Plant taxonomists have the difficult task of organizing a quarter of a million species into a classification system. Classification systems are important because they allow the human mind to work with, find patterns in, and understand diversity. The objective is to divide flowering plants into groups that express historic evolutionary linkages—that is, the classification should be phylogenetic (see Chapter 18). Recall, however, that the fossil record is very meager in information about how flowering plants evolved: They seem to spring fully developed into the record. So taxonomists must *infer* evolutionary relationships, based on deductions about which traits are primitive and which are advanced. Since inferences depend on assumptions, taxonomists differ in the conclusions they draw about evolutionary relationships among flowering plants. Consequently, different classification schemes exist. Typically, however, Magnoliophyta is divided into two classes, **Magnoliopsida** (the dicots) and **Liliopsida** (the monocots).

Phylogenetic Relationships Are Based on Vegetative and Reproductive Traits

One system that found much favor early in the 20th century was developed by Charles Bessey, and it is still widely followed. Bessey developed 26 criteria for determining degrees of evolutionary advancement from what he considered the most primitive angiosperm groups (now called the subclasses Magnoliidae and Ranunculidae).

Some of Bessey's criteria assume that more advanced flowering plants will indicate the following trends: (1) shortening of the floral axis; (2) change from a spiralled arrangement of floral parts to a whorled arrangement; (3) decreasing number of stamens and carpels; (4) a tendency for stamens and carpels to become less free and instead to be attached together or to petals; (5) reduction of flowers to incomplete (lacking petals or sepals) or imperfect (unisexual) status; (6) irregular symmetry; and (7) an inferior ovary.

More recent taxonomists, particularly Armen Takhtajan, Arthur Cronquist, Robert Thorne, and Rolf Dahlgren, have amended and expanded Bessey's criteria to include trends in wood anatomy, plant growth form, stomatal structure, types of inflorescence, pollination vectors, details of the female gametophyte, seed structure, and fruit type. Takhtajan, for example, assumes that as plants advance they change along these lines: (1) from woody to herbaceous growth forms; (2) from simple, alternate, entire leaves with pinnate venation to compound, opposite, lobed leaves with other kinds of venation (see Chapter 7); (3) from xylem with few vessels to xylem with many vessels; (4) from a terminal, cymelike infloresence with few

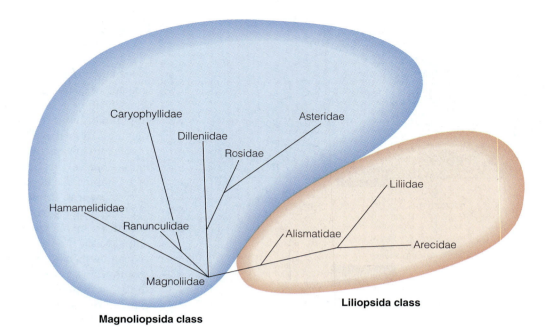

Figure 25.8 Takhtajan's phylogenetic arrangement of classes and subclasses within the Magnoliophyta. (Redrawn from A. Takhtajan, "Outline of the Classification of Flowering Plants (Magnoliophyta)." *Botanical Review* 46 (1980): 225–359.)

flowers to a lateral, umbel-like inflorescence with many flowers (see Chapter 13); (5) from pollen grains with an unornamented outer wall to grains with elaborate sculpturing; and (6) from medium-sized seeds with a small embryo and abundant endosperm to large or small seeds with a large embryo and a small amount of residual endosperm.

Takhtajan uses his criteria to divide the Magnoliopsida into seven subclasses and the Liliopsida into three subclasses (Fig. 25.8). His system shows the Magnoliidae as being the evolutionary base for both classes, which is in agreement with Bessey's earlier scheme.

As new investigations use new techniques of plant analysis, results are added into the definition of what is primitive or advanced. These techniques include the analysis of protein structure, comparison of DNA sequencing, and model building based on sophisticated cladistic statistics. It is certain that the classification scheme in Figure 25.8 will be modified someday by the conclusions of studies not yet done. However, our current state of knowledge is based on Takhtajan's subclasses. The rest of this section deals with some of these subclasses. They are discussed in reference to their order in Figure 25.8.

The Magnoliidae Is the Most Primitive Subclass

The Magnoliidae subclass contains tropical and warm-temperate trees with simple leaves, pinnate venation, large flowers with many stamens and pistils, and relatively simple wood anatomy. Magnolia (see Fig. 25.2) is a good example. Other members of this subclass and of its close relative the Ranunculidae include nutmeg (*Myris-*

tica), cinnamon (*Cinnamomum*), sassafras (*Sassafras*), avocado (*Persea*), bay laurel (*Laurus*), black pepper (*Piper*), water lily (*Nymphaea, Nelumbo, Victoria*), pitcher plant (*Nepenthes* and *Sarracenia*), opium poppy (*Papaver*), many ornamental flowers in the buttercup family (Ranunculaceae, Fig. 25.9), and the tropical plant *Chondrodendron*, which contains the drug curare. The flowers of plants in these two subclasses have regular symmetry, with many separate parts and a superior ovary (Fig. 25.9). They are often large, white, and lack nectar—and are typically pollinated by beetles.

The Alismatidae and Arecidae Are Primitive Monocots

The Liliopsida (monocots) diverged early in evolutionary time from the Magnoliidae (dicots). The most primitive Liliopsida are in the flowering rush family (Butomaceae) and the water plantain family (Alismataceae), both in the subclass Alismatidae. Water plantains number nearly 100 species, chiefly from the northern hemisphere. They are herbaceous, aquatic plants with vessels only in their roots. Arrowhead, or wapeto (*Sagittaria*, Fig. 25.10), is a member of this family. It has floating leaves, arising from a rhizome, and a root system embedded in lake bottom sediments. Several species are cultivated for their edible tubers in Hawaii, Malaya, and Kashmir, as well as by Native Americans in North America.

A somewhat more advanced subclass is the Arecidae. The largest family in the Arecidae is the Arecaceae, the palms: There are almost 2700 species distributed throughout the tropical latitudes of the Old and New Worlds. A few species enter the temperate zone, including California fan palm (Fig. 25.11). Palms have a distinctive growth

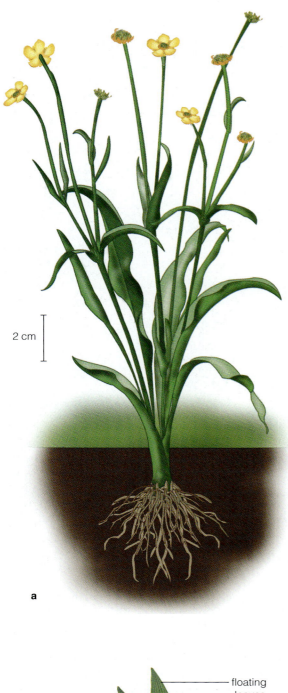

2 cm

a

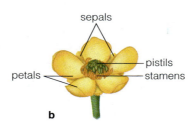

sepals

petals

pistils
stamens

b

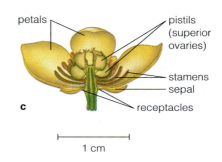

petals

pistils
(superior
ovaries)

stamens
sepal

receptacles

c

1 cm

Figure 25.9 The primitive subclass Ranunculidae is represented by buttercup (*Ranunculus alismaefolius*). (**a**) Entire plant. (**b**) One flower, showing regular symmetry. (**c**) Section through one flower, showing multiple, separate (not fused) parts, and superior ovary position. Primitive traits include regular symmetry of flowers, many flower parts on an elongated receptacle, a superior ovary position, a pistil made from a single carpel, and simple leaves.

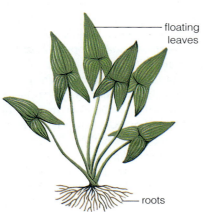

floating
leaves

roots

Figure 25.10 Arrowhead (*Sagittaria*), a primitive monocot (class Liliopsida) in the subclass Alismatidae.

Figure 25.11 The subclass Arecidae is represented by California fan palm (*Washingtonia filifera*), which grows in desert oases of southern California.

Table 25.1 The 20 Largest Families of Flowering Plants

Rank	Family (common name)	Class/Subclass	Number of Species	Distribution
1	Asteraceae (sunflower)	Magnoliopsida/Asteridae	21,000	Worldwide
2	Orchidaceae (orchid)	Liliopsida/Liliidae	17,500	Worldwide
3	Fabaceae (bean)	Magnoliopsida/Rosidae	16,400	Worldwide
4	Rubiaceae (madder)	Magnoliopsida/Asteridae	10,700	Worldwide
5	Poaceae (grass)	Liliopsida/Liliidae	7,950	Worldwide
6	Euphorbiaceae (spurge)	Magnoliopsida/Rosidae	7,750	Worldwide
7	Lamiaceae (mint)	Magnoliopsida/Asteridae	5,600	Worldwide
8	Melastomataceae (melastoma)	Magnoliopsida/Rosidae	4,750	Tropical, warm, esp. South America
9	Liliaceae (lily)	Liliopsida/Liliidae	4,550	Worldwide
10	Scrophulariaceae (snapdragon)	Magnoliopsida/Asteridae	4,500	Worldwide
11	Acanthaceae (acanthus)	Magnoliopsida/Asteridae	4,350	Tropical
12	Myrtaceae (eucalyptus)	Magnoliopsida/Rosidae	3,850	Tropical, warm, esp. Australia
13	Cyperaceae (sedge)	Liliopsida/Liliidae	3,600	Worldwide
14	Ericaceae (heather)	Magnoliopsida/Dilleniidae	3,350	Worldwide
15	Apiaceae (carrot)	Magnoliopsida/Rosidae	3,100	Worldwide
16	Rosaceae (rose)	Magnoliopsida/Rosidae	3,050	Temperate, warm northern hemisphere
17	Brassicaceae (mustard)	Magnoliopsida/Dilleniidae	3,000	Worldwide
18	Moraceae (mulberry)	Magnoliopsida/Dilleniidae	2,975	Tropical, subtropical, temperate
19	Araceae (arum)	Liliopsida/Arecidae	2,950	Tropical, subtropical
20	Asclepiadaceae (milkweed)	Magnoliopsida/Asteridae	2,900	Tropical, warm, esp. Africa

Data on number of species and distribution from D. J. Mabberley, *The Plant-Book*. New York: Cambridge University Press, 1987.

form: an unbranched trunk, a terminal tuft of compound or dissected leaves, and a coconutlike fruit. All palms are sensitive to frost; if their single apical meristem is killed, the entire plant dies. One advantage of an unbranched morphology is minimal wind resistance for maximum survival of hurricane or typhoon winds.

Palms are among the most important tropical families. Coconut palm (*Cocos nucifer*) yields edible endosperm from its huge seeds, oil pressed from dried coconut meat (copra), and cordage (coir) from its stringy outer husk. Dates come from *Phoenix dactylifera*; carnauba waxes from *Copernicia*; and palm cooking oils from *Elaeis* species. Many species are favorite ornamentals in warm climates.

Other notable members of this subclass are the smallest flowering plants, the duckweeds (family Lemnaceae); the taro plant (*Colocasia*), which has an edible root; and the popular ornamentals *Philodendron*, *Anthurium*, and calla lily (*Zantedeschia*). The latter three are in the family Araceae, which is among the 20 largest families of flowering plants (Table 25.1).

The Liliidae Are the Most Advanced Monocots

The most advanced subclass of monocots, the Liliidae, is also very diverse. One line (which includes orchids, lilies, and bromeliads) features colorful flowers that are highly coevolved with insect pollinators. The other line (which includes grasses and sedges) features flowers reduced to small, colorless, wind-pollinated forms. Four of the 20 largest families are found in this subclass (Table 25.1), families that are enormously important to humans. All of our grains—barley (*Hordeum*), corn (*Zea*), millet, oats (*Avena*), rice (*Oryza*), rye (*Secale*), wheat (*Triticum*)—belong to the grass family (*Poaceae*). All of the turf grasses in landscaping and the bamboos in tropical buildings are in this family. Other families yield the food plants pineapple (*Ananas*); banana (*Musa*); onion, garlic, leek, and chive (various *Allium* sp.); asparagus (*Asparagus*); and yam (*Dioscorea*). Also included are the spices ginger (*Zingiber*), cardamom (*Elettaria*), and turmeric (*Curcuma*). The ancient paper of Egypt was made from a sedge, *Papyrus* (family Cyperaceae). Many ornamental house and garden plants also belong to this subclass: iris

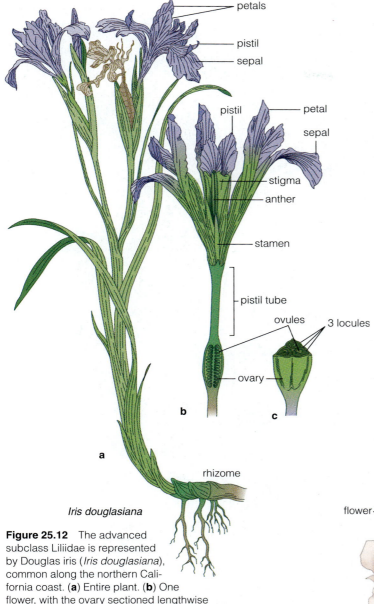

petals

pistil

sepal

pistil

petal

sepal

stigma

anther

stamen

pistil tube

ovules

3 locules

ovary

ovary

b

c

a

rhizome

Iris douglasiana

Figure 25.12 The advanced subclass Liliidae is represented by Douglas iris (*Iris douglasiana*), common along the northern California coast. (**a**) Entire plant. (**b**) One flower, with the ovary sectioned lengthwise to show many ovules. (**c**) Cross section of the ovary, showing three locules that represent three fused carpels. Advanced traits include inferior ovary position, irregular flower shape, reduction in the number of flower parts, and a pistil made up of several fused carpels.

Figure 25.13 The large subclass Rosidae includes many species in the Rosaceae family with edible fruits, such as boysenberry (*Rubus ursinus* var. *loganbaccus*). (**a**) Portion of a stem, showing compound leaves, a mature flower, and young fruits. (**b**) Section of one flower. This subclass has a mixture of primitive and advanced traits. Primitive traits include flowers with regular symmetry and many flower parts; advanced traits include compound, toothed leaves and some degree of fusion of petals with sepals at their bases.

(*Iris*, Fig. 25.12), bird-of-paradise (*Strelitzia*), water hyacinth (*Eichornia*), tulip (*Tulipa*), lily (*Lilium*), daffodil (*Narcissus*), and of course the orchids (family *Orchidaceae*).

The Hamamelididae and Rosidae Are Primitive Dicots

The Hamamelididae is a relatively small subclass of temperate-zone woody plants, including some important hardwoods of the eastern deciduous forest (see Chapter 27). The most familiar examples are sycamore (*Platanus*), witch hazel (*Hamamelis*), sweet gum (*Liquidambar*), birch (*Betula*), alder (*Alnus*), hazelnut (*Corylus*), beech (*Fagus*), oak (*Quercus*), chestnut (*Castanea*), and the she-oaks of Australia (*Casuarina*). Commercial cork is derived from the bark of the Mediterranean cork oak (*Quercus suber*).

The Rosidae is the most diverse subclass of dicots. Among the many families, the largest in the temperate zone are bean (Fabaceae), spurge (Euphorbaceae), eucalyptus (Myrtaceae), rose (Rosaceae, Fig. 25.13), and carrot (Apiaceae, Fig. 25.14). The tropical family Melastomataceae also is one of the 20 largest families. Economically important uses include fruits, nuts, vegetables, ornamentals, timber trees, spices, dyes, and drugs. Botanical oddities range from the small insectivorous Venus's-flytrap to

compound leaf

young fruit

flower

stigma

ovary (future achene)

receptacle

stamen

petal

sepal

a

b

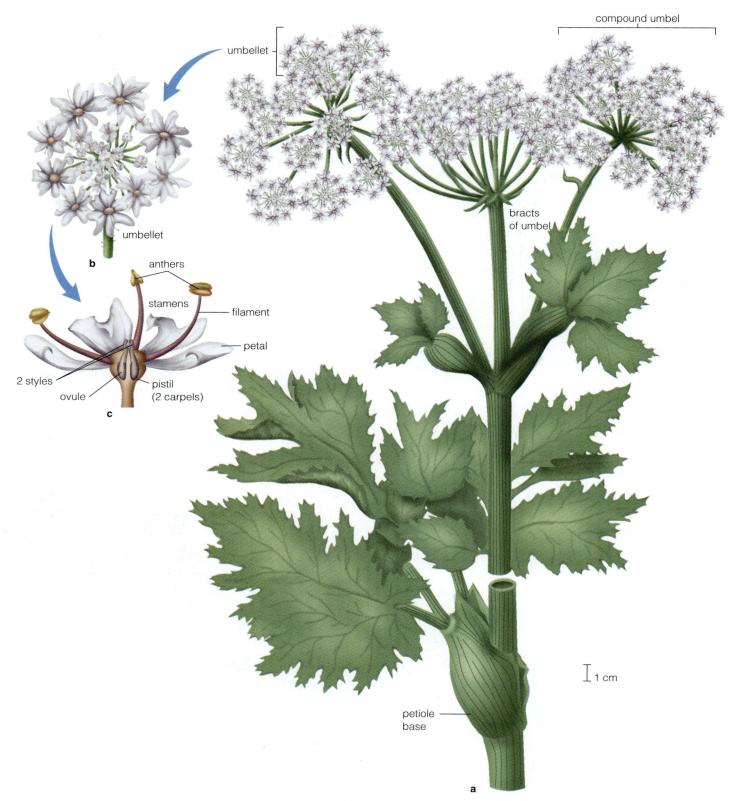

compound umbel

umbellet

umbellet

b

anthers

stamens

filament

petal

2 styles

ovule

pistil
(2 carpels)

c

bracts
of umbel

1 cm

petiole
base

a

Figure 25.14 Another large family in the subclass Rosidae is the carrot family, Apiaceae, represented here by cow parsnip (*Heracleum lanatum*). (**a**) Upper portion of a large plant, about 1 m (3 ft) tall, showing deeply lobed leaves and several inflorescences called *umbels* or *compound umbels*. (**b**) One portion of an umbel (an umbellet), made up of a cluster of two dozen flowers. (**c**) One flower, sectioned to show the inferior ovary. Advanced traits include reduced flower size, reduced number of flower parts, inferior ovary, fusion of two carpels into a pistil, and deeply lobed and toothed leaves.

parasitic mistletoes to the 1-m-diameter (3-ft) flowers of *Rafflesia*.

Just a sample of food plants in this group would list apple (*Malus*); almond, cherry, apricot, nectarine, peach, and plum (*Prunus*); blackberry, loganberry, and raspberry (*Rubus*); gooseberry and currant (*Ribes*); snap, string, green, kidney, lima, and mung beans (*Phaseolus*); soybean (*Glycine*); peanut (*Arachis*); guava (*Psidium*); cassava (*Manihot*); grape (*Vitis*); cashew (*Anacardium*); pistachio (*Pistacia*); mango (*Mangifera*); walnut (*Juglans*); carrot (*Daucus*); and lemon, lime, orange, and grapefruit (*Citrus*). Spices and scents include fennel (*Foeniculum*), dill (*Anethum*), caraway (*Carum*), clove (*Syzygium*), pimento (*Pimenta*), ginseng (*Panax*), frankincense (*Boswellia*), and myrrh (*Commiphora*).

Common ornamentals in this group are roses (*Rosa*), *Hydrangea*, *Daphne*, dogwood (*Cornus*), holly (*Ilex*), *Impatiens*, nasturtium (*Tropaeolum*), and ivy (*Hedera*). Commercial timber comes from mangroves, *Eucalyptus*, *Terminalia*, sandalwood (*Santalum*), maple (*Acer*), mahogany (*Swietenia*), and lignum vitae (*Guaiacum*). The famous dye indigo comes from *Indigofera* plants, rubber from *Hevea* trees, and the drug cocaine from the coca shrub (*Erythroxylum*).

The Caryophyllidae, Dilleniidae, and Asteridae Are Advanced Dicots

The largest families in the subclass Caryophyllidae include iceplants (Aizoaceae), cacti (Cactaceae, Fig. 25.15), pinks (Caryophyllaceae), and beets (Chenopodiaceae), each with about 2000 species. Plants in this subclass with important human uses include sugar beet (*Beta*), spinach (*Spinacea*), purslane (*Portulaca*), rhubarb (*Rheum*), buckwheat (*Fagopyrum*), and the South American quinoa (*Chenopodium*) and *Amaranthus*. Landscape ornamentals include tropical *Bougainvillaea*, temperate carnations (*Dianthus*), sea lavender (*Limonium*), and many cacti and iceplants.

The Dilleniidae subclass has many families, but only a few are large and well known in temperate latitudes: heather (Ericaceae), mulberry (Moraceae), and mustard (Brassicaceae), each with about 3000 species; and elm (Ulmaceae), with 2000 species. The world's main source of hardwood lumber comes from a tropical group of genera commonly called dipterocarps (*Dipterocarpus*, *Hopea*, *Shorea*, *Vatica*). Other timber trees include linden and basswood (*Tilia*), balsa (*Ochroma*), elm (*Ulmus*), and ebony (*Diospyros*).

Cotton (*Gossypium*, Fig. 25.16), jute (*Corchorus*), and hemp (*Cannabis*) provide economically important fibers from this subclass.

Ornamentals in the Dilleniidae include peonies (*Paeonia*), *Camellia*, *Hibiscus*, violets (*Viola*), rockrose (*Cistus*), *Tamarix*, *Begonia*, willows (*Salix*), *Rhododendron*, *Cyclamen*, and *Primula*.

Edible products include figs (*Ficus*); mulberry (*Morus*); Brazil nut (*Bertholletia*); passion fruit (*Passiflora*); papaya (*Carica*); caper (*Capparis*); huckleberry, cranberry, blue-

Figure 25.15 Cactus flower, in the subclass Canyophyllidae, showing relatively primitive large flower with multiple parts.

berry, and bilberry (*Vaccinium*); persimmon (*Diospyros*); cantaloupe, honeydew, and cucumber (*Cucumis*); pumpkin, squash, and various gourds (*Cucurbita*, Fig. 25.17); watermelon (*Citrullus*); turnip, kale, cabbage, mustard, Brussels sprouts, broccoli, and cauliflower (*Brassica*); tea (*Camellia*); and cocoa (*Theobroma*).

The Asteridae—the most advanced dicot subclass—includes tomatoes, mints, and sunflowers. Advanced traits include irregular flower symmetry, inferior ovary, small flowers, and herbaceous habit. This subclass has some of the most widely distributed and well-known families. The largest of all flowering plant families is the sunflower family, Asteraceae (21,000 sp., see Table 25.1). Other large families include madder (Rubiaceae), mint (Lamiaceae) (Fig. 25.18), snapdragon (Scrophulanaceae) milkweed (Asclepiadaceae), tomato (Solanaceae) (Fig. 25.19), and forget-me-not (Boraginaceae).

Important garden plants from this subclass include gentians (*Gentiana*), periwinkle (*Vinca*), oleander (*Nereum*), jasmine (*Jasminum*), lilac (*Syringa*), *Phlox*, heliotrope (*Heliotropium*), forget-me-not (*Myosotis*), *Lantana*, lavender (*Lavendula*), snapdragon (*Antirrhinum*), *Penstemon*, African violet (*Saintpaulia*), *Lobelia*, *Gardenia*, honeysuckle (*Lonicera*), *Gerbera*, cosmos (*Bidens*), daisy (*Aster*), and *Dahlia*.

No major global food plants are found in this subclass, but components and flavors of regional diets are provided from olive (*Olea*); potato (*Solanum*); tomato (*Lycopersicon*); peppers, paprika, and chilies (*Capsicum*); tomatillo (*Physalis*); sweet potato (*Ipomoea*); elderberry (*Sambucus*); lettuce (*Lactuca*); endive and chicory (*Cichorium*); and artichoke (*Cynara*). Spices include mint (*Mentha*), oregano (*Origanum*), sage (*Salvia*), thyme (*Thymus*), basil (*Ocimum*), and sesame (*Sesamum*).

Other economically important products are coffee (*Coffea*), madder dye (*Rubia*), safflower oil (*Carthamus*), sunflower oil (*Helianthus*), tobacco (*Nicotiana*), teak (*Tectona*), and the drugs digitalin from foxglove (*Digitalis*) and quinine from the tropical tree *Cinchona*.

We conclude this chapter with a brief look at the global distribution patterns of plant families.

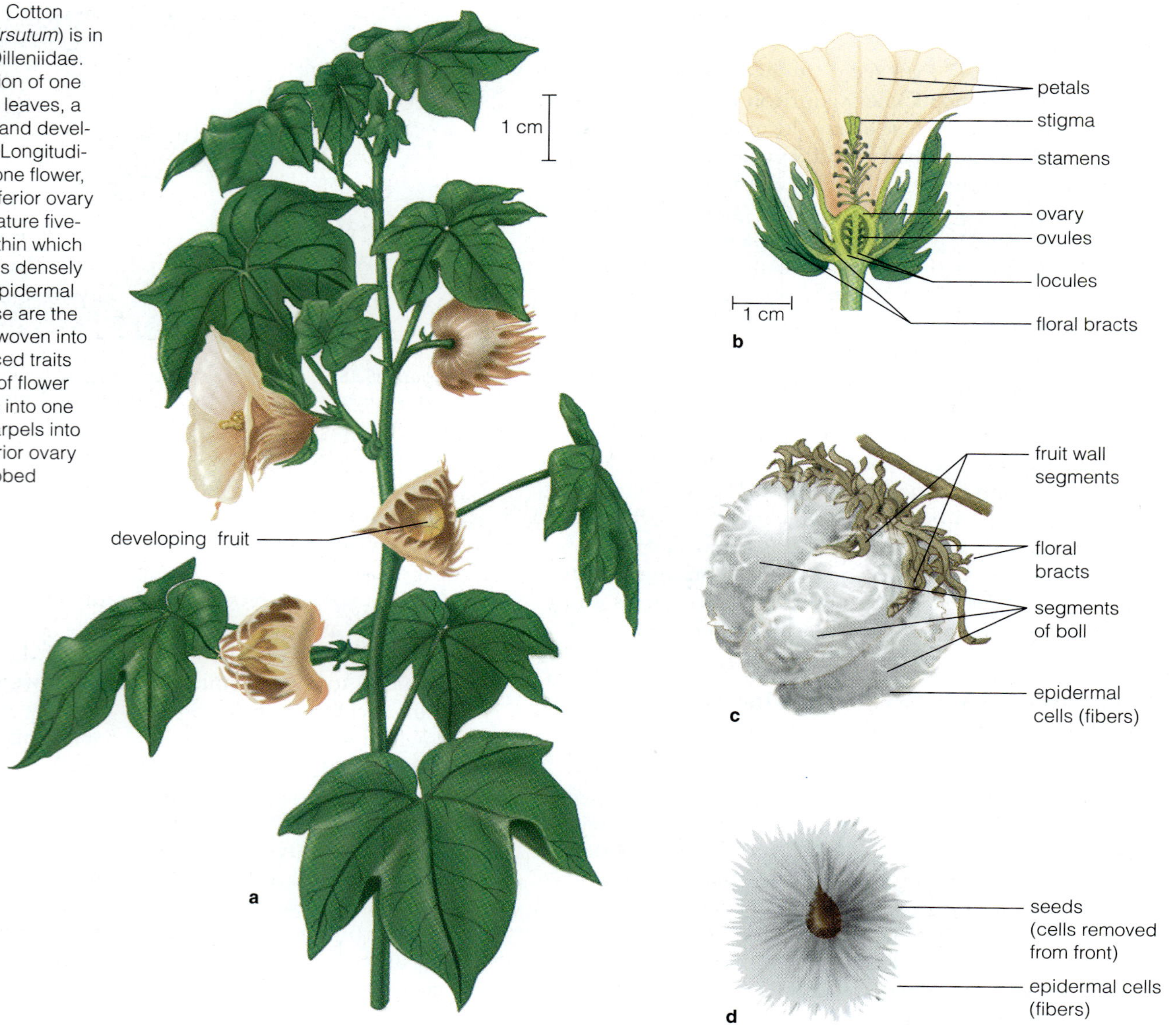

Figure 25.16 Cotton (*Gossypium hirsutum*) is in the subclass Dilleniidae. (**a**) Upper portion of one plant, showing leaves, a mature flower, and developing fruit. (**b**) Longitudinal section of one flower, showing the inferior ovary position. (**c**) Mature five-parted fruit, within which are many seeds densely covered with epidermal fibers (**d**). These are the fibers that are woven into cotton. Advanced traits include fusion of flower parts (stamens into one column, five carpels into one pistil), inferior ovary position, and lobed leaves.

1 cm

developing fruit

a

petals
stigma
stamens
ovary
ovules
locules
floral bracts

1 cm

b

fruit wall segments
floral bracts
segments of boll
epidermal cells (fibers)

c

seeds (cells removed from front)
epidermal cells (fibers)

d

25.6 PLANT GEOGRAPHY

Plant geography is the branch of plant biology that describes the distribution of plant taxa (orders, families, genera, and species) over the surface of the world. It also tries to explain how, why, or when these patterns of distribution occurred. Generally, seed plants—and particularly flowering plants—are the focus. Rather than classification theories or technological breakthroughs, studies of plant distribution require only field trips conducted by energetic botanists with keen observational skills.

Plant geography was stimulated by voyages of exploration in the 17th, 18th, and 19th centuries. Plants new to science were brought back to botanical gardens in Europe alive and whole, as seeds, or as pressed herbarium specimens (see sidebar, "Botanical Gardens, Smuggling, and Colonialism," p. 446). Important botanists who tried to make sense of this cascading mountain of information included Carl von Willdenow, Alexander von Humboldt, Johannes Schouw, August Grisebach, Alphonse de Condolle, Oscar Drude, Adolf Engler, George Marsh, Asa Gray, and Charles Darwin.

Humboldt, for example, led an expedition to Central and South America at the start of the 19th century. For five years he traveled in Cuba, Venezuela, Peru, Mexico, and the Orinoco and Amazon River basins. He walked through steamy lowland rain forest, semiarid thorn scrub, dry deserts, and cold alpine elevations as high as 6000 m (19,700 ft) atop Mt. Chimborazo. His expedition took along the best equipment of the day for measuring elevation, location, and weather. More than 60,000 plant specimens were collected. On his return to Europe,

Figure 25.17 The gourds (represented by *Cucurbita pepo*) are also in the subclass Dilleniidae. (**a**) Terminal portion of a creeping stem, showing leaves, dioecious flowers, and a fruit. (**b**) Male flower. (**c**) Female flower, with ovary sectioned to show ovules. (**d**) Young fruit in cross section, showing that the pistil is made of several fused carpels. Advanced traits include fusion of flower parts, inferior ovary, reduction in number of flower parts, unisexual flowers, and lobed, toothed leaves.

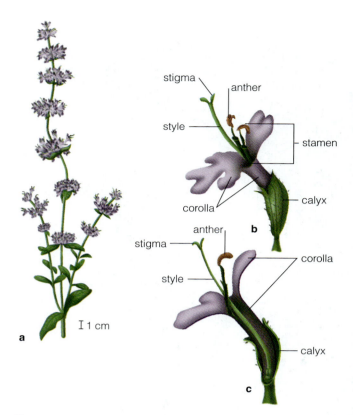

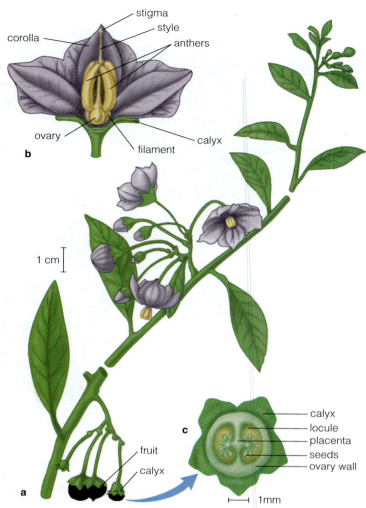

Figure 25.18 The subclass Asteridae contains mint and sage (represented by *Salvia mellifera*). (**a**) Upper part of one plant, showing opposite leaves, square stems, and lateral clusters of flowers. (**b**) One flower. Note fused petals and sepals. (**c**) Longitudinal section of a flower. Note that stamens are fused to the petals. In this species, some stamens do not fully develop and are small pegs attached to the corolla near the throat. Advanced traits include irregular symmetry, small flower size, fused petals and calyx, and nonterminal inflorescences.

Figure 25.19 The subclass Asteridae also contains the tomato family, here represented by nightshade (*Solanum parishii*). (**a**) Upper portion of one plant, showing flower buds, mature flower, and developing fruit. (**b**) One flower, with some stamens removed to show the superior ovary. (**c**) Cross section of a young berry fruit, showing two locules and developing seeds. Advanced traits include small flower size and fused petals, sepals, and stamens.

Humboldt wrote a monumental 30-volume work summarizing the expedition; the first 14 volumes were devoted to botany. He is generally credited with being the founder of the science of plant geography. In addition, his keen observations, records, and interpretations were important contributions to the new discipline of ecology.

It has become clear that the world's flora is not uniform. Although a dozen or so very large families of flowering plants are cosmopolitan, commonly found on every continent except Antarctica (see Table 25.1), most of the 380-plus families have a regional flavor or a habitat bias in their distribution. Entire families and groups of families are restricted to certain continents or regions. As a result, plant geographers have divided the world into floristically homogeneous units. Ronald Good's scheme

(Fig. 25.20) recognizes more than 30 global units. Each is characterized by its own endemic families and unique mixes of the more cosmopolitan families.

Studies by plant geographers have also shown that the vegetation of one particular kind of climate sometimes looks very similar wherever that climate recurs around the world—even though plants at each location may belong to unrelated taxa. For example, desert vegetation with cactuslike growth forms exists in the New and Old Worlds; the New World species, however, belong to the Cactaceae, whereas the Old World species are in the Euphorbiaceae (see Fig. 18.10). A similar, shrubby vegetation with a dense, stiff, small-leaved canopy occurs in five different places with a Mediterranean-type climate; yet the families of plants are different. Tropical rain

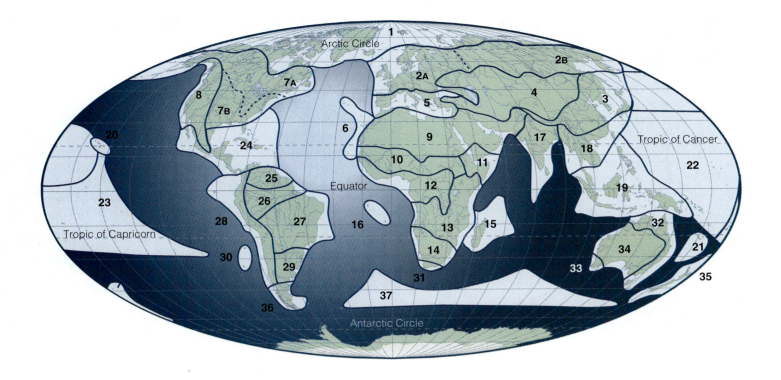

1. Arctic and Sub-arctic
2. Euro–Siberian
   A. Europe
   B. Asia
3. Sino–Japanese
4. W. and C. Asiatic
5. Mediterranean
6. Macaronesian
7. Atlantic North American
   A. Northern
   B. Southern
8. Pacific North American
9. African–Indian Desert
10. Sudanese Park Steppe

11. N.E. African Highland
12. W. African Rain-forest
13. E. African Steppe
14. South African
15. Madagascar
16. Ascension and St. Helena
17. Indian
18. Continental S.E. Asiatic
19. Malaysian
20. Hawaiian
21. New Caledonia
22. Melanesia and Micronesia
23. Polynesia
24. Caribbean

25. Venezuela and Guiana
26. Amazon
27. South Brazilian
28. Andean
29. Pampas
30. Juan Fernandez
31. Cape
32. N. and E. Australian
33. S.W. Australian
34. C. Australian
35. New Zealand
36. Patagonian
37. S. Temp. Oceanic Islands

Figure 25.20 Floristic regions of the world, according to Good. (Redrawn from R. Good, *The Geography of the Flowering Plants*. New York: John Wiley, 1961.)

BOTANICAL GARDENS, SMUGGLING, AND COLONIALISM

Botanical gardens are collections of living plants. They began as places where plants that were new to science (collected on voyages of discovery in the 16th and 17th centuries) could be propagated and studied. The Royal Botanic Garden at Padua, Italy, was the first to be established, in 1545. Other Italian gardens at Pisa, Florence, and Bologna soon followed. In the 17th century, gardens were constructed in France, the Netherlands, England, Germany, Sweden, Scotland, and Japan. Early botanical gardens were small and formally laid out; the English were the first to plant larger, more naturally landscaped gardens.

The Royal Botanic Gardens at Kew, England, are the finest botanical gardens in the world. They began on 4 hectares (10 acres) in 1759, but they have expanded to occupy more than 100 hectares (250 acres) today. Over 28,000 species of plants from around the world are growing in the garden (Fig. 1). Major sections include an arboretum (trees), tropical plants maintained in warm greenhouses, a rock garden, alpine plants in refrigerated greenhouses, and a grass collection. Libraries, research laboratories, collections of botanical art, and an herbarium of 6 million pressed specimens are also on the grounds.

Kew Gardens exemplifies several objectives common to most botanical gardens. They conduct basic taxonomic research on living specimens; investigate plants for potential economic value; propagate horticultural plants that have ornamental value; educate the public; provide the public with a place to enjoy nature; and provide a secure location to maintain rare plants.

Sometimes the collection of plants for botanical gardens has involved questionable activities. For example, by the mid-19th century, rubber had become a valuable, desired plant product. The industrial process of vulcanization had enormously expanded the use of rubber. Rubber comes from tropical plants of the New World, and the best rubber trees (*Hevea brasiliensis*) were native to Brazil. The Brazilian government definitely wished to maintain its monopoly and prohibited the export of rubber tree cuttings or seeds.

The British India Office, however, was determined to start rubber plantations in the new British colony of Ceylon. The India Office believed that Ceylon soils and climate were similar enough to those of Brazil to provide a suitable new home for rubber production. In 1876 the India Office secretly commissioned an English rubber worker named H. A. Wickham to smuggle out rubber tree seeds, agreeing to pay him $50 per thousand. Wickham proved to be very efficient: he collected 70,000 seeds, packing them in covered baskets labeled simply, "Botanical Specimens for Her Majesty's Gardens at Kew." He eventually found a

Figure 1   Kew Gardens, England.

willing ship captain with available space who slipped the seeds past the customs office at Belem without incident.

Once in England, the seeds were turned over to botanists at Kew Gardens, who carefully germinated and nurtured 2400 seedlings from the 70,000 seeds. Most of these plants were shipped within a year to plantations that had been made ready in Ceylon. By 1912, these plantations were producing a princely income of $45,000 per hectare (2.5 acres) per year. Not surprisingly, Wickham was knighted in 1920. Today, 95% of the world's rubber comes from Southeast Asia, and plantations have also been established in Africa.

Many other economically valuable plants were moved to new homes during the imperialist and colonialist times of the 17th to 19th centuries. Trade in plants was one of the driving economic forces during those centuries. For example, profits in brokering spice between Southeast Asia and Europe were every bit as attractive to investors as trade in gold, cotton, or slaves. Cornering the market of a spice or any other important plant species was a reward for aggressive imperialism. Consequently, coffee (*Coffea*) was taken from Africa to Central and South America, to Indonesia, and to Hawaii; tea (*Thea*) was moved from China to India; chili peppers (*Capsicum*) were transplanted from Mexico to Europe; pineapple (*Ananas*) and sugarcane (*Saccharum*) were hijacked from South America to Hawaii; rice (*Oryza*) was taken from Asia to Texas and California; oranges (*Citrus*) were brought from China to Florida. In all these cases, modern centers of production are far distant from the places where the plants were once native and restricted. Truly, humans are homogenizing the world's edible landscape.

Figure 25.22 Desert vegetation in Baja California, showing boojum trees (*Idria*), century plants (*Yucca*), and cardon cactus (*Pachycereus*). No other desert in the world has a similar collection of growth forms.

Figure 25.21 Alpine vegetation in African mountains. (**a**) Giant lobelia (*Lobelia rhynchorpetalum*). (**b**) Giant senecio (*Senecio keniodendron*).

forests of Africa, Australasia, and South America share a similar climate, a similar array of plant growth forms, and a similar forest architecture, but in each continent the assemblage of important families is different.

Sometimes, however, the environmental conditions, the isolation of a place, and the genetic potential of plants growing there combine to create unique and bizarre vegetation found nowhere else in the world. An excellent example is the flora and vegetation of alpine zones on tropical African mountains such as Mt. Kilimanjaro in Kenya. Tree-size plants in the normally herbaceous genera *Senecio* and *Lobelia* dominate the landscape (Fig. 25.21). The plants have clustered leaves at the ends of irregularly formed branches, and they may live for a century.

The desert of Baja California is another place where unique taxa and growth forms exist. The boojum tree (*Idria columnaris*), giant cacti, arborescent yuccas, and trees with succulent stems (*Bursera, Pachycormus*) give the horizon an otherworldly aspect (Fig. 25.22).

The process of evolution has not been equal for all taxa of flowering plants. Some orders and families have many species throughout the world but nevertheless have a limited range of growth forms. Other families have fewer species that are less widespread, yet they exhibit a wide range of growth forms. Still others are narrowly restricted and have very few species. Apparently, some families come with the genetic potential to be numerous, aquatic, succulent, woody, frost tolerant, or salt

tolerant; to evolve flowers pollinated by bats, or fruits dispersed by birds; or to accumulate herbivore-inhibiting chemicals. Others are not so genetically equipped. No group of plants has the potential to fill every terrestrial niche, but some groups can fill fewer niches than others. We also see that environmental stresses or problems are solved by different plants in different ways.

Perhaps future plant biologists will be able to explain why evolution has proceeded along the exact route it has taken among the flowering plants. For now, we can only marvel at the diverse results, treasure this biotic heritage of the past, and protect it for the future.

SUMMARY

1. Angiosperms are seed plants. The ovule and the seed that develops from the ovule are enclosed within an ovary, rather than borne naked as in the gymnosperms. The ovary is part of a new organ called a carpel or pistil, and the carpel is part of a new complex structure called the flower. Synonymous names for *angiosperm* include flowering plant, Anthophyta, and Magnoliophyta.

2. The fossil record of Magnoliophyta begins in the Cretaceous period of the Mesozoic era, about 130 million years ago. Paleobotanists theorize that the real beginnings of this division extend further back in time, to the Triassic period, and that the Magnoliophyta is linked to gymnosperms of that time (such as seed ferns or Bennettitales). Flowering plant fossils dating back to this period may be missing because they most likely originated in an

upland habitat where erosion would have prevented the sedimentation necessary for fossil formation.

3. The flower is thought to represent a modified leafy shoot.

4. Flowering plants dominated world vegetation throughout the Cenozoic era, despite major geographic, climatic, and vegetational changes. The temperate zone of North America showed a trend to a progressively cooler and drier climate. The last 2 million years of the Cenozoic era are the Quaternary period, which includes the Ice Age (Pleistocene) and Recent (Holocene) epochs. During the Pleistocene, ice sheets advanced and retreated several times. Temperate-zone vegetation is still recovering from the last glacial retreat, which ended 10,000 years ago. The past 10,000 years—the Recent epoch—may merely be another interglacial stage.

5. Humans have affected plant evolution during the Pleistocene and Recent epochs by selecting plants for food, managing the landscape with fire, and introducing exotic species as they wandered.

6. Novel aspects of the angiosperm life cycle include: (1) further reduction (when compared with gynmosperms) in the size and complexity of the gametophyte generation; (2) location of the ovule within another structure (the ovary); (3) a two-step process of fertilization, with one of those steps producing endosperm tissue; (4) dispersal of the seed within a fruit. In comparison with gymnosperms, these innovations further adapt angiosperms to life on land, and they help conserve plant food reserves; thus these adaptations have ecological consequences and may explain why angiosperms supplanted gymnosperms as the dominant terrestrial plant life-form.

7. About 250,000 species of flowering plants exist on earth today. Phylogenetic classifications of these plants are still subjective; therefore, it's not clear which classification scheme—among several recently proposed—is the best. The most widely accepted model assumes that the primitive condition is represented by *Magnolia*: woody plants whose xylem has few vessels; leaves that are simple and entire; inflorescence that is terminal and few-flowered; a pollen grain wall that is nonornamented; flowers that are large, regularly symmetrical, and with many free parts; an ovary that is superior; pollination by beetles; and seed that contains abundant endosperm and a small embryo.

8. The Magnoliophyta contains two classes, Liliopsida (monocots) and Magnoliopsida (dicots). The presumed most primitive subclass in the Liliopsida is the Alismatidae, which includes water plantains. The presumed most primitive subclass in the Magnoliopsida is the Magnoliidae. The most advanced subclasses are the Liliidae among monocots and the Asteridae among dicots.

9. Plant geography is the study of distribution patterns of plants over the world's surface. Flowering plants are usually the focus. Some plant groups are cosmopolitan, but others are more narrowly distributed and characterize floristic regions of the Earth. It is clear that each plant group has its own genetic limitations and cannot evolve to fit in all of the niches available. It is also clear that the same environmental stresses or problems are solved by different plant groups in different ways.

Questions

1. Diagram the evolutionary steps necessary to convert an ovuliferous pine scale into an ovary of a flowering plant. According to the fossil record as we understand it today, when did the first flowering plant appear?

2. Compare the size, anatomical complexity, and degree of independence of a fern gametophyte, a pine female gametophyte, and a magnolia female gametophyte. Which one is the most insulated and protected from the environment?

3. How can a fossil assemblage of flowering plants in a given location be used to estimate the climate at that location at that time? What assumptions are made in order to reconstruct the past climate?

4. According to the botanist Takhtajan, what angiosperm traits are primitive (representing early evolutionary groups) and what traits are advanced (representing most recently evolved groups? Do we have extensive fossil evidence to support Takhtajan's criteria?

5. In your opinion, which subclass of angiosperms is most important to the most people? Give some examples of important plants within that subclass. Are the families in this subclass rich in species? How are they distributed in the world? Do they represent a primitive or an advanced evolutionary state?

6. How can widely separated places in the world that share similar environments have similar vegetation—when the taxa of flowering plants that make up the vegetation are completely different from place to place?

Further Readings

Buchman, S. L., and G. P. Nabhan. 1996. *The Forgotten Pollinators.* Covelo, Calif: Island Press. The authors combine stories from the field with discussions of ecology, botany, and crop science for a fascinating account of plant–pollinator relationships. Examples come from the Galapagos, Panama, Malaysia, and the arid Southwest.

Burns, R. M., and B. H. Honkala, eds. 1990. *Hardwoods.* Vol. 2 of *Silvics of North America.* U.S. Department of Agriculture. Agriculture Handbook 654. Descriptions of 128 species of hardwoods that are major components of the landscapes of the United States. Each description discusses the biology, distribution, ecology, economic uses, and growth requirements of the species.

Cronquist, A. 1988. *The Evolution and Classification of Flowering Plants.* 2d ed. New York: New York Botanic Garden. The most complete modern text available on angiosperms. More technical and not as well illustrated as Heywood's *Flowering Plants of the World.*

Doyle, J. J. 1993. "DNA, Phylogeny, and the Flowering of Plant Systematics." *BioScience* 43: 380–89. A concise review of the kinds of molecular evidence currently being used as evidence of phylogenetic relationships, especially among flowering plants.

Heywood, V. H., ed. 1993. *Flowering Plants of the World.* New York: Oxford University Press. A reader-friendly description of more than 300 families of flowering plants. Each family is illustrated with color and black-and-white drawings and is described in terms of habitats, global distribution, evolutionary relationships, number of species, and economic uses.

Hill, C. R., and P. R. Crane. 1982. "Evolutionary Cladistics and the Origin of Angiosperms. Pages 269–361 in *Problems of Phylogenetic Reconstruction*. Edited by K. A. Joysey and A. E. Friday. New York: Academic Press. A technical summary of the evidence that angiosperms originated as far back as the Carboniferous, that angiosperms arose from gymnosperms, and that all angiosperms evolved from a common ancestor (that is, that angiosperms are monophyletic).

Jones, D. L. 1995. *Palms Throughout the World*. Washington, D.C.: Smithsonian Books. A complete survey of 800 species of palms, including a photo of each and notes on cultivation, natural distribution, and economic importance.

Parry, J. W. 1969. *Spices: The Story of Spices and the Spices Described*. New York: Chemical Publishing. An impressive historical and botanical book that shows how closely linked the colonial politics and geography of the 1500s–1800s were to the location of spices and their trade routes. The biology and cultivation of 37 major spices are also described, from allspice to turmeric. Ends with an excellent bibliography.

Perlin, J. 1989. *A Forest Journey*. Cambridge, Mass.: Harvard University Press. The history of the exploitation of Europe's forests, from the epic of Gilgamesh to industrial England. We have always lived in a Wood Age, the author concludes.

Sauer, J. D. 1988. *Plant Migration*. Berkeley: University of California Press. Readable summaries of plant introductions, plant eliminations due to pathogens or herbivores, and unusual natural plant distributions around the world—all written by a master plant geographer and storyteller.

Simpson, B. B., and M. C. Ogorzaly. 1995. *Economic Botany*. 2d ed. New York: McGraw-Hill. A well-organized, reader-friendly survey of major economically important plants throughout the world: plants used for building materials, food, beverages, fabric, and pharmaceuticals.

Takhtajan, A. L. 1980. "Outline of the Classification of Flowering Plants (Magnoliophyta)." *Botanical Review* 46: 225-359. The essence—the short version—of Takhtajan's classification model, including a full description of his criteria and a synopsis of each subclass with a complete list of all families.

Williams, M. 1989. *Americans and Their Forests*. New York: Cambridge University Press. The story of American exploitation of forests, from Native American burning to the clear-cutting of the eastern deciduous forest and of conifer forests in the Great Lakes states, the Southeast, and the West.

26 ECOLOGY, ECOSYSTEMS, AND PLANT POPULATIONS

1. Each organism has one of three roles in any ecosystem: producer, consumer, or decomposer. (a) Producers are green plants or protists that manufacture their own carbohydrate food from inorganic water and carbon dioxide; (b) consumers are animals that obtain food by ingesting other animals or plants; and (c) decomposers are nongreen protists or prokaryotes that digest dead organic remains of producers and consumers. The actual transfer of energy among these three categories is only a fraction of the potential energy; thus food chains or food webs have limited complexity.

2. The population is a local group of organisms belonging to the same species. Each population is genetically distinct. Most species have hundreds to thousands of unique populations, each population subtly adapted to variations in the local environment; thus the population is the basic ecological unit of any species.

3. Every species can be assigned to one of a relatively small number of life history patterns. Each pattern represents a unique budget of time, activities, and resources that allows a population to continue in existence from generation to generation. Some of the activities budgeted are germination, growth, and reproduction.

4. The distribution of a population is affected by abiotic factors such as soil nutrient level, soil moisture availability, intensity of solar radiation, or the incidence of wildfire. Every plant must solve a zero-sum budget for dissipating incoming solar radiation energy into reflection, reradiation, convection, metabolism, storage, and transpiration.

5. Distribution is also determined by biotic interactions with other species, such as competition (in which a substance is removed from the environment), amensalism (in which a substance is added to the environment), herbivory (consumption by an animal), and mutualism (cooperative behavior in which the probability of survival is increased for both interacting populations).

Figure 26.1 A year-old conifer seedling near the base of its parent tree. During its life span of several hundred years, the resulting plant must successfully budget time and resources to satisfy simultaneous demands of growth, maintenance, reproduction, competition, and herbivore defense.

26.1 ECOSYSTEMS AND BIOMASS PYRAMIDS

Plant ecologists seek to find an underlying order to the pattern of plant distribution over the world's surface. What threads link plants to each other and to their environment? How flexible are those threads, and how intertwined? How do plants solve the problems of germination, acquisition of energy and nutrients, competition, avoidance of herbivores, reproduction, and dissemination of seeds? Are there as many ways to allocate time and energy during a life cycle as there are species, or are there only a small number of successful "budgets" that many species share (Fig. 26.1)? How do plants withstand natural disturbances such as fire, flood, or storm? What can plants tell us, by their presence, vigor, or abundance, about the past and future history of a habitat? Can plants be used as a scientific tool to analyze the environment, to test hypotheses about the course of evolution, or to lead us to wise management of the land?

All these questions, and more, are being investigated by plant ecologists. The word *ecology* was coined more than a century ago by the German zoologist Ernst Haeckel. It comes from the Greek roots *oikos*, meaning "home," and *logos*, meaning "study of." Therefore, ecology is the study of organisms in their home, the environment. More generally, **ecology** is the study of organisms in relation to their natural environment.

Environment is the sum of all biotic (living) and abiotic (nonliving) elements that surround and influence an organism. *Environment* is synonymous with the terms **habitat** and **ecosystem**. Examples of biotic elements in an ecosystem are neighboring plants, animals, and soil microbes; examples of abiotic elements in an ecosystem are temperature, moisture, wind, sunlight, soil nutrients, and episodic fires. Depending on the organisms being studied, a habitat or ecosystem can be a few square centimeters of bare rock being colonized by mosses, a small lake filling with sediment, or an entire mountain range.

All the hundreds of species within any one ecosystem can be conveniently placed in a few trophic (nutritional) categories. Photosynthetic plants, protists, and prokaryotes are considered **producers** because they generate hydrocarbon food from sunlight and inorganic chemicals alone. Herbivorous animals and plant parasites are **primary consumers**. Carnivorous animals are **secondary consumers**, and carnivores that consume secondary consumers are **tertiary consumers**. Carrion feeders, nonphotosynthetic protists, and microbes are **decomposers**. Consumers and decomposers need organic hydrocarbons

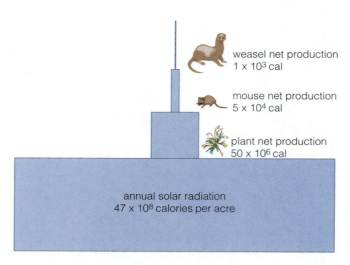

weasel net production
1×10^3 cal

mouse net production
5×10^4 cal

plant net production
50×10^6 cal

annual solar radiation
47×10^8 calories per acre

Figure 26.2 A typical ecosystem biomass/energy pyramid for an acre of grassland in southern Michigan. Notice how steeply each trophic level's biomass content declines. The decline is a result both of inefficiency in gathering energy at each level and of the use of most of that energy for respiration rather than net growth. In this particular grassland, solar radiation reaching the ground totals 47×10^8 calories over the course of one year. Plants are able to fix only 1% of that radiation in photosynthesis, and they use up most of the fixed calories in respiration. At the end of the growing season, the total plant biomass contains 50×10^6 calories. Of this, mice—the major primary consumer—eat about 1%. The mice use up much of that food in respiration, producing a net biomass at the end of the year of only 5×10^4 calories. Weasels are the major secondary consumer, and they feed on mice, consuming 10% of mouse biomass. Weasel net growth at the end of a year is 1×10^3 calories. Thus the net retention of energy through only three trophic levels (from solar radiation to weasel growth) is only 0.00002%.

Figure 26.3 An aspen (*Populus tremuloides*) population. Aspen is the most widespread tree species in North America. It is capable of spreading asexually. This population of many individuals could all belong to the same genotype, but most populations consist of a mixture of genotypes.

to survive, and these hydrocarbons are ultimately manufactured by producers.

Each trophic category can be summarized in terms of the number of species that belong to it, in terms of the tissue weight of all the members (the **biomass**), and in terms of energy content (usually expressed in terms of calories, where an average gram dry weight of plant or animal tissue is equivalent to 5000 cal). The biomass or the energy content of each trophic level is linked to other levels in a predictable manner: Each level is no more than 1/10 that of the previous level. For example, 100 g (5 oz) of producers (plants) per square meter (square yard) of ground can support at most 10 g (0.5 oz) of primary consumers; those in turn can support at most 1 g of secondary consumers; and those in turn can support at most 0.1 g (0.05 oz) of tertiary consumers. The reason for the 1/10 efficiency is that most of the energy at any one level must go into the growth, maintenance, and reproduction of the organisms within that level. If all the biomass were harvested by the next trophic level, the ecosystem would crash. Instead, measurements show that only 10% of the energy or biomass is harvested.

If a secondary consumer is a large bird of prey, weighing 4 kg (about 9 lb)—instead of 1 g—it will have to forage over an area of 4000 m² (about 1.5 acres). In fact, the foraging area would have to be many times larger because the

bird does not feed on all herbivore species, instead choosing only some of them; furthermore, in nesting season the foraging area must be large enough to support two parents and their chicks. This relationship is called a **biomass pyramid** (Fig. 26.2). The pyramid declines steeply because of this 10-to-1 transfer factor between trophic levels.

26.2 THE POPULATION: THE BASIC ECOLOGICAL UNIT

Organisms are usually not studied as individuals by ecologists; rather, they are grouped together into populations. A **population** is a group of freely interbreeding individuals of the same species occupying the same habitat. A plant population might consist of thousands of individuals, especially if they are small; or it might include only a few individuals, if they are rare. Not all individuals in a population actually interbreed every year, sharing pollen and genes, but they all grow close enough together so that they could *potentially* interbreed. Enough of them do interbreed to maintain a relatively homogeneous mix of genes throughout the population.

In special circumstances, a population can consist of many genetically identical individuals. Aspen (*Populus tremuloides*) reproduces not only sexually but also by underground roots and sucker shoots. A grove of aspen (Fig. 26.3) may have hundreds of trees, all connected underground, representing a single clone.

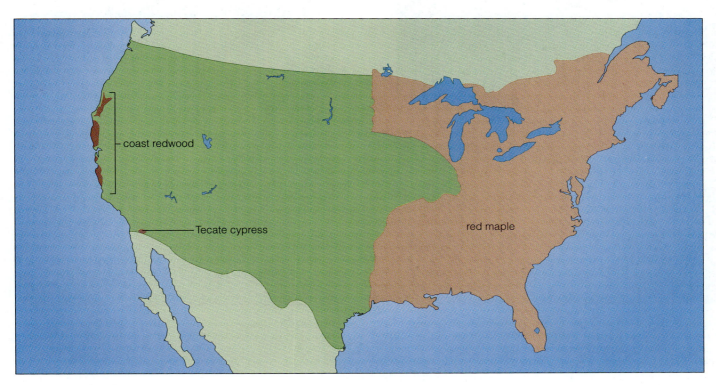

Figure 26.4 Distribution limits of a common species, red maple (*Acer rubrum*); an uncommon species, California coast redwood (*Sequoia sempervirens*); and a rare species, Tecate cypress (*Cupressus forbesii*). (Redrawn from U.S. Dept. of Agriculture, *Silvics of Forest Trees in the United States*. Handbook No. 271.)

Although a population consists of only a single species, most species are made up of many populations. Each population occupies the same kind of habitat, scattered over a landscape or region. Some widespread species have thousands of populations, which range over many degrees of latitude or longitude and through hundreds of meters of elevation. In such a case, populations at the extremes of the species' territory are isolated from each other and have very little opportunity to share genes. Over time, these outlier populations may evolve different structural or functional traits that allow the species to exploit new habitats. Red maple (*Acer rubrum*) is a good example of a widespread species: It ranges throughout the entire eastern United States (Fig. 26.4) into habitats that vary from wet to dry. An uncommon species, such as California coast redwood (*Sequoia sempervirens*), has narrower habitat requirements than red maple. As a result it has a smaller range and only several populations (Fig. 26.4), though the number of individuals in each population is large. The rarest species with the narrowest habitat requirements may have only one or a few small populations. For instance, the closed-cone conifer Tecate cypress (*Cupressus forbesii*) occurs in only four little groves, all in southern California (Fig. 26.4).

Ecologists who study single populations and environmental factors that affect the population are engaged in **population ecology**. In contrast, **community ecology** or **ecosystem ecology** is the study of groups of different populations that coexist in the same habitat and of the environmental factors that affect them. We will discuss population ecology in this chapter and community and ecosystem ecology in the next chapter.

26.3 LIFE HISTORY PATTERNS

Every plant population has a problem to solve: how to allocate time and energy during the life cycle so as to maximize the probability of successful reproduction. Every plant has limits to its life span, its size, and the resources available to it; therefore, the plant's budget of time and energy is not as open-ended as, say, the national debt.

Ecological research now indicates that there are only a few basic kinds of life cycle budgets and that each kind is shared by many species. We can think of each budget as a life history pattern: a collection of inherited traits and behaviors that permits a plant population to survive. Life spans, growth forms, timing of reproduction, and sexuality are the major components of life history patterns.

Life Spans Are Annual, Biennial, or Perennial

Annual plants go from seed to seed in less than one year. They are usually small and inhabit open areas with short, unpredictable growing seasons. Desert annuals are good

Figure 26.5 Desert ephemerals in flower. These annuals actually have a very flexible life span, which may be as short as three months or as long as eight months, depending on temperature and rainfall.

Figure 26.6 Jack-in-the-pulpit (*Arisaema triphylla*), an herbaceous perennial that is dioecious. Male and female plants differ in size.

examples (Fig. 26.5). *Boerhaavia repens* (desert four o'clock) of the Sahara is a small annual that can go from seed to seed in only 10 to 14 days, but most annuals have a life span of three to eight months. Perhaps the term **ephemeral** is more appropriate than *annual*, which implies a full year of life.

Winter annuals in the Mojave Desert of California usually germinate in November, with the onset of winter rains. They grow slowly during the winter, mainly developing a root system and a whorl of basal leaves. Warming temperatures in March and April bring rapid stem elongation and flower production. Seed maturity occurs by May, and the parent plant dies in June, killed by both the exhaustion of energy reserves that went into reproduction and by the stresses of summer heat and drought. The total life span is about eight months. However, the same winter annuals can complete their life cycle in only three to four months if winter rains are delayed until March. The plants then skip the winter phase of growth and go on to the reproductive stage; they flower as much smaller individuals but nevertheless complete seed set before dying on schedule in June.

Biennials live for two years. In the first year, growth takes place in the root system and a basal rosette of leaves. Cold temperatures during the first winter induce hormonal changes, so that during the second spring the shoot elongates and produces flowers. After setting seed, the plant dies. Carrot (*Daucus*) and lettuce (*Lactuca*) are examples of biennials. Most biennials can be induced experimentally to live longer than two years, so we know that they are not genetically programmed to die in that span of time. If a biennial is kept warm over the winter, it

remains in the vegetative state longer. Alternatively, biennials can be sprayed with gibberellin and experimentally induced to flower during the first year.

Perennials live for many years and usually flower repeatedly. Their growth form may be herbaceous or woody. **Herbaceous perennials** die back to underground parts each winter. Only their bulbs, corms, roots, or rhizomes are perennial; the aboveground leaves and stems are annual. *Iris, Gladiolus*, onion (*Allium*), potato (*Solanum*), the ferns, many wild grasses, and most spring-flowering herbs of the eastern deciduous forest (Fig. 26.6) are herbaceous perennials. They often occupy moderately harsh environments such as desert, seasonally dry grassland, shaded forest floors, or alpine regions.

It's difficult to determine the age of an herbaceous perennial because a woody stem with growth rings is absent. The perennial organs may retain a scar from each year's leaf system, however; then the scars can be counted to determine plant age. Other perennials spread outward vegetatively from rhizomes or runners, and their age can be estimated by knowing the annual rate of spread. From these sorts of evidence, we think that many herbaceous perennials live 20 to 30 years.

Woody perennials accumulate aboveground stem tissue year after year. What makes them woody is that the stems have secondary growth. The size and shape of the stems determine whether the plants are subshrubs, shrubs, vines, or trees.

Subshrubs are multibranched and genetically dwarfed, seldom becoming taller than 30 cm (1 ft). At the end of each growing season, their stems die back partially, but not all the way to the soil surface. They have short life spans,

Figure 26.7 Creosote bush (*Larrea tridentata*), a common shrub of the warm deserts of North America. Individuals can live for more than 100 years. Individuals are able to reproduce vegetatively, producing a continually widening circle of offspring. Some of these clones are estimated to be several thousand years old.

similar in length to those of herbaceous perennials. They often invade disturbed sites and are later overtopped and outcompeted by taller, longer-lived shrubs and trees. Rabbitbrush (*Chrysothamnus* species) is a common subshrub of the intermountain Great Basin.

Shrubs are also multibranched, but they do not die back annually. Sometimes an entire branch may die, but the tips of all living branches continue to grow taller. Shrubs have life spans that may exceed a century. Some reproduce asexually with underground stems, and an entire clone can attain a very great age. Creosote bush (*Larrea tridentata*), for example, is a desert shrub capable of spreading vegetatively from the base. Some clones are old enough to form a ring several meters in diameter (Fig. 26.7), estimated to be several thousand years old.

Vines (called **lianas** in the tropics) have weak, single trunks. They require additional support from neighboring shrubs or trees, obtaining that support by twining about the host's trunk or literally sprawling on top of the host's leafy canopy. Common vines in the eastern deciduous forest include Virginia creeper (*Parthenocissus quinquefolia*) and Dutchman's pipe (*Aristolochia durior*). Vines have secondary wood, but typically that wood is light because it contains many more vessels than tracheids and fibers. Such wood is metabolically inexpensive to build, leaving more energy to be allocated to the plant's growth in height. As a consequence, vines have a relatively rapid growth rate. The life span of a vine is similar to that for subshrubs and shrubs.

Trees have strong, single trunks and an elevated branch system that does not die back annually. As with shrubs, individual branches or even a portion of the trunk may die, but the rest of the tree continues to grow in height and girth. **Broadleaf trees**, such as oak (*Quercus* species), are typically angiosperms; they are also called **hardwoods** because of the density of their secondary wood. **Needle-leaf trees**, such as pine (*Pinus* species), are typically gymnosperms; they are called **softwoods**.

Hardwoods and softwoods can be either evergreen or deciduous. If they are **evergreen**, each individual leaf has a life span longer than 1 year; if they are **deciduous**, all leaves have a life span less than 1 year and fall synchronously (at the same time), leaving the tree bare for part of the year. Leaf life spans mirror the pattern for tree life spans, being longer for softwood trees than for hardwood trees. Broadleaf evergreens, such as some southern and western oaks, have leaf life spans of only 2 years; but needle-leaf evergreens, such as some mountain pines, have leaf life spans of 5 to 30 years.

Evergreen leaves, also called **sclerophylls** (hard leaves), are energetically more expensive to manufacture than deciduous leaves. They are thick, tough in texture, and bounded by a well-developed cuticle; they also have few air spaces in the mesophyll, exhibit a low surface-to-volume ratio, and contain high concentrations of metabolic by-products. Collectively, these features mean that it takes more carbohydrates to build an evergreen leaf than a deciduous leaf. Evergreen leaves are more drought tolerant than deciduous leaves, but at the same time they have a lower maximum rate of photosynthesis. Ecologists think that the long life of sclerophylls and their efficient use of water are compensations for their expensiveness and low photosynthetic rates.

Reproduction Can Be Semelparous or Iteroparous

Timing of reproduction is another component of a life history pattern. Annuals, biennials, and some herbaceous perennials reproduce sexually only once, at the end of their life spans. This is called **semelparous reproduction** (from the Latin *semel*, meaning "once," and *parere*, to "bear"). An advantage of semelparous reproduction is that a single burst of reproduction, delayed to the time of maximum plant size, yields so many seeds that some are likely to germinate and live long enough to produce the next generation. The California chaparral plant our Lord's candle (*Yucca whipplei*, Fig. 26.8), is a semelparous perennial that flowers at the end of its 20-year life span. Semelparous plants spend 25% of their stored caloric resources on their single, massive reproductive event. It is possible that they die by exhausting their food reserves in setting so many flowers and seeds in such a short time.

In contrast, **iteroparous** plants reproduce many times during their lives (*itero* means "repeat"). Each reproductive event uses 10 to 15% of the plant's food reserves. Iteroparous plants go through a juvenile period before the first reproductive event. The length of the juvenile period depends on the species; in general, it accounts for 10% of

Figure 26.8 Our Lord's candle (*Yucca whipplei*), a semelparous perennial. It flowers only once, at the end of its 20-year life. This species grows in coastal hills of central and southern California.

the entire life span. In commercial fruit trees, the juvenile phase lasts four to six years, but wild plants often have a much longer juvenile period. Some iteroparous species reproduce annually once the juvenile period is over. Others, such as oak, pine, and fir, reproduce every few years or reproduce more abundantly some years than others. The years of high reproduction are called **mast** years.

Sexual Identity Is Not Always Fixed

The distribution of sexual identity by individuals is yet another component of life history patterns. Most species of angiosperms are monoecious, with male and female flower parts (stamens and pistils) on the same individual. Other species, which have only male or only female flower parts on one individual, are dioecious. We can imagine that the advantage of being dioecious is that cross-pollination is absolutely required; thus genetic recombination and variation are high. The advantage of being monoecious is that a single isolated individual can reproduce successfully. Most weedy, invasive species are monoecious.

In dioecious species, the sex of the individual plant is usually genetically determined, but not by X and Y chromosomes, as in humans. Sometimes sex is not under any genetic control but is a consequence of plant size or habitat conditions. Jack-in-the-pulpit (see Fig. 26.6) is an herbaceous perennial in which the sex of an individual is environmentally controlled. Very young individuals go through a juvenile phase during which they grow vegetatively and become larger but do not reproduce. Plants exhibiting sexuality are older and larger in size (as measured by total area of leaves produced and volume of perennial underground parts) than the juveniles. Of these adult plants, the largest—but not necessarily the oldest—are female. Sexuality in this case is actually a function of

size, not of age. If a female plant is manipulated so that it does not accumulate much stored food one year, it will produce fewer leaves the next year, and the size of its underground system will shrink. If the loss in size is significant, the plant will become male. As it accumulates stored food over the years, it will become female again. If male plants are similarly starved, the following year they may become nonreproductive, like juveniles. As they grow, they will become male again.

Some species segregate sex by habitat. Female plants occupy moister or nutritionally richer soils, whereas males occupy drier, sterile sites. Female plants require better sites because so much stored food must go into seeds and fruits, but male plants can occupy marginal sites because less energy is needed to produce pollen. Dioecious species thus subdivide the habitat and take advantage of all sites. Another advantage of the dioecious state is that seed predation by animals is reduced. Female plants—with their energy-rich seeds and fruits—are mixed with males across the landscape; thus, an herbivore that finds a female plant may not encounter another female plant nearby, making it unlikely that the herbivore will devour all the seeds in a locale.

Patterns of sexuality in other species are less obvious. Red maple (*Acer rubrum*) individuals, for example, seem to change sex randomly from year to year, very few remaining constantly male or female. We still have much to learn about the allocation of sexuality in plants of dioecious species.

Life History Patterns Range from r to K

When we put life span, growth form traits, timing of reproduction, and allocation of sexuality together, some concrete life history patterns emerge. It is useful to recognize two extremes because all other types can be arranged between the extremes. Those extremes are called *r* and *K* life history patterns (Table 26.1).

Typically, **r-selected species** are annuals in open habitats. They have a rapid growth rate, deciduous leaves, and a short juvenile period; they are semelparous and allocate considerable caloric energy to reproduction. **K-selected species** are usually perennials with a larger body size. These plants inhabit more stable habitats, covered by a more continuous plant canopy. They may be evergreen or deciduous. Compared with r-selected species, they have a slower growth rate and a longer juvenile stage; they expend less energy on reproductive events and are iteroparous. Basically, r-selected species maximize reproduction in unstable habitats at the expense of long life and large body size, whereas K-selected species maximize long-term occupation of a site and competitive abilities at the expense of rapid growth and early reproduction. Most species lie somewhere along a continuum between these extremes.

Another classification of life history patterns recognizes three extremes (Fig. 26.9) instead of two. In this sys-

Table 26.1 Some Traits of r- and K-selected Species

Trait	r-selected	K-selected
Habitat (growing conditions)	Unpredictable from year to year	More constant or predictable
Population size	Variable from year to year, usually below the carrying capacity of the habitat	More constant, often at or near the carrying capacity
Survivorship	Type I or II (see Fig. 26.13)	Typically, Type III
Competition	Variable and usually not an important factor	Usually keen
Life span and reproduction	Usually short life span with semelparous reproduction, which has a high reproductive effort	Long (perennial) life span with iteroparous reproduction, low reproductive effort each time
Growth pattern	Rapid growth and development, small body size at maturity, short juvenile period	Slower growth and development, larger body size at maturity, long juvenile period

tem, the **R type** (abbreviation for *ruderal*, or *roadside*) is equivalent to the r in the r–K continuum we just described. The **C type** (for *competitive*) is equivalent to the K. The **S type** (*stress-tolerant*) is a new category, representing species that grow in stressful habitats such as salt marshes, rock outcrops, moving sand, and deserts. Plant density in such habitats is low; thus, a tolerance of biotic stresses (such as competition) is less important than a tolerance of abiotic stresses.

R, C, and S species can sometimes grow in close association. For example, rock outcrops in the Appalachian Mountains often are occupied by islands of vegetation (Fig. 26.10). At the outermost edge, growing on bare rock, are the mosses *Grimmia*, *Polytrichum*, and *Bryum*. These slow-growing plants are S types. Further into the island of vegetation, growing on top of a thin layer of soil that mosses have accumulated, occur herbaceous R-type plants such as hair grass (*Deschampsia flexuosa*). In the center, with deepest soil, grow C-type shrubs and trees such as *Rhododendron* and chestnut oak (*Quercus prinus*). Most ecologists agree that this tripartite division is an improvement in our understanding of life history patterns.

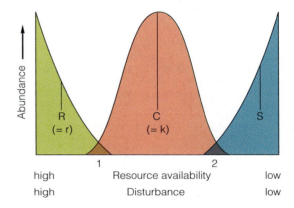

Figure 26.9 Diagram of the abundance of ruderal or roadside (R), stress-tolerant (S), and competitive (C) plants along environmental gradients of resource availability and frequency of disturbance. At point (1) disturbance is rare enough to favor C types over R types. At point (2) resources become so limiting that S types are favored over C types. (Redrawn from J. P. Grime, *American Naturalist* 111 (1977):1169–94.)

26.4 PLANT DEMOGRAPHY: POPULATION AGE STRUCTURE OVER TIME

A population usually contains individuals of all ages: seedlings, juveniles, reproductive adults, and senescent oldsters. The proportion of the population in each age category tells us something about the history and future of that population. For example, Figure 26.11 is a summary of the ages of all pine and hardwood individuals in a forest near Gainesville, Florida. (Actually, ages are not shown; rather, the diameter at breast height (dbh) of the trunks is indicated. However, tree age is generally pro-

Figure 26.10 C-, S-, and R-selected plants may occur in the same local habitat, separated only by microenvironmental factors. Rock outcrops in the Appalachian Mountains show a concentric zonation of plants, with S types at the outermost fringe, R types in a middle zone, and C types in the very center.

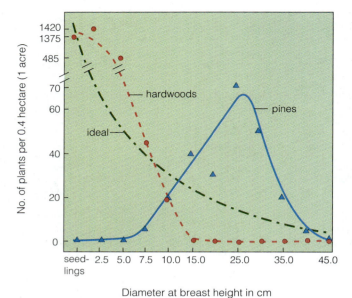

Figure 26.11 Age distribution (based on trunk diameters at breast height, dbh) for pines and hardwoods in a forest near Gainesville, Florida. The pine population is senescent, and the hardwood population is invading. The green line shows the expected age distribution of a stable population, in contrast.

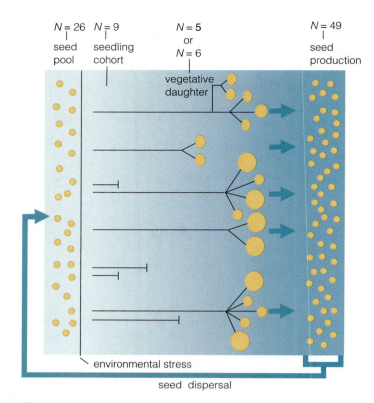

Figure 26.12 A model of one population's life history. Time increases along the horizontal line, from left to right. This species is semelparous and flowers only once, at the end of its life span. It is capable of vegetative production during this life span. Note that not all seeds germinate and that not all seedlings reach maturity. N is the number of living individuals in the cohort at the time. Because the seed output is larger than the initial seed pool, this population is growing. (Redrawn from J. L. Harper, *Population Biology of Plants*. New York: Academic Press, 1977.)

portional to dbh. A tree 25 cm (10 in) dbh is about 100 years old in this forest; one 45 cm (18 in) dbh is about 200 years old.) Notice that the pine population in the figure has many mature trees 15 to 30 cm (6 to 12 in) dbh, but it has no seedlings, saplings, or young trees smaller than 5 cm (2 in) dbh. From this, we conclude that the pine population has not been reproducing itself for some decades. The hardwood population, in contrast, has only seedlings and saplings, with no large, mature individuals. This suggests that the hardwood population is a recent invader of the forest.

Putting the two population trends together, we can predict that the pine forest will become a hardwood forest. The present overstory pines will eventually reach their natural life span limits, of about 200 years, and die. As they leave the canopy, they will be replaced by the young hardwoods growing beneath them. In time, the forest will become a hardwood forest, and pines will be rare or absent. When that happens, we can predict that the distribution of ages is likely to follow the green curve shown in Figure 26.11: There will be many young hardwoods and fewer and fewer old ones. Such an age structure will be stable over time because as each old tree dies, the probability is high that youngsters of the same population are already growing beneath it and that they will grow up to maintain the same overstory composition.

Plant demography is the study of changes in population age structure over time. By examining age structure, a demographer can create a mathematical model for predicting how long an average individual will live, when it will enter and leave reproductive age, how many seeds it will shed in its lifetime, and how many of those seeds will survive to germinate. A diagram of such a model appears in Figure 26.12. Let's tease apart the model.

On the left, represented by black dots, are all the seeds currently in the ground; this is the **seed pool**. These seeds are 26 in number ($N = 26$). Environmental and physiological stresses (such as being eaten by animals, parasitized by soil microbes, or kept dormant by some internal mechanism) take their toll, so that only a fraction of those seeds (9 out of 26) germinate this year. The 9 that germinate together are part of a single **cohort**, a segment of the population that is the same age.

Of the 9 germinated seeds, environmental stresses—such as heat, drought, or disease—kill 4 before they can reproduce (represented in Fig. 26.12 by short, truncated lines). The remaining 5 plants of the cohort all reach sexual maturity and produce flowers (indicated as open circles in Fig. 26.12). One member of the cohort reproduces vegetatively before this event, producing an additional sixth plant, which also flowers. Notice that not all the

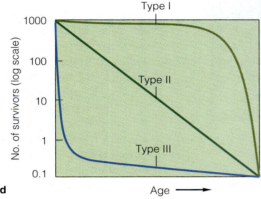

Figure 26.13 Three different survivorship curves. (**a**) A Type I curve is characteristic of annuals, (**b**) Type II of plants in extreme environments, (**c**) and Type III of large perennials. (**d**) Notice that the vertical scale is logrithmic, meaning that each population unit changes by an order of 10. (Redrawn from E. S. Deevey, Jr., *Quarterly Review of Biology* 22 (1947): 283–314.)

plants produce the same number or sizes of flowers. Their reproductive effort probably correlates with the particular patch of ground each plant occupies.

This particular model is for a semelparous species because all 6 plants die upon release of seeds (the horizontal lines stop). Notice that the number of seeds released (49) is much greater than the number we started with (26). Therefore, we might predict that this population will grow in numbers. Because about 35% of the original seed pool germinated ($9 \div 26 \times 100 = 35\%$), we can predict that next year's cohort of seedlings should total about 17 (35% of 49 seeds).

Another way of presenting changes in population age structure over time is in the form of a survivorship curve (Fig. 26.13). Plant and animal populations have three basic kinds of survivorship curves. A Type I curve is characteristic of r-selected (or R-type) populations, such as annual plants. There is high survivorship (low mortality) for most of the life span; then all the individuals die within a relatively short time. A Type III curve is characteristic of K-selected (or C-type) populations, such as forest trees. There is high mortality early in life but then very low mortality once the individuals reach maturity. A Type II curve is characteristic of S-type populations such as alpine herbaceous perennials. There is constant mortality throughout the life span because of the overwhelming importance of abiotic environmental stress at all stages of the life cycle.

26.5 POPULATION INTERACTIONS WITH THE ENVIRONMENT

We have already divided the environment into biotic and abiotic components, but an additional division is useful. The **macroenvironment** is that part of the environment determined by the general climate, elevation, and latitude of the region. Weather bureau data on rainfall, wind speed, and temperature are measurements of the macroenvironment. Such measurements are taken at a standard height above the ground, in cleared areas well away from buildings or trees.

The **microenvironment** is that part of the environment determined by conditions close to the surface of a plant, animal, structure, or the ground. The microenvironment is modified by nearby surfaces. For example, bare soil tends to absorb heat; consequently, on a clear day the temperature just above or below the soil surface is much higher than air temperature. Prostrate plants with basal leaves will experience microenvironmental temperatures much warmer than those recorded by the weather bureau 1.5 m (5 ft) above the ground. Microenvironmental air as far as 10 mm (0.4 in) from the surface of a leaf on a still day is less turbulent, higher in humidity, and warmer than free air farther from the leaf. Light quality and quantity are much different for herbs in the microenvironment beneath a forest canopy than for the leaves of the canopy itself. Herbs on the forest floor sense

Figure 26.14 An epiphytic cactus in a semitropical forest.

perhaps only 1 to 5% of the light received at the top of tree canopies. Moreover, the light that passes through the canopy is relatively enriched in green and far-red wavelengths because photosynthetic pigments in the canopy leaves absorb in the blue and red portions of the spectrum. Plants growing in shallow depressions experience frost more frequently than those growing on higher ground because cold air flows downhill. Soils may also be wetter and more anaerobic in depressions. Populations differ in their range of tolerance to both microenvironmental factors and macroenvironmental factors.

Water and Soil Are Important Environmental Factors

Probably no single abiotic factor is as significant in the distribution of plant populations as the supply of water. So important are plant–water relationships that species have been formally classified into **xerophytes** (plants able to grow in very dry places), **mesophytes** (those that grow best in moist soil), **helophytes** (those able to grow in saturated soil such as marshes and bogs), and **hydrophytes** (those able to grow rooted, submerged, emergent, or floating in standing water).

Plants in each category have unique morphological, anatomical, and physiological traits that adapt them to a particular moisture regime. Xerophytes, for example, may have small, hard leaves; an epidermis with thick cuticle, light color, and hairlike trichomes; stomata that are sunken in epidermal depressions and that close early during the day; and, sometimes, **succulent** tissue (the cells of which are large and contain water-filled vacuoles). Or leaves may be permanently absent or seasonally absent. If

they are permanently absent, green stems conduct most of the annual photosynthesis; if they are seasonally absent, they fall during extended rainless periods, rather than in cold periods; such a plant is **drought deciduous**. Root systems may be extensive.

Xerophytes can live in humid areas as well as dry ones, provided that their roots are in dry microenvironments. Some xerophytes live in tropical rain forests as **epiphytes**—plants that grow on tree trunks or branches with their roots twining through relatively dry bark; they obtain all their water by trapping rainfall in leaf axils. Many of them have succulent leaves and stems and specialized water-storing tissue on root surfaces. One common group of tropical epiphytes belong to the Cactaceae, or cactus, family (Fig. 26.14). Cacti probably evolved in the tropics, later moving into dry deserts as those habitats appeared in the late Cenozoic. Other important epiphytic families are the Bromeliaceae, to which the pineapple (*Ananas*) belongs, and the Orchidaceae, the orchids.

Other xerophytes live on soil whose physical or chemical properties modify the amount of free water available to plant roots. Very coarse soils, high in sand or gravel, permit most rainwater to percolate rapidly, leaving an island of aridity within a climate that might seem wet and adequate for luxuriant plant growth. Similarly, shallow soils or rock outcrops store little water in the root zone. Soils high in clay can potentially store much water, but the rate at which they can absorb water is slow; if rains come in torrential bursts, much of the precipitation will run off, rather than percolate. Saline soils are said to be physiologically arid because dissolved salts lower the water potential of soil water, making the water less available. Again, local dry islands of vegetation will be the result.

The seasonal timing of precipitation can be as critical to vegetation as the annual amount. In the tropical latitudes, annual precipitation is generally more than 200 cm (80 in). When that precipitation is evenly divided among 12 months, a three-storied, evergreen forest results; but if the same annual precipitation is divided into wet and dry seasons, a less complex forest with many deciduous trees is the result. The form of precipitation is also important. Through the midelevations of California mountains, annual precipitation changes little, averaging 150 cm (60 in). Below an elevation of 2000 m (6600 ft), however, most of that 150 cm falls as rain. Above 2000 m elevation, most of it falls as snow, building a winter snowpack 3 to 4 m (10 to 13 ft) deep. As a result, the species of trees, shrubs, and herbs are dramatically different in the two zones, even though total and seasonal precipitation are the same.

Solar Radiation Is Another Environmental Factor

SOLAR RADIATION Solar radiation strikes the outer limits of our atmosphere with an energy content of about 2 calories per square centimeter per minute, a value called

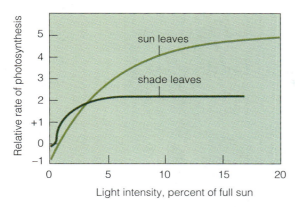

Figure 26.15 The rate of photosynthesis in sun leaves and in shade leaves at different light intensities. Leaves are from beech (*Fagus sylvatica*). (Redrawn from P. Boysen-Jensen and D. Muller, *Jahr. Wiss. Bot.* 70 (1929): 493.)

the **solar constant**. Solar wavelengths of the electromagnetic spectrum range from 300 nm to more than 10,000 nm. Technically, **light** is the portion of the solar spectrum visible to human eyes, or wavelengths from 400 nm (violet light) to 740 nm (red light). Plants can use the same range of wavelengths in photosynthesis (see Chapter 10).

Only half of the solar constant reaches the lower atmosphere and vegetation within it because certain wavelengths are absorbed or reflected back to space. In particular, a significant amount of shortwave, ultraviolet radiation is removed by ozone in the stratosphere.

Solar radiation is further depleted if it passes through foliage. Below a forest canopy, sunlight is reduced to 5% of its level just above the foliage. Many shade-tolerant herbs, shrubs, and tree saplings can still grow in this level of sunlight.

Long-lived plants must be very flexible in their light requirements because, as they grow, they experience different levels of light intensity. As saplings they may grow in dense shade, but as adults they are exposed to full sun. Furthermore, adult trees are so large that not all the leaves experience the same amount of sunlight. Leaves high in the canopy—sun leaves—receive high amounts of solar radiation, whereas those lower in the canopy—shade leaves—receive less. Shade leaves exhibit a different morphology, anatomy, and physiology from those that develop in sunlight, even when both are attached to the same plant. Shade leaves are larger, thinner, contain less chlorophyll per gram of tissue, and have less well-defined palisade and spongy mesophyll layers. They reach their maximum rate of photosynthesis at much lower light intensities than sun leaves (Fig. 26.15).

Eastern hemlock (*Tsuga canadensis*) is a good example of a long-lived, shade-tolerant tree species: It can grow in deep shade as a sapling but continues growth in full sun as an adult. Mature trees live for 1000 years. Saplings grow very slowly while shaded. Saplings only 2 m (7 ft) tall and 2 to 3 cm (1 in) dbh may have a ring count indicating an age of 60 years. Juveniles can remain alive in deep shade for as long as 400 years. If the canopy opens

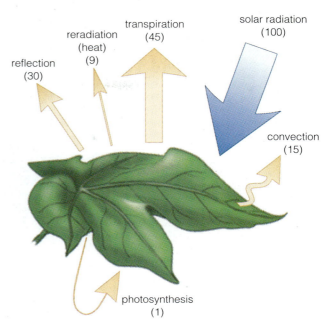

Figure 26.16 Energy budget of a leaf. Incoming solar radiation is given a relative value of 100 energy units. The leaf gives back 99 units in the forms of reflection, reradiation, transpiration, and convection. It assimilates 1 unit in photosynthesis.

up because of the death of a mature tree, the increased light releases a juvenile into a fast-growth mode, and it soon fills the canopy gap. Many other trees, especially pines, are not shade tolerant, and their juveniles can grow only in open habitats.

Solar radiation is also depleted when it passes through water. At about 170 m (560 ft) below the surface of clear ocean water, sunlight is reduced to 0.5% of its level at the surface. Light level sets the **compensation depth** for most aquatic plants—the depth at which they can just maintain positive net photosynthesis.

BALANCING INCOMING AND OUTGOING SOLAR RADIATION At the present time, the Earth is neither cooling down nor heating up. Therefore, we can conclude that what comes in also goes out; that is, the solar **energy budget** equals zero. This zero-sum budget holds not only for the Earth itself, but for every organism and for every part of an organism on earth. Let's examine the energy budget of an imaginary plant leaf as an example (Fig. 26.16).

Some of the solar radiation that reaches the leaf will be reflected. Perfectly white leaves reflect all radiation, but typical green leaves reflect only 30% of the energy. The energy that is absorbed is converted to longwave energy (wavelengths greater than 740 nm), which is felt as **heat**; that is, the leaf warms up. This heat is dissipated in four ways: transpiration, convection, reradiation, and metabolism (photosynthesis).

Transpiration is the conversion of water into gas and its loss from the leaf (see Chapter 11). When water changes state to gas, it uses up a significant amount of

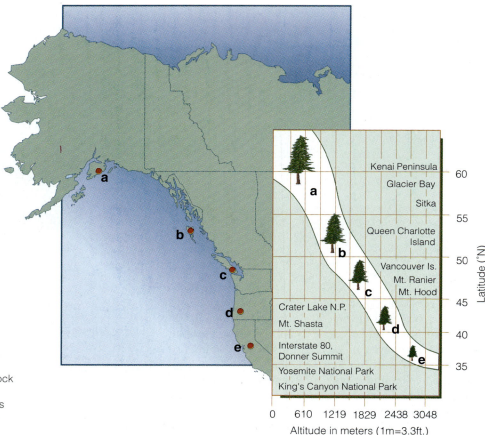

Figure 26.17 Distribution of mountain hemlock (*Tsuga mertensiana*) along the west coast of North America from latitude 62° N to 38° N. As latitude increases, the elevational zone of the tree drops.

energy. For our imaginary leaf, with well-hydrated tissues and open stomata, transpiration accounts for 45% of the energy budget. In a desert, transpiration would account for less of the energy budget. **Convection** is the transfer of heat energy back to air. For our imaginary large, entire leaf, convection accounts for only 15% of the energy budget. In a desert plant, with smaller leaves, convection would account for more of the energy budget. **Reradiation** is the loss of longwave radiation back to space; its effect is best seen at night, when the temperature of the leaf falls. Under the cover of a forest canopy, reradiation accounts for only 9% of our imaginary leaf's heat budget. In a desert, under a cloudless sky, reradiation would account for more of the energy budget. The overall remainder of the solar radiation, 1%, is converted into chemical energy by photosynthesis. Photosynthesis does not vary much, regardless of the plant or habitat. This particular energy budget formula is:

incoming solar radiation
     (100 units)

= reflection + transpiration + convection + reradiation + photosynthesis
  (30 units)   (45 units)   (15 units)   (9 units)   (1 unit)

This equation does balance (100 units incoming = 100 units outgoing; there is zero gain or loss).

SOLAR RADIATION'S EFFECT ON TEMPERATURE As noted above, heat is a consequence of solar radiation.

Longwave radiation is experienced as heat, and it is an important environmental factor. Heat can be measured with thermometers and expressed in many ways, temperature being just one.

Latitude and topography greatly influence solar radiation and heat. As we move north or south from the equator, the angle at which solar radiation strikes the ground becomes less direct and more oblique; this means that radiation travels through more of the earth's atmosphere and that more of it is absorbed or reflected. Light at the poles in summer may last 24 hours a day, but it is a weak light compared with the intensity at the equator. One consequence is that polar summer days are not very warm.

Because temperature is a limiting factor for plant life, different species prosper at different latitudes and elevations. For instance, some plant species are restricted to certain belts of elevation in mountains, and the elevation of those belts changes with latitude. Plants that grow in mountains respond to the diminishing amount of heat away from the equator by shifting their habitat lower and lower. Mountain hemlock (*Tsuga mertensiana*), for example, grows at high subalpine elevations (2700 m) in the Sierra Nevada of California, but its habitat drops toward the north until it reaches sea level along the Alaskan coast (Fig. 26.17).

The direction a slope faces also affects solar radiation and heat. In the northern hemisphere, south-facing slopes receive more sunlight than north-facing slopes; consequently, they have higher temperatures and drier soils. These microenvironmental differences affect plant population distribution and behavior. A canyon gorge running east-west through an Indiana forest illustrates the magnitude of this effect. The gorge is 65 m (215 ft) wide at the top and 45 m (149 ft) deep; its sides support scattered trees, shrubs, and herbs. Microenvironmental instruments were placed 15 cm (6 in) above and below the soil surface midway down each side. In spring, the south-facing slope had a larger daily range of both soil and air temperature, a higher average air temperature, a higher rate of soil water evaporation, lower soil moisture, and lower relative air humidity than the north-facing slope. Of nine spring-flowering species present on both slopes, the flowering time averaged six days earlier on the south-facing side. To gain a similar difference in flowering time on level land running north-to-south, one would have to pick sites 180 km (110 mi) apart.

Some species are sensitive to brief extremes of high or low temperature, and their distributions are determined more by the extremes of temperature than by long-term average temperatures. Most species of palms and cacti, for example, are sensitive to frost and do not occur in regions that experience yearly frost, even if the duration of freezing temperature is short.

Other species are sensitive to variation in temperature over seasons or days, and it is this amplitude of temperature that is more important than the average. The difference between maximum daytime temperature and minimum nighttime temperature within a 24-hour cycle is called a **thermoperiod**. California coast redwood (*Sequoia sempervirens*) grows best under the almost constant day–night and winter–summer temperatures of its habitat, near the temperature-buffering Pacific Ocean and bathed by summer fogs (refer back to Fig. 26.4). Experiments with seedling redwoods in growth chambers have shown that the optimum thermoperiod for coast redwood is less than 4°C (7°F). In contrast, conifers that grow inland in the Sierra Nevada have optimum thermoperiods greater than 13°C (23°F).

Fire Can Be a Natural Part of the Environment

In the 20th century, fire has been rediscovered as a natural environmental factor in North America. We say "rediscovered" because Native Americans were well aware of its occurrence and its effects, and they learned how to use it for their own purposes. Early Euro-American explorers and colonists invariably commented on the frequency of fires in forests and grasslands, but the pervasiveness of natural fire was ignored by land use managers until recently.

Most natural fires are started by lightning strikes unaccompanied by rain, which occur after a prolonged spell

Figure 26.18 Fire in the southeastern coastal plain.

of dry weather. Climates that have such conditions are said to be **fire-type climates**, and they are widespread throughout North America. Lightning accounts for 5000 fires in the United States each year, two-thirds of all our wildfires. Human-caused fires are in the minority. Grasslands, chaparral scrub, the northern boreal forest, southeastern forests (Fig. 26.18), some mountain conifer forests, and even desert palm oases owe their presence in large measure to climates with episodic natural fires.

When humans suppress fires in the mistaken notion that all fires are unnatural and catastrophic, they begin to change the mix of species and the shape of the landscape (see sidebar, "The Natural Fire Cycle in the Southeastern Pine Savanna," p. 464). This is because many species not only tolerate fire but depend on it, either to complete their life cycle or to maintain dominance over other species. Despite our current understanding of fire's importance and the creation of programs to use it in national parks and forests, only a small fraction of the land that should burn each year actually does burn. Thus, the accumulated magnitude of environmental change gets larger year by year, and we have not returned North America to the landscapes seen by the first explorers.

Each Population Is Ecologically and Genetically Unique

We now realize that each population interacts with its local environment so that over evolutionary time it becomes genetically better adapted to its microenvironment. Each population within a given species, then, is probably genetically different from all others. The genetic differences can be expressed as plant morphology, timing of life cycle events (such as flowering or leaf drop), or plant metabolism.

Not so long ago, botanists did not understand that species could contain such genetic diversity. In the 1920s,

THE NATURAL FIRE CYCLE IN
THE SOUTHEASTERN PINE SAVANNA

One of the first vegetation types shown to be dependent on fire for its maintenance in this country was the pine savanna of the southeastern coastal plain (Fig. 1). Tall, scattered loblolly, slash, shortleaf, and longleaf pines dominate the region, and a thick growth of grasses (mainly *Andropogon* and *Aristida*), with some broad-leaved herbs, cover the ground beneath. Euro-American settlers found the pines to be valuable for lumber and turpentine, and the grass for grazing livestock. Wildfires were thought to be dangerous, so during the 19th century some of this vegetation was protected from fire. Gradually, however, the pines gave way to oaks, and in time a dense oak forest with little grass replaced the pine savanna. Ecological studies in the 1930s then revealed the importance of fire to the pines, especially to longleaf pine (*Pinus palustris*).

Longleaf pine is tolerant of fire and is dependent on it. Its seeds germinate in the fall, soon after dropping to the ground from cones. During their first year of growth, the seedlings are very sensitive to even the slightest fire. However, during the next several years, longleaf pine seedlings are in a "grass" stage (see Fig. 1). Food reserves are shunted to the root system, the stem remains stunted, and a dense cluster of long needles surrounds the apical meristem, which is located at the ground surface. If a fire sweeps through the area at this time, the apical bud is insulated, and the seedling remains alive. A fire actually improves seedling survival because the high temperatures kill a common fungus that otherwise parasitizes the needles.

After the grass stage, longleaf pine saplings enter a 4–5 year period of rapid stem elongation. By the time the sapling is 8–9 years old, its canopy is high enough above the ground that it is beyond the reach of ground-fire flames. In addition, the bark is thick enough to insulate the cambium from damage.

Pine savanna systems maintain themselves naturally so long as ground fires sweep through every several years. However, if fires are suppressed for as long as 15 years, hardwood seeds carried into the area by animals will germinate and produce young saplings. Hardwood saplings are very sensitive to ground fire; they would not be able to invade an area where burns occurred every few years. In the absence of fire, however, they compete well with grasses and pines, and they are tolerant of shade from the overstory pines. As the oaks grow, they create more shade, which inhibits pine seedling establishment, growth, and survival. When overstory pines die, they are replaced by hardwoods, not by pines. In time, an oak forest exists where a pine savanna used to be.

Besides the pines, many herbaceous plant species decline in abundance because fire no longer keeps down competing grass and smothering litter. Other species disappear because their life cycles somehow require fire: Their seeds can germinate only after the seed coat is scarified by high temperatures, or they flower profusely only after fire has released a

a

Figure 1 Longleaf pine (*Pinus palustris*) savanna, North Carolina. (**a**) Aspect of the savanna. Overstory trees are widely spaced and average about 50 years old; beneath them is a well-developed understory of grasses and broad-leaved herbs. (**b**) Close-up of *Pinus palustris* in the "grass" stage.

b

pulse of nutrients into the soil. In addition, certain grazing animals, birds, and insects that live best in open savanna will be absent from this oak forest. In other words, a whole community of organisms will change in the absence of fire.

Standard land management practice now is to set a prescribed ground fire purposely every 4 years if a natural fire has not visited the area. Under this system, the pine savanna maintains itself indefinitely, just as it would with natural wildfires.

The use of natural or prescribed fire as a management tool is still not widely practiced because it affects air quality, is expensive, and can be conducted only during narrow windows of time when there is a low probability that a fire will escape. The infamous Yellowstone fire of 1988 is a good example of a natural fire that was allowed to burn but that got out of control when the weather changed. It caused a great deal of economic damage to surrounding communities and resulted in a congressional investigation of the whole concept of fire management. The congressional report ratified the concept that natural fire is beneficial, but it recommended additional restrictions as to when natural fires would be allowed to burn.

the botanist Gote Turesson collected seeds of species that had a wide range of habitats—from lowland, southern, and central Europe to northeastern Russia in the Ural Mountains—and he germinated them all in a garden in Akarp, Sweden. He was the first to demonstrate that members of a widespread species were not genetically homogeneous. Despite the uniform environment of the common garden, plants that grew from seeds collected in warm, lowland sites often were taller and flowered later in the year than those that grew from seeds collected in cold, northern sites. They differed in frost tolerance, leaf traits, and onset of dormancy. Yet these variants all had the same flower traits and all could be cross-pollinated; so they are members of the same species. Turesson called these variants within species **ecotypes**, and today we assume that every population is its own ecotype.

One example of ecotypes involves alpine sorrel (*Oxyria digyna*), a small perennial herb that grows in rocky places above timberline. Some populations grow at high elevations in mountains; these are **alpine** populations (Fig. 26.19a). Others grow in the far north at low elevations; these are the **arctic** populations (Fig. 26.19b). Both populations experience certain environmental stresses in common but differ in others. For example, they both experience long periods of freezing winter weather, but during summer days alpine plants receive more solar radiation. Researchers have demonstrated that the alpine and arctic populations are separate ecotypes and that they differ in such physiological ways as the amount of light that is optimum for photosynthesis (Fig. 26.19c). Alpine plants photosynthesize best at 50% full sun, whereas arctic plants do best in lower light at 25% full sun.

The ecotype concept is very important in forestry. After logging, sites may be planted with tree seedlings. The seedlings come from seeds that were collected as close (in distance and elevation) to the logged area as possible. In mountainous terrain, seed sources must be within 100 m (330 ft) of elevation of the planting site. The Forest Service recognizes that the probability of seedling survival will be greatest for a population that is already genetically adapted to the local region.

26.6 ## INTERACTIONS AMONG NEIGHBORING POPULATIONS

Very rarely does a single population, to the exclusion of all others, occupy a habitat. Usually, several plant, animal, and microbe populations—representing as many different species—coexist. Individuals of the various populations intermingle, growing next to each other, and this proximity allows them to affect each other. Sometimes the results are positive, with the organisms showing enhanced growth or survival; sometimes the results are negative, with the organisms suffering a decline. There are four basic kinds of interaction: competition, amensalism, herbivory, and mutualism.

a   b

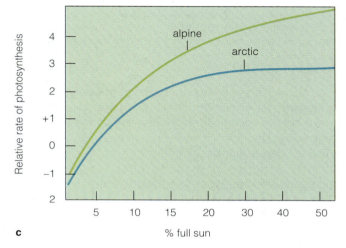

c

Figure 26.19 Ecotypes of alpine sorrel (*Oxyria digyna*). (**a**) Alpine ecotype. (**b**) Arctic ecotype. (**c**) Photosynthesis of each in different light intensities. (**c** from H. A. Mooney and W. D. Billings, *Ecological Monographs* 31 (1961):1-29.) (Courtesy of W. D. Billings)

Competition Creates Stress by Removing a Limiting Resource

Competition may be defined as the decreased growth of two populations because of an insufficient supply of some necessary resource such as light, moisture, space, nutrients, or pollinators. Sometimes the limitation lies with a single resource, but often several are lacking. Competition may be the most important biotic factor affecting plant distribution. Many populations restricted to saline, dry, or nutritionally poor soils, for example, would actually grow better on normal soil if other populations were first removed. These restricted populations are more tolerant of stress but are poor competitors in comparison with the plants that populate normal soil. When planted together, the restricted populations have slower root and shoot growth rates; consequently, they obtain less soil water and sunlight, and fewer of them survive to produce seeds. In time they are completely eliminated from all but the most stressful sites, where plants from normal soils cannot maintain themselves.

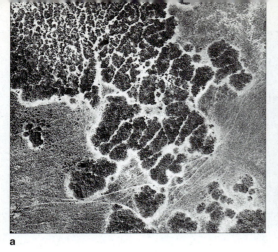 

a   b

Figure 26.20 Possible amensalism between sage shrubs (*Salvia* and *Artemisia*) and grasses in coastal areas of southern California. (**a**) Aerial photograph of a patch of sage; note the bare soil beneath the shrubs and for some distance around them. (**b**) View at the edge of a shrub patch, showing a zone of bare ground and stunted grasses extending for several meters away from the shrubs. Some researchers speculate that a volatile chemical exuded by the shrubs prevents the grasses from germinating or growing well.

The intensity of competition lessens in a process called **divergent evolution**. Over time, two populations of different species become more different in such traits as the time of germination or flowering, tolerances for soil aridity or depth, nature of pollination or seed-dispersal vectors, or degree of shade tolerance. The portion of the microenvironment each population uses is called its **niche**. Over time, each population's niche becomes more distinct and separate from the niches of nearby populations. It is theoretically impossible for two populations to have the same niche because competition would be too intense; hence the expression, "One niche, one population."

Amensalism Creates Stress by Adding a Resource

Amensalism may be defined as the inhibition of one population by another through the *addition* of something to the environment. The added material can be a metabolic by-product exuded from a living root or the decomposition products from dead litter. It can be a solid that accumulates beneath the parent plant, a liquid carried into the soil by percolating rainwater, or a volatile molecule carried off by the wind. A striking pattern of avoidance by two species in nature is often taken as evidence of amensalism (Fig. 26.20).

Amensalism may be very common because plants are leaky systems, passively contributing all sorts of chemicals into the environment. One investigator grew seedlings representing 150 different flowering plant species in a nutrient culture. Water around the roots contained several radioactive elements, as markers. The plants took up the isotopes through their roots and transported them to all organs, including the leaves. The plants were then exposed to a mist, and the water that condensed and ran off the leaves was collected for analysis. He found that 14 elements, 7 sugars, 23 amino acids, and 15 organic acids—all radioactive—had been leached from the plants.

Another study showed that roots are similarly leaky. Root tips were collected from mature trees in a New Hampshire forest and rinsed in distilled water. Analysis of the water revealed sugars, amino acids, organic acids, and various cations and anions. Birch (*Betula alleghaniensis*) exuded five times the amount of substances as maple (*Acer saccharum*). In nature, such substances from leaves, roots, or litter would accumulate in the soil, where they might affect microbial decomposers, soil pH, soil physical structure, and the growth of nearby plants.

Herbivory is the Consumption of Plant Biomass by Animals

Herbivory is the harvesting by an animal of vegetative or reproductive parts of a plant; it may leave the plant still alive or kill it. Herbivory can play a striking role in plant distribution. A glance along a pasture fence shows that certain palatable species are absent inside the pasture yet common outside it. Grazing animals are responsible for the difference.

Although plants are often thought of as passive organisms, unable to move (as animals do), they are able to defend themselves against herbivores. The defenses are slowly elaborated over evolutionary time. One defense is the dispersal of population members over space, making it difficult for herbivores to locate and damage all the individuals. Dispersal can be achieved by vegetative runners or rhizomes, by explosive fruits, or by the use of wind as a vector. Another defense is the dispersal of population members over time. This can be accomplished by seeds with complicated dormancies or by reproduction that is irregular from year to year (as with masting or semelparous flowering). A third defense is the development of physical barriers such as sclerophylls, spines, thick bark, and hard fruits or seed coats.

A fourth defense is amensal: the manufacture and accumulation of metabolic by-products that are distasteful or that otherwise inhibit herbivores from feeding. There has been much recent research into **chemical herbivore defenses**. The type of chemical a plant manufactures varies with its life history pattern: r-selected plants tend to manufacture toxins, while K-type plants tend to accumulate tannins or terpenes. Toxins repel herbivores by interfering with nerve and muscle activity, hormone function, or liver and kidney metabolism. Toxins are small, relatively energetically inexpensive molecules of less than 500 molecular weight. They are effective in small concentrations, and they account for less than 2% of leaf dry weight. In contrast, tannins and terpenes repel herbivores by their bitter, unpleasant taste. They are large, complex molecules of 500 to 3000 molecular weight and may account for 6% of leaf dry weight. Animals attempting to feed on plant tissue that is rich in tannins or terpenes move away to other plants.

Some chemical herbivore defenses are induced in response to an attack by an herbivore, parasite, or pathogen. Such chemicals are called **phytoalexins.** The plant's chemical defense is analogous to antibody formation in animals.

Humans are herbivores, too, and therefore plant toxins can and do poison humans as well as grazing animals or insects. Humans, however, have learned to use some poisons as medicines or psychotropic drugs, and we cultivate the plants that manufacture them. The nicotine that accumulates in tobacco (*Nicotiana*) leaves inhibits herbivores, but humans use nicotine as a drug. Cocaine functions as an herbivore deterrent in coca shrubs (*Erythroxylon*), but Andean Indians learned to chew the leaves for stimulation while doing strenuous activity at high elevations.

Mutualism Increases the Success of Both Populations

The previous examples of interactions were all negative, causing depressed growth or reproductive success in one or both interacting populations. **Mutualism** is an interaction that benefits both partners; further, it is essential, in the sense that the success of both partners is reduced when one of the partners is absent.

We have already described several examples of mutualism earlier in this book. Animal partners form mutualistic relationships with plants. Bees, moths, and beetles—which cross-pollinate flowers while themselves feeding on nectar, pollen, or flower parts—are familiar examples (see Chapter 13). Ants, birds, and mammals that consume fruits while dispersing seeds are other examples (see Chapter 14). Sea anemones, which mutualistically harbor algae in their cells, provide yet another example. Lichens are mutualistic associations of fungi and algae (see Chapter 21), mycorrhizae are mutualistic associations of fungi and higher plants (see Chapter 5), and nitrogen-fixing bacteria live symbiotically in the root nodules of legumes (see Chapter 11).

If the benefits are not shared by both interacting populations (if the interaction is not obligate), the relationship is not mutualistic but **commensal**. Epiphytes growing in a tropical tree, for example, form a commensal relationship with the tree. The host tree gains no benefit from the epiphyte. At the same time, the host tree is not usually hindered by the epiphyte because the epiphyte is not a parasite, nor is it so large that it shades the tree or breaks off limbs. The tree neither benefits nor loses by the association, whereas the epiphyte is promoted.

SUMMARY

1. Ecology is the study of organisms in relation to their natural environment. The environment contains all the biotic and abiotic elements that surround and influence an organism. The macroenvironment reflects the general, regional climate; the microenvironment reflects conditions near the surface of organisms or objects. Ecosystems consist of organisms and their environment. These organisms participate in trophic functions within the ecosystem: They produce food, they consume it, or they decompose it.

2. The basic ecological unit is the population. Every species typically has many populations, each one genetically distinct and adapted to its particular habitat. A population may be equivalent to an ecotype, unique in its morphology, physiology, or behavior. Population ecology is the study of a population, together with the environmental factors affecting that population.

3. A life history pattern is the budget of time and energy that carries a population through all phases of its life span. Plant life spans are annual (ephemeral), biennial, or perennial. Herbaceous perennials die back to below-ground organs each year. They occupy stressful habitats and commonly live for 20 to 30 years. Woody perennials include subshrubs, shrubs, vines, and trees. Trees include broadleaf angiosperm and needle-leaf gymnosperm categories. Broadleaf trees are shorter lived, have shorter-lived leaves, and build harder wood than needle-leaf trees.

4. Semelparous populations reproduce only once, at the end of a plant's life span. Iteroparous populations reproduce repeatedly. Each pattern has its own advantages. Sexuality is sometimes determined by environmental factors or plant size; it is not necessarily under genetic control. The dioecious condition promotes cross-pollination, allows a population to occupy a wider array of sites, and reduces the loss of seeds to herbivores. The monoecious condition permits weedy, aggressive plants to disperse into new habitats, even when only a single individual has arrived.

5. Species with an r-selected life history pattern are typically small, fast-growing annuals adapted for open habitats; they are semelparous and allocate a great deal of caloric energy to reproduction. K-selected species are typically slow-growing but ultimately large and long-

lived forest trees in more stable habitats; they are iteroparous and allocate less energy to reproductive events. Most species lie somewhere along a continuum between these extremes.

6. Plant demography is the study of population age structure over time. It uses mathematical models to summarize the life span of individuals and cohorts in a population, their reproductive potential, their mortality rate over time, and future changes in the size of the population.

7. The intensity and quality of solar radiation change as the radiation passes through the atmosphere, plant leaves, and water. Light (the wavelengths of solar radiation visible to humans) is used by plants in photosynthesis. Large, long-lived, K-selected plants successively live through different light environments and therefore must be very flexible in their light requirements. This flexibility is enhanced by an ability to produce sun leaves and shade leaves and to be shade tolerant when young.

8. Every organism and organ has an energy budget with a zero sum. Solar radiation striking a leaf, for example, is first partitioned into reflected and absorbed energy. Absorbed energy is dissipated as transpiration, convection, reradiation, and photosynthesis. Mesophytes expend most solar radiation by transpiration; xerophytes by convection, reradiation, and reflection.

9. Fire caused by dry lightning strikes is a natural occurrence in many parts of North America. When natural fires are suppressed, the distribution of many plant populations is affected. Open pine savannas of the southeastern coastal plain, for example, change into dense oak forests if fires are not allowed to burn every 3 to 15 years. Currently, our use of prescribed fire is too limited to restore all North American landscapes to conditions before fire suppression.

10. Neighboring populations in the same habitat interact and affect each other through competition, amensalism, herbivory, and mutualism. Competition, amensalism, and herbivory are negative interactions, with one or both partners suffering a decline. Competition is caused by a deficient amount of an essential resource. Amensalism is caused by the addition of a deleterious substance into the microenvironment. Herbivory is the consumption of vegetative or reproductive plant material by an animal. Plants have evolved herbivore defenses in the form of: (a) dispersal over space and time, (b) repulsive physical structures, and (c) toxic or unpalatable metabolic by-products. The latter mechanism is a form of amensalism. Mutualism is a positive interaction, with the partners showing enhanced growth or survival.

Questions

1. A grassy acre in southern Michigan receives 47×10^8 calories of solar radiation over the course of one year. Plants fix, in photo- synthesis, about 58×10^6 calories of that radiation. They retain 50×10^6 calories, and of this mice consume 25×10^4 calories. The mice retain 5×10^4 calories, and of this weasels consume 5×10^3 calories. Weasel net growth is 1×10^3 calories. If a young weasel must gain 2000 grams of weight in that year, and each gram is equivalent to 5×10^3 calories, how many acres of grassland must be its range?

2. Can a species be rare and in danger of extinction if its only population has thousands of individuals? Can a species be rare and in danger of extinction if it has thousands of populations, each population of which has only a few individuals?

3. Why do semelparous plants die after their single reproductive event? What are possible advantages to reproduction in this way? Shrubs and trees are rarely semelparous; do you know of any in your region?

4. Is a single habitat inhabited by only r-selected or only K-selected species, or can they coexist in the same habitat?

5. What is the definition of *plant demography*? Try to diagram the demography of a human population using the same model shown for a plant population in Fig. 26.12.

6. In an irrigated, dark green pasture in Texas, what fraction of annual solar radiation do you guess would be dissipated by transpiration, by reflection, by convection, and by photosynthesis? If that same pasture were allowed to revert to semiarid, nonirrigated natural grassland, what might those figures be?

7. How did Turesson show that species were not ecologically and genetically homogeneous? Is ecotype a synonym for population?

8. Describe the difference between competition and amensalism. Why are both of these interactions called negative, whereas mutualism is said to be positive?

Further Readings

Barbour, M., J. Burk, and W. Pitts. 1998. *Terrestrial Plant Ecology*. 3d ed. Menlo Park, Calif.: Benjamin/Cummings. A very readable general survey of all modern aspects of the science of plant ecology, with examples of research from throughout the United States.

Bertness, M. 1992. "The Ecology of a New England Salt Marsh." *American Scientist* 80: 260–68. Excellent illustration of how important species interactions are to the structure of a complex ecosystem.

Carrier, J. 1989. *Summers of Fire: the Great Yellowstone Fires of 1988*. Layton, Utah: Peregrine Smith. A detailed description of the Yellowstone wildfire, which began as a "let-burn" ground fire and escalated into an extensive crown fire that threatened our modern fire management policy.

Chambers, J., and J. MacMahon. 1994. "A Day in the Life of a Seed." *Annual Review of Ecology and Systematics* 25: 263–92. An imaginative summary of seed dispersal and seed survival in major world biomes.

Larcher, W. 1995. *Physiological Plant Ecology*. 3d ed. New York: Springer-Verlag. A technical but masterful summary of energy budgets, life history patterns, and plant response to various environmental stresses.

Silvertown, J. W., and J. L. Doust. 1993. *Introduction to Plant Population Biology*. 3d ed. Cambridge, Mass.: Blackwell. An authoritative survey of plant demography, population genetics, competition, the role of disturbance on population age structure, sex allocation, and life history patterns.

27 ECOLOGY AND PLANT COMMUNITIES

1. The plant community is a group of recurring species that inhabit a characteristic habitat and collectively have unique architecture, species richness, biomass, productivity, and cycles of nutrients and energy. North America contains thousands of plant communities.

2. A vegetation type is composed of many communities that differ only in the identity of dominant and/or associated species but that otherwise share a similar physiognomy and environment. Two-thirds of North America is covered by only three major vegetation types: boreal forest, grassland, and tundra.

3. Successional plant communities change over time; climax plant communities do not show directional change, though they may fluctuate from year to year.

4. Conservation biology is a relatively new science that investigates ways to preserve, restore, and maintain biotic diversity in the face of human exploitation of natural ecosystems.

Figure 27.1   A red spruce (*Picea rubens*)–Fraser fir (*Abies fraseri*) forest community in the Appalachian Mountains of North Carolina.

27.1   THE NATURE OF PLANT COMMUNITIES

Most plant populations do not grow in isolation. Single populations do not usually monopolize a habitat to the exclusion of all other organisms. Normally, there is a mixture of coexisting plant and animal populations, as well as of many less visible fungal, protist, and moneran populations.

Wherever a particular habitat repeats itself within a region, many of the same species recur. The species composition does not replicate itself completely, but there is a nucleus of species that do repeat. These clustered species are said to be associated with each other and to be members of a biotic **community**. For simplicity, we can break down any community into its complement of animal, plant, or microbial species. Since this is a plant biology textbook, we will discuss communities from the standpoint of the plants they contain; that is, we will discuss the plant community.

Every plant community is named after its **dominant species** and is characterized by its own roster of associated species and their combined architecture. For example, high peaks in the Great Smoky Mountains of North Carolina all have red spruce (*Picea rubens*) and Fraser fir (*Abies fraseri*) trees as the tallest layer of vegetation. The conifers are rather dense, and their canopies usually touch, creating a deep shade (Fig. 27.1). Beneath this needle-leaf canopy is a second, much more open layer of vegetation of scattered mountain ash trees (*Sorbus americana*). Mountain ash is a deciduous broadleaf species that sheds its leaves in fall. A final layer of vegetation carpets the ground and consists mainly of ferns, mosses, and broad-leaved herbaceous perennials.

The spruce and fir populations are said to dominate the community because they contain the largest individuals and contribute the most biomass. By being largest, they create the microenvironment within which smaller associated species live.

The architecture of this red spruce–Fraser fir community can be summarized as having three canopy layers: The uppermost is almost continuous and is made up of needle-leaf evergreens; the second is open and is made up of shorter, broadleaf trees; and the third is a continuous ground layer of herbs. The habitat is high-elevation, north-temperate, winter-cold, with well-drained soils. Wherever this habitat appears in the Appalachian Mountains, the spruce–fir community recurs.

There are many other plant communities in the Appalachian Mountains. They are dominated by different species with different growth forms—such as shrubs, grasses, broad-leaved evergreen trees, or broad-leaved herbs. If we expand our vision to the entire North American continent, we could identify thousands of different plant communities. In addition, we could discern blendings or mixtures in **ecotones**, where two or more communities and environments grade into each other.

Each Plant Community Has Unique Attributes

Each plant community has features that transcend a mere list of member species. These features, or *attributes*, have to do with community architecture, species richness, the spatial patterns in which individuals are arranged, the efficiency with which they trap sunlight and cycle energy or nutrients through the community, and the stability of the associated species in the face of environmental stress or change. These attributes are summarized in Table 27.1; we will discuss just a few of them now.

PHYSIOGNOMY   We have already described the architecture of an Appalachian spruce–fir community. A tech-

Table 27.1 Some Attributes of Plant Communities

Physiognomy (architecture)
      Canopy cover and leaf area index
      Growth forms of dominant species
      Spatial pattern
      Timing of life cycle events (germination, bud break, flowering, leaf drop)
Species diversity
      Number of species
      Relative importance (abundance) of each species
Productivity
      Biomass
      Efficiency
      Allocation of biomass (roots, stems, leaves, reproduction)
Nutrient cycling
      Nutrient demand
      Location and size of storage pools
      Efficiency
Change over time (succession)
      Primary vs. secondary
      Progressive vs. retrogressive

a

b

Figure 27.2 Communities with different physiognomies. (**a**) Desert scrub in southern Nevada, dominated by creosote bush (*Larrea tridentata*), has a single open canopy layer. (**b**) A tropical rain forest in northern Argentina has several tree canopy layers.

nical synonym for community architecture is *physiognomy*. **Physiognomy** is the external appearance of the community, its vertical structure, and the growth forms that dominate each canopy layer. A desert community, consisting of only a single canopy layer of widely scattered shrubs (Fig. 27.2a), might have 10% **canopy cover** or less (that is, 10% of the ground is directly beneath the foliage of the shrubs; 90% is open and unshaded). A tropical rain forest, with several overlapping tree layers (Fig. 27.2b), has 100% canopy cover.

Another attribute of the canopy has to do with its thickness—that is, the number of layers of overlapping leaves through which light must pass on its way to the ground. Canopy cover does not measure this attribute, but **leaf area index (LAI)** does. LAI is the total area of leaf surface (one side only) for all leaves that project over a given area of ground. If the leaf and ground areas are measured in the same units (so many square meters of leaf surface per square meter of ground), then LAI becomes a dimensionless number. The desert community

Figure 27.3 The importance of a species to its community can be measured by its density, canopy cover, and frequency. A rectangular frame outlines a sample of one desert scrub community, as seen from above. *Density* is the number of individuals of a species within the sample. *Cover* is the fraction of the sample shaded by each species. *Frequency* is a measure of the dispersion of members of a species, and cannot be estimated from one quadrat. A species present in three out of four quadrats would have a 75% frequency. Dotted lines show canopy edge, rounded out.

we mentioned earlier has an LAI of 1, and the tropical rain forest has an LAI of 10. The optimal LAI for such crops as corn is 4. Planting corn so densely that the LAI is higher creates too much shading for lower leaves.

SPECIES RICHNESS **Biotic diversity**—a term often used by the news media—simply means the total number of species that occur in a given community or region. In this sense, diversity is synonymous with **species richness**. Communities differ in the number of associated species they contain. Tropical rain forests appear to have the highest diversity of plant species, up to 365 per 1000 m² (1/3 acre). Temperate forest and woodland communities have more moderate diversities of 50 to 100 species in a similar area, and desert communities have fewer than 50 species in the same area.

Not every species in a community is equally important. A few species are represented by many individuals (such species are common), a few other species are rare, and most are intermediate in abundance. It is the relative scale of importance of each species that collectively gives the community its unique physiognomy. The importance of each species can be quantified by counting its density of individuals (the number per unit area), its canopy cover, its biomass, or its frequency of occurrence in the community. A widely dispersed population has individuals that would be frequently encountered, whereas individuals that are clumped together would be encountered infrequently (Fig. 27.3).

BIOMASS AND PRODUCTIVITY Communities differ in the amount of biomass they have above and below ground. Tropical rain forests, for example, may have up to 500,000 kg (1.1 million pounds) of aboveground biomass per hectare (abbreviated ha, equivalent to 2.5 acres), whereas desert communities have only 1% of that amount. *Allocation* of biomass above and below ground

also varies with the community. Forest communities typically have five times as much biomass above ground as below ground. Another way of saying this is that the **root-to-shoot biomass ratio** of a forest is 0.2. In contrast, scrub communities have equal amounts of biomass above and below ground (the ratio is 1.0), and grassland communities have more biomass below ground than above ground (the ratio is 3.0).

Communities also differ in the amount of biomass they produce each year, their **productivity**. Those communities with the highest biomass usually are the most productive because they have a higher LAI with which to trap solar radiation. High-biomass communities also tend to live where the growing season is longest, another factor contributing to productivity. Thus, tropical rain forest communities have an annual productivity of 20,000 kg per ha (18,000 lbs per acre), compared to only 2000 kg per ha (1800 lbs per acre) for deserts.

NUTRIENT CYCLING Communities differ in their demand for mineral nutrients from the soil, in the locations where such nutrients travel and accumulate, and in the rate at which they are returned to the soil or are lost from the ecosystem by erosion. This process of nutrient uptake, use, and return is called a **nutrient cycle**.

The nitrogen cycle (Fig. 27.4), for example, moves nitrogen from inorganic to organic forms and back again. Nitrogen gas (N_2) in the air is ordinarily not available to plants, which can absorb nitrogen only as ammonium (NH_4^+) or nitrate (NO_3^-). The biological process of nitrogen fixation (see Chapter 11), accomplished by certain bacteria, converts nitrogen gas into ammonium. Volcanic action also creates ammonium, as does lightning; this ammonium is then brought to the ground (and to plant roots) as ions dissolved in rainwater. Certain bacteria are capable of converting ammonium into nitrate. Once in plants, the ammonium or nitrate is metabolically altered,

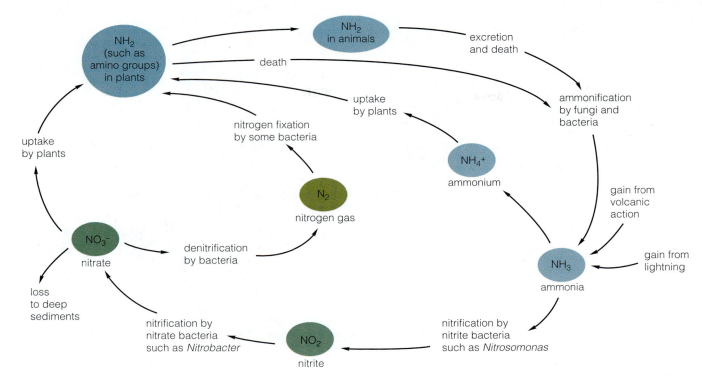

Figure 27.4 The nitrogen cycle. Blue color indicates reduced, organic forms of nitrogen; green indicates oxidized forms.

moved onto amino acids, and then incorporated into proteins, nucleic acids, alkaloids, and many other important molecules.

Some of the nitrogen returns to the soil when leaves are shed or when the plant dies. Certain soil microbes convert the organic forms of nitrogen in litter back into ammonium, and others change the ammonium into nitrite (NO_2) and then into nitrate. About 10% of the nitrogen in a plant is passed on to the next generation in the form of stored food in fruits and seeds; another 10% is ingested by grazing animals and becomes part of their metabolism. When they excrete waste or die, their nitrogen is converted into ammonium, nitrite, and nitrate by the same microbes that decompose plant remains.

Similar nutrient cycle diagrams could be prepared for sulfur, carbon (see Fig. 10.11), water, and phosphorus. The details of each cycle, the rates of nutrient movement within it, and its overall efficiency are different from community to community. Tropical forest communities, for example, contain broadleaf trees with a high demand for nitrogen and other nutrients. The forest trees store nutrients in trunks (the wood of Fig. 27.5a) for long periods of time, and they return only a small fraction of the nutrients to the soil each year via leaf shedding and decomposition. The largest pool of nutrients in this particular ecosystem is in the wood. Decomposition, from litter to humus to soil, is very rapid because of continuously warm temperatures and adequate moisture. Most of the nutrients released to the soil are rapidly taken back up by a dense network of fine roots in the topsoil. As a conse-

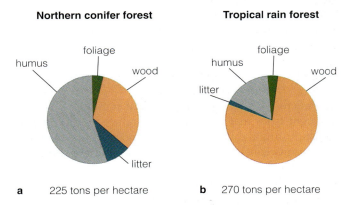

Northern conifer forest   **Tropical rain forest**

a   225 tons per hectare   b   270 tons per hectare

Figure 27.5 Distribution of carbon in (**a**) a northern conifer forest community and (**b**) a tropical rain forest community. Units are in metric tons (2200 lbs) per hectare (2.5 acres).

quence, tropical soils tend to be low in residual nutrients. There is very little loss of nutrients from erosion because the soil is covered with vegetation and anchored in place by dense roots. Most minerals are said to have a long *residence time* in the ecosystem, and the cycling of nutrients is *tight*.

Northern forest communities, in contrast, contain conifer trees with a low demand for nutrients. These trees have a smaller pool of nutrients in their wood, and they return more nutrients to the soil in an annual rain of litter. The rate of litter decomposition is slow because of cold temperatures, however, so litter accumulates in a thick layer. The largest pool of nutrients in this ecosystem is in

Figure 27.6 Secondary succession in abandoned fields in the Piedmont region of North Carolina. (**a**) Field abandoned for one year. Dominant plants include short-lived herbs and grasses. (**b**) Ten-year-old field, showing perennial herbs and grasses, and 7–8 year-old pines in the background. (**c**) Pine stand abandoned 50 years prior to this photograph. Old furrows are still obvious in the picture. Understory dominated by broad-leaved tree seedlings and saplings. (**d**) Mature deciduous forest.

humus (Fig. 27.5b). Soils tend to be low in residual nutrients because they are acidic. Conifer foliage contains high amounts of organic acids, which dissolve cations in the soil, allowing them to be leached from the root zone. The cycling of nutrients in such a community is *loose*.

Plant Communities Change over Time

The microenvironment within a plant community is very different from that outside. Temperature, humidity, soil moisture, and light are all affected and modified. A stable community consists of K-selected species whose seedlings can survive to maturity in this unique microenvironment (see Chapter 26). Seedlings of species from other communities are at a disadvantage and cannot normally invade.

CHARACTERISTICS OF SUCCESSION If the stable community is removed by some disturbance—such as landslide, storm, canopy fire, or logging—then the soil surface is exposed, and the microenvironment has changed. The first species to colonize the site after the disturbance are usually not those of the old community. They are seedlings of r-selected species adapted to open sites. Only after a passage of time, when the biomass of invasive species has altered the microenvironment, will species of the old, stable community gradually return. The process of community change at one place over time is called **plant succession**.

Succession has defining characteristics. It occurs at population and community scales. Thus, the death of one individual and its replacement by another is not succession. We must instead look to change over a larger area than the space taken up by individuals for evidence of

succession. Succession also has a *maximum* area limit because it must occur within a uniform macroenvironment. When we begin to cross soil boundaries or zones of elevation or latitude, the area is no longer homogeneous, so the changes that we see from place to place are not the result of succession. Succession generally occurs over an area from 1 hectare (2.5 acres) to several square kilometers (one square mile).

Succession is usually measured over the course of several years to several hundred years. Plant changes from season to season are not succession; they merely reflect life cycle phases. Succession does not occur over time periods shorter than a year. There is also a maximum time—certainly shorter than 1000 years—because succession must occur within a constant macroenvironment. Once we reach scales of time that incorporate global climate change, the community changes we see are not part of succession. Furthermore, we all know that the environment is not constant from year to year. In good years, the biomass, productivity, and species richness of a community will be different from those in poor years. These random fluctuations in a community over relatively short periods of time are not part of succession.

STAGES OF SUCCESSION Succession proceeds from pioneer to climax phases. For example, in parts of the southeastern United States, oak–hickory forest seems to be stable. If such a forest is cleared and then abandoned, the path of succession is as follows: The first plants to invade are annual and perennial herbs that grow well in high light intensity, such as horseweed (*Conyza canadensis*), *Aster pilosus*, and broomsedge (*Andropogon virginicus*). These make up a **pioneer stage** (Fig. 27.6a).

Table 27.2 Comparison of Some Community Traits During Early and Late Stages of Progressive Succession

Trait	Early Stages	Later Stages
Biomass	Small	Large
Architecture	Simple	Complex
Nutrient pool	Soil	Vegetation
Mineral cycling	Loose	Tight
Productivity	High	Low
Stability	Low	High
Species diversity	Low	High
Life history patterns	r-selected	K-selected
Site quality	Extreme, microenvironment not well developed	Moderated, microenvironment well developed

Pine seeds blow in from large, old pines that grow as scattered individuals in the surrrounding oak–hickory forest. They are more readily dispersed than the heavier seeds of oak (*Quercus*) and hickory (*Carya*). They also germinate and grow well in semiopen habitats. Within ten years of the disturbance, many short pine saplings are visible (Figure 27.6c). As the pines grow, their emerging canopies begin to alter the microenvironment. Some of the pioneer herbs do not maintain their populations and disappear from the community or become very rare. Within 30 years of the disturbance, a pine savanna has grown up (Fig. 27.6b).

Examination of the herb layer beneath the pines shows many oak and hickory seedlings and very few pine seedlings, however. Pine seedlings grow poorly in shade and under conditions of root competition for moisture from other trees, but those of oak and hickory are more tolerant of shade and competition.

As we saw in the last chapter, if fire sweeps through the pine savanna community frequently enough, the oaks and hickories are selectively eliminated. But if fire does not visit this community, succession continues. In this case, within 50 years of disturbance, a well-defined understory of hardwood saplings and young trees exists beneath the pines (Fig. 27.6b). Whenever a pine dies, it is replaced in the upper canopy by a maturing oak or hickory. Within 200 years of the disturbance, the forest again consists of mature oak and hickory trees with an occasional old pine. The shrubs and herbs associated with oak and hickory are also present, and the dense grass understory of the pine savanna has disappeared (Fig. 27.6d). The stable oak–hickory forest community is the end point of succession and is called the **climax stage**.

The succession we have just described is an example of **secondary succession**, which takes place on vegetated land. **Primary succession** occurs on newly exposed ground not previously occupied by plants. Examples of new land include mobile sand dunes, volcanic lava flows, mud ex-

posed by a drop in lake water level, bare rock scraped clean by a retreating glacier, or the in-filling of a small lake. The rate of succession is slower in primary succession, for several reasons. First, the parent materials may be bedrock, coarse sand, or wind-blown silt—all of which lack clay particles and thus essential nutrients. Second, there is no bank of plant seeds, bulbs, or rhizomes already in the soil, left over from a predisturbance community. Primary successions often take several hundred years, whereas secondary successions may take only decades.

Many community traits change during succession, and these are summarized in Table 27.2. Basically, these changes make the community more complex and massive, the cycles of energy and nutrients more efficient, and the microenvironment less stressful. These changes are the result of **progressive succession**.

But not all succession is progressive. For example, Figure 27.7 is a summary of primary succession on Alaskan floodplains left dry when a river meanders and changes form. At first, progressive succession (clockwise in Fig. 27.7) leads to increasing herb cover; then, to invasion by willow (*Salix*) and alder (*Alnus*) shrubs. Alders can fix nitrogen in their roots, so soils become richer. Biomass increases, shading the site. K-selected white spruce (*Picea glauca*) then invades the shade and slowly grows through the shrub canopy. As the spruce creates an overarching tree canopy, the r-selected alders decline in its shade.

Now **retrogressive succession** begins, reversing the process, so that the community becomes simpler and less massive, cycles of energy and nutrients less efficient, and the microenvironment more severe. The mature spruce canopy creates a cold, dense shade. One consequence is that *Sphagnum* moss invades the ground surface; this moss has the capacity to retain large amounts of water, and it literally raises the water table. The shade and thick layer of *Sphagnum* cools the soil, allowing a permanently

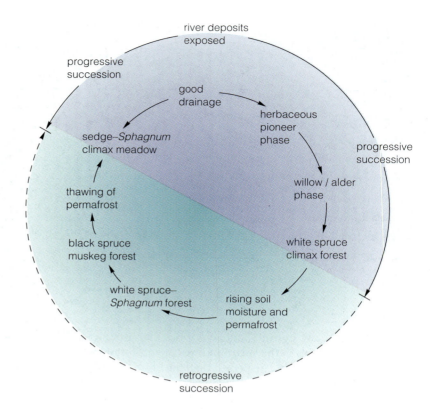

river deposits
exposed

progressive
succession

good
drainage

herbaceous
pioneer
phase

progressive
succession

sedge–*Sphagnum*
climax meadow

willow / alder
phase

thawing of
permafrost

white spruce
climax forest

black spruce
muskeg forest

white spruce–
Sphagnum forest

rising soil
moisture and
permafrost

retrogressive
succession

Figure 27.7 Progressive (purple) and retrogressive (blue) succession on an Alaskan floodplain. (Redrawn from W. H. Drury, Jr., *Contributions from the Gray Herbarium* 178 (1956): 1–30.)

frozen layer (**permafrost**) to rise toward the surface. White spruce requires well-drained soil and a deep water table, so it begins to die. Scattered, stunted black spruce (*Picea mariana*) trees replace the declining white spruce. Black spruce woodland permits more light to reach the ground, heating the soil and melting back the permafrost, creating an even wetter habitat. As a consequence, the soil becomes shallow and soggy, and an open bog or muskeg replaces the spruce forest. Retrogressive succession ends in a meadow with dwarf shrubs, grasses, and sedges as the dominant growth forms.

The next section describes some of the most widespread vegetation types in North America.

27.2 VEGETATION TYPES

Plant communities blend into each other. Every species has its range limits, and at those limits each can be replaced by some other species that may have a similar growth form and niche. Therefore, the name of the plant community will change as we cross the species range limits, but the architecture of the vegetation and the general habitat might stay the same. So long as the architecture and environment remain constant, several sequential communities all belong to a single **vegetation type**.

The term **vegetation** refers to the dominant growth form, not to the dominant species. Each vegetation type has a two-part name that describes the dominant growth form and the habitat; the name does not include any information

about species. Thus, we have upland conifer forest, tropical rain forest, alpine tundra, eastern deciduous forest, tidal marsh, and desert scrub (scrub = shrub) vegetation types, among many others. The Appalachian red spruce–Fraser fir community fits within a more widespread vegetation type called subalpine conifer forest, which includes spruce–fir communities from the Rocky Mountains and the Cascade Range, as well as pine–hemlock communities from the Sierra Nevada.

There are, then, many fewer vegetation types than communities. North America has thousands of communities but fewer than a dozen major vegetation types (Fig. 27.8). Most of the world's major vegetation types are represented in North America, but extremely arid deserts and several tropical grassland, savanna, and forest types are missing. We will describe nine of North America's major vegetation types, beginning in the north and moving clockwise through the continent. The vegetation, of course, is simply one component of the regional ecosystem; animals, microbes, and environmental factors are other components. These large, regional, climatically controlled ecosystems are often called **biomes**.

Tundra Vegetation Occurs Beyond Timberline

As one travels north toward the pole, trees gradually become scattered, stunted, and less common. They finally disappear completely in a zone called **timberline**. At the same time, low shrubs, perennial herbs, grasses, and

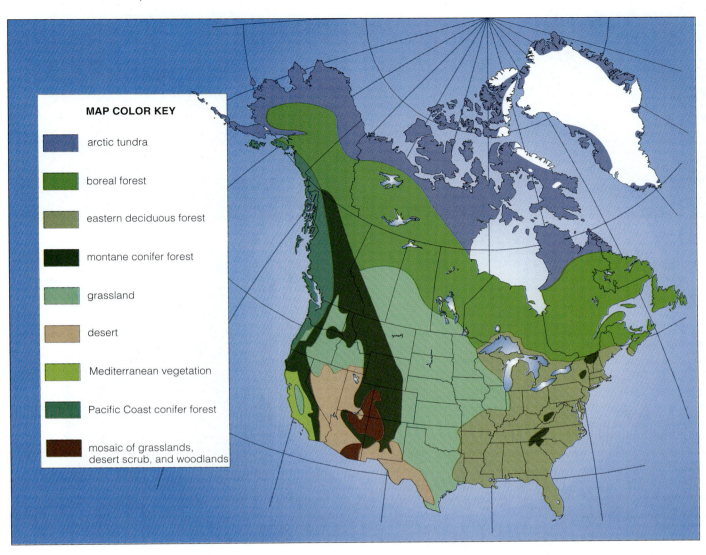

Figure 27.8 Major vegetation types of North America. (Redrawn from M. G. Barbour and N. Christensen, "*Vegetation of North America*," in vol.1 of *Flora of North America*. New York: Oxford University Press.)

MAP COLOR KEY

- arctic tundra
- boreal forest
- eastern deciduous forest
- montane conifer forest
- grassland
- desert
- Mediterranean vegetation
- Pacific Coast conifer forest
- mosaic of grasslands, desert scrub, and woodlands

sedges become dominant. Most of the biomass is below the soil surface; the root-to-shoot ratio is about 4. Shrubs are dwarfed, gaining normal height only in the protection of the lee of boulders or small hills. The winter wind, carrying drying air and ice, acts like a sandblast, pruning back stems wherever they project above the top of snow or beyond a protecting object. Perennials produce many large flowers, but the most successful reproduction is by rhizomes. Annuals are rare. This is **tundra**, a word that comes from Finnish or Lapp and means "marshy hill". Vegetation ecologists, however, use the word to denote vegetation (Fig. 27.9a), not topography. This vegetation type covers 19% of North America's land area.

Tundra is a one-layered vegetation with nearly 100% cover at its southern extent, but dropping to 5 to 50% cover at its northern limits. A similar vegetation occurs at lower latitudes in mountains, at elevations above timber-line (Fig. 27.9b). The climates of the two tundra regions—arctic and alpine—have both similarities and differences. In both regions the warmest month has an average daily mean temperature of 10°C (50°F) or less. The growing season is short, only two to three months of the year having average daily temperatures above freezing. During the growing season the top 30 cm (1 ft) of soil thaws, and roots may freely penetrate it; below this depth, soil and water are frozen year-round in an impermeable, meters-thick permafrost layer. Annual precipitation is less than 25 cm (10 in); thus, winter snowpacks are shallow—just deep enough to cover and insulate the low-growing vegetation. Few plants have evolved to tolerate this environment. Only 700 species in North America have ranges that extend into the arctic tundra.

The alpine tundra differs from the arctic tundra in that summertime solar radiation and temperatures near

Figure 27.9 Tundra vegetation. (**a**) Arctic tundra at high latitudes in Alaska, north of timberline. The antler is from a caribou. (**b**) Alpine tundra at high elevation above timberline, here at 3500 m (11,500 ft) elevation in southern Colorado.

the soil surface are higher; also, thermoperiods have a broader range than in arctic tundra. An alpine plant could experience leaf temperatures of 30°C (86°F) at midday and −5°C (23°F) at night, for a 35°C (63°F) thermoperiod. During the same 24-hour period, an arctic plant would experience a weak sun that never sets and temperature extremes of only 15°C (59°F) maximum and 5°C (41°F) minimum, for only a 10°C (18°F) thermoperiod. Alpine tundra plants share many of the same environmental stresses felt by desert shrubs thousands of meters lower in elevation: strong winds, high solar radiation, short growing season, low soil moisture, and widely fluctuating temperatures.

Figure 27.10 Boreal forest in northern Canada, adjacent to a clear-cut so that its architecture is revealed.

Boreal Forest Is the Taiga of North America

South of the arctic tundra lies a broad belt of low-elevation conifer forest, the **boreal forest**. It covers 28% of North America. At the forest's northern limit, trees meet and mingle with the tundra, and they can be shrublike and slow growing. The critical environmental factor determining the location of timberline is the amount of heat received during the growing season. The growing season in the boreal forest is three to four months long, and temperatures can be much warmer than in the tundra. Annual precipitation is 30 to 90 cm (12 to 36 in), most of it falling in summer; so water is not limiting. Winter temperatures, however, can be even lower than they are in the tundra because oceans—with their moderating effect on temperature—are more distant. Soils are relatively young, acidic, and leached of nutrients. Some soils beneath the northern portion of the boreal forest have a permafrost layer.

This vegetation has a two-layered architecture. Trees are slender, short (15 to 20 m tall, or 50 to 65 ft), and relatively short-lived (less than 300 years); but they are densely packed (Fig. 27.10). The Eurasian term for this

vegetation, **tataiga**, means "dense forest." Carpeting the ground beneath the overstory canopy is a continuous layer of bryophytes, seedless vascular plants, and herbaceous perennial angiosperms.

Disturbance by storms and wildfires recurs every 200 or more years, affecting large areas and setting secondary succession in motion. Black spruce (*Picea mariana*), one of the most widespread trees in the boreal forest, is a closed-cone conifer well-adapted to fire. As an intense fire sweeps through a black spruce forest, killing all aboveground vegetation, the millions of sealed cones are melted open, and for days following the burn a quiet rain of seeds falls on an ash-rich seed bed. A new black spruce forest will replace the old one.

The Eastern Deciduous Forest Has a Complex Physiognomy

Boreal softwoods give way, at about 45° N latitude, to deciduous hardwoods as the growing season increases to six months and annual precipitation climbs above 100 cm (40

Figure 27.11 The eastern deciduous forest. (**a**) The leafy overstory canopy changes color dramatically in autumn, just before leaf drop. (**b**) Active spring growth of perennial herbs occurs before the new overstory completely leafs out.

in). Winters are cold but are neither as intensely cold nor as long as in the boreal forest. Soils are richer in nutrients and less acidic. This is the **eastern deciduous forest**, a vegetation type dominated by a diversity of broadleaf, winter-deciduous tree species (Fig. 27.11a). Along the southern limits of the forest, at about 30° N latitude, winters are mild enough for evergreen broadleaf trees to mix—as a minor element—with deciduous trees. Precipitation declines to the west, where an arid grassland exists as a boundary to the forest. Sometimes the transition from forest to grassland is abrupt; sometimes it is many kilometers wide. This important ecotone between woody and herbaceous vegetation meanders north to south along the line of 95° W longitude.

The eastern deciduous forest, which covers 11% of North America, is a complex mix of communities. Very little of this forest has escaped human impact. Native Americans influenced its structure for thousands of years by burning the understory. They maintained an open forest to encourage populations of birds and grazing animals used for food. Before the forest changes brought about by European colonists, forest buffalo and grouse grazed as far east as New York State. In 200 to 300 years, Euro-Americans modified the forest further by clearing, selectively cutting, grazing, or accidentally introducing foreign pathogens. The modern forest no longer resembles the descriptions by early explorers. The forest overstory of Ohio and Pennsylvania, for example, was once dominated by 400-year-old oaks (*Quercus*), sugar maples (*Acer saccharum*), and chestnuts (*Castanea*) 1 to 2 m (3 to 7 ft) dbh, with straight trunks rising 25 m (65 ft) before the first side branch. Black walnuts (*Juglans*), shagbark hickories (*Carya*), and cotton-

woods (*Populus*) grew in flood plains near rivers and were big enough to be made into dugout canoes 20 m (65 ft) long and more than 1 m (3 ft) across. Today, only a few forest reserves contain remnants of this ancient forest.

Seasonality is a striking feature of eastern deciduous forests. Overstory leaves turn brilliant yellow, red, and orange in autumn. After a winter dormancy, the first green of spring comes from a diversity of herbaceous perennials on the forest floor (Fig. 27.11b). These species take advantage of a narrow time window when light and temperature permit rapid growth. As the overstory canopy completes its expansion of new leaves, these spring-flowering herbs shed seeds and enter dormancy. Two intermediate canopy layers—between the ground herbs and the overstory trees—consist of scattered shrubs (many in the heath family) and small trees such as dogwood (*Cornus*). In addition, many vines (such as Virginia creeper, *Parthenocissus quinquefolia*), grow up through all of the tree and shrub canopies.

Grasslands Cover One-Fifth of North America

Across an east-west expanse of perhaps 100 km (60 mi), we pass along a gradient of increasing aridity and through a series of vegetation types. At the western edge of the deciduous forest, the overstory tree canopy becomes discontinuous, and herbaceous plants begin to form a continuous ground cover. Forest becomes woodland and then savanna, steppe, and finally grassland. **Woodland** is grassland with overtopping trees whose canopies cover 30–60% of the ground; **savanna** is grassland with overtopping

Figure 27.12 Grassland on rolling hills in North Dakota.

Figure 27.13 The Great Basin desert vegetation has a uniform sagebrush shrub (*Artemisia tridentata*) architecture. It is cold in winter, with snow and hard frost. Warm desert vegetation is more diverse and includes succulents, trees, and drought-deciduous shrubs.

trees that are regularly present but whose canopies cover less than 30% of the ground; **steppe** is grassland interspersed with shrubs. **Grassland** (synonym, **prairie**), finally, is vegetation dominated by herbaceous plants growing in a climate too dry for trees (Fig. 27.12). Trees may be present, but they are restricted to special, localized topography, such as along waterways or on rocky ridgelines with thin soil. Perennial and annual grasses dominate the biomass, but broad-leaved dicot herbs (**forbs**) dominate in terms of numbers of species.

Some plant geographers have concluded that there is no such thing as a grassland climate. Their position is that grasslands replace woodlands for reasons other than gradients in climate. A certain annual precipitation, say 50 cm (20 in), could support grassland, savanna, woodland, or forest. Grassland is favored in those places where variation in rainfall from year to year is high, creating years of drought that alternate with years of above-average precipitation. Grassland is also favored in fire-type climates, where the probability of wildfire revisiting the same hectare of land every one to three years is high. Additionally, grassland is favored where growing temperatures are high enough and air humidity low enough to promote transpiration. In forests, the ratio of incoming precipitation to outgoing transpiration is greater than 1, but in grasslands the ratio falls to 1 or just below. Finally, grassland is favored where soils are deep and texture is loamy to clayey.

Grasslands once covered 21% of North America. The largest area of grassland occurs across central North America, from Manitoba to Texas and from Iowa to the Rocky Mountains. It consists of tall-grass, mixed-grass, and short-grass prairies. Other major grasslands are: (1) along the edge of warm deserts in Texas, New Mexico, and Arizona; (2) scattered through the intermountain Great Basin (with a finger extending into the Palouse area of southeastern Washington and outliers in the Willamette Valley of Oregon and north coastal California); and (3) within the Great Central Valley of California.

All of our grasslands have been significantly modified by Euro-Americans in the past 200 years. Most of the central grasslands have been cleared and plowed, converted into farmland. The desert grasslands have been overgrazed and fire has been suppressed; as a result, desert shrubs have invaded. The intermountain, Palouse, and California grasslands have been overgrazed, and invaded by aggressive annuals from Eurasia (for example, cheatgrass, *Bromus tectorum*) as well as having had many hectares converted to farmland, pasture, and urban sprawl.

Desert Scrub Is Dominated by Shrubs

Scrub is any vegetation dominated by shrubs. It occurs where either precipitation or the water storage capacity of the soil is too low to support grassland. Thornscrub, chaparral, and desert scrub are examples. **Desert scrub** occurs where annual rainfall is less than 25 cm (10 in) and a pronounced dry season exists every year. The annual precipitation-to-transpiration ratio drops to 0.1. Another feature of desert climate is the high variation in rainfall from year to year, so that the concept of an average rainfall is meaningless. Desert vegetation covers only 5% of North America.

Deserts are generally warm to hot during the summer, but they may be quite cold in winter. The high-elevation intermountain desert, for example, regularly receives snow and hard frost every winter; this stress reduces species and growth form diversity (Fig. 27.13). Low-elevation and more southerly Sonoran desert scrub has almost no frost; it has a greater diversity of species and growth forms. The two other North American deserts—the Mojave (of southern California and Nevada) and the Chihuahuan (of Texas, New Mexico, and Mexico)—are intermediate in elevation and temperature regimes.

Associated with shrubs in the warm deserts may be succulent cacti, green-stemmed trees, subshrubs, herba-

Figure 27.14 Coachwhip (*Fouquieria splendens*), a drought-deciduous plant that is able to shed and remake several crops of leaves a year. (**a**) Plants with leaves, soon after heavy rains. (**b**) Same plants, leafless during an extended dry period.

ceous perennials, and ephemerals. Total ground cover may be 50% at a maximum and 5% at a minimum. Other parts of the world have more extremely arid climates than North America, and in those places plant cover may be less than 1%.

All plants in desert scrub must be adapted to survive extended droughts. There are five basic techniques for drought tolerance or avoidance. One is the **phreatophyte** syndrome, a suite of traits shared by woody plants that have deep roots in permanent contact with groundwater. These plants have green stems and leaves that are winter deciduous. Their leaves are well supplied with water during the hot summer, and therefore they are under stress (of cold temperature) only in wintertime, which is when the leaves are shed.

Other shrubs with shallower roots retain their leaves only during the wet season and drop them during dry seasons; this is the **drought-deciduous** syndrome. Drought-deciduous leaves are thin and energetically inexpensive; they can be cast off and remade several times a year (Fig. 27.14).

A few shrubs, the **true xerophytes**, have evergreen leaves. They are able to tolerate the desert dry season because their metabolism proceeds at a slow rate the entire year. Under prolonged drought, some of the leaves may be shed; but if all fall, the plant dies. Their leaves are typically small and have many anatomical features that retard transpiration.

Succulents, such as cacti, store water in the vacuoles of large cells. They typically exhibit crassulacean acid metabolism, which features stomata that open by night and close by day, thereby lowering transpiration. Their leaves and bodies also minimize the amount of surface for a given volume or mass, another feature that reduces transpiration. Cacti have shallow root systems, capable of absorbing moisture from even light rains. Succulent plants appear to avoid drought, rather than tolerate it.

Ephemerals exhibit a fifth syndrome: They live for six weeks to six months and complete their life cycle during the wettest, least stressful part of the year. They avoid the drought season by passing through it in a state of dormancy, as seeds. Although woody plants dominate the desert biomass, ephemerals contribute the most species.

Mediterranean Vegetation Includes Chaparral and Woodland

Mediterranean climates and vegetation are found in five locations throughout the world: (1) the Mediterranean rim of southern Europe, the Middle East, and northern Africa; (2) the Cape region of South Africa; (3) southern and southwestern Australia; (4) central Chile; and (5) California. All these regions lie between 40 and 32° N or S latitude. They occupy western or southwestern edges of continents, receive 27 to 90 cm (11 to 36 in) of annual precipitation (mainly winter rainfall), have minimal frost, and experience episodic wildfire. Mediterranean climates are fire-type climates, basically with hot, dry summers and cool, wet winters.

Vegetation is remarkably similar in these five areas, even though different families of plants predominate in each. **Mediterranean vegetation** ranges from forest (in the wettest locations) to woodland to scrub (in the driest locations). All these vegetation types are dominated by broad-leaved, evergreen, woody flowering plants, with a relatively high percentage of canopy cover.

Mediterranean scrub is called **chaparral**, from the Spanish word for "low-growing." Chaparral is dense and one-layered, about 1 to 3 m (3 to 10 ft) tall; it is composed of rigidly branched shrubs with small, hard leaves and an extensive root system (Fig. 27.15a). In contrast to open desert scrub, chaparral has 100% cover and a leaf area index twice that of desert scrub. Herbs are uncommon.

Wildfire recurs every 20 to 50 years, consuming all the aboveground vegetation and producing very high temperatures. Chaparral shrubs respond to fire in several ways. Some are capable of sprouting from the stem base, which is buried beneath the soil surface and thus insulated from high temperatures during the fire. Others have hard-coated seeds that lie dormant until cracked by

Figure 27.15 Mediterranean vegetation types. (**a**) Chaparral. (**b**) Foothill woodland.

Figure 27.16 Pacific Coast conifer forest (also known as the temperate rain forest). (**a**) Cathedral Grove Provincial Park, Vancouver Island, British Columbia. The dominant trees are 2 m (7 ft) dbh and more than 65 m (200 ft) tall. Note human figure (orange jacket) for scale. (**b**) Because of the high rainfall and high humidity, tree trunks and branches support a rich assortment of epiphytic mosses and lichens.

moderately high temperatures. By both re-sprouts and seedlings, the chaparral community recovers its preburn cover and species composition within half a dozen years.

Chaparral grows on steep slopes with coarse, shallow soils, at elevations below 1000 m (3300 ft). Nearby, within the same macroenvironment, are **foothill woodlands**, dominated by oak (*Quercus*) species and underlain by grasses and forbs (Fig. 27.15b). Tree canopy cover is 30 to 60%. Woodland and chaparral vegetation types are spatially separated according to microenvironmental factors, not macroenvironmental ones. The woodland occupies gentler slopes (with deeper, less coarse soil) as well as moist north-facing or east-facing slopes (which burn less frequently than the drier chaparral sites).

Before the arrival of Euro-Americans, the woodland understory consisted mainly of perennial bunchgrasses. Because of the introduction of aggressive, weedy annuals in the past 150 years, the understory today consists mainly of annual grasses. These woodlands once supported the highest population densities of Native Americans in all of North America. Oak acorns were the major resource that supported them. Acorns were easily collected every fall, leached of tannin, ground into flour, stored, and used the entire year—much as grains are used in other cultures.

Mediterranean vegetation covers only 1% of North America, but many people are familiar with it because California landscapes are often in the news—witness the ever-recurring catastrophes of wildfires, floods, landslides, earthquakes, and droughts—and because they serve as backdrops to many popular films and television programs.

Pacific Coast Conifer Forests Are the Most Massive in the World

The conifer forests of the Pacific coast are the most luxuriant, most productive, most massive vegetation in the world. They are dominated by a rich diversity of big, long-lived tree species underlain by equally rich canopies of shrubs, herbs, bryophytes, and epiphytes (Fig. 27.16). The **Pacific Coast conifer forests** are situated in a low-elevation strip of coastline that extends from Cook Inlet, Alaska, south to Monterey, California. The climate is mild, buffered by the nearby ocean and summer fog banks; so thermoperiods are only 6 to 10°C (11 to 18°F). Hard frosts are uncommon. Annual precipitation is very high—80 to 300 cm (32 to 120 in)—most of which falls in wintertime. Because of the climate, this type of forest is often called a **temperate rain forest**.

Dominant tree species include coast redwood (*Sequoia sempervirens*), Douglas fir (*Pseudotsuga menziesii*), lowland white fir (*Abies grandis*), Sitka spruce (*Picea sitchensis*), western hemlock (*Tsuga heterophylla*), and western red cedar (*Thuja plicata*). These trees all commonly live for 400 to 1200 years and attain heights of 100 m (330 ft). Annual productivity is as high as 25 metric tons per hectare (12 tons per acre), and standing biomass is typically 85 metric tons per hectare (42 tons per acre); in one case, it is as high as 230 metric tons per hectare (110 tons per acre). Tropical rain forest vegetation has a similar productivity, but the biomass is several times smaller.

Although this vegetation type covers only 3% of North America, its timber volume and value have been highly significant to both local human communities and distant corporations during the past century. Logging has had a heavy impact on the region. Conservationists, using the endangered northern spotted owl as a symbol of logging threats to biotic diversity, have managed to decrease the intensity of logging in the United States in recent years. Canada has not yet reduced its harvest of coastal forests in British Columbia, however, and clear-cuts in British Columbia, are said to be among the few human artifacts visible from orbiting spacecraft.

Upland Conifer Forests Have a Wide Range

Coniferous forests clothe the slopes of the higher Appalachian peaks, the Rocky Mountains, the Cascade-Sierra Nevada axis, the Coast Ranges of Washington and Oregon, and the Transverse and Peninsular Ranges of southern California. These upland forests are called **montane conifer forests**. They range from 65 to 19° N latitude (well into the Sierra Madre of Mexico), and they cover 7% of North America. Annual precipitation rises from 60 cm (24 in) at lowest elevations to over 200 cm (80 in) at higher elevations. Precipitation in winter is in the form of snow, and deep snowpacks can accumulate. Summers are warm and relatively dry. Montane conifer forests of the West are in a fire-type climate.

The structure, diversity, and productivity of montane conifer forests are intermediate between the two other conifer-dominated vegetation types we have already described, the boreal forest and the Pacific Coast forest. Many of our most popular national parks, the jewels of the park system, are located in this vegetation type: Glacier, Great Basin, Great Smoky Mountains (uppermost elevations), North Cascades, Rocky Mountain, Yellowstone, and Yosemite National Parks (Fig. 27.17).

Zonation of forest communities along elevation gradients is a common phenomenon. **Lower montane** (low-elevation) **forests** tend to be rather open savannas or woodlands, intermingled with species from adjacent grasslands, Mediterranean woodlands and chaparral, or deserts. Frequent wildfires seem to be essential to the

Figure 27.17 Midmontane conifer forest, Yosemite National Park, California, about 2000 m (6600 ft) elevation.

maintenance of some of these communities, and they have been significantly degraded by overgrazing and changes in fire frequency over the past 150 years. Pinyon pines (*Pinus monophylla* and others), ponderosa pine (*P. ponderosa*), and junipers (*Juniperus scopulorum* and others) are common woodland trees.

Midmontane (intermediate-elevation) **forests** are typically rich in overstory species, such as Douglas fir (*Pseudotsuga menziesii*), white fir (*Abies concolor*), ponderosa pine (*Pinus ponderosa*), and many regionally limited tree species. A variety of shrubs in the heath and rose families are also common, as are seasonally present herbaceous perennials. These are complex, four-layered forests that often require wildfire to maintain their structure; they need ground fires that burn relatively cool, consuming only litter, shrubs, and young trees. When fire is kept out by fire suppression management, these forests change their physiognomy, becoming denser. Then, when a ground fire does start (and the presence of a fire-type climate assures us that it *will* start), the ground fire will *ladder up* the younger trees and become a raging, destructive crown fire (Fig. 27.18).

The midmontane elevation belt contains many other vegetation types besides forest. Perhaps half the area is interrupted by meadow or scrub vegetation. **Meadow** is grassland that occurs within a climate capable of supporting forest vegetation. Forbs often dominate meadows, grasses being reduced to associate status. Forest is absent here because of local topography or soil that creates seasonally wet conditions. **Montane scrub** occurs on rocky ridges, on south-facing slopes, or as a temporary successional stage following wildfire. In the latter case, forests will slowly reclaim the site by growing through the scrub and shading it out.

Upper montane and **subalpine** (highest-elevation) **forests** are densest and simplest and experience the deepest snowpacks. Fir (*Abies*), hemlock (*Tsuga*), pine (*Pinus*), and spruce (*Picea*) are the dominant genera. This is the ele-

ground fire   crown fire

Figure 27.18 Forest fires can be ground fires (left) or crown fires
(right). Recurrent natural wildfires keep many montane forests
open, so that fires remain on the ground and consume only litter,
shrubs, and young trees. Fire temperatures are moderate, and mature trees survive. Fire suppression, however, allows more trees of
all sizes to become established; thus, when a fire occurs it ladders
from the ground up to the crown, becomes hotter, and kills the
trees. (Drawing by M. Yuval from M. Barbour et al., *California's
Changing Landscapes*. Sacramento: California Native Plant
Society, 1993.)

vation zone where the oldest individual plants in the
world exist, bristlecone pines (*Pinus longaeva*, Fig. 27.19).
Above the subalpine forest, the amount of heat during the
growing season is too little to support the growth of trees,
and we enter an alpine, high-elevation type of tundra.

Wetlands and Aquatic Ecosystems Are Very Productive

MARINE ECOSYSTEMS Wetlands are terrestrial sites
where the upper soil is saturated by saline or fresh water
for at least a few weeks of the year. **Tidal wetlands**, or
salt marshes, are coastal meadows subject to periodic
flooding by the sea (Fig. 27.20). They occupy nearly level
shores that receive only low-energy waves, such as
are found along estuaries or tidal flats behind barrier islands or sandy peninsulas. Environmental stresses include flooding—bringing mechanical disturbance and
anaerobic conditions—and salinity. As the ground slopes
up landward, however, these two stresses decline in frequency and intensity.

The vegetation is usually a single, low-growing, nearly
closed layer of perennial herbs. The soil beneath is crowded
with rhizomes and roots. The flora is rather simple, with
only a handful of species coexisting in a given local wetland. Tidal wetlands along the entire Atlantic and Gulf
coasts of the United States support fewer than 350 vascular
plant species, and those along the entire Pacific coast, from
Point Barrow to the tip of the Baja California peninsula,
contain fewer than 80 species. Apparently, not many
species have the genetic capacity to tolerate tidal marsh

Figure 27.19 Bristlecone pine (*Pinus longaeva*) in the subalpine
zone of the White Mountains of California. This species occurs on
high peaks throughout the Great Basin, and individuals can attain
ages of 5000 years.

Figure 27.20 Tidal wetland (salt marsh) vegetation, Morro Bay,
California.

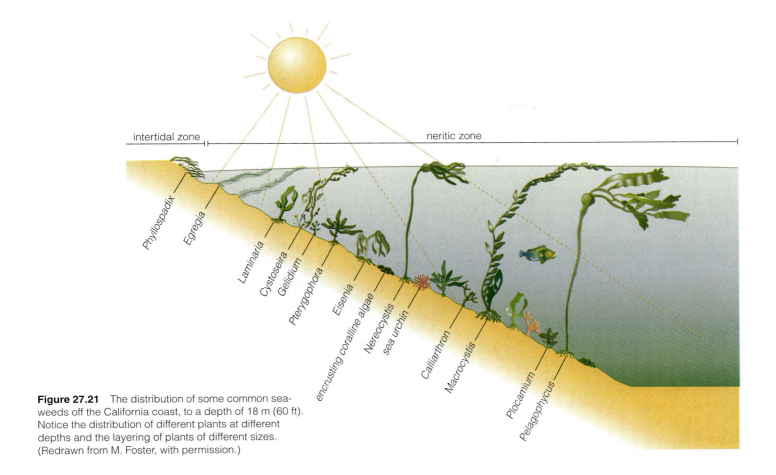

intertidal zone   neritic zone

Phyllospadix
Egregia
Laminaria
Cystoseira
Gelidium
Pterygophora
Eisenia
encrusting coralline algae
Nereocystis
sea urchin
Calliarthron
Macrocystis
Plocamium
Pelagophycus

Figure 27.21 The distribution of some common sea-weeds off the California coast, to a depth of 18 m (60 ft). Notice the distribution of different plants at different depths and the layering of plants of different sizes. (Redrawn from M. Foster, with permission.)

stresses. A few have enormous ranges that extend through nearly all coastlines of North America: arrow-grass (*Triglochin maritimum*), cordgrass (*Spartina alterniflora*), pickleweed (*Salicornia virginica*), saltbush (*Atriplex patula*), and salt grass (*Distichlis spicata*). Some of the traits these successful species exhibit are succulence, asexual reproduction by rhizomes, and aerenchymna tissue in stems and roots (which channels oxygen down into the anaerobic soil).

Enormous losses of acreage in tidal wetlands have occurred during the 20th century. Tidal wetlands have been diked, drained, filled, and converted to farmland, ports, and cities. We have discovered their ecological value rather late. One ecological function these wetlands serve is as a biological filter for runoffs from the land, which contain pollutants and excessive nutrients. The marsh and its soil act as a sieve, cleaning the water before passing it on to the ocean. Another ecological function is service as a nursery for the young of many aquatic animals, including commercially valuable fish. Although salt marsh vegetation has a low profile, its annual productivity is as high as that of a tropical rain forest. The plants are herbaceous, however; so this tremendous amount of new plant biomass does not accumulate as woody tissue. Instead, it is shed each year into the water, where it fuels an extensive food chain through microbes, algae, plankton, invertebrates, fish, and humans.

Along exposed coasts that receive the full brunt of wave action, a **rocky intertidal** ecosystem replaces the salt marsh. Flowering plants are few in the rocky intertidal zone, but they include surf grass (*Phyllospadix*), a relative of other sea grasses more commonly found in quieter harbor waters. Many seaweeds and a few kelps are also attached to rocks in this habitat, each at its own particular depth within the intertidal zone. The different depths at which they are distributed probably reflect different degrees of adaptation to the stresses of exposure.

Below the intertidal zone is the **neritic zone**. This aquatic ecosystem is a rocky shelf always covered by water but shallow enough to admit adequate sunlight for attached algae to grow along the bottom. Many kelps grow in the neritic zone, especially in cool-temperate oceans. As in the intertidal zone, different species of algae dominate at different depths (Fig. 27.21). The pattern of distribution probably reflects differing tolerances among algal species to low light. Both the quality and quantity of light change with passage through water. The compensation depth, where positive growth is no longer possible, is usually shallower than 170 m (560 ft). In deep oceanic bodies of water, the habitat between the surface and the compensation depth is dominated by phytoplankton. One exception is the Sargasso Sea off Bermuda, where large, floating masses of *Sargassum* seaweed are present.

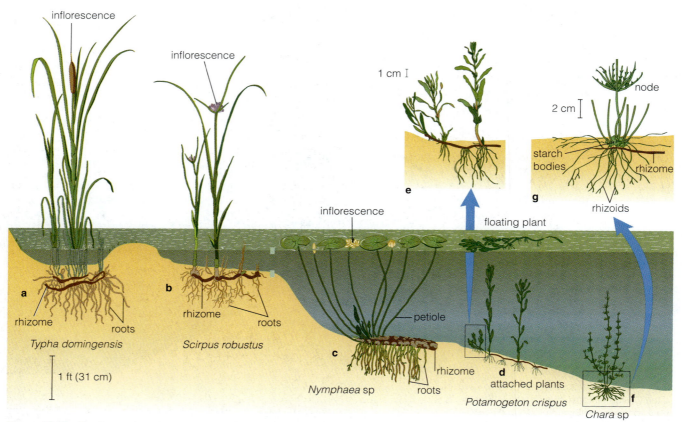

inflorescence

inflorescence

1 cm

node

2 cm

starch
bodies

rhizome

inflorescence

floating plant

rhizoids

e

g

a

rhizome

roots

Typha domingensis

b

rhizome

roots

Scirpus robustus

petiole

c

rhizome

Nymphaea sp

roots

attached plants

d

Potamogeton crispus

Chara sp

f

1 ft (31 cm)

Figure 27.22   Freshwater ecosystems support flowering plants and algae. (**a**) Cattail (*Typha domingensis*). (**b**) Sedge (*Scirpus robustus*), more tolerant of flooding than cattail. (**c**) Water lily (*Nymphaea*), with floating leaves and flowers. (**d**) Submerged and floating segments of pondweed (*Potamogeton crispus*). (**e**) Enlargement. (**f**) The stonewort *Chara*, a complex alga differentiated into rhizoid, "rhizome," and "stem" regions. (**g**) Enlargement.

This seaweed may play an important role in the survival of young sea turtles, giving them a place to feed and hide from predators until they become large adults.

FRESHWATER ECOSYSTEMS   Freshwater wetlands are found along the shorelines of lakes, rivers, sinks, and springs. Trees, shrubs, and herbs that occupy these habitats must be tolerant of occasional flooding, and they must be resilient to the physical disturbance of floodwaters. They often have fast growth rates, produce abundant wind-distributed seed, and are capable of vegetative reproduction. This fringe of **riparian** (bank) **vegetation** is ecologically important as a filter of eroded soil and nutrients that otherwise would enter the aquatic ecosystems and degrade them. The filter function is very important in the modern landscape because of intensive agricultural practices that add fertilizers and pesticides to the land.

At shallow margins of freshwater aquatic ecosystems (Fig. 27.22) are flowering plants such as sedge (*Scirpus, Carex, Cyperus*), cattail (*Typha*), pondweed (*Potamogeton*), and water lily (*Nymphaea*). These are called *emergent plants*

because although they are rooted under water, some part of the plant body extends above the water surface. Associated with them are algae that grow attached either to the muddy surface or epiphytically to other plants: mainly diatoms, green algae, and golden algae. In somewhat deeper water, submerged or floating flowering plants are common, associated with larger algae such as stonewort (*Chara*).

If lakes are deep enough, their water may stagnate into three zones. An upper zone, the **epilimnion**, is relatively warm and sunlight is intense enough to support large phytoplankton populations. Oxygen levels in the epilimnion are high. Below the epilimnion is a narrow transition zone where temperature drops rapidly with depth (on the order of 4°C per meter—2°F per foot). This is the **thermocline**, and it serves as a barrier to any mixing between the upper epilimnion and the lower **hypolimnion**, which extends to the bottom of the lake. Phytoplankton may occur in the hypolimnion, but only at low densities.

We conclude this chapter with some thoughts about how this diversity of vegetation might be maintained into the future.

27.3 CONSERVATION BIOLOGY

Each year, the extent of natural vegetation grows smaller. Vast areas of some vegetation types, such as tundra and boreal forest, still remain as wilderness; but many of our deciduous forests, grasslands, deserts, woodlands, montane conifer forests, and tidal wetlands are endangered. They are threatened by timber harvest, land conversion, water diversion, and the spread of weeds and pathogens—and even by the recreational impact of too many people in too small a place.

Only a generation ago, *conservation* meant a rate of natural resource consumption that would result in sustained, continued existence of that resource far into the future. We are beginning to understand that humans have great difficulty in staying within boundaries that would sustain a particular resource. Our species has had a sad, consistent history of overshooting these balance points. Our farming, fishing, and forestry activities have consistently led to overexploitation and decline of natural resources. Consequently, *conservation* now usually means the restricted use, nonuse, or preservation of natural resources.

Conservation biology is a relatively new science that studies the impact of human societies on the nonhuman landscape. Conservation biologists ask the question: Can a growth-oriented, technological culture coexist with its surrounding natural systems? We know that nontechnological cultures, such as those of Native Americans, did coexist with their surrounding natural systems for thousands of years, but we recognize that their demands on nature were much different from ours.

Conservation biologists are investigating ways to measure sustainability: How do we know when a plant or animal population is maintaining itself? How do we measure biotic diversity? How do we design parks so that the probability of extinction for any rare population is as low as possible (see sidebar, "Nature in Flux or Nature in Balance?" p. 488)? How do we restore degraded habitats and their plant and animal communities? How do our technological activities interweave with the biosphere in unexpected ways to magnify into global stresses (acid rain, ozone depletion, climate change, pollutants carried through food chains), and how might we best modify these technological acitivties to reduce the stress?

Conservation biology is an exciting field, but its contributions will be limited if the human population continues to rise. In the middle of the 20th century we numbered 3 billion, and we are expected to reach 6 billion by the start of the 21st century. Sustainability is an unattainable myth in the face of such climbing population pressure on the earth. Without a doubt, the control of human population growth has been and remains the greatest challenge our species must solve.

SUMMARY

1. A plant community consists of a cluster of associated species that repeats wherever a particular habitat repeats itself. Plant communities are named after their dominant or characteristic species.

2. Plant communities have a characteristic architecture, or physiognomy, which is a combination of the external appearance of the community, its vertical structure, and the growth forms of each canopy layer.

3. Other community attributes include percent cover, leaf area index, species richness, productivity, biomass, allocation of biomass (to roots, woody tissue, leaves, reproductive organs), rate of nutrient cycling, and relative stability.

4. Plant communities change over time in a process called *succession*. Technically, succession is cumulative, directional change in a homogeneous area over several years to several hundred years. Secondary succession takes place on already vegetated land that is disturbed; primary succession takes place on new land not previously occupied by plants. Primary succession is much slower than secondary succession.

5. Progressive succession leads to increasingly complex and massive communities, in which the cycles of energy and nutrient flow become tighter and more efficient and the microenvironment becomes less stressful. Retrogressive succession exhibits the reverse trends.

6. Plant communities blend gradually into each other over space because every species has its own range limits. However, the architecture and habitat may remain constant. All plant communities that share the same architecture and habitat belong to the same vegetation type. Vegetation types are named after location and dominant growth form, not after the dominant species. North America has thousands of plant communities, but only about a dozen major vegetation types.

7. The major vegetation types of North America differ profoundly in their productivity, biomass, physiognomy, and habitat traits—and in the portion of North America that they occupy.

8. Two-thirds of North America is covered by boreal forest, grassland, and tundra. Tundra is herbaceous vegetation that occurs where the growing season is too cool and short for trees. Arctic tundra is at low elevations in far northern latitudes, whereas alpine tundra is at high elevations in mountain chains at more southern latitudes. Grassland vegetation lacks trees because of periodic droughts, a low precipitation-to-evaporation ratio, recurring wildfire, and a fine soil texture. Boreal forest (taiga) is dense, short, and relatively low in biotic diversity. It occurs in a wide band just south of arctic tundra, where the growing season is three to four months long and the dif-

NATURE IN FLUX OR NATURE IN BALANCE?

A "paradigm" is any widely accepted viewpoint. In science, a paradigm is made up of a family of related theories that seem to explain some behavior in nature. Until recently, one classical paradigm in ecology was the "equilibrium paradigm," which saw nature as dominated by stable, long-undisturbed climax ecosystems. An old-growth montane forest, one acre of central prairie, or a strip of undisturbed salt marsh was each said to be in equilibrium, or in balance, with its environment.

This classical paradigm and its metaphor, "the balance of nature," suggest that any unit of nature is conservable; thus, any unit of a landscape can be an adequate nature reserve. The paradigm also suggests that reserves isolated from direct human disturbance will maintain themselves. That is, people and their activities are not a part of natural systems.

An example of the application of this paradigm is Mettler's Woods, the last remaining uncut upland forest in central New Jersey. The forest, which had been protected since colonial days, and eventually became a part of Rutgers University, is now known as the Hutcheson Memorial Forest Center (HMFC). The forest was considered to be climax and thus self-perpetuating. No human disturbance was to be permitted. Visitors were restricted to a single trail and had to be accompanied by a qualified guide.

Reality has shown that merely protecting the forest from humans is not sufficient to maintain it. The old oaks (*Quercus* species), the average age of which is 245 years, are senescing and becoming sensitive to the combined stresses and disturbances that naturally occur. The overstory is now quite open, the mid-layer dogwoods (*Cornus florida*) have succumbed to a regional epidemic of anthracnose (a fungal disease), and many species of shade-tolerant wildflowers have become rare. After severe defoliation of the oaks caused by gypsy moth caterpillars in the early 1980s, several sun-loving shrubs and herbs invaded the forest understory and have maintained large populations. Using insecticides to manage the gypsy moth population was considered to be inappropriate in a nature reserve, and so was not done. Although the oaks do produce large seed crops, there is an almost complete failure of seedling oaks to survive into the sapling category. It is now understood that ground fires had been a regular part of the history of the forest before fire protection policies were adopted in 1711. Fire scars preserved in the trunks of the oldest oaks document that light ground fires used to occur every 10 years. Perhaps these fires cleared away competing herbs and shrubs and thereby permitted seedlings to survive. Thus, the paradigm—and the conservation strategy derived from it—failed to preserve the forest.

Failures of the equilibrium paradigm became so numerous and well-known by 1980 that ecologists developed a new paradigm, the "nonequilibrium paradigm." First, it accepts natural systems as being open, meaning that they must be put into the context of their surroundings. Second, it recognizes that natural systems are subject to physical disruption from a wide range of natural forces and events such as fire, drought, windstorm, earth movement, volcanism, herbivore outbreak, disease epidemic, and so on. Third, the new paradigm permits the inclusion of humans in the scope of conservation. Once the importance of natural disturbances is recognized, it is a short logical step to include humans as just another agent of change. A metaphor for this new viewpoint is "nature in flux."

The simplest translation of the contemporary paradigm into conservation biology would state that conservation cannot always be passive, but must often involve active management. If natural disturbance had been a characteristic that preserved a particular ecosystem, then managed distur-

---

ference between winter and summer temperatures is relatively large.

9. Other North American forest types include the eastern deciduous forest, dominated by winter-deciduous broadleaf trees; the Pacific Coast temperate rain forest, the most massive vegetation type in the world; and montane conifer forests, home to famous national parks in the Rocky Mountains, the Cascade Range, and the Sierra Nevada.

10. Nonforest vegetation types include Mediterranean scrub and woodlands, desert scrub, and wetlands. Mediterranean climate occurs in only five small parts of the world; in North America it dominates California. Characteristic vegetation includes a dense scrub called *chaparral* and an open woodland of evergreen and drought-deciduous small trees. Desert scrub plants are more dispersed than chaparral shrubs, and they grow in a more arid environment. All desert plants either tolerate or avoid drought, and they do so by being phreatophytes, drought deciduous, true xerophytes, succulents, or ephemerals. Wetlands are very productive vegetation types, and they play important filtering roles as a buffer between terrestrial and aquatic ecosystems. Wetlands may be saline or freshwater. Saline wetlands include salt marshes and rocky intertidal habitats; freshwater wetlands include riparian vegetation along the banks of rivers, lakes, springs, and sinks.

11. Conservation biology is a relatively new science that investigates ways to preserve, restore, and maintain biotic diversity in the face of human exploitation of natural ecosystems. Given the current global population (projected to reach 6 billion early in the 21st century), it will be difficult to develop a resource management plan capable of sustaining our natural ecosystems.

bances that mimic the natural ones must be created. Managed disturbances might include setting a controlled ground fire, thinning and mulching of young trees, biological control of certain insects or pathogens, weeding out invasive plants, or permitting seasonal flooding. The size of the reserve must be large enough to permit such episodic disturbances to run their course within at least a portion of the reserve. Also, the location of the reserve must be chosen in appropriate relation to the land use practices, politics, and social attitudes within adjoining parcels. All pieces of an ecosystem are not equally capable of preservation, and certainly not without active management. The conservation of isolated species or habitats is not likely to be successful. We need to demarcate entire landscapes as parks, preferably with surrounding buffer lands and corridors to other parks, allowing biota to move more widely. The complexities of compromise and management required in such situations, of course, are quite challenging.

This approach to such regional, ecological park planning is being attempted in southern California, where decades of residential development have displaced and fragmented a major regional vegetation type called coastal sage scrub (Fig. 1). A number of rare plant and animal species are at risk of extinction if development continues in the same manner. Furthermore, small preserves are not sustainable because preserves must be large enough to permit episodic wildfires to burn some patches and to permit the movement of displaced animals to adjacent, unburned areas.

As an alternative to continued future habitat and biotic diversity loss, a consortium of federal, state, and local government agency planners, developers, environmentalists, and citizens of several counties joined together in a multiple-year activity. Their objectives have been to locate the best and largest remnants of the scrub, to suggest

Figure 1 Coastal sage scrub in southern California. Approximately 80% of its original range has been displaced and fragmented by urban development in the past half century.

optimal boundary shapes, to design corridors between preserves, to designate appropriate human activities within buffer lands next to the corridors and preserves, and to identify developable areas and transportation routes within the context of this mosaic of regional open space.

Portions of this essay were condensed from S. T. A. Pickett, V. T. Parker, and P. L. Fiedler, "The New Paradigm in Ecology: Implications for Conservation Biology Above the Species Level." Pages 65–88 in *Conservation Biology*, edited by P. L. Fiedler and S. K. Jain. New York: Chapman and Hall, 1992.

Questions

1. Some ecologists believe that plant communities are very tightly organized, with the associated species being somehow interdependent on one another. How could you test this idea, either by observation or experimentation?

2. What traits do communities exhibit that individual plants, or plant populations, do not?

3. When Mount St. Helens erupted in the 1980s, the explosion blew down forest trees and covered the ground with deep volcanic ash. Succession has been studied since the eruption, and it is proceeding slowly. Is this primary or secondary succession, or both?

4. In the boreal forest, storms often uproot and blow down every tree in a stand. Over time, herbs and shrubs invade and cover the site; then aspen may dominate, and conifers reinvade the site under the shade of the aspen. In time, the conifers overtop the aspen, and a boreal forest is reestablished. Explain why this succession is either progressive or retrogressive.

5. Why is it critically important for tundra plants to be low-growing perennial herbs?

6. Describe the physiognomy (architecture and dominant growth forms) of an eastern deciduous forest.

7. Consider a desert phreatophyte, a cactus, and an evergreen shrub all growing near each other in a hectare of desert. Which one do you think would grow the least in one year, and why?

8. Why is wetland vegetation important to an adjacent aquatic ecosystem?

9. Do you think it's possible for humans to do all four of the following simultaneously: continue economic growth, continue population growth, maintain biological diversity, and attain sustainability in natural resource use? If not, how many of these four could be simultaneously achieved?

Further Readings

Archibold, W. O. 1995. *Ecology of World Vegetation*. New York: Chapman and Hall. An outstanding, easily read, and richly illustrated review of the world's major biomes.

Bakeless, J. 1961. *The Eyes of Discovery*. New York: Dover. A classic synopsis of America as the early explorers saw it, with an emphasis on vegetation and landscapes.

Barbour, M. G., and W. D. Billings, eds. 1998. *North American Terrestrial Vegetation*. 2d ed. New York: Cambridge University Press. A detailed account of the vegetation of all regions of North America, including outliers in the Caribbean, Mexico, Central America, and Hawaii. Each chapter is written by regional experts, and each chapter includes sections on successional relationships, conservation activities, and important topics for future research—in addition to summaries of major plant communities and vegetation types.

Barbour, M. G., and N. L. Christensen. 1993. "Vegetation of North America." Pages 97–131 in vol. 1 of *Flora of North America North of Mexico*. New York: Oxford University Press. A concise summary of the major vegetation types of North America.

Brownson, J. M. J. 1995. *In Cold Margins*. Missoula, Mont.: Northern Rim Press. An excellent integration of human culture, vegetation, environment, and geography within the boreal forest region of the north temperate hemisphere. Chapters on Europe, Siberia, Northeast Asia, and North America. Very reader-friendly.

Fiedler, P. L., and Subodh K. J., eds. 1992. *Conservation Biology*. New York: Chapman and Hall. A collection of 18 essays by outstanding vegetation and wildlife ecologists, who describe the new science of conservation ecology and reach important general conclusions about future research directions and the practical matters of restoration ecology.

Ludwig, D. 1993. "Environmental Sustainability: Magic, Science, and Religion in Natural Resource Management." *Ecological Applications* 3: 555–58. A wonderful essay that concludes that humans have never achieved sustainability in any of the ecosystems they know best—fisheries, farms, forests—so what makes us think we can ever achieve it globally? Magic is founded on the belief that "hope cannot fail nor desire deceive," and Ludwig concludes that the current practice of resource management fits this definition of magic; it certainly is not based on experience or science.

GLOSSARY

Technical Terms Including Their Origin

Abbreviation	Meaning
A.S.	Anglo-Saxon
D.	Dutch
dim.	diminutive
F.	French
Gr.	Greek
It.	Italian
L.	Latin
Lapp.	Lapland
M.E.	Medieval English
M.L.	Medieval Latin
N.L.	New Latin
O.F.	Old French
O.E.	Old English
R.	Russian
Sp.	Spanish

The glossary serves two purposes, (a) to define terms used in the text and (b) to give their derivation. It should be noted that in carrying out the first purpose, with few exceptions, only the meanings actually used in the text are given. Reference to dictionaries will give additional meanings, while other textbooks may define the same words in slightly different ways.

Scientific language and slang have much in common. They are both vital, growing, and changing phases of modern English. The same words will have different meanings in different parts of the country.

Scientific language grows in a number of ways. Some botanical terms, such as "seed," are used by scientist and layman alike and can be traced back to Anglo-Saxon days, when the word *sed* was used to designate anything sown in the ground, or, in our terms, a seed. In other instances, words from Greek or Latin have been taken into the scientific language with their identical original meaning; for instance, the Greeks called

wood *xylon*, and we use this term, changed to xylem, to designate the woody conducting tissue of plants. Other words have a more complex history. *Metabolos* is the Greek word "to change." When the scientists of Europe wrote in Latin they took this Greek word, made a Medieval Latin word from it, and used it to describe the changes undergone by some insects. It was used in 1639 to mean changes in health, and in 1845 the German scientist who elaborated the cell doctrine made a German word of it and first used it as we use it here, designating the changes taking place within a cell.

In using the glossary, pay as much attention to derivation as you do to the definition. To memorize the definition alone is to learn, parrot-fashion, only one word. To learn the derivation is to understand the word and possibly to introduce yourself to a whole family of new words. Pay particular attention to such combining forms as *hetero*, *auto*, *micro*, *phyll*, *angio*, *plast* or *plasm*, *spore*, and the like.

ABA abscisic acid

Abscisic acid a hormone variously inducing abscission, dormancy, stomatal closure, growth inhibition, and other responses in plants

Abscission zone (L. *abscissus*, cut off) zone of delicate thin-walled cells extending across the base of a petiole, the breakdown of which disjoins the leaf or fruit from the stem

Absorb (L. *ab*, away + *sorbere*, to suck in) to suck up, to drink up, or to take in; in plant cells materials are taken in (absorbed) in solution

Absorption spectrum a graph relating the ability of a substance to absorb light of various wavelengths

Accessory bud a bud located above or on either side of the main axillary bud

Accessory pigment a pigment that absorbs light energy and transfers energy to chlorophyll *a*

Achene simple, dry, one-seeded indehiscent fruit, with seed attached to ovary wall at one point only

Acid (F. *acide*, from L. *acidus*, sharp) a substance that can donate a hydrogen ion; most typical acids are sour and are compounds of hydrogen with another element or elements

Acid-growth hypothesis a hypothesis to explain the stimulation of growth of plant cells by auxin: states that the main effect of auxin is to cause cells to secrete acid (H^+ ions) and that the acid stimulates the changes in plasticity

Actinomorphic (Gr. *aktis*, ray + *morphe*, form) said of flowers of a regular or star pattern, capable of bisection in two or more planes into similar halves

Action spectrum (F. *acte*, a thing done) a graph relating the degree of physiological response (e.g., phototropism, photosynthesis) caused by different wavelengths of light

Activation energy the increase in potential energy of a complex of two or more molecules required for a chemical reaction

Active site a region of an enzyme whose shape permits the binding of substrates and where catalytic activity takes place

Active solute uptake the intake of dissolved materials by cells against a concentration gradient and requiring an expenditure of metabolic energy

Adaptation (L. *ad*, to + *aptare*, to fit) adjustment of an organism to the environment

Adenine a purine base present in nucleic acids and nucleotides

Adenosine triphosphate (ATP) a substance formed in metabolism from ADP and inorganic phosphate; the most prominent and universal molecule that acts as a carrier of energy in metabolism

Adhesion (L. *adhaerere*, to stick to) a sticking together of unlike things or materials

Adnation (L. *adnasci*, to grow to) in flowers, the growing together of two or more whorls to a greater or lesser extent; compare **Adhesion**

ADP adenosine diphosphate

Adsorption (L. *ad*, to + *sorbere*, to suck in) the concentration of molecules or ions of a substance at a surface or an interface (boundary) between two substances

Advanced (M.E. *advaunce*, to forward) said of a taxonomic trait thought to have evolved late in time from some more primitive trait

Adventitious (L. *adventicius*, not properly belonging to) referring to a structure arising from an unusual place: buds at other places than leaf axils, roots growing from stems or leaves

Aerate to supply or impregnate with common air, such as by bubbling air through a culture solution

Aerobe (Gr. *aer*, air + *bios*, life) an organism living in the presence of molecular oxygen and using it in its respiratory process

Aerobic respiration respiration involving molecular oxygen

Agar (Malay *agaragar*) a gelatinous substance obtained mainly from certain species of red algae

Aggregate fruit (L. *ad*, to + *gregare*, to collect; to bring together) a fruit developing from the several separate carpels of a single flower; for example, a strawberry

Akinete (Gr. *a*, not + *kinein*, to move) enlarged thick-walled nonmotile reproductive cell produced by some cyanobacteria

Alcohol (M.L. from Arabic *al-kuhl*, a powder for painting eyelids; later applied, in Europe, to distilled spirits that were unknown in Arabia) a product of the distillation of wine or malt; any one of a class of compounds analogous to common alcohol; the ending *ol* designates a member of this class of compounds

Aleurone layer (Gr. *aleurone*, flour) the outer most cell layer of the endosperm of wheat and other grains

Alfisol modified podsol soil, typical of the northern part of the deciduous forest

Alga (plural, **algae**) (L. *alga*, seaweed) a member of the large group of thallus plants containing chlorophyll and thus able to synthesize carbohydrates

Algin a long-chain polymer of mannuronic acid found in the cell walls of the brown algae

Alkali (Arabic *alqili*, the ashes of the plant saltwort) a substance with marked basic properties

Allele (Gr. *allelon*, of one another, mutually, each other) variant form of a gene; **multiple-**, several alleles at a single locus; **pseudo-**, factors that recombine within a single locus

Alpine (L. *Alpes*, the Alps Mountains) meadowlike vegetation at high elevation, above tree line

Alternate referring to bud or leaf arrangement in which there is one bud or one leaf at a node

Alternation of generations the alternation of haploid (gametophytic) and diploid (sporophytic) phases in the life cycle of many organisms; the phases (generations) may be morphologically quite similar or very distinct, depending on the organism

Amensalism (L. *a*, not + *mensa*, table) a form of biological interaction in which one organism is inhibited by another, but the other is neither inhibited nor stimulated

Amino acid (Gr. *Ammon*, from the Egyptian sun god, in N.L. used in connection with ammonium salts) an acid containing the group NH_2; one of the building blocks of a protein

Ammonification (*Ammon*, Egyptian sun god, near whose temple ammonium salts were first prepared from camel dung + L. *facere*, to make) decomposition of amino acids, resulting in the production of ammonia

Amoeboid (Gr. *amoibe*, change) eating or moving by means of temporary cytoplasmic extensions from the cell body

Amyloplast (L. *amylum*, starch + *plastos*, formed) cytoplasmic organelle specialized to store starch; abundant in roots and in storage organs such as tubers

Anabolism (Gr. *ana*, up + metabolism) the constructive phase of metabolism, in which more complex molecules are built from simpler substances

Anaerobe (Gr. *a*, without + *aer*, air + *bios*, life) an organism able to respire in the absence of free oxygen, or in greatly reduced concentrations of free oxygen

Anaphase (Gr. *ana*, up + *phais*, appearance) that stage in mitosis in which half chromosomes or sister chromatids move to opposite poles of the cell

Androecium (Gr. *andros*, man + *oikos*, house) the aggregate of stamens in the flower of a seed plant

Aneuploid (Gr. *aneu*, without + *ploid*) the condition in which the number of chromosomes differs from the normal by less than a full set (*n*), for example, $3n + 1$, $2n - 1$, $2n + 4$; compare **Polyploid**

Angiosperm (Gr. *angion*, a vessel + *sperma* from *speirein*, to sow, hence a seed or germ) literally a seed borne in a vessel, thus a group of plants whose seeds are borne within a matured ovary

Angstrom (Å) (after A. J. Angstrom, a Swiss physicist, 1814–1874) a unit of length equal to 0.0001 of a micron (0.1 nanometer), 10,000 angstroms = 1 micron = 0.001 millimeter

Anions (Gr.) a negatively charged ion, as hydroxide, chloride, or a protein with a net excess of negatively charged carboxylate functional groups

Anisogamy (Gr. *an*, prefix meaning not + *isos*, equal + *gamete*, spouse) the condition in which the gametes, though similar in appearance, are not identical

Annual (L. *annualis*, within a year) a plant that completes its life cycle in one year and then dies

Annual ring in wood, a layer of growth formed during one year and consisting of spring- and summerwood

Annular vessels (L. *annularis*, a ring) vessels with lignified rings of secondary wall material

Annulus (L. *anulus* or *annulus*, a ring) in ferns, a row of specialized cells in a sporangium, of importance in opening of the sporangium; in mosses, thick-walled cells along the rim of the sporangium to which the peristome is attached

Anther (M.L. *anthera*—from the Gr. *anthros*, meaning flower—a medicine extracted from the internal whorls of flowers by medieval pharmacists; confined to pollen-producing parts by herbalists in 1700) pollen-bearing portion of stamen

Antheridium (anther + Gr. *idion*, dim. ending, thus a little anther) male gametangium or sperm-bearing organ of plants other than seed plants

Anthocyanin (Gr. *anthros*, a flower + *kyanos*, dark blue) a blue, purple, or red vacuolar pigment

Antibiotic (Gr. *anti*, against or opposite + *biotikos*, pertaining to life) a natural organic substance that retards or prevents the growth of organisms; generally used to designate substances formed by microorganisms that prevent growth of other microorganisms

Antibody (Gr. *anti*, against + body) a protein produced in an organism, in response by the organism to a contact with a foreign substance, and having the ability of specifically reacting with the foreign substance

Anticlinal cell division (Gr. *anti*, against + *klinein*, incline) cell division where the newly formed cell wall is perpendicular to the axis of the organ surface

Antipodal (Gr. *anti*, opposite + *pous*, foot) generally, referring to one object situated opposite another object; specifically, referring to cells or nuclei at the end of the embryo sac opposite that of the egg apparatus

Apex (L. *apex*, a tip, point, or extremity) the tip, point, or angular summit of anything: the tip of a leaf, that portion of a root or shoot containing apical and primary meristems

Apical dominance the inhibition of lateral buds or meristems by the apical meristem

Apical meristem a mass of dividing cells at the very tip of a shoot or root

Apomixis (Gr. *apo*, away from + *mixis*, a mingling) the production of offspring in the usual sexual structures without the mingling and segregation of chromosomes

Apoplast (Gr. *apo*, away from + *plastides*, formed) the region of a plant tissue between the cells, outside of the plasma membrane of the cells that form it; contrast to **Symplast**

Apothecium (Gr. *apotheke*, a storehouse) a cup-shaped or saucer-shaped open ascocarp

Archaebacteria (Gr. *arche*, beginning + bacteria) a class or kingdom of primitive Monera including methane-producing, halophilic, and hot-acid-dwelling forms

Archegonium (L. dim. of Gr. *archegonos*, literally a little founder of a race) female gametangium or egg-bearing organ, in which the egg is protected by a jacket of sterile cells

Aril (M.L. *arillus*, a wrapper for a seed) an accessory seed covering formed by an outgrowth at the base of the ovule in *Taxus*

Ascocarp (Gr. *askos*, a bag + *karpos*, fruit) a fruiting body of the Ascomycetes, generally either an open cup, a vessel, or closed sphere lined with special cells called asci; see **Ascus**

Ascogenous hyphae hyphae arising from the ascogonium, after the formation of *n+n* paired nuclei; the hymenial layer of the ascocarp develops from the ascogenous hyphae

Ascogonium the oogonium or female gametangium of the Ascomycetes

Ascomycetes (Gr. *askos*, a bag + *mykes*, fungus) a large group of true fungi with septate hyphae producing large numbers of asexual conidiospores and meiospores called ascospores, the latter in asci

Ascospore (Gr. *askos*, a bag + spore) meiospore produced within an ascus

Ascus (plural, **asci**) (Gr. *askos*, a bag) a specialized cell, characteristic of the Ascomycetes, in which two haploid nuclei fuse, immediately after which three (generally) divisions occur, two of which constitute meiosis, resulting in eight ascospores still contained within the ascus

Asepsis (Gr. *a*, not + *septos*, putrid) the condition of being germ-free

Asexual (Gr. *a*, without + L. *sexualis*, sexual) any type of reproduction not involving the union of gametes or meiosis

Aspect (L. *aspectus*, appearance) the direction of slope of a surface, as a hillside with a south-facing aspect

Assimilation (L. *assimilare*, to make like) the transformation of food into protoplasm

Atoms (F. *atome*, from the Gr. *atomos*, indivisible) the smallest particles in which the elements combine either with themselves or with other elements, and thus the smallest quantity of matter known to possess the properties of a particular element; a unit of matter consisting of a dense, central nucleus surrounded by a number of negatively charged electrons that are in constant motion: the nucleus consists of several positively charged protons and uncharged neutrons

ATP see **Adenosine triphosphate**

Auricles (L. *auricula*, dim. of *auris*, ear) earlike structures; in grasses, small projections that grow out from the opposite side of the sheath at its upper end where it joins the blade.

Autoecious (Gr. *auto*, self + *oikia*, dwelling) having a complete life cycle on the same host

Autoradiograph (Gr. *auto*, self + L. *radiolus*, a ray + Gr. *graphe*, a painting) a photographic print made by a radioactive substance acting upon a sensitive photographic film

Autotetraploidy the condition in which the doubling of the chromosome number occurs in one cell or between cells on the same plant

Autotrophic (Gr. *auto*, self + *trophein*, to nourish with food) pertaining to a plant that is able to manufacture its own food

Auxin (Gr. *auxein*, to increase) a plant growth-regulating substance regulating cell elongation

Axil (Gr. *axilla*, armpit) the upper angle between a petiole of a leaf and the stem from which it grows

Axillary bud a bud formed in the axil of a leaf

Bacillus (L. *baculum*, a stick) a rod-shaped bacterium

Backcross the crossing of a hybrid with one of its parents; when the genes of an individual are to be tested, the crossing of that individual with an individual that is homozygous recessive for all genes being tested

Bacteria (Gr. *bakterion*, a stick) common name for the kingdom Monera

Bacteriochlorophyll a light-harvesting pigment found in certain bacteria; the molecular structure is similar to that of chlorophyll a, but certain side groups and the absorption spectrum are unique

Bacteriology (bacteria + Gr. *logos*, discourse) the science of bacteria

Bacteriophage (bacteria + Gr. *phagein*, to eat) literally, an eater of bacteria; a virus that infects specific bacteria, multiples therein, and usually destroys the bacterial cells

Bark (Swedish *bark*, rind) the external group of tissues, from the cambium outward, of a woody stem or root

Base a substance that can accept a proton (H^+); also the purine and pyrimidine groups in nucleic acids and nucleotides are collectively called bases

Base pair the nitrogen bases that pair in the DNA molecule, adenine with thymine, and guanine with cytosine

Base sequence a sequence of nucleotide bases in DNA or RNA, potentially containing information used to direct the synthesis of a protein

Basidiomycetes (M.L. *basidium*, a little pedestal + Gr. *mykes*, fungus) group of fungi characterized by the production of meiospores on special cells, the basidia

Basidiospore (M.L. *basidium*, a little pedestal + spore) type of meiospore borne by basidia in the Basidiomycetes

Basidium (plural, **basidia**) (M.L. *basidium*, a little pedestal) a specialized reproductive cell of the Basidiomycetes in which nuclei fuse and meiosis occurs. It may be a special club-shaped cell, a short filamentous cell, or a short four-celled filament

Berry a simple fleshy fruit, the ovary wall fleshy and including one or more carpels and seeds

Biennial (L. *biennium*, a period of two years) a plant that requires two years to complete its life cycle. Flowering is normally delayed until the second year.

Bifacial leaf (L. *bis*, twice + *facies*, face) a leaf having distinctly different upper and lower surfaces

Binomial (L. *binominis*, two names) two-named; in biology each species is generally indicated by two names: the genus to which it belongs and its own species name

Bioassay (Gr. *bios*, life + L. *exagere*, to weigh or test) to test for the presence or quantity of a substance by using an organism's response as an indicator.

Biological barrier a barrier to crossing (hybridization) of plants caused by differences in pollination vector or timing in flower opening, in contrast to physiological barriers (incompatibility of pollen with stigma or style) or ecological barriers (habitats too far apart)

Biological clock an endogenous timing mechanism inferred to exist in cells to explain various cyclical behaviors

Biology (Gr. *bios*, life + *logos*, word, speech, discourse) the science that deals with living things

Biosystematics (Gr. *bios*, life + *synistanai*, to place together) a field of taxonomy that emphasizes breeding behavior and chromosome characteristics

Biotic (Gr. *biokitos*, relating to life) relating to life

Biotin a vitamin of the B complex

Bladder (O.E. *bladre*, a blister) a gas-filled sac whose buoyancy keeps some aquatic plants upright

Blade typically the thin expanded portion of a leaf; in some algae, the leaflike frond

Bloom (Gr. *blume*, flower; also Indo-European *bhlo*, to spring up) a increase in phytoplankton density sufficient to color bodies of fresh water

Bolting the rapid formation of a flowering shoot from the apex of a plant with a rosette habit

Bordered pit a pit in a tracheid or vessel member having a distinct rim of the cell wall overarching the pit membrane

Botany (Gr. *botane*, plant, herb) the science dealing with plant life

Bract (L. *bractea*, a thin plate of precious metal) a modified leaf, from the axil of which arises a flower or an inflorescence

Bud (M.E. *budde*, bud) an undeveloped shoot, largely meristematic tissue, generally protected by modified scale-leaves; also a swelling on a yeast cell that will become a new yeast cell when released

Bud scale a modified protective leaf of a bud

Bud scar a scar left on a twig when the bud or bud scales fall away

Bulb (L. *bulbus*, a modified bud, usually underground) a short, flattened, or disk-shaped underground stem, with many fleshy scale-leaves filled with stored food

Bundle scar scar left where conducting strands passing out of the stem into the leaf stalk were broken off when the leaf fell

Bundle sheath sheath of parenchyma cells that surround the vascular bundles of leaves, sometimes called border parenchyma

^{13}C, ^{14}C isotopes of carbon which have additional neutrons; only ^{14}C is radioactive

C_3 **cycle** the Calvin Benson cycle of photosynthesis, in which the first products after CO_2 fixation are three-carbon molecules

C_4 **cycle** the Hatch-Slack cycle of photosynthesis, in which the first products after CO_2 fixation are four-carbon molecules

Callose (L. *callum*, thick skin + *ose*, a suffix indicating a carbohydrate) an amorphous carbohydrate deposited around pores in sieve-tube members and in other areas in cell walls

Callus (L. *callum,* thick skin) mass of large, thin-walled cells, usually developed as the result of wounding

Calorie (L. *calor,* heat) the amount of heat needed to raise the temperature of 1 g of water 1°C (usually from 14.5 to 15.5°C), also called gram-calorie; 1000 calories = 1 kilocalorie

Calyptra (Gr. *kalyptra,* a veil, covering) in bryophytes, an envelope covering the developing sporophyte, formed by growth of the venter of the archegonium

Calyx (Gr. *kalyx,* a husk, cup) sepals collectively; outermost flower whorl

Cambium (L. *cambium,* one of the alimentary body fluids supposed to nourish the body organs) a layer, usually regarded as one or two cells thick, of persistently meristematic tissues, giving rise to secondary tissues, resulting in growth in diameter

Canopy (Gr. *kanopeion,* a cover over a bed to keep off gnats) the leafy portion of a tree or shrub

Capillaries (L. *capillus,* hair) very small spaces, or very fine bores in a tube

Capillary forces forces that pull water into the narrow space between hydrophilic surfaces; includes adhesive force beween water and the surfaces and cohesive force between water molecules.

Capsule (L. *capsula,* dim. of *capsa,* a case) a simple, dry, dehiscent fruit, with two or more carpels

Carbohydrate (chemical combining forms, *carbo,* carbon + *hydrate,* containing water) a food composed of carbon, hydrogen, and oxygen, with the general formula $C_nH_{2n}O_n$

Carbon fixation the enzymatic reaction in which CO_2 is attached to a receiver compound such as ribulose bisphosphate, thereby adding to the supply of organic carbon; occurs chiefly in photosynthesis

Carboxysome (F. *carbone,* carbon + Gr. *oxys,* acidic + Gr. *soma,* body) polyhedral bodies rich in RuBP carboxylase, found in some photosynthetic bacteria and cyanobacteria

Carcinogenic (Gr. *karkinoma,* cancer + L. *genitalis,* to beget) that which induces cancer

Carotene (L. *carota,* carrot) a reddish-orange plastid pigment

Carpel (Gr. *karpos,* fruit) a floral leaf bearing ovules along the margins

Carrageenan a polysaccharide found in the walls of red algae that reacts with milk proteins to make a stable, creamy, thick solution or gel

Caruncle (L. *caruncula,* dim. of *caro,* flesh, wart) a spongy outgrowth of the seed coat, especially prominent in the castor bean seed

Caryopsis (Gr. *karyon,* a nut + *opsis,* appearance) a simple, dry, one-seeded, indehiscent fruit, with pericarp firmly united all around to the seed coat

Casparian strip suberized strip that impregnates the radial and transverse wall of endodermal cells

Catalyst (Gr. *katelyein,* to dissolve) a substance that accelerates a chemical reaction but that is not used up in the reaction

Cation exchange (Gr. *kata,* downward) the replacement of one positive ion (cation) by another, as on a negatively charged clay particle

Cations (Gr.) positively charged ions, such as sodium, potassium, and hydronium ions, and proteins with a net positive charge.

Catkin (literally a kitten, apparently first used in 1578 to describe the inflorescence of the pussy willow) a type of inflorescence, really a spike, generally bearing only pistillate flowers or only staminate flowers, which eventually fall from the plant entire

Caulescent (Gr. *kaulos,* a plant stem) a plant whose stem bears leaves separated by visibly elongated internodes, as opposed to a rosette plant

Cell (L. *cella,* small room) a structural and physiological unit composing living organisms, in which take place the majority of complicated reactions characteristic of life; it is surrounded by a plasma membrane, contains a metabolic system, and has a store of DNA

Cell cycle the repeating sequence of events involved in the reproduction of a eukaryotic cell, including G_1, S (DNA synthesis), G_2, and M (mitosis and cytokinesis) phases

Cell plate a structure that forms at the equatorial plane of the cell at right angles to the spindle fibers during cytokinesis; the precursor of the middle lamella

Cell theory the theory that states that the cell is the fundamental unit of living matter and that organisms are formed from a complex of one or more cells

Cellulose (cell + *ose,* a suffix indicating a carbohydrate) a complex carbohydrate occurring in the cell walls of the majority of plants; cotton is largely cellulose; it is composed of hundreds of simple sugar molecules, glucose, linked together in a characteristic manner

Cell wall the somewhat rigid organelle surrounding the protoplast of a bacterial, fungal, protist, or plant cell

Cenozoic (Gr. *kainos,* recent + *zoe,* life) the geologic era extending from 65 million years ago to the present

Chalaza (Gr. *chalaza,* small tubercle) the region on a seed at the upper end of the raphe where the funiculus spreads out and unites with the base of the ovule

Chaparral (Sp. *chaparro,* low, woody vegetation) a vegetation type characterized by small-leaved, evergreen shrubs growing together into a nearly impenetrable scrub

Chemoautotroph (Gr. *chymeia,* to pour, later alchemy, chemistry + *autos,* self + *trophein,* to feed) bacteria that oxidize reduced inorganic compounds such as H_2S to obtain energy and that use CO_2 as a carbon source

Chemoheterotroph (Gr. *chymeia,* to pour + *heteros,* other + *trophein,* to feed) bacteria that oxidize or reduce a variety of organic compounds to obtain both energy and carbon

Chemotropism (Gr. *chymeia,* to pour + Gr. *tropos,* a turning) influence of a chemical substance on the direction of growth

Chiasma (Gr. *chiasma,* two lines placed crosswise) the cross formed by breaking, during prophase 1 of meiosis, of two nonsister chromatids of homologous chromosomes and the rejoining of the broken ends of different chromatids

Chitin (Gr. *chiton*, a coat of mail) a polymer in which the monomer unit is the modified sugar *N*-acetyl glucosamine; it is the principal stiffening material in the cell walls of most fungi and in the exoskeleton of insects and crustaceans

Chlamydospore (Gr. *chlamys*, a horseman's or young man's coat + spore) a heavy-walled resting asexual spore

Chlorenchyma (Gr. *chloros*, green + *enchyma*, a suffix meaning tissue) parenchyma tissue possessing chloroplasts

Chlorophyll (Gr. *chloros*, green + *phyllon*, leaf) the green pigment found in the chloroplast, important in the absorption of light energy in photosynthesis

Chloroplast (Gr. *chloros*, green + *plastos*, formed) specialized cytoplasmic body, containing chlorophyll, in which occur important reactions of starch or sugar synthesis

Chlorosis (Gr. *chloros*, green + *osis*, diseased state) failure of chlorophyll development, because of a nutritional disturbance or because of an infection of virus, bacteria, or fungus

Chromatid (chromosome + *id*, L. suffix meaning daughters of) the half-chromosome during prophase and metaphase of mitosis, and between prophase 1 and anaphase 11 of meiosis

Chromatin (Gr. *chroma*, color) substance in the nucleus which readily takes artificial staining; also, that portion which bears the determiners of hereditary characters made up of DNA

Chromatin bodies bodies in bacteria that give some of the histochemical reactions that are associated with the chromosomes of higher organisms

Chromatophores (Gr. *chromo*, color + *phorus*, a bearer) in algae, bodies bearing chlorophyll; in bacteria, small bodies, about 100 nanometers in diameter, containing chlorophyll, protein, and a carbohydrate

Chromoplast (Gr. *chroma*, color + *plastos*, formed) specialized plastid containing yellow or orange pigments

Chromosome (Gr. *chroma*, color + *soma*, body) a nuclear body containing genes in a linear order and undergoing characteristic division stages; one of the threads or rods of condensed chromatin visible during cell division

Cilia (singular, **cilium**) (F. *cil*, an eyelash) protoplasmic hairs that, by a whiplike motion, propel certain types of unicellular organisms, gametes, and zoospores through water

Cisterna (plural, **cisternae**) (L. *cistern*, a reservoir) generally referring to sections of the endoplasmic reticulum that appear as parallel membranes, each about 5 nanometers thick bounding a space about 40 nanometers wide

Citric acid cycle a system of reactions that contributes to the catabolic breakdown of foods in respiration and that provides building materials for a number of anabolic pathways; also called the Krebs cycle and the tricarboxylic acid (TCA) cycle

Cladode (Gr. *kladodes*, having many shoots) a cladophyll

Cladophyll (Gr. *klados*, a shoot + *phyllon*, leaf) a branch resembling a foliage leaf

Class (L. *classis*, one of the six divisions of Roman people) a taxonomic group below the division level but above the order level

Clay soil particles less than 2 microns in diameter, composed mainly of aluminum (Al), oxygen (O), and silicon (S)

Cleistothecium (plural, **cleistothecia**) (Gr. *kleistos*, closed + *thekion*, a small receptacle) the closed, spherical ascocarp of the powdery mildews

Climax community the last stage of a natural succession; a community capable of maintaining itself as long as the climate does not change

Clone (Gr. *klon*, a twig or slip) the aggregate of individual organisms produced asexually

Closed bundle a vascular bundle lacking cambium

Coal Age the Carboniferous period, beginning 345 million years ago and ending 280 million years ago

Coalescence (L. *coalescere*, to grow together) a condition in which there is union of separate parts of any one whorl of flower parts; synonyms are **Connation** and **Cohesion**

Coccus (plural, **cocci**) (Gr. *kokkos*, a berry) a spherical bacterium

Codominance concerning alleles of a gene, sharing in influence; the trait produced by two codominant alleles is intermediate between (or different from) that produced by either alone; also incomplete dominance

Codon a sequence of three bases (nucleotides) along the RNA molecule that code for a single amino acid

Coenobium (Gr. *koinois*, common + Gr. *bios*, life) a colony of unicellular organisms surrounded by a common membrane

Coenocyte (Gr. *koinos*, shared in common + *kytos*, a vessel) a plant or filament whose protoplasm is continuous and multinucleate and without any division by walls into separate protoplasts

Coenogamete (Gr. *koinos*, shared in common + gamete) a multinucleate gamete, lacking cross walls

Coenzyme a substance, usually nonprotein and of low molecular weight, necessary for the action of some enzymes

Cohesion (L. *cohaerere*, to stick together) union or holding together of parts of the same materials; the union of floral parts of the same whorl, as petals to petals

Coleoptile (Gr. *koleos*, sheath + *ptilon*, down, feather) the first leaf in germination of grasses which sheaths the succeeding leaves

Coleorhiza (Gr. *koleos*, sheath + *rhiza*, root) sheath that surrounds the radicle of the grass embryo and through which the young root bursts

Collenchyma (Gr. *kolla*, glue + *enchyma*, a suffix, derived from parenchyma and denoting a type of cell tissue) a tissue composed of cells that fit rather closely together and with walls thickened at the angles of the cells; found in young stems and petioles

Colloid (Gr. *kolla*, glue + *eidos*, form) referring to matter composed of particles, ranging in size from 0.0001 to 0.000001 millimeter, dispersed in some medium; milk and mayonnaise are examples

Colony (L. *colonia*, a settlement) a growth form characterized by a group of closely associated, but poorly differentiated, cells; sometimes filaments can be associated together in a colony (as in *Nostoc*), but more typically unicells are associated in a colony

Community (L. *communitas*, a fellowship) all the populations within a given habitat; usually the populations are thought of as being somewhat interdependent

Companion cell cell associated with sieve-tube members

Compensation depth (L. *compensare*, to counterbalance) that depth, in a body of water, at which light intensity is so low that photosynthesis of floating or submerged plants just equals respiration

Compensation point (L. *compensare*, to counterbalance) the light intensity (light compensation point) or the carbon dioxide concentration (CO_2 compensation point) at which photosynthesis just equals respiration

Competition (L. *competere*, to strive together) a form of biological interaction in which both organisms (at least initially) decline in growth or success because of the insufficient supply of some necessary factor(s)

Complete flower a flower having four whorls of floral leaves: sepals, petals, stamens, and carpels

Compound (M.F., *compondre*, put together) a chemically distinct substance produced by the union of two or more elements in definite proportion by weight; formed from specific molecules

Compound leaf a leaf whose blade is divided into several distinct leaflets

Conceptacle (L. *conceptaculum*, a receptacle) a cavity or chamber of a frond (of *Fucus*, for example) in which gametangia are borne

Condensation reaction a chemical or physical reaction from which one product is water

Conduction (L. *conducere*, to bring together) act of moving or conveying a substance through the plant; generally the movement of water through the xylem or food through the phloem

Cone (Gr. *konos*, a pine cone) a fruiting structure composed of modified leaves or branches, which bear sporangia (microsporangia, megasporangia, pollen sacs, or ovules), and frequently arranged in a spiral or four-ranked order; for example, a pine cone

Cone scale the flat, woody parts of pine cones that spiral out from the central axis and bear the ovules (and later seeds) on their upper surfaces; each is subtended by a sterile bract

Conidia (singular, **conidium**) (Gr. *konis*, dust) asexual reproductive cells of fungi, arising by fragmentation of hyphae, by the cutting off of terminal or lateral cells of special hyphae, or by being pushed out from a flask-shaped cell

Conidiophore (conidia + Gr. *phoros*, bearing) conidium-bearing branch of hypha

Conidiosporangium (Gr. *konis*, dust + sporangium) sporangium formed by being cut off from the end of a terminal or lateral hypha

Conidiospore (conidia + spore) spore formed as described for conidia

Conifer (cone + L. *ferre*, to carry) a cone-bearing tree

Conjugation (L. *conjugatus*, united) process of sexual reproduction involving the fusion of isogametes

Conk the fruiting body of a bracket fungus

Connation (L. *connatus*, to be born together) condition in a flower where there is a union of similar parts of any one whorl of appendages; synonym of **Coalescence**

Conservative (L. *conservare*, to keep) said of a taxonomic trait whose expression is not modified to any great extent by the external environment; a trait that is constant unless its genetic base is changed

Convergent evolution process of successive progeny, originally of quite distinct parents, coming to appear more and more alike through time because of selection pressure

Cork (L. *quercus*, oak) an external, secondary tissue impermeable to water and gases

Cork cambium the cambium from which cork develops

Corm (Gr. *kormos*, a trunk) a short, solid, vertical, enlarged underground stem in which food is stored

Corolla (L. *corolla*, dim. of *corona*, a wreath, crown) petals, collectively; usually the conspicuous colored flower whorl

Cortex (L. *cortex*, bark) region of primary tissue in a stem or root bounded externally by the epidermis and internally in the stem by the phloem and in the root by the pericycle

Cotyledon (Gr. *kotyledon*, a cup-shaped hollow) seed leaf; two, generally storing food in dicotyledons; one, generally a digestive organ in the monocotyledons

Covalent bond a chemical bond between two atoms formed by shared electrons, that is, electrons in bonding orbitals shared by the atoms

Cristae (L. *crista*, a crest) crests or ridges, used here to designate the infoldings of the inner mitochondrial membrane

Crossing-over the exchange of corresponding segments between chromatids of homologous chromosomes

Cross-pollination the transfer of pollen from a stamen to the stigma of a flower on another plant, except in clones

Crustose lichens that are firmly attached, along their entire lower surface, to a substrate

Cuticle (L. *cuticula*, dim. of *cutis*, the skin) waxy layer on outer wall of epidermal cells

Cutin (L. *cutis*, the skin) waxy substance that is but slightly permeable to water, water vapor, and gases

Cutinization impregnation of cell wall with a substance called cutin

Cyclosis (Gr. *kyklosis*, circulation) cytoplasmic streaming in a cell

Cyme (Gr. *kyma*, a wave, a swelling) a type of inflorescence in which the apex of the main stalk or the axis of the inflorescence ceases to grow quite early, relative to the laterals

Cystocarp (Gr. *kystos*, bladder + *karpos*, fruit) a peculiar diploid spore-bearing structure formed after fertilization in certain red algae

Cytochrome (Gr. *kytos*, a receptacle or cell + *chroma*, color) a class of several electron-transport proteins serving as carriers in mitochondrial oxidations and in photosynthetic electron transport

Cytokinesis (Gr. *kytos*, a hollow vessel + *kinesis*, motion) division of cytoplasmic constituents at cell division

Cytokinin (Gr. *kytos*, a receptacle or cell + *kinetos*, to move) a class of growth hormones important in the regulation of nucleic acid and protein metabolism, in cell division, delaying senescence, and organ initiation

Cytology (Gr. *kytos*, a hollow vessel + *logos*, word, speech, discourse) the science dealing with the cell

Cytoplasm (Gr. *kytos*, a hollow vessel + *plasma*, form) all the protoplasm of a protoplast outside the nucleus

Cytosine a pyrimidine base found in DNA and RNA

Cytoskeleton the framework of protein filaments, including microfilaments, microtubules, and intermediate filaments, in the cytoplasm

Day-neutral plants varieties of plants whose flowering does not depend on day length; contrast **Long-day plants** and **Short-day plants**

Deciduous (L. *deciduus*, falling) referring to trees and shrubs that lose their leaves in the fall

Decomposer (L. *de*, from + *componere*, to put together) an organism that obtains food by breaking down dead organic matter into simpler molecules

Decomposition (L. *de*, to denote an act undone + *componere*, to put together) a separation or dissolving into simpler compounds; rotting or decaying

Dehiscent (L. *dehiscere*, to split open) opening spontaneously when ripe, splitting into definite parts

Deletion (L. *deletus*, to destroy, to wipe out) used here to designate an area, or region, lacking from a chromosome

Denitrification (L. *de*, to denote an act undone + *nitrum*, nitro, a combining form indicating the presence of nitrogen + *facere*, to make) conversion of nitrates into nitrites, or into gaseous oxides of nitrogen, or even into free nitrogen

Deoxyribose nucleic acid (DNA) hereditary material; the DNA molecule carries hereditary information

Desert shrub (M.E. *schrubbe*, shrub) a vegetation type characterized by evergreen or drought-deciduous shrubs growing together rather openly, generally in an area with annual precipitation below 25 cm

Determinate (L. *determinare*, to limit) generally, having defined limits; in plant development, a morphogenetic process that ends with a cessation of cell division and growth

Determined in plant development, a tissue having a limited number of developmental possibilities

Detritus (L. *detritus*, worn away) particulate organic matter released in the processes of decomposition of dead organisms or parts of organisms (such as plant litter)

Deuterium or heavy hydrogen; a hydrogen atom, the nucleus of which contains one proton and one neutron; it is written as ^{2}H; the common nucleus of hydrogen consists only of one proton

Development (F. *developper*, to unfold) developmental changes of a cell, tissue, or organ leading to the presence of features that equip that cell, tissue, or organ for performing specialized functions

Diastase (Gr. *diastasis*, a separation) a complex of enzymes that brings about the hydrolysis of starch with the formation of sugar

Diatom (Gr. *diatomos*, cut in two) member of a group of golden brown algae with silicious cell walls fitting together much as do the halves of a pill box

Diatomite (Gr. *diatomos*, cut in two) fossil deposits of diatom cell walls; currently mined for such commercial purposes as filters, extenders, and stabilizers

Dichotomy (Gr. *dicha*, in two) the forking of an axis into two branches

Dicotyledon (Gr. *dis*, twice + *kotyledon*, a cup-shaped hollow) a plant whose embryo has two cotyledons

Dictyosome (Gr. *diktyon*, a net + *some*, body) one of the component parts of the Golgi apparatus; in plant cells a complex of flattened double lamellae

Differentially permeable referring to a membrane through which different substances diffuse at different rates; some substances may be unable to diffuse through such a membrane

Differentiation (L. *differre*, to carry different ways) development from one cell to many cells, accompanied by a modification of the new cells for the performance of particular functions

Diffuse porous wood with an equal and random distribution of large xylem vessel members throughout the growth season

Diffusion (L. *diffusus*, spread out) the movement of molecules, and thus a substance, from a region of higher concentration of those molecules to a region of lower concentration

Digestion (L. *digestio*, dividing, or tearing into pieces, an orderly distribution) the processes of rendering food available for metabolism by breaking it down into simpler compounds, chiefly through actions of enzymes

Dihybrid cross (Gr. *dis*, twice + *hybrida*, the offspring of a tame sow and a wild boar, a mongrel) a cross between organisms differing in two characters

Dikaryon (Gr. *di*, two + *karyon*, nut) the $n+n$ paired nuclei, each usually derived from a different parent one male, one female

Dioecious (Gr. *dis*, twice + *oikos*, house) unisexual; having the male and female elements in different individuals

Diploid (Gr. *diploos*, double + *oides*, like) having a double set of chromosomes, or referring to an individual containing a double set of chromosomes per cell; usually a sporophyte generation

Disease (L. *dis*, a prefix signifying the opposite + M.E. *aise*, comfort, literally the opposite of ease) any alteration from state of metabolism necessary for the normal development and functioning of an organism

Divergent evolution process of successive progeny, originally of quite similar parents, coming to appear more and more different through time because of isolation and selection pressure

Division a major portion of the plant kingdom; equivalent to **Phylum**

DNA see **Deoxyribose nucleic acid**

Dominant (L. *dominari*, to rule) referring, in ecology, to species of a community that receive the full force of the macroenvironment; usually the most abundant of such species, and not all of them; referring, in heredity, to that gene (or the expression of the character it influences) that, when present in a hybrid with a contrasting gene, completely dominates in the development of the character; in peas, tall is dominant over dwarf

Donor one who gives

Dormant (L. *dormire*, to sleep) being in a state of reduced physiological activity such as occurs in seeds, buds, etc.

Dorsiventral (L. *dorsum*, the back + *venter*, the belly) having upper and lower surfaces distinctly different, as a leaf does

Double bond a covalent bond that involves four electrons

Double fertilization in the embryo sac, the fusion of the egg and sperm and the simultaneous fusion of the second male gamete with polar nuclei

Drupe (L. *drupa*, an overripe olive) a simple, fleshy fruit, derived from a single carpel, usually one-seeded, in which the exocarp is thin, the mesocarp fleshy, and the endocarp stony

Ecology (Gr. *oikos*, home + *logos*, discourse) the study of plant life in relation to the environment

Ecosystem (Gr. *oikos*, house + *synistanai*, to place together) an inclusive term for a living community and all the factors of its nonliving environment

Ecotype (Gr. *oikos*, house + *typos*, the mark of a blow) genetic variant within a species that is adapted to a particular environment yet remains interfertile with all other members of the species

Ectomycorrhiza (Gr. *ektos*, outside + *mykos*, fungus + *riza*, root) a type of mycorrhiza that produces a mantle of hyphae around the outside of a young root; hyphae do not penetrate cortex cells

Edaphic (Gr. *edaphos*, soil) pertaining to soil conditions that influence plant growth

Egg (A.S. *aeg*, egg) a female gamete

Elater (Gr. *elater*, driver) an elongated, spindle-shaped, sterile, hygroscopic cell in the sporangium of a liverwort sporophyte

Electron (Gr. *elektron*, gleaming in the sun, by way of L. *electrum*, a bright alloy of gold and silver, and finally amber, from which the first electricity was produced by friction) an elementary particle of matter bearing a unit of negative electrical charge; low in mass and in constant motion, electrons surround the atom's positively charged nucleus and define the size and chemical properties of the atom or molecule

Electronegativity (L. *electrum*, amber + negativity) the power of an atom to attract electrons

Electron microscope a microscope that uses a beam of electrons rather than light to produce a magnified image

Electron transport chain a membrane-bound series of electron carriers that controls the flow of electrons from reduced to oxidized compounds, so that some of the energy carried by the electrons is used to form ATP; the chain consists of several compounds (carriers) that alternately accept and donate electrons; found in mitochondria and chloroplasts

Electrophoresis (Gr. *elektron*, amber + *phora*, motion + *esis*, drive) the process of causing charged molecules (e.g., proteins) to move between positively and negatively charged poles

Element (L. *elementa*, the first principles; according to one system of medieval chemistry as recent as 1700, there were four elements composing all material bodies: earth, water, air, and fire) in modern chemistry, a substance that cannot be divided or reduced by any known chemical means to a simpler substance; 92 natural elements are known, of which gold, carbon, oxygen, and iron are examples; several, including plutonium, have been formed in atomic piles

Embryo (Gr. *en*, in + *bryein*, to swell) a young sporophytic plant, while still retained in the gametophyte or in the seed

Embryo sac the female gametophyte of the angiosperms; generally a seven-celled structure; the seven cells are two synergids, one egg cell, three antipodal cells (each with a single haploid nucleus), and one endosperm mother cell with two haploid nuclei

Emulsion (L. *emulgere*, to milk out) a suspension of fine particles of a liquid in a liquid

Endocarp (Gr. *endon*, within + *karpos*, fruit) inner layer of fruit wall (pericarp)

Endodermis (Gr. *endon*, within + *derma*, skin) the layer of living cells, with various characteristically thickened walls (Casparian strips) and no intercellular spaces, which surrounds the vascular tissue of certain plants and occurs in nearly all roots and certain stems and leaves

Endogenous (Gr. *endon*, within + *genos*, race, kind) developed or added from outside the cell

Endomembrane system the organelles in a cell that exchange patches of membranes, including the nuclear envelope, endoplasmic reticulum, and Golgi apparatus

Endomycorrhiza (Gr. *endon*, within + *mykos*, fungus + *riza*, root) a type of mycorrhiza in which fungal hyphae penetrate cortex cells; an external fungal mantle is absent, though hyphae do extend out the root into the soil

Endoplasmic reticulum (Gr. *endon*, within + *plasma*, anything formed or molded; L. *reticulum*, a small net) originally, a cytoplasmic net adjacent to the nucleus, made visible by the electron microscope; now, any system of paired membranes within the cytoplasm; frequently abbreviated to ER

Endosperm (Gr. *endon*, within + *sperma*, seed) the nutritive tissue formed within the embryo sac of seed plants; it is often consumed as the seed matures, but remains in the seeds of corn and other cereals

Endosperm mother cell one of the seven cells of the mature embryo sac, containing the two polar nuclei and, after reception of a sperm cell, giving rise to the primary endosperm cell from which the endosperm develops

Endospore (Gr. *endon*, within + *spora*, seed) a thick-walled resting spore, formed within the cells of certain bacteria after a complex process of nuclear division and cytoplasmic reorganization

Enzyme (Gr. *en*, in + *zyme*, yeast) a protein of complex chemical constitution produced in living cells, which, even in very low concentration, speeds up certain chemical reactions but is not used up in the reaction

Epicotyl (Gr. *epi*, upon + *kotyledon*, a cup-shaped hollow) the upper portion of the axis of embryo or seedling, above the cotyledons

Epidermis (Gr. *epi*, upon + *derma*, skin) a superficial layer of cells occurring on all parts of the primary plant body: stems, leaves, roots, flowers, fruits, and seeds; it is absent from the root cap and not differentiated on the apical meristems

Epigeal (Gr. *epi*, upon + *ge*, the earth) type of germination where cotyledons rise above the ground

Epigyny (Gr. *epi*, upon + *gyne*, woman) the arrangement of floral parts in which the ovary is embedded in the receptacle so that the other parts appear to arise from the top of the ovary

Epilimnion (Gr. *epi*, upon + *limne*, marsh) an upper, aerated, warm zone of water that lies above a lower, less aerated, cold zone (the hypolimnion); common in large bodies of fresh water such as deep lakes

Epiphyte (Gr. *epi*, upon + *phyton*, a plant) a plant that grows on another plant, yet is not parasitic

Episome a piece of DNA that can replicate either autonomously or as part of a main chromosome; in bacteria, includes plasmids and certain viral genomes

Equatorial rain forest vegetation with several tree strata; characteristic of warm, wet regions; synonymous with tropical rain forest

ER see **Endoplasmic reticulum**

Ergot (F. *argot*, a spur) a fungus disease of cereals and wild grasses in which the grain is replaced by dense masses of purplish hyphae, the ergot

Erosion (L. *e*, put + *rodere*, to gnaw) the wearing away of land, generally by the action of water

Ethylene C_2H_4, a growth hormone regulating fruit ripening, various aspects of vegetative growth, and also important in the abscission process

Etiolation (F. *etioler*, to blanch) a condition involving increased stem elongation, poor leaf development, and lack of chlorophyll, found in plants growing in the absence, or in a greatly reduced amount, of light

Eubacteria (Gk. *eu*, good, true + bacteria) a class or kingdom of relatively advanced Monera; includes the majority of bacterial species now present on Earth

Eukaryote (L. *eu*, true + *karyon*, a nut, referring in modern biology to the nucleus) any organism characterized by having the cellular organelles, including the nucleus, bounded by membranes

Eutrophication (Gr. *eu*, good, well + *trephein*, to nourish) pollution of bodies of water resulting from slow, natural, geological, or biological processes such as siltation or encroachment of vegetation or accumulation of detritus, also called natural eutrophication

Evapotranspiration (L. *evaporare*, *e*, out of + *vapor*, vapor + F. *transpirer*, to perspire) the process of water loss in vapor form from a unit surface of land both directly and from leaf surfaces

Evolution (L. *evolutio*, an unrolling) the development of a race, species, genus, or other larger group of plants or animals

Exine (L. *exterus*, outside) outer layer of pollen grain wall

Exocarp (Gr. *exo*, without, outside + *karpos*, fruit) outermost layer of fruit wall (pericarp)

Exogenous (Gr. *exe*, out, beyond + *benos*, race, kind) produced outside of, originating from, or due to external causes

F_1 first filial generation in a cross between any two parents

F_2 second filial generation, obtained by crossing two members of the F_1, or by self-pollinating the F_1

Facultative (L. *facultas*, capability) referring to an organism having the power to live under a number of certain specific conditions; for example, a facultative parasite may be either parasitic or saprophytic

Family (L. *familia*, family) in plant taxonomy, a group of genera; families are grouped in orders

Fascicle (L. *fasciculus*, a small bundle) a bundle of pine or other needle-leaves of gymnosperms

Fascicular cambium cambium within vascular bundles

Fat (A.S. *faett*, fatted) one of the three major types of foods (the other two are carbohydrates and proteins); nonpolar, containing carbon, hydrogen, and small amounts of oxygen; rich in energy; used synonymously with **lipids**

Feedback inhibition the inhibition of an enzyme by its product or the product of the metabolic pathway of which it is a part

Fermentation (L. *fermentum*, a drink made from fermented barley beer) an oxidative process in foods in which molecular oxygen is not involved, such as the production of alcohol from sugar by yeasts

Ferredoxin an electron-transferring protein containing iron, involved in photosynthesis and in the biological production and consumption of hydrogen gas

Fertilization (L. *fertilis*, capable of producing fruit) that state of a sexual life cycle involving the union of egg and sperm and hence the doubling of chromosome numbers

Fiber (L. *fibra*, a fiber or filament) an elongated, tapering, thick-walled strengthening cell occurring in various parts of plant bodies

Fiber-tracheid xylem elements found in pine that are structurally intermediate between tracheids and fibers

Fibril (dim. of L. *fibra*, fiber) submicroscopic threadlike units of cellulose found in cell walls

Field capacity the amount of water retained in a soil (generally expressed as percent by weight) after large capillary spaces have been drained by gravity

Filament (L. *filum*, a thread) stalk of stamen bearing the anther at its tip; also, a slender row of cells (certain algae)

Fission (L. *fissilis*, easily split) asexual reproduction involving the division of a single-celled individual into two new single-celled individuals of equal size

Flagellum (plural, **flagella**) (L. *flagellum*, a whip) a long, slender whip of protoplasm (in eukaryotes) or proteins (in bacteria)

Flora (L. *floris*, a flower) an enumeration of all the species that grow in a region; also, the collective term for all the species that grow in a region

Floret (F. *fleurette*, a dim. of *fleur*, flower) one of the small flowers that make up the composite flower or the spike of the grasses

Flower (F. *fleur*, L. *flos*, a flower) floral leaves grouped together on a stem and adapted for sexual reproduction in the angiosperms

Foliose (L. *folium*, a leaf) lichens that are leaflike, attached to the substrate only along part of their surface

Follicle (L. *folliculus*, dim. of *follis*, bag) a simple, dry, dehiscent fruit, with one carpel, splitting along one suture

Food (A.S. *foda*) any organic substance that directly furnishes energy and building materials for vital processes

Food chain the path along which caloric energy is transferred within a community (from producers to consumers to decomposers)

Foot (O.E. *fot*, foot) that portion of the sporophyte of bryophytes and lower vascular plants that is sunk in gametophyte tissue and absorbs food parasitically from the gametophyte

Form genus a scientific name given to an organism from the fossil record, when only a portion of the entire plant has been recovered and is known

Fossil (L. *fossio*, a digging) any impression, natural or impregnated remains, or other trace of an animal or plant of past geological ages that has been preserved in the earth's crust

Free energy the internal thermodynamic potential of a portion of matter; change in free energy during a chemical reaction determines the direction of the reaction

Fret (O.F. *frette*, latticework) flattened membrane sacs that connect grana in chloroplasts

Frond (L. *frons*, branch, leaf) a synonym for a large divided leaf, especially a fern leaf; also the leaflike blades of some algae

Fruit (L. *fructus*, that which is enjoyed, hence product of the soil, trees, cattle, etc.) a matured ovary; in some seed plants other parts of the flower may be included; also applied, as fruiting body, to reproductive structures of other groups of plants

Frustule (L. *frustulum*, little piece) a diatom cell, composed of two overlapping halves (valves)

Fruticose (L. *frutex*, a shrub) lichens that are erect and shrublike or pendant from branches; may have radial symmetry

Fucoxanthin (Gr. *phykos*, seaweed + *xanthos*, yellowish brown) a brown pigment found in brown algae

Fungus (plural, **fungi**) (L. *fungus*, a mushroom) a thallus plant unable to make its own food, exclusive of the bacteria

Funiculus (L. *funiculus*, dim. of *funis*, rope or small cord) a stalk of the ovule, containing vascular tissue

Fusiform initials (L. *fusus*, spindle + form) meristematic cells in the vascular cambium that develop into xylem and phloem cells composing an axial system

G₁ period of cell cycle preceding DNA synthesis

G₂ period of cell cycle preceding mitosis

Gametangium (Gr. *gametes*, a husband, *gamete*, a wife + *angeion*, a vessel) organ bearing gametes

Gamete (Gr. *gametes*, a husband, *gamete*, a wife) a protoplast that fuses with another protoplast to form the zygote in the process of sexual reproduction

Gametophyte (gamete + Gr. *phyton*, a plant) the gamete-producing plant

Gel (L. *gelare*, to freeze) jellylike, colloidal mass

Gemma (plural, **gemmae**) (L. *gemma*, a bud) a small mass of vegetative tissue; an outgrowth of the thallus

Gene (Gr. *genos*, race, offspring) a group of base pairs in the DNA molecule in the chromosome that determines or conditions one or more hereditary characters

Generation (L. *genus*, birth, race, kind) any phase of a life cycle characterized by a particular chromosome number, as the gametophyte generation and the sporophyte generation

Genetic code the relationship between codons (sequences of three bases) in DNA or messenger RNA and the amino acids they specify in a protein

Genetics (Gr. *genesis*, origin) the science of heredity

Genome (gene + chromosome) the collection of all genes in an organism

Genotype (gene + type) the assemblage of genes in an organism

Genus (plural, **genera**) (Gr. *genos*, race, stock) a group of structurally or phylogenetically related species

Germination (L. *germinare*, to sprout) the beginning of growth of a seed, spore, bud, or other structure

Gibberellins a group of growth hormones (not identical with auxins), the most characteristic effect of which is to increase the elongation of stems in a number of kinds of higher plants

Gill (M.E. *gile*, a lip, probably due to its resemblance in shape and arrangement to gills of fishes) in mushrooms, thin spore-bearing plates on the underside of the cap

Girdle (O.E. *gyrdel*, enclosure, girdle) that region of a stem from which a ring of bark extending to the cambium has been removed

Glucose (Gr. *glykys*, sweet + *ose*, a suffix indicating a carbohydrate) a simple sugar, grape sugar, $C_6H_{12}O_6$

Glume (L. *gluma*, husk) an outer and lowermost bract of a grass spikelet

Glycogen (Gr. *glykys*, sweet + *gen*, of a kind) a carbohydrate related to starch but found generally in the liver of animals

Glycolysis (Gr. *glykys*, sweet + *lysis*, a loosening) decomposition of sugar compounds without involving free oxygen; early steps of respiration

Glyoxisome (Gr. *glykys*, sweet + *soma*, body) a type of microbody that contains enzymes involved in the conversion of fats to carbohydrates during germination of fat-storing seeds

Golgi body (Italian cytologist Camillo Golgi 1844–1926, who first described the organelle) in animal cells, a complex perinuclear region thought to be associated with secretion; in plant cells, a series of flattened bladders

Gram's stain (after Hans C. M. Gram) a method for the differential staining of bacteria; distinguishes species on the basis of cell wall composition, which determines whether stain is retained

Grana (singular, **granum**) (L. *granum*, a seed) structures within chloroplasts, seen as green granules with the light microscope and as a series of parallel lamellae with the electron microscope

Gravitropism (L. *gravis*, heavy + Gr. *tropos*, turning) a growth curvature induced by gravity

Ground meristem (Gr. *meristos*, divisible) a primary meristem that gives rise to cortex, mesophyll, and pith

Grow (A.S. *growan*, probably from Old Teutonic *gro*, from which grass is also derived) of living bodies generally; to increase gradually in size by natural development

Growth retardant a chemical (such as cycocel, CCC) that selectively interferes with normal hormonal promotion of growth, but without appreciable toxic effects

Guanine a purine base found in DNA and RNA

Guard cells specialized epidermal cells found on young stems and leaves; between each pair of guard cells is a small pore through which gases enter or leave; a pair of guard cells plus the pore constitute a stoma

Guttation (L. *gutta*, drop, exudation of drops) exudation of water from plants, in liquid form

Gynoecium (Gr. *gyne*, woman + *oikos*, house) the aggregate of carpels in the flower of a seed plant

Habitat (L. *habitare*, inhabit, dwell) the place or natural environment where an organism naturally grows

Hallucinogenic (L. *hallucinari*, to mentally wander + *genitalis*, to beget) that which induces hallucinations

Haploid (Gr. *haploos*, single + *oides*, like) having a single complete set of chromosomes, or referring to an individual or generation containing such a single set of chromosomes per cell; usually a gametophyte generation

Haptera (singular, **hapteron**) fingerlike, tubular projections that make up the holdfast of certain kelps

Haustorium (plural, **haustoria**) (M.L. *haustrum*, a pump) a projection of hyphae that acts as a penetrating and absorbing organ

Head an inflorescence; typical of the composite family, in which flowers are grouped closely on a receptacle

Heartwood wood in the center of old secondary stems that is plugged with resins and tyloses and is not active

Helix (Gr. *helix*, anything twisted) anything having a spiral form; here, quite generally refers to the double spiral of the DNA molecule

Hemicellulose (Gr. *hemi*, half + cellulose) a class of polysaccharides of the cell wall, built of several different kinds of simple sugars linked in various combinations

Herb (L. *herba*, grass, green blades) a seed plant that does not develop woody tissues

Herbaceous (L. *herbaceus*, grassy) referring to plants having the characteristics of herbs

Herbal (L. *herba*, grass) a book that contains the names and descriptions of plants, especially those that are thought to have medicinal uses

Herbarium (L. *herba*, grass) a collection of dried and pressed plant specimens

Herbicide (L. *herba*, grass or herb + *cidere*, to kill) a chemical used to kill plants, frequently chemically related to a hormone (as the herbicide 2,4-D is related to the hormone IAA); an herbicide may have narrow or wide selectivity (range of target organisms)

Herbivory the ingestion of plants by animals

Heredity (L. *hereditas*, being a heir) the transmission of morphological and physiological characters of parents to their offspring

Hermaphrodite flower (Gr. *hermaphroditos*, a person having the attributes of both sexes, represented by Hermes and Aphrodite) a flower having both stamens and pistils

Heterobasidiomycotina (Gr. *heteros*, other + Basidiomycete) a subclass of Basidiomycetes with variable basidia, never club-shaped cells

Heterocyst (Gr. *heteros*, different + *cystis*, a bag) an enlarged colorless cell that may occur in the filaments of certain blue-green algae

Heteroecious (Gr. *heteros*, different + *oikos*, house) referring to fungi that cannot carry through their complete life cycle unless two different host species are present

Heterogametes (Gr. *heteros*, different + gamete) gametes dissimilar from each other in size and behavior, like egg and sperm

Heterogamy (Gr. *heteros*, different + *gamos*, union or reproduction) reproduction involving two types of gametes

Heterophyte (Gr. *heteros*, different + *phyton*, a plant) a plant that must secure its food ready-made

Heterosis (Gr. *heteros*, different + *osis*, suffix indicating a state of) the state of a genotype having a large degree of heterozygosity

Heterospory (Gr. *heteros*, different + spore) the condition of producing microspores and megaspores

Heterothallic (Gr. *heteros*, different + thallus) referring to species in which male gametangia and female gametangia are produced in different filaments or by different individual plant bodies

Heterotrichy (Gr. *heteros*, different + *trichos*, a hair) in the algae, the occurrence of two types of filaments, erect and prostrate

Heterotrophic (Gr. *heteros*, different + *trophein*, to nourish with food) referring to a plant obtaining nourishment from outside sources

Heterozygous (Gr. *heteros*, different + *zygon*, yoke) having different genes of a Mendelian pair present in the same cell or organism; for instance, a tall pea plant with genes for tallness (*T*) and dwarfness (*d*)

Hexose (Gr. *hexa*, six + *ose*, suffix indicating, in this usage, carbohydrate) a carbohydrate with six carbon atoms

Hilum (L. *hilum*, a trifle) scar on seed, which marks the place where the seed broke from the stalk

Histology (Gr. *histos*, cloth, tissue + *logos*, discourse) science that deals with the microscopic structure of animal and vegetable tissues

Holdfast an anchoring organ in certain seaweeds; not a true root because it lacks vascular tissue; furthermore, most absorption occurs elsewhere on the thallus

Homobasidiomycetidae (Gr. *homo*, the same + Basidiomycete) a subclass of Basidiomycotinia with a typical club-shaped cell as a basidium

Homologous chromosomes (Gr. *homologos*, the same) members of a chromosome pair; they may be heterozygous or homozygous

Homospory (Gr. *homos*, one and the same + spore) the condition of producing one sort of spore only

Homothallic (Gr. *homos*, one and the same + thallus) referring to species in which male gametangia and female gametangia are produced in the same filament or by the same individual plant body

Homozygous (Gr. *homos*, one and the same + *zygon*, yoke) having similar genes of a Mendelian pair present in the same cell or organism; for instance, a tall pea plant with genes for tallness (*TT*) only

Hormone (Gr. *hormaein*, to excite) a specific organic product, produced in one part of a plant or animal body, and transported to another part where, effective in small amounts, it controls or stimulates another and different process

Host (L. *hospes*, host, guest) an organism on or in which another organism lives

Humidity, relative (L. *humidus*, moist) the ratio of the weight of water vapor in a given quantity of air, to the total weight of water vapor that quantity of air is capable of holding at the temperature in question, expressed as percent

Humus (L. *humus*, the ground) decomposing organic matter in the soil

Hybrid (L. *hybrida*, offspring of a tame sow and a wild boar, a mongrel) the offspring of two plants or animals differing in at least one Mendelian character; or the offspring of plants or animals differing in many characters

Hybrid vigor (heterosis) (Gk. *heterosis*, alteration fr. *heteros*, other) a greater capacity for growth frequently observed in crossbred animals or plants as compared to those resulting from inbreeding

Hydathode (Gr. *hydro*, water + O.E. *thoden*, stem or *thyddan*, to thrust) a structure, usually on leaves, that releases liquid water during guttation

Hydrocarbon chain a linear molecule formed from hydrogen and carbon

Hydrogen acceptor a substance capable of accepting hydrogen atoms or electrons in the oxidation-reduction reactions of metabolism

Hydrogen bond a weak bond between a hydrogen atom attached to a strongly electronegative atom and another strongly electronegative atom (e.g., oxygen, nitrogen)

Hydroid (Gr. *hydro*, water + *eidos*, a shape) a water-conducting cell found in some mosses

Hydrolysis (Gr. *hydro*, water + *lysis*, loosening) union of a compound with water, attended by decomposition into less complex compounds; usually controlled by enzymes

Hydrophilic (Gr. *hydro*, water + *philos*, loving) soluble in water, capable of forming weak associations with water molecules

Hydrophobic (Gr. *hydro*, water + *phobos*, fearing) insoluble in water

Hydrophobic bond the tendency of hydrophobic molecules in an aqueous environment to stick together, occurring because of the high free energy required to force them into solution

Hydrophyte (Gr. *hydro*, water + *phyton*, a plant) a plant that grows wholly or partly submerged in water

Hymenium (Gr. *hymen*, a membrane) spore-bearing tissue in various fungi

Hypantheum (L. *hypo*, under + Gr. *anthos*, flower) fusion of calyx and corolla partway up their length to form a cup, as in many members of the rose family

Hypersensitive response a reaction of plant cells to the presence of pathogenic agents, involving a loss of ions, the synthesis of reactive oxygen species (superoxide, hydrogen peroxide), and eventual death

Hypertonic (Gr. *hyper*, above, over + *tonos*, to stretch) a solution having a concentration high enough so that water will move into it across a membrane from another solution

Hypertrophy (Gr. *hyper*, over + *trophein*, to nourish with food) a condition of overgrowth or excessive development of an organ or part

Hypha (plural, **hyphae**) (Gr. *hyphe*, a web) a fungal thread or filament

Hypocotyl (Gr. *hypo*, under + *kotyledon*, a cup-shaped hollow) that portion of an embryo or seedling between the cotyledons and the radicle or young root

Hypogeal (Gr. *hypo*, under + *ge*, the earth) type of germination where cotyledons remain below ground

Hypogyny (Gr. *hypo*, under + *gyne*, female) a condition in which the receptacle is convex or conical, and the flower parts are situated one above another in the following order, beginning with the lowest: sepals, petals, stamens, carpels

Hypolimnion (Gr. *hypo*, under + *limne*, marsh) a lower, cold, relatively nonaerated zone of water that lies below a warmer zone (the epilimnion); common in large bodies of fresh water, such as deep lakes

Hypothesis (Gr. *hypothesis*, foundation) a tentative theory or supposition provisionally adopted to explain certain facts and to guide in the investigation of other facts

Hypotrophy (Gr. *hypo*, under + *trophein*, to nourish with food) an underdevelopment of an organ or part

IAA see **Indoleacetic acid**

Imbibition (L. *imbibere*, to drink) the absorption of liquids or vapors into the ultramicroscopic spaces or pores found in such materials as cellulose or a block of gelatin; an adsorption phenomenon

Imperfect flower a flower lacking either stamens or pistils

Imperfect fungi fungi reproducing only by asexual means

Inclusion body a body found in the cells of organisms with a virus infection

Incomplete dominance see **Codominance**

Incomplete flower a flower lacking one or more of the four kinds of flower parts

Indehiscent (L. *in*, not + *dehiscere*, to divide) not opening by valves or along regular lines

Indeterminate generally, having no defined limits; in plant development, a morphogenetic process that produces new organs or sections of an organ while preserving a meristem in which new cells are formed to continue the process indefinitely

Indicator species a species that has a narrow range of tolerance for one or more environmental factors so that, from its occurrence at a site, one can predict these factors at that site (e.g., nutrient availability or summer temperatures)

Indoleacetic acid a naturally occurring growth regulator, an auxin

Indusium (plural, **indusia**) (L. *indusium*, a woman's undergarment) membranous growth of the epidermis of a fern leaf that covers a sorus

Infect (L. *infectus*, to put into, to taint with morbid matter) specifically to produce disease by such agents as bacteria or viruses

Inferior ovary an ovary partially or completely united with the calyx

Inflorescence (L. *inflorescere*, to begin to bloom) a flower cluster

Inheritance (O.F. *enheritance*, inheritance) the reception or acquisition of characters or qualities by transmission from parent to offspring

Inorganic referring in chemistry to compounds that do not contain carbon

Integument (L. *integumentum*, covering) external layer of ovule that later develops into the seed coat

Inter a prefix, from the Latin preposition *inter*, meaning between, in between, in the midst of

Intercalary (L. *intercalare*, to insert) descriptive of meristematic tissue or growth not restricted to the apex of an organ, that is, growth at nodes

Intercellular (L. *inter*, between + cells) lying between cells

Interfascicular cambium (L. *inter*, between + *fasciculus*, small bundle) cambium that develops between vascular bundles

Intermediary metabolism the collection of all the metabolic pathways in a cell

Intermediate (L. *inter-*, between + *medius*, middle) a compound in a metabolic pathway that occurs before the final product

Internode (L. *inter*, between + *nodus*, a knot) the region of a stem between two successive nodes

Interphase (L. *inter*, between + Gr. *phasis*, appearance) the period of preparation for cell division; state between two mitotic or meiotic cycles

Intertidal zone the strip of coastal land that is alternately inundated and exposed as tides rise and fall

Intine (L. *intus*, within) the innermost coat of a pollen grain

Intra a prefix from the Latin preposition *intra*, meaning on the inside, within

Intracellular (L. *intra*, within + cell) lying within cells

Introgressive hybridization (L. *intro*, to the inside + *gress*, walk; *hybrida*, half-breed) backcrossing between complete or partial hybrids and the original parental stock

Involucre (L. *involucrum*, a wrapper) a whorl or rosette of bracts surrounding an inflorescence

Ion (Gr. *ienai*, to go) a charged particle formed by the breakdown of substances able to conduct a electric current

Ionic bond a chemical bond formed between ions of opposite charge

Ion pump a protein in a cellular membrane that catalyzes the transport of an ion from one side of the membrane to the other; pump implies that the transport is active, requiring the expenditure of metabolic energy and potentially occurring from a lower concentration of the ion to a higher concentration

Irregular flower a flower in which one or more members of at least one whorl are of different form from other members of the same whorl; zygomorphic flower

Isobilateral leaf (Gr. *isos*, equal + L. *bis*, twice, two-fold + *lateralis*, pertaining to the side) a leaf having the upper and lower surfaces essentially similar

Isodiametric (Gr. *isos*, equal + diameter) having diameters equal in all directions, as a ball

Isogametes (Gr. *isos*, equal + gametes) gametes similar in size and behavior

Isogamy (Gr. *isos*, equal + *gamete*, spouse) the condition in which the gametes are identical

Isomers (Gr. *isos*, equal + *meros*, part) two or more compounds having the same molecular formula, for example, glucose and fructose are both $C_6H_{12}O_6$

Isotonic (Gr. *isos*, equal + *tonos*, to stretch) having equal osmotic concentration

Isotope (Gr. *isos*, equal + *topos*, place) any of two or more forms of an element having the same or very closely related chemical properties

Karyogamy (Gr. *karyon*, nut + *gamos*, marriage) the fusion of two nuclei

Karyon (Gr. *karyon*, nut) term used in conjunction with the nuclei in cells of Ascomycetes and Basidiomycetes; dikaryon, two nuclei per cell, each derived from a different parent, $n+n$; heterokaryon, the situation in which members of a dikaryon pair carry different alleles

Kelp (M.E. *culp*, seaweed) a collective name for any of the large brown algae

Kinetic energy energy associated with moving objects

Kinetin (Gr. *kinetikos*, causing motion) a purine that acts as a cytokinin in plants but probably does not occur in nature

Kinetochore (Gr. *kinein*, to move + *chorein*, to move apart) specialized body of a chromosome, which seems to direct its movement

Krebs cycle see **Citric acid cycle**

K selection natural selection that favors long-lived, late-maturing individuals that devote a small fraction of their resources into reproduction; many tree species are K strategists

Lamella (plural, **lamellae**) (Gr. *lamin*, a thin blade) cellular membranes, frequently those seen in chloroplasts

Lamina (L. *lamina*, a thin plate) blade or expanded part of a leaf

Lateral bud a bud that grows out of the side of a stem

Laterite (L. *later*, a brick) a soil characteristic of rain forest vegetation; color is red from oxidized iron in the A horizon; synonymous with oxisol, a soil order

Latex (L. *latex*, juice) a milky secretion

Leaf (O.E. *leaf*) lateral outgrowth of stem axis, which is the usual primary photosynthetic organ, and in the axil of which may be a bud

Leaf axil angle formed by the leaf stalk and the stem

Leaf gap the region composed of parenchyma that is located in the primary vascular cylinder above the point of departure of the leaf vascular tissue

Leaflet separate part of the blade of a compound leaf

Leaf primordium (L. *primordium*, a beginning) a lateral outgrowth from the apical meristem, which will become a leaf

Leaf scar characteristic scar on stem axis made after leaf abscission

Leaf trace the vascular bundle extending from the stem to the base of a leaf

Legume (L. *legumen*, any leguminous plant, particularly bean) a simple, dry dehiscent fruit with one carpel, splitting along two sutures

Lemma (Gr. *lemma*, a husk) lower bract that subtends a grass flower

Lenticel (M.L. *lenticella*, a small lens) a structure of the bark that permits the passage of gas inward and outward

Leptoid (Gr. *leptos*, small, thin + *eidos*, a shape) a sugar-conducting cell found in some mosses

Leukoplast (Gr. *leuk-*, white + *plastid*) a colorless plastid

Liana (F. *liane* from *lier*, to bind) a plant that climbs upon other plants, depending upon them for mechanical support; a plant with climbing shoots

Lichen (Gr. *leichen*, thallus plants growing on rocks and trees) a composite plant consisting of a fungus living symbiotically with an alga

Lignification (L. *lignum*, wood + *facere*, to make) impregnation of a cell wall with lignin

Lignin (L. *lignum*, wood) an organic substance or group of substances impregnating the cellulose framework of certain plant cell walls

Ligule (L. *ligula*, dim. of *lingua*, tongue) in grass leaves, a outgrowth from the upper and inner side of the leaf blade where it joins the sheath

Line transect a method of sampling vegetation by stretching a tape along a straight line and measuring the canopy cover of plants beneath that line or which cut through a vertical plane described by that line

Linkage the grouping of genes on the same chromosome

Linked characters characters of a plant or animal controlled by genes grouped together on the same chromosome

Lipid (Gr. *lipos*, fat + L. *ides*, suffix meaning son of; now used in sense of having the quality of) any of a group of fats or fatlike compounds insoluble in water and soluble in fat solvents

Lipid body lipid storage organelle found in seeds

Lipopolysaccharide a substance containing lipid and a carbohydrate polymer; specifically, the outer membrane of a Gram-negative bacterium

Liverwort (liver + M.E. *wort*, a plant; literally, a liver plant, so named in medieval times because of its fancied resemblance to the lobes of the liver) common name for the class Hepaticae of the Bryophyta

Loam (O.E. *lam* or Old Teutonic *lai*, to be sticky, clayey) a particular soil texture class, referring to a soil having 30 to 50% sand, 30 to 40% silt, and 10 to 25% clay

Lobed leaf (Gr. *lobos*, lower part of the ear) a leaf divided by clefts or sinuses

Locule (L. *loculus*, dim. of *locus*, a place) a cavity of the ovary in which ovules occur

Long-day plants plants that are induced to flower only when the length of the day exceeds a certain value

Longevity (L. *longaevus*, long-lived) length of life

Lumen (L. *lumen*, light, an opening for light) the cavity of the cell within the cell walls

Lysis (Gr. *lysis*, a loosening) a process of disintegration and cell destruction

Macroenvironment (Gr. *makros*, large + O.F. *environ*, about) the environment due to the general, regional climate; traditionally measured some 4 ft above the ground and away from large obstructions

Macronutrient (Gr. *makros*, large + L. *nutrire*, to nourish) an essential element required by plants in relatively large quantities

Map distance on a chromosome, the distance in crossover units between designated genes

Medulla (L. *medulla*, marrow) the filamentous center of certain lichens and kelp blades and stipes

Megapascal (MPa) a measure of pressure and water potential

Megaphyll (Gr. *megas*, great + *phyllon*, leaf) a leaf whose trace is marked with a gap in the stem's vascular system; megaphylls are thought to represent modified branch systems

Megasporangium (Gr. *megas*, large + sporangium) sporangium that bears megaspores

Megaspore (Gr. *megas*, large + spore) the meiospore of vascular plants, which gives rise to a female gametophyte

Megasporocyte (Gr. *megas*, large + *spora*, seed + L. *cyta*, vessel) a diploid cell in which meiosis will occur, resulting in four megaspores; synonymous with megaspore mother cell

Megasporophyll (Gr. *megas*, large + spore + Gr. *phyllon*, leaf) a leaf bearing one or more megasporangium

Meiocyte (meiosis + Gr. *kytos*, currently meaning a cell) any cell in which meiosis occurs

Meiosis (Gr. *meioun*, to make smaller) two special cell divisions occurring once in the life cycle of every sexually reproducing plant and animal, halving the chromosome number and effecting a segregation of genetic determiners

Meiospore (meiosis + spore) any spore resulting from the meiotic divisions

Membrane (L. *membrana*, skin, parchment) generally, a thin, soft, pliable sheet; specifically, a limiting surface, within or surrounding a cell, formed from phospholipids, glycolipids, or other hydrophobic compounds

Meristem (Gr. *meristos*, divisible) undifferentiated tissue, the cells of which are capable of active cell division and differentiation into specialized tissues

Meristoderm (meristem + epidermis) the outer meristematic cell layer (epidermis) of some Phaeophyta

Mesocarp (Gr. *mesos*, middle + *karpos*, fruit) middle layer of fruit wall (pericarp)

Mesophyll (Gr. *mesos*, middle + *phyllon*, leaf) parenchyma tissue of leaf between epidermal layers

Mesophyte (Gr. *mesos*, middle + *phyton*, a plant) a plant avoiding both extremes of moisture and drought

Mesosome (Gr. *mesos*, middle + *soma*, body) one of a series of paired membranes occurring in many bacteria

Mesozoic (Gr. *mesos*, middle + *zoe*, life) a geologic era beginning 225 million years ago and ending 65 million years ago

Metabolic pathway a set of compounds related by enzymatically catalyzed chemical reactions, such that the first compound is transformed into the second, the second into the third, etc.

Metabolism (M.L. from the Gr. *metabolos*, to change) the process, in an organism or a single cell, by which nutritive material is built up into living matter, or aids in building living matter, or by which protoplasm is broken down into simple substances to perform special functions

Metabolite (Gr. *metabolos*, changeable + *ites*, one of a group) a chemical that is a normal cell constituent capable of entering into the biochemical transformations within living cells

Metamorphic rock (Gr. *meta*, change + *morphe*, shape or form) one of three major categories of rock; rocks whose original structure or mineral composition has been changed by pressures or temperatures in the earth's crust

Metaphase (Gr. *meta*, after + *phasis*, appearance) stage of mitosis during which the chromosomes, or at least the kinetochores, lie in the central plane of the spindle

Metaxylem (Gr. *meta*, after + *xylon*, wood) last formed primary xylem

Microcapillary space exceedingly small spaces, such as those found between microfibrils of cellulose

Microenvironment (Gr. *mikros*, small + O.F. *environ*, about) the environment close enough to the surface of a living or nonliving object to be influenced by it

Microfibrils (Gr. *mikros*, small + fibrils, dim. of fiber; literally, small little fibers) the translation of the name expresses the concept very well; microfibrils are exceedingly small fibers visible only with the high magnifications of the electron microscope

Microfilament a intracellular organelle; thread-like, formed of actin subunits, and often participating in the movement of other organelles around the cell

Microfossil (Gr. *mikros*, small + L. *fossilis*, dug up) fossils of microscopic organisms, only visible when thin sections of rock are examined

Micrometer (Gr. *mikros*, small + *metron*, measure) one millionth (10^{-6}) of a meter, or 0.001 millimeter; also called a micron, and abbreviated μm

Micronutrient (Gr. *mikros*, small + L. *nutrire*, to nourish) an essential element required by plants in relatively small quantities

Microphyll (Gr. *mikros*, small + *phyllon*, leaf) a leaf whose trace is not marked with a gap in the stem's vascular system; microphylls are thought to represent epidermal outgrowths

Micropyle (Gr. *mikros*, small + *pyle*, orifice, gate) a pore leading from the outer surface of the ovule between the edges of the two integuments down to the surface of the nucellus

Microsporangium (plural, **microsporangia**) (Gr. *mikros*, little + sporangium) a sporangium that bears microspores

Microspore (Gr. *mikros*, small + spore) a spore that, in vascular plants, gives rise to a male gametophyte

Microsporocyte (Gr. *mikros*, small + *spora*, seed + L. *cyta*, vessel) a diploid cell in which meiosis will occur, resulting in four microspores; synonymous with microspore mother cell

Microsporophyll (Gr. *mikros*, little + spore + Gr. *phyllon*, leaf) a leaf bearing microsporangia

Microsurgical dissections (Gr. *mikros*, small + surgical) surgical experiments done on individual cells or groups of cells, such as a shoot apex

Microtubule (Gr. *mikros*, small + *tubule*, dim. of tube) a tubule 25 nm in diameter and of indefinite length, occurring in the cytoplasm of many types of cells

Middle lamella (L. *lamella*, a thin plate or scale) original thin membrane separating two adjacent protoplasts and remaining as a distinct cementing layer between adjacent cell walls

Millimeter the 0.001 part of a meter, equal to 0.0394 inch

Mitochondrion (plural, **mitochondria**) (Gr. *mitos*, thread + *chondrion*, a grain) a small cytoplasmic particle associated with intracellular respiration

Mitosis (plural, **mitoses**) (Gr. *mitos*, a thread) nuclear division, involving appearance of chromosomes, their longitudinal duplication, and equal distribution of newly formed parts to daughter nuclei

Mitospore (mitosis + spore) a spore forming after mitosis

Mixed bud a bud containing both rudimentary leaves and flowers

Molecular biology a field of biology which emphasizes the interaction of biochemistry and genetics in the life of an organism

Molecule (F. *môle*, mass + *cule*, dim.; literally, a little mass) a unit of matter, the smallest portion of an element or a compound that retains chemical identity with the substance in mass; the molecule usually consists of a union of two or more atoms; some organic molecules contain a very large number of atoms

Mollisol (L. *mollis*, soft + *solum*, soil, solid) one of the 11 world soil orders, characterized by containing more than 1% organic matter in the top 17.5 cm and associated with grassland vegetation; synonymous with chernozem

Monocotyledon (Gr. *monos*, solitary + *kotyledon*, a cup-shaped hollow) a plant whose embryo has one cotyledon

Monoecious (Gr. *monos*, solitary + *oikos*, house) having the reproductive organs in separate structures, but borne on the same individual

Monohybrid (Gr. *monos*, solitary + L. *hybrida*, a mongrel) a cross involving one pair of contrasting characters

Monomer (Gr. *monos*, single + *meros*, part) a single subunit that can be used to form a larger complex; e.g., glucose, a simple sugar, is a monomer of a cellulose chain

Monophyletic (Gr. *mono*, single + *phyle*, tribe) said of organisms having a common, but sometimes quite ancient, ancestor

Monostromatic (Gr. *monos*, single, solitary + *stroma*, a bed, currently meaning a supporting framework) referring to a thallus, one cell in thickness

Morphogenesis (Gr. *morphe*, form + L. *genitus*, to produce) the structural and physiological events involved in the development of an entire organism or part of an organism

Morphology (Gr. *morphe*, form + *logos*, discourse) the study of form and its development

Moss (L. *muscus*, moss) a bryophytic plant

Multiciliate (L. *multus*, many + F. *cil*, an eyelash) having many cilia present on a sperm or spore or other type of ciliated cell

Multiple fruit a cluster of matured ovaries produced by separate flowers; for example, a pineapple

Mutant (L. *mutare*, to change) an individual containing a gene that differs from those of its parents

Mutation (L. *mutare*, to change) a sudden, heritable change appearing in an individual as the result of a change in genes or chromosomes

Muton (mutation + *tron*, or *on*, from the last syllable of proton, electron, and other such words indicating an elementary particle) the smallest element, the alteration of which can cause a mutation; possibly a single nucleotide pair

Mutualism (L. *mutuus*, reciprocal) a form of biological interaction in which both organisms must associate together for the continued success of both

Mycelium (Gr. *mykes*, mushroom) the mass of hyphae forming the body of the fungus

Mycobiont (Gr. *mykes*, mushroom + *bios*, life) the fungal partner in a lichen

Mycology (Gr. *mykes*, mushroom + *logos*, discourse) the branch of botany dealing with fungi

Mycorrhiza (Gr. *mykos*, fungus + *riza*, root) a symbiotic association between a fungus and, usually, the root of a higher plant

Myxomycophyta (Gr. *myxa*, mucus + *mykes*, mushroom + *phyton*, plant) a division comprising the "slime fungi"

NAD nicotinamide adenine dinucleotide, a coenzyme capable of being reduced

NADH reduced NAD

NADP nicotinamide adenine dinucleotide phosphate, a coenzyme capable of being reduced

NADPH reduced **NADP**

Naked bud a bud not protected by bud scales

Nanometer (Gr. *nanos*, small) one billionth (10^{-9}) of a meter, abbreviated nm

Natural classification a classification scheme that is based on the phylogenetic nature of the organisms classified; contrasts with an artificial classification, which separates organisms on the basis of convenient traits, but fails to show the evolutionary relationships among the organisms

Natural selection the effect of the environment in channeling the genetic variation of organisms down particular pathways

Nectar (Gr. *nektar*, drink of the gods) a fluid rich in sugars secreted by nectaries, which are often located near or in flowers

Nectar guide a mark of contrasting color or texture that may serve to guide pollinators to nectaries within the flower

Nectary (Gr. *nektar*, the drink of the gods) a nectar-secreting gland, found in flowers

Neritic zone a subtidal but relatively shallow offshore zone, often dominated by large kelps

Net productivity the arithmetic difference between calories produced in photosynthesis and calories lost in respiration

Net radiation the arithmetic difference between incoming solar radiation and outgoing terrestrial radiation

Net venation veins of leaf blade visible to the unaided eye, branching frequently and joining again, forming a network

Neutron (L. *neuter*, neither) an uncharged particle found in the atomic nucleus of all elements except hydrogen; the helium nucleus has two protons and two neutrons; mass of a neutron is equal to 1.67×10^{-24} gram

Niche (It. *nicchia*, a recess in a wall) the functional position of an organism in its ecosystem

Nitrate reduction the enzymatic reaction that converts nitrate ion, NO_3^-, to nitrite ion, NO_2^-

Nitrification (L. *nitrum,* nitro, a combining form indicating the presence of nitrogen + *facere,* to make) change of ammonium salts into nitrates through the activities of certain bacteria

Nitrogen cycle a continuous series of natural processes by which nitrogen passes through successive stations in air, soil, and organisms, involving principally ammonification, nitrogen fixation, nitrification, nitrate reduction, and denitrification

Nitrogen fixation the process of reducing N_2 gas into ammonia and incorporating it into the protoplast; accomplished only by certain prokaryotes

Node (L. *nodus,* a knot) slightly enlarged portion of the stem where leaves and buds arise, and where branches originate

Nodule (L. *nodulus* dim. of *nodus,* a knot) knot or swelling on a root, especially one containing nitrogen-fixing bacteria

Nonpolar having a uniform distribution of electric charge

Nonseptate descriptive of hyphae lacking cross walls

Nucellus (L. *nucella,* a small nut) tissue composing the chief part of the young ovule, in which the embryo sac develops; megasporangium

Nuclear envelope the double membrane surrounding the nucleus of a eukaryotic cell

Nucleic acid an acid found in all nuclei, first isolated as part of a protein complex in 1871 and separated from the protein moiety in 1889; all known nucleic acids fall into two classes, DNA and RNA; they differ from each other in the sugar, in one of the nitrogen bases, in many physical properties, and in function

Nucleoid the region in a bacterial cell containing the principal chromosome; differs from a eukaryotic nucleus in not being surrounded by a nuclear envelope

Nucleolus (L. *nucleolus,* a small nucleus) dense protoplasmic body in the nucleus

Nucleosides components of nucleic acids consisting of a nitrogen base and a sugar; in DNA, the sugar is deoxyribose, and in RNA, ribose; adenine, guanine, and cytosine occur in both DNA and RNA, thymine occurs in DNA, and uracil occurs in RNA

Nucleotides components of nucleic acid: nucleoside (nitrogen base + sugar) + phosphoric acid

Nucleus (L. *nucleus,* kernel of a nut) a dense protoplasmic body essential in cellular synthetic and developmental activities; present in all eukaryotic cells except mature sieve-tube members

Nut (L. *nux,* nut) a dry, indehiscent, hard, one-seeded fruit, generally produced from a compound ovary

Nyctinastic movement (Gr. *nyct,* night + *nastos,* close-pressed, firm) a movement of plant parts (e.g., leaf petioles) associated with diurnal changes in temperature or light intensity

Obligate anaerobe an organism obliged to live in the absence of oxygen

Obligate parasite an organism obliged to live strictly as a parasite

Obligate saprophyte an organism obliged to live strictly as a saprophyte

Ontogeny (Gr. *on,* being + *genesis,* origin) the development of an individual organism or part

Oogamy (Gr. *oion,* egg + *gamete,* spouse) the condition in which the gametes are different in form and activity, that is, sperm and eggs

Oogonium (L. dim. of Gr. *oogonos,* literally, a little egg layer) female gametangium of egg-bearing organ not protected by a jacket of sterile cells, characteristic of the thallophytes

Oospore (Gr. *oion,* egg + spore) a resistant spore developing from a zygote resulting from the fusion of heterogametes

Operculum (L. *operculum,* a lid) in mosses, cap of sporangium

Opposite referring to bud or leaf arrangement in which there are two buds or two leaves at a node

Orbital a solution of the Schrödinger wave equation describing a possible mode of motion of a single electron in an atom or molecule

Order (L. *ordo,* a row of threads in a loom) a taxonomic category below class and above family

Organ (L. *organum,* an instrument or engine of any kind, musical, military, etc.) a part or member of an animal or plant body or cell adapted by its structure for a particular function

Organelle a specialized region within a cell, such as the mitochondrion or dictyosome

Organic referring in chemistry to the carbon compounds, many of which have been in some manner associated with living organisms

Osmometer (Gr. *osmos,* pushing + *meter,* measure) a devise for measuring the magnitude of osmotic force

Osmosis (Gr. *osmos,* a pushing) diffusion of a solvent through a differentially permeable membrane

Osmotic potential the maximum theoretical suction that can be developed in a solution as a result of osmosis when the solution is placed in an osmometer surrounded by pure water; it is a measure of the concentration of the solute(s)

Osmotic pump a mechanism of forcing solution through a pipe (e.g., xylem) using osmosis to generate pressure at one end

Ovary (L. *ovum,* an egg) enlarged basal portion of the pistil, which becomes the fruit

Ovulate referring to a cone, scale, or other structure bearing ovules

Ovule (F. *ovule,* from L. *ovulum,* dim. of *ovum,* egg) a rudimentary seed, containing, before fertilization, the female gametophyte, with egg cell, all being surrounded by the nucellus and one or two integuments

Ovuliferous (ovule + L. *ferre,* to bear) referring to a scale or sporophyll bearing ovules

Oxidation (F. *oxide,* oxygen + *tions,* suffix denoting action) to increase the positive valence or decrease the negative valence of an element or ion; loss of an electron by an atom

P_{fr} and P_r abbreviations for the far-red (FR) or red (R) absorbing form of phytochrome (P)

P_i an abbreviation for inorganic orthophosphate (a mixture of $H_2PO_4^{-1}$ and HPO_4^{-2}, depending on the pH)

Palea (L. *palea*, chaff) upper bract that subtends a grass flower

Paleobotany (Gr. *palaios*, ancient, prehistoric + *botane*, a plant) the study of fossil plants

Paleoecology (Gr. *palaios*, ancient) a field of ecology that reconstructs past vegetation and climate from fossil evidence

Paleozoic (Gr. *palaios*, ancient + *zoe*, life) a geologic era beginning 570 million years ago and ending 225 million years ago

Palisade parenchyma elongated cells, containing many chloroplasts, found just beneath the upper epidermis of leaves

Palmately veined (L. *palma*, palm of the hand) descriptive of a leaf blade with several principal veins spreading out from the upper end of the petiole

Panicle (L. *panicula*, a tuft) an inflorescence, the main axis of which is branched, and whose branches bear loose racemose flower clusters

Pappus (L. *pappus*, woolly, hairy seed or fruit of certain plants) scales or bristles representing a reduced calyx in composite flowers

Paradermal section (Gr. *para*, beside + *derma*, skin) a section cut parallel to a flat surface, such as a leaf section cut parallel to the surface of the blade

Parallel venation type of venation in which veins of a leaf blade that are clearly visible to the unaided eye are parallel to each other

Paraphysis (plural, **paraphyses**) (Gr. *para*, beside + *physis*, growth) a slender, multicellular hair (*Fucus*, etc.); one of the sterile branches or hyphae growing beside fertile cells in the fruiting body of certain fungi

Parasexual cycle (Gr. *para*, beside) a sexual cycle involving changes in chromosome number, differing in time and place from the usual sexual cycle; occurring in fungi in which the normal cycle is suppressed or apparently absent

Parasite (Gr. *parasitos*, one who eats at the table of another) an organism deriving its food from the living body of another plant or an animal

Parenchyma (Gr. *parenchein*, an ancient Greek medical term meaning to pour beside and expressing the ancient concept that the liver and other internal organs were formed by blood diffusing through the blood vessels and coagulating, thus designating ground tissue) a tissue composed of cells that usually have thin walls of cellulose, and that often fit rather loosely together, leaving intercellular spaces

Parent material the original rock or depositional matter from which the soil of a region has been formed

Parietal (F. *pariétal*, attached to the wall, from L. *paries*, wall) belonging to, connected with, or attached to the wall of a hollow organ or structure, especially of the ovary or cell

Parietal placentation a type of placentation in which placentae are on the ovary wall

Parthenocarpy (Gr. *parthenos*, virgin + *karpos*, fruit) the development of fruit without fertilization

Parthenogenesis (Gr. *parthenos*, virgin + *genesis*, origin) the development of a gamete into a new individual without fertilization

Pathogen (Gr. *pathos*, suffering + *genesis*, beginning) an organism that causes a disease

Pathology (Gr. *pathos*, suffering + *logos*, account) the study of diseases, their effects on plants or animals, and their treatment

Peat (M.E. *pete*, of Celtic origin, a piece of turf used as fuel) any mass of semi-carbonized vegetable tissue formed by partial decomposition in water

Pectin (Gr. *pektos*, congealed) a white amorphous substance which, when combined with acid and sugar, yields a jelly; the substance cementing cells together; the middle lamella

Pedicel (L. *pediculus*, a little foot) stalk or stem of the individual flowers of an inflorescence

Peduncle (L. *pedunculus*, a late form of *pediculus*, a little foot) stalk or stem of a flower that is borne singly; or the main stem of an inflorescence

Penicillin an antibiotic derived from the mold *Penicillium*

Pentose (Gr. *pente*, five + *ose*, a suffix indicating a carbohydrate) a five-carbon-atom sugar

Peptide (Gr. *pepton*, cooked or digested, a substance remaining after the digestion of proteins) two or more amino acids linked end to end

Peptide bond the bond between carbon and nitrogen that unites two amino acid residues in a polypeptide chain or protein

Peptidoglycan a particular macromolecule that makes up the walls of bacteria and cyanobacteria

Perennial (L. *perennis*, lasting the whole year through) a plant that lives more than two years

Perfect flower a flower having both stamens and pistils; hermaphroditic flower

Perianth (Gr. *peri*, around + *anthos*, flower) the petals and sepals taken together

Pericarp (Gr. *peri*, around + *karpos*, fruit) fruit wall, developed from ovary wall

Periclinal cell division cell division where the newly formed cell wall is parallel to the axis of the organ

Pericycle (Gr. *peri*, around + *kyklos*, circle) tissue, generally of root, bound externally by the endodermis and internally by the xylem and phloem

Periderm (Gr. *peri*, around + *derma*, skin) protective tissue that replaces the epidermis after secondary growth is initiated; consists of cork, cork cambium, and phelloderm

Peridium (plural, **peridia**) (Gr. *peridion*, a little pouch) external covering of the hymenium of certain fungi; in Myxomycota, the hardened envelope that covers the sporangium

Perigyny (Gr. *peri*, about + *gyne*, a female) a condition in which the receptacle is more or less concave, at the margin of which the sepals, petals, and stamens have their origin, so that these parts seem to be attached around the ovary; also called half-inferior

Peristome (Gr. *peri*, about + *stoma*, a mouth) in mosses, a fringe of teeth about the opening of the sporangium

Perithecium (Gr. *peri*, around + *theke*, a box) a spherical or flask-shaped ascocarp having a small opening

Permafrost (L. *permanere*, to remain + A.S. *freosan*, to freeze) soil that is permanently frozen; usually found some distance below a surface layer that thaws during warm weather

Permanent wilting the condition of wilting of a plant when it can no longer obtain moisture from the soil

Permanent wilting percentage the maximum amount of water (expressed as % of the dry weight of the soil) that a soil can hold that is unavailable to a plant

Permeable (L. *permeabilis*, that which can be penetrated) said of a membrane, cell, or cell system through which substances may diffuse

Peroxysome an organelle of the microbody class that contains enzymes capable of making and destroying hydrogen peroxide, glycolic oxidase, and catalase

Petal (Gr. *petalon*, a flower leaf) one of the flower parts, usually conspicuously colored

Petiole (L. *petiolus*, a little foot or leg) stalk of leaf

PGA see **Phosphoglyceric acid**

pH (Fr. *p(ouvoir)*, to be able + *h(ydrogené)*, hydrogen) a symbol for the degree of acidity (values from 0 to 7) or alkalinity (values from 7 to 14); values representing the relative concentration of the hydrogen ion in solution

Phage (Gr. *phago* to eat) a virus infecting bacteria; originally bacteriophage

Phelloderm (Gr. *phellos*, cork + *derma*, skin) a layer of cells formed in the stems and roots of some plants from the inner cells of the cork cambium

Phellogen (Gr. *phellos*, cork + *genesis*, birth) cork cambium, a cambium giving rise externally to cork and in some plants internally to phelloderm

Phenotype (Gr. *phaneros*, showing + type) the external visible appearance of an organism

Phloem (Gr. *phloos*, bark) food-conducting tissue, consisting of sieve-tube members with companion cells or sieve cells, phloem parenchyma, and fibers

Phosphoenolpyruvate carboxylase the enzyme responsible for the fixation of inorganic CO_2 into oxaloacetic acid in a dark reaction of the C_4 photosynthesis cycle

Phosphoglyceric acid (PGA) a three-carbon compound formed by the interaction of carbon dioxide (CO_2) and a five-carbon compound, ribulose bisphosphate; the reaction yields two molecules of PGA for each molecule of the ribulose bisphosphate; the first step in the C_3 carbon cycle of photosynthesis

Phospholipid a complex lipid compound, composed of fatty acids, glycerol, phosphate, and one additional hydrophilic residue (e.g., choline, serine, inositol); a major constituent of biological membranes

Phosphorylation a reaction in which phosphate is added to a compound, for example, the formation of ATP from ADP and inorganic phosphate

Photoautotroph (Gr. *photos*, light + *autos*, self + *trophein*, to feed) bacteria that use light as an energy source and CO_2 as a carbon source

Photoheterotroph (Gr. *photos*, light + *heteros*, other + *trophein*, to feed) bacteria that use light as an energy source and various organic compounds as a carbon source

Photon a quantum of light; the energy of a photon is proportional to its frequency; $E = \hbar v$, where E is energy; $\hbar$, Planck's constant, 6.62×10^{-27} erg-second; and v is the frequency

Photoperiod (Gr. *photos*, light + period) the optimum length of day or period of daily illumination required for the normal growth and maturity of a plant

Photophosphorylation a reaction in which light energy is converted into chemical energy in the form of ATP produced from ADP and inorganic phosphate

Photoreceptor (Gr. *photos*, light + L. *receptor*, a receiver) a light-absorbing molecule involved in converting light into some metabolic (chemical energy) form, for example, chlorophyll and phytochrome

Photosynthesis (Gr. *photos*, light + *syn*, together + *tithenai*, to place) a process in which carbon dioxide and water are brought together chemically to form a carbohydrate, the energy for the process being radiant energy

Phototropism (Gr. *photos*, light + *tropos*, turning) a growth curvature in which light is the stimulus

Phycobiliproteins pigments found in the red algae and cyanobacteria, similar to bile pigments and always associated with proteins

Phycobilisomes (Gr. *phykos*, seaweed + L. *bilis*, relating to greenish bile + Gr. *soma*, body) rods or discs of accessory pigments that are attached to photosynthetic membranes in cyanobacteria; they absorb light in the green to orange part of the spectrum

Phycobiont (Gr. *phykos*, seaweed + *bios*, life) the algal partner in a lichen

Phycocyanin (Gr. *phykos*, seaweed + *kyanos*, blue) a blue phycobilin pigment occurring in cyanobacteria

Phycoerythrin (Gr. *phykos*, seaweed + *erythros*, red) a red phycobilin pigment occurring in red algae

Phycology (Gr. *phykos*, seaweed + *logos*, word, thought) the science of the study of algae

Phycomycetes (Gr. *phykos*, seaweed + *mykes*, mushroom or fungus) a class of fungi that approaches the algae in some characters

Phylogenetic classification see **Natural classification**

Phylogeny (Gr. *phylon*, race or tribe + *genesis*, beginning) the evolution of a group of related individuals

Phylum (Gr. *phylon*, race or tribe) a primary division of the animal or plant kingdom; synonymous with division

Physiology (Gr. *physis*, nature + *logos*, discourse) the science of the functions and activities of living organisms

Phytobenthon (Gr. *phyton*, a plant + *benthos*, depths of the sea) attached aquatic plants, collectively

Phytochrome a reversible pigment system of protein naturally found in the cytoplasm of green plants; it is associated with the absorption of light that affects growth, development, and differentiation of a plant, independent of photosynthesis, for example, in the photoperiodic response

Phytoplankton (Gr. *phyton*, a plant + *planktos*, wandering) free-floating plants, collectively; usually algae

Pigment (L. *pingere*, to paint) a substance that absorbs visible light; hence, appears colored

Pileus (L. *pileus*, a cap) umbrella-shaped cap of fleshy fungi

pilus (L. *pilus*, hair) a structure on the surface of a bacterial cell resembling a hair

Pinna (plural, **pinnae**) (L. *pinna*, a feather) leaflet or division of a compound leaf (frond)

Pinnately veined (L. *pinna*, a feather + *vena*, a vein) descriptive of a leaf blade with single midrib from which smaller veins branch off, somewhat like the divisions of a feather

Pinocytosis (Gr. *pinein*, to drink + *kytos*, hollow vessel) the process by which a cell may take in food or other material by forming an invagination of the plasma membrane that is pinched off into the cytoplasm

Pioneer community the first stage of a succession

Pistil (L. *pistillum*, a pestle) central organ of the flower, typically consisting of ovary, style, and stigma

Pistillate flower a flower having pistils but no stamens

Pit a minute, thin area of a secondary cell wall

Pith the parenchymatous tissue occupying the central portion of a stem

Placenta (plural, **placentae**) (L. *placenta*, a cake) the tissue within the ovary to which the ovules are attached

Placentation (L. *placenta*, a cake + *tion*, state of) manner in which the placentae are distributed in the ovary

Plankton (Gr. *planktos*, wandering) free-floating aquatic plants and animals, collectively

Plasma membrane (Gr. *plasma*, anything formed + L. *membrana*, parchment) a delicate cytoplasmic membrane found on the outside of the protoplast adjacent to the cell wall

Plasmid (Gr. *plasma*, to form) a piece of extra-chromosomal DNA, found in some bacteria

Plasmodesma (plural, **plasmodesmata**) (Gr. *plasma*, something formed + *desmos*, a bond, a band) fine protoplasmic thread passing through the wall that separates two protoplasts

Plasmodium (Gr. *plasma*, something formed + mod. L. *odium*, something of the nature of) in Myxomycetes, a slimy mass of protoplasm, with no surrounding wall and with numerous free nuclei distributed throughout

Plasmogamy (Gr. *plasma*, anything molded or formed + *gamos*, marriage) the fusion of protoplasts, not accompanied by nuclear fusion

Plasmolysis (Gr. *plasma*, something formed + *lysis*, a loosening) the separation of the cytoplasm from the cell wall due to removal of water from the protoplast

Plastid (Gr. *plastis*, a builder) the cellular organelle in which carbohydrate metabolism is located

Plastoquinone a quinone, one of a group of compounds involved in the transport of electrons during photosynthesis in chloroplasts

Plumule (L. *plumula*, a small feather) the first bud of an embryo or that portion of the young shoot above the cotyledons

Pneumatophores (Gr. *pneuma*, breath + *phore*, F. *pherein*, to carry) extensions of the root systems of some plants growing in swampy areas; in contrast to most roots, they are negatively geotropic and grow out of the water and thus assure adequate aeration

Polarity (Gr. *pol*, an axis) the observed differentiation of an organism, tissue, or cell into parts having opposed or contrasted properties or form

Polar transport the directed movement within plants of compounds (usually hormones) predominantly in one direction; polar transport overcomes the tendency for diffusion in all directions

Pollen (L. *pollen*, fine flour) the germinated microspores or partially developed male gametophytes of seed plants

Pollen mother cell see **Microsporocyte**

Pollen profile a diagrammatic summary of the sequence and abundance of pollen types that have been chronologically trapped in sediments

Pollination the transfer of pollen from a stamen or staminate cone to a stigma or ovulate cone

Pollinium (L. *pollentis*, powerful or *pollinis*, fine flour + *ium*, group) a mass of pollen that sticks together and is transported by pollinators as a mass; present in orchids and milkweeds

Polygenes many genes influencing the development of a single trait; results in continuous variability; compare **Allele**

Polymer (Gr. *polys*, many + *meros*, part) a compound formed by repeating structural units; e.g., cellulose is a polymer of glucose, a simple sugar

Polymerization the chemical union of monomers, such as glucose, or nucleotides to form starch or nucleic acid

Polynomial (Gr. *polys*, many + L. *nomen*, name) scientific name for an organism, composed of more than two words; see **Binomial**

Polynucleotides (Gr. *polys*, much, many) long-chain molecules composed of units (monomers) called nucleotides; nucleic acid is a polynucleotide

Polypeptide chain a linear polymer formed from amino acids held together by peptide bonds

Polyphyletic (Gr. *polys*, many + *phyle*, tribe) referring to organisms having more than one common ancestor

Polyploid (Gr. *polys*, many + *ploos*, fold) referring to a plant, tissue, or cell with more than two complete sets of chromosomes, for example, 4*n*, 6*n*

Polyribosome (Gr. *polys*, many + ribosomes) an aggregation of ribosomes, frequently simply called polysome

Polysaccharides (Gr. *polys*, much, many + *sakcharon*, sugar) long-chain molecules composed of units (monomers) of a sugar; starch and cellulose are polysaccharides

Pome (L. *pomum,* apple) a simple fleshy fruit, the outer portion of which is formed by the floral parts that surround the ovary

Population (L. *populus,* people) a group of closely related, interbreeding organisms

Potential energy energy associated with position; concerning a chemical reaction, includes the energy of a set of reactants associated with the position of their constituent atoms and the shape of their bonding orbitals

p-protein a mass of protein material formerly called slime found in sieve-tube members

Prairie (L. *pratum,* meadow) grassland vegetation, with trees essentially absent; often considered to have more rainfall than does the steppe

Predation (L. *predatio,* plundering) a form of biological interaction in which one organism is destroyed (by ingestion); parasitism, carnivory, and seed herbivory are forms of predation

Primary (L. *primus,* first) first in order of time or development

Primary endosperm cell a cell of the embryo sac after fertilization, generally containing a nucleus resulting from fusion of the two polar nuclei with a sperm nucleus; the endosperm develops from this cell

Primary meristems meristems of the shoot or root tip giving rise to the primary tissues

Primary pit field thin areas of primary cell walls

Primary tissues those tissues, epidermis, xylem, phloem, and ground tissues, that form from primary meristems

Primary wall the first formed cell wall layer formed during cell expansion

Primitive (L. *primus,* first) referring to a taxonomic trait thought to have evolved early in time

Primordium (L. *primus,* first + *ordiri,* to begin to weave; literally beginning to weave, or to put things in order) the beginning or origin of any part of an organ

Procambium (L. *pro,* before + cambium) a primary meristem that gives rise to primary vascular tissues and, in most woody plants, to the vascular cambium

Producer (L. *producere,* to draw forward) an organism that produces organic matter for itself and other organisms (consumers and decomposers) by photosynthesis

Proembryo (L. *pro,* before + *embryon,* embryo) a group of cells arising from the division of the fertilized egg cell before those cells that are to become the embryo are recognizable

Prokaryotes (L. *pro,* before + Gr. *karyon,* a nut, referring in modern biology to the nucleus) primitive organisms and bacteria, that do not have the DNA separated from the cytoplasm by an envelope

Prophase (Gr. *pro,* before + *phasis,* appearance) an early stage in nuclear division, characterized by the shortening and thickening of the chromosomes and their movement to the metaphase plate

Proplastid (Gr. *pro,* before + plastid) a type of plastid, occurring generally in meristematic cells, that will develop into a chloroplast

Protease (protein + *ase,* a suffix indicating an enzyme) an enzyme breaking down a protein

Protein (Gr. *proteios,* holding first place) naturally occurring complex organic substances (egg albumen, meat) composed of amino acids, which are connected by peptide bonds, and which acquire specific three-dimensional shapes required for their biological functions

Protein body protein storage organelle in seeds, sometimes called aleurone grain

Proterozoic (Gr. *protera,* before in time + *zoe,* life) the earliest geologic era, beginning about 4.5 to 5 billion years ago and ending 570 million years ago; also called Precambrian era

Prothallium (Gr. *pro,* before + *thallos,* a sprout) in ferns, the haploid gametophyte generation

Protochlorophyll (Gr. *protos,* first + *chloros,* green + *phyllos,* leaf) one of the precursors of chlorophyll; it accumulates in dark-grown and potentially green tissue

Protochlorophyllide holochrome a light-sensitive compound or complex composed of photochlorophyll and a protein; absorption of light converts the protochlorophyll part to chlorophyll

Protoderm (Gr. *protos,* first + *derma,* skin) a primary meristem that gives rise to epidermis

Proton (Gr. *protos,* first) the nucleus of a hydrogen atom is a single positively charged particle, the proton; the nucleus of all other elements consists of protons and neutrons; the mass of a proton is 1.67×10^{-24} gram

Protonema (plural, **protonemata**) (Gr. *protos,* first + *nema,* a thread) an algal-like filamentous growth; an early stage in development of the gametophyte of mosses

Protoplasm (Gr. *protos,* first + *plasma,* something formed) living substance

Protoplast (Gr. *protoplastos,* formed first) the organized living unit of a single cell; a cell from which the cell wall has been removed

Protoxylem (Gr. *protos,* first + *xylon,* wood) first formed primary xylem

Pseudopodium (Gr. *psuedes,* false + *podion,* a foot) in Myxomycetes, an armlike projection from the body by which the plant creeps over the surface

Purine a group of organic bases having a double-ring structure, one five-carbon, the other six-carbon

Pyrenoid (Gr. *pyren,* the stone of a fruit + L. *oïdes,* like) a denser body occurring within the chloroplasts of certain algae and liverworts and apparently associated with starch deposition

Pyrimidine an organic base having a single-ring structural formula

Quadrat (L. *quadrus,* a square) a frame of any shape which, when placed over vegetation, defines a unit sample area within which the plants may be counted or measured

Quantum (L. *quantum,* how much) an elemental unit of energy; its energy value is $\hbar v$, where $\hbar$, Planck's constant, is 6.62×10^{-27} erg-second and v is the frequency of the vibrations or waves with which the energy is associated

Quiescent center (L. *quiescere*, to rest) disk-shaped region of root apex containing slowly dividing cells

Raceme (L. *racemus*, a bunch of grapes) an inflorescence in which the main axis is elongated but the flowers are born on pedicels that are about equal in length

Rachilla (Gr. *rhachis*, a backbone + L. dim. ending *illa*) shortened axis of spikelet

Rachis (Gr. *rhachis*, a backbone) main axis of spike; axis of fern leaf (frond) from which pinnae arise; in compound leaves, the extension of the petiole corresponding to the midrib of an entire leaf

Radicle (L. *radix*, root) portion of the plant embryo that develops into the primary or seed root

Radioactive decay the disintegration of the nucleus of an atom through the emission of a high energy alpha or beta particle

Random plant distribution a distribution pattern of a plant population within an area such that the probability of finding an individual at one point is the same for all points

Raphe (Gr. *rhaphe*, seam) ridge on seeds, formed by the stalk of the ovule, in those seeds in which the funiculus is sharply bent at the base of the ovule

Raphides (Gr. *rhaphis*, a needle) fine, sharp, needlelike crystals

Ray initials (L. *radius*, a beam or ray) meristematic cells in the vascular cambium that develop into xylem and phloem cells composing the ray system

Ray system system of cells in secondary tissues that are oriented perpendicular to the long axis of the stem, formed from ray initials of the vascular cambium

Receptacle (L. *receptaculum*, a reservoir) enlarged end of the pedicel or peduncle to which other flower parts are attached

Recessive character that member of a pair of Mendelian characters which, when both members of the pair are present, is subordinated or suppressed by the other, dominant character

Recombinant DNA a molecule of DNA formed by the joining of segments from different sources

Recombination (L. *re*, repeatedly + *combinatus*, joined) the mixing of genotypes that results from sexual reproduction

Redox reaction (*red*uction + *ox*idation) a chemical reaction in which one or more reactants is reduced (gains electrons) and one or more reactants is oxidized (loses electrons)

Red tide a coloring of offshore marine water caused by dense phytoplankton populations; often accompanied by toxic by-products

Reduction (F. *reduction*, L. *reductio*, a bringing back) originally "bringing back" a metal from its oxide, that is, iron from iron rust or ore; any chemical reaction involving the removal of oxygen from or the addition of hydrogen or an electron to a substance; energy is required and may be stored in the process as in photosynthesis

Regular flower a flower in which the corolla is made up of similarly shaped petals equally spaced and radiating from the center of the flower; star-shaped flower; actinomorphic flower

Replication the production of a facsimile or a very close copy; here used to indicate the production of a second molecule of DNA exactly like the first molecule

Reproduction (L. *re*, repeatedly + *producere*, to give birth to) the process by which plants and animals give rise to offspring

Reproductive isolation the separation of populations in time or space so that genetic flow between them is cut off

Residual meristem meristematic region near shoot tip that remains after differentiation of the pith and cortex

Resin duct (L. *resina*, rosin + *ductus*, led) resin canal; in conifers, continuous tubes lined with secretory cells that run through the sap-wood; they function as repositories for metabolic by-products

Respiration (L. *re*, repeatedly + *spirare*, to breathe) a chemical oxidation controlled and catalyzed by enzymes that in living protoplasm break down carbohydrate and fats, thus releasing energy to be used by the organism in doing work

Restriction enzyme an enzyme that hydrolyzes DNA at a particular base sequence (specific to the type of enzyme); also called restriction endonuclease

Reticulum (L. *reticulum*, a small net) a small net

Rhizoid (Gr. *rhiza*, root + L. *oïdes*, like) one of the cellular filaments that perform the functions of roots

Rhizome (Gr. *rhiza*, root) an elongated, underground, horizontal stem

Rhizophores (Gr. *rhiza*, root + *phoros*, bearing) leafless branches that grow downward from the leafy stems of certain Lycophyta and give rise to roots when they come into contact with the soil

Rhizosphere soil zone immediately outside a root containing micro-organisms

Ribonucleic acid a nucleic acid containing the sugar ribose, phosphorus, and the bases adenine, guanine, cytosine, and uracil; present in all cells and concerned with protein synthesis in the cell

Ribose a five-carbon sugar, a component of RNA

Ribosomes (ribo, from RNA + Gr. *somatos*, body) small particles 10 to 20 nanometers in diameter, containing RNA

Ribulose bisphosphate carboxylase (F. *carbone*, carbon + Gr. *oxys*, acidic) the enzyme responsible for the fixation of inorganic CO_2 into organic compounds in the dark reactions of the C_3 photosynthesis cycle

Ring porous wood wood with large xylem vessel members mostly in early wood; compare with **Diffuse porous wood**

Ripening (A.S. *rifi*, perhaps related to reap) changes in a fruit that follow seed maturation and that prepare the fruit for its function of seed dispersal

RNA see **Ribonucleic acid**

Root (A.S. *rot*) the descending axis of a plant, normally below ground, serving to anchor the plant and absorb and conduct water and mineral nutrients

Root cap A thimblelike mass of cells covering and protecting the apical meristems of a root, also site of gravity perception

Root hairs epidermal projections of root cells in region of maturation that increase the absorptive surface of the root

Root pressure pressure developed in the root as the result of osmosis and inducing guttation

Rootstock an elongated, underground, horizontal stem

Rosette (dim. of L. *rose*, rose) a shoot with a very short stem, composed of several unelongated internodes but with fully expanded leaves

r selection natural selection that favors short-lived, early-maturing individuals that devote a large fraction of their resources to reproduction; annual herbs are r strategists

Runner a stem that grows horizontally along the ground surface

Salt a compound formed from a defined mixture of cations and anions

Samara (L. *samara*, the fruit of the elm) simple, dry, one- or two-seeded indehiscent fruit with pericarp bearing a wing-like outgrowth

Sand soil particles between 50 and 2000 microns in diameter

Saprophyte (Gr. *sapros*, rotten + *phylon*, a plant) an organism deriving its food from the dead body or the nonliving products of another plant or animal

Sapwood peripheral wood that actively transports

Savannah (Sp. *sabana*, a large plain) vegetation of scattered trees in a grassland matrix; originally limited to tropical regions

Scalariform vessel (L. *scala*, ladder + form) a vessel with secondary thickening resembling a ladder

Schizocarp (Gr. *schizein*, to split + *karpos*, fruit) dry fruit with two or more united carpels that split apart at maturity

Sclereids (Gr. *skleros*, hard) cells of different shapes having heavily lignified cell walls

Sclerenchyma (Gr. *skleros*, hard + *enchyma*, a suffix denoting tissue) a strengthening tissue composed of cells with heavily lignified cell walls

Scrub (A.S. *scrob*, a shrub) vegetation dominated by shrubs; described as thorn forest in areas with moderate rainfall, or as chaparral or desert in areas with low rainfall

Scutellum (L. *scutella*, a dim. of *scutum*, shield) single cotyledon of grass embryo

Seaweed large marine algae, generally reds and browns

Secondary tissue those tissues, xylem, phloem, and periderm, that form from secondary meristems

Secondary wall wall material deposited on the primary wall in some cells after elongation has ceased

Sedimentary rock (L. *sedere*, to sit) rock formed from material deposited as sediment, then physically or chemically changed by compaction and hardening while buried in the earth's crust

Seed (A.S. *sed*, anything that may be sown) popularly, as originally used, anything that may be sown; that is, "seed" potatoes, "seeds" of corn, sunflower, etc.; botanically, a seed is the matured ovule without accessory parts

Self-pollination transfer of pollen from the stamens to the stigma of either the same flower or flowers on the same plant

Seminal root the root or roots forming from primordia present in the seed

Senescence (L. *senescere*, to grow old) the phase of plant growth that extends from full maturity to death and is characterized by a breakdown of functional cellular components, accumulation of metabolic products, and (often) a increase in respiratory rate and synthesis of ethylene

Sepals (M.L. *sepsium*, a covering) outermost flower structures that usually enclose the other flower parts in the bud

Septate (L. *septum*, fence) divided by cross walls into cells or compartments

Septicidal dehiscence (L. *septum*, fence + *caedere*, to cut; *dehiscere*, to split open) the splitting open of a capsule along the line of union of carpels

Septum (L. *septum*, fence) any dividing wall or partition; frequently a cross wall in a fungal or algal filament

Serpentine (L. *serpens*, a serpent) referring to soil derived from metamorphic parent material characterized, among other things, by low calcium (Ca), high magnesium (Mg), and a greenish-gray color

Sessile (L. *sessilis*, low, dwarf, from *sedere*, to sit) sitting, referring to a leaf lacking a petiole or a flower or fruit lacking a pedicel

Seta (plural, **setae**) (L. *seta*, a bristle) in bryophytes, a short stalk of the sporophyte, which connects the foot and the capsule

Sexual reproduction reproduction that requires meiosis and fertilization for a complete life cycle

Sheath part of leaf that wraps around the stem, as in grasses

Shoot (derivation uncertain, but early referring to new plant growth) a young branch that shoots out from the main stock of a tree, or the young main portion of a plant growing above ground

Shoot tip portion of the shoot containing apical and primary meristems and early stages of differentiation

Short-day plants plants that are induced to flower only when the length of the day drops below a certain value

Sibling species species morphologically nearly identical but incapable of producing fertile hybrids

Sieve cells a long and slender sieve element without a companion cell, with relatively unspecialized sieve areas, and with tapering end walls that lack sieve plates

Sieve plate wall area in a sieve-tube member containing a region of pores through which pass strands connecting sieve-tube protoplasts

Sieve tube a series of sieve-tube members forming a long cellular tube specialized for the conduction of food materials

Sieve-tube members portion of a sieve tube composed of a single protoplast and separated from other sieve-tube members by sieve plates

Silique (L. *siliqua*, pod) the fruit characteristic of Brassicaceae (mustards); two-celled, the valves splitting from the bottom and leaving the placentae with the false partition stretched between

Silt soil particles between 2 and 50 microns in diameter

Simple pit pit not surrounded by an overarching border, in contrast to bordered pit

Soil (L. *solum,* soil, solid) the uppermost stratum of the earth's crust, which has been modified by weathering and organic activity into (typically) three horizons: an upper A horizon which is leached, a middle B horizon in which the leached material accumulates, and a lower C horizon, which is unweathered parent material

Soil texture refers to the amounts of sand, silt, and clay in a soil, as sandy loam, loam, or clay texture

Solute (L. *solutus,* from *solvere,* to loosen) a dissolved substance

Solution (M.E. *solucion,* from O.F. *solucion,* to loosen) a homogeneous mixture, the molecules of the dissolved substance (e.g., sugar), the solute, being dispersed between the molecules of the solvent (e.g., water)

Solvent (L. *solvere,* to loosen) a substance, usually a liquid, having the properties of dissolving other substances

Soredium (plural, **soredia**) (Gr. *soros,* a heap) asexual reproductive body of lichens, consisting of a few algal cells surrounded by fungus hyphae

Sorus (plural, **sori**) (Gr. *soros,* a heap) a cluster of sporangia

Species (L. *species,* appearance, form, kind) a class of individuals usually interbreeding freely and having many characteristics in common; a population is a subset of a species

Sperm (Gr. *sperma,* the generative substance or seed of a male animal) a male gamete

Spermatophyte (Gr. *sperma,* seed + *phyton,* plant) a seed plant

Spike (L. *spica,* an ear of grain) an inflorescence in which the main axis is elongated and the flowers are sessile

Spikelet (L. *spica,* an ear of grain + *let,* dim. ending), the unit of inflorescence in grasses; a small group of grass flowers

Spindle (A.S. *spinel,* an instrument employed in spinning thread by hand) referring in mitosis and meiosis to the spindle-shaped intracellular structure in which the chromosomes move

Spirillum (L. *spira,* a coil) a bacterial cell that has a spiral shape

Spodosol (Gr. *spodos,* wood ashes; R., *pod,* under + *zola,* ashes) one of the 10 world soil orders, characterized by an ashy, sandy, bleached acidic A_2 horizon and associated mainly with coniferous forest vegetation; synonymous with podzol

Sporangiophore (sporangium + Gr. *phore,* a root of *phorein,* to bear) a branch bearing one or more sporangia

Sporangium (spore + Gr. *angeion,* a vessel) spore case

Spore (Gr. *spora,* seed) a reproductive cell that develops into a plant without union with other cells; some spores, such as meiospores, occur at a critical stage in the sexual cycle, but others are asexual in nature

Sporidium (dim. meaning a little spore) the basidiospore of smut fungi

Sporophore (spore + Gr. *phorein,* to bear) the fruiting body of fleshy and woody fungi, which produces spores

Sporophyll (spore + Gr. *phyllon,* leaf) a spore-bearing leaf

Sporophyte (spore + Gr. *phyton,* a plant) in alternation of generations, the plant in which meiosis occurs and which thus produces meiospores

Stamen (L. *stamen,* the standing-up things or a tuft of thready things) flower structure made up of an anther (pollen-bearing portion) and a stalk or filament

Staminate flower a flower having stamens but no pistils

Starch (M.E. *sterchen,* to stiffen) a complex insoluble carbohydrate, the chief food storage substance of plants, that is composed of several hundred hexose sugar units and that easily breaks down on hydrolysis into these separate units

Statolith (Gr. *statos,* standing + *lithos,* stone) a starch grain that moves to its position in a cell as a result of gravity, thus providing an initial sensing of, or orientation to, gravity by a cell

Stele (Gr. *stele,* a post) the central vascular cylinder, inside the cortex, of roots and stems of vascular plants

Stem (O.E. *stemn*) the main body of the portion above ground of tree, shrub, herb, or other plant; the ascending axis, whether above or below ground, of a plant, in contradistinction to the descending axis or root

Steppe (R. *step,* a lowland) an arid grassland vegetation, usually with scattered shrubs

Stereid (Gr. *stereos,* solid) thick-walled cells at or near the epidermis of certain mosses

Sterigma (plural, **sterigmata**) (Gr. *sterigma,* a prop) a slender, pointed protuberance at the end of a basidium, which bears a basidiospore

Sterilization (L. *sterilis,* barren) the process of making something germ-free

Sterols a class of lipids, including (in animals) cholesterol and (in plants) ergosterol, often contributing to the structure of biological membranes

Stigma (L. *stigma,* a prick, a spot, a mark) receptive portion of the style to which pollen adheres

Stipe (L. *stipes,* post, tree trunk) the stem portion of a kelp, to which are attached bladders and blades

Stipule (L. *stipula,* dim. of *stipes,* a stock or trunk) a leaflike structure from either side of the leaf base

Stolon (L. *stolo,* a shoot) a stem that grows horizontally along the ground surface

Stoma (plural, **stomata**) (Gr. *stoma,* mouth) a minute opening plus two guard cells in the epidermis of leaves and stems, through which gases pass

Strobilus (Gr. *strobilos,* a cone) a number of modified leaves (sporophylls) or ovule-bearing scales grouped together on an axis

Stroma (Gr. *stroma,* a bed or covering) a mass of protecting vegetative filaments; the background substance of chloroplasts, probably the location of the carbon cycle of photosynthesis

Style (Gr. *stylos*, a column) slender column of tissue that arises from the top of the ovary and through which the pollen tube grows

Suberin (L. *suber*, the cork oak) a waxy material found in the cell walls of cork tissue

Substrates the reactants of an enzymatically catalyzed chemical reaction

Succession (L. *successio*, a coming into the place of another) a sequence of changes in time of the species that inhabit an area, from an initial pioneer community to a final climax community

Succulent (L. *sucus*, juice) a plant having juicy or watery tissues

Sucrose (F. *sucre*, sugar + *ose*, ending designating a sugar), cane sugar ($C_{12}H_{22}O_{11}$)

Superior ovary an ovary completely separate and free from the calyx

Suspensor (L. *suspendere*, to hang) a cell or chain of cells developed from a zygote whose function is to place the embryo cells in an advantageous position to receive food

Suture (L. *sutura*, a sewing together; originally the sewing together of flesh or bone wounds) the junction, or line of junction, of contiguous parts

Symbiosis (Gr. *syn*, with + *bios*, life) an association of two different kinds of living organisms involving benefit to both

Sympetaly (Gr. *syn*, with + *petalon*, leaf) a condition in which petals are united

Symplast (Gr. *syn*, together + *plastides*, formed) the region of a plant tissue inside of the plasma membrane of the cells that form it, especially when the cell interiors are connected by plasmodesmata; contrast to **Apoplast**

Synandry (Gr. *syn*, with + *andros*, a man) a condition in which stamens are united

Syncarpy (Gr. *syn*, with + *karpos*, fruit) a condition in which carpels are united

Synergids (Gr. *synergos*, toiling together) the two nuclei at the upper end of the embryo sac, which, with the third (the egg), constitute the egg apparatus

Synsepaly (Gr. *syn*, with + *sepals*) a condition in which sepals are united

Taiga (Teleut, *taiga*, rocky mountainous terrain) a broad northern belt of vegetation dominated by conifers; also, a similar belt in the mountains just below alpine vegetation

Tannin a substance that has an astringent, bitter taste

Tapetum (Gr. *tapes*, a carpet) nutritive tissue in the sporangium, particularly an anther

Taxon (plural, **taxa**) (Gr. *taxis*, order) a general term for any taxonomic rank, from subspecific to divisional

Taxonomy (Gr. *taxis*, arrangement + *nomos*, law) systematic botany; the science dealing with the describing, naming, and classifying of plants

Teliospore (Gr. *telos*, completion + spore) resistant spore characteristic of the Heterobasidiomycetidae, in which karyogamy and meiosis occur and from which a basidium develops

Telium (plural, **telia**) (Gr. *telos*, completion) a sorus of teliospores

Telophase (Gr. *telos*, completion + phase) the last stage of mitosis, in which daughter nuclei are reorganized

Template a pattern or guide used in manufacturing; a sequence of DNA bases used to specify the sequence of synthesis of complementary strand of DNA or RNA

Tendril (L. *tendere*, to stretch out, to extend) a slender coiling organ that aids in the support of stems

Terminal bud a bud at the end of a stem

Testa (L. *testa*, brick, shell) the outer coat of the seed

Test cross the mating between a recessive homozygote and the corresponding dominant to determine whether the latter is homozygous or heterozygous.

Tetrad (Gr. *tetradeion*, a set of four) a group of four, usually referring to the meiospores immediately after meiosis

Tetraploid (Gr. *tetra*, four + *ploos*, fold) having four sets of chromosomes per nucleus

Tetraspores (Gr. *tetra*, four + spores) four spores formed by division of the spore mother cell, used particularly for meiospores in certain red algae

Tetrasporine line (tetraspore + L. *ine*, suffix meaning like) a line of evolutionary development in the algae in which mitosis is directly followed by cytokinesis, resulting in a filament, thallus, or complex plant body of varied form

Thallophytes (Gr. *thallos*, a sprout + *phyton*, plant) all plants whose body is a thallus, that is, lacking roots, stems, and leaves: algae, fungi, liverworts

Thallus (Gr. *thallos*, a sprout) plant body without true roots, stems, or leaves

Thermocline (Gr. *therme*, heat + *klino*, slope) a zone of rapidly changing temperature that separates the epilimnion from the hypolimnion in large bodies of fresh water

Thylakoid (Gr. *thylakos*, sack + N.L. *oid*, a thing that is like) one of membranes in the chloroplasts

Thymidine a nucleoside incorporated in DNA, but not in RNA

Thymidine ^{3}**H** tritiated or radioactive thymidine

Thymine a pyrimidine occurring in DNA, but not in RNA

Tiller (O.E. *telga*, a branch) a grass stem arising from a lateral bud at a basal node; tillering is the process of tiller formation

Tissue a group of cells of similar structure that performs a special function

Tonoplast (Gr. *tonos*, stretching tension + *plastos*, molded, formed) the cytoplasmic membrane bordering the vacuole; so-called by de Vries, as he thought it regulated the pressure exerted by the cell sap

Totipotent (L. *totus*, whole + *potens*, being able) capable of development along any of the lines inherently possible to cells of its species

Toxin (L. *toxicum*, poison) a poisonous secretion of a plant or animal

Tracheid (Gr. *tracheia*, windpipe) an elongated, tapering xylem cell with lignified pitted walls, adapted for conduction and support, and without open end walls

Tracheophytes (Gr. *tracheia*, windpipe + *phyton*, plant) vascular plants

Trait a distinctive definable characteristic; a mark of individuality

Transcription (L. *trans*, across + *scribere*, to write) the process of RNA formation from a DNA code

Transfer cells specialized cells modified by their cell wall projections that may facilitate short distance transport

Transformation mutation or change in genes of bacteria by the direct intervention of extracellular DNA

Translation generally, rendering from one language into another; specifically, the use of a sequence of RNA bases to specify a sequence of amino acids in the process of protein synthesis

Translocation (L. *trans*, across + *locare*, to place) the transfer of food materials or products of metabolism; in genetics, the exchange of chromosome segments between nonhomologous chromosomes

Transmit to pass or convey something from one person, organism, or place to another person, organism, or place

Transpiration (F. *transpirer*, to perspire), the giving off of water vapor from the surface of leaves

Trichogyne (Gr. *trichos*, a hair + *gyne*, female) receptive hairlike extension of the female gametangium in the Rhodophyta and Ascomycetes

Trichome (Gr. *trichoma*, a growth of hair) a short filament of one or more cells extending from the epidermis

Triglyceride a lipid compound composed of three fatty acids plus glycerol; usually functions as a form of stored carbon and energy

Triose (Gr. *treis*, three + *ose*, suffix indicating a carbohydrate) any three-carbon sugar

Trisomic (Gr. *treis*, three + *soma*, body) a plant containing one additional chromosome, $2n + 1$

Tritium a hydrogen atom, the nucleus of which contains one proton and two neutrons; it is written as ^3H; the more common hydrogen nucleus consists only of a proton

Tropism (Gr. *trope*, a turning) movement or curvature due to an external stimulus that determines the direction of movement

Tuber (L. *tuber*, a bump, swelling) a much-enlarged, short, fleshy underground stem

Tundra (Lapp, *tundar*, hill) meadowlike vegetation at low elevation in cold regions that do not experience a single month with average daily maximum temperatures above 50° F

Turgid (L. *turgidus*, swollen, inflated) swollen, distended; referring to a cell that is firm due to water uptake

Turgor pressure (L. *turgor*, a swelling) the pressure within the cell resulting from the absorption of water into the vacuole and the imbibition of water by the protoplasm

Tylose (plural, **tyloses**) (Gr. *tylos*, a lump or knot) a growth of one cell into the cavity of another

Type specimen the herbarium specimen selected by a taxonomist to serve as a basis for the naming and description of a new species

Umbel (L. *umbella*, a sunshade) an inflorescence, the individual pedicles of which all arise from the apex of the peduncle

Unavailable water water held by the soil so strongly that root hairs cannot readily absorb it

Unicell (L. *unus*, one + cell) an organism consisting of a single cell; generally used in describing algae

Uniseriate (L. *unus*, one + M.L. *seriatus*, to arrange in a series) said of a filament having a single row of cells

Uracil a pyrimidine found in RNA but not in DNA

Uredium (plural, **uredinia**) (L. *uredo*, a blight) a sorus of uredospores

Uredospore (L. *uredo*, a blight + spore) a red, one-celled summer spore in the life cycle of the rust fungi

Vacuole (L. dim. of *vacuus*, empty) a watery solution of various substances forming a portion of the protoplast distinct from the protoplasm

Vascular (L. *vasculum*, a small vessel) referring to any plant tissue (xylem and phloem) or region consisting of or giving rise to conducting tissue; for example, bundle, cambium, ray

Vascular bundle a strand of tissue containing primary xylem and primary phloem (and procambium if present) and frequently enclosed by a bundle sheath of parenchyma or fibers

Vascular cambium cambium giving rise to secondary phloem and secondary xylem

Vector (L. *vehere*, to carry) an organism, usually an insect, that carries and transmits disease-causing organisms; a DNA molecule used to transmit genes in a transformation procedure

Vegetation (L. *vegetare*, to quicken) the plant cover that clothes a region; it is formed of the species that make up the flora, but is characterized by the abundance and life form (tree, shrub, herb, evergreen, deciduous plant, etc.) of certain of them

Venation (L. *vena*, a vein) arrangement of veins in leaf blade

Venter (L. *venter*, the belly) enlarged basal portion of an archegonium in which the egg cell is borne

Ventral canal cell the cell just above the egg cell in the archegonium

Ventral suture (L. *ventralis*, pertaining to the belly) the line of union of the two edges of a carpel

Vernalization (L. *vernalis*, belonging to spring + *izare*, to make) the promotion of flowering by naturally or artificially applied periods of extended low temperature; seeds, bulbs, or entire plants may be so treated

Vesicle (L. *vesicula*, small bladder) a small, membrane-enclosed cavity in the cytoplasm of a cell

Vessel (L. *vasculum*, a small vessel) tube of determinate length composed of vessel members joined end-to-end by opened perforation plates. The end walls of a vessel are closed

Vessel member a portion of a vessel derived from a single cell of the vascular cambium or procambium

Vibrio (L. *vibrare*, to shake, vibrate) a genus of short rigid motile bacteria having one or more polar flagella, being typically shaped like a comma or an *S*

Virion (L. *virulentus*, full of poison) infectious virus particle as it exists outside a host

Virulence (L. *virulentia*, a stench) the relative infectiousness of a bacteria or virus, or its ability to overcome the resistance of the host metabolism

Virus (L. *virus*, a poisonous or slimy liquid) a disease principle that can be cultivated only in living tissues, or in freshly prepared tissue brei

Vitamins (L. *vita*, life + amine) naturally occurring organic substances necessary in small amounts for the normal metabolism of plants and animals

Volva (L. *volva*, a wrapper) cup at base of stipe or stalk of basidiomycete fruiting body

Water potential refers to the difference between the activity of water molecules in pure distilled water at atmospheric pressure and 30°C (standard conditions) and the activity of water molecules in any other system; the activity of these water molecules may be greater (positive) or less (negative) than the activity of the water molecules under standard conditions

Weed (A.S. *wēod*, used at least since 888 in its present meaning) generally a herbaceous plant or shrub not valued for use or beauty, growing where unwanted, and regarded as using ground or hindering the growth of more desirable plants

Whorl a circle of flower parts, or of leaves

Whorled referring to bud or leaf arrangement in which there are three or more buds or three or more leaves at a node

Wild-type in genetics, the gene normally occurring in the wild population, usually dominant

Wood (M.E., *wode*, *wude*, a tree) a dense growth of trees, or a piece of a tree, generally the xylem

Xanthophyll (Gr. *xanthos*, yellowish brown + *phyllon*, leaf) a yellow chloroplast pigment

Xerophyte (Gr. *xeros*, dry + *phyton*, a plant) a plant very resistant to drought, or that lives in very dry places

Xylem (Gr. *xylon*, wood) a plant tissue consisting of tracheids, vessel members, parenchyma cells, and fibers; wood

Zoology (Gr. *zoon*, an animal + *logos*, speech) the science having to do with animal life

Zoosporangium (Gr. *zoon*, an animal + sporangium) a sporangium bearing zoospores

Zoospore (Gr. *zoon*, an animal + spore) a motile spore

Zygomorphic (Gr. *zygo*, yoke, pair + *morphe*, form) referring to bilateral symmetry; said of organisms, or a flower, capable of being divided into two symmetrical halves only by a single longitudinal plane passing through the axis

Zygospore (Gr. *zygon*, a yoke + spore) a thick-walled resistant spore developing from a zygote resulting from the fusion of isogametes

Zygote (Gr. *zygon*, a yoke) a protoplast resulting from the fusion of gametes (either isogametes or heterogametes)

ANSWERS TO GENETICS PROBLEMS

1. Two plants with red flowers may produce 25% white-flowered progeny if both the parents are heterozygous. Under the assumptions of the problem, white-flowered plants are homozygous and cannot produce red-flowered progeny.

2. If *ein* is recessive and the tall plant is mutant in *ein*, then the tall plant must be homozygous.

3. 50% of the seeds will be yellow; 50% will be green.

4. She expected 25 of the 100 seeds to be green. She would not necessarily expect a second plant to have green seeds because it might be homozygous for yellow seeds.

5. The genotypes of the parents were *ccPP* and *CCpp*.

6. Red, normal (*RrShsh*); yellow, normal (*rrShsh*); red, shrunken (*Rrshsh*); and yellow, shrunken (*rrshsh*)—all in equal proportions (25%).

7. Alleles *B* and *C* (and *b* and *c*) are linked. Gene *A*(*a*) is not linked to either gene *B*(*b*) or *C*(*c*).

8. The homologous chromosomes containing the *R*(*r*) and *W2*(*w2*) genes looked like this: _____ *R*_____*w2*_____ and _____ *r*_____*w2*_____ . The proportion of recombinant pollen was $(20 + 28)/(20 + 172 + 180 + 28) = 0.12$.

9. Assuming that the starting strains were homozygous, 100% of the progeny of cross 1 would have the *Rv* allele and so be resistant. All the progeny from this cross (and further crosses) that are resistant would be heterozygous; so 50% of the progeny of cross 2 would be resistant, and 50% of progeny from all further crosses (including cross 10). 100% of the progeny of cross 1 would have unlinked wild tomato alleles. For cross 2, 50% would have wild tomato alleles for any gene unlinked to *Rv* (the other 50% would have domestic alleles). For cross 3, because 50% of the parents (progeny from cross 2) have wild tomato alleles, $50\% \times 50\% = (0.5)^2 \times 100 = 25\%$ of the progeny would have unlinked alleles; by extension, for cross 10, the number would be $(0.5)^9 \times 100\%$.

10. If the trait were controlled by a nuclear gene, it would be inherited according to Mendelian laws. Taking this as a hypothesis, the fact that all progeny from the first cross were sterile implies that the male-sterile allele must be dominant and that the progeny from the first cross must be heterozygous. But if that were true, some of the progeny from the second cross should be male-fertile. Because they were not, the trait could not be controlled by a single Mendelian gene. The results are consistent with a maternally inherited gene. (There are other possibilities: What if the male-sterile mutant had dominant alleles in many genes, any one of which could cause male sterility? The probability of finding a male-fertile progeny in the second cross could become very small.)

CREDITS AND ACKNOWLEDGMENTS

Chapter Opener Photo Credits
Ch. 1, © James Carmichael, Jr./NHPA;
Ch. 2, © Stuart Westmorland/Tony Stone
Images; Ch. 3, © C. W. Brown/Photo
Researchers; Ch. 4, © Robert Brons/
BPS/Tony Stone Images; Ch. 5, © Walter
H. Hodge/Peter Arnold, Inc.; Ch. 6, ©
Nigel J. Dennis/ NHPA; Ch. 7, © Scott
Camazine & Sue Trainor/Photo
Researchers; Ch. 8, © Greg Vaughn/Tony
Stone Images; Ch. 9, © Angelina
Lax/Photo Researchers; Ch. 10, ©
Charles D. Winters/Photo Researchers;
Ch. 11, © David Cavagnaro/ Peter
Arnold, Inc.; Ch. 12, © Biophoto Associ-
ates/Photo Researchers; Ch. 13, © L. S.
Stephanowicz/Visuals Unlimited; Ch.
14, © Inga Spence/Tom Stack & Associ-
ates; Ch. 15, © Ed Reschke/Peter Arnold,
Inc.; Ch. 16, © Matt Meadows/ Peter
Arnold, Inc.; Ch. 17, © Holt Studios
International/Inga Spence/Photo Re-
searchers; Ch. 18, © John Cancalosi/
Peter Arnold, Inc.; Ch. 19, © Sharon
Gerig/Tom Stack & Associates; Ch. 20, ©
David M. Dennis/Tom Stack & Associ-
ates; Ch. 21, © M. I. Walker/NHPA;
Ch. 22, © Tim Hauf/ Visuals Unlimited;
Ch. 23, © Gregory G. Dimijian/Photo Re-
searchers; Ch. 24, © Jeremy Burgess/
Science Photo Library/Photo Re-
searchers; Ch. 25, © Art Wolfe/Tony
Stone Images; Ch. 26, © John Gerlach/
Visuals Unlimited; Ch. 27, © Art Wolfe/
Tony Stone Images.

Chapter 1
1.1 (a) NASA Page 5 Phillip Grime/
University of Sheffield

Chapter 2
2.6 Art by Raychel Ciemma / 2.11
Richard Sessions/University of Bristol /
2.13 Micrograph Biophoto
Associates/Science Source/Photo
Researchers / 2.14 (b) Art by Raychel
Ciemma

Chapter 3
3.1 (a) National Library of Medicine;
(b) James M. Bell/Photo Researchers /
3.2 (c–d) Jeremy Pickett-Heaps, School of
Botany, University of Melbourne / 3.3
Micrograph M. C. Ledbetter, Brookhaven
National Laboratory; art by Raychel
Ciemma / 3.4 Art by Raychel Ciemma /
3.6 P. A. Roelofsen / 3.8 (a-left) Don W.
Fawcett/Visuals Unlimited; (a-right)
A. C. Faberge, *Cell and Tissue Research,*
151:403–415, 1974 / 3.10 Micrograph
Gary W. Grimes; art by Robert Demarest
after a model by J. Kephart / 3.11 (a) Art
by Raychel Ciemma; (b) L. K. Shumway;
(f) A. R. Spurr and W. H. Harris, *Amer. J.
Bot.* 55, 1210. © 1968 by Botanical Society
of America / 3.12 (a) Art by Raychel
Ciemma; (b) Micrograph Keith R. Porter
/ 3.13 Robert Seagull / 3.17 (a–j) Dr. S.
Wick, *Jour. Cell Biol.* 89 (1981): 685–90.
New York: © The Rockefeller University
Press / 3.19 Andrew S. Bajer, University
of Oregon

Chapter 4
4.5 J. S. Pate and B. E. S. Gunning, *Protoplasma* 68, 140 / 4.10 (c–d) Steve
Mauseth, University of Texas / 4.12
Sonia Cook, University of California,
Davis / 4.15 (a–b) Professor V. Fran-
ceschi, Washington State University /
4.16 (a–b) Professor Richard Falk, University of California, Davis / 4.23 (a)
David R. Frazier/Photo Researchers; (c)
G. I. Bernard/NHPA

Chapter 5
5.5 (a) J. W. Perry Photo Library / 5.9
Professor Margaret McCully, McMaster
University, Carlton, Ontario, Canada /
5.10 (a) J. W. Perry Photo Library / 5.12
(a) Art by Raychel Ciemma; (c–d) art by
Leonard Morgan / 5.13 Zankowski et al.,
Lindleyana 2 (1987): 1–7. Used with per-
mission of the authors / 5.19 (b) Dr. Tom
Lanini, University of California, Davis /
5.20 (a) Photo by Adrian P. Davies/Bruce
Coleman Ltd. / 5.23 Dr. Judith Jernstedt,
University of California, Davis

Chapter 6
6.23 (a–c) Susan Larson, University of
California, Davis / 6.26 (b) D. Cav-
agnaro/Visuals Unlimited / 6.30 George
Hunter/Tony Stone Images

Chapter 7
7.1 & 7.2 Art by D. & V. Hennings / 7.5
Art by Raychel Ciemma / 7.12 (a) Dr. J.
Lin / 7.13 Professor S. Poethig, Univer-
sity of Pennsylvania, Philadelphia / 7.16
After John G. Torrey, Development in
Flowering Plants, by permission of
Macmillan Publishing Company, copy-
right © 1967 by John G. Torrey / Page
117 (figure 1) Tim Metcalf; (figures 2–3)
Masters of Linen/USA / 7.18 (b) William
J. Weber/Visuals Unlimited; (c) Jerome
Wexler/Photo Researchers

Chapter 8
8.3 Richard Sessions, University of
Bristol

Chapter 9
9.9 (b–c) Art by Raychel Ciemma

Chapter 10
10.1 Photograph Carolina Biological Supply Company; art by Raychel Ciemma /
10.2 Micrograph David Fisher; art by
Raychel Ciemma / 10.10 (a) Art by Raychel Ciemma / 10.12 Art by Leonard
Morgan

Chapter 11
11.9 (c) John Troughton and L. A.
Donaldson

Chapter 12
12.1 (a) Ronald W. Hoham, Dept. of Biology, Colgate University; (b) Ripon
Microslides Inc. / 12.5 Art by Raychel
Ciemma / 12.7 Photograph D. J. Patter-
son/Seaphot Limited: Planet Earth Pictures; art by Raychel Ciemma

Chapter 13
13.4 Rob Preston and Professor Maureen
Stanton, University of California, Davis
/ 13.9 (a–e) Dr. Joseph Lin, Davis, CA /

Page 210 (figure 1) Thomas Eisner, Cornell University; (figure 3) Edward S. Ross; (figure 4) Dr. R. Norris, University of California, Davis / 13.18 Grady L. Webster

Chapter 14
14.15 (a–b) J. W. Perry / 14.25 (a–b) J. W. Perry / 14.29 J. W. Perry / 14.31 (a) Inga Spence © 1995/Tom Stack & Assoc.; (b) John D. Cunningham/Visuals Unlimited / 14.32 (a–b) Robert S. Boyd, Auburn University

Chapter 15
15.9 Christina McWhorter / 15.11 A. Lang / 15.14 A. C. Leopold and M. Kawase, *American Journal of Botany*, 51:294–298, 1964 /15.15 J. Goeschel and H. Pratt / 15.16 Illustration Services / 15.17 Illustration Services / 15.20 (a) Jan Zeevart; (b) Plant Industry Station, Crops Research Division, Agricultural Research Service, U.S. Department of Agriculture

Chapter 16
16.10 Pioneer Hi-Bred International, Inc., © 1996

Chapter 17
17.8 Keith V. Wood / 17.9 Susan Wessler, University of Georgia / 17.11 (a–b) Monsanto Company; (c–d) Calgene, Inc.

Chapter 18
Page 294 Gregory G. Dimi/Photo Researchers / 18.2 (a–b) scanning electron micrographs by John L. Bowman / 18.10 Tim Metcalf / 18.12 Redrawn with permission of M. L Wheelis from Figure 1 in M. L. Wheelis, O. Kandler, and C. R. Woese "On the nature of global classification." *Proc. Natl. Acad. Sci. USA* 89:2930–2934 (April 1992).

Chapter 19
19.1 (a) Photo L. Santo; (b) art by Palay Beaubois / 19.2 (b) Dr. David Dressler and Dr. Huntington Potter, Harvard Medical School / 19.3 Charles C. Brinton, Jr. and Judith Carnahan, University of Pittsburg / 19.4 art by Raychel Ciemma / 19.5 (b) J. W. Costerton, reproduced with permission from *Ann. Rev. Microbiol.* 33:459 (1979). Copyright 1979 by Annual Reviews, Inc. / 19.6 (a) J. Pangborn (b-left) M. L. DePamphilis and Julius Adler / 19.8 (a) W. C. Trentini and

the American Society of Microbiology, from *J. Bact.* 93:1699 (1967); (b) *Biochemistry of Chloroplasts*, vol. 1, T. W. Goodwin, ed., 1966. Courtesy of W. Menke and Academic Press, Inc.; (c) Stanley W. Watson, International Journal of Systematic Bacteriology, 21:254–270, 1971 / 19.9 T. J. Beveridge, University of Guelph/BPS / 19.10 Hans Reichenbach, Gesellschaft für Biotechnologische Forschung, Braunschweig, Germany / 19.12 Biological Photo Services / 19.13 Lawrence Migdale © 1985 / 19.16 Dr. O. Bradfute/Peter Arnold; (b) D. L. D. Caspar, from *Science* 227:773 (1985). Copyright 1985 by the AAAS / 19.17 (b) Kenneth Corbett (c) Brecks, Peoria, IL

Chapter 20
20.2 Introductory Mycology, 3rd Edition, by C. J. Alexopolulos and C. W. Mims, © John Wiley & Sons, Inc., 1979. Reprinted by Permission of John Wiley & Sons, Inc. / 20.3 (a) J. W. Perry / 20.7 Art by Raychel Ciemma / 20.4 R. M. Thornton / 20.6 R. M. Thornton /20.8 Photo by John Webster / 20.13 (a–e) J. W. Perry / 20.16 (a–b) J. W. Perry / 20.11 (d) redrawn from E. A. Gäuman, *The Fungi.* © 1952 by Hafner Publishing Co. and E. A. Gäuman, *Comparative Morphology of Fungi.* © 1928 by McGraw-Hill Book Co. Reprinted with permission of the publisher.

Chapter 21
21.1 (c) N. Lang / 21.4 Holt Studios International/Nigel Cattlin/Photo Researchers / Page 352 J. W. Perry / 21.11 (c) J. W. Perry / 21.12 (a) J. W. Perry; (b) H. Hoops / 21.13 (a) J. W. Perry / 21.14 (a) Biological Photo Service; (b) Gerald & Buff Corsi/Tom Stack & Assoc.; (c) © Claire Ott/Photo Researchers, Inc.

Chapter 22
22.3 (a) Copyright © 1997 by Stephen P. Rae (www.musci.com); (b) copyright © 1997 by Stephen P. Rae (www.musci.com) /22.4 W. Russell / 22.6 J. W. Perry / 22.7 J. W. Perry / 22.8 (a–b) J. W. Perry / 22.12 Photograph Jan Burton/Bruce Coleman Ltd.; art by Raychel Ciemma / 22.14 (a) J. W. Perry / 22.17 (a–c) J. W. Perry / 22.18 (b) J. W. Perry / 22.19 (a–c) W. Russell and D. Hess

Chapter 23
23.1 (a–b) J. W. Perry; (c) Doug Sokell/Visuals Unlimited / 23.4 (b) Dr. Judith Jernstedt, University of California, Davis / 23.8 Field Museum of Natural History / 23.9 Art by Raychel Ciemma / 23.11 J. W. Perry / 23.12 (a) J. W. Perry / 23.18 (a) J. Michael Elchelberger/Visuals Unlimited; (b) James W. Richardson/Visuals Unlimited / 23.20 (a) Philipe E. Keenan/Visuals Unlimited; (b) Ed Reschke/Peter Arnold, Inc.; (c) © R. Calentine/Visuals Unlimited; (d) Laurie Campbell/NHPA / 23.21 (a) John D. Cunningham/Visuals Unlimited; (b) Ken Wagner/Visuals Unlimited / 23.22 (a) J. W. Perry; (b) Jack M. Bostrack/Visuals Unlimited; (c) John D. Cunningham/Visuals Unlimited

Chapter 24
24.4 American Museum of Natural History / 24.5 J. W. Perry / 24.11 J. W. Perry / 24.14 (a–c) J. W. Perry / 24.15 (a–c) J. W. Perry / 24.17 (a–b) J. W. Perry / 24.19 J. W. Perry / 24.20 (d) J. W. Perry

Chapter 25
Page 244 David Parker/Science Photo Library/Photo Researchers / 25.21 (a–b) G. J. James/Biological Photo Service

Chapter 26
26.7 © 1987 Dan Suzio/Photo Researchers / 26.13 (b) John Shaw/NHPA; (c) George Ranalli/Photo Researchers / 26.14 John D. Cunningham/Visuals Unlimited / 26.18 Paige Martin/Archbold Biological Station / 26.19 (a) Laurie Campbell/NHPA; (b) © Steve Krasemann/Photo Researchers / 26.20 (a–b) C. Muller

Chapter 27
27.6 (a–d) Norm Christensen, Duke University / 27.9 (a) James McGraw / 27.17 Greg Vaughn/Tom Stack & Assoc.

Among the many computer illustrations rendered by the artists at Illustrious Interative, Page Two, and Precision Graphics, a few examples are listed below:
Illustrious Interactive—figures 3.18, 4.1, 5.11, 25.12
Page Two—figures 12.8, 22.5, 23.15, 25.14
Precision Graphics—figures 2.15, 10.6, 12.4, 16.3

INDEX

Boldface indicates pages on which primary discussion appears.

fertilization, **190**, 262
  in angiosperms, 432–433
  in conifers, 417–418
  double, **202**
fertilizers, 173, 358
fibers, **54**, 55, 58, 62, 115,
  116–117
fiber-tracheids, **412**
fibrous root system, **70**, 72
Ficus (fig), 115, 224, 230, 428,
  439
  bengalensis (banyan), 72, 73
  carica, 225
  pumila (climbing fig), 72, 73
fiddleheads, **399**
field bindweed. *See Convolvu-*
  lus arvensis
fig. *See Ficus*
filament, **198**
fire, 461, 462, 482
fire-type climates, **461**
flagella, 40, **312**
flannel bush. *See Fremontoden-*
  dron californicum
flax. *See Linum*
Fleming, Alexander, 339
florigen, 257
flower buds, **99**
flowering plants. *See*
  angiosperms
flowers, 197–211
  architecture, 203–205
  in clusters, 205–206
  development, 202–203
  incomplete or complete, **203**
  parts, 197–202
  pollination, 206–211
Foeniculum (fennel), 439
foliose lichens, **354**, 355
follicle fruits, **226**
food chains, **9**
foothill woodlands, **480**
foraminiferans, 8
forbs, **478**
form families, **338**
form genera, **338**
fossil fuels, 387–390
fossils, **295**–296, 301, 302
  algae, 348, 354
  angiosperm, 426, 427, 428
  bryophte, 367
  lower vascular plant, 384,
    385
  mycorrhizae, 326
founder effect, **300**
Fouquieria splendens (coachwhip), 479
foxtail. *See Hordeum hispida*
Fragaria (strawberry), 101,
  102, 184, 185, 224, 228
fragmentation, 190, **359**

free energy, **126**–127
Fremontodendron californicum
  (flannel bush), 233
freshwater wetlands, **484**
fronds, **398**, 399
fruits, **223**–230
  aggregate, **224**, 230
  compound, **224**, 230
  development, 224
  follicle, **226**
  improving, 288
  key to, 223
  multiple, **224**, 225, 230
  ripening, 142, 245, 252–253
  simple, **224**–229
  without seeds, 230
fruticose lichens, **354**, 355
Fuchsia, 200
fucoxanthin, **348**
Fucus (rockweed), 192–193,
  347, 358, 361
Funaria, 374
functional groups, **19**
fungi
  Ascomycota, **328**, 330–334
  Basidiomycota, **328**,
    335–338
  bracket, **336**
  classification, 328–329
  Deuteromycota, **328**, 331,
    338–339
  garden of, 331
  in lichens, 354–356
  mycelium, **326**–327
  reproduction, 327–328,
    329–330, 332–334,
    337–338, 338–339, 356
  resistance to, 253, 274
  roles, 326
  traits, 325–326
  Zygomycota, **328**, 329–330,
    331
  See also Fungi kingdom
fungi imperfecti, **328**
Fungi kingdom, 3, 8, 295, 305,
  325–339. *See also* fungi
funiculus, 218
Fusarium, 274
fusiform initials, **92**, 93

G

gametangia, **190**–191, **329**
gametes, **187**
gametic life cycle, **192**–193,
  360–361
gametophytes, **188**
  asexual reproduction by,
    190–191
  female, **202**
  male, **198**

Ganoderma applanatum, 337
Gardenia, 439
Gelidium, 358
gemmae cups, **370**, 371, 376
generative cells, **198**
genes, 24, **261**
  and chromosomes, 262, 263
  endosymbiosis, 297
  expression of, 240–241
  linked or unlinked, 269,
    270, **271**, 272–273
  mutated, 296–297
  and natural selection, 299
  recombinations of, 297,
    338–339
  regulating cell cycle, 42–43
  transmission of, 262–271
genetic code, **24**
genetic drift, **300**
genetic engineering, 280–289
  applications, 286–289
  safety, 289
genetic information, **24**
genetics, 7, 261–276
  and plant breeding, 271,
    274–276
  population, **299**
genome, **262**
genotype, **265**
Gentiana (gentian), 439
genus, **293**, 295
geographic isolation, **300**
Geranium (cranesbill), 231,
  234. *See also Pelargonium*
Gerbera, 439
germination, **199**, 219–222
  epigeal, **220**
  and gibberellins, 248–249
  hypogeal, **220**, 221
Gibberella fujikuroi, 247
gibberellins, 243, 244, **245**, 246,
  247–249, 257
Ginkgo biloba (maidenhair
  tree), 406, 407, 418–419
Ginkgophyta, **418**–419
ginseng. *See Panax*
  quinquefolium
girdling, 99
Gladiolus, 101, 102, 452
Gleditsia (honey locust), 107
global warming, 5, 161–162
Gloeotrichia, 356
glycerol, 22
Glycine (soybean), 82, 439
glycolipids, **22**
glycolysis, **133**–134, 135
glyoxylate cycle, 138–139
glyoxysomes, **40**
Gnetophyta, **419**–421
Gnetum, 420
Golgi apparatus, **37**

Gonium, 347, 353
Gonyaulax, 356
Good, Ronald, 442, 443
gooseberry. *See Ribes*
Gossypium hirsutum (cotton),
  41, 187, 288, 439, 440
  flowers, 201, 202
  seed, 215, 216
Gracilaria, 358
grains, 217, **219**, 226–227
Gram-negative bacteria, **311**
Gram-positive bacteria, **311**
grandfather's beard. *See*
  Ramalina reticulata
granum (grana), **149**
grapefruit. *See Citrus*
grape. *See Vitis*
grasses, 89–90, 107, 179–180,
  209, 219
grasses of the sea. *See*
  phytoplankton
grassland, 477–**478**
gravitropism, **244**, 246
gravity, 167
Gray, Asa, 440
green algae, 31, 192, 318, 346,
  351, 353–354
greenhouse effect, 162
Grew, Nehemiah, 29
Grime, Phillip, 5
Grimmea (rock moss), 366, 455
Grisebach, August, 440
ground meristem, **65**, 76,
  86–87
ground tissue, 51, **89**
ground tissue system, **49**,
  51–54
growth, 62–63, 238–258
  coordination of, 241–242
  and hormones, 242–253
  and light, 254–258
  patterns of, 238–239
growth regulators, **243**
Guaiacum (lignum vitae), 439
guard cells, **60**, 63, 109,
  171–172
guttation water, 179–180
Gymnodinium, 356
gymnosperms, 49, **406**–422
  Cycadophyta, **418**, 419
  Ginkgophyta, **418**–419
  Gnetophyta, **419**–421
  importance of, 421–422
  in Mesozoic era, 406–407
  Pinophyta, **407**–418
  sieve cells in, 58, 59
  tracheids in, 57, 412
  traits, 406
  wood of, 96–97
gynoecium, 197, **198**, 200–201,
  205

conifer, 416–417
  means of, 207–211
  study of, 206
  types of, 206
pollination drop, **417**
pollination syndromes, **207**
polymers, **19**
polypeptide chains, **19**
polyploidy, **275–276**, 300, 301
Polypodium (polypody fern), 399
polyribosomes, **36**
polysaccharides, **21–22**
Polysiphonia, 342
Polytrichum (hair cap moss), 373, 374, 376, 377, 455
pomes, **229**
pondweed. *See Potamogeton*
popcorn, 219
poppy. *See Papaver*
population ecology, **451**
population genetics, **299**
populations, **292**, **450–465**
  and environment, 457–463
  evolution in small, 300
  interactions among, 463–465
  life history patterns of, 451–455
  over time, 455–457
Populus (cottonwood), 107, 428, 477
  tremuloides (aspen), 184, 185, 450
pore fungi, **336**
Porella, 371, 372
Porphyra, 347, 357, 358
Portulaca (purslane), 439
positional information theory, **66**
Potamogeton (pondweed), 211, 484
potato blight, 345–346
potato. *See Solanum*
potential energy, **121–123**
P-protein, **57**, 59
prairie, **478**
preprophase band (PPB), **44**
prickles, **101**
Priestley, Joseph, 146
primary cell walls, 35, 55
primary consumers, **449**
primary endosperm nucleus, **202**
primary growth, of stem, 86–90
primary meristems, 63, 64, **65**, 76, **86–88**
primary phloem, **57**, 59, 88
primary root, **72**
primary succession, **473**

primary thickening meristem (PTM), **90**
primary xylem, **55**, 88
primitive character states, **303**
Primula, 200, 439
Principal Control Point Hypothesis, 42
procambium, **65**, 76, 78
  residual, **88**
  stem, 86, 87–88
Prochloron, 318
producers, **449**
productivity, **159–162**, **470**
products, **122**
progressive succession, **473**, 474
progymnosperms, **406**
Prokaryota, **308**
prokaryotic cells, **8**
prolamellar bodies, 37
prophase, 43, **44**, 45
prop roots, **72**
Prosopis juliflora (mesquite), 232
proteins, **19–21**
  motor, 40
  in plasma membrane, 33
  synthesis, 24–25, 36
  transport, 36–37, 40
Protista kingdom, **3**, 8–9, 295, 304, 305, **342–363**
  algae. *See* algae
  Chytridiomycota (chytrids), 325, **343–344**
  Myxomycota (slime molds), **344–345**
  Oomycota, **345–346**
protoagriculture, **432**
protoderm, **65**, 76, **86**
protonema, **373**
proton pumps, **33**
protons, **12**
protophloem, **80**
protoplasts, **35**, 286
protoxylem, **78–80**
protozoans, 8
Prunus (plum), 223, 439
  amygdalus (almond), 224, 227, 228
  avium (cherry), 188–190, 228, 251
  persica (peach), 223, 294
Prymnesium parvum, 356
Pseudomonas
  cepacia E-37, 317
  syringae, 253, 319
Pseudosporochnus nodosus, 386
Pseudotsuga, 409
  menziesii (Douglas fir), 409, 422, 481
Psidium (guava), 439

Psilocybe, 336
Psilophyta, **383**, 390–392
Psilotum (whisk fern), 383, 390–392
Pteridium (bracken fern), 399
Pterophyta, **383**, 390, 398–402
Puccinia (rust fungus), 338
puffballs, **336**
purslane. *See Portulaca*
Pyracantha (fire thorn), 102
Pyrrhophyta, 346, **349–350**, 356, 357
Pyrus (pear), 54, 229

Q

Quaternary period, 428, 429, 432
Quercus (oak), 29, 83, 209, 227, 428
  alba (white oak), 95, 96
  as angiosperms, 86, 90, 96, 437, 453
  leaves, 107, 110
  in plant communities, 473, 477, 480, 486
  prinus (chestnut oak), 455
  rubra (red oak), 93, 94
  suber (cork oak), 98, 99, 437
quiescent center (QC), **73–75**
quillwort. *See Isoetes*

R

rabbitbrush. *See Chrysothamnus*
racemes, **205**
rachis, **399**
radial sections, **92**, 93, 94, 96
radiation
  adaptive, **300–301**
  electromagnetic, **151**
  solar, 458–461
radicle, **72**
radioactive carbon, 12, 13, **155**
radioactive decay, 13
radish. *See Raphanus sativus*
Rafflesia, 439
ragweed. *See Ambrosia*
Ramalina reticulata (grandfather's beard), 355
ramie. *See Boehmeria nivea*
Ranunculus (buttercup), 53, 79, 434
  alismaefolius, 435
  aquatilis, 115
Raphanus sativus (radish), 74, 77, 80, 205
raphe, 218
ray initials, **92**, 93

rays, **92**, **412**
ray system, **92**, 93
reactants, **122**
reactions. *See* chemical reactions
receptacle, **197**
receptor, **241**
recessive alleles, **265**
recombinant DNA, **281–282**
recombination, **265**, 266, **271**, 297, 338–339
recycling, 104
red algae, 318, 346, **350**, 358
red–far-red response, 254–255
red tides, 350, **356**
reduction, **14**
reduction division, **188**
redwoods, 409–411. *See also Metasequoia; Sequoia; Sequoiadendron*
region of elongation, **75–76**
region of maturation, **76**
regular symmetry, **203**, 204
Reinke, J., 148–149
Renalia, 385
reproduction
  differential, **298**
  iteroparous, **453–454**
  semelparous, **453**
  See also asexual reproduction; sexual reproduction
reproductively isolated, **300**
reradiation, **460**
residual procambium, 81, **88**
resin ducts, **97**, **412**
resistance to disease, 253–254, 287–288
resolution, **30**
respiration, **131–143**
  aerobic, **132**, 136–138, 143
  alternate pathways of, 138–139
  anaerobic, **132**, 134, 136
  and cell structure, 139–141
  energy transfer in, 132–133
  and environment, 141–143
  functions of, 132
  as oxidation-reduction process, 132
  reactions of, 133–139
  as series of reactions, 132
restriction endonucleases, **281**
restriction enzymes, **281**
retrogressive succession, **473–474**
retting, 117
Rheum (rhubarb), 439
Rhizobium, 82, 173, 177, 308, 319, 320
rhizoids, **368**

Z

Zamia (cycad), 406, 418, 419
Zantedeschia (calla lily), 436
Zea mays (corn, maize), 38, 44, 159, 165, 275, 286, 338
  classification, 86, 436
  flower, 203, 208, 211
  genetics, 261, 262, 263, 274, 275
  leaf, 107, 110, 112
  photosynthesis in, 157, 158
  root system, 70, 73, 75, 184
  seed, 215, 220, 222, 224
Zingiber (ginger), 436
Zostera (eelgrass), 211
Zygomycota, **328**, 329–330, 331
zygosporangium, **329**
zygospore, **328**
zygote, **190**, **202**
zygotic life cycle, **192**, 193, 327–328, 359–360